Air Pollution Modeling and Its Application XVI

Previous Volumes in this Mini-Series

Air Pollution Modeling and Its Application XVI

Edited by

Carlos Borrego
University of Aveiro
Aveiro, Portugal

and

Selahattin Incecik
Istanbul Technical University
Istanbul, Turkey

Springer Science+Business Media, LLC

Library of Congress Cataloging-in-Publication Data

Air pollution modeling and its application XVI/edited by Carlos Borrego and Selahattin Incecik.
 p.   cm.
Includes bibliographical references and index.
ISBN 978-0-306-48464-3   ISBN 978-1-4419-8867-6 (eBook)
DOI 10.1007/978-1-4419-8867-6
  1. Air—Pollution—Mathematical models—Congresses. 2. Atmospheric diffusion—
Mathematical models—Congresses. I. Borrego, C. II. Incecik, Selahattin.
III. NATO/CCMS International Technical Meeting on Air Pollution Modeling and Its
Application (26th: 2003: Istanbul, Turkey)

TD881.Aa47254 2004
628.5'3'015118—dc22
                                                                  2004042430

CCMS

FRONT COVER

The cover depicts a view of Bosphorus—Istanbul, separating the two continents; Asia on the right, and Europe on the left. Bosphorus is the strait between the Black Sea and the Sea of Marmara. There are two suspension toll bridges on this Strait: The first one is Bogazici Bridge and the second one is Fatih Sultan Mehmet Bridge (seen in the photo). At the bottom on the left of the photograph is Rumelihisari (Rumeli Fortress), built in only four months during the 1390s.

Istanbul was inhabited in 3000 BC. In the early 100's BC, Istanbul became part of the Roman Empire and in AD 306, Emperor Constantine the Great, made Byzantium capital of the entire Roman Empire. The city was known as Constantinople. Ottoman Turks conquered Constantinople in 1453. The city became the third and last capital of the Ottoman Empire. It was renamed Istanbul after World War I.

Proceedings of 26th NATO/CCMS International Technical Meeting on Air Pollution Modeling and Its Application, held May 2003, at Istanbul Technical University in Istanbul, Turkey

© Springer Science+Business Media New York
Originally pulished by Kluwer Academic / Plenum Publishers, New York in 2004

http://www.wkap.nl/

10 9 8 7 6 5 4 3 2 1

A C.I.P. record for this book is available from the Library of Congress

PREFACE

In 1969 the North Atlantic Treaty Organization (NATO) established the Committee on Challenges of Modern Society (CCMS). The subject of air pollution was from the start, one of the priority problems under study within the framework of various pilot studies undertaken by this committee. The organization of a periodic conference dealing with air pollution modelling and its application has become one of the main activities within the pilot study relating to air pollution. The first five international conferences were organized by the United States as the pilot country; the second five by the Federal Republic of Germany; the third five by Belgium; the next four by The Netherlands; and the next five by Denmark; and with this one the last two by Portugal.

This volume contains the papers and posters presented at this 26th NATO/CCMS International Technical Meeting on Air Pollution Modelling and Its Application held in Istanbul Technical University, Istanbul, Turkey during 26-30 May 2003. The key topics distinguished at this ITM included: Role of Atmospheric Models in Air Pollution Policy and Abatement Strategies; Integrated Regional Modelling; Global and Long-Range Transport; Aerosols as Atmospheric Contaminants; New Developments; and Model Assessment and Verification.

94 participants representing 33 countries from Europe, North America and Asia attended the 26th ITM. The Conference was jointly organized by the University of Aveiro, Portugal (Pilot Country) and by Istanbul Technical University, Turkey (Host Country). The total 57 oral and 24 poster papers were presented during the conference.

Invited papers by E.L.Genikhovich of Russia (Statistical Approach to Detereministic Modelling), D.Syrakov of Bulgaria (Advection Schemes), M.Z.Boznar of Slovenia (Neural-Network Technique), P.Thunis of Italy (CityDelta), A.Ryaboshapko of Russia (Modeling of Atmospheric Mercury Transport) and R.Bornstein of USA (Urbanization of Numerical Mesoscale Models) were presented.

We wish to thank Prof. Sema Topcu who made important contribution in preparing this volume and Miss Ceyhan Kahya who has also worked as a technical assistance. This includes the technical contributors and presenters, we wish to thank the Scientific Committee members, and the Istanbul Technical University and University Aveiro staff involved in coordinating the meeting and preprint volume.

On behalf of the Scientific Committee and as Organizers and Editors, we would like to express our gratitude to all participants who made the meeting successful. The efforts of the chairpersons and rapporteurs were appreciated.

Special thanks are due to the sponsoring institutions:

Istanbul Technical University, Turkey
University of Aveiro, Portugal
EURASAP: European Association for the Sciences of Air Pollution
GRICES: Office for International Relations in Science and Higher Education, Portugal
NATO/CCMS: Committee on the Challenges of Modern Society
TUBITAK: The Scientific and Technical Research Council of Turkey
TUBITAK, Marmara Research Center, Turkey

The next conference in this series will be held in 2004 in Bannf (Canada).

Selahattin INCECIK   Carlos BORREGO
Local Conference Organizer   Scientific Committee Chairperson
Turkey   Portugal

THE MEMBERS OF THE SCIENTIFIC COMMITTEE FOR THE 26TH NATO/CCMS INTERNATIONAL TECHNICAL MEETINGS ON AIR POLLUTION MODELLING AND ITS APPLICATION

HISTORY OF NATO/CCMS AIR POLLUTION PILOT STUDIES

**Pilot Study on Air Pollution: International Technical Meetings (ITM)
on Air Pollution Modelling and Its Application**

Dates of Completed Pilot Studies:

1969 - 1974 Air Pollution Pilot Study (Pilot Country United States)
1975 - 1979 Air Pollution Assessment Methodology and Modelling (Pilot Country Germany)
1980 - 1984 Air Pollution Control Strategies and Impact Modelling (Pilot Country Germany)

Dates and Locations of Pilot Study Follow-Up Meetings:

Pilot Country – United States (R.A. McCormick, L.E. Niemeyer)

Feb 1971 – Eindhoven, The Netherlands
First Conference on Low Pollution Power Systems Development

Jul 1971 – Paris, France
Second Meeting of the Expert Panel on Air Pollution Modelling

All of the following meetings were entitled NATO/CCMS International Technical
Meetings (ITM) on Air Pollution Modelling and Its Application

Oct 1972 – Paris, France Third ITM
May 1973 – Oberursel, Federal Republic of Germany Fourth ITM
Jun 1974 – Roskilde, Denmark- Fifth ITM

Pilot Country – Germany (Erich Weber)

Sep 1975 – Frankfurt, Federal Republic of Germany – Sixth ITM
Sep 1976 – Airlie House, Virginia, USA – Seventh ITM
Sep 1977 – Louvain-La-Neuve, Belgium – Eighth ITM
Aug 1978 – Toronto, Ontario, Canada – Ninth ITM
Oct 1979 – Rome, Italy – Tenth ITM

Pilot Country – Belgium (C. De Wispelaere)

Nov 1980 – Amsterdam, The Netherlands – Eleventh ITM
Aug 1981 – Menlo Park, California, USA – Twelfth ITM
Sep 1982 – Ile des Embiez, France – Thirteenth ITM
Sep 1983 – Copenhagen, Denmark – Fourteenth ITM
Apr 1985 – St. Louis, Missouri, USA – Fifteenth ITM

Pilot Country – The Netherlands (Han van Dop)

Apr 1987 – Lindau, Federal Republic of Germany – Sixteenth ITM
Sep 1988 – Cambridge, United Kingdom – Seventeenth ITM
May 1990 – Vancouver, British Columbia, Canada – Eighteenth ITM
Sep 1991 – Ierapetra, Crete, Greece – Nineteenth ITM

Pilot Country – Denmark (Sven-Erik Gryning)

Nov 1993 – Valencia, Spain .Twentieth ITM
Nov 1995 – Baltimore, Maryland, USA – Twenty-First ITM
Jun 1997 – Clermont-Ferrand, France – Twenty-Second ITM
Sep 1998 – Varna, Bulgaria – Twenty-Third ITM
May 2000 – Boulder, Colorado, USA – Twenty-Fourth (Millennium) ITM

Pilot Country – Portugal (Carlos Borrego)

Oct 2001 – Louvain-la-Neuve, Belgium – Twenty-Fifth ITM
May 2003 – Istanbul, Turkey – Twenty-Sixth ITM

CONTENTS

ROLE OF ATMOSPHERIC MODELS IN AIR POLLUTION POLICY AND ABATEMENT STRATEGIES

INTEGRATED REGIONAL MODELLING

GLOBAL AND LONG-RANGE TRANSPORT

AEROSOLS AS ATMOSPHERIC CONTAMINANTS

NEW DEVELOPMENTS

MODEL ASSESSMENT AND VERIFICATION

POSTER SESSION

ROLE OF ATMOSPHERIC MODELS IN AIR POLLUTION POLICY AND ABATEMENT STRATEGIES

Chairperson: D. Anfossi

Rapporteur: C. Hogrefe

STATISTICAL APPROACH TO DETERMINISTIC MODELLING OF AIR POLLUTION AND ITS APPLICATIONS IN RUSSIA

Eugene L. Genikhovich

1. INTRODUCTION

Mathematical modeling is used in atmospheric sciences as a powerful tool for improving the scientific knowledge and supporting the decision-making. It is widely believed that the more physics is included in the models and more sophisticated they are the more precise are the results of their application. It was shown, however, that any models could not predict meteorological processes over the time span more, say, than several weeks (on topic of the limit of predictability see for example Shukla, 1985). In other words, generally speaking, there is no sense in "asking" the model to predict the weather that will happen at a certain place in two months. This result could be considered as warning that not every thinkable question is sensible enough to be asked from the atmospheric models. The goal of this paper is to demonstrate that there is another limitation on performance of atmospheric models that could be of importance, in particular, in connection with dispersion modeling. This limitation follows from the stochastic nature of quantities predicted by the models and deterministic predictions generated by the models.

It is well known that atmospheric processes have both, deterministic and stochastic, components and that usual short-time averaging does not completely filter out this stochastic component. In relation to concentrations from a single point source, it was proven by Gifford (1959) who derived their probability distribution function (PDF) using certain simplifications. In this situation, deterministic dispersion models can be efficiently used for predictions of the air pollution only if the variance of the stochastic component is comparatively small. It is not the case, for example, for dispersion from the point source, and, as a result, the agreement between calculated and measured concentrations is usually rather poor here. The models, however, can perform much

Eugene L. Genikhovich, Voeikov Main Geophysical Observatory,
194121 St. Petersburg, Russia

Air Pollution Modeling and Its Application XVI, Edited by
Borrego and Incecik, Kluwer Academic/Plenum Publishers, New York, 2004

3

better, if they are aimed to predict certain statistics of concentrations rather than their "individual" values. Since 1960th the corresponding approach has been systematically implemented in Russia to development of regulatory dispersion models. That is why certain conclusions in this paper will be supplemented with discussion related to these models.

2. STOCHASTIC VARIABILITY OF CONCENTRATIONS

Primarily characteristics of the air pollution are concentrations of pollutants averaged over the certain time interval, usually from 20 – 30 min in Russia to 1 hour in many western countries. Stochastic features of these mean concentrations were found in numerous field experiments and are accounted for in modern techniques of validations of dispersion models (Irwin, 1999).

Statistical properties of concentrations in the plume at given meteorological conditions were theoretically studied by Gifford (1959). He indicated, in particular, that centerline concentration is a stochastic variable. Using certain physical assumptions for short-term concentrations, valid mainly on average, Gifford derived an analytical expression for the probability density of centerline concentrations. Empirical PDFs of the centerline concentrations were studied by Irwin and Lee (1997). Processing the Kincaid data set (see Ohlesen, 1997), Genikhovich and Filatova (2001) found that, having stratified the sample of measure concentration accordingly to the distances from the source and meteorological conditions and having removed several outliers corresponding the lowest measured values, one can approximate PDFs of the centreline ground-level concentrations with the log-normal distribution which corresponds to the following probability density:

$$p(c) = \exp[-\ln^2(c/m)/(2s^2)]/(sc\sqrt{2\pi}),$$

(1)

where $\ln(m)$ is the mean value ("mathematical expectation") of logarithms of concentrations c (in other words, m is the geometric mean of concentrations), and s is the standard deviation of these logarithms of concentrations ("logarithmic standard deviation"). Such an approximation will be used in this paper too, but the results obtained can be easily reformulated for other PDFs.

The magnitude of s is critical for selection the "strategy" of modeling the air pollution. If s is small enough, the process of the air pollution could be considered as essentially deterministic, and one can expect only small discrepancies between model predictions and observations, if this model properly reproduces the main physical features of this process. The large is s, however, the more are these discrepancies. As a measure of the scatter one can use, for example, the normalized mean square error (NMSE) that is defined as follows:

$$NMSE = < (M - P)^2 > /(< P > \cdot < M >),$$

(2)

where M and P are measured and predicted concentrations, respectively, and brackets symbolize the procedure of averaging. A graph of the relative error (the square root from

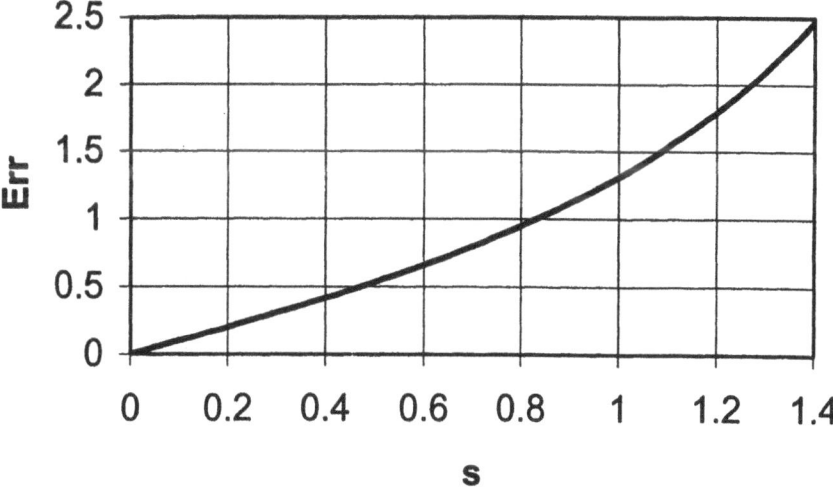

Figure 1. Relative error as a function of s.

NMSE) , Err, as a function of s is given on Fig. 1; it is assumed here that we have at our disposal a perfect model (see Genikhovich, 2002) that predicts exactly the mean value <c> of concentrations distributed accordingly (1).

Genikhovich and Filatova (2001) found, however, that the Kincaid data correspond to s in Eq. (1) that is varying in the interval from 0.6 to 1.2, depending on the distance from the source and governing meteorological conditions. It corresponds, accordingly to Fig. 1, to relative errors from 70% to 180%. These errors cannot be reduced by any improvements of the models; they are inherent to the stochastic nature of the atmospheric diffusion. As a result of the scatter, measured concentrations can be either higher or smaller than <c>. It can be easily derived from (1) that measured concentrations are large than <c> for approximately from 30% to 40% of all measurements, if s is inside the aforementioned interval. In all these cases the perfect model, adjusted to predicting the mean concentrations <c>, underestimates real concentrations. Therefore, decisions made with the use of predictions of this model and aimed for preventing the possible hazard for population and environment could be not satisfactory in 30 to 40 percent of all cases. It seems to be more preferable in these circumstances to tune the models to predicting, for example, upper percentiles of PDFs or other statistical characteristics.

3. ATMOSPHERIC DISPERSION OF INDUSTRIAL EMISSIONS

The problem of atmospheric diffusion is usually formulated as a part of the problem of geophysical hydrodynamics that is aimed to description of mean and turbulent characteristics of the airflow. It can be based on solution of the system of equations that looks like

$$\partial f_i / \partial t + U_k \partial f_i / \partial x_k = \partial(-\overline{u_k f_i}) / \partial x_k + S_i;$$

(3)

where summation is assumed over repeated indices. Here x_k are Cartesian coordinates (notations x, y, z also are used in this paper; and x-axis is oriented along the wind direction near the ground, y-axis is perpendicular to x-axis in the horizontal plane and z-axis is always directed in the vertical direction). Depending on selection of the source term S_i, functions f_i can be equated with components U_1, U_2 or U_3 of the mean wind speed, the air temperature T, humidity q, kinetic energy of turbulent fluctuations of the wind speed E, dissipation rate ε, as well as concentrations of different pollutants C_j. The system (3) is supplemented with the continuity equation and initial and boundary conditions.

Turbulent fluxes in (3) are frequently approximated using the first- or second-order turbulent closure. Special precautions should be taken with horizontal fluxes of pollutants because corresponding approximations should account for the Richardson effect, *i.e.*, for the dependence of the horizontal eddy diffusivity, K_y, on the spatial scale of the dispersing cloud or plume. When modeling the stationary diffusion from a single point source, this effect can be simulated, if one assumes that $K_y = \sigma_\theta U x$ with σ_θ and U determined as the standard deviation of horizontal fluctuations of the wind direction and the wind speed profile, correspondingly. Then the multiple-source case can be handled using superposition of concentration fields from individual sources.

In certain instances (flat terrain, for example) the fields of the wind speed and eddy diffusivities are known from field and laboratory measurements, and Eq. (3) can be reduced to the advection-diffusion equation. Its solution can be found analytically (in the simplest cases) or numerically as soon as all coefficients and initial and boundary conditions are specified. The number of governing parameters that determine this solution is limited. In the case of the stationary point source with the constant emission rate M, one can assume that the concentration field depends on the time "parametrically", *i.e.*, varies in time accordingly to variations of these governing parameters. Thus, concentration of the pollutant, c, can be represented as

$$c = M \cdot G(x_1, x_2, x_3, \Omega);$$

(4)

where $\Omega = (\omega_1, \omega_2, \omega_3.......)$ are governing meteorological parameters and G implicitly depends also on technical parameters of the source (stack height and diameter, effluent temperature and volume rate and so on). Assuming also that $\Phi(\Omega)$ is a joint PDF of governing parameters ω_i with the density $\varphi(\Omega)$, one can find PDF of modeled concentrations, Fm(c), using the following expression (Pugachev, 1979):

$$Fm(c, x_1, x_2, x_3) = \int_{M \cdot G(x_1, x_2, x_3, \Omega) < c} \varphi(\Omega) d\Omega;$$

(5)

where $d\Omega = d\omega_1 d\omega_2 d\omega_3 \ldots$. The upper (1-δ) quantile of modeled concentrations, $c_{1-\delta}$, can be found as a solution of the equation Fm(c, x_1, x_2, x_3) = 1- δ. It follows from Eq. (5) that $c_{1-\delta} = M \cdot \chi(x_1, x_2, x_3, \delta)$ where function χ does not depend on actual meteorological parameters but is rather governed by characteristics of their regime "hidden" in $\Phi(\Omega)$. The field $c_{1-\delta}$ is referred to as the "majorant" or "upper-limit" concentration field

6

corresponding to the upper $(1-\delta)$ quantile. It should be noted that Fm(c) does not reproduce the influence of stochastic fluctuations of concentrations described in Eq. (1) so that Fm(c) is, generally speaking, should be different from PDF of experimentally measured concentrations, F(c). However, meandering of the plume, which is mainly responsible for these stochastic fluctuations, should not reduce highest measured concentrations because their occurrence corresponds to "direct hit" of the monitoring point by the plume. That is why one can expect a reasonably good agreement between measured and calculated majorant concentration fields.

A more practical approach to estimating the majorant concentrations is based on separation of the phase space, S, of meteorological parameters Ω in two parts, S_δ and S/S_δ where S_δ is a subset of Ω corresponding to the highest modelled concentrations that occurred with probability δ, and S/S_δ is the rest of the elements of S. Then $c_{1-\delta}$ can be estimated as a conditional extremum of modelled concentrations c over the subset S/S_ξ. This approach was used, in particular, to derive formulae of the Russian national regulatory dispersion model (Berlyand et al., 1987). It is a multiple-source model, which is used for calculation of dispersion from point, line and area sources. The model accounts for technical parameters of the sources, characteristics of effluents, terrain and building features, and so on. Main steps in development of this model will be illustrated here considering the simplest case of ground-level concentrations from a single point source located in flat terrain.

Having in mind that input information for dispersion models in Russia is mainly available from the standard meteorological observations inside the surface layer, coefficients in the advection-diffusion equation (3) were first parameterized using the wind speed U at the wind-vane level z_v, the wind direction φ, the roughness parameter z_0, the depth of the surface layer h, standard deviation of horizontal fluctuations of the wind direction σ_θ, and the stratification parameter λ, which is determined as a ratio of the vertical eddy diffusivity, K_z, to the product of the wind speed U(z) and corresponding height z (this ratio should be calculated at the height $z = z_1$ with $z_1 = 1$ m). The last parameter is closely correlated with the Rchardson number Ri, the Monin-Obukhov length scale L and other quantitative characteristics of the atmospheric stability. In particular, σ_θ can be expressed via λ and h can be expressed via λ and the Coriolis parameter. Using dimensional analysis, one can show from Eq. (3) that ground-level concentrations could be represented in the following form:

$$c|_{z=0} = c_{m\lambda U} L_s(x/x_{m\lambda U}, \lambda, z_0/h) \exp[-0.5 y^2/(\sigma_\theta^2 x^2)];$$

(6)

where L_s is normalized with the condition $L_s(1, \lambda, z_0/h) = 1$ and $c_{m\lambda U}$ is the maximum ground-level concentration reached at the distance $x_{m\lambda U}$ from the source. It is follows from calculations that $c_{m\lambda U}$ and $x_{m\lambda U}$ can be approximated by the expressions

$$c|_{m\lambda U} = M F_1(H_e/h, \lambda)/(U H_e^2);$$

(7)

$$x_{m\lambda U} = H_e F_2(H_e/h, \lambda);$$

(8)

where F_1 and F_2 are dimensionless functions and H_e is the effective stack height.

The extremum of concentrations in Eq. (7) is related to critical values of the stratification parameter λ_m and wind speed U_m. It should be noted, however, that this extremum is conditional and, when looking for maximum of $c_{m\lambda U}$ in the domain S/S_δ, one should introduce a certain correlations between U and λ that, actually, separate S/S_δ from S_δ. In particular, it is accounted for that the boundary $\lambda \sim a_2 \hat{U} /U$ separates S_δ at "small" U and "large" λ with a constant a_2 depending on climatic conditions at the location considered. Another limitation follows from the fact that the depth of the surface layer h usually is not exceed $100 - 150$ m. The resulting conditional extremum (7) over λ at the given value U, c_{mU}, is approximated by the following expression:

$$c_{mU} = \Gamma z_1^{1/3} M / (uH_e^{7/3});$$

(9)

where Γ is a function of the wind speed.

The critical wind speed, $U_{m\lambda}$, is determined here as the value of the wind speed U that corresponds to the extremum of concentrations in Eq. (9). The expression for $U_{m\lambda}$ can be derived from (7), if one assumes a certain model for the initial plume rise. Some conclusions about the critical wind speed at near-neutral conditions could be obtained even from dimensional analysis. Let us consider a single point source characterized by the following parameters: stack height H, stack diameter D, effluent velocity W and effluent temperature T_s. It is known (see, for example, Briggs, 1984) that trajectories of plumes and jets are governed by the "fluxes" of momentum, F_m, and buoyancy, F_b, where $F_m = 0.25 W^2 D^2 T_a/T_s$ and $F_b = 0.25 g W D^2 (T_s - T_a)/T_s$. Here, T_a is the ambient air temperature and g is the gravitational acceleration. It should be noted that dimensions of F_m and F_b are $L^4 T^{-2}$ and $L^4 T^{-3}$, where L and T are length and time units. When considering centerline plume trajectories in shear-free flow, the wind speed at the stack top, U_H, is also used as a governing parameter. Dimensional analysis suggests that buoyancy-dominated plumes should be scaled with the length scale $l_b = F_b/U_H^3$, and that momentum-driven jets should be scaled with $l_m = (F_m)^{1/2}/U_H$.

When considering the critical wind speed, U_{cr}, the stack height H should be used as a governing parameter instead of U_H (indeed, U_{cr} should correspond to the "most efficient" transport of pollutants from the source level to the ground). Thus, dimensional analysis for the buoyancy-dominated plumes suggests that the wind speeds should be scaled with the following velocity scale V_b:

$$V_b = (F_b / H)^{1/3}.$$

(10)

Similarly, for momentum-dominated jets the scale for the wind speed is provided by the following expression:

$$V_m = \sqrt{F_m} / H.$$

(11)

Note here that the ratio of these two scales might be considered as a dimensionless measure of the relative importance of buoyancy and momentum effects for the source under consideration. Actually, in the regulatory model introduced by Berlyand et al. (1987) such a measure, f, is determined by the ratio $f \sim (V_m/V_b)^3$, or $f = a_1 F_m^{1.5}/(F_b H^2)$, where a_1 is a dimensionless constant (for the sake of convenience it is introduced in such

8

a way that the transition from buoyancy- to momentum-dominated plume rise occurs at f = 100). Obviously, f = 0 corresponds to buoyancy-dominated plumes, and f tends to infinity for momentum-dominated jets. In the intermediate case, where both buoyancy and momentum fluxes are of importance, the scale for the critical wind speed, U_{scale}, may be written as

$$U_{scale} = V_b \Phi(f),$$

(12)

where $\Phi(f)$ is a dimensionless function normalized by the conditions $\Phi(0) = 1$ and $\Phi(f) \sim (f/a_1)^{1/3}$ when f tends to infinity (this equation can be also rewritten in a different form with V_m appeared in the right-hand side as a scaling parameter). Using this velocity scale, one can represent the critical wind speed in the following way:

$$U_{cr} = U_{scale} G_1,$$

(13)

where G_1 is a dimensionless factor which can depend only upon dimensional arguments. In the simplest case of shear-free flow, one could expect that G_1 is simply a constant or, at least, does not depend upon source parameters. Therefore, U_{cr} should be proportional to $H^{-1/3}$ for buoyant plumes and to H^{-1} for jets. In other words, both $(U_{cr})^{-3}$ for buoyant plumes and $(U_{cr})^{-1}$ for jets should be proportional to H (for confirmation of this conclusion see Genikhovich, 2001). In the intermediate case the dependence is more complex and will not be analyzed here.

The resulting extremum c_m of the maximum ground-level concentrations c_{mU} corresponds to the critical wind speed. In the case of the flue gas emitted at the temperature T_g, which is much higher than the temperature of the ambient air T_a, c_m can be calculated here as follows:

$$c = c_2 AMm / \{H^2 [\beta V_1 (T_g - T_a)]^{1/3}\};$$

(14)

where β is the buoyancy parameter, A is a dimensionless coefficients describing the climatic conditions of turbulent diffusion in the region considered, V_1 is the volume rate, c_2 is a dimensionless constant for conversion of mass units of emission into mass units of concentrations, and m is a dimensionless function of f.

Similar technique was applied to other terms in Eq. (6). The resulting expressions can be used to calculate the majorant concentration field from a single point source. Using the superposition principle, these expressions are applicable also in the cases of multiple point sources, line- and area sources, and so on. Numerous validation tests have proven that these expressions predict the field of the upper 98th percentiles of concentrations with an error inside 25%. An example of such validation taken from Genikhovich et al. (1999) is shown on Fig. 2. It represents the sample estimate of PDF of ratios of concentrations measured in field experiments to calculated ones. It was calculated as an average of nine individual PDFs each corresponding to a field experiment. These experiments were carried out in 1970th - 80th around aluminium plants located in different geographic regions of the fUSSR, and concentrations of HF were measured at the distances from 1 km up to 10 km from the plants. The total volume of the sample used to plot this figure is more than 44600 values of concentrations. Each measured concentration was compared with its upper 98th percentile calculated accordingly to the technique described. Thus, one can expect that the ratios of measured to calculated concentrations should be mainly relatively small. If these ratios are always significantly less than one, it means that the technique for predicting upper percentiles is too conservative, *i.e.*, that upper percentiles were overpredicted; if a noticeable fraction of the ratios is significantly large than one, it means that the upper percentiles were underpredicted and, therefore, decisions made with the use of this technique could be not sufficient for protecting human health and environment

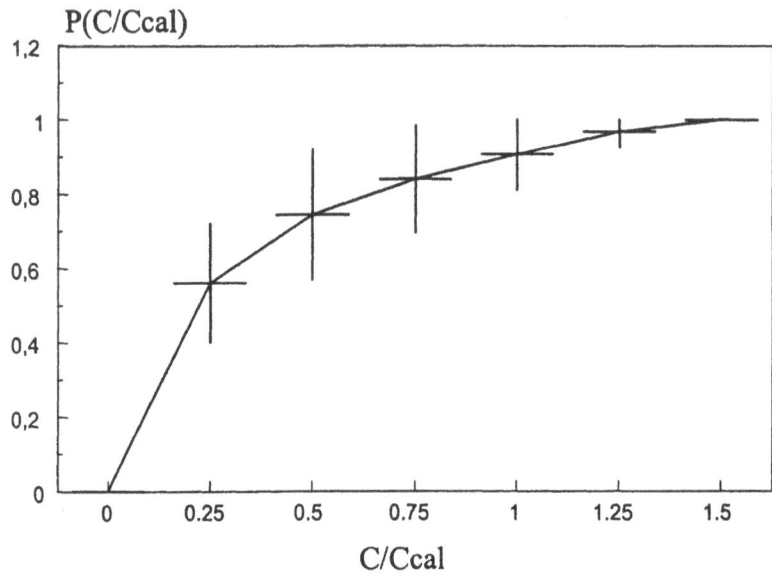

Figure 2. Probability distribution function P of ratios of concentrations measured in nine field experiments to calculated using the majorant dispersion model. Horizontal lines represent mean values of nine individual PDFs, and vertical bars characterise standard deviations of individual PDFs relative to their mean values.

4. MAJORANT APPROACH TO MODELLING OF LONG-TERM AVERAGES

Accordingly to its definition, the long-term averaged atmospheric concentration of the pollutant at the given receptor point (x,y,z) is determined as the integral over the averaging time, τ_Σ, from corresponding values of short-term averaged concentrations at this receptor point divided by τ_Σ. Therefore, one can model long-term averages using an "imitational" dispersion model, which simulates time series of short-term concentrations at the receptor points and then calculates their average values. It has been mentioned in this paper, however, that even the purely stochastic component of the error of simulation of individual short-term concentrations is very high (roughly speaking, about 100% for centreline ground-level concentrations and at least an order of magnitude more for concentrations at given receptor points). When calculating their average value, this stochastic error is decreased proportionally to the factor of $N^{-1/2}$ where N is the number of "independent concentrations" in the sample used for averaging. It seems to be plausible that, if one would use 8760 hourly values of concentrations over the year, than the stochastic error will be reduced in approximately 94 times. It is well known, however, from the empirical data analysis (*e.g.*, Genikhovich *et al.*, 2003) that short-term concentrations are usually correlated, especially, due to the influence of daily variations. If one would apply daily averaging to filter out this correlation, the volume of sample will be reduced to 365 and, therefore, the factor of reduction of errors would drop to approximately 19. So, when using this approach, the expected stochastic error of calculated mean annual concentrations should be not less than 50%. Actually, such an error should be significantly more because the plume only a faction of the whole year influences each receptor point (it depends, in particular, on the probability distribution of wind directions, i.e., on the wind rose).

An alternative approach (see for example Genikhovich et al., 2000) is based no integration of the expression for short-term concentrations over the phase space S of governing meteorological parameters Ω with the weight $p(\Omega)$ which is the probability density (p.d.) of observation of Ω. Then S is decomposed in a sum of subsets S_j, each of them corresponding to a "reduced" number of governing parameters. Such a decomposition simplifies the task of determining the weights $p_j(\Omega)$ (*i.e.*, "projections" $p(\Omega)$ in corresponding subsets) from empirical data sets; the same goal can be achieved, if some of parameters ω_i forming Ω are independent so that $p(\Omega)$ is decomposed in the product of probability densities lesser dimensions.

It is shown by Genikhovich et al. (2000) that the first term in decomposition of the long-term averaged concentration from a single point source, <c>, can be represented in the following way:

$$< c > = p_1(\varphi) \iint p_2(U,\lambda) c'(r,U,\lambda) dU d\lambda / r,$$

(15)

where (r,φ) are polar co-ordinates of the receptor point corresponding (x,y), $p_1(\varphi)$ and $p_2(U,\lambda)$ are p.d. of the azimuth of the plume centreline and joint p.d. of the wind speed and stability parameter, and c' is the crosswind integrated concentration from the point source (could be obtained, for example, from Eq. 6)..Here, $p_1(\varphi)$ is expressed via the wind rose $p_1(\varphi)$ by its normalizing and reversing relative to its origin ($p_1(\varphi)$ (because northern wind directions correspond to southern orientations of the plume). Next terms in the aforementioned decomposition represent effects of scavenging of the pollutants and other physical effects not accounted for in Eq. (15).

Eq. (15) was validated by comparison of calculated concentrations with those measured in field conditions (see for example Genikhovich et al., 2000). It is used in the draft of the Russian regulatory guideline on regulatory dispersion modelling that is now goes through procedures of official approval. Here, this expression is written to demonstrate that the long-term averages depend on the meteorological regime characterised by p.d. like $p_1(\varphi)$ and $p_2(U,\lambda)$. When calculating for example mean annual concentrations for the specific year, these p.d. should be estimated from the meteorological data sets corresponding to this year. Therefore, they are stochastic functions varying from year to year, and it results in variability of estimated mean annual concentrations. As it follows from, Eq. (15), <c> is more sensitive to variations of p_1

because, unlike p_1, p_2 is placed under the integral and, therefore, is averaged with the weight c'.

The variability of $<c>$ can be significantly reduced, if p_1 and p_2 are estimated with the use of the multi-annual sample of governing meteorological parameters (it corresponds to calculation of the "climatologically averaged" mean annual concentrations). It should be noted, however, that the "actual" values of annual averages corresponding to given years are distributed relative to their climatological mean value (CMV) accordingly to the Gaussian distribution. In other words, about 50% of actual mean annual concentrations should be higher than their CMV. Thus, the policy measures aimed to guarantee that CMVs are less than corresponding ambient air quality standards or that assessed via CMV corresponding risk is less than a tolerance level result in not satisfying these criteria "each second year". More correct approach here should be based on calculations of certain upper percentiles of PDF of mean annual concentrations. This approach is used, in particular, in the draft of a new Russian guideline on regulatory dispersion modelling where the tolerance level is selected as 90%.

5. DISPERSION MODELLING OF ACCIDENTAL RELEASES

Accidental releases of noxious chemicals in the atmosphere unfortunately happen frequently enough to be considered as a serious problem for modern civilization. The number of victims of these accidents could be as high as several thousand like after the Bhopal tragedy. The main sources of accidents are industrial enterprises and storage facilities where poisonous chemicals are generated or stored or used as a part of technological processes (transportation of the noxious chemicals is also a possible reason for accidental releases). For these enterprises and facilities, the potential sources of accidental releases are usually well localized, and possible amount of pollutants, which could be discharged in the case of accidents, is known in advance. For the emergency preparation, it is required in Russia to estimate the size of the zone (also called "the scale of contamination") around an enterprise, which could be affected by harmful pollutants in the case of accidental release at such level that certain precautions are to be taken (for example, people could be evacuated from this zone in the case of emergency). A corresponding regulatory model was introduced in Russia in the beginning of 1990th (Berlyand et al., 1990), and its description was given by Berlyand, Genikhovich, Gracheva (1994). It is based on the "dose approach" (it is often referred to as the "exposure approach"), i.e., on calculation of the doses, D, of different noxious pollutants, corresponding to 1-hour exposure time, τ, and comparison those with criteria limiting the permissible level of doses. It should be noted here that doses are monotonous functions of τ; that is why one can be sure limiting 1-hour doses with a certain criterion that all doses corresponding to lesser exposure times will be also limited with the same criterion.

An equation for doses was derived by integration Eq. (3) over the time. It was assumed that the non- in stationarity (3) results from the non-stationary accidental discharge of pollutants into the atmosphere. This equation, similar to those for the stationary concentration distribution, was solved numerically and the results, analytically approximated, are used in computations. The model could be used for different kinds of sources (discharge of gases from tanks and pipes, evaporation of liquid spills and so on) as well as for preparation of emergency preparedness scenarios ("prognostic mode") and for calculations in actual emergency response situations ("diagnostic mode"). For the sake of simplicity, in this paper let us limit ourselves with the case of the gaseous emission.

The characteristic parameters of the accidental release like its height, orientation of impulse, and volume rate of the source are usually not known well, especially in the cases when the model works in the prognostic mode. It seems to be natural in this situation to use those combinations of these parameters that result in the highest level of impact of the release. The worst orientation of the jet emitted from the source, in particular, could be directed downward, and this situation could be simulated with the ground-level source of emission. The spatial distribution of pollutants from the ground source, which is an obvious majorant to those from the elevated source (it should be stressed out, however, that at large enough distances from the source, the influence of the source height could become negligible), can be given in the following form:

$$D = \frac{0.94\psi Q}{\lambda^{3/2} U x^2} \cdot \exp(-\frac{1.8 y^2}{\lambda x^2}),$$

(16)

where U is the wind speed at the vane level and ψ is a function of $(x/U\tau_0)$ with the time scale τ_0 expressed via the Coriolis parameter. Here, ψ is normalized with a condition $\psi(0) = 1$ and $\psi \sim (x/U\tau_0)^{1/2}$ when x tends to infinity.

As in the previous section, Eq. (15) is used for constructing a majorant dose field. An accidental release, however, usually does not continue for the whole year; that is why the approach described in the previous section and aimed to determining the extremum of concentrations over the whole range of variations of thermal stratifications cannot be directly transferred here. It seems to be possible, instead, to introduce several "characteristic modes" of the thermal stratification of the atmosphere and derive formulae for majorant concentration fields corresponding to these modes and given values of the wind speed. Critical conditions here obviously correspond to low winds and most stable stratifications. That is why these conditions are assumed when running the model in the prognostic mode. In diagnostic mode, however, the actual meteorological conditions during the accidental release should be taken into account. In doing so, the value of the stratification parameter λ as a certain low quantile of the probability distribution function of λ corresponding to the chosen stratification mode. Similarly, the wind speed should be taken from the "lower side" of their distribution corresponding to its actually measured value.

It is very important also to take majorant over the lateral spread of the plume of doses. It should account for uncertainties of location of the plume centerline because of plume meandering as well as for differences in wind directions measures at the meteorological station and existing at the location of the accidental release. It is mentioned by Nieuwsatdt and Van Dop (1982) that the standard deviation $\sigma^2_{\theta A - \theta B}$ of the differences $\theta_A - \theta_B$ of wind directions, measured in a big city at locations A and B, can be approximated as $\sigma^2_{\theta A - \theta B} = (15° + 5.7 \ln x_{AB})^2$, where x_{AB} is a distance from A to B. Averaged over the area, the standard deviation of wind direction, $\sigma^2_{\theta, AV}$ depends on the wind speed as follows: $\sigma^2_{\theta, AV} = (5°)^2 + (60°/U)^2$. For a normally distributed stochastic variable with a mean value m and standard deviation σ, the value 4σ could be considered as the uncertainty interval corresponding to the tolerance limit of 10% (because the probability of been confined inside the interval (m-2σ, m+2σ) is approximately equal to 0.1). The aforementioned formulae indicate, in particular, that at the wind speed $U = 1$

m/s the uncertainty interval for the wind direction is equal to 241°. In other words, at low wind speed the plume, emitted at the arbitrarily located point of release, could be oriented in practically any direction, independently on the wind direction measured at the monitoring station.

Because the spatial distribution of the doses is similar to those for the stationary concentration field, the model was validated using experimental data from the "classical" data sets like, for example, Prairie Grass and from wind-tunnel simulations. In addition, the field experiments were organized in Uzbekistan on the special test side near the natural-gas producing field. The natural gas with high content of H_2S was used as a tracer. A reasonably good agreement was found between calculations and measurements.

6. CONCLUSION

The existence of the stochastic component influencing meteorological variables was discussed by meteorologists for several decades (see for example Obukhov, 1988). As a way to handle corresponding problems, the ensemble forecasts were introduced in the weather predicting practice. Probabilistic approaches are widely used in forecasting precipitations and other "spotty" meteorological events. It seems to be reasonable to account for stochastic nature of concentration field in modelling the air pollution too. The methodology in use in Russia gives an example of direct computation of the majorant concentration fields for regulatory purposes. The computed results are widely used in engineering and public practice including decision making, designing, and so on. Similar models are developed in the USA and called SCEEN-type models.

Majorant concentration fields can be used in solution of different environmental problems where the maximal level of impact of pollutants should be limited. More detailed results can be obtained, if one calculates PDFs of concentrations (indeed, in this case one could estimate different percentiles of concentrations, their mean values and so on). When the dispersion model at hand (*e.g.* the Gaussian one) is simple enough to simulate the long-term time series of short-term averaged concentrations at the given receptor point, one can construct the PDF as the frequency distribution of simulated concentrations. It is shown in this paper, however, that the individual short-term concentrations are calculated with extremely high errors. Thus, one can expect high errors in the estimated PDF too, especially in its high and low percentiles.

If an analytical expression for PDF is known, its parameters can be found using corresponding values of the mean concentrations and their upper percentiles (one can use here the approaches described in sections 3 and 4). For example, if the log-normal distribution is applicable, its two parameters can be derived from known 98th percentile and mean annual value. In general case, when concentrations are generated from the numerical solution of Eq (3), an ensemble approach seems to be an efficient way to estimating the expected distributions of concentrations. It should be noted, however, that, using this approach, one should be sure that the governing equations properly describe the components of the turbulent spectra of concentrations corresponding to all influencing frequencies. Obviously, it might be expected in some cases, for example, if the meteorological driver of the dispersion model includes procedures of data assimilation. On the contrary, if characteristics of the wind flow are generated exceptionally by dynamic equations solved in a limited computational domain, one hardly could expect any proper description of the low-frequency part of spectra corresponding to the length scales larger than the size of this domain (it could be also a

14

factor limiting applicability of results obtained with LES techniques). An additional limitation of the ensemble approach follows from the fact that, due to computational restrictions, the number of individual realisations in the ensemble is rather limited. Thus, to obtain here more-or-less reliable estimates of PDFs, one should have *a priory* information about their functional form. Corresponding techniques are still under development.

6. REFERENCES

Berlyand, M.E., Gasilina, N.K., Genikhovich, E.L., Onikul, R.I., Glukharev, V.A. (Ed) 1987 Method for Calculation of Concentrations of Air Pollutants the Industrial Emission Contains. National Regulatory Document OND - 86. Hydrometeorological Publishers, L., 92 p.(in Russian)

Berlyand M.E., Genikhovich E.L., Gracheva I.G. (1994) Modeling and forecast of air pollution under accidents. Proc. 4th Intern. Conf. On Atmospheric Sciences and Application to Air Quality, Seul, Korea,p. 91 – 93

Berlyand, M.E., Suldin, Yu.I., Genikhovich, E.L., Gracheva, I.G., Malyshev, V.P., Isaev, V.S., Chicherin S.S., Onikul, R.I., Eliseev, V.S., Zachek, V.S., Korzunov, S.N., Semenov, V.I. (1990) Method for Forecasting of Scales of Contamination With Poisonous Toxic Pollutant From Accidental Releases at Chemically Dangerous Units and Transport. RD 52.04.253-90. Hydrometeorological Publishers, Leningrad, 23 p. (in Russian)

Briggs, G.A., (1984) Plume rise and buoyancy effects. In: Atmospheric Science and Power Production, D. Randerson (Ed.), US Department of Energy/TIC 27601.

Genikhovich, E.L. (2001) Critical wind speed for a single point source located either in flat terrain or near buildings. In: PHYSMOD2001. Proc. of the International Workshop on Physical Modelling of Flow and Dispersion Phenomena, September 3 – 5, 2001, Hamburg, Germany. – 6 p.

Genikhovich, E. (2002) Indicators of performance of dispersion models and their reference values. Proceedings of 8th Intern. Conference on Harmonization Within Atmospheric Dispersion, p. 40 – 47

Genikhovich, E., Berlyand M., Onikul R. 1999 Developing the theory of atmospheric diffusion as a basis for decision making in the air pollution prevention. In: Contemporary Investigations at the Main Geophysical Observatory to Its 150th Anniversary, v. 1, p. 99 - 126

Genikhovich, E., Filatova, E. (2001) A PDF approach to processing the data of tracer experiments for validation of dispersion models. Proc. 7th Int. Conf. On Harmonization Within Atmospheric Dispersion Modelling for Regulatory Purposes, Belgirate, Italy. EC JRC, p. 33 – 36

Genikhovich,E.L., Gracheva, I.G., Groisman, P.Ya., Khurshudyan, L.G. 2000 A new Russian regulatory dispersion model MEAN for calculation of mean annual concentrations and its meteorological preprocessor, *Int. J. of Environment and Pollution*, v.14, Nos. 1-6, p. 443 – 452

Genikhovich, E., Ziv, A., Iakovleva, E., Palmgren, R., Berkowicz, R. (2003) Joint analysis of air pollution in street canyons in St. Petersburg and Copenhagen. (submitted to *Atmosperic Environment*)

Gifford, F. (1959) Statistical plume model. In: Atmospheric Diffusion and Air Pollution (Ed. F.N. Frenkiel, P.A. Sheppard), Acad. Press, NY, pp. 143 – 164 (translated into Russian in 1962).

Irwin, J.S. (1999) Effects of concentration fluctuations on statistical evaluations of centerline concentrations estimated by atmospheric dispersion models. Proceedings of 6th Conf. Harmonization Atmospheric Dispersion Models, p. 38 – 47

Irwin, J.S., Lee, R. (1997) Comparative evaluation of two air quality models: within-regime evaluation statistic. *Int. J. Environment. Pollution*, vol. 8, Nos 3-6, pp. 346-355

Nieuwsatdt, F.T.M., Van Dop, H. (Ed.) (1982) Atmospheric Turbulence and Air Pollution Modelling. D. Reidel Publ. (transated into Russian in 1985, 350 p.)

Obukhov, A.M. (1988) Weather and turbulence. In: Turbulence and Atmospheric Dynamics, Hydrometeorological Publishers, p. 281 – 288

Olesen, H.R. (1997) 'Data sets and protocol for model validation', *Int. J. Environment. Pollution*, vol. 5, Nos. 4-6, p. 693-701

Pugachev, V.S. (1979) Probability Theory and Mathematical Statistics. Nauka Publ., M., 495 p. (in Russian)

Shukla, J. (1985) Predictability. In: Advances in Geophysics, v. 28, Part B (Ed. S. Manabe), Academic Press, p. 96 – 130 (translated into Russian in 1988)

DISCUSSION

D. STEYN

I think all air pollution meteorologists will accept that a stochastic approach to air pollution concentrations is needed. I am concerned however that administrators and regulators will not appreciate this – they may not even understand what a stochastic variable is.

E. GENIKHOVICH

Fortunately, we did not have in Russia a long history of applications of the Gaussian model and, as a result, our administrators and regulators had no psychological problems with accepting such a stochastic approach. Speaking seriously, I understand your point but think that it is our responsibility to help the lawmakers, users etc. to switch to more adequate descriptions of the processes in question.

J. BARTNICKI

Do you assume purely deterministic emissions in your approach?

E. GENIKHOVICH

The stochastic variability of concentrations is governed by the turbulent properties of the atmosphere and by stochastic variations of the emissions. I was trying to demonstrate here that the high level of "atmospheric" stochasticity results in the siginificant noise. Certainly, variability of emissions should add a lot to the resulting noise. In such circumstances it is even more important to use dispersion models for calculating statistically stable characteristics (like upper percentiles) rather than concentrations at given meteorological conditions.

D. ANFOSSI

I have a comment: I agree that it is very important to consider the influence of natural stochasticity in the concentration measurements but I hope that this will not lead us to the conclusion that it is no more useful to try to improve our models.

E. GENIKHOVICH

No doubts, it is very important to work on improvement of our dispersion models and to make them more efficient in predicting the deterministic component of air pollution. My talk, however, was about taking into account the fact that any deterministic model cannot reproduce the stochastic component of air pollution, about concequenses of this fact and ways to handle them.

16

S.-E. GRYNING

You did not consider the effect of the height of the mixing layer (both in stable and unstable atmospheric conditions) on the critical wind speed nor in the discussion on accidental release. Why did you omit this effect?

E. GENIKHOVICH

For the sake of illustration, I referred to dispersion models that used meteorological input data from observations in the surface layer. The rest of the atmospheric boundary layer was parameterised in these models via governing parameters in use and it included parameterisations of the mixing height. In Eq. (16), in particular, its influence is represented by ψ.

RISK ASSESSMENT OF AIRBORNE SULPHUR SPECIES IN POLAND

Katarzyna Juda-Rezler

1. INTRODUCTION

Risk which airborne pollutants cause on the environment is always connected with amounts of the pollutants emitted as well as with environment sensitivity to the pollution. The emissions of air pollutants were substantially reduced during last fourteen years in Poland. However, Poland is still among the countries with the highest emissions of sulphur dioxide. The 1999 emissions amounted 1,719 million tonnes, which represented almost 10% of European emissions. Such significant amounts stem from the unique structure of the energy supply system, which, unlike elsewhere in the world, is largely dominated by coal combustion (65% of primary energy consumption). This results in quite high SO_2 and SO_4^{2-} concentrations in the air as well as high total sulphur deposition on ecosystems. Both sulphur species in the air and sulphur deposition may cause harmful adverse effects on human health and the environment.

In the paper an integrated assessment model ROSE (**R**isk **O**f airborne **S**ulphur species on the **E**nvironment), which facilitates complex evaluation of the sulphur species threat to the environment and human health in Poland, is presented.

The POLSOX-I numerical air pollution model was applied for the calculation of the current (1999 emission data) sulphur species concentration and total sulphur deposition fields in the area of Poland. Environment sensitivity to the airborne sulphur species was assessed by calculating *critical values* fields for the country. The following protective measures were applied: critical levels for SO_2 and SO_4^{2-} as well as conditional critical load for sulphur.

Maps of calculated ratios of current levels/loads to critical ones for Poland, are presented and discussed. The concept of *assessment thresholds* has been applied for the specification of the areas at risk.

Katarzyna Juda-Rezler, Institute of Environmental Engineering Systems, Warsaw University of Technology, Nowowiejska 20, 00-653 Warsaw, Poland.

Air Pollution Modeling and Its Application XVI, Edited by
Borrego and Incecik, Kluwer Academic/Plenum Publishers, New York, 2004

2. THE ROSE MODEL

The ROSE (Risk Of airborne Sulphur species on the Environment) model belongs to the family of Integrated Assessment Models (IAMs) which provide a framework for bringing together disparate information related to a particular environmental problem. IAMs were used to help formulate the protocols to the Geneva Convention on Long-range Transboundary Air Pollution - the 1994 so called *second sulphur protocol* as well as the 1999 so called *multi pollutant-multi-effect protocol* (UN-ECE, 2003). In both protocols reductions of sulphur emissions aim at protecting human health and the environment from adverse effects, in particular *acidifying effects*. Such an effect-based approach is used in RAINS model (e.g. Schöpp *et al.*, 1999), which was widely used in Europe and Asia. IAMs were also used to derive abatement strategies for the specific regions - e.g. for heavy polluted *Black Triangle* region of Europe (Loweles et al., 1998). For that region two IAMs were developed - one of them used critical loads as the ideal deposition that the model should try to attain, while the second one estimated a region's sensitivity to pollution based on critical levels of SO_2.

The ROSE model is an effect-based IAM developed for the area of Poland. In the model *both critical levels and critical loads* are used to indicate the state of required environmental protection. The aim of developing the model was to investigate all possible environmental impacts of sulphur emissions. These are invoked by:

1. Sulphur dioxide in the air, which directly affects people's health (mainly respiratory diseases) and which is also a phytotoxic pollutant; advances have been made in demonstrating the significance of very low concentrations on growth and yield and on changing plant sensitivity to other environmental stresses (WHO, 2000).
2. Particulate sulphate, formed in the air from sulphur dioxide and belonging to the secondary particles. It directly affects people's health, reduces visibility and is also a component of the acid mists, which have a direct impact on vegetation, especially in mountain regions.
3. Total sulphur deposition, which is one of the principal contributors to acidification, especially affecting aquatic ecosystems and forest soil and having indirect effects on vegetation and human health.

The objectives of each IAM are twofold. Primarily, the purpose of the "scenario analysis" mode is to assess the environmental and/or population risk of a given control strategy. Secondly, the optimization mode allows for environmental/population risk minimisation in optimal way. In the ROSE model the optimisation module, based on evolutionary computation techniques, is built in. This allows formulating multi-criterion goal function as well as non-linear constrains. The system generates scenarios for risk minimizations and seeks the optimal solutions.

However, the first step in such an integrated modelling is to investigate and analyse the current threat to the environment. The results of such investigations are presented herein.

In the beginning the state of the environment resulting from the pollution pressure have to be recognised. To this end the POLSOX-I, numerical two-dimensional Eulerian grid (K-theory) model, developed at the Warsaw University of Technology (WUT) was applied (see Juda-Rezler and Abert, 1997; Juda-Rezler, 2003). The model code combines transport, diffusion, chemistry and deposition processes (dry and wet) in the study area. For the POLSOX-I modelling purposes the so called outer grid, covering the region of 1050 km x 900 km including Poland and the neighbouring central European countries was adopted. Such a grid allows incorporation of emission sources from the Czech Republic and eastern part of Germany influencing the air quality in Poland. The proper ROSE model computational grid is smaller (900 km x 750 km). Both grids constitute a part of the EMEP model (EMEP, 2001) grid, only with finer spatial resolution (30 km x 30 km).

The calculations were performed for 1999 emission data. Annual mean values of SO_2 and SO_4^{2-} concentrations as well as annual total S deposition were obtained. The model estimates were verified by measurements. A very good agreement was obtained between observations and modelled values (Juda-Rezler, 2003).

Next the sensitivity of the receiving media (humans, vegetation, soil, water) to sulphur species had to be assessed for the area of Poland. Critical values concept was applied to this end. The methods used are described below.

Finally, calculated current concentrations and depositions were compared to the critical values. By subtracting the critical values from the present values the so called *critical values exceedances* were obtained, which are direct estimates of environmental and population risk.

3. CRITICAL VALUES DISTRIBUTION IN POLAND

3.1. The Methods of Calculations

The statistical model (ASPECT), based on cluster and regression analyses, was developed for calculating 5th percentile critical levels for SO_2 - 5%CLev(SO_2) - in the adopted grid covering Poland. The model concept, implementation and results have been presented by Juda-Rezler and Matuszewski (2003). Applied computational algorithm takes into account land-cover of separate grid cells (European Land Use Database of RIVM was used) as well as WHO recommendations of critical SO_2 levels for a given ecosystem type (WHO, 2000).

As the WHO annual SO_2 target value (50 µg/m3) is much bigger than critical levels for ecosystems (15-30 µg/m3), obtained grided critical levels for Poland protect also humans from direct adverse effects.

Mapping the sensitivity of humans or vegetation to particulate sulphate is more difficult. First of all neither WHO nor EU do not set any target or limit value for sulphate aerosol - which makes mapping protective measures for humans impossible at present. On the other hand it was recognised that mists can contain solute concentrations up to ten times those of rain, and can thus have a direct impact on vegetation (WHO, 2000). Since mists and clouds occur most frequently at high altitudes, and are intercepted with particular efficiency by forest, mountain forests are at risk of injury by acid mist. The author believes that, taking into account forest decline in Poland, setting critical level for sulphate is very important. As measurements or modelling of sulphate concentration in cloud water or mist are sparse, the critical values should be based on the equivalent sulphate concentration in air. Cape (1993) showed that critical levels for exposure of forest trees to cloud water would correspond to particulate sulphate concentration in the air in the range 1-3.3 µg (S) /m^3. Also WHO recommended a critical level of particulate sulphate as an annual mean for trees where ground level cloud is present 10% or more of the time. In the author's opinion, however, in the WHO guidelines a mistake occurs when presenting a guideline - 1 µg (SO$_4$) /m^3 instead of 1 µg (S) /m^3.

In the ROSE model the sensitivity of mountain trees to acid mist is incorporated by adopting critical sulphate level - CLev (SO$_4^{2-}$) according to Cape - 1µg (S)/m^3 (3µg (SO$_4$) /m^3) for the grids where such a vegetation exists.

Mapping critical loads for sulphur acting alone required cooperation with CCE/RIVM. Since the deposition of both sulphur and nitrogen can contribute to the acidification of an ecosystem, no unique acidity critical load can be defined, but the combinations of nitrogen and sulphur deposition not causing "harmful effects" lie on the so-called *critical load function* of the ecosystem defined by the three quantities: maximum critical load for sulphur - CL$_{max}$(S), maximum critical load fort nitrogen - CL$_{max}$(N) and minimum critical load for nitrogen - Cl$_{min}$(N). However, if one is interested in reductions of only one from two pollutants, a unique critical load can be derived which is consistent with the above formulation. If emission reduction deal with S only, a unique critical load of S for a fixed N deposition (N$_{dep}$) can be derived from the critical load function (Posch and Hettelingh, 1997). Thus conditional critical load of S - CL(S|N$_{dep}$) - is computed for a given year.

The 5-percentile conditional critical loads of sulphur for Poland in 1999 (5% CL(S|N99$_{dep}$) were calculated at RIVM (Max Posch, personal communications). The latest national data were used (Mill and Schlama, 2001) as well as EMEP calculations of N deposition for 1999. Calculations were performed for the "new" EMEP grid with 50 km x 50 km resolution and afterwards interpolated using GIS methodology to the ROSE computational grid.

3.1. The Results

The results obtained from calculations presented above are given in Figures 1 - 3. In all maps the areas with the smallest critical levels/loads are the most sensitive ones. Fig. 1 presents 5% critical SO$_2$ levels in the area of Poland. The most sensitive areas are situated mainly in south-eastern and north-western parts of Poland. The areas for which the critical level for SO$_4^{2-}$ were applied are shown in Fig. 2. The mountain forests - receptors being the most sensitive to acid mist - are situated in Poland on its southern boundary. Fig. 3 presents 5% conditional critical loads of S in Poland. The most sensitive areas are situated mainly in western and central-northern parts of the country.

Presented maps demonstrate clearly, that critical values distribution in Poland, and also vegetation sensitivity to different forms of sulphur pollution, vary essentially within the country.

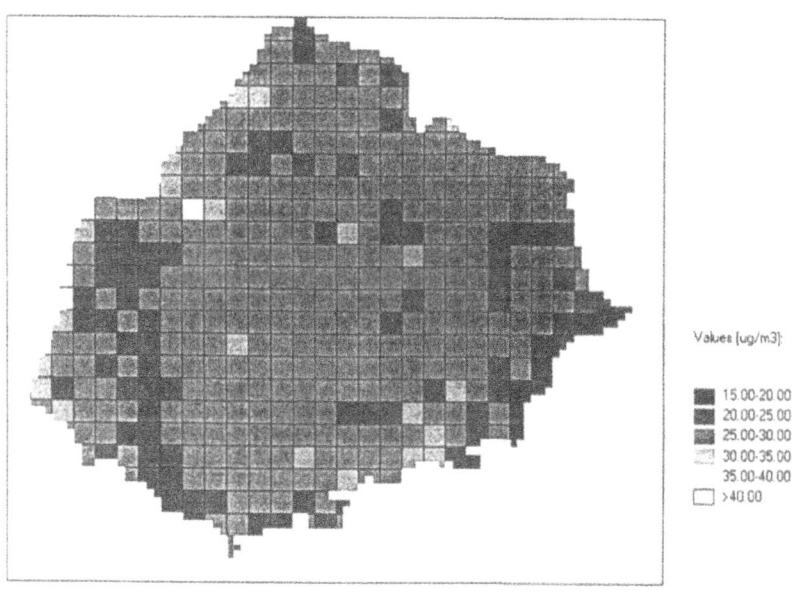

Figure 1. Calculated 5% critical levels for SO₂ in Poland .

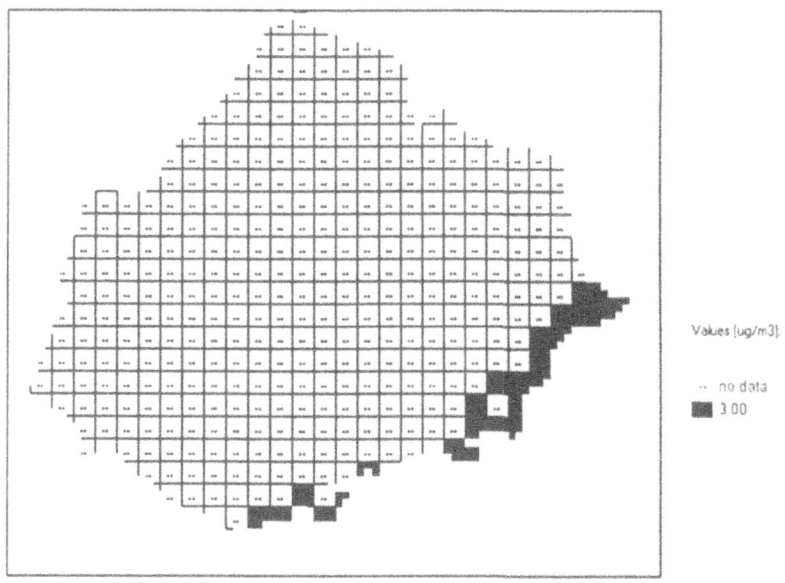

Figure 2. Areas in Poland for which the critical level for SO₄²⁻ was adopted.

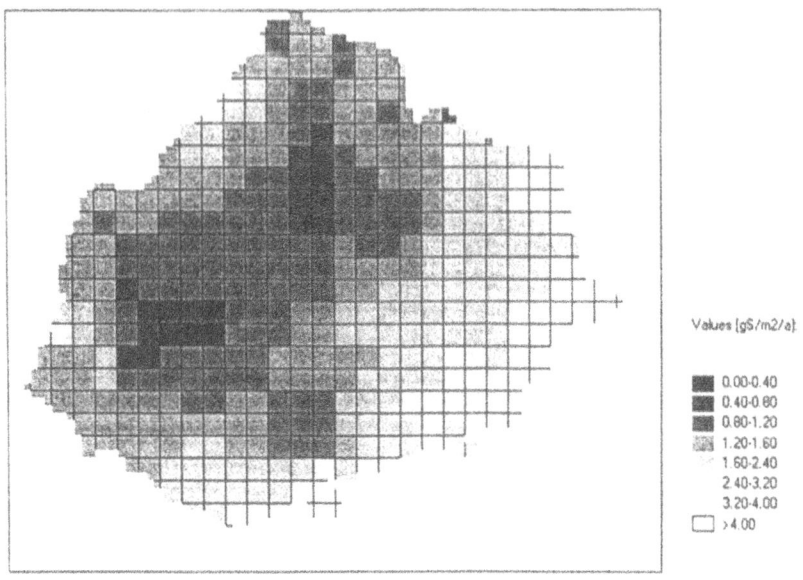

Values [gS/m2/a]:

- 0.00-0.40
- 0.40-0.80
- 0.80-1.20
- 1.20-1.60
- 1.60-2.40
- 2.40-3.20
- 3.20-4.00
- >4.00

Figure 3. Conditional 5% critical loads of S in Poland calculated for 1999 N deposition.

4. EXCEEDANCES OF THE CRITICAL VALUES

In order to assess current (1999 data) threat of sulphur species to the environment in Poland the *critical values exceedances* were calculated. Calculated exceedances are expressed in the units of a given pollutant and show the areas where critical values are exceeded and the areas where they are not exceeded. Such presentation has two disadvantages. First of all direct comparison of individual maps is difficult, secondly the areas with pollution levels very close to critical levels are not identified. To overcome above disadvantages the results presented herein are given in terms of *current-to-critical values ratios (CCR)*. Additionally the concept of *"lower" and "upper" assessment thresholds* used in EC Directives relating to limit values for air pollutants in ambient air (e.g. Council Directive 1999/30/EC) has been applied. According to that, the following division of the areas in the country is proposed:

1. The secure areas - where current level/load of pollution is below lower assessment threshold, i.e. below 40% of critical value; CCR≤0.4.
2. The areas at low risk - where current level/load of pollution is between lower and upper assessment threshold, i.e. between 40% and 60% of critical value; 0.4<CCR≤0.6.
3. The areas at increased risk - where current level/load of pollution is between upper assessment threshold and critical value, i.e. between 60% and 100% of critical value; 0.6<CCR≤1.0.
4. The areas at high risk - where current level/load of pollution is above critical value, i.e. areas where exceedances occurs; CCR>1.0.

The results obtained are presented in Figures 4-6 for SO_2, SO_4^{2-} and S, respectively.

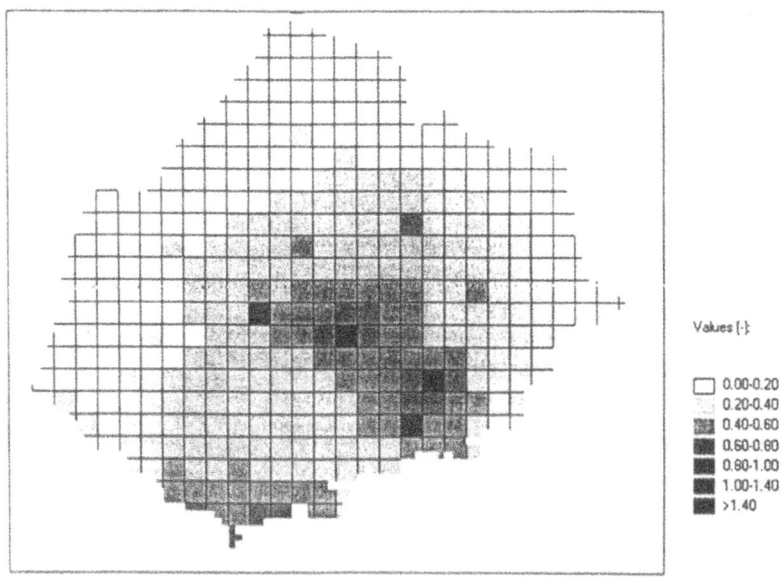

Figure 4. Ratios (CCR) of the current SO_2 concentrations to the critical SO_2 levels in Poland for 1999.

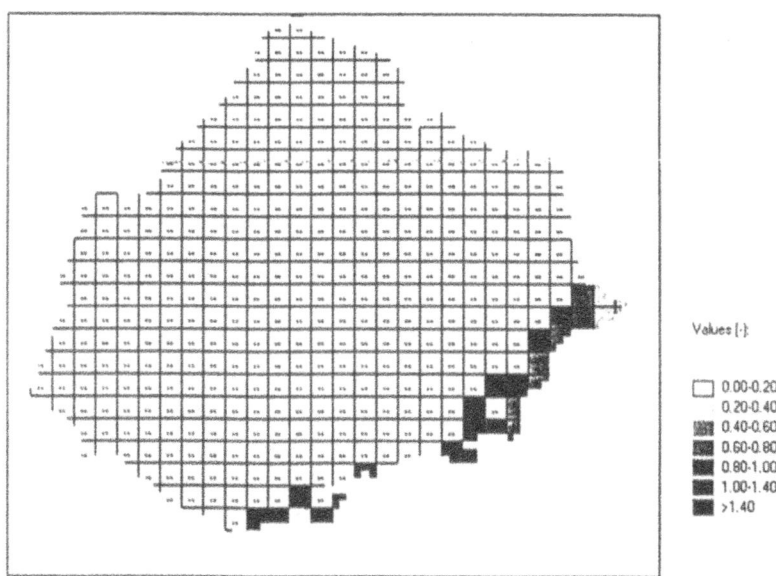

Figure 5. Ratios (CCR) of the current SO_4^{2-} concentrations to the critical SO_4^{2-} levels in Poland for 1999.

It can be observed that distribution of increased and high risk areas determined by the use of specific protective measures is varying for Polish territory. The southern regions are endangered mostly by acid mist, while central-southern parts (with Silesia industrial region) of the country as well as the Black Triangle region are still at increased risk of SO_2 pollution.

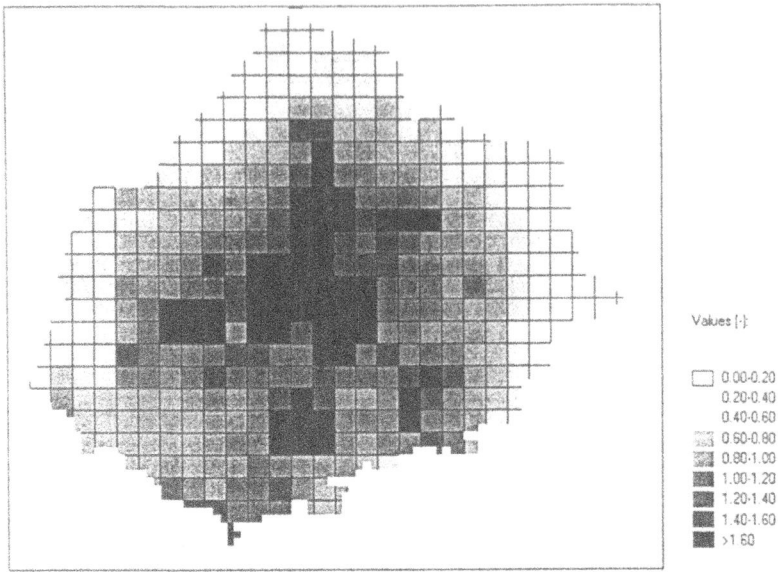

Values [-]:

☐ 0.00-0.20
0.20-0.40
0.40-0.60
0.60-0.80
0.80-1.00
1.00-1.20
1.20-1.40
1.40-1.60
>1.60

Figure 6. Ratios (CCR) of the current S depositions to the 5% conditional critical loads of S in Poland for 1999.

Central part of the country is endangered by sulphur acidifying deposition. The high risk area connected with acidification has the widest range. For some regions the exceedances are substantial.

5. CONCLUSIONS

The first results of ROSE Integrated Assessment Model developed for a regional scale have been presented. The critical levels for SO_2 and SO_4^{2-} as well as conditional critical loads of S for Poland were calculated and mapped in grid compatible with EMEP model grid. The risk to ecosystems was assessed by comparing these values with current (1999) national levels/loads data. The results demonstrate that area sensitivity to the respective sulphur species varies across Polish territory and that areas at risk invoked by particular pollutants are distributed differently. Some regions of Poland are still under substantial non-sustainable stress, which could lead to harmful effects on ecosystems.

The results simultaneously show that, although the high risk areas connected with acidifying effects have a wider range in Poland, still certain areas are at risk invoked by direct adverse effects of sulphur species. Therefore the scenarios for emission reduction prepared for the individual countries should be based on the complex evaluation of the pollutant threat to human health and the environment, as it was done for sulphur species in this work.

7. REFERENCES

Cape, J.N., 1993, Direct damage to vegetation caused by acid rain and polluted cloud: Definition of critical levels for forest trees, *Environmental Pollution*, **82**: 167.

EMEP, 2001, Summary Report 2001, *Transboundary acidification, eutrophication and ground level ozone in Europe - Summary Report 2001*, EMEP Report 1/2001, Oslo 2001.

Juda-Rezler, K., 2003, Current sulphur concentration levels in Poland - model estimates and comparison to observations, in: *Environmental Engineering Studies. Polish Research on the Way to the EU*, L. Pawłowski, M.R. Dudzińska, A. Pawłowski, eds., Plenum/Kluwer, N. York, in press.

Juda-Rezler, K. and Abert, K., 1997, Results from the POLSOX model, in: *Emission Abatement Strategies and the Environment (EASE) - Final Report*, H.M. ApSimon, ed., Imperial College of Science, Technology and Medicine, London, pp. 387-398.

Juda-Rezler, K. and Matuszewski, A., 2003, Critical levels of sulphur dioxide in Poland and their exceedances, in: *Environmental Engineering Studies. Polish Research on the Way to the EU*, L. Pawłowski, M.R. Dudzińska, A. Pawłowski, eds., Plenum/Kluwer, N. York, in press.

Lowles, I., ApSimon, H., Juda-Rezler, K., Abert, K., Brechler, J., Holpuch, J. and Grossinho, A., 1998, Integrated assessment models - tools for developing emission abatement strategies for the *Black Triangle* region, *Journal of Hazardous Material*, **61**:229.

Mill, W.A. and Schlama, A., 2001, National Focal Centre Reports - Poland, in: *Modelling and Mapping of Critical Thresholds in Europe. Status Report 1997*, M. Posch, P.A.M. de Smet, J.-P. Hettelingh, and R.J. Downing, eds., CCE/RIVM, Bilthoven, pp. 159-162.

Posch, M and Hettelingh, J.-P., 1997, Remarks on critical load calculations, in: *Calculation and Mapping of Critical Thresholds in Europe. Status Report 2001*, M. Posch, J.-P. Hettelingh, P.A.M. de Smet and R.J. Downing, eds., CCE/RIVM, Bilthoven, pp. 25-28.

Schöpp, W., Amann, M., Cofala, J., Heyes, Ch., Klimont, Z., 1999, Integrated assessment of European air pollution emission control strategies, *Environmental Modelling & Software*. **14(1)**:1.

UN-ECE, 2003,Geneva (January 7, 2003); http://www.unece.org/env/lrtap/status/lrtap_s.htm

WHO, 2000, *Air quality guidelines for Europe. Second Edition*, World Health Organization, Copenhagen, 273 pp.

DISCUSSION

V.P. ANEJA   To calculate the critical value how do you calculate the deposition velocity?

K. JUDA-REZLER   The deposition velocities were used for calculations of the current S deposition. The POLSOX-I numerical air pollution model was applied for the calculation of total sulphur deposition fields in the area of Poland. Simplified deposition submodel was connected with the main model. It calculates SO_2 dry deposition velocities, dry deposition velocity for SO_4^{2-} was kept constant. Variable (space and time dependent) SO_2 dry deposition velocities were calculated for the constant flux layer height, using resistance analogy formula and the Monin-Obukhov similarity theory for the surface layer. The Businger-Dyer integrated stability functions were applied. The basic $V_d(SO_2)$ referring to the height $z = 1$ m, equal to 0.7 cm/s was accepted.

E. GENIKHOVICH   Did you check applicability for the criteria introduced your model like $1 \ \mu g/m^3$ for SO_4^-?

K. JUDA-REZLER   The critical level for SO_4^{2-} equal to $1 \mu g \ (S)/m^3$ applied in my model is adopted from Cape (1993). Cape, based on experimental exposure of plants to simulated acidic rain, fog or mist, showed that critical levels for exposure of forest trees to cloud water would correspond to particulate sulphate concentration in the air in the range $1-3.3 \ \mu g \ (S)/m^3$. Also WHO recommended a critical level of particulate sulphate as an annual mean for trees where ground level cloud is present 10% or more of the time.

M. SOFIEV   Could you please compare your integrated assessment model ROSE with RAINS model from IIASA?

K. JUDA-REZLER   Yes. First of all, the ROSE model is a regional scale model developed for the area of Poland, while the RAINS model is a continental scale model (Europe, Asia). So, more detailed input data can be introduced to the ROSE model. Secondly, in the RAINS model risk of sulphur species on the environment is connected *only* with acidifying effects (critical loads exceedances). In the ROSE model *one pollutant-multi-effect* approach was employed; exceedances of both critical levels for

SO_2 and SO_4^{2-} as well as conditional critical load for sulphur are used as measures of sulphur emissions threat to the environment. Thirdly, the optimisation modules in both models are different. In the ROSE model, evolutionary computation techniques are employed. This allows formulating multi-criterion goal function as well as non-linear constrains.

APPLICATION OF MODELS-3/CMAQ IN AIR POLLUTION POLICY STRATEGIES DURING HIGH PARTICULATE EPISODES OVER EASTERN NORTH AMERICA

J. Sloan, R. Bloxam, A. Chtcherbakov, S. Wong, P.K. Misra, M. Pagowski, and X. Lin[*]

1. INTRODUCTION

Atmospheric models can provide information to assist in the development and evaluation of policies for reducing air pollution. Before applying models for assessing abatement strategies, the models need to be evaluated to demonstrate that they are credible. This study presents results of an evaluation of the US EPA air quality modelling system MODELS-3/CMAQ/MM5 during one winter and two summer high particulate matter episodes over eastern North America and some emission reduction sensitivity tests with implications for policy and abatement strategies.

The performance of the modelling system was evaluated during high particulate episodes, which occurred over eastern North America on July 9-17, 1995, February 5-13, 1998, and July 13-19, 1999. The Community Multiscale Air Quality (CMAQ) is a third generation chemical transport model for urban to regional scale simulation of ozone, acid deposition, and fine particulate matter. The meteorological driver for CMAQ is MM5 – the Pennsylvania State University/NCAR meso-meteorological model.

Emissions from US and Canadian inventories were processed for a modelling domain with 36 km resolution with 30 vertical levels used for model evaluation runs. The domain covers the eastern part of North America.

Modelling results for $PM_{2.5}$, PM_{10}, secondary aerosols and precursor gases (SO_2, NO_2) were compared with observed data derived from a number of networks of monitoring stations operated in the USA and Canada during the episodes. Model evaluation techniques included scatter plots, calculation of temporal and spatial correlation coefficients, bias, index of agreement, RMSE, gross error and visual analysis.

[*] J. Sloan, University of Waterloo, Waterloo, Canada, R. Bloxam, A. Chtcherbakov, S. Wong, P.K. Misra, Ontario Ministry of Environment, Toronto, Canada, M. Pagowski, Cooperative Institute for Research in the Atmosphere, Boulder, USA, X. Lin Kinectrics, Toronto, Canada

Air Pollution Modeling and Its Application XVI, Edited by
Borrego and Incecik, Kluwer Academic/Plenum Publishers, New York, 2004

This modelling system was used to explore the relative effects of reducing primary PM and precursor gas emissions on $PM_{2.5}$ and PM_{10}. The base case model runs and the emission reduction sensitivity tests were also assessed for the spatial extent of the concentration changes, seasonal differences in contributions to $PM_{2.5}$ and PM_{10} and the degree of non-linearity in the concentration changes.

2. DESCRIPTION OF EPISODES

The February 1998 episode extended over southern Ontario, Quebec and the north-eastern US. It was dominated by a high-pressure system under stagnant atmospheric conditions. During this episode the surface temperatures were above 0°C and no precipitation was observed. Circulation at 850 mb brought warm air northward increasing static stability in the lower troposphere, reducing mixing in the boundary layer. The episode ended with the passage of a cold front on February 13. The multi-day build-up of the stagnation episode began approximately on February 7 and continued through February 11.

From July 7 to July 10, 1995, the eastern part of the continent was dominated by an elongated trough, which slowly moved north-eastward towards the Atlantic with surface pressure rising in its wake. On July 12, a Bermuda High extended over eastern Canada and the US leading to conditions that are favourable for development of summer particulate episodes: weak winds, high temperatures, vertical descent, clear skies and low inversions. This high-pressure system persisted for the next 2-3 days until a cold front from Northern Ontario swept through the area bringing precipitation and clean, cool northern air.

For the July 1999 episode, an area of high pressure off the Atlantic coast extended inland over the mid-Atlantic states resulting in south-westerly winds through eastern Canada and the north-eastern U.S. states. Over the south and central U.S. states wind speeds were very light. Over eastern Canada and the northern U.S. states, light to moderate winds from the south west to west persisted from July 14 to 17. On July 18 and 19 a cold front advanced from the north west bringing much cleaner air to the region.

3. DESIGN OF MODEL RUNS

In this study we used MM5 as the meteorological driver, MEPPS (Models-3 Emission Processing and Projection System) for emission processing and CMAQ as the chemical transport model.

In MM5, the parameterization of the boundary layer was done using the Blackadar scheme, with the Kain-Fritsch scheme coupled with a mixed phase explicit scheme used for moist processes (*MM5 Modeling System Version 3, NCAR*, 2002). 36-hour forecasts were calculated for each day of the episodes with a spin-up period of 12 hours.

The Canadian emission inventory supplied with the Models-3 database was improved significantly for this study. The improvements included:

- New, annual average emission inventories for point, area and mobile sources for 1995 were put into MEPPS.

- Canadian area and point source temporal allocation factors were added to the temporal lookup table.
- Information on the distribution of crop and forest species in Canada was integrated with land use data to provide better input to the BEIS-2 model.
- For eastern Canada, MOBILE5a input files were replaced with province specific 1995 data.
- Population and dwelling data by census enumeration area were put into MEPPS to replace the much coarser census division data originally in the database.

CMAQ's set-up used mostly the default modules along with the RADM-2 chemical mechanism including aerosol and aqueous chemistry (*D.W. Byun and J.K.S. Ching*, 1999). Observed data were used to set up time dependent inflow boundary conditions.

4. MODEL EVALUATION

The results of CMAQ runs were evaluated against extensive observed data, derived from several US and Canadian networks. Evaluation of the model was performed in a reduced domain starting 3 cells (108 km) inside each boundary of the model domain. The model was run with a 2-day "warm up" period, during which the results were not analyzed.
Detailed description of model evaluation (1995 and 1998 episodes) could be found in (*An. Chtcherbakov et. al.*, 2002).

February 1998 evaluation

Ammonia emission rates have no seasonal variability in MEPPS. This causes a significant overestimation in February since manure spreading, the largest source of ammonia emissions, is not occurring in north-eastern North America at that time. This is very important for the winter episode, which is dominated by formation of ammonium nitrate. A model run with emissions of NH_3 reduced by 50 % shows improved comparisons with observed data. Fig. 1 shows scatter plots for NO_3 and NH_4 for the base case NH_3 emissions and half NH_3 emissions. In both cases NO_3 is high, but the magnitude is much better with half ammonia emissions. NH_4 corresponds well with observations.

PM_{10} is slightly low, and PM_{25} is high by comparison with the observations. Modelled sulphate concentrations correlate with observed data but are somewhat lower than the measurements. Scatter plots for SO_2 and NO_2 give good correlations between modelled and observed concentrations, but the modelled SO_2 is high by about 50%. Modelled NO_2 concentrations are highly correlated with measured data and show very little bias.

The time series correlation coefficients for sulphate and ammonium are between 0.5 and 0.9 at a large majority of the sites. Sites in southern Ontario had time correlation coefficients greater than 0.8 with sites in Quebec having values up to 0.75. The area of maximum correlation for ammonia is located just to the north of Lake Huron. The magnitude of the correlation smoothly decreases in the east direction.

Very good correspondence can be seen between CASTNET weekly observations and modelled data both in the pattern of the contours and the magnitude for NO_3, NH_4 and

SO$_2$. Areas with maximum concentrations have basically the same locations. For the SO$_2$ concentration field, the modelled concentrations are somewhat high. The modelled maximum concentrations for NO$_3$ and NH$_4$ are shifted to the east against observed data.

July 1995 evaluation

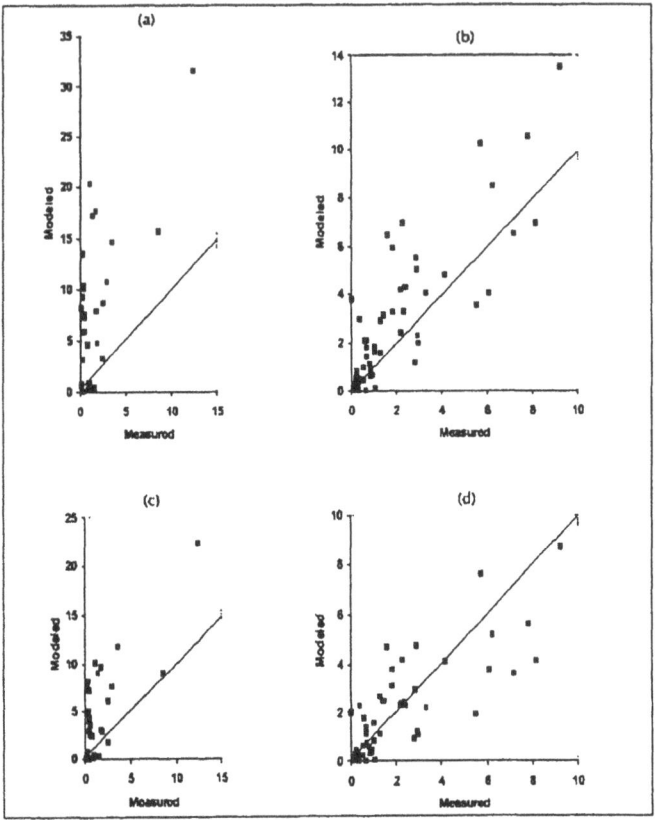

Figure 1. Scatter plots of modelled vs. observed surface concentrations (μg/m^3) for February 5-13, 1998: a) NO$_3$ and b) NH$_4$ with base case emissions and c) NO$_3$ and d) NH$_4$ with NH$_3$ emission rates reduced by 50%

For this episode, the modelled sulphate concentrations were strongly correlated with observed data and showed very little bias. PM$_{10}$, PM$_{2.5}$ and ammonium were also strongly correlated with observed data, but the modelled results were slightly lower than the observed data on average. Modelled nitrate was higher then observed data however a number of the nitrate measurements could be biased low because of volatilization losses. Similar analyses of some of the precursor gases show nitric acid predictions to be strongly correlated to measured data but biased high relative to the data. Measured and modelled concentrations of SO$_2$ and NO$_2$ were correlated with modelled SO$_2$ results tending to be biased slightly high and NO$_2$ biased slightly low.

Modelled time series for sulphate and ammonium were strongly correlated with observed data at a large majority of the sites (correlation coefficients are more then 0.75).

Both species demonstrate the same spatial pattern with maximum correlations in Quebec and a gradual decrease in the west direction.

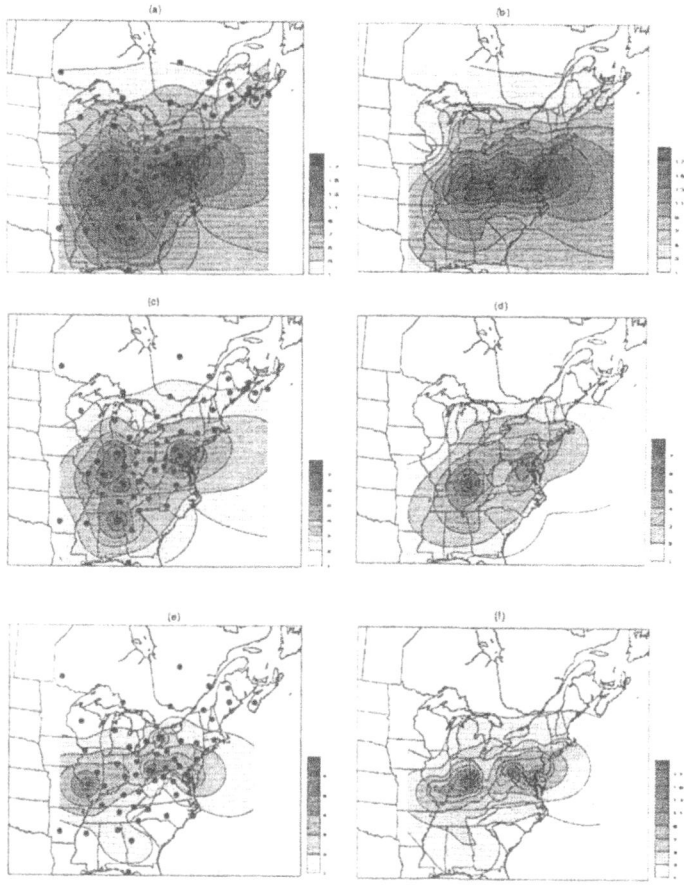

Figure 2. Measured (a, c, e) and modelled (b, d, f) concentrations for SO_4 (a, b), NH_4 (c, d) and SO_2 (e, f) averaged over the period from 13:00 GMT July 11, 1995 to 22:00 GMT July 17, 1995

The comparison of CASTNET data (weekly observations) versus averaged modelled data for SO_4, NH_4 and SO_2 (see fig. 2) shows an almost exact match in the locations of maxima for all species, with SO_4 and NH_4 matching measured values. The modelled SO_2 concentrations are an order of magnitude high, but the spatial correlation is excellent.

July 1999 evaluation

The July 1999 episode shows almost the same features as the July 1995 episode. PM concentrations are dominated by sulphate formation; there is very good correspondence between modelled and observed SO_4, PM_{25}, and PM_{10}; and the time series correlation coefficients for sulphate are very high.

Fig. 3 shows the spatial distribution of the weekly averaged SO_4 modelled and observed concentrations. As it was in the 1995 episode there is almost exact match in the maximum locations and their values, which stay the same and are in the range 15-16 $\mu g/m^3$.

In general, the model demonstrates reasonably accurate predictions for particulate matter during the episodes, but some secondary aerosols and precursor gases were either higher or lower than the measured data (up to 50% for SO_2). The time correlation coefficients for most species were very high and the model also captures the spatial distributions of the surface concentrations, the timing of the evolution of the episodes, and the chemical composition of the aerosols.

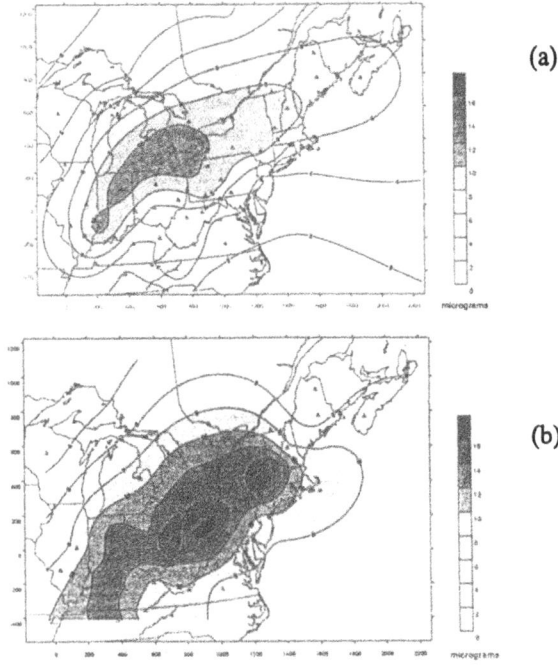

Figure 3. Observed (a) and modelled (b) concentrations for SO_4 averaged over the period from July 13 to July 19, 1999

5. IMPACT OF DOMAIN WIDE REDUCTIONS IN PRIMARY PM AND PRECURSOR GAS EMISSIONS ON PM CONCENTRATIONS

For the February 1998 and the July 1995 episodes, the domain wide anthropogenic emissions of primary particulate matter, SO_2, NOx and VOCs were reduced by 50%. The effects of these reductions on $PM_{2.5}$ are described here. Some factors which could affect policies and abatement strategies are the seasonal contribution of each emitted species to the $PM_{2.5}$ mass, the spatial extent of the concentration changes and possible non-linear responses to emission reductions.

The modelled base case compositions of $PM_{2.5}$ are very different in the winter and summer. On days having high concentration, the February 1998 episode showed a large contribution from ammonium nitrate (Fig. 1) while the summer episode had high sulphate concentrations (Fig. 2 and 3). The percentage of the modelled $PM_{2.5}$ due to emission of primary PM was larger in the winter episode. Primary PM impacts were much higher in urban centres than in rural areas for both the summer and winter. Secondary organic aerosols (SOAs) due to anthropogenic VOC emissions were a small contributor to modelled $PM_{2.5}$ concentrations on both summer and winter days during the episodes.

Fig. 4 and 5 show the sulphate decreases resulting from a 50% reduction in SO_2 emissions. Sulphate concentrations were reduced by 40 to 50% over most of the domain on July 14th. The fixed concentrations at the inflow boundaries result in much smaller percentage reductions in these areas.

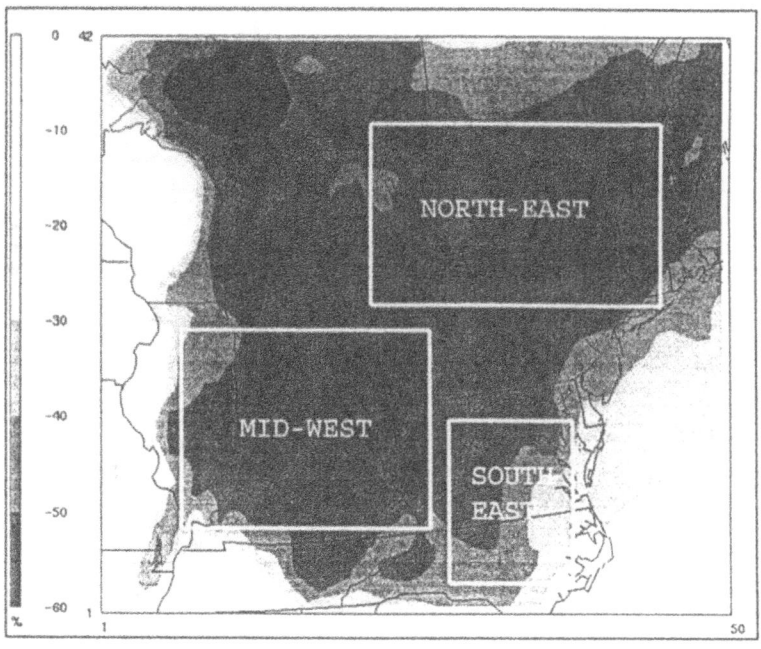

Figure 4. Subdomains and percent changes of SO_4 concentrations due to a 50% reduction of SO_2 emissions, July 14, 1995, 3:00 GMT

Fig. 5 presents time series of percentage changes in sulphate for 3 subdomains where

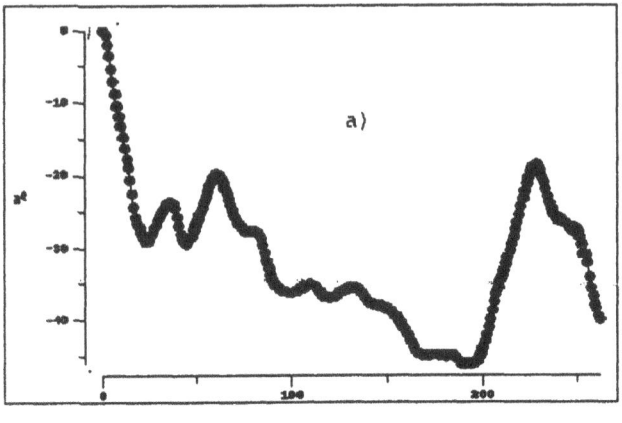

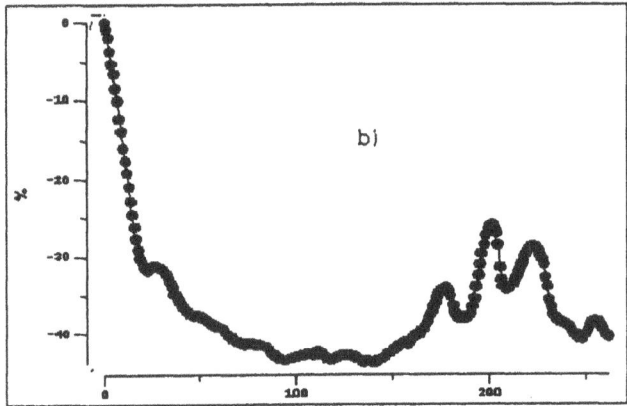

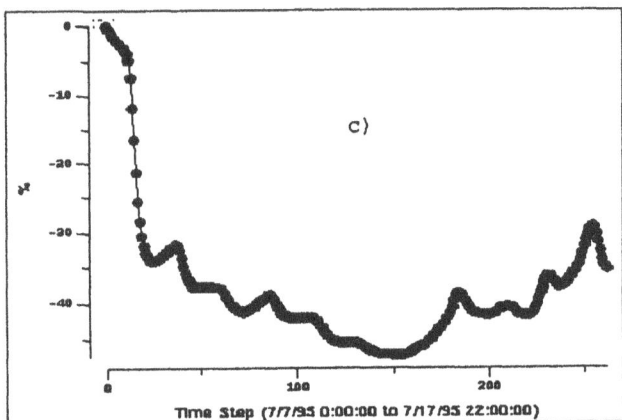

Figure 5. Time series of percent changes of SO4 concentrations averaged over sub-domains for July 7-17, 1995, as a result of 50 % domain-wide cut off in SO₂ emissions: a) north-east, b) mid-west, c) south-east

the first 2 days are model warm up. The modelled response to the emission reductions was nearly linear except for cleaner days in the midwest and northeast where the sulphate reductions were in the 25 to 40% range. Nitrate concentrations increased when SO_2 emissions were reduced since more ammonia was available to form ammonium nitrate. Although the percentage increases were very high on episode days, the base case concentrations were generally low on those days.

Reducing NOx emissions by about 50% for the summer 1995 episode gave nearly linear decreases in nitrate on cleaner days but the concentrations were reduced by only 25 to 40%. An additional modelled benefit was a reduction of 5 to 20% in sulphate concentrations. These sulphate reductions could be due to reductions in oxidant concentrations. On cleaner days, south and central U.S. states showed smaller decreases in sulphate while the model predicted increases in sulphate concentrations for eastern Canada and the northeast U.S. states.

In both the summer and winter, reductions of 50% in primary particulate matter emissions gave a reduction of 50% in modelled primary PM concentrations. The effects of primary PM emissions were very localized.

Decreasing anthropogenic VOC emissions by about 50% resulted in a nearly linear decrease in SOA concentrations for both seasons. Reductions in VOC emissions also affected sulphate and nitrate concentrations for the winter episode. Decreases in nitrate concentrations varied across the domain with most regions showing 1 to 3% decreases but some sub-regions had up to 10% decreases in nitrate. Sulphate concentrations also decreased with the largest reductions reaching 10 to 20%.

The February 1998 episode was dominated by very stagnant conditions for most of the period; transport began only near the end of the episode. The highest modelled sulphate concentrations also occurred near the end of the episode. The modelled response to a 50% reduction in SO_2 emissions was non-linear. Sulphate concentrations decreased by 10 to 30% in the northern part of the domain with 20 to 30% reductions in the south. Nitrate concentrations generally increased. The increases in nitrate were up to 20% but varied across the domain and over time. In the north-eastern U.S. and eastern Canada nitrate concentrations increased by 0 to 5%.

Decreasing NOx emissions by about 50% gave very non-linear reductions in nitrate. In the northern portion of the domain, nitrate was reduced by 10 to 15% on cleaner days and up to 20% during the high PM episodes. For the southern states, nitrate was reduced by 10 to 15% on cleaner days and 15 to 30% on days with higher nitrate concentrations. On days with high PM, the NOx emission reduction resulted in 40 to 60% increases in modelled sulphate concentrations. This further decreased the impact of NOx emission reductions on $PM_{2.5}$ concentrations.

6. CONCLUSIONS

Reducing SO_2 would result in the largest and most wide-spread reductions in $PM_{2.5}$ concentrations during summer episodes of high PM. Model runs with reduced NOx emissions resulted in lower $PM_{2.5}$ in the summer but more reductions were found for the winter episode. The response of $PM_{2.5}$ to either NOx or SO_2 cuts on high concentration winter days was very non-linear with modelled increases in sulphate or nitrate concentrations, respectively, partially offsetting reductions in $PM_{2.5}$.

Reducing primary $PM_{2.5}$ emissions had much larger effects in urban centres than in rural areas, with the modelled reduction higher in the winter than in the summer. Although modelled SOA concentrations due to anthropogenic VOC emissions were small, some reductions in winter $PM_{2.5}$ were found due to decreases in nitrate and sulphate concentrations.

7. REFERENCES

Byun, D.W. and Ching, J.K.S. 1999, Science Algorithms of the EPA Models-3 Community Multiscale Air Quality (CMAQ) Modeling System. EPA/600/R-99/030.

PSU/NCAR Mesoscale Modeling System Tutorial Class Notes and Users' Guide: *MM5 Modeling System Version 3, NCAR*, 2002.

Chtcherbakov, An., Pagowski, M., Sloan, J., Soldatenko, S., Lin, Xuide, Bloxam, R., Wong, S., and Misra P.K., Models-3/CMAQ Evaluation During High Particulate Episodes Over Eastern North America in Summer 1995 and Winter 1998, in: *Proceedings of Eighth International Conference on Harmonization Within Atmospheric Dispersion Modelling For Regulatory Purposes, Sofia, Bulgaria*, October 14-17, 2002, pp. 345-349.

DISCUSSION

S.A.-AKSOYOGLU When you say that the agreement between model and measurements is good or bad, how do you define good, not bad, etc...

J. SLOAN To carry out this study, it was first necessary to replace the approximate Canadian emissions inventories contained in the original release of the *Models-3* system with correct values. Then we carried out a series of model runs to assess the performance of the model by comparison with results from several observational networks. Thus it is important to remember that the uncertainties in the observational results must also be considered in the evaluation procedure.

For our detailed analyses of the results, the techniques included scatter plots, temporal and spatial correlation coefficients, bias, index of agreement, RMSE, gross error and visual analysis by mapping. The numerically-based comparisons were used in our detailed "in-house" evaluation of the results, which were provided to the policy assessment group at the Ontario Ministry of the environment, along with the results of other studies not published here. For the purpose of this presentation, we used visual comparisons *via* contour maps because these are the most rapid and effective method to convey the overall quality of the agreement between the modelled and measured values for the base case studies we have reported here. The designations of "good" and "bad" agreement are simply meant to convey our opinion of the qualitative agreement presented visually in the maps.

R.BORNSTEIN Do you think the results of your Figure 2 could be due to the K-theory formulation in MM5, which may have under estimated horizontal diffusion and thus has produced a too large SO_2 concentration and (2) a too narrow real extent of the regional plume, while conserving SO_2 mass?

J. SLOAN This discrepancy occurred in several different runs. We originally thought that the cause of the narrow SO_2 spatial distributions was related to advection, but closer inspection showed that this is not the case. The sources of the SO_2 are very tall stacks, which also emit CO and NO. The advection of the CO and NO is normal, so we conclude that transport is not the cause of the disagreement between the modelled and measured SO_2. Thus it appears that this difference has a chemical or microphysical cause, possibly related to cloud scavenging or cloud chemistry.

O.SELAND   Do you treat scavenging?

J. SLOAN   Yes. The scavenging scheme is discussed in Chapter 11 of the Models-3 Science Documentation, which can be found at http://www.epa.gov/asmdnerl/models3/doc/science/ch11.pdf.
Briefly, scavenging is calculated by two methods, depending upon whether the pollutant participates in the cloud chemistry and on the liquid water content. (1) For those pollutants that are absorbed into the cloud water and participate in the cloud chemistry (and provided that the liquid water content is > 0.01 g/m3), the amount of scavenging depends on Henry's law constants, dissociation constants, and cloud water pH. (2) For pollutants which do not participate in aqueous chemistry (or for all water-soluble pollutants when the liquid water content is below 0.01 g/m3), the model uses the Henry's Law equilibrium equation to calculate ending concentration and deposition amounts. The detailed equations for these processes and also for related wet deposition processes can be found at the above website.

V.ANEJA   How good is the ammonia emission inventory during winter and summer in Canada?

J. SLOAN   The ammonia emissions inventories are not well known in either Canada or the US. Efforts are currently being made to address this problem, but they still remain poorly documented in both jurisdictions. A more serious problem, and the one responsible for the discrepancy mentioned in the presentation, is the fact that the model has no variation in the ammonia temporal factors. While this might be reasonable for temperate climates, it is inappropriate for cold climates like that of Canada, where ammonia emissions from agricultural activities are high in the summer and very low in the winter.

D.MICHELANGELI 1. How did you treat the size distribution for the emissions of $PM_{2.5}$ and PM_{10}?

2. Did you do any other comparisons to observations on different time scales (daily or hourly)

J. SLOAN

1. The aerosol component of the CMAQ is derived from the Regional Particulate Model (RPM) which in, turn, is based upon the paradigm of the Regional Acid Deposition Model (RADM). The particles are divided into two groups – fine particles and coarse particles. These groups generally have separate source mechanisms and chemical characteristics. The fine particles result from combustion processes and chemical production of material that then condenses upon existing particles or forms new particles by nucleation. The coarse group is composed of material such as wind-blown dust and marine particles (sea salt). The anthropogenic component of the coarse particles is most often identified with industrial processes.

The approach chosen for implementation in CMAQ is to model the particles as a superposition of lognormal sub-distributions called modes (See Whitby, K. T., The physical characteristics of sulfur aerosols, *Atmos. Environ.*, 12, 135-159, 1978). The modes are PM2.5 (particles with diameters less than 2.5 μm) and PM10 (particles with diameters less than 10 μm). Note that PM10 includes PM2.5. Thus coarse particles are those with diameters between 2.5 and 10 μm. Then, the mass of the coarse particles is the difference between the masses in PM10 and PM2.5.

To model the aerosol size distribution, PM2.5 is treated as two interacting sub-distributions or modes. The coarse particles form a third mode. Within the fine group, the smaller (nuclei or Aitken) mode represents fresh particles either from nucleation or from direct emission, while the larger (accumulation) mode represents aged particles. Primary emissions may also be distributed between these two modes. The two modes interact with each other through coagulation. Each mode may grow through condensation of gaseous precursors; each mode is subject to wet and dry deposition. Finally, the smaller mode may grow into the larger mode and partially merge with it. The chemical species treated in the aerosol component are fine species, sulfates, nitrates, ammonium, water, anthropogenic and biogenic organic carbon, elemental carbon, and other unspecified material of anthropogenic origin. The coarse-mode species include sea salt, wind-blown dust, and other unspecified material of anthropogenic origin. Because atmospheric transparency or visual range is an important air quality related value, the aerosol component also calculates estimates of visual range and aerosol extinction coefficient.

2. Yes. The CMAQ modeled results were compared with all available observed data, which had a variety of different timescales (hourly and daily as well as the weekly comparisons discussed in the presentation). For example we used AIRS (Aerometric Information Retrieval System) network data (http://www.epa.gov/airs/), which contains hourly and daily-observed datasets. The detailed description of the comparison of the CMAQ modeling results against the observed data recorded by several networks can be found in An. Chtcherbakov, M. Pagowski, J. Sloan, S. Soldatenko, X. Lin, R. Bloxam, S. Wong and P.K. Misra, "Models-3/CMAQ Evaluation During High Particulate Episodes Over Eastern North America in Summer 1995 and Winter 1998" *Proceedings of Eighth International Conference on Harmonization Within Atmospheric Dispersion Modelling For Regulatory Purposes*, Sofia, Bulgaria, 14-17 October 2002, p.345-349

INTEGRATED REGIONAL MODELLING

Chairperson: M. Schatzmann

J. Baldasano

N. Chaumerliac

S.-E. Gryning

Rapporteur: M. Kaasik

E. Batchvarova

P. Suppan

A. Ryaboshapko

MODELLING OF PARTICULATE AIR POLLUTION IN NORTHERN ITALY

S. Andreani-Aksoyoglu[1], J. Keller[1], A.S.H. Prévôt[1], J. Dommen[1],
U. Baltensperger[1], J.P. Putaud[2]

1. INTRODUCTION

The northern part of the Po basin is one of the densely populated and strongly industrialized areas in Europe. The PIPAPO experiment (Pianura Padana Produzione di Ozono) was designed to investigate the photooxidant production in the Milan metropolitan region of Italy in summer 1998, within the frame of the EUROTRAC-2 project LOOP. Photooxidant production and limitation of ozone formation during two intensive observation periods (IOP1: May 11-14, IOP2: June 1-10) were studied using numerous measurements and models (Dommen et al., 2002; Neftel et al., 2002; Ritter et al., 2002; Martilli et al., 2002). However, modelling studies were mainly focused on the gas-phase chemistry. In this work, the Eulerian photochemical model CAMx was applied for the first time including the aerosol chemistry. The main objective is to investigate the capability of the model to predict the formation and spatial and temporal distribution of secondary organic and inorganic aerosols.

2. MODELLING

The Comprehensive Air Quality Model with extensions (CAMx) is an Eulerian photochemical dispersion model that allows for integrated assessment of gaseous and particle air pollution over many scales (Environ, 2002). The chemical mechanism used to invoke the aerosol chemistry is a version of CBM-IV which includes condensible organic gas species and a second olefin species to account for the biogenic olefins. The CAMx aerosol chemistry routine is a hybrid of the linear aqueous sulfate chemistry and parameterized

[1] Laboratory of Atmospheric Chemistry, Paul Scherrer Institut, CH-5232 Villigen PSI, Switzerland
[2] Joint Research Centre, Environmental Institute, European Commission, I-21020 Ispra, Italy

Air Pollution Modeling and Its Application XVI, Edited by
Borrego and Incecik, Kluwer Academic/Plenum Publishers, New York, 2004

47

sulfate/nitrate/ammonium equilibrium. In this mechanism, first of all the photochemistry calculates the transformation of SO_2 to sulfate via the homogeneous gas-phase reaction, the production of nitric acid and the production of condensible organic carbon and then they are supplied to the aerosol routine. The equilibrium between sulfate, nitrate and ammonium is handled in two modes; ammonia-lean conditions and ammonia-rich conditions (Pilinis and Seinfeld, 1987). The aerosol species calculated by CAMx include sulfate, nitrate, ammonium, organic carbon, sodium, chlorine, and primary inert PM. No particle size distribution is currently modeled, however the user can choose a representative size range and density for each aerosol species. In this study, the particle size range for these secondary aerosol species was chosen as 0.04 - 2.5 μm, as indicated by experimental results (Baltensperger, 2001).

CAMx requires inputs to describe photochemical conditions, surface characteristics, initial and boundary conditions, emission rates and various meteorological fields over the entire modelling domain. Information about the preparation of the input parameters and the meteorological model SAIMM is described in detail in Ritter et al. (2002). In addition to the basic emissions, NH_3 and monoterpene emissions were included in this study, in order to calculate the secondary aerosols. The procedure to obtain NH_3 emissions is given by Martilli et al. (2002). The model domain consists of 47x54 grid cells with a resolution of 3km x 3km. It covers the Lombardy region and some part of Southern Switzerland (Figure 1). There are 8 layers in a terrain-following coordinate system, the first being 50 m above ground. The model top is set at 3000 m above ground. Photolysis rates are derived for each grid cell assuming clear sky conditions as function of five parameters: solar zenith angle, altitude, total ozone column, surface albedo and atmospheric turbidity. Simulations started on May 11 at 1200 central European summer time (CEST) and ended on May 13 at 2400 CEST. The first 12 hours were used as initialization of the model.

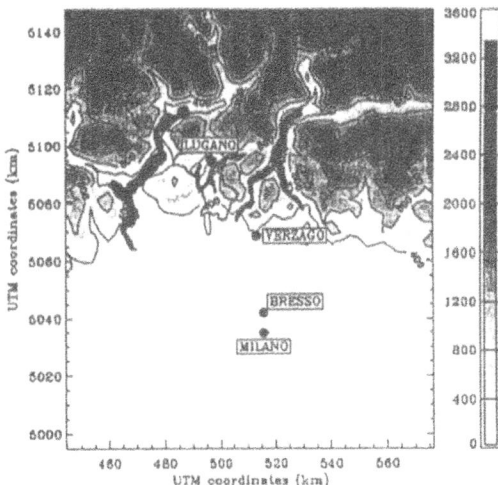

Figure 1 The topography of the model domain (masl).

3. RESULTS

3.1. Gaseous Species

The photochemically active plume formed near Milan moved to the north and ozone mixing ratio reached about 195 ppb at 1500 CEST near Verzago which is about 30 km north of Milan. Ozone mixing ratios outside the plume were between 100 and 120 ppb. Figure 2 shows the comparison of model results with ground-level measurements during May 12-13 at the semi-rural site Verzago. The predicted ozone peak on May 13 is delayed by about one hour. The model reproduced the mixing ratios in the afternoon well except for SO_2 both for Verzago (Figure 2) and for the urban site Bresso (not shown). Sensitivity tests indicated that the SO_2 emissions might be overestimated.

3.2. Aerosol Species

Inorganic aerosol components were measured on-line during May 12-13 at Verzago and are shown in Figure 3 together with the model predictions. At the urban site Bresso, the measurements are also available except for NH_4^+ (Figure 4). The model predicted the nitrate concentrations very well especially on May 13 at both urban and semi-rural sites. Calculated sulfate and ammonium concentrations are comparable to measured values. The correlations between measured NH_4^+ and $NO_3^- + 2SO_4^{2-}$ suggest that enough NH_3 was available to neutralize the aerosol to yield particulate NH_4NO_3 and $(NH_4)_2SO_4$ (Putaud et al., 2002).

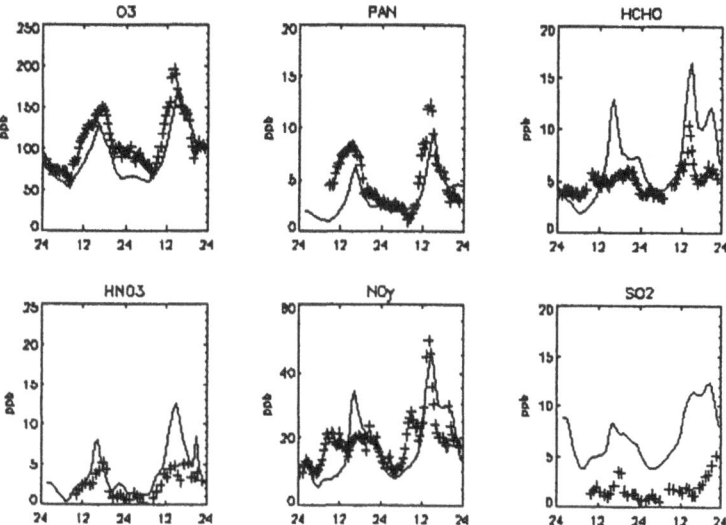

Figure 2 Diurnal variation of measured (+) and predicted (−) mixing ratios (ppb) for gaseous species during May 12-13 at the semi-rural site Verzago.

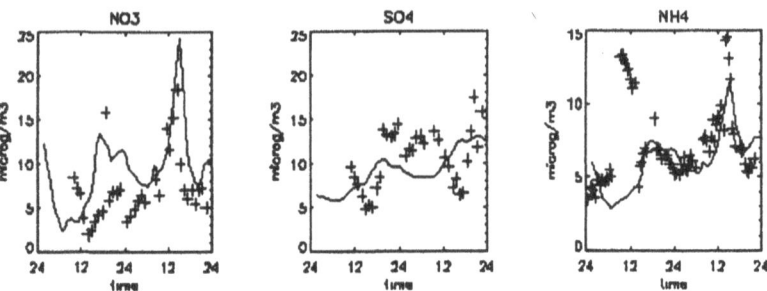

Figure 3 Diurnal variation of measured (+) and predicted (−) concentrations of particulate NO_3^-, SO_4^{2-} and NH_4^+ (µg m^{-3}) during May 12-13 at the semi-rural site Verzago.

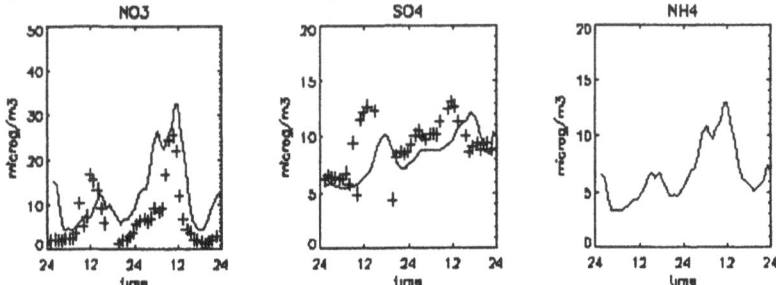

Figure 4 Diurnal variation of measured (+) and predicted (−) concentrations of particulate NO_3^-, SO_4^{2-} and NH_4^+ (µg m^{-3}) during May 12-13 at the urban site Bresso.

Baltensperger et al. (2002) showed that the measured maximum nitrate/sulfate ratios occur at 1100 and 1530 CEST in Bresso and Verzago, respectively. The peak nitrate/sulfate ratio in Verzago coincides with the highest O_3 mixing ratio due to the arrival of the Milan plume in the afternoon. The diurnal variation of predicted nitrate/sulfate mixing ratio by CAMx in Bresso and Verzago is shown in Figure 5. The model reproduced the same feature of a delayed peak of the NO_3/SO_4 ratio in Verzago which occurs at the same time of peak O_3.

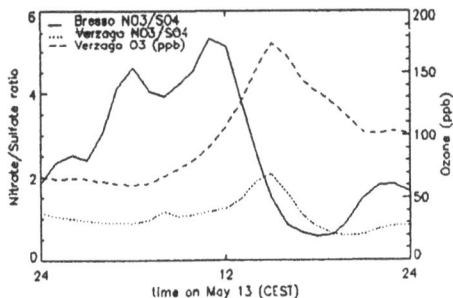

Figure 5 Diurnal variation of the predicted hourly nitrate/sulfate mixing ratio in Bresso and Verzago on May 13 1998. The peak of the nitrate/sulfate ratio in Verzago is delayed compared to Bresso and coincides with the highest ozone mixing ratio at this site.

50

There were a few measurements of total organic carbon (TOC) and black carbon (BC) at Bresso and Verzago. The measured TOC may contain both primary (POC) and secondary (SOC) organic carbon while model results refer only to SOC due to lack of emission data. Therefore, a direct comparison of model SOC results with the measured TOC concentrations cannot be done. However, BC, which is solely primary, can be used as a tracer of POC assuming that BC and POC have the same sources and that there is a constant ratio of POC/BC for the primary aerosol. In Figure 6, the relationship between the TOC and BC measured in Bresso during the measurement campaign is shown. The minimum value of TOC/BC ratio may be used as ratio for the primary aerosol (Castro et al., 1999). Using this ratio derived from Figure 6 and the measured TOC and BC concentrations, the amount of SOC in both sites was estimated (Table 1). SOC concentrations predicted by CAMx at both sites (Figure 7) are in the same range as the values estimated from the measurements. The concentrations at Verzago are higher and the maximum levels are reached later than at Bresso. Considering the high uncertainties (about 50 %) in the measurements (Putaud et al., 2002), estimation of the primary fraction of the measured TOC and unknown uncertainties in the biogenic emissions which play an important role in the SOC formation, the agreement between the measured and predicted values can be considered as satisfactory. The total secondary particle mass concentration (i.e. the sum of NO_3^-, SO_4^{2-}, NH_4^+, SOC) is positively correlated with secondary pollutants such as ozone and formaldehyde (Figure 8).

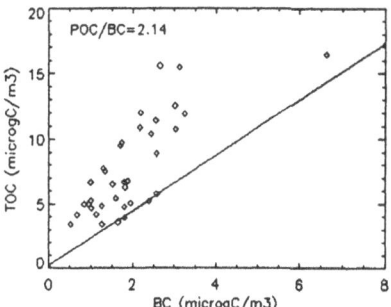

Figure 6 The relationship between black carbon and total organic carbon measured in Bresso. Minimum value of TOC/BC ratio is assumed to represent POC/BC ratio.

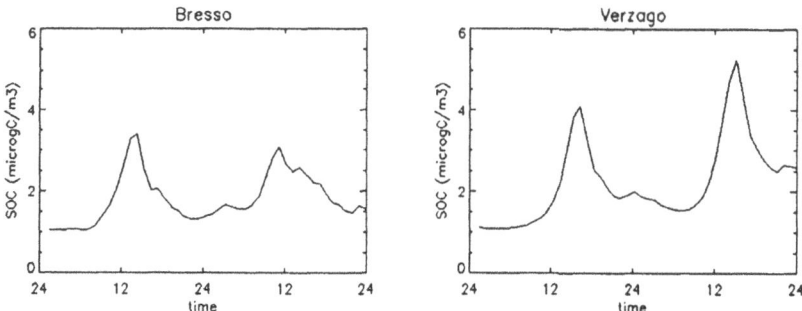

Figure 7 Diurnal variation of the predicted SOC concentrations at Bresso (left) and Verzago (right) on May 12-13.

Table 1 Estimation of SOC (secondary organic carbon) from the TOC (total organic carbon) and BC measurements (TOC= POC+SOC) (Putaud et al. 2002).

site	period	POC/BC	TOC (μgCm^{-3})	BC (μgCm^{-3})	POC (μgCm^{-3})	SOC (μgCm^{-3})
Bresso	morning		10.9	3.6	7.7	3.2
	afternoon	2.14	6.0	1.8	3.9	2.2
	night		7.1	2.1	4.5	2.6
Verzago	morning		6.1	1.5	3.2	2.9
	afternoon	2.14	6.3	1.0	2.1	4.2
	night		6.1	0.8	1.7	4.4

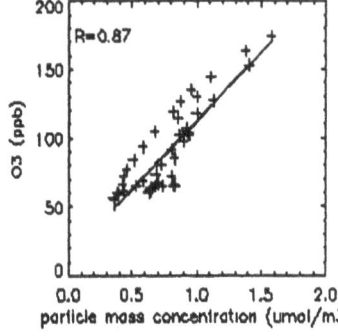 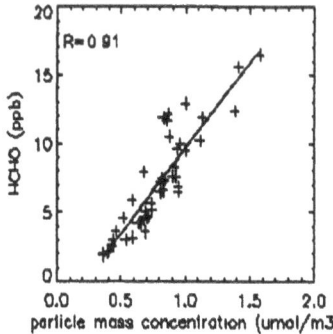

Figure 8 Correlation between predicted mixing ratios of gaseous species and total secondary (i.e. sum of NO_3^-, SO_4^{2-}, NH_4^+, SOC) particle mass concentration ($\mu mol\ m^{-3}$) in Verzago during May 12-13. Solid line is the linear regression.

3. 3. Effects of emission reductions

Two simulations were performed with reduced NO_x and VOC emissions, each 35% separately. The change in ozone in Bresso and Verzago due to the reductions of NO_x and VOC emissions are shown in Figure 9. At both stations ozone production is mainly sensitive to VOC emissions and therefore peak ozone concentrations decrease with VOC reductions, but increase with NO_x reductions. Secondary organic carbon aerosols show a similar behaviour. On the other hand, particulate NH_4^+ and NO_3^- concentrations decrease with decreasing NO_x emissions and are not affected by VOC emission reductions.

3. 4. Effects of biogenic emissions on SOC production

Biogenic hydrocarbons such as terpenes are among the most reactive gaseous compounds in the atmosphere, and they are important precursors of secondary organic aerosol (Raes et al., 2000). In this study, the difference in SOC concentrations in two simulations which were carried out with and without monoterpene emissions showed that monoterpenes contribute up to about 25% of the SOC produced in the northern part of the model domain which is mostly a forested area (Figure 10).

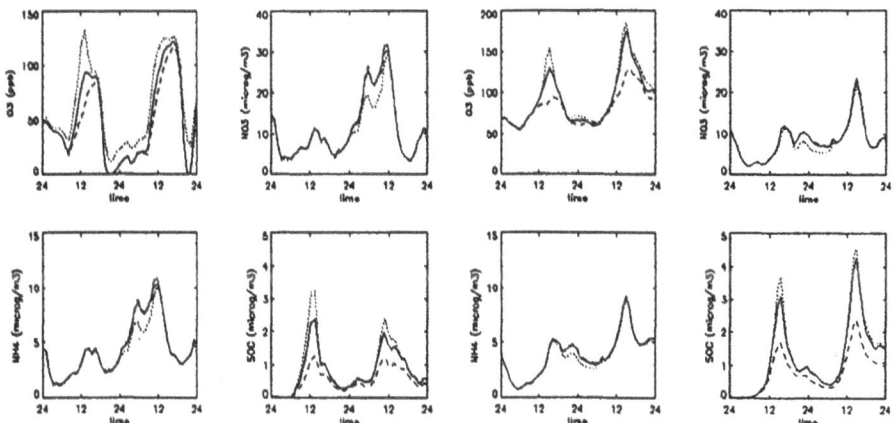

Figure 9 Ozone and particulate NO$_3^-$, NH$_4^+$, and SOC concentrations for the base case (–) and for the cases with NO$_x$ (...) and VOC (- -) emission reductions at Bresso (left) and Verzago (right).

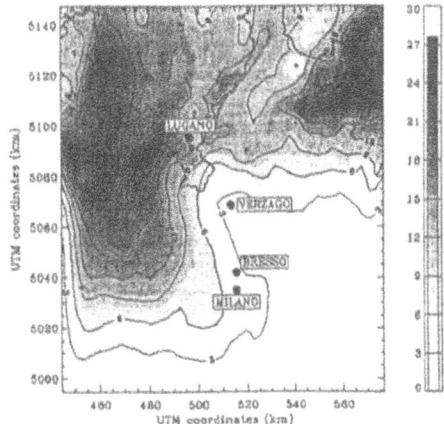

Figure 10 Increase (%) in SOC concentrations when monoterpene emissions are included.

4. CONCLUSIONS

The performance of aerosol modules of the three-dimensional eulerian model CAMx was evaluated for the first time in a domain covering the Lombardy region in northern Italy. The secondary aerosol species such as particulate NO$_3^-$, NH$_4^+$, SO$_4^{2-}$, and SOC (secondary organic carbon) were calculated for the particle size below 2.5 μm and compared with the measurements available from the PIPAPO field campaign which took place in May-June, 1998. Model results were comparable to the measurements in the afternoon for well mixed conditions both in urban and semi-rural stations. The total secondary particle mass concentration is positively correlated with secondary gaseous pollutants such as ozone and formaldehyde. In most of the model domain, particulate NO_3^- and NH_4^+ concentrations decreased with reduced NO_x emissions. On the other hand, the levels of secondary organic carbon aerosols decreased with reduced VOC emissions but increased with reduced NO_x emissions, similarly to ozone. Monoterpene emissions were predicted to contribute about 25% of the SOC produced in the northern part of the model domain which is mostly a forested area. In the southern part of the domain which is mainly urban, the contribution is much less. In general, inspite of the high uncertainties concerning NH_3 emissions, model results can be considered as satisfactory. Improvement of emission data and inclusion of particle size distribution in the model will be considered for the future work. To evaluate a model in more detail, more temporally and spatially resolved measurements of the aerosol composition is needed.

Acknowledgements

We would like to thank the LOOP community for providing us with various data.

5. REFERENCES

Baltensperger U., 2001, The art of aerosol measurement and modelling with within AEROSOL, Eurotrac-2 Symposium 2000, Garmisch-Partenkirchen, Germany, Springer-Verlag Berlin Heidelberg, pp. 39-44.

Baltensperger U., Streit N., Weingartner E., Nyeki S., Prévôt A.S.H., Van Dingenen R., Virkkula A., Putaud J.-P., Even A., ten Brink H., Blatter A., Neftel A., Gäggeler H.W., 2002. Urban and rural aerosol characterization of summer smog events during the PIPAPO field campaign in Milan, Italy, *J. Geophys. Res.*, 107:8193, doi:10.1029/2001JD001292

Castro L.M., Pio C.A., Harrison R.M., Smith D.J.T., 1999, Carbonaceous aerosol in urban and rural European atmospheres: estimation of secondary organic carbon concentrations, *Atmos. Environ.*, 33:2771.

Dommen J., Prévôt A.S.H., Neininger B., Bäumle M., 2002, Characterization of the photooxidant formation in the metropolitan area of Milan from aircraft measurements, *J. Geophys. Res.*, 107:8197,doi:10.1029/2000JD000283

Environ, 2002, User's Guide, Comprehensive Air Quality Model with Extensions (CAMx), Version 3.10, Environ International Corporation, California.

Martilli A., Neftel A., Favaro G., Kirchner F., Sillman S., Clappier A., 2002, Simulation of the ozone formation in the Northern Part of the Po Valley, *J. Geophys. Res.*, 107: 8195, doi:10.1029/2001JD000534

Neftel A., Spirig C., Prévôt A.S.H., Furger M., Stutz J., Vogel B., Hjorth J., 2002, Sensitivity of photooxidant production in the Milan Basin: An overview of results from a EUROTRAC-2 Limitation of Oxidant Production field experiment, *J. Geophys. Res.*, 107: 8188 doi:10.1029/2001JD001263

Pilinis C. and Seinfeld J.H., 1987, Continued development of a general equilibrium model for inorganic multicomponent atmospheric aerosols, *Atmos. Environ.*, 21:2453.

Putaud J.-P., Van Dingenen R., Raes F., 2002, Submicron aerosol mass balance at urban and conurban sites in the Milan area (Italy), *J. Geophys. Res.*, 107: 8198, doi:10.1029/2000JD000111

Raes F., Van Dingenen R., Vignati E., Wilson J., Putaud J.-P., Seinfeld J.H., Adams P., 2000, Formation and cycling of aerosols in the global troposphere. *Atmos. Environ.*, 34:4215.

Ritter N., Prévôt A.S.H., Dommen J., Andreani-Aksoyoglu S., Keller J., 2002, Model study with UAM-V in the Milan area (I) during PIPAPO: Simulations with changed emissions compared to ground and airborne measurements. Submitted to *Atmos. Environ.*

DISCUSSION

Z.KLAIC   How did you determine turbulent exchange coefficient fields (K) which are necessary as a CAMx input?

S.A.-AKSOYOGLU   Meteorological data were calculated in another project for the same domain and period. Turbulent exchange coefficients were derived in the meteorological model SAIMM from the turbulent kinetic energy.

MEASUREMENTS AND MODELING OF REGIONAL AIR QUALITY IN THREE SOUTHEAST UNITED STATES NATIONAL PARKS

Daiwen Kang, Viney P. Aneja, Rohit Mathur, and John D. Ray[*]

1. INTRODUCTION

Since the passage of the 1970 Clean Air Act (CAA), regulatory efforts to comply with the 0.12-ppmv National Ambient Air Quality Standard (NAAQS) for O_3 have proved inadequate [National Research Council (NRC), 1991; Dimitriades, 1989]. O_3 nonattainment continues to be a problem, especially in the southeast United States. This is attributed to the oxidation of NO_X in the presence of excessive amounts of biogenically emitted VOCs such as isoprene [Trainer et al., 1987; Chameides et al., 1988]. The new 8-hour O_3 National Ambient Air Quality Standard (NAAQS) (0.08 ppm) is likely to bring more suburban and rural locations into noncompliance [Chameides et al., 1997]. Biogenic VOCs emitted by vegetation [Fuentes et al., 2000; Fehsenfeld et al., 1992; Lamb et al., 1993] and anthropogenic VOCs emitted by human activities are both widely present in rural areas [Kang et al., 2001; Hagerman et al., 1997]. Previous studies indicate that the influence of these VOCs on important aspects of atmospheric chemistry such as O_3 production can be significant [Trainer et al., 1987; Chameides et al., 1988]. Clearly, if O_3 concentrations are to be successfully controlled by implementation of control on primary pollutant emissions, the roles of both natural and anthropogenic VOCs in these rural areas must be thoroughly understood. However, our understanding of O_3 and VOC budgets in rural areas is still very limited. Emissions of biogenic VOCs as well as the roles of both biogenic and anthropogenic VOCs in O_3 production in rural areas are largely uncharacterized [Guenther et al., 2000].

[*] Daiwen kang, NERL, U.S. Environmental Protection Agency, RTP NC, 27711, USA. Viney P. Aneja, Department of Marine, Earth, and Atmospheric Sciences, North Carolina State University, Raleigh NC, 27695. Rohit Mathur, Carolina Environmental Program, University of North Carolina, Chapel Hill, NC 27599-1105, USA. John D. Ray, Air Resources Division, National Park Service, Denver CO, 80225, USA.

Air Pollution Modeling and Its Application XVI, Edited by
Borrego and Incecik, Kluwer Academic/Plenum Publishers, New York, 2004

57

In order to further investigate impacts of hydrocarbons and O_3 production in rural areas, this study focuses on a modeling analysis of O_3 and VOCs in three southeast United States national parks. Our previous study [Kang et al., 2001] presented a comprehensive analysis of data collected from the same three parks. Even though previous studies [Hagerman et al., 1997] claim that the rural areas of interest in this study are NOx-limited for the formation of O_3, our study indicates a significant contribution from local VOCs. In order to evaluate the model, we compare model predictions with measured values. We also examine VOC emission-perturbation scenarios in the context of model process budgets to develop insights into the role of VOCs on O_3 concentrations at these rural sites.

2. THE MODELING SYSTEM AND MEASUREMENT DATA

2.1. Overview of MAQSIP

The Multiscale Air Quality Simulation Platform (MAQSIP) [Odman and Ingram, 1996] is a comprehensive Eulerian grid model that has also served as a prototype for the U.S. EPA's Models-3 concept [Byun et al., 1999]. The anthropogenic emissions inventory is the Ozone Transport Assessment (OTAG) inventory for 1995, which in turn derived from the earlier national emissions inventory for 1990 [Houyoux et al., 1996]. Biogenic emissions are calculated using the USEPA Biogenic Emission Inventory System 2 (BEIS2). The Carbon Bond Mechanism IV (CB4) used in MAQSIP is a modified version of that proposed by Gery et al. (1989). Modifications reflect our increased understanding of atmospheric chemistry involving the organic peroxy radicals which form inert organic nitrates [Kasibhatla et al., 1997] and isoprene chemistry.

2.2. Domain selection of the modeling system

The modeling domain was chosen so as to adequately represent conditions at sampling sites in three national parks [Kang et al., 2001]: Shenandoah National Park, Big Meadows (SHEN) located at 38°31'21" N, 78°26'09"W with an elevation of 1073 m, Great Smoky Mountains National Park Cove Mountain (GRSM) located at 35°41'48"N, 83°36'35"W with an elevation of 1243 m, and Mammoth Cave National Park (MACA) at 37°13'04"N, 86°04'25" W with an elevation of 219 m. As Figure 1 shows, the domain of this modeling system consists of 34x42 cells using a 36-km horizontal resolution. The vertical domain varying from the surface to 100 mb is discretized using 22 layers of variable resolution; the lowest layer has a depth of 38 m. Since this study focuses on concentration field at the surface, only concentration fields at the lower 12 layers of the vertical domain are extracted from the model output.

The time period for the model exercise selected is from 12:00 pm July 14 to 12:00 pm July 29, 1995 to comply with measurements that were made during July when photochemical activity was at a maximum [Kang et al., 2001]. Time-varying lateral boundary conditions for various model species were derived from previous model simulations conducted over the eastern U.S. for the study period.

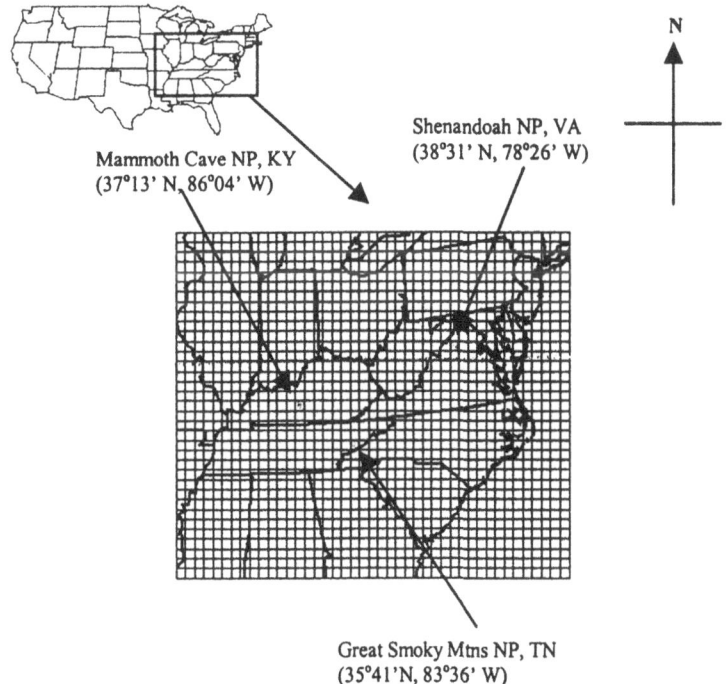

Figure 1. Map of the model domain (36 km grid size). Locations of the three monitoring sites in national parks are indicated with the shaded squares.

2.3. Measurement description

One-hour integrated ambient-air samples for determination of VOCs were collected (a single measurement at local noon for each day) at the three U.S. national parks described above during June through July for the year 1995. In the analysis of VOC samples, standards, both for retention time and quantification, were run routinely, and four internal standards were added to every chromatographic run to verify performance of the analytical system [Farmer et al., 1994]. Identification of the target compounds was confirmed by mass spectrometer analysis. The method detection limit is 0.10 parts per billion carbon (ppbC), with an uncertainty of ±20%. Any target compounds not detected in the sample above 0.10 ppbC were reported as not detected (ND) and were not included in any statistical analysis. The quality of the data is further guaranteed through correlation and ratio analysis. Further details on the measurement and data analysis can be found in Kang et al. (2001).

2.4. Organization of simulations

Table 1 gives the simulation details. Since our primary concerns in these rural areas are the characteristics of biogenic hydrocarbons represented primarily by isoprene (the contribution of terpenes to PAR and OLE can be neglected) in the model versus anthropogenic hydrocarbons, in the simulation design the emissions perturbation factor of isoprene is changed more often than that of other VOCs. Each simulation is assigned a name for use in the subsequent discussion. The perturbation factors represent net reduction or increase in emissions with respect to the Base Scenario designated by CS0. Further details on emissions in the Base Scenario are presented in Houyoux et al. (2000). The factors are uniformly applied to all cells in the modeling domain.

Table 1. Simulation details and motivations

Simulation	Emissions perturbation factors						Motivation
Designation	ISOP	PAR	ETH	OLE	TOL	XYL	
CS0	1	1	1	1	1	1	Base scenario
CS1	0	0	0	0	0	0	Assess effect of boundary inflow
CS2	0.5	0.5	0.5	0.5	0.5	0.5	Assess effect of reduced emissions
CS3	1.5	1.5	1.5	1.5	1.5	1.5	Assess effect of increased emissions
CS4	0	1	1	1	1	1	Assess effect of anthropogenic emissions
CS5	1	0	0	0	0	0	Assess effect of biogenic emissions
CS6	2	2	2	2	2	2	Assess effect of doubled emissions
CS7	3	1	1	1	1	1	Assess effect of increased biogenic emissions only
CS8	5	1	10	10	10	100	Match the measurement

3. OBSERVED VOCS AND MODEL PREDICTIONS

3.1. Concentrations

Figure 2 shows the time series of both model predicted and observed concentrations for 4 model species at different locations. The observed values stand for hourly integrated mean concentrations from 12:00 – 13:00 local time and the model predictions are the mean concentrations at hour 12:00 and 13:00 (local time) on the same days when observations are available. The observed isoprene concentrations are between the values predicted by CS7 and CS8 except the one on July 23 at GRSM; it is a good match between observed concentrations and the predicted values by CS8 at MACA even though most of the observed concentrations are still higher than the predicted ones by this scenario. In general, the base scenario underpredicts isoprene concentrations with a factor of about 3-5 at the three locations. The model predicted and observed distributions of anthropogenic VOC concentrations vary both in species and in location. The single-bonded one atom surrogate, PAR, is the species whose predictions have closer agreement with observations compared to the other species, and the difference between predications and observations are within a factor of 2 at GRSM and within a factor of 2 for more than half of the points and a factor of 3 to 5 for the rest at MACA (the largest increase in PAR emissions is a factor of 2). The base scenario significantly underpredicts all anthropogenic species except PAR. The predictions for OLE by CS8 are a good match at

GRSM, but are still too low at MACA for 4 of the 12 data points. Notice that the variation pattern of PAR is very similar to OLE at MACA and the higher concentrations on July 19 through July 22 may be indicative of local pollution events. The molecular surrogate TOL is overpredicted by CS8 at GRSM, but it is a good match at MACA.

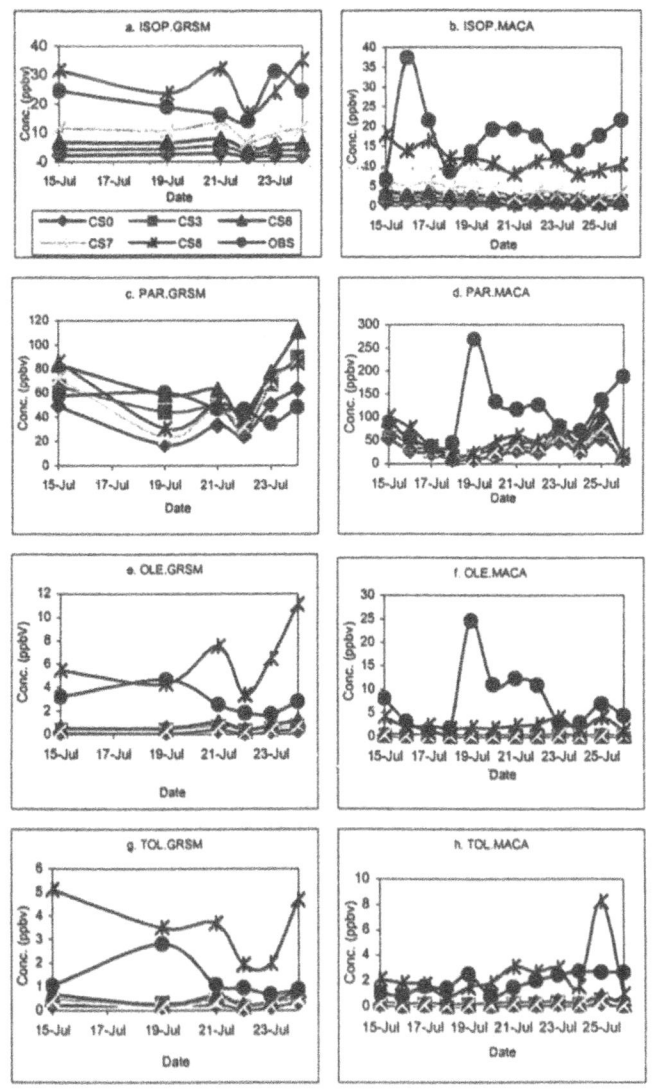

Figure 2. Time series of model predicted and observed lumped species
CS0, CS3, CS6, CS7, and CS8 are simulation scenarios and OBS is observation.
The title of each figure contains the name of the lumped species and the name of
the location separated by a dot.

3.2. Reactivity

Mathur et al. (1994) proposed a reactivity weighted organic (RWOG) scale. In this study we have extended the RWOG to evaluate the impact of hydrocarbon reactivity on ozone production. As Figure 3 indicates, with the increase of hydrocarbon reactivity, mean daily maximum O_3 concentrations increase initially, then reach a maximum point, and decrease after that point. The observed values fit well on the simulation trends for both locations. Also note that the observed values at both national park locations are on the decrease phase of the variation pattern, suggesting that hydrocarbon concentrations are at such a level that helps suppress O_3 production at these two locations.

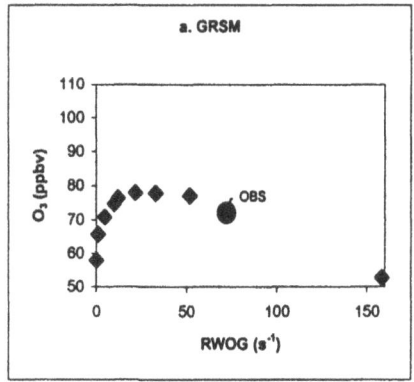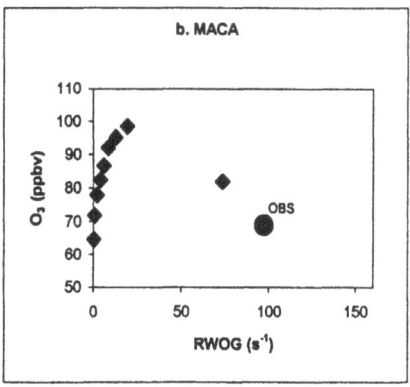

Figure 3. Variations of mean daily predicted and observed maximum O3 concentrations with reactivity-weighted organic (RWOG). OBS is observed values. Others are model simulations.

4. PROCESS ANALYSIS

A comprehensive process analysis reveals that chemistry processes are important to both O_3 and VOCs local budgets in the areas studied. Even O_3 transported to the area is also produced through photochemistry processes in the atmosphere. VOCs are predominantly emitted locally, however, about half are consumed by chemistry, and about 33 to 50% are exported to other areas. Most of the exported VOCs are removed by various chemical and photochemical processes. This analysis reveals that deposition of VOCs is small. Therefore, it is necessary to further analyze the chemistry processes in detail. The magnitude of O_3 chemistry budget represents the net ozone is produced through local photochemical processes. During the photochemically active period of the day (10:00 am – 5:00 pm), the value is always positive. The magnitude of VOC chemistry values represents how much VOCs has been removed through local chemical and photochemical processes. The value is always negative, meaning that VOCs are consumed.

The relationship between net O_3 production (P_{O3}) and VOC loss (L_{VOC}) due to chemistry at the three locations are presented in Figure 4. The last scenario, CS8, is not included because this scenario significantly reduces O_3 production at all three locations. Note that while VOCs are consumed, O_3 is produced in the model O_3 chemistry (For convenience's sake, the negative sign of the VOC chemistry term is dropped). It is clear that the consumption of VOCs contributes to an approximately linear production of O_3 at all three locations, but their actual relationship varies from location to location.

We define VOC potential for O_3 production (VPOP) as:

$$VPOP = \frac{d[O_3 \ chemistry]}{d[VOC \ chemistry]} \quad (1)$$

VPOP is a measure of how much O_3 is produced per unit of VOCs consumed or, in other words, the O_3 production efficiency of VOC chemistry. It represents the slope of the best-fit regression line through the data in Figure 4.

From Figure 4, the VPOP values are 0.065 (GRSM), 0.36 (MACA), and 0.11 (SHEN) ppbv O_3 per ppbv VOCs. Thus, one ppbv VOCs consumed produces 5.5 times more O_3 at MACA than at GRSM. It is interesting to note that, if VOC chemistry is zero, values for O_3 production due to chemistry are similar at all three locations (3.1 to 4.4 ppbv/hr). These may represent background O_3 production from inorganic reactions such as CO and CH_4 and long-range transported stratospheric O_3.

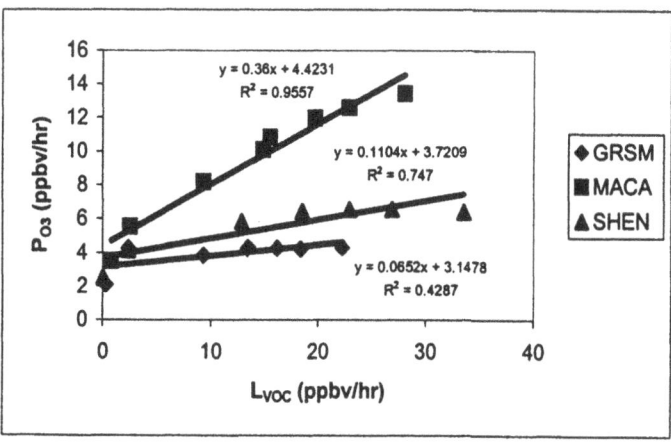

Figure 4. O_3 chemistry and VOC chemistry (the negative sign for L_{VOC} is dropped)

5. CONCLUSONS

Limited comparisons between predicted and measured VOCs indicate that the predictions of VOC species concentrations are less successful. Out of six nonmethane hydrocarbon species, PAR is best predicted with a factor of 2 for uncertainty in most cases, followed by isoprene with an uncertainty factor of 3 to 5. Most other species are predicted at about one order of magnitude lower than observed values. However, the relative reactivity of the various VOC species is captured properly by MAQSIP and the observed total reactivity-weighted organic (RWOG) values fit well on the model simulated variation trends. The comparison of the observed and model predicted RWOG suggests that the model Base Scenario underestimates RWOG by a factor of 3 to 10 for the three locations studied.

The VOC potential for O_3 production (VPOP), which is 3 and 6 times greater at MACA than at SHEN and GRSM, respectively, measures the efficiency of O_3 production due to the chemistry process of VOCs. This metric can be used to quantitatively evaluate the relative strength of chemistry process and O_3 production at a location.

6. REFERENCES

Byun, D.W. and J.K.S. Ching, Eds.,: Science algorithms of the EPA Models-3 Community Multi-scale Air Quality (CMAQ) modeling system, EPA/600/R-99/030, Office of Research and Development, U.S. Environmental Protection Agency, 1999.

Chameides, W.L., R. W. Lindsay, J. Richardson and C.S. Kiang, The role of biogenic hydrocarbons in urban photochemical smog: Atlanta as a case study, *Science,* 241, 1473-1474, 1988.

Chameides, W.L., R.D. Saylor, and E.B. Cowling, Ozone pollution in the rural united states and the new NAAQS, Science, 276, 916, 1997.

Dimitriades, B., Photochemical oxidant formation: overview of current knowledge and emerging issues. In Atmospheric Ozone Research and its Policy Implications, ed. T. Schneider et al. Elsevier, Amsterdam, 1989.

Fehsenfeld, R.M., J. Calvert, R. Fall, P. Goldan, A. Guenther, B. Lamb, S. Liu., M. Trainer, H. Westberg and P. Zimmerman, Emissions of volatile organic compounds from vegetation and the implications for atmospheric chemistry, *Global Biogeochem. Cycles,* 6, 389-430, 1992.

Fuentes, J.D., M. Lerdau, R. Atkinson, D. Baldocchi, J.W. Bottenheim, P. Ciccioli, B. Lamb, C. Geron, L. Gu, A. Guenther, T.D. Sharkey, and W. Stockwell, Biogenic hydrocarbons in the atmospheric boundary layer: A review, *Bulletin of the American Meteorological Society,* 81, 1537-1575, 2000.

Gery, M. W., G. Z. Whitten, J. P. Killus, and M. C. Dodge, A photochemical kinetics mechanism for urban and regional scale computer modeling, *J. Geophys. Res.,* 94, 12,925-12,956, 1989.

Guenther, A., C. Geron, T. Pierce, B. Lamb, R. Harley, and R. Fall, Natural emissions of non-methane volatile organic compounds, carbon monoxide, and oxides of nitrogen from north America, *Atmos. Environ.,* 34, 2205-2230, 2000.

Hagerman, L. M., V. P. Aneja, and W. A. Lonneman, Characterization of non-methane hydrocarbons in the rural Southeast United States. *Atmos. Environ.,* 31, 4017-4038, 1997.

Houyoux, M.R., C.J. Coats, A. Eyth, and S.C.Y. Lo, Emissions modeling for SMRAQ: A seasonal and regional example using SMOKE, paper presented at AMWA Computing in Environmental Resources and Management Conference, Research Triangle Park, North Carolina, Dec. 2 to Dec. 4, 1996.

Kang, D., V. P. Aneja, R. G. Zika, C. Farmer, and J. D. Ray, Nonmethane hydrocarbons in the rural southeast United States national parks, *J. Geophys. Res.,* 106, 3133-3155, 2001.

Kasibhatla, P., W. L. Chameides, B. Duncan, M. Houyoux, C. Jang, R. Mathur, T. Odman, and A. Xiu, Impact of inert organic nitrate formation on ground-level ozone in a regional air quality model using the carbon bond mechanism 4, *Geophys. Res. Lett.,* 24, 3205-3208, 1997.

Lamb, B., D. Gay, H. Westberg, and T. Pierce, A biogenic hydrocarbon emission inventory for the USA using a simple forest canopy model, *Atmos. Environ.,* 27, 1709-1713, 1993.

Mathur, R., K. L. Schere, and A. Nathan, Dependencies and sensitivity of tropospheric oxidants precursor concentrations over the Northeast United States: A model study, *J. Geophys. Res.,* 99, 10,535-10,552, 1994.

National Research Council, Rethinking the Ozone Problem in Urban and Regional Air Pollution. National Academy Press, Washington DC., 1991.

Odman, T. and C. L. Ingram, Multiscale air quality simulation platform (MAQSIP): Source code documentation and validation, MCNC Technical Report, ENV-96TR002-v1.0, 1996.

Trainer, M., E. J. Williams, D. D. Parrish, M. P. Buhr, E. J. Allwine, H. H.Westberg, F.C. Fehsenfeld, and S. C. Liu, Models and observations of the impact of natural hydrocarbons on rural ozone, *Nature,* 329, 705-707, 1987.

DISCUSSION

S. A. AKSOYOGLU	When you compare model results with observed lumped voc species, you probably use each single voc measured, and calculate the particular species such as PAR or OLE, etc. Can you cover all VOCs contributing to PAR, are all measured?
D. KANG	No. But since most of the identified VOC species by GC/MS are included in this calculation, I would say more than 90% of the VOC species are taken into consideration.
D. STEYN	I am concerned about the statistical details of your evaluation procedure with 600 points, all constrained to the upper night quadrant, you are almost bound to achieve magnificant correlations even if the model performance is not particularly useful. There are two approaches possible: 1) Contrast your statistics with the same statics for 600 points randomly distributed in the upper right quadrant. 2) Reduce the number of points by analysing the data in subsets weather type; day-of-week; or other external variables.
D. KANG	Thank you. This is a good suggestion for our further analysis.
R. SAN JOSE	What is the time step > m need to solve the chemistry in MM5-CHEM?
D. KANG	We didn't run the model MM5-CHEM that is developed and run by Georg Grell for US NOAA's Forecast System Laboratory. But as far as I know, the time step is 1 hour for the output of O_3 concentration field.
P. BUILTJES	Did you investigate whether the forecast models are better than persistence?
D. KANG	Yes, we did. Among the three models, only MAQSIP can beat the persistence.
R. BORNSTEIN	An interesting and important additional statistics you could calculate would be the missed episode (=a/ (a+b)) which are important from a public health point of view.
D. KANG	In fact, we do have the metrics for a/(a+b), and it is False Alarm Rate (FAR).

65

MODELING OF AIR POLLUTION AND DEVELOPMENT OF AIR QUALITY MANAGEMENT PLAN FOR ISKENDERUN REGION IN TURKEY

M.Tahir Chaudhary and Aysel T. Atimtay[*]

1. INTRODUCTION

Clean air is considered to be a basic requirement for human health and their well-being. Air pollution has become an alarming problem with industrialization, and protection of air quality turned into a topic of great interest since early 1960's. The United Nations Conference on Environment and Development (UNCED) held in 1992 at Rio de Janeiro, Brazil, adopted the Framework Climate Convention (Agenda 21) which underlined the need of air pollution control. Also, the declaration of Habitat II (United Nations Conference on Human Settlements held in June 1996, at Istanbul, Turkey) emphasized the "sustainable development" and "sustainable human settlements" for the protection of environment. In sustainable human settlements clean air was one of the most important considerations put forward among the other environmental issues.

Industry is a major source of pollution when proper controls of the emissions are not made. Consequently, industrial operations can affect the health of workforces, the environment and the health of nearby and some times very far located populations. Iron and steel industry is regarded as one of the basic industries for the development of any country. On the other hand, iron and steel industries are one of the major sources of air pollution, which is included in the list of industries with significant health impacts (WHO, 2000).

1.1. Objectives of the Study

The main objectives of this study are:

---

[*] Middle East Technical University, Environmental Engineering Department, 06531 Ankara, Turkey. Tel: +90-312-210-5879, Fax: +90-312-210-1260, E-mail: aatimtay@metu.edu.tr

Air Pollution Modeling and Its Application XVI, Edited by
Borrego and Incecik, Kluwer Academic/Plenum Publishers, New York, 2004

- To prepare an emission inventory, and to find the contributions of several industries to the air pollution in the region,
- To estimate the ground level concentrations of pollutants by dispersion modeling and to prepare ground level concentration maps.
- To list suggestions based on modeling results for the development of a "clean air plan" for the Iskendern Region, which is necessary according to Turkish Regulation.

1.2. Location and Physical Characteristics of the Region

The Iskenderun Region is located between the Mediterranean Sea and Amanos mountain range where peaks reach up to 1700m high from sea level. This gulf region forms a narrow coastal area between the sea and mountains. There are 415,000 inhabitants residing in the study area (SIS, 2002). Major population centers are İskenderun, Dörtyol and Payas with populations of 160,000, 54,000 and 32,000, respectively. The rest of 169,000 people live in several sub-urban areas and villages scattered in the study area. Iskenderun Gulf region is the most industrialized region of the Southeast Turkey. This region is also regarded as the citrus depot of Turkey. During winter most of the dwellings use coal for domestic heating. There are two major inter-city highways passing through the study area, the Iskenderun-Adana Motorway and the E-5 Highway.

Industrial complexes are located at about 17 km north of Iskenderun city. Among these industries, the largest one is ISDEMIR which is an integrated plant with a production capacity of 2,200,000 tons/year and it is the 2nd largest integrated iron and steel works of Turkey. In Turkey, 37% of the total integrated iron and steel production capacity is installed at ISDEMİR.

There are three industrial zones in the area along with ISDEMIR, two in the north and the other in the south of ISDEMIR. Re-rolling steel mills are concentrated in these Organized Industrial Estates. These industries use fossil fuels, iron ores (hematite, magnetite, pyrite) and steel scrap for their production and were suspected to emit large amounts of pollutants including SO_2, NO_x, CO, CO_2, VOC and particulate matter. ISDEMIR produces 45% of the steel production in the region.

2. MATERIALS AND METHODS

2.1. Emission Inventory

An emission inventory of the study area has been prepared in this work by taking into account all the possible emission sources. These sources include all the industrial and residential sources (rural and urban), ISDEMIR and all other industries in the Organized Industrial Estates of Iskenderun, Dörtyol and Payas. Moreover, emissions from traffic sources on the Iskenderun-Adana motorway, Iskenderun-Adana highway and urban roads in the cities were also included. Four major pollutants, namely, PM, SO_2, NO_x and CO were included in this study. Emissions from industries were measured at sources, while emissions from domestic heating and traffic were estimated by using the CORINAIR emission factors (Corinair, 1999).

2.2. Meteorological Data

The air quality modeling of the Iskenderun region in this study is based on the emission inventory data for the year 2001. Therefore, meteorological data of the same year was used in this modeling work. The meteorological data of the Iskenderun station for year 2001 was acquired from the State Department of Meteorology (SDM). This data consisted of sequential hourly records of values for temperature, atmospheric pressure, sunshine hours, cloud cover, mixing height, precipitation, and wind speed and wind direction. The hourly meteorological data is required by the dispersion model (ISCST3) to be used. In addition to these parameters, data for morning and afternoon mixing heights measured at Adana, the nearest synoptic station to Iskenderun, were also taken from the SDM. In order to prepare the meteorological data input files from the raw data the PCRAMMET program was used. The annual wind rose for Iskenderun for the year 2001 is given in Figure 1.

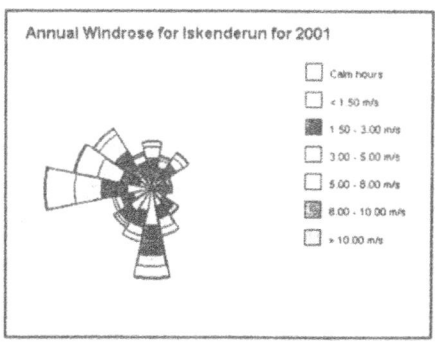

Figure 1. Annual wind rose for Iskenderun for year 2001

2.3. Air Quality Modeling

Industrial Source Complex Short Term Version 3 (ISCST3), a model developed by the U.S. EPA (1995) was used to model the air pollution of the study area. The ISCST3 model provides options to model emissions from a wide range of sources that might be present at a typical industrial complex and area surrounding it.

3. RESULTS AND DISCUSSION

3.1. Emission Inventory

The emission inventory prepared for the study area by Chaudhary M.T. in his Ph.D. research work was used as input to the air quality modeling. The results of emission inventory are given in the following tables. Table 1 shows the total emissions from Industrial activities in the Iskenderun Region. Table 2 and 3 give contributions of several sources to the total emission and annual emissions from all sources in the study area.

Table 1. Annual industrial emissions

INDUSTRY	PM (ton/y)	SO₂ (ton/y)	NOₓ (ton/y)	CO (ton/y)	NMVOC (ton/y)
ISDEMIR	19,009	32,836	6,490	79,585	208
All Other Industries	94	6,802	804	143	10
Total	19,103	39,638	7,295	79,728	218
ISDEMIR	99.5%	82.8%	89.0%	99.8%	96.0%
All Other Industries	0.5%	17.2%	11.0%	0.2%	4.0%

Table 2. Contributions of several sources in the annual pollution load in the study area

SOURCES	PM (ton/y)	SO₂ (ton/y)	NOₓ (ton/y)	CO (ton/y)	NMVOC (ton/y)
Industrial	19,103	39,638	7,295	79,728	218
Domestic Heating	696	1,195	140	42	114
Traffic	152		3329	30,168	5333
Total	19,951	40,833	10,764	109,938	5665

Table 3. Annual emissions from all sources

SOURCES	PM, %	SO₂, %	NOₓ, %	CO, %	NMVOC, %
Industrial	95.75	97.07	67.77	72.52	3.85
Domestic Heating	3.49	2.93	1.30	0.04	2.01
Traffic	0.76		30.93	27.44	94.14

3.2. Air Quality Predictions

The area of interest in this study is the Iskenderun Region covering an area with a width of 25 km and length of 50 km. This area contains emission sources as point, area and line sources. Industrial sources are modeled as point sources. They include stacks of ISDEMIR and other industries located in the organized industrial estates of Iskenderun, Dortyol and Payas. Area sources contain the dwellings in the city of Iskenderun, town of Dortyol, and other rural residential areas. Emissions from traffic in urban areas and also on the highway and motorway were considered as line sources.

PM

The annual average ground level concentrations of PM estimated from all sources including industrial, residential heating and traffic sources are shown in Figure 2a. As can be seen from the figure, the annual average PM concentration in most of the region is below 127 μg/m³. The long term limit in the Turkish Air Quality Protection Regulation (TAQPR, 1986) is 150 μg/m³. In the southeastern parts of Iskenderun City, in towns of Dortyol and Payas PM concentration is between 40 to 70 μg/m3. This concentration is also below the EC Regulations limit of 80 μg/m³. The proposed revision of TAQPR suggests reducing the long-term limit for PM to 100 μg/m³. Even in this case the limits are not exceeded in urban areas. Almost all of the rural areas have average annual PM concentration between 10 and 40 μg/m³. Further investigations by running the model with discrete receptors revealed that the contributions of domestic heating sources in the

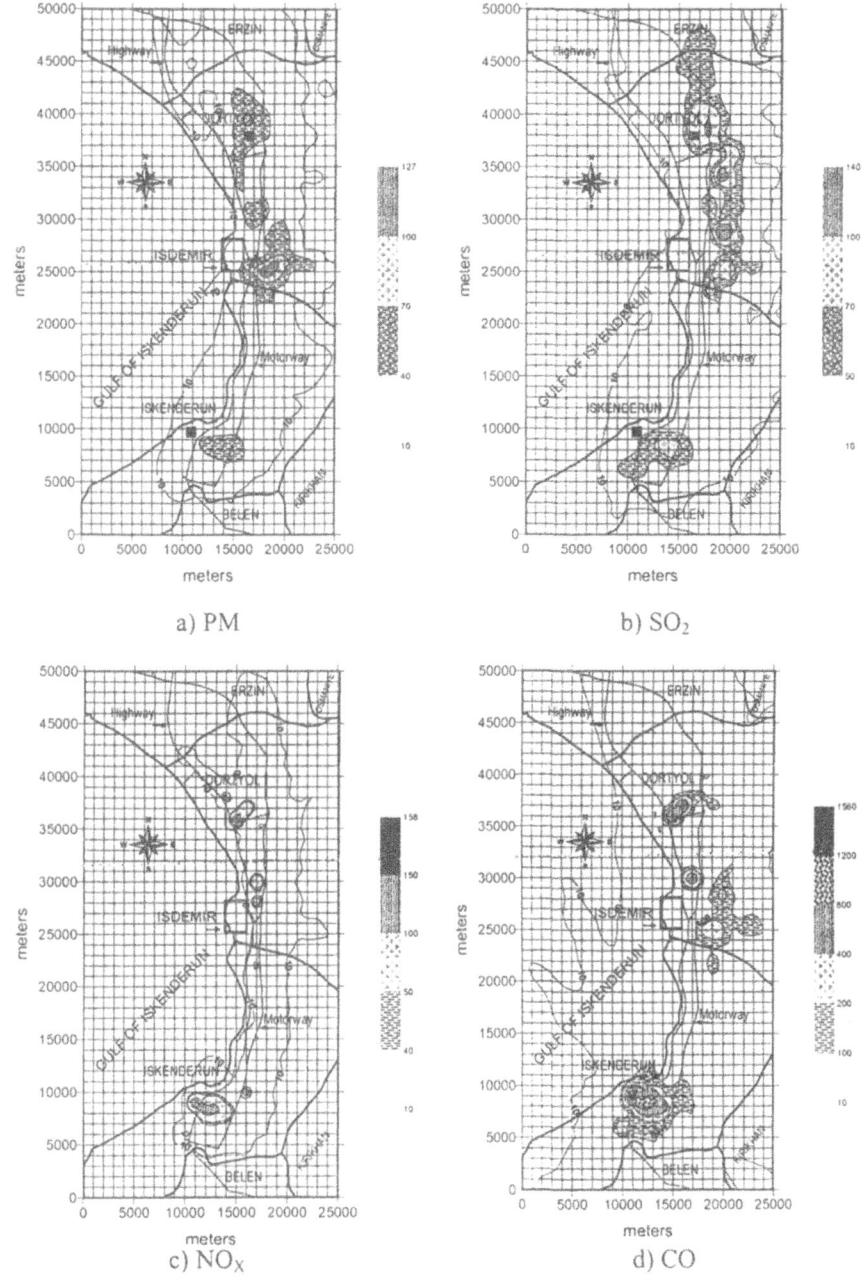

a) PM

b) SO$_2$

c) NO$_X$

d) CO

Figure 2. Annual average ground level concentrations ($\mu g/m^3$) from all sources

annual average PM concentrations at city centers of Iskenderun, Payas and Dortyol were 84%, 59% and 77%, respectively. For the same locations ISDEMIR was responsible for

71

2.8%, 37% and 19% of annual average concentrations of PM. The PM concentrations due to other industries were 1%.

In the areas to the ESE of ISDEMIR the PM concentration rises to 127 $\mu g/m^3$ in a small area due to dominant westerly winds. The areas surrounding this small section have concentration levels of 70-100 $\mu g/m^3$. However, these high concentration areas are located at altitudes of 300-600 m and even 900 m above sea level in the mountains. Moreover, there are no population centers located in this high concentration area. ISDEMIR accounts for 88% of the annual average PM concentrations in this zone of high concentration. This result can be attributed to the presence of tall stacks of ISDEMIR and dispersion of PM to long distances.

SO2

In Figure 2b the lowest annual average ground level concentration of SO_2 is between 10 to 50 $\mu g/m^3$. However, there are some regions where the concentration is reaching up to 140 $\mu g/m^3$. The long-term limit in the TAQPR is 150 $\mu g/m^3$. The concentration reaches up to 140 $\mu g/m^3$ in a very small area to the ENE of ISDEMIR where altitude is 100 to 200m above sea level. Near this location there are villages and the average concentration of SO_2 is in the range of 70-100 $\mu g/m^3$. The effects of high SO_2 concentrations on the people living in these villages, on forest and other vegetation are not known. It will be interesting to make a study on this subject. In the city of Iskenderun, SO_2 concentration rises to 50-70 $\mu g/m^3$. The concentration reaches up to 70-100 $\mu g/m^3$ in a small area of about 3 km^2 in the eastern parts of Iskenderun. In the north almost all parts of the town of Dortyol have SO_2 concentration of 50-70 $\mu g/m^3$, which rises to 70-100 $\mu g/m^3$ in the central and eastern parts of the town due to dominant westerly winds. Comparison of results from different source groups have shown that contributions of domestic heating in annual average SO_2 concentrations at city centers of Iskenderun, Payas and Dortyol were 98%, 85% and 73%, respectively.

The limit of SO_2 in WHO Guidelines and EC Regulation is 50 $\mu g/m^3$. It is seen from Figure 2b that in some parts of these cities, SO_2 concentration exceeds this limit. Therefore, SO_2 concentrations pose a danger for human health in many areas.

NO$_x$

Figure 2c shows the annual average ground level concentrations of NO_X. As can be seen from this figure, the annual average ground level concentration reaches up to 158 $\mu g/m^3$ in the city center of Iskenderun. Mostly the concentration in Iskenderun city lies between 50-150 $\mu g/m^3$. In Dortyol and Payas maximum concentrations are 150 and 100 $\mu g/m^3$, respectively. The long-term limits for NO_2 and NO are defined as 100 and 200 $\mu g/m^3$, respectively, in TAQPR. The major nitrogen oxide in the atmosphere is NO_2. The annual average NO_2 concentrations exceed this limit at the locations mentioned above except Payas. Around these high concentration locations and along the highway and the motorway, there are some pockets where the NO_X concentration is in the range of 100 to 200 $\mu g/m^3$. These regions are above the limits set by TAQPR for NO_2. The annual average concentrations of NO_X in the study area except the ones mentioned above are from 10 to 50 $\mu g/m^3$. The long-term limits of NO_2 set by EC Regulation and WHO Guidelines are 50 and 40 $\mu g/m^3$, respectively. A very characteristic result is seen in Figure 2c. As can be seen in the figure there is a belt of NO_X concentration areas along the highway and motorway where the concentration is between 10 and 40 $\mu g/m^3$. Although the limits of TAQPR are satisfied in most of the study area except in cities of Iskenderun and Dortyol, the NO_X concentrations are above the limits according to the EC

Regulation and the WHO Guidelines. The sources of NO_X emissions are industry, traffic and partly residential places. The contribution of each in total annual NO_X emissions was 68%, 31% and 1%, respectively. According to the emissions measurements and the emission inventory it was found that 60% of the annual NO_X emissions in the study area were due to ISDEMIR. In order to find the contribution of ISDEMIR on the annual average ground level NO_X concentrations the model has been run to see its effect separately also. The results have shown that the contribution of industry on the annual average NO_X concentrations is small, up to 20 $\mu g/m^3$ in two small areas. Moreover, it was revealed that almost all of the ground level NO_x concentration in the study area was due to traffic sources on inter-city (26%) and urban roads (65%). 8% was due to residential heating in Iskenderun.

CO

Figure 2d shows the annual average ground level concentrations of CO from all sources. The CO concentration does not exceed 1560 $\mu g/m^3$ anywhere in the study area. The maximum CO concentration is found toward ESE of ISDEMIR in the mountains at an altitude of about 200-300 m, near a village. The long-term limit set by TAQPR is 10,000 $\mu g/m^3$, and the calculated CO concentration is very small as compared to this limit. WHO Air Quality Guidelines define the CO limit as 30,000 $\mu g/m^3$ and 10,000 $\mu g/m^3$ for 1hr and 8hr exposure times, respectively. However, they have not defined any value for annual average CO concentration. The annual average CO concentration in most of the study area is in the range of 10-100 $\mu g/m^3$

Further analysis of results showed that the traffic is the major contributor to the annual average CO concentration in the study area. The maximum CO concentration of 1560 $\mu g/m^3$ is found in the city of Iskenderun. The CO concentration around this area changes between 800 and 1200 $\mu g/m^3$. Outside these areas and most parts of Dortyol receive a CO concentration of 10 to 100 $\mu g/m^3$ due to ISDEMIR. The CO concentrations due to traffic are followed by the concentrations from Isdemir. However, when traffic sources are taken into account by itself in the calculations, it is seen that an area where CO concentration is between 10 and 100 $\mu g/m^3$ stretches along the two major highways.

Although the contribution of traffic sources in the total annual CO emissions in the study area is only 27.4%, however, due to very close to ground level release of exhaust gases from vehicles, the dispersion of CO could not take place effectively. Therefore, the annual average ground level CO concentrations due to traffic along the motorway and the highway were found to be between 10 to 400 $\mu g/m^3$, which in urban areas it will reach above 1000 $\mu g/m^3$ at various places.

3.3. Model Performance in PM and SO₂ Predictions

In order to evaluate the model performance in PM and SO_2 predictions, the daily average predicted and observed ground level concentrations for January 2001 were used. The statistical analyses were carried out and the results of statistical analyses indicate that the overall performance of model is good. The value of index of agreement was found to be 0.63 and 0.67 for PM and SO_2, respectively. The relevant analysis of RMSE shows that there is small error in the model predictions for SO_2 as compared to measured values.

These methods of model evaluations have been used in several air pollution dispersion modeling studies all over the world. In a study in Jamshedpur India, which is rich is steel industries like the study area under consideration in this work, 68% accuracy in NO_X predictions by ISCST3 was reported (Sivacoumar et al., 2001). In a similar study in Izmir, Turkey the accuracy of ISCST3 prediction of SO_2 concentrations was found to be 72% (Elbir, 2002). The comparison of the model accuracy found in this study with those in the previous studies shows a good agreement.

4. CONCLUSIONS

The result of the dispersion modeling showed that the Iskenderun City, the largest residential area in the region was least affected by industrial emissions due to the prevailing wind directions from S and W, and location of industries at least 15 km north of this city. The annual average ground level PM concentration of 40-70 $\mu g/m^3$ found in some parts of residential areas were basically due to domestic heating activities in winter in these cities. These concentrations are below the limits set by TAQPR and EC Regulations. During winter, the average PM concentrations exceed the limit. This is because of usage of coal for heating.

The SO_2 concentration reaches up to 70-100 $\mu g/m^3$ in the eastern parts of Iskenderun. In the north almost all parts of the town of Dortyol have SO_2 concentration of 50-70 $\mu g/m^3$. In some parts of these cities, SO_2 concentration exceeds the EC limit, which is 50 $\mu g/m^3$. Therefore, SO_2 concentrations pose a danger for human health in many areas.

Although the limits of TAQPR are satisfied in most of the study area except in cities of Iskenderun and Dortyol, the NO_X concentrations are above the limits according to the EC Regulation and the WHO Guidelines. The results have shown that the contribution of industry on the annual average NO_X concentrations is small. Moreover, it was revealed that almost all of the ground level NO_x concentration in the study area was due to traffic sources on inter-city (26%) and urban roads (65%). Analysis of results showed that the traffic is the major contributor to the annual average CO concentration in the study area.

4.1. Acknowledgement

The help provided by ISDEMIR, SIS, SDM, Municipalities of Iskenderun, Dortyol, Payas, and other government organizations during this study is greatly appreciated.

5. REFERENCES

CORINAIR, 1999, Emission Inventory Guidebook-3rd Edition, EEA (European Environment Agency), Copenhagen.

Elbir, T., 2002, "Air Quality Modeling and Decision Making Studies for Izmir," Ph.D. Thesis, Dokuz Eylul University, Izmir, Turkey.

Sivacoumar, R., Bhanarkar, A.D., Goyal, S.K., Gadkari, S.K., Aggarwal, A.L., 2001, Air pollution modelling for an industrial complex and model performance evaluation, *Environmental Pollution*. 111: 471.

State Institute of Statistics (SIS), 2002, Population statistics, Administrative Division, Ankara (D.I.E.).

Turkish Air Quality Protection Regulation (TAQPR), 1986, Official Gazette, No.16269, Ankara, Turkey.

U.S. Environmental Protection Agency, 1995, User's Guide for the Industrial Source Complex (ISC3) Dispersion Model, Volume I and II., Research Triangle Park, North Carolina: Office of Air Quality Planning and Standards Emission, Monitoring and Analysis Division.

World Health Organization (WHO), 2000, Air Quality Guidelines for Europe, 2nd Ed., WHO Regional Office for Europe, Copenhagen, Denmark.

DISCUSSION

R.BORNSTEIN   Where was the meteorological station located?

A.ATIMTAY   It is located in the northern part of the Iskenderun City on the sea cost.

E.GENIKHOVICH   Is it possible that in your calculations the receptor points mostly did not hit the highway and were located mainly at some distance from this source? It could result in underestimation of highest concentrations of NO_x and CO due to the highway traffic.

A. ATIMTAY   It could be possible. However, not for most of the receptor points. The program calculates the concentration at the grid nodes. Therefore, if the road goes between the grid nodes, the concentration calculated at the nearby grid points will be less than the concentration on the road itself.

LONG-RANGE TRANSPORT OF LARGE PARTICLES RELEASED DURING A NUCLEAR ACCIDENT

Jerzy Bartnicki, Brit Salbu, Jørgen Saltbones, Anstein Foss, and Ole Christian Lind[*]

1. INTRODUCTION

In case of a nuclear accident such as an explosion or a fire in a nuclear power plant, a significant part of the refractory radionuclides is emitted to the atmosphere in the form of particulate matter. Existing dispersion models take only partly this fact into account, assuming that only small size (diameter of ~1 μm) particles are subject to long-range transport in the air. Large particles (diameter   10 μm) are usually not included in most long-range transport models, assuming that they are deposited relatively close to the sources. A typical assumption in these models is the assignment of one dry deposition velocity to each radionuclide. This means that each radionuclide is transported and deposited as a separate small particle, whereas in reality several radionuclides are imbedded in particles and can be transported as large particles.

The above assumptions are not entirely confirmed by observations. Available data showed that large particles emitted from the Chernobyl accident were found all over Europe (Pöllänen *et al.*, 1997). These particles, often referred to as "hot particles", are quite frequently found far away from the source and sometimes their activity may be so high that even a single particle can cause severe health effects. Fuel particles containing several radionuclides, as well as ruthenium particles have been found in Scandinavia (Fig. 1), more than 2000 km from the release site (Salbu, 1988). Fuel particles released during the explosion (deposited to the West of the reactor) and fuel particles released during the fire (deposited to the North) varied with respect to structure and density (Salbu *et al.*, 2001).

---

[*] Jerzy Bartnicki, Jorgen Saltbones, Anstein Foss, Norwegian Meteorological Institute, P.O. Box 43, Blindern, N-0313 Oslo, Norway. Brit Salbu, Ole Christian Lind, Agricultural University of Norway, P.O Box 5026, NO-1432 Aas, Norway.

Air Pollution Modeling and Its Application XVI, Edited by
Borrego and Incecik, Kluwer Academic/Plenum Publishers, New York, 2004

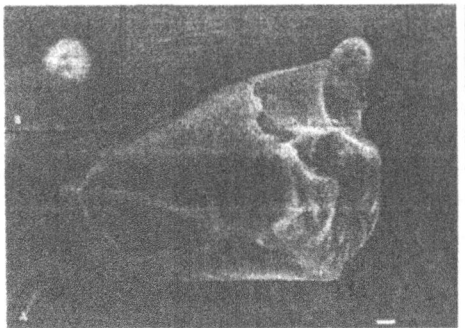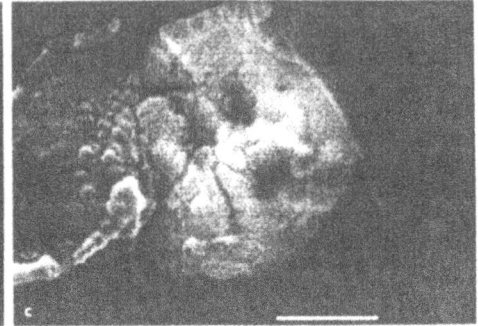

Figure 1. [106]Ru particle (with spherical structure) attached to a larger Si particle. This particle was collected in Valdres, Norway in 1987. Both horizontal bars represent 1 μm.

In order to estimate the range and importance of transport and deposition of large particles, several simulations of the Chernobyl accident have been made with the modified version of the operational dispersion model SNAP developed at the Norwegian Meteorological Institute (Saltbones *et al.*, 1995).

These modifications of the operational version of SNAP (Bartnicki *et al.*, 2001) included: (1) new source term parameterisation reflecting a complicated structure and large particles emitted from the Chernobyl reactor, (2) implementation of gravitational settling velocity – essential for large particles, (3) modified parameterisation of vertical advection and diffusion, taking into account the influence of gravitational settling velocity. Spherical shape of large particles was assumed in the calculations of the gravitational settling velocity. This simplification could underestimate the range of the transport.

The modified version of the SNAP model was run for 12 different classes of particles (three densities and four sizes). For each class there was a different source term resulting in different atmospheric transport and deposition pattern. These runs were performed with meteorological data from the NORLAM Weather Prediction Model, available at the Norwegian Meteorological Institute.

2. RELEASE OF PARTICLES DURING THE CHERNOBYL ACCIDENT

Noble gases, gases of volatile elements, aerosols, fuel particles and fuel fragments were released to the atmosphere during 10 days of release from the Chernobyl Accident. General description of the release can be found in the OECD Report (OECD, 1996). Gaseous forms, such as krypton and xenon escaped completely from the fuel material, while approximately 50-60% of gaseous iodine was released. Some of the iodine, as well as cesium and tellurium were attached to aerosols with a typical size of 0.5-1.0 μm. One of the features of the source term was a release of fuel material in form of hot particles of varying size. An early estimate for fuel material released to the atmosphere was $3\pm1.5\%$ (IAEA, 1986). The revised estimate (Bedyaev *et al.*, 1991), $3.5\pm0.5\%$, corresponds to the six tonnes emitted as fragmented fuel.

Following Buzulakow and Dobrynin (1993), evolution of the release pattern is shown in Figure 2. Large release at the beginning of the period was caused mainly by the mechanical fragmentation of the fuel during the explosion. The second large release, between day 7 and day 10, was associated with the high temperature reached in the core melt and the subsequent fire. Rapid cooling of the fuel probably caused the drop of release after 10 days.

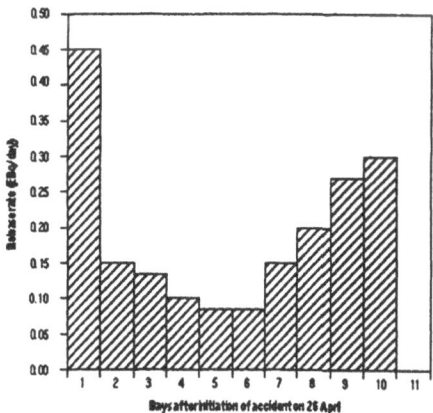

Figure 2. The daily release rate of radioactive substances into the atmosphere (not including noble gases). The range of uncertainty for all releases is ±50%.

Based on particle characterization of such as size, structure and density (Salbu et al., 2001), emission rates for particles were calculated, separately for each size class and density class, for 10 days of the Chernobyl Accident (Bartnicki *et al.*, 2001). These emission rates, used for the model simulations, are summarized in Table 1. The assumed effective release heights, for day 1, days 2-6 and days 7-10 were 1000 m – 2000 m, 50 m – 1000 m and 1000 m – 1500 m, respectively.

3. RESULTS OF THE MODEL SIMULATIONS

Following the specification in Table 1, the modified version of the SNAP model, which included parameterisation of gravitational settling velocity also for large Reynolds numbers, was run for 12 different classes of particles.

The results of both the operational and modified version of the SNAP model are on-line displayed on the screen during the simulation as a cloud of particles superimposed on the map of the model domain. An example of what can be seen during the model simulation is shown in Fig. 3. All classes of particles still remaining in the air on the 7th May, approximately 11 day after the release started are shown in this figure. Figure 3 illustrates the transport pattern from the release of days 7-10, which reached Norway from the South, as a second wave. It includes different particles from the entire period of 10 days emission in Chernobyl.

Table 1. Source term data (based on experimental information on size, structure and density) used in the model simulations of the Chernobyl accident. M_e is the emitted mass. Densities: $U - 19$ g cm^{-3}, $UO_2 - 11$ g cm^{-3}, $U_3O_8 - 8.3$ g cm^{-3}.

| | 1. class: 10% of M_e (100μm - 500μm) $d_p = 200$μm | 2. class: 30% of M_e (30μm - 100μm) $d_p = 50$μm | 3. class: 45% of M_e (1μm - 30μm) $d_p = 10$μm | 4. class: 15% of M_e (0.1μm - 1μm) $d_p = 0.2$μm |
Day				
1	Total emission: 60 kg Rate: 60 kg/day Divided in: U: 30 kg/day UO₂: 30 kg/day U₃O₈: 0 kg/day	Total emission: 180kg Rate: 6180 kg/day Divided in: U: 90 kg/day UO₂: 90 kg/day U₃O₈: 0 kg/day	Total emission: 270kg Rate: 270 kg/day Divided in: U: 135 kg/day UO₂: 135 kg/day U₃O₈: 0 kg/day	Total emission: 90 kg Rate: 90 kg/day Divided in: U: 45 kg/day UO₂: 45 kg/day U₃O₈: 0 kg/day
2-6	Total emission: 60 kg Rate: 12 kg/day Divided in: U: 6 kg/day UO₂: 6 kg/day U₃O₈: 0 kg/day	Total emission: 180kg Rate: 36 kg/day Divided in: U: 18 kg/day UO₂: 18 kg/day U₃O₈: 0 kg/day	Total emission: 270kg Rate: 54 kg/day Divided in: U: 21.6 kg/day UO₂: 21.6 kg/day U₃O₈: 0 kg/day	Total emission: 90 kg Rate: 18 kg/day Divided in: U: 9 kg/day UO₂: 9 kg/day U₃O₈: 0 kg/day
7-10	Total emission: 80 kg Rate: 20 kg/day Divided in: U: 0 kg/day UO₂: 0 kg/day U₃O₈: 20 kg/day	Total emission: 240kg Rate: 60 kg/day Divided in: U: 0 kg/day UO₂: 60 kg/day U₃O₈: 60 kg/day	Total emission: 360kg Rate: 90 kg/day Divided in: U: 0 kg/day UO₂: 0 kg/day U₃O₈: 90 kg/day	Total emission: 120kg Rate: 30 kg/day Divided in: U: 0 kg/day UO₂: 0 kg/day U₃O₈: 30 kg/day

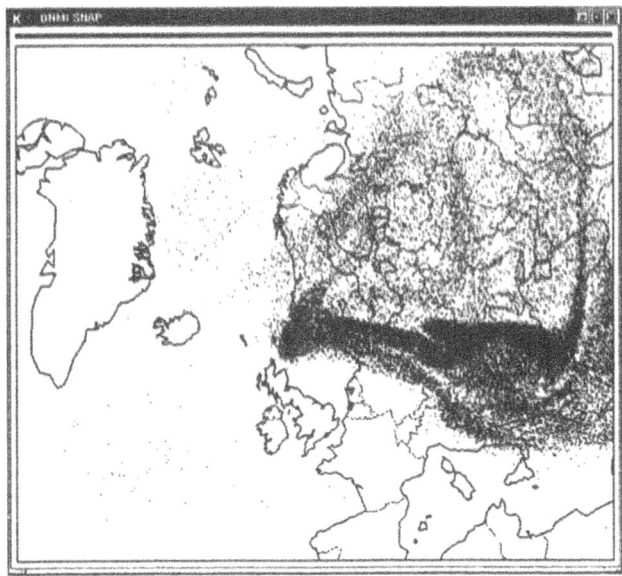

Figure 3. Results of the SNAP model simulation for the 7th of May 1986, 272 hours after the release start. All particles released during the Chernobyl Accident and still remaining in the air are shown in the figure.

The maps of accumulated total (wet + dry) deposition for each of the two size classes, 0.2 μm and 10 μm, and the two density classes, 8.3 g cm^{-3} and 19 g cm^{-3}, are shown in Figures 4 to 7.

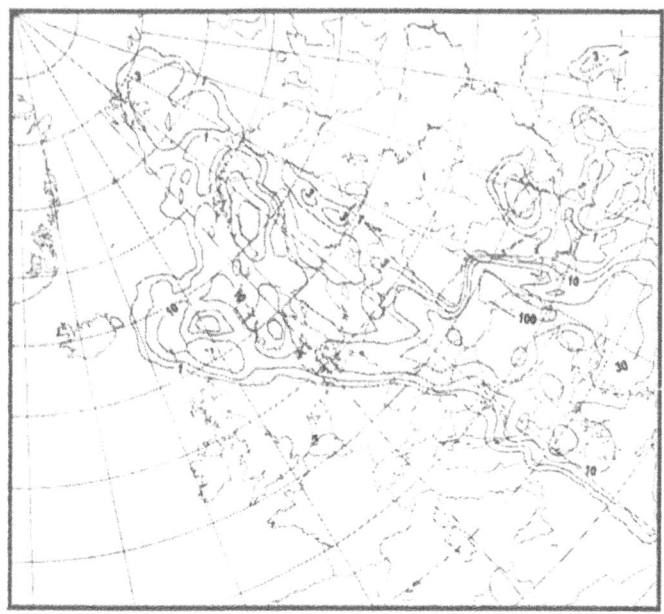

Figure 4. Results of the SNAP model simulation for the 11th of May 1986, 372 hours after the release start. Isolines of accumulated total (wet + dry) deposition for particle size 0.2 μm and density 8.3 g cm^{-3} are shown for the entire period of simulation. Units: ng m^{-2}.

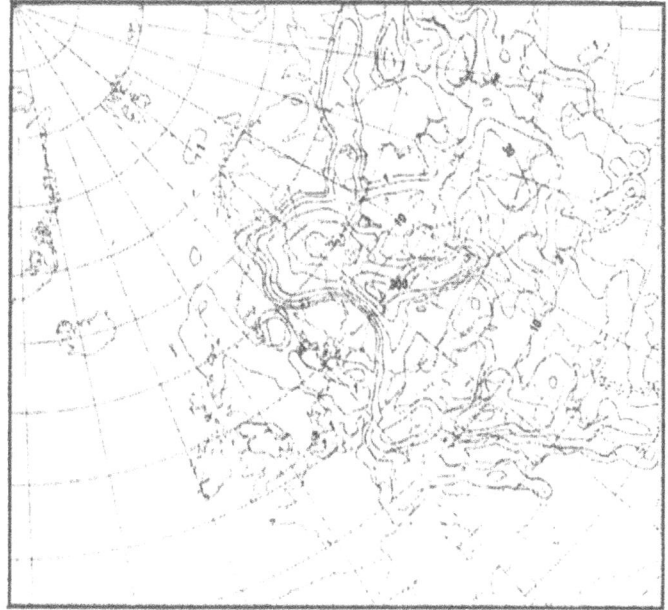

Figure 5. Results of the SNAP model simulation for the 11th of May 1986, 372 hours after the release start. Isolines of accumulated total (wet + dry) deposition for particle size 0.2 μm and density 19 g cm^{-3} are shown for the entire period of simulation. Units: ng m^{-2}.

81

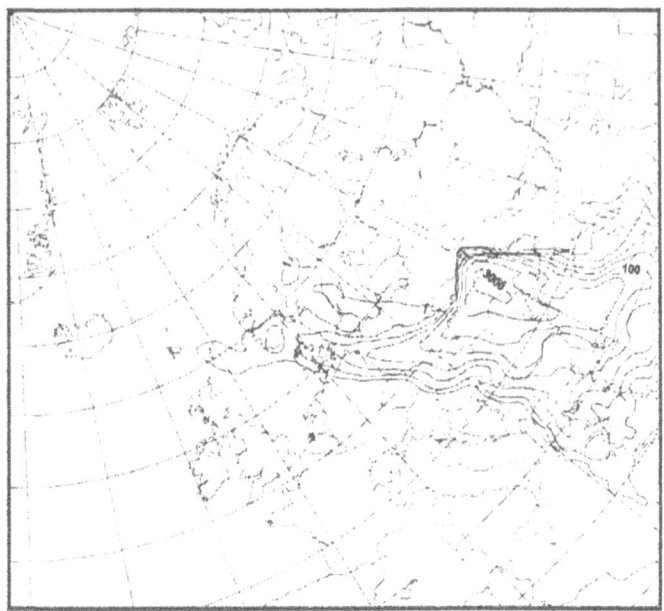

Figure 6. Results of the SNAP model simulation for the 11th of May 1986, 372 hours after the release start. Isolines of accumulated total (wet + dry) deposition for particle size 10 μm and density 8.3 g cm^{-3} are shown for the entire period of simulation. Units: ng m^{-2}.

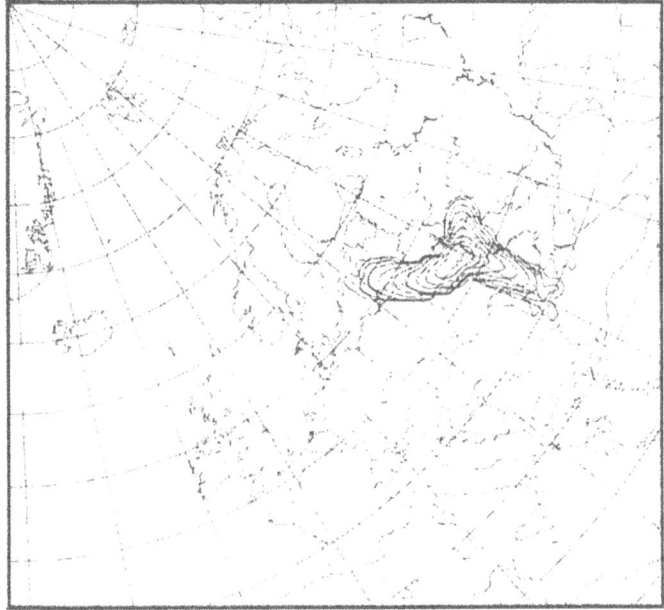

Figure 7. Results of the SNAP model simulation for the 11th of May 1986, 372 hours after the release start. Isolines of accumulated total (wet + dry) deposition for particle size 10 μm and density 19 g cm^{-3} are shown for the entire period of simulation. Units: ng m^{-2}.

82

There are significant differences in the deposition patterns for particles with a diameter of 10 μm and below, and for particles with a diameter of 50 μm and above. Smaller particles are transported for long distances (more than 1000 km), whereas larger particles are deposited relatively close to the source (less then 100 km).

There is relatively small difference in the transport range for two density classes of the smallest particles (0.2 μm), but this difference is quite important for distant countries and Norway in particular (Figs. 4 and 5), because they were assumed to be released at different phases of the accident.

All three density classes of the smallest particles reached Norway, but from different directions and in different periods. Two types of particles; U metal and UO$_2$ particles with densities 19 g cm^{-3} and 11 g cm^{-3}, respectively, came to Norway mainly from the East at the beginning of the simulation period (emissions from the first day, see Table 1). U$_3$O$_8$ particles with density 8.3 g cm^{-3} reached Norway from the South, and at the end of the simulation (emissions from days 7-10, see Table 1). Transport and deposition of these particles are not much different in the operational and modified version of the SNAP model, because gravitational settling velocity in this case is small and not very important for the effectiveness of the dry deposition process. In addition, for such small particles, dry deposition is definitely less effective than wet deposition.

For 10 μm particles, differences between density classes (and related emission periods) are even more pronounced than for the smallest particles (Figs. 6 and 7). In this case, U$_3$O$_8$ and UO$_2$ particles with densities 8.3 g cm^{-3} and 11 g cm^{-3}, respectively, also came to Norway in different directions and in different periods. The third class, the heaviest U particles with density 19 g cm^{-3}, did not reach Norway at all, according to these simulations.

4. CONCLUSIONS

Main conclusions from the SNAP model simulations can be summarized as follows:
o Parameterization and proper description of the source term is very important for modelling long-range transport of radioactive particles.
o Large particles have to be taken into account in the long-range transport models, because as proved by observations (Figure 1), they can travel long (more than 1000 km) distances.
o The modified version of the SNAP model was able to simulate transport of 10 μm particles released during the Chernobyl accident to Norway. This means that the model was able to reproduce observations, since deposition of such particles was observed in Norway after the Chernobyl accident.
o The border line (concerning particle size and structure) between long-range and local scale particle deposition remains unclear. The answer to this question may be found after more systematic model simulations with particle size/shape/density classes. The answer can be significantly influenced by implementation of deep convection and correction factor for non-spherical particles in the dispersion model.

6. REFERENCES

Bartnicki J., Salbu B., Saltbones J., Foss A., and Lind O.C., 2001, Gravitational settling of particles in dispersion model calculations using the Chernobyl Accident as a test case. DNMI Research Report No. 131. Norwegian Meteorological Institute, Oslo, Norway.

Buzulakow Y.P., and Dobrynin Y.L., 1993, Release of radionuclides during the Chernobyl accident. In: The Chernobyl Papers 1 (Eds: S.E. Merwinand M.I. Balonov). Research Enterprise, Richland, WA, pp. 3-21.

IAEA (1986) Summary Report on the Post-Accident Review Meeting on the Chernobyl accident. Safety Series 75, INSAG-1, IAEA, Vienna, Austria.

OECD, 1996, Chernobyl. Ten years on. Radiological and health Impact. OECD NEA Report.

Valkama I., and Toivonen H., 1997, Transport of radioactive particles from the Chernobyl Accident. *Atmospheric Environment* 31, 3557-3590.

RAFF, 1999, Properties of nuclear fuel particles and release of radionuclides from carrier matrix. RAFF Final Report. A research programme carried out with a financial support from the Commission of the EC - DG XII. Contract No. FI4CCT960007.

Salbu B., 1988, In: *Radionuclides associated with colloids and particles in rainwaters*, H. von Philipsborn, and F. Steinhäuser, Eds. (Bergbau - und Industrimuseum, Theuern) 16, 83.

Salbu, B., Krekling, T., Lind, O. C., Oughton, D. H., Drakopoulos, M., Simionovici, A., Snigireva, I., Snigirev, A., Weitkamp, T., Adams, F., Janssens, K. & Kashparov, V.A., 2001, High energy X-ray microscopy for characterisation of fuel particles. *Nucl. Instr. and Meth. A*, 467 (21), 1249-1252.

Saltbones J., Foss A., and Bartnicki J., 1995, SNAP: Severe Nuclear Accident Program. Technical description. DNMI Research Report No. 15. Norwegian Meteorological Institute, Oslo, Norway.

DISCUSSION

A.RYABOSHAPKO What height of the source do you use in your model? Do you take in to account vertical spreading of the source just after the explosion?

J.BARTNICKI Assumed release height for day 1, days 2-6, and days 7-10 was 1000m–2000m, 50m–1000m, and 1000m 1500m, respectively.

A.L. NORMAN Could the source flux be improved by inverse modelling?

J.BARTNICKI Probably yes, in case of cesium and iodine, because of the relatively large number of measurements available. Probably no, in case of large fuel particles, because there are not enough measurements.

D.KORACIN How did you treat turbulence and diffusion and their link in such relatively coarse resolution for large scale transport?

J.BARTNICKI Diffusion in the model is parameterized by random diplacements of particles in the horizontal and vertical directions. The displacements are proportional to the lenght scales of horizontal and vertical turbulent motions.

E.GENIKHOVICH In the diffusion limit, there are two major factors to be accounted for when simulating the dispersion of large particles. The first one is the gravitational sedimentation with the terminal velocity and the second one is the reduction in the effective eddy diffusivity with increase of the terminal velocity due to decorrelation of particles initially located inside the same turbulent eddy. Did you take the Judine effect into account?

J.BARTNICKI This effect is not taken into account in the model.

B.TERLUIC You didn't mention any radioactive reaction in your presentation. Did you take into account those reactions in your calculations?

J.BARTNICKI Radioactive reactions are not included in the model equations, however, we take into account radiactive decay.

CHEMICAL MECHANISMS IN TWO PHOTOCHEMICAL MODELLING SYSTEMS: A COMPARISON PROCEDURE

Joana Ferreira, Anabela Carvalho, Ana C. Carvalho, Alexandra Monteiro, Helena Martins, Ana I. Miranda and Carlos Borrego[*]

ABSTRACT

The main purpose of this study is the evaluation of different chemical mechanisms included in two Eulerian photochemical models, MARS and CAMx, comparing simulation results. Both mechanisms (KOREM and EMEP) included in MARS were applied, whereas concerning CAMx the tested mechanisms were the default Carbon Bond IV and the revised radical termination reactions for the same mechanism.

These two models were applied to a Portuguese coastal region, Aveiro, where a field campaign was carried out during a summer period, from 25th June to 2nd July 2001. The understanding of the influence of mesoscale meteorological phenomena on the ozone production over the region of interest was the major objective of this field campaign, which was designed considering photochemical modelling evaluation. Therefore, a 48 hours simulation was performed using a 200 x 140 km^2 domain. Meteorological inputs to both photochemical models were obtained by the application of the mesoscale meteorological model MM5 using its nesting capabilities. Meteorological and ozone concentration simulation results were evaluated using the data collected during the field campaign. Preliminary results concerning the different chemical schemes applied, KOREM, EMEP and CB-IV, present a good skill for ozone peak simulations. All mechanisms may be applied for population alert situations. Although, it is important to notice that KOREM scheme offers an interesting and simple operative solution for the study region.

1. INTRODUCTION

Atmospheric chemistry-transport models are useful tools for the understanding of pollutant dynamics in the atmosphere and are the best available way to explain ozone

---

[*] Department of Environment and Planning, University of Aveiro, 3810-193 Aveiro, Portugal.

Air Pollution Modeling and Its Application XVI, Edited by
Borrego and Incecik, Kluwer Academic/Plenum Publishers, New York, 2004

episodes. Nevertheless, the chemical mechanisms included in the Eulerian air quality models may be numerically expensive in terms of computing time. Due to this, the chemical scheme is often simplified in a small limited number of chemical reactions (Aumont et al., 1997). Additionally to the kinetic module, a photochemical system should be able to reproduce daily ozone variations due to horizontal turbulence effects, vertical mixing, local removal by nitrogen oxide (NO) and response to fast changing emissions (Hogrefe et al., 2001). For that, it should be taken into account meteorological effects, emissions, transport, chemical transformation and removal processes at surface of ozone concentration (Rao et al., 2000).

To evaluate the chemical mechanism influence on air quality modelling and to define their performance relative to each other, four different chemical mechanisms included in two photochemical mesoscale modelling systems, were tested and applied to the Aveiro study area. The CAMx model includes the Carbon Bond – IV and the CB – IV with extensions whereas the MARS model considers the KOREM and the EMEP mechanisms, all of them condensed concerning molecular structure reactivity.

The region of interest is located in the Atlantic coast of Portugal, where high levels of photochemical pollutants, like ozone, have been measured. This coastal region is characterised by a high concentration of human activities, which produces significant anthropogenic atmospheric emissions. On the other hand, the complex topography and some favourable synoptical situations imply the occurrence of mesoscale circulations, namely sea/land breezes and anabatic/katabatic flows.

In order to study the atmospheric mesoscale circulations importance on regional ozone transport and production, an air quality field campaign was designed and carried out in the Aveiro region, from 25th June to 2nd July 2001 (Evtiouguina et al., 2002)

2. THE SELECTED CHEMICAL MECHANISMS

A short description of the chemical mechanisms included in the Eulerian photochemical models under evaluation will be given. Both photochemical modelling systems apply a numerical Arakawa C grid.

2.1. CAMx model: CB-IV mechanisms

CAMx is a 3D Eulerian model that contains five chemical mechanisms, four of them based on the Carbon Bond Mechanism version 4 (CB-IV) and the other one on the SPARC99 (ENVIRON, 2002). For this study, two of the CB-IV revised mechanisms were applied: CB-IV as implemented in the EPM UAM-IV (extensively used at the University of Aveiro, Borrego et al., 2000b; Monteiro et al., 2002) and CB-IV with updated isoprene and reactive chlorine chemistry.

Concerning the VOC speciation, CB-IV is a lumped structure mechanism used for both urban and regional scale modelling. In this lumped structure approach, organics are divided into smaller reaction elements based on the types of carbon bonds existing in each of the 12 VOC species (alkanes, 1-alkenes, ethene, toluene, monoalkybenzene, dialkybenzenes, trialkybenzenes, formaldehyde, other aldehydes isoprene, methylglyoxal and unsaturated dicarbonyls).

The differences between the two selected mechanisms relay on the number of chemical reactions (91 and 110 respectively), primary and secondary photolysis reactions and gas species (24 and 34 respectively).

2.2. MARS model: KOREM and EMEP mechanisms

The MARS model, a 3D Eulerian well-tested model in Europe, numerically simulates photo-oxidants formation considering the chemical transformation process of pollutants together with its transport in the atmospheric boundary layer (Moussiopoulos et al., 1995). The EMEP (Simpson et al., 1993) mechanism describes the tropospheric gas-phase chemistry with 66 species and 139 photochemical reactions, including 34 photolysis reactions; and the KOREM mechanism, which is simpler, includes 39 chemical reactions and 20 reactive pollutants. KOREM results from the combination of inorganic reactions of the CERT mechanism (Atkinson et al., 1982) and organic reactions of the compact mechanism of Bottenheim and Strausz (1982). Volatile organic compounds (VOCs) are lumped in five classes: methane, alkanes, alkenes, aromatics and aldehyds On the other hand, the EMEP mechanism considers the following VOC speciation: methane, ethane, n-butane, ethane, propene, o-xylene, formaldehyde, acetaldehyde, methylethylketone, methanol, ethanol and isoprene (Moussiopoulos, 1992).

3. METHODOLOGY

During the performed field campaign, measurements of all the main meteorological parameters and ozone and its precursors concentrations were taken at surface and in altitude. Three air quality stations were installed at Aveiro, Sangalhos and Covelo (Figure 1). Their location was chosen taking into account the most frequent wind direction (from N-NW) in the west Portuguese coast during summer season. This wind direction can be associated with different synoptical and mesoscale meteorological situations, the Azores anticyclone oriental border, Iberian Peninsula thermal low and sea-breezes circulations. The data used in MM5 validation were acquired at the meteorological stations located at Aveiro, Sangalhos and Anadia (near Sangalhos).

To obtain the resolution required by air quality models, the anthropogenic emissions data (NO_x, CO and VOC) were disaggregated (spatial and temporally) from the national emissions values of the most recent CORINAIR inventory using adequate statistical indicators (top-down methodology, Borrego et al., 2002). Biogenic emissions (isoprene and monoterpene) from forest were calculated using local and detailed dataset and emission factors (bottom-up methodology, Borrego et al, 1998). The VOC emissions were split in different species, according to each chemical mechanism, using adequate profiles. Therefore, the only difference between VOC emissions considered by each mechanism is related with VOC profiles both for biogenic and anthropogenic emissions.

The meteorological model MM5 (Grell et al., 1995) was used to drive the meteorological fields needed by both photochemical modelling systems. MM5 is a wide spread community model applied worldwide, namely by several research institutions of the Iberian Peninsula, as a research tool or operational (URL 1).

Using MM5 capability of doing multiple nestings, the meteorological model was applied with the one way nesting option and for two nests: (i) a large domain covering the

Iberian Peninsula (45 km resolution), (ii) a first nest covering the Atlantic Coast of Iberian Peninsula (15 km resolution), (iii) and a second nest for the North part of Portugal (5 km resolution). The grid sizes are 82 x 50, 67 x 67, and 43 x 31 grid points, respectively. MM5 simulations were initialised from the gridded NCEP/AVN reanalysis data. All modelling domains have the same vertical structure with 23 unequally spaced σ levels, with first level height at 22 m. The 5 km grid is governed by the Reisner graupel moisture scheme, Grell cumulus scheme, and Mellor-Yamada scheme (used into the Eta model) for the boundary layer parameterisations.

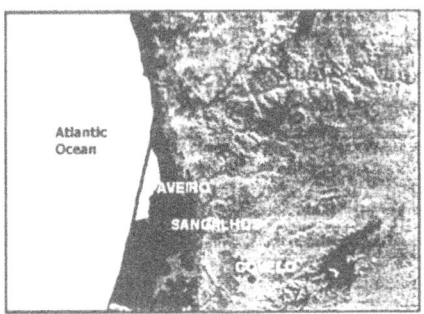

Figure 1. Location of meteorological/air quality stations in the photochemical simulation domain (200 km x140 km).

Carvalho *et al.* (2001) developed and implemented an interface regarding MM5-MARS compatibility considering the meteorological parameters calculation in different numerical grid types. In conformity with MARS model turbulence schemes, vertical turbulent diffusion coefficients were calculated based on turbulent kinetic energy fields simulated by MM5. Both photochemical modelling systems assumed a initial state of clean air for initial and boundary conditions.

4. SIMULATION PERIOD CHARACTERISATION

To evaluate the two photochemical model systems over Aveiro region, the period between 27th and 28th of June 2001 was selected. During these days the Iberian Peninsula was suffering the influence of an anticyclone (a common weather pattern over Portugal) on an intensifying process, favouring the development of mesoscale meteorological phenomena (URL 2, URL3).

Table 1. Differences between maximum and minimum temperature values.

	$\Delta T=$ Tmax-Tmin (°C) 27th of June	$\Delta T=$ Tmax-Tmin (°C) 28th of June
Aveiro	5.3	10.1
Anadia	11.6	14.5
Covelo	8.9	16.8

90

The amplitude in temperature-measured values (Table 1) is bigger in the inner station (Covelo) and that values increase on the 28th of June. However, the highest temperature was registered on June 28th at Anadia (27.2 °C) followed by Covelo (27.1 °C). All the meteorological stations show the mesoscale circulation overlapping the synoptic pattern during the afternoon for both days – the wind speed increases during the afternoon, temperature values drop for a small period of time and relative humidity increases at the same time.

The highest values of ozone measurements were detected at Covelo during June 28th (Table 2), were it can be observed the maximum and minimum measured ozone values, and their time of occurrence. The presented values indicate the importance of mesoscale phenomena in air pollutants transport and further ozone production. Except for Aveiro on June 27th, the maximum values have occurred latter in time and enhanced, as the stations location is more inland. Ozone minimum values vary between 8 and 36 µg m[-3] and occur between 3 and 8 a.m., later at Covelo. Once again, Aveiro does not present a regular behaviour, which can indicate that air quality station detects more ozone from urban origin than regional one. Greater amplitudes in ozone concentration could be directly correlated to greater amplitudes in temperature.

Table 2. Maximum and minimum measured ozone concentration ($\mu g.m^{-3}$) and corresponding hour of occurrence (local standard time).

Location	\multicolumn{4}{c}{27[th] of June}				\multicolumn{4}{c}{28[th] of June}			
	Min O_3	Hour	Max O_3	Hour	Min O_3	Hour	Max O_3	Hour
Aveiro	19.8	9	79	21	15.8	4	71.1	12
Sangalhos	19.8	4	65.2	15	7.9	4	88.9	14
Covelo	35.6	6-7	73.1	18-19	15.8	7-8	104.7	17

5. RESULTS AND DISCUSSION

The MM5 skill is evaluated through the application of quantitative error analysis introduced by Keyser and Anthes (1977).

$$E = \left(\sum_{i=1}^{N} (\phi_i - \phi_{iobs})^2 \Big/ N \right)^{1/2} \qquad E_{UB} = \left(\sum [(\phi_i - \phi_0) - (\phi_{iobs} - \phi_{0obs})]^2 \Big/ N \right)^{1/2}$$

$$S = \left(\sum_{i=1}^{N} (\phi_i - \phi_0)^2 \Big/ N \right)^{1/2} \qquad S_{obs} = \left(\sum_{i=1}^{N} (\phi_{iobs} - \phi_{0obs})^2 \Big/ N \right)^{1/2}$$

The parameter E is the root mean square error (rmse), E_{UB} the rmse after the removal of a certain deviation and S and S_{obs} the standard deviation of the modelled and observed data. If ϕ_i and ϕ_{iobs} are individual modelled data and observed in the same mesh cell, respectively; ϕ_0 and ϕ_{0obs} the average of ϕ_i and ϕ_{iobs} for some sequence in study, and N the number of observations, then the simulation presents an acceptable behaviour when $S \approx S_{obs}$, $E < S_{obs}$ and $E_{UB} < S_{obs}$.

Regarding parameter S/S_{obs} (Table 3), the results are considerably close to unity, except for Aveiro concerning temperature and for Anadia regarding wind speed and direction. These values indicate the model capability to better simulate temperature natural variability at Anadia, although this variability is better resolved for the wind speed and direction at Aveiro and Sangalhos. A preliminary conclusion is that advective processes are well represented on the photochemical modelling systems.

Table 3. Statistical analysis for temperature, wind speed and direction at the three meteorological stations.

		Aveiro	Anadia	Sangalhos
Temperature	S/S_{obs}	0.50	0.95	1.24
	E/S_{obs}	0.62	0.35	0.39
	E_{UB}/S_{obs}	1.81	0.31	0.36
Wind speed	S/S_{obs}	0.80	1.61	0.93
	E/S_{obs}	0.54	2.08	1.00
	E_{UB}/S_{obs}	0.62	1.23	0.51
Wind direction	S/S_{obs}	0.93	0.66	0.99
	E/S_{obs}	0.98	1.22	1.33
	E_{UB}/S_{obs}	1.08	1.03	1.23

Concerning photochemical models assessment, each mechanism is evaluated through the application of the USEPA (1991) guideline. The statistical analysis was applied rejecting the first 8 hours of simulation, to consider the spin up of the model. The selected quality indicators were focused on models ability to predict the peak ozone values, unpaired highest-prediction accuracy (A_u), normalized bias test (D), average station peak prediction accuracy ($\bar{A}$), bias of all station peaks (D_{peak}) and fractional bias for peak concentration (mean and standard deviation, F_m and F_S, respectively). The values obtained for A_u parameter should be around ± 15-20% and D in the range of ±0.05-0.15. F_m and F_S vary from –2 to +2, negative values indicate over prediction and positive values indicate under prediction.

$$A_u = 100\left(\frac{C_{max\ obs} - C_{max\ pred}}{C_{max\ pred}}\right)$$

$$D = \frac{1}{N_T}\sum_{i=1}^{N}\sum_{j=1}^{H_i}\frac{C_{obs}(i,j) - C_{pred}(i,j)}{C_{obs}(i,j)}$$

$$\bar{A} = \frac{100}{N}\sum_{i=1}^{N}\frac{C_{peak\ obs}(i,t_i) - C_{pred}(i,t_i)}{C_{peak\ obs}(i,t_i)}$$

$$D_{peak} = \frac{1}{N}\sum_{i=1}^{N}\left(C_{peak\ obs}(i,t_i) - C_{pred}(i,t_i)\right)$$

$$F_m = 2\cdot\frac{m_{max\ obs} - m_{peak\ pred}}{m_{max\ obs} + m_{peak\ pred}}$$

$$F_S = 2\cdot\frac{S_{max\ obs} - S_{peak\ pred}}{S_{max\ obs} + S_{peak\ pred}}$$

Table 4 presents the statistical analysis results from the comparison between measured and simulated ozone values. No differences were verified on the comparative temporal variation of CB-IV and CB-IV with extensions simulations results, for the 3 stations. Therefore, the obtained statistical parameters are the same for these two mechanisms. As can be seen, all the four chemical mechanisms simulated reasonably well the surface ozone concentration.

According to the number and location of air quality stations available for this study, the highest ozone level inside the domain (represented by A_u) is well simulated by both model systems. Nevertheless, this parameter presents better results for the first day of simulation and for KOREM and CB-IV mechanisms. Concerning statistical parameters that evaluate ozone peaks the results present good skill for all mechanisms and for both days.

Table 4. Statistical analysis for the 3 different model applications, at the 3 stations.

	27 June						28 June					
	Au (%)	D	Ä (%)	D_{peak}	F_m	F_s	Au (%)	D	Ä	D_{peak}	F_m	F_s
CAMx CB-IV	9.7	-0.25	5.3	4.36	-0.02	0.88	24.2	-0.56	9.5	10.1	-0.02	0.72
MARS KOREM	3.2	0.11	13.2	10.9	0.15	-0.53	23.5	-0.45	18.9	19.1	0.02	0.63
MARS EMEP	10.1	0.28	20.0	15.4	0.26	-0.46	31.2	-0.29	37.8	33.10	0.13	0.98

Despite its simpler formulation, parameter D shows that KOREM mechanism has the best performance for the first simulated day. Worst results of this parameter can be justified by the worst fitting of the simulated and measured concentrations during the night period at the inner stations (Sangalhos and Covelo). Both modelling systems cannot reproduce the removal process occurring during the night. This can be related to several assumptions, namely NO_x emissions underestimation during the night, incorrect tuning of the boundary layer parameterisations, not time dependent boundary conditions and the fact that all air quality stations are located in the southwest part of the domain.

6. CONCLUSIONS

Both modelling systems presented a reasonable performance for ozone peak levels simulation at the air quality stations location. Concerning the chemical mechanisms, preliminary conclusions point to a non-significant improvement of results associated to a more complete description of photochemical reactions. Considering the two CAMx mehanisms application over the study region, no differences were detected. It is not possible to conclude which mechanism is more adequate, but the KOREM results show that a more sophisticated mechanism does not mean more accurate results. Previous applications of MARS model lead to the same conclusions (Miranda et al., 2002). These studies should be regarded as a first attempt to compare model performance using different chemical mechanisms and this kind of study requires a representative air quality network over the simulation domain, one of the weak points of this analysis.

7. ACKNOWLEDGEMENTS

The authors wish to thank to the financial support of the 3rd EU Framework Program and the Portuguese Ministério da Ciência e do Ensino Superior, for the PhD grants of J. Ferreira

(SFRH/BD/3347/2000), A.C. Carvalho (PRAXIS XXI/BD/21474/99), A. Carvalho (SFRH/BD/10882/2002), and A. Monteiro (SFRH/BD/10922/2002); and for the research grant of H. Martins in the scope of QUIMERA Project (POCTI/34346/CTA72000).

8. REFERENCES

Atkinson, R.; Lloyd, A. C. and Winges, L., 1982, An updated chemical mechanism for hydrocarbon/NO$_x$/SO$_2$ photo-oxidations suitable for inclusion in atmospheric simulation models, *Atmospheric Environment* 16, 1341-1355.

Aumont, B., Jaecker-Voirol, A., Martin, B., Toupance, G., 1997, Tests of some reduction hypotheses made in photochemical mechanisms, *Atmospheric Environment* 30, 2061-2077.

Borrego, C; N. Barros, M. Lopes; M. Conceição; M. J. Valinhas, O. Tchepel; C. Ferreira, M. Coutinho and S. Lemos, 1998, Emission inventory for simulation and validation of mesoscale models. *Symp98, EUROTRAC-2, Transport and Chemical Transformation in the Troposphere*. Garmisch-Partenkirchen.

Borrego, C.; Tchepel, O.; Barros, N. e Miranda, A.I., 2000, Impact of road traffic emissions on air quality of the Lisbon region, *Atmospheric Environment*, 34, Pergamon, pp. 4683-4690.

Borrego, C.; Tchepel, O.; Monteiro, A.; Barros, N. and Miranda, A., 2002), Influence of traffic emissions estimation variability on urban air quality modelling, *Water, Air and Soil Pollution: Focus 2*: 487-4999. Kluwer Academic Publishers.

Bottenheim, J. W. and Strausz, O. P., 1982, Modelling study of a chemically reactive power plant plume, *Atmospheric Environment*, 16, pp 85-106.

Carvalho, A.C.; Carvalho, A.; Monteiro, A.; Borrego, C.; Miranda, A.I.; Gelpi, I.; Pérez-Muñuzuri, V.; Méndez, M.R.; Souto, J.A., 2001, Air quality study over the Atlantic coast of Iberian Peninsula. *Air Pollution Modelling and its Application XV, Eds Carlos Borrego and Guy Schayes*, Kluwer Academic/ Plenum Publishers, New York, pp.59-66.

ENVIRON International Corporation, 2002, Comprehensive Air Quality Model with Extensions – CAMx. Version 3.10. User's guide.

Evtiouguina, M., Nunes, T. e Pio, C., 2002, Temporal variations and lateral profiles of volatile organic compounds (VOC) in the breeze front of the Portugal west coast during summer period. *European Geophys. Soc. XXVII General Assembly*. Nice, *Geophysical Res. Abst., Vol 4*.

Grell, G., Dudhia, J., and Stauffer, D., 1995, A Description of fifth-generation Penn State/NCAR Mesoscale Model (MM5), NCAR Tech. Note NCAR/TN-398+STR, Boulder, pp. 122.

Hogrefe, C; Rao, S. Trivikrama; Kasibhatla, P; Hao, W.; Sistla, G; Mathur, R. and McHenry, J., 2001, Evaluating the performance of regional-scale photochemical modelling systems: Part II – ozone predictions, *Atmospheric Environment*, 35, pp 4175-4188.

Keyser, D. and Anthes, R.A., 1977, The applicability of a mixed-layer model of the planetary boundary layer to real-data forecasting. Mon. Weather Rev., Vol. 105, pp. 1351-1371.

Miranda, A.I.; Martins, H.; Monteiro, A.; Ferreira, J.; Carvalho, A.C. and Borrego, C., 2002, Evaluation of two mesoscale modeling systems using different chemical mechanisms. *Proceedings of 4th Symposium on the Urban Environment, Joint Session*, eds. American Meteorological Society, Boston, USA, pp J77 – J78

Monteiro, A.; Lopes, M.; Borrego, C. and Miranda, A.I., 2002, Contribution of air pollution to the management of carbon cycle on a Portuguese coastal region. *Coastal Environmental, Environmental Problems in Coastal Regions IV*, eds C. A. Brebbia, WITpress Southampton, pp. 395-404.

Moussiopoulos, N., 1992, MARS (Model for the Atmospheric Dispersion of Reactive Species): Technical Reference, Aristotle University, Thessaloniki.

Moussiopoulus, N., Sahm,. P. and Kessler, Ch., 1995, Numerical simulation of photochemical smog formation in Athens, Greece – A case study, *Atmospheric Environment* 29, 3619-3632.

Rao, S. T.; Hogrefe, C.; Mao, H.; Biswas, J.; Zurbenko, I.; Porter, P. S.; Kasibhatla, P. and Hansen, D. A., 2000, *24th Int. Tech. Meeting of NATO/CCMS on Air Pollution Modelling and its Application*, Gryning & Schiermeier, eds, Kluwer Academic/Plenum Press, pp. 25-34.

Simpson, D.; Andersson-Sköld Y. A. and Jenkin M. E., 1993, Updating the chemical scheme for the EMEP MSC-W Note 2/93. The Norwegian Meteorological Institute. Oslo, Norway.

USEPA, 1991, Guideline for Regulatory Application of the Urban Airshed Model. EPA-450/4-91-013.

URL1: http://rediberiacamm5.uib.es/

URL2: http://www.wetterzentrale.de/topkarten/fsavneur.html

URL3: http://www.infomet.am.ub.es/arxiu/mapes_fronts/

DISCUSSION

S.A.- AKSOYOGLU   1. Korem is very simple, only with 5 VOC classes, Therefore results cannot be very reliable, species with very different reaction rate constants are described in the same class.

2. Comparison of differrent mechanisms should not be done only for ozone, but also for other species.

J.FERREIRA   1. Actually, KOREM is a very simple mechanism compared to EMEP and CB-IV, they consider 13 and 12 VOC classes respectively. Of course, this has to be taken into account when analysing photochemical simulation results. Also the kind of input data should be considered in this process, namely the fact that the available emission inventories do not disaggregate VOC. Currently the used VOC classes came from literature disaggregation schemes. On the other hand, the objective of this study was, precisely, the comparison of different chemical machanisms with different number of reactions and VOC speciations.

2. Other species, for example NO_x, were analysed in this comparison procedure, but only ozone results were presented.

Z. KLAIC   Why did you employ different numbers of vertical layers in CAMx and MARS simulations?

J.FERREIRA   MARS model considers 8 vertical layers. Concerning CAMx, it is advisable to consider a minimum of 13 vertical layers in simulations. Although, the 8 first layers were the same for both MARS and CAMx applications.

R.BORNSTEIN   It appears that Korem-chem is more rapid than CAMx-chem, thus significant O_3 forms closest the coast with Korem.

J.FERREIRA   Once KOREM mechanism has a simpler formulation it performs ozone production faster than CB-IV mechanism. This is evident on the ozone concentration fields at 36 hours of simulation obtained for KOREM and CB-IV mechanisms applications and shown during the presentation.

K.J. REZLER

1. Have you drawn your conclusions only from one 48h simulation (presented here) or from bigger number of episodic simulation?

2. How these conclusion can be dependent on specific meteorological / terrain conditions?

J.FERREIRA

1. Previous studies with MARS model were carried out with the purpose of evaluating photochemical simulation performance for ozone episodes. Thus, 48 hours simulations were done and the same conclusions were drawn – KOREM presented the best performance.

2. The various applications were performed for the same region within different meteorological patterns, namely anticiclone and thermal low conditions. In both conditions the KOREM mechanism presented better results.

WINTERTIME AND SUMMERTIME EVALUATION OF THE REGIONAL PM AIR QUALITY MODEL AURAMS

V.S. Bouchet[1], M.D. Moran[2], L-P. Crevier[1], A. P. Dastoor[1], S. Gong[2], W. Gong[2], P.A. Makar[2], S. Menard[1], B. Pabla[2] and L. Zhang[2*]

1. INTRODUCTION

Three dimensional air quality models have established themselves as valuable tools for the development of emission control strategies. They give scientists the ability to investigate the consequences of multiple emission reduction scenarios as well as research optimum sets of pollutant and/or precursor emission reductions under various meteorological conditions. This capability is becoming crucial as advances in atmospheric science in the past decades have revealed numerous links between ground-level air pollution problems such as ozone, acid rain and particulate matter (PM) (Hidy et al., 1998). With the recognition of PM, especially, as a severe human health concern (Samet et al., 2000; Brook et al., 2002), efforts have lately focused on addressing this new issue without compromising any achievement in reducing ozone and acid rain. As demonstrated by Meng et al. (1997), decreasing VOCs, which are ozone precursors, could free nitrogen oxides and results, under certain conditions, in an increase in PM mass. The so-called 'one atmosphere' models are therefore desirable to study PM issues.

AURAMS, A Unified Regional Air quality Modelling System, developed by the Meteorological Service of Canada, is such a model. It is designed to study interactions between nitrogen oxides (NO_x), volatile organic compounds (VOCs), ammonia (NH_3), ozone (O_3) and, primary and secondary PM through aqueous, gaseous and heterogeneous reactions. Before being used for policy applications however, large air quality models need to undergo thorough evaluation to assess their performance under different meteorological and chemical regimes. As an episodic model, AURAMS has to be evaluated on relatively short multi-day simulations. As a first step, two cases were

---

1   Véronique S. Bouchet, Louis-Philippe Crevier, Ashu P. dastoor and Sylvain Ménard, Meteorological Service of Canada, CMC, 2121 Route TransCanadienne, Dorval, Québec, H9P 1J3, Canada

2   Michael D. Moran , Sunling Gong, Wanmin Gong, Paul A. Makar, Balbir Pabla and Leiming Zhang, Meteorological Service of Canada, 4905 Dufferin Street, Downsview, Ontario, M3H 5T4, Canada

selected in this study: a summertime and a wintertime episode. Results from a preliminary performance evaluation are presented in the following.

2. MODEL DESCRIPTION

AURAMS is an episodic, regional scale, air quality modelling system (Moran et al., 1998). It is composed of 5 different programs run in sequence, as detailed in Makar et al. (these proceedings). The Canadian Emission Processing System -CEPS- (Scholtz et al., 1999) and the meteorological driver, either the Canadian forecast model GEM (Côté et al, 1998) or the Canadian mesoscale community model MC2 (Benoit et al., 1997), provide the corresponding input fields. The chemical transport model itself uses a non-oscillatory semi-Lagrangian advection scheme (Pudykiewicz et al., 1997) to describe the transport of up to 145 individual chemical tracers. Gas-phase and aqueous reactions are simulated with the ADOM-II mechanisms (Lurmann et al., 1986), with a significant update to allow for secondary organic aerosol (SOA) formation (Odum et al., 1996), while heterogeneous reactions are described, in the version of AURAMS used here, by the thermodynamic equilibrium module ISORROPIA (Nenes et al., 1998). The dynamic of internally mixed aerosol is represented with the Canadian Aerosol Module (CAM), which uses a sectional approach (Gong et al., 2002). Deposition of both gases and particles was updated and is described in Zhang et al. (2001) and Zhang et al. (2002).

For the present study, AURAMS was run at 40km resolution on its eastern domain, covering most of eastern Canada and the United-States (Figure 1). Aerosols were represented using 12 size bins, ranging from 0.01 to 40.96 μm in diameter, and 5 chemical species (sulphate, nitrate, ammonium, sea-salt and SOA). Elemental carbon, crustal material and aerosol-bound water were not included for these runs. All gas and aqueous phase reactions included in AURAMS were taken into account, while heterogeneous chemistry was limited to the SO_4^{2-} -HNO_3-NH_3-H_2O system. A typical 2-day warm-up period was used for each simulation, and gases and aerosol mass concentrations were output every hour (AURAMS internal time step was 900s here). Post-processing of the chemical output is required to produce fields comparable to the observations as will be described in the next section. Finally, despite some obvious limitations, emissions input are based on the 1985 NAPAP emission inventory (US EPA, 1989), which did not include any information regarding primary PM emissions. These episodes will be simulated again, once the updated 1995 emission inventory is processed for input in AURAMS.

3. EPISODE AND DATA DESCRIPTION

July 9 to 14, 1995 and February 4 to 13, 1998 were chosen as evaluation periods. The first event is a typical summertime regional episode associated with a slow-moving high pressure system over most of North-Eastern America (Ryan et al., 1998). Peak ground-level ozone concentrations exceeding 120 ppbv, and as high as 179 ppbv, were observed at more than 35 sites throughout the eastern U.S. seaboard between July 12 and 15, 1995 (Ryan et al., 1998). High ozone concentrations extended into Canada, with southern

Ontario and southern Quebec experiencing levels above 80 ppbv during the same period. Levels of $PM_{2.5}$ were also elevated through, with 24 hr averaged values exceeding 30 $\mu g/m^3$ in the U.S. on July 12th and reaching Canada on the 14th (Figure 2). This episode was selected by the North American Research Strategy for Tropospheric Ozone (NARSTO) as an intercomparison period for oxidant models, meaning that ozone but not PM data were exploited. The second episode is an extreme wintertime PM event that affected the same areas in February 1998. Very high levels of $PM_{2.5}$ were observed in Vermont and Quebec: 24hr average mass concentration exceeded 30 $\mu g/m^3$ on February 9, 10 and 11th throughout all the northeastern States, and 1hr levels exceeded 100 $\mu g/m^3$ for on February 11th in Québec. These unusual circumstances sparked an effort across state and provincial agencies to gather all available PM measurements for this regional episode.

Prior to 1998, network PM measurements were relatively sparse both in time and frequency. Measurements are typically made with non continuous filter samplers, usually twice a week (CAPMoN being the exception), and with a minimum time of exposure of 24 hr. In addition, a lot of variability exists between networks operating procedures. A summary of the differences between the network data used for the July 1995 and February 1998 comparisons is given in Table 1. However, due to the disparities in operating procedures, PM data density is highly variable from day to day. The observations maps for July 12 and 14, 1995, in Figure 2, illustrate the contrast in the amount of information available on those two days (non continuous instruments only).This is a serious impairment for evaluating a PM model such as AURAMS. The situation is slightly better for the 1998 episode, as more PM monitors were starting to be installed on various networks, especially continuous ones. However, early TEOMs measurements, and especially those made during the winter, are not as reliable due to nitrate evaporation problems (T. Dann, personal communication). The preliminary evaluations presented here will focus on 24 hr filter measurements only.

Table 1: Characteristics of the measurement networks included in this study. OP. Cond. stands for operating conditions, TEC for elemental carbon and TOC for organic carbon. CAPMoN is the Canadian air and precipitation monitoring network, CAAMP the Canadian air aerosols monitoring program, operating in 1995 but not in 1998, GAViM the Guelph aerosol and visibility monitoring program, IMPROVE the interagency monitoring network of protected visual environment, and NAPS the (Canadian) national air pollution surveillance network.

Network	Kind	Frequency	Op. Cond.	Start time	Size cut	Species measured
CAPMoN	24 hr filters	daily	STP	8:00 local	none	$SO_4^=$, NO_3^-, NH_4^+
CAAMP*	24 hr filters	daily	ambient	0:00 local	2.5 µm	$PM_{2.5}$, PM_{10}, $SO_4^=$, NO_3^-, NH_4^+
GAViM	24 hr filters	1/3 days	ambient	0:00 local	2.5 µm	$PM_{2.5}$, $SO_4^=$
IMPROVE	24 hr filters	1/3 days	ambient	0:00 local	2.5 µm	$PM_{2.5}$, PM_{10}, $SO_4^=$, NO_3^-, NH_4^+, EC, OC
NAPS	24 hr filters	1/6 days	STP	0:00 local	0-2.5 µm & 2.5-10 µm	$PM_{2.5}$, $SO_4^=$, NO_3^-, NH_4^+

As illustrated in Table 1, some networks measure PM mass (NAPS, IMPROVE), while other concentrate on particle composition (CAPMoN). All relevant information is being used for AURAMS evaluation. Reconciling model output with observations requires careful manipulations of the model fields to accommodate for the differences in operating start time, measurement conditions (STP versus ambient) and the inlet size-cuts. It is handled through AURAMS post-processor.

4. EPISODE ANALYSES

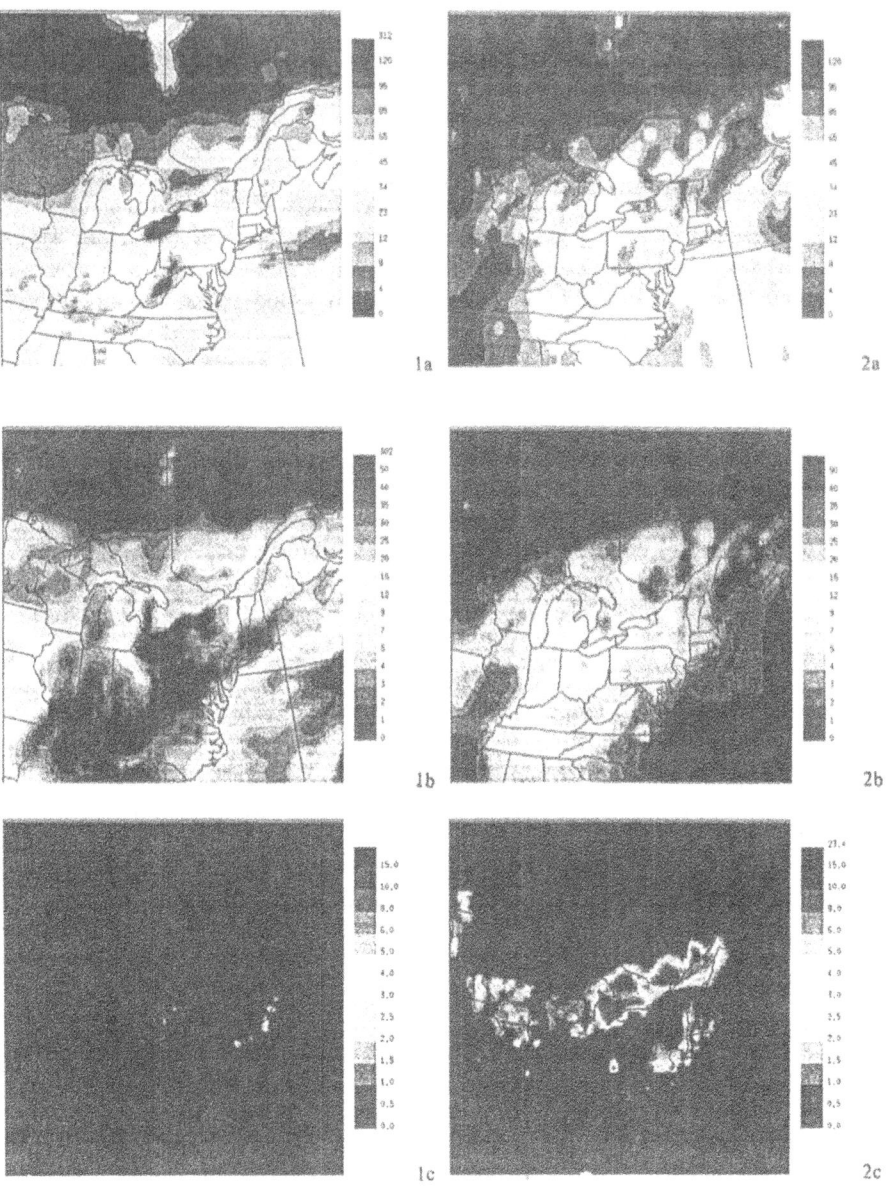

Figure 1: AURAMS surface fields for July 12, 1995 (1) and February 11, 1998 (2) in $\mu g/m^3$. (a) $PM_{2.5}$; (b) SO_4^-; (c) NO_3 ;

Figure 1 presents the 24hr averaged PM$_{2.5}$, fine particulate SO$_4^-$ and fine particulate NO$_3^-$ surface fields, as simulated by AURAMS for July 12, 1995 and February 11, 1998. High PM$_{2.5}$ levels (> 45 µg/m^3), driven by the sulphate component, are simulated for most of eastern U.S. and southern Canada during the summer event. Nitrate levels are extremely low with only a few grid points exceeding 2 µg/m^3. Ammonia levels (not shown) vary between 1 to 3 µg/m^3 through most of the domain. In strong contrast, winter spatial patterns show a much stronger contribution from nitrate to PM$_{2.5}$ mass, with almost equal proportions of fine SO$_4^-$ and fine NO$_3^-$ in following Lake Ontario and the St-Lawrence river. It reflects the change in the chemistry of the SO$_4^-$-HNO$_3$-NH$_3$, where formation of ammonium nitrate particles is favored at colder temperatures and lower sulphate levels (resulting from the decrease in photochemical conversion of SO$_2$). Simulated ammonium levels for February 1998 are however similar to the ones predicted for the summer case. Ammonia emissions in the 1985 NAPAP inventory are known to be underestimated (M. Moran, personal communication). This could explain the low ammonium levels simulated by AURAMS, as well as limit the amount of particulate nitrate that AURAMS can form.

Actual comparisons with observations are presented in Figures 2 to 4 for PM$_{2.5}$, sulphate and nitrate mass. The summer maps (Figures 2 and 3) clearly show an overprediction of PM$_{2.5}$ levels in the U.S. and to a lesser extent in Canada. This is due to a corresponding overprediction in SO$_4^-$ levels (Figure 3). Simulated nitrate and ammonium levels (not shown) are much lower than observed during that summer period.

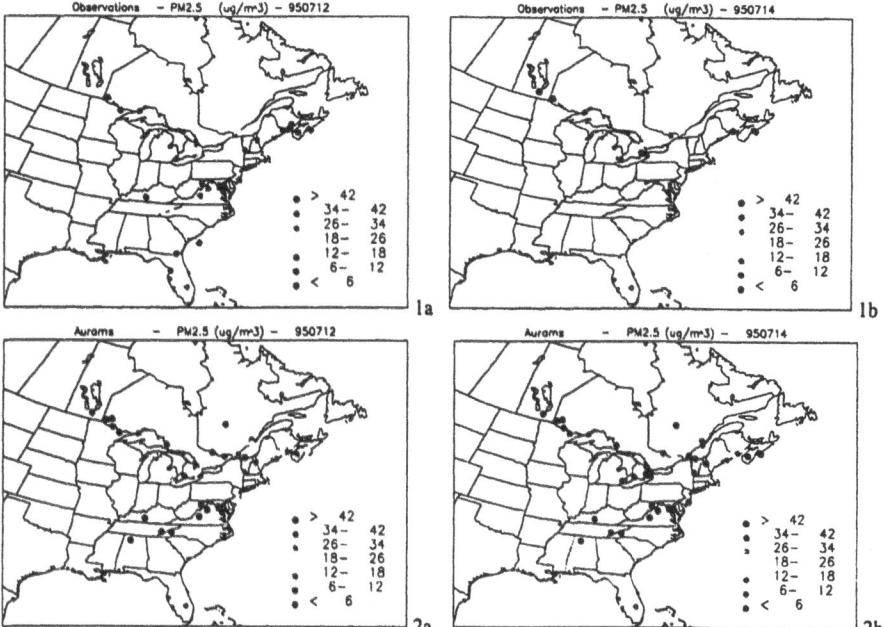

Figure 2: PM$_{2.5}$, 24 hr average mass concentrations at various station locations for July 12 (a) and 14 (b), 1995. (a): observed values; (b) modelled values (at the nearest grid point front station location). Modelled values are shown at all possible locations within AURAMS domain, while (1a) and (1b) show the actual available measurement for the corresponding dates.

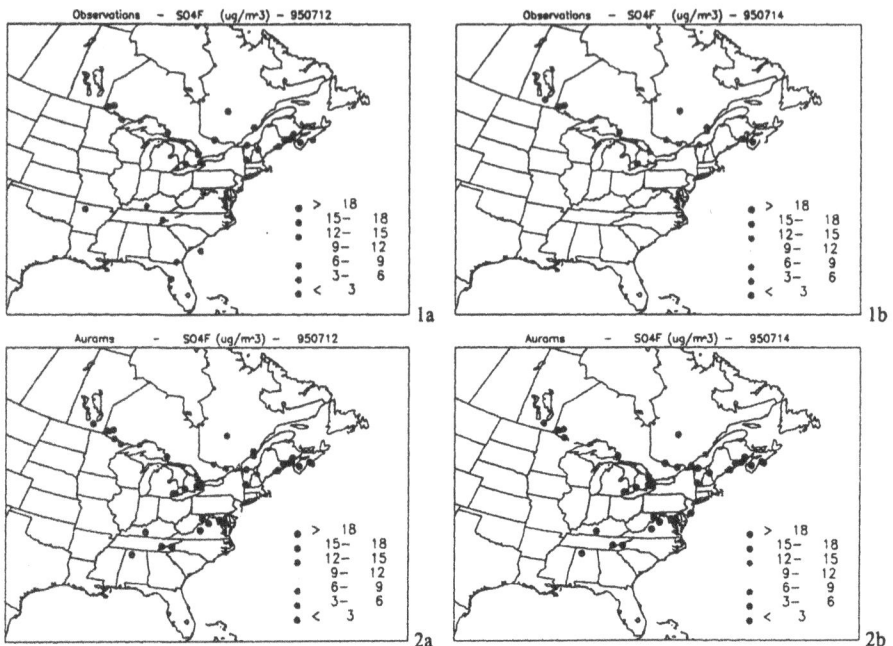

Figure 3: SO_4^-, 24 hr average mass concentrations at various station locations for July 12 (a) and 14 (b), 1995. (a): observed values; (b) modelled values (at the nearest grid point front station location). Different size-cut measurements are shown here, modelled values are treated accordingly.

Figure 4 show the comparison between observed and modelled levels of sulphate for February 11 and nitrate for February 10, 1998. Again AURAMS overpredicts SO_4^- mass concentrations in the U.S. Ohio Valley, but it does captures the change in chemical regime due to the winter conditions and SO_4^- levels are approximately 1/3 of the summer ones. As the 1985 NAPAP emission inventory contains some small, but direct emissions of particulate SO_4^-, areas of overprediction correspond to U.S. regions with major emission sources. On the Canadian side, the model seems to be underpredicting, on the contrary, most of the high level sites except along Lake Erie. Overall, winter SO_4^- levels are much closer to the observed ones than for the summer case. Nitrate mass concentrations (Figure 4 (b)) are also in much better agreement, with observation raging from 3 to 11 $\mu g/m^3$ and modelled values varying from 2 to 27 $\mu g/m^3$ in the Windsor-Quebec City corridor. Correspondingly, ammonium levels (not shown) in the corridor follow actual measurements, which is in contradiction with having low ammonia emissions in the inventory. Based on these preliminary comparisons, AURAMS performance is generally better for the winter case than the summer one. Further examination will however be needed to identify the reasons for the different behavior.

5. CONCLUSION AND FUTURE WORK

A preliminary evaluation of AURAMS was performed for two PM episodes characterised by drastically different meteorological conditions. For the summer case, AURAMS generally overpredicted sulphate mass and underpredicted nitrate and ammonium mass in the particle phase. As a result, $PM_{2.5}$ levels were overpredicted over most of the model domain. The situation was completely different for the winter case, where modelled values for nitrate and ammonium were in much better agreement with observations and sulphate levels only slightly overpredicted. At this point in time, only a preliminary evaluation was conducted and a thorough comparison will be underway shortly, including reruns of the episodes with an updated emissions inventory and an analysis of the meteorological inputs. In addition, hourly PM measurements and U.S. filter measurements from the AIRS database will also be included in the more detailed comparison. This preliminary study did illustrate, however, that evaluations under various meteorological and chemical regimes are critical to characterise a model's performance.

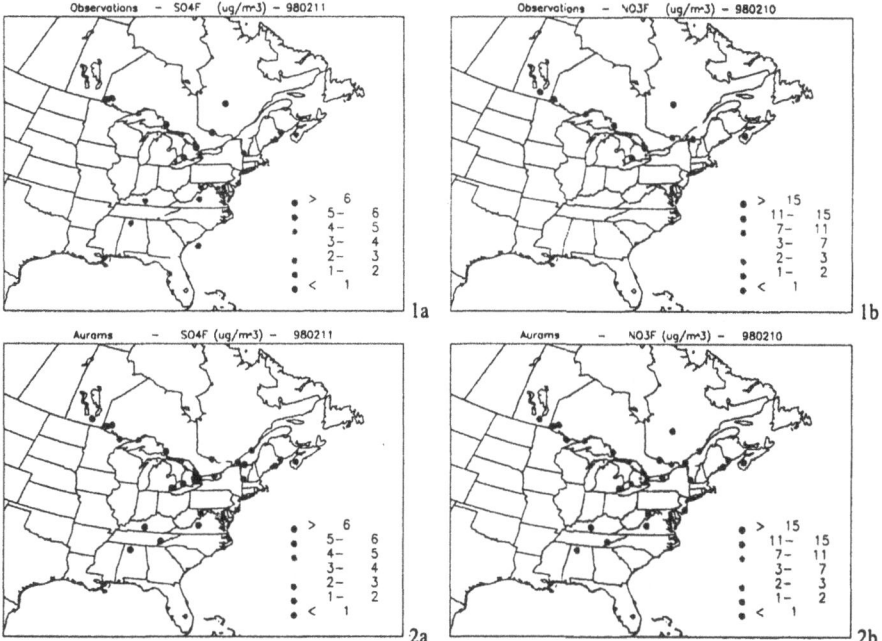

Figure 4: 24 hr average mass concentrations of (a) SO_4^-, February 11, 1998, and (b) NO_3^-, February 10, 1998, at various station locations. (1): observed values; (2) modelled values (at the nearest grid point front station location). Different size-cut measurements are shown here, modelled values are treated accordingly.

REFERENCES

Benoit, R.,M. Desgagné, P. Pellerin, S. Pellerin, Y. Chartier and S. Desjardins, 1997. The Canadian MC2: a semi-Lagrangian, semi-implicit wideband atmospheric model suited for finescale process studies and simulation. *Mon. Wea. Rev.*, 125, 2382-2415.

Brook, R.D., J.R. Brook, B. Urch, R. Vincent, S. Rajagopalan, and F. Silverman, 2002. Inhalation of fine particulate air pollution and ozone causes acute arterial vasoconstriction in healthy adults. *Circulation*, 105, 1534-1536.

Côté, j., J.-G. Desmarais, S. Gravel, A. Méthot, A. Patoine, M. Roch and A. Staniforth, 1998. The operational CMC/MRB Global Environmental Multiscale (GEM) model. Part I: Design considerations and formulation. *Mon. Wea. Rev.*, 126, 1373-1395.

Gong et al., *J.Geophys.Res., in press*, 2002.

Hidy, G.M.; P.M. Roth, J.M. Hales, and R. Scheffe, 1998. White paper for *1998 North American Research Strategy for Tropospheric Ozone (NARSTO) Critical Review Series*, June 1998.

Lurmann, F.W., A.C. Lloyd, and R. Atkinson, 1986. A chemical mechanism for use in long-range transport/ acid deposition computer modelling, *J. Geophys. Res.*, 91, 10905-10936.

Makar, P.A., V. Bouchet, L.P. Crevier, S. Gong, W. Gong, S. Menard, M. Moran, B. Pabla, S. Venkatesh, 2003. AURAMS runs during the Pacific2001 time period – a Model/Measurement Comparison., this Proceeding.

Meng, Z., D. Dabdub, and J.H. Seinfeld, 1997. Chemical coupling between atmospheric ozone and particulate matter. *Science*, 277, 116-119.

Moran, M. D., A. P. Dastoor, S.-L. Gong, W. Gong and P. A. Makar, 1998. Conceptual design for the AES unified regional air quality modelling system. Air Quality Research Branch, Meteorological Service of Canada, Downsview, Ontario M3H 5T4, Canada.

Nenes, A., C. Pilinis, and S.N. Pandis, 1998. ISORROPIA: a new thermodynamic equilibrium model for multiphase multicomponent marine aerosols, *Aquatic Geochem.*, 4, 123-152.

Odum, J.R., T. Hoffman, F. Bowman, D. Collins, R.C. Flagan, and J.H. Seinfeld, 1996. Gas/particle partitioning and secondary aerosol formation. *Environ. Sci. Technol.*, 30, 2580-2585.

Pudykiewicz, J.A., A. Kallaur and P.K. Smolarkiewicz, 1997. Semi-Lagrangian modelling of tropospheric ozone, *Tellus*, 49B, 231-248.

Ryan, W. F., B.G. Doddridge, R.R. Dickerson, R.M. Morales, K.A. Hallock, P.T. Roberts, D.L. Blumenthal, J.A. Anderson, and K.L. Civerolo, 1998. Pollutant transport during a regional O_3 episode in the Mid-Atlantic States. *Air & Waste Manage. Assoc.*, 48, 786-797.

Samet, J.M., Scott L.Z., Dominici F., Curriero F., Coursac I., Dockery D.W., Schwartz J., and A. Zanobetti, 2000. The national morbidity, mortality and air pollution study. Part II: Morbidity, mortality and air pollution in the united States. *Heath Effects Institute*, 94, part II, June 2000.

Scholtz et al., 1999. Proc. *9th AWMA Emission Inventory Symp., Oct 26-29,Raleigh, North Carolina*, 456-468.

U.S. EPA, 1989. The 1985 NAPAP emission inventory (version 2). Development of the annual data and modeler's tapes. Rep. EPA-600/7-89-012a, 692pp., Natl. Tech. Info. Serv., Springfield, Va.

Zhang, L., M.D. Moran, P.A. Makar, J.R. Brook, and S. Gong, 2001. Asize-segregated particle dry deposition scheme for an atmospheric aerosol module. Atmos. Environ., 35, 549-560.

Zhang, L., M.D. Moran, P.A. Makar, J.R. Brook, and S. Gong, 2002. Modelling gaseous dry deposition in AURAMS: a unified regional air-quality modelling system. Atmos. Environ., 36, 537-560.

DISCUSSION

S.A.-AKSOYOGLU   With such sophisticated aerosol model and speciated aerosol measurements, one expects perfect results. Where do the problems come from?

W.GONG   More sophisticated model does not guarantee better results. As far as aerosol measurements are concerned, we do have some speciated information from the various monitoring networks. However the information on size (or speciated and size-segregated information) is lacking. There is a problem with time resolution and availability with the network observations as they tend to be collected at different frequency with different averaging time. There may also be issues concerning QA/QC and different standards from different networks. From the modelling side of things, we know that for this particular model run presented here we had some serious problems (some due to coding errors). In general, the sophistication in model representation of various physical and chemical processes increases as the advancement of our understanding and the increase in computing power. More sophisticated model allows us to look into the various processes and their interaction in more details, but, at the same time, it also imposes a demand for more specialized and detailed measurement in order to evaluate the model representations of the various processes.

P.BUILTJES   Can you explain why the sulphate calculated values are much too high? Sulphate modelling is not so complicated.

W.GONG   In this case we now know that we had some serious problems (including errors in emission) with this run. The corrected model run is being conducted at the moment, and hopefully we will see some improvement. At least we do see better comparison in sulphate from our Pacific 2001 simulation as will be presented later (from an updated model run as opposed to the simulation presented in the manuscript).

APPORTIONMENT OF POLLUTANT S IN AN URBAN AIRSHED: CALGARY, CANADA, A CASE STUDY

A.L. Norman[*], H.R. Krouse and J. M. MacLeod

ABSTRACT

Sources of pollutant sulphur in the city of Calgary, in western Canada were identified using natural stable sulphur and oxygen isotope abundance ratios, and a simple apportionment model was applied to quantify pollutant emissions. Three sources of isotopically distinct sulphur found in Calgary air were vehicle exhaust at +9 ‰, oil recycling plant emissions near –8 ‰, and sour gas flaring and processing emissions advected to the city near +20 ‰. Based on the lack of $\delta^{34}S$ values lower than +9 ‰ for SO_2 and aerosol sulphate, emissions from the oil recycling plant were deemed an insignificant contributor to the pollutant sulphur load in Calgary. A simple apportionment model showed that, on average 77% of the SO_2, 64% of aerosol sulphate, and 55 % of precipitation sulphate in Calgary was derived from emissions from the oil and gas industry. A model of 20% primary sulphate mixed with secondary sulphate from heterogeneous SO_2 oxidation in the presence of dilute Fe^{3+} was consistent with measured oxygen isotope compositions for precipitation sulphate. An association of higher $\delta^{34}S$ and $\delta^{18}O$ values in both precipitation and aerosol sulphate suggest primary sulphate was largely from sour gas flaring activities. In contrast, secondary sulphate was derived from both vehicle exhaust a sour gas processing. Upwind at a relatively pristine mountain site, sulphate in snow had $\delta^{34}S$ values near +5 ‰ but at concentrations lower than that for Calgary precipitation, confirming the input of atmospheric sulphur from oil and gas and vehicle exhaust to the city.

1. INTRODUCTION

Airborne pollutants in urban environments have an adverse effect on human health. Elevated concentrations of sulphur dioxide and fine aerosol sulphate, alone and in conjunction with other airborne compounds have been linked to increased hospital admissions (Aunan, 1996: Burnett et al., 1995, 1998). Understanding the sources, characteristics, transport, and deposition of sulphur compounds within the urban arena is necessary in order to implement appropriate mitigation technologies.

---

[*] The University of Calgary, Department of Physics & Astronomy, 2500 University Dr. N.W., Calgary, Alberta T2N 1N4
Ph: 403-220-5405, Fax: 403-220-7773, annlisen@phas.ucalgary.ca

Air Pollution Modeling and Its Application XVI, Edited by
Borrego and Incecik, Kluwer Academic/Plenum Publishers, New York, 2004

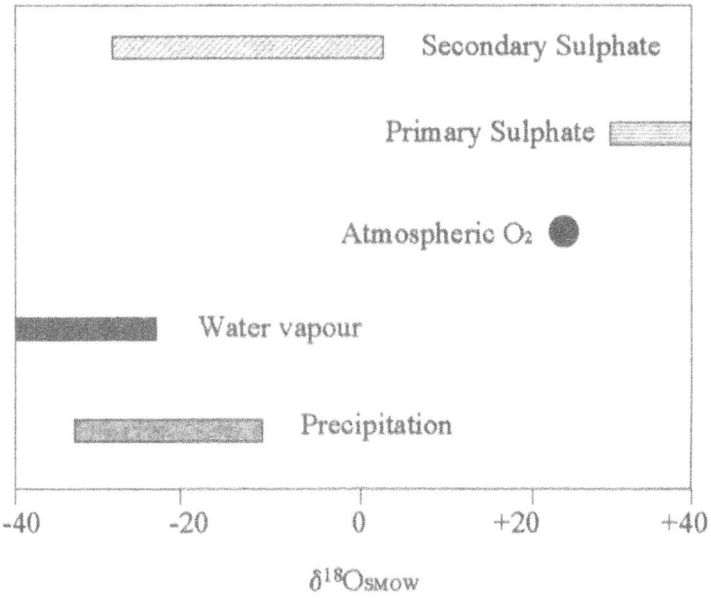

$\delta^{18}O_{SMOW}$

Figure 2. Oxygen isotope variation in precipitation and aerosol sulphate for primary and secondary oxidation.

2. METHODS

2.1. Sample Collection

2.1.1. Aerosol Sulphate and Sulphur Dioxide

Aerosol sulphate and SO_2 were sampled using Sierra Miscu High Volume samplers modified to collect particulate matter on a glass fiber filter (EMP 2000) followed by a high flow filter (SS410) impregnated with a potassium carbonate or triethanolamine and glycerol mixture to trap sulphur dioxide (the isotope composition for SO_2 trapped on TEA and K_2CO_3 impregnated filters was the same). High volume samplers were located at the University of Calgary weather station in 1989 and 1990 to sample well-mixed air. A second suite of samples was collected only during westerly winds at a site west of the city in 2000. Samplers were shut off manually in 1989 and 1990 during rain events and were set up with a computer interface to shut off for relative humidity above 98 % in 1999 and 2000. Additional sampling was conducted at a series of locations throughout the city in 1999 to characterize sulphur emisons from local industrial sites and vehicle exhaust using a high volume sampler fitted with a cascade impactor. Equipment was set up on an elevated site (e.g. rooftops of buildings) at each site.

The isotope composition of aerosol sulphate and sulphur dioxide were used in this study to determine the proportion of pollutants from isotopically distinct emission sources. The heavy to light isotope ratio in a sample was compared to that for a standard material. The S isotope composition was referenced to VCDT and the oxygen isotope composition was referenced to V-SMOW. The relative isotope abundance was expressed using delta notation; for example

$$\delta^{34}S = ((^{34}S/^{32}S)_{sample}/(^{34}S/^{32}S)_{VCDT} - 1) \times 10^3 \tag{1}$$

Calgary, Alberta, Canada is a city of about a million people situated to the east of the Rocky Mountains and its industry largely revolves around the production and processing of oil and gas in rural areas of the province. Sour gas, natural gas that contains ~ 5 % hydrogen sulphide, is processed at a number of sites surrounding the city and removes sulphur from the gas stream. The final product that is sold for commercial and residential use contains less than 1 ppb (< 0.0001 %) H_2S. The processing facilities, or sour gas plants, convert S in the gas stream to elemental sulphur via the Claus, or other extraction process with an efficiency of 98 % or greater. The remaining sulphur fraction is combusted to sulphur dioxide at temperatures that typically reach 400 °C or more. Stack control technology ensures that very little primary sulphate is emitted from the stacks at the sour gas plants under normal operating conditions. However, when upsets to the gas stream or facility occur, the gas stream may be diverted and raw or partially stripped gas is flared. Under these conditions, primary sulphate is produced. Primary sulphate as well as sulphur dioxide can also be expected from flaring associated with oil and gas well start-up and/or operating procedures. To test the volume of gas in a reservoir, raw gas is flared until a measureable pressure drop occurs. For larger gas wells, this type of test can last up to two weeks. Gases from oil reservoirs are also often flared if it is not commercially viable to divert the gas stream to an appropriate pipeline for processing.

Emissions from sour gas processing and operations may impact sites distant from the source because the residence time for sulphur dioxide (several days to a week) and sulphate (several weeks) makes for efficient dispersal. The magnitude of the emissions from gas processing are large. For example, a single sour gas processing site 40 kilometers north of the city emits approximately ten tonnes of SO_2 per day (Rowe *et al.*, 1977). Within the city of Calgary, large point source emitters of pollutant sulphur are not present but sulphur dioxide and sulphate concentrations are not negligible, reaching ~ 60 μgS/m^3 and 8.7 μgS/m^3 during peak pollution periods. On average sulphur dioxide and sulphate concentrations within the city are 3.2 μgS/m^3 and 2.7 μgS/m^3 (Myrick and Hunt, 1998). In this study, we have examined the sources of pollutant sulphur within the city and its vicinity, its oxidation in the atmosphere, and its deposition.

Atmospheric sulphur budgets typically rely on emission source data averaged over an annual cycle. On a regional scale they are difficult to confirm experimentally using concentration measurements which can be highly variable over time and depend on boundary layer height, source strength, chemical conversion, and local meteorology. Further, with concentration measurements alone, local biogenic and anthropogenic sources are indistinguishable. Natural isotope abundance ratios can be used to distinguish two or more sources that are isotopically distinct. In Alberta, the isotope composition of sulphur from sour gas processing is isotopically heavier than that in soils and plants (Figure 1). Plants assimilate sulphur from soil with little to no isotope fractionation (Krouse and Herbert, 1988). Upon decay, sulphide gases such as H_2S and DMS are released into the atmosphere and reflect the isotope composition of plant and ultimately the parent soil from which their sulphur was derived. Biogenic sulphur can also be released to the atmosphere through gases that escape from shallow anoxic sediments in which bacterial sulphate reduction occurs. In this process sulphate is consumed by sulphate reducing bacteria. These organisms preferentially break the ^{32}S bond, leaving sulphate enriched in ^{34}S. The product of this reaction is hydrogen sulphide that is depleted in ^{34}S relative to sulphate. In shallow aquatic environments a significant proportion of the hydrogen sulphide produced in the sediments is able to escape the water column unoxidized. Once in the atmosphere, sulphides are oxidized to sulphur dioxide and sulphate. Sulphur isotope fractionation during oxidation is a confounding factor but previous studies have found little fractionation (Saltzmann *et al.*, 1983). Therefore, the isotope composition of sulphate can be used to distinguish biogenic and anthropogenic sulphur emissions. In this paper, we refer to the biogenic source as "background" or "natural" sulphur dioxide and sulphate. Biogenic emissions are likely maximum in summer when anoxic conditions prevail in the sediments of shallow sloughs and ponds.

Information about the oxidation and transport of sulphur through the atmosphere can also be inferred from its isotope composition: for sulphate the oxygen isotopes are most useful. Once formed, sulphate does not exchange oxygen atoms with its environment (Holt, 1991). Other investigators (Holt *et al.* 1982, Holt and Kumar, 1984) showed that oxygen isotopes for primary sulphate (S oxidized prior to release to the atmosphere) from combustion sources were isotopically much heavier than those for secondary sulphate (S oxidized after release) which is oxidized in the presence of water or water vapour (Holt *et al.*, 1981, 1983). Figure 2 shows the range in oxygen isotope composition for sulphate as a function of the isotope composition of liquid water. In this study we have used both oxygen and sulphur isotopes in sulphate to study the atmospheric sulphur budget in the city of Calgary.

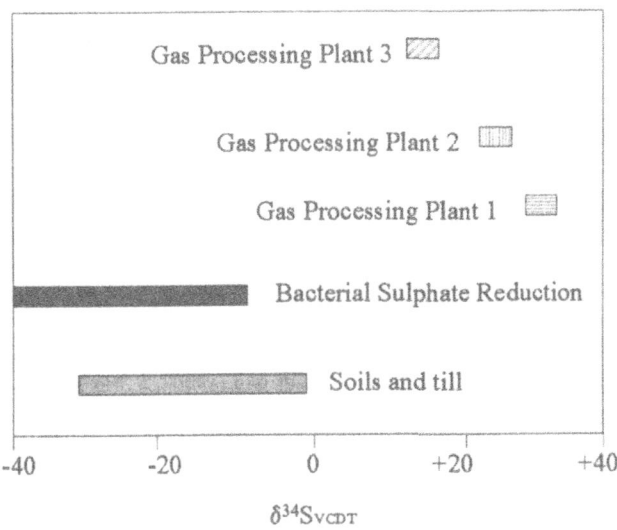

Figure 1. S isotope variations for atmospheric S precursors near Calgary Alberta.

For reference, a site was sampled to the west of the city high on a mountain in the Kananaskis with westerly prevailing winds. This site was remote from industrial emissions and roadways but was near to a ski lodge where a diesel generator was used to provide local power.

2.1.2 Precipitation

Precipitation in the city of Calgary was collected in the summer of 1989 at a downtown location near a very busy intersection using a large polyethylene sheet rinsed with distilled deionized water (DDH$_2$O) and at the University of Calgary using a large rain collector with a retractable cover designed to open only during precipitation events.

Snow at the Kananaskis site was collected intermittently throughout the late winter months. In all cases, the snow appeared to be relatively recent and was separated from earlier snowfall by a well-defined crust. Snow markers were attached to a tree at the site to avoid sampling a mixture of events. On the 3rd of February, 1990, both the top and bottom layers of the snow were collected since the marker was well beneath the two top snow layers.

2.2. Sample Analysis

Aerosol sulphate was extracted from the filters by sonication for half an hour in a dilute acid solution. After initial filtration to remove filter fibres, sulphate in solution was precipitated as BaSO$_4$ by the addition of excess 10 % by weight BaCl$_2$ solution. It was acidified to pH < 3 and heated to reduce the volume to be filtered. The remainder of the particulate filters were retained and any remaining oxidized form of aerosol sulphur (called insoluble sulphur) was reacted under nitrogen with Thode Reduction mixture (HI, HNO$_3$, HCl) to form H$_2$S which was subsequently trapped in a cadmium aetate solution as CdS. The CdS was converted to Ag$_2$S using silver nitrate solution.

Sulphur dioxide filters were treated in a similar manner those for aerosol sulphate except the filters were left in an 0.3 % H$_2$O$_2$ solution overnight prior to extraction to oxidize the SO$_2$ to sulphate.

Sulphur in fuels was extracted by combusting 1g aliquots of fuel in a Parr bomb pressurized to 28 Atmospheres with O$_2$ to convert all sulphur forms to sulphate. The bomb was rinsed and BaCl$_2$ was added to filtered rinsate to form BaSO$_4$ precipitate.

Precipitation sulphate was concentrated using ion exchange resin (DOWEX 8-1) and after elution from the resin it was precipitated as BaSO$_4$ as described above. All precipitates were filtered onto pre-wieghed 0.45 mm ashless nucleopore filters and either carefully scraped from the filter (BaSO$_4$ for ^{18}O analysis and Ag$_2$S) or the filter and precipitate were separated by combusting the filter at 800 °C for 1.5 hours.

The sulphur isotope composition of Ag$_2$S and BaSO$_4$ were analyzed on an EA coupled to a Prism (Giesemann et al., 1994), or were analyzed off-line by dual inlet using a modified Nuclide instrument. Data from the two mass spectrometers were standardized by using laboratory references spanning the range of isotope compositions in the samples ($BaSO_4$ and/or Ag_2S).

For oxygen isotope measurements, $BaSO_4$ was mixed with graphite and heated to 1000 °C in a platinum crucible. CO from the decomposition of $BaSO_4$ was converted to CO_2 in a disproportionation reaction and cryogenically purified and sealed in a glass ampule for introduction to a modified Nuclide 903 for oxygen isotope analysis.

Sulphur concentrations were measured gravimetrically.

3. RESULTS

3.1. Aerosol Sulphate, insoluble S, and SO_2

Table 1 summarizes the S concentration and the $\delta^{34}S$ values for aerosol sulphur (sulphate and insoluble sulphur) and SO_2 from the University of Calgary, Fortress Mountain and the site west of the city. Also shown are the data for aerosol sulphate less than 0.24 micrometers in diameter and sulphur dioxide at several sites within the city including the sewage treatment plant, the airport, the city centre, and an oil recycling plant. Data for a single sample that was collected directly behind an idling vehicle in the parking lot at the University of Calgary are also included.

The range in isotope composition for sulphur dioxide is quite large, from –4.5 ‰ at the oil recycling plant to +28 ‰ at the University of Calgary. Sulphur dioxide from the city center and the airport are identical at +12.8 ‰. SO_2 from vehicle exhaust and the sewage treatment plant are lighter, at +9.3 ‰ and +10.7 ‰ respectively.

Sulphate $\delta^{34}S$ values lie between +9.1 and +28.3 ‰, while $\delta^{18}O$ values range from +6.6 to +13.9 ‰. Aerosol sulphate less than 0.23 um in diameter is similar to sulphur dioxide in that similar isotope compositions are observed at the airport and the city center, while the lightest $\delta^{34}S$ value is for the oil recycling plant. It is also worthy to note that the fine aerosol sulphate and sulphur dioxide composition at almost every site nearby a sulphur source are very nearly the same. The two exceptions are the sewage treatment plant where the sulphate isotope composition is 3 ‰ heavier than the sulphur dioxide, and at the oil recycling plant, where sulphate is 13 ‰ heavier.

Insoluble sulphur is isotopically lighter than both sulphur dioxide and sulphate with an average $\delta^{34}S$ value of +6.7 ‰ and the concentration of insoluble sulphur was much lower than for sulphate and sulphur dioxide in every case.

Sulphur dioxide was present at higher concentrations than sulphate and insoluble sulphur in almost every instance. An extreme case, on February 14, 1990, had SO_2 concentrations of 62 $\mu gS/m^3$ while the corresponding sulphate concentration was 0.01 $\mu gS/m^3$. It is interesting to note that exceptions to this are mainly associated with $\delta^{18}O$ values of ~ +13 ‰ or more.

113

The isotope composition of sulphate, sulphur dioxide and insoluble sulphur at the Kananaskis site was +14.6 ‰, +10.2 ‰, and +1.5 ‰ respectively: quite similar to the $\delta^{34}S$ values for Calgary.

Table 1. Isotope composition of airborne sulphur compounds and their concentration.

Site & Date (mm/dd/yy)	$\delta^{34}S$ SO_2 (‰)	$[SO_2]$ (μgS/m^3)	$\delta^{34}S$ SO_4 (‰)	$[SO_4^{2-}]$ (μgS/m^3)	$\delta^{34}S$insol (‰)	[insol.] (μgS/m^3)	$\delta^{18}O$ SO_4 (‰)
University of Calgary							
06/21/89	+15.0	1.8	+28.3	0.38	-	-	-
02/14/90	+27.6	61.9	+11.9	0.01	+5.8	0.07	-
03/06/90	+13.1	0.50	+12.4	0.32	+12.0	0.17	-
03/07/90	+17.5	0.31	+11.7	0.47	+4.3	0.08	+7.7
03/26/90	+18.7	2.57	+15.2	0.31	+3.6	0.02	-
03/30/90	+23.4	2.60	-	0.41	+3.4	0.02	
06/18/90	+16.3	1.43	+15.2	0.52	+12.1	0.03	-
06/21/90	+14.6	1.31	+11.3	0.28	-	-	-
06/22/90	+17.4	2.29	+15.9	0.24	+5.4	0.02	-
06/26/90	+17.1	0.89	+17.0	0.19	+9.4	0.04	-
06/28/90	+20.4	0.20	-	0.06	-	-	-
07/02/90	+20.2	1.06	+21.0	0.49	-	-	+12.8
07/04/90	+20.7	0.56	+22.4	0.99	-0.5	0.08	+13.8
07/05/90	+22.7	4.32	+15.8	1.65	+7.4	-	-
09/13/90	+18.1	0.46	+18.6	0.51	+15.0	0.08	-
09/14/90	+18.2	0.29	+17.4	1.19	+12.8	0.08	+13.9
09/15/90	+20.2	0.69	+19.0	0.37	+9.8	0.02	+6.9
09/19/90	+19.8	2.75	+16.1	0.32	+4.5	0.01	+10.5
09/26/90	+14.6	0.38	+19.5	0.69	+1.9	0.08	+13.4
09/27/90	+17.4	1.22	+15.6	0.70	-	0.08	-
09/28/90	+19.1	0.98	+18.8	0.63	-	0.04	+12.8
09/30/90	+19.1	2.41	+15.8	0.93	6.25	-	+13.7
West of Calgary							
04/10/00	+16.5	2.44	-	-	-	-	-
06/05/00	+9.3	0.69	+17.4	0.02	-	-	-
06/19/00	+6.4	1.00	+15.4	0.74	-	-	-
06/26/90	+4.8	1.81	+9.1	-	-	-	-

07/04/90	+19.5	0.64	+19.9	-	-	-	-
07/17/00	+20.4	1.31	+14.6	0.09	-	-	-
07/24/00	+19.2	-	-	-	-	-	-
08/08/00	+12.0	1.06	+12.6	0.63	-	-	-
12/04/00	+17.4	3.67	+12.9	0.34	-	-	-
Kananaski s							
08/14/90	+14.6	-	+10.2	-	+1.5	-	+13.5
Source Character- isation							
Vehicle exhaust	+9.3	104.5	+10.0	0.37*	-	-	-
Airport	+12.8	0.25	+13.0	0.08*	-	-	-
Sewage Plant	+10.7	1.16	+14.4	0.39*	-	-	-
Downtown	+12.7	0.56	+12.1	0.18*	-	-	-
Oil Recycling Plant	-4.5	1.63	+9.1	0.62	-	-	-

* size fraction 50% cutoff <0.24 micrometers

3.2. Precipitation Sulphate

The concentration and oxygen and sulphur isotope composition for sulphate for each precipitation event where sufficient sample was present for analysis are shown in Table 2. Sulphate concentrations in Calgary ranged from 0.7 to 9.5 mgSO$_4$/l while the sulphur isotope composition varied from +8.4 to +20.4 ‰; similar to the values that were found for sulphur dioxide and aerosol sulphate. However, the oxygen isotope composition of precipitation sulphate covered a wider range of values and were more negative than those for aerosol sulphate.

Sulphate in snow at the Kananaskis site had lighter δ^{34}S and δ^{18}O values than for snow in Calgary and was present at much lower concentration.

Table 2. Concentration and isotope composition of sulphate in precipitation.

Site and/or Date (mm/dd/yy)	Rain or Snow	δ^{34}S SO$_4$ (‰)	[SO$_4^{2-}$] (mg SO$_4^{2-}$/l)	δ^{18}O SO$_4$ (‰)	δ^{18}O SO$_4$precip (‰)
Downtown					

115

06/07/89-AM	rain	+12.3	-	+6.5	-17.3
06/07/89-PM	rain	+18.5	1.88	+11.0	-13.8
06/16/89-AM	rain	+11.9	-	+7.6	-18.0
06/16/89-PM	rain	+11.7	-	+5.2	-16.7
06/17/89	rain	+13.7	2.46	-	-
01/25/90	snow	+12.5	1.50	+5.6	-28.2
08/17/90	rain	+13.3	0.71	+11.9	-13.8
University of Calgary					
01/25/90	snow	+16.3	2.46	+4.4	-26.0
02/01/90	snow	+17.9	4.39	+1.0	-30.3
03/08/90	snow	+20.4	6.52	+6.0	-22.1
04/26/90	snow	+16.9	2.78	-3.0	-
05/08/90	snow	+8.4	4.97	+7.0	-19.1
06/09/90	rain	+15.6	-	-	-
06/10/90	rain	+15.0	2.55	-	-13.4
06/16/90	rain	+17.8	2.30	+13.2	-16.6
06/20/90	rain	+10.9	0.80	-	-15.8
06/24/90	rain	+18.5	4.45	+12.3	-14.0
06/28/90	rain	+20.3	7.10	+6.2	-12.9
06/30/90	rain	+9.3	3.49	+9.2	-9.6
07/02/90	rain	+15.7	1.14	+14.3	-13.0
07/06/90	rain	+15.6	0.98	+11.0	-17.7
07/23/90	rain	+14.8	9.49	+12.3	-10.7
07/27/90	rain	+18.2	6.54	+13.5	-9.9
09/27/90	rain	+17.6	1.08	+14.0	-13.9
Kananakis					
01/28/90	snow	+5.5	0.1	-1.8	-27.9
02/03/90-top	snow	+10.5	0.3	+6.5	-26.4
02/03/90-bot.	snow	+7.7	0.4	+6.2	-25.2
07/02/90	snow	+7.2	0.2	+9.2	-26.1
22/03/90	snow	+6.6	0.5	+6.6	-28.1

3.3. Gasoline, Diesel and Oil

Over twenty samples of gasoline and diesel fuel were sampled within Calgary to examine the extent and variability of its isotope composition. Measurements were also made to assess the sulphur isotope composition of engine additives that contribute to vehicle emissions on combustion. Gasoline and diesel were similar isotopically, with $\delta^{34}S$ values of +8.9 +/- 0.6 ‰, and +9.9 +/- 2 ‰, respectively. Three oil samples were isotopically lighter than the fuel and had $\delta^{34}S$ values of +1.6 +/- 1.5 ‰. Diesel engines use an engine additive, molybdenum disulphide (MoS_2) as an anti-knock agent, and a sample from a truck stop was found to have an isotope composition of +5.5 ‰.

4. DISCUSSION

4.1. Isotopically Distinct Sources

Within the city of Calgary, large point source industrial emissions of sulphur compounds are largely absent. This is reflected in the relatively low concentrations for SO_2 and sulphate relative to cities with a larger industrial component. However, vehicles are used extensively within the city for private and industrial transport and an oil recycling facility in the southeast is a potential source of atmospheric sulphur pollution.

High temperature oxidation of fuel containing trace sulphur produces sulphur dioxide, sulphur trioxide and sulphate. Hot stack or tailpipe emissions rise and mix with ambient air where they disperse and oxidize. Dispersal efficiency is a function of local meteorological conditions and the temperature of the emissions. Primary sulphate may be released as a gas or as a component of solid or liquid aerosols and fine sulphate aerosols form as emissions cool to ambient temperature. The lifetime of fine aerosols <0.1 um is only a few minutes, so this aerosol fraction is best used to define the isotope composition of nearby sulphate-sulphur emissions.

The isotope composition for sulphur dioxide and fine aerosol sulphate in vehicle exhaust (Table 1) are +9.3 and +10 ‰ respectively and confirm that there is little to no isotope fractionation in the formation of sulphur dioxide and sulphate from the fuel source. Downtown, where emissions from vehicle exhaust, building ventilation and heating systems are mixed with sulphur dioxide transported into the city center, the isotope composition is slightly higher at +12.8 and +12.1 ‰, for SO_2 and fine aerosol sulphate respectively. Airplane fuel is very low in sulphur but emissions from vehicles, hangars, and heating, would likely contribute to local emissions. Oxidation of sulphur dioxide in the immediate vicinity of the sampler would form a city "background" of fine aerosol sulphate on which nearby emissions would be superimposed. The isotope composition of both fine aerosol sulphate and sulphur dioxide at the airport and in the city center are likely the sum of several sources, and the widespread and uniform value for fine aerosol sulphate and sulphur dioxide suggests that pollution in the air is very well mixed.

In contrast to the downtown and airport samples, the isotope composition for fine aerosol sulphate and sulphur dioxide at the sewage treatment plant differed. Sulphur dioxide was isotopically lighter, with a $\delta^{34}S$ value of +10.7 ‰, while fine aerosol sulphate was +14.5 ‰. This suggests one of three things:

- isotopically distinct emissions of sulphate and sulphur dioxide at the site
- isotopically lighter sulphur dioxide and heavier sulphate superimposed on the city "background" value near +12 ‰ or
- isotope fractionation in the formation or deposition process.

The sewage treatment plant is located in the river valley toward the southern border of the city. Nearby roads are major traffic routes and sewage treatment ponds are aerated. Double the concentration of sulphate and sulphur dioxide were found at this site compared to downtown and quadruple that at the airport. A higher proportion of sulphur dioxide from vehicle exhaust or the oxidation of reduced sulphur from sewage might be responsible for the additional sulphur dioxide at this site.

Distinct isotope compositions for sulphur dioxide and sulphate were observed at the oil recycling plant but the difference was much more pronounced than at the sewage treatment plant (Table 1). Here, sulphur dioxide concentrations were high, 1.6 $\mu gS/m^3$, and its isotope composition was negative, -4.5 ‰. The fine aerosol sulphate concentration was the highest of the source samples, 0.62 $\mu gS/m^3$, but resembled vehicle exhaust isotopically ($\delta^{34}S$ = +9.1 ‰). These results can be explained as the sum of city "background" sulphur dioxide with values near +12 ‰ mixed with isotopically lighter stack emissions. Stack controls remove primary sulphate but sulphur dioxide that is isotopically lighter than –4.5 ‰ (-7.6 ‰ using the airport as "background) is emitted and a portion is oxidized to fine aerosol sulphate. If we use the airport to represent city "background", the mass balance calculations for sulphate and sulphur dioxide suggest the $\delta^{34}S$ for the emissions are about –8.1 +/- 0.5 ‰.

In summary, three isotopically distinct sulphur sources of anthropogenic sulphur that are superimposed on city "background" at +12 ‰ can be used for identification;

- vehicle exhaust +9 ‰
- oil recycling –8 ‰
- oil and gas ~ +18 ‰ or more

Because the $\delta^{34}S$ values for the oil recycling plant overlap with those expected for oxidation of reduced sulphur from the sediments of shallow ponds, care must be taken in the interpretation of the data from site to site and from one season to the next for apportionment calculations.

4.2. Isotope Fractionation and Source Apportionment

Figure 3a is used to examine the relationship between sulphur dioxide and these isotopically distinct sources. With the exception of two samples from the west of the city during the summer when sulphur dioxide from sloughs may have influenced the results, the isotope composition of sulphur dioxde at low concentration falls between +9 and +20 ‰. As concentrations increase, the isotope composition trends toward +20 ‰ or more. Values intermediate to these fall within a triangular region that is bounded by a two-source mixing line. The mixing line shown describes a constant source of SO_2 with a $\delta^{34}S$ value of +9 ‰, consistent with vehicle exhaust, mixed with a variable source of SO_2 having an isotope composition of +28 ‰. The latter value is taken from the extreme value measured in February 14, 1990. From these results it is possible to identify sulphur dioxide in Calgary as a mixture of vehicle exhaust and oil and gas processing emissions at low concentration, and as coming from a sour gas processing facility with emissions characterized by values near +28 ‰, as they are for the Crossfield sour gas processing plant, periodically at high concentration.

It is interesting to note that are no trends in a similar plot for aerosol sulphate, even though the isotope composition falls within the same range (Figure 3b). However, it is clear that the source of sulphur dioxide and sulphate are frequently the same from a plot of $\delta^{34}S$ values for SO_2 versus those for sulphate (Figure 4a), that no systematic isotope fractionation occurs on oxidation (since the values fall along rather than above or below the 1:1 line), and that as the fraction of sulphate (Figure 4b) increases $\delta^{34}S$ values for SO_2 near +18 ‰ are more common.

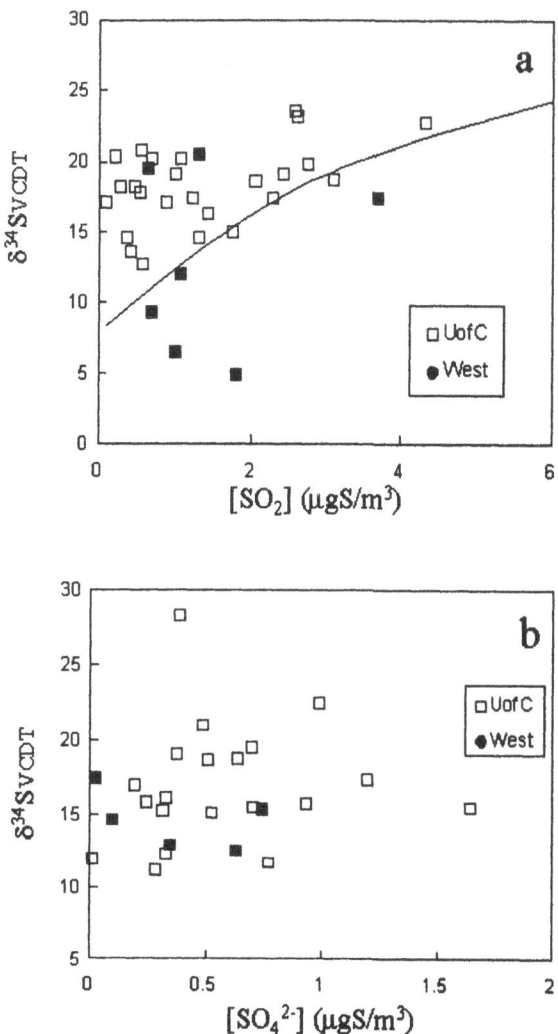

Figure 3. Sulphur isotope composition versus concentration for SO_2 (a) and sulphate (b) at the University of Calgary within the city (U of C), and a site northwest of the city core (West).

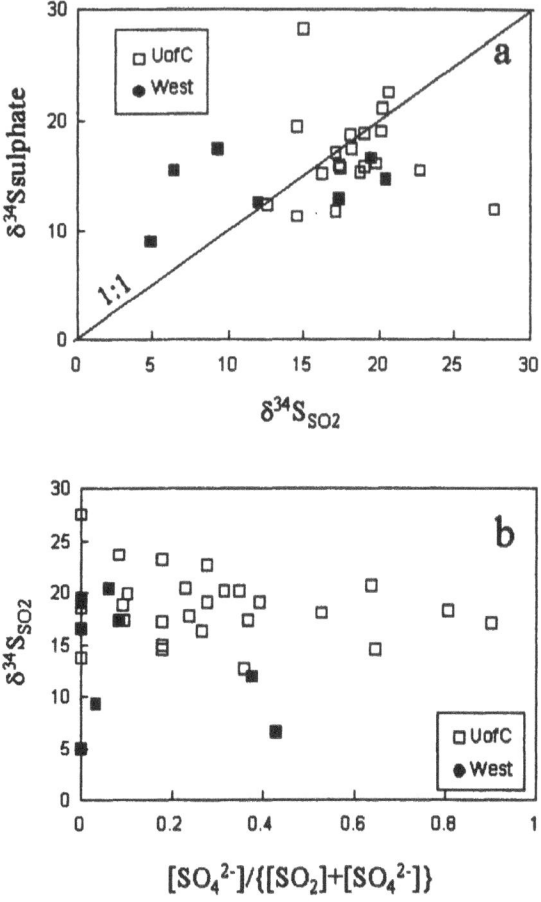

Figure 4. Similar $\delta^{34}S$ values for SO_2 and sulphate are apparent in 4a. The isotope composition of SO_2 as a fraction of atmospheric sulphur as sulphate is shown in 4b.

$\delta^{34}S$ values for precipitation sulphate trend toward more positive values with increasing sulphate concentration, similar to the results for SO_2 and suggest sulphate in precipitation is at least partially derived from SO_2 oxidation. The source of the isotopically heavy sulphate in rain and snow can found as intercepts from plots of $\delta^{34}S$ versus inverse sulphate concentration (Figure 5a). The intercept for the upper range of data on the graph for both rain and snow is near +21 ‰, as could be expected for a mixture of sulphur from two sour gas plants to the north of the city.

It is worth noting that $\delta^{34}S$ values for sulphate at the upwind site at Kananaskis were isotopically lighter than sulphate in Calgary snow and sulphate was present only at low concentration. If this is considered to be representative of natural background sulphate in the region, then mass balance calculations could be performed to remove this "source". However, diesel generators and woodsmoke were present in the vicinity of the sampling site and sulphate concentrations at that site are considered influenced by some local emissions, so these calculations were not performed.

Information on the oxidation pathway is inferred from the oxygen isotope composition of sulphate and its relationship to precipitation water. Heterogeneous oxidation of SO_2 during precipitation events makes up a significant fraction of the precipitation sulphate measured since a strong relationship between the oxygen isotope composition of precipitation and sulphate is evident (Figure 5b). Holt (1991) mapped the relationship between the oxygen isotopes of precipitation and sulphate for both heterogeneous and homogeneous oxidation mechanisms and the region described by these reactions is shown in Figure 5 for comparison. The straight line fit to the data in Figure 5 are described by the relationship:

$$\delta^{18}O_{SO4} = (0.64 +/- 0.08) \, \delta^{18}O_{liquid} + (21.4 +/- 1.8 \text{ ‰}) \qquad r = 0.91$$

(2)

This is well below the line for high temperature oxidation (primary sulphate: slope 0.06 intercept +37.5 ‰ for water vapour) and slightly above lines describing secondary oxidation. From 19 reactions studied by Holt and colleagues the best fit to the data is for heterogeneous oxidation of sulphur dioxide in the presence of dilute Fe^{3+} (Holt's reaction used 7.5×10^{-5} M at 25 °C: slope 0.78 intercept +12 ‰). Assuming the fractionation between liquid water and water vapour as 11 ‰, then the proportion of secondary sulphate (a) can be found from the equation

$$\delta^{18}O_{SO4} = (0.06 + 0.72 \, a) \, \delta^{18}O_{liquid} + (37.5 - 25.5a) \text{ ‰}$$

(3)

The observed slope is 0.64 so a = 0.8 and

$$\delta^{18}O_{SO4} = 0.64\delta^{18}O_{liquid} + 17.1 \text{ ‰}$$

(4)

The intercept from this model of 20% primary sulphate mixed with 80% secondary sulphate agrees reasonably well with the observations. The presence of iron can be reasonably assumed since iron concentrations in precipitation from two sites in Alberta, Rocky Mountain House to the northwest, and Edmonton 200 km north of the city of Calgary, were measured by Klemm and Gray in 1978 and found to be 1 and 5 µg/l respectively.

Sources of primary and secondary sulphate in Calgary are examined using a plot of the $\delta^{34}S$ versus $\delta^{18}O$ for sulphate in Figure 6. More positive $\delta^{18}O$ values and greater proportions of primary sulphate, are associated with $\delta^{34}S$ values between +14 and +22 ‰. Those with lower $\delta^{18}O_{SO4}$ values are found for $\delta^{34}S$ values between +8 and +16 ‰. This

121

suggests primary sulphate in Calgary largely originates from the oil and gas industry rather than from vehicle exhaust, while secondary sulphate is formed from a mixture of both these sources.

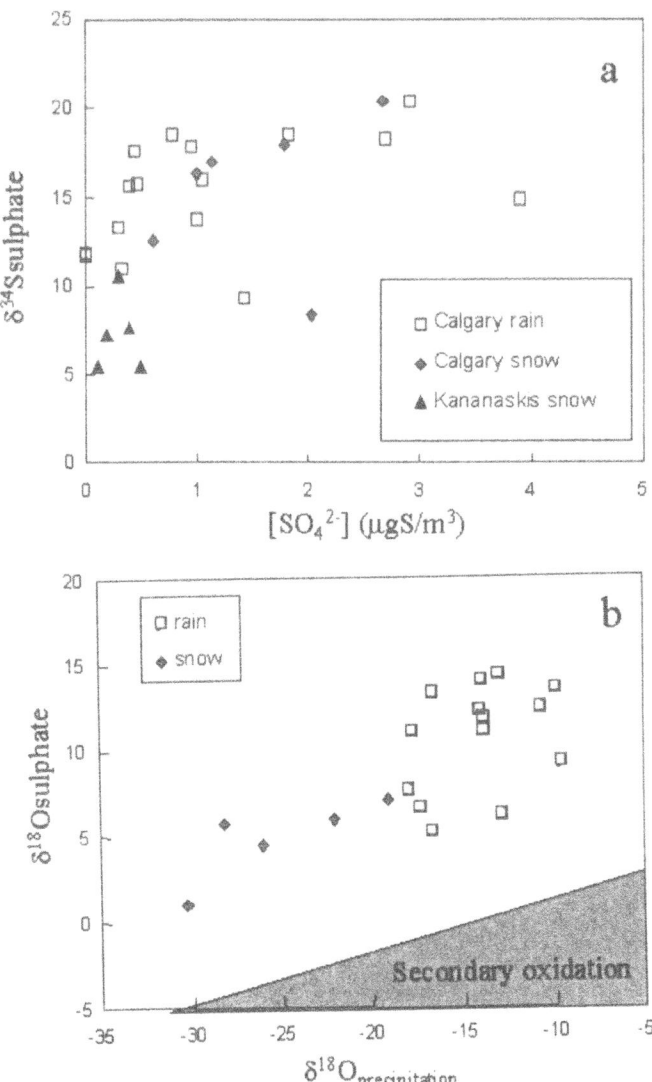

Figure 5. The sulphur (a) and oxygen (b) isotope composition of precipitation are shown as a function of concentration and the isotope composition of precipitation water respectively.

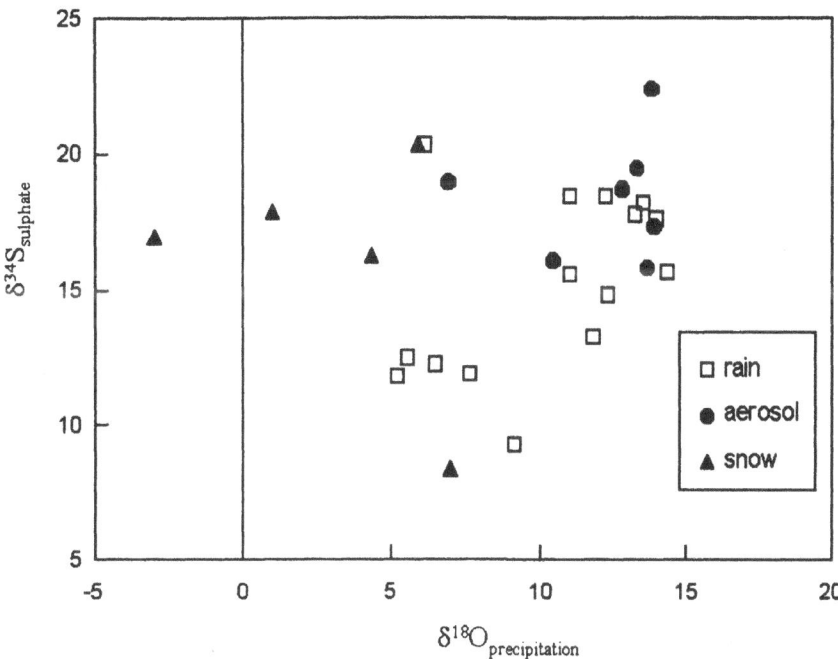

Figure 6. Sulphur versus oxygen isotope composition for aerosol and precipitation sulphate.

In the following discussion emissions from the oil recycling plant, to the east of the downtown core, are considered insignificant relative to vehicle exhaust and sour gas processing. This is not unreasonable given the lack of samples approaching the isotope composition of its emissions (near –8 ‰).

Given the above assumptions that sulphate from upwind sites and the oil recycling plant are insignificant in Calgary relative to vehicle exhaust and oil and gas industry emissions, it is possible to calculate the proportion of sulphur dioxide and sulphate from the latter two sources. The fraction of sulphur dioxide as well as aerosol and precipitation sulphate can be calculated for a scenario in which the $\delta^{34}S$ value for emissions from oil and gas are assumed to be +20 ‰. On average 77 +/- 41 % of the sulphur dioxide, 65 +/- 36 % of aerosol sulphate, and 55 +/- 30 % of precipitation sulphate in Calgary over the period of study were derived from flaring and or processing emissions from the oil and gas industry surrounding the city. A rough calculation of the emissions from vehicle exhaust in the city based on the average sulphur content for diesel and gasoline of 0.008 %, an average mileage of 10 litres per 100km for personal vehicles, approximately 500,000 vehicles traveling 10,000 km/yr, gives a value of 40,000 kg S/yr as SOx. If we further assume that emissions from vehicles are half SO_2 and half sulphate, then according to the apportionment calculations above, oil and gas industry emissions can be considered a source of approximately 67,000 kg S/yr and 37,000 kg S/yr SO_2 and aerosol sulphate respectively, or 104,000 kg S/yr. Compared to the emissions of 3,630,000 kg S/yr from a single gas plant north of the city in 1990, this seems large. However, over 20 such sour gas processing plants are located within a 100 km radius of

123

the city, mainly to the northwest.

5. CONCLUSIONS

Sources of sulphate and sulphur dioxide in Calgary, Alberta were distinguished and used to estimate the contribution to pollutant sulphur on the basis of differing sulphur isotope abundance ratios. Three isotopically distinct pollutant sulphur sources were identified: an oil recycling plant with emissions near -8 ‰, vehicle exhaust at $+9$ ‰, and oil and gas emissions from outside the city that contribute to the pollutant sulphur load. Further, isotope fractionation was not observed in the formation of fine aerosol sulphate relative to SO_2, an important feature if the isotope composition of sulphate is to be used for apportionment studies. This was also reflected in a large number of the SO_2 and total aerosol sulphate samples that had similar $\delta^{34}S$ values, suggesting common sources for both compounds. As the fraction of total sulphur as sulphate increased, $\delta^{34}S$ values for SO_2 converged, indicating a source of SO_2 with $\delta^{34}S$ values near $+18$ ‰, consistent with emissions from a sour gas processing plant just north of the city. In contrast $\delta^{34}S$ values for aerosol sulphate varied between $+12$ and $+22$ ‰ with increasing sulphate fraction and imply the addition of multiple sources of sulphate with pollution transport. $\delta^{34}S$ values for sulphate in precipitation in Calgary converged at high concentration near $+21$ ‰ whereas those for snow from a site upwind in the relatively pristine mountain region west of the city were much lower, around $+5$ ‰. These results confirm that sulphur from oil and gas emissions contribute significantly to pollutant sulphur loads in Calgary. Application of a simple two-source mixing model showed that, on average, vehicles contribute 40,000 kgS/yr as SOx to the city while the influx of emissions from oil and gas processing are on the order of 104,000 kgS/yr.

The proportion and source of primary and secondary sulphate was also derived in this study. Oxygen isotopes in precipitation sulphate were used to show that roughly 20 % was primary and that the remaining secondary sulphate best fit a model for heterogeneous oxidation of SO_2 in the presence of iron. Primary sulphate had $\delta^{34}S$ values associated with oil and gas processing, while secondary sulphate had $\delta^{34}S$ values that indicate an origin from both vehicle exhaust and sour gas.

5.1. Acknowledgements

Funding for this study was provided by an NSERC operating grant.

6. REFERENCES

Aunan, K., 1996 Exposure-response functions for health effects of air pollutants based on epidemiological findings. Risk Analysis, 16, 693-709.

Burnett R.T., Cakmak S. and Brook J.R., 1998 The effect of the urban ambient air pollution mix on daily mortality rates in Canadian Cities. Canadian J. Public Health, 89(3),152-156.

Burnett R.T., Dales R.E., Krewski D., Vincent R., Dann T. and Brook J.R.,1995: Associations between ambient particulate sulfate and admissions to Ontario hospitals for cardiac and respiratory diseases, American Journal of Epidemiology, 142, 15-22.

Calhoun, J. A., T. S. Bates, and R. J. Charlson, 1991 Sulphur isotope measurements of submicrometer sulphate aerosol particles over the Pacific Ocean. *Geophys. Res. Lett.* **18**, 1877-1880.

Giesemann, A., Jαger, H.J., Norman, A.L., Krouse, H.R. and Brand, W.A., 1994, On-line sulfur isotope determination using an elemental analyzer coupled to a mass spectrometer. *Anal. Chem.* **66**, 2816-2819.

Holt, B.D. (1991) The isotopic analysis of sulphur and oxygen isotopes: oxygen isotopes. In: Stable isotopes in the Assessment of Natural and Anthropogenic Sulphur in the Environment. Krouse, H.R. and Grinenko, V.A. (eds.), John Wiley & Sons, N.Y., 55-64.

Holt, B.D. and Kumar, R. (1984) Oxygen-18 study of high-temperature air oxidation of SO_2. Atmospheric Enviornment, 18, 2089-2094.

Holt, B.D., Cunningham, P.T., and Engelkemeir, A.G. (1976) Oxygen isotopy in atmospheric sulfate formation. Chemical Engineering Division: Environmental Chemistry Annual Report, July 1975-June 1976. U.S. Energy Research and Development Administration ANL-76-107. 27-33.

Holt, B.D., Kumar, R., and Cunningham, P.T. (1981) Oxygen-18 study of the aqueous phase oxidation of sulfur dioxide. Atmospheric Environment, 15, 557-566.

Holt, B.D., Cunningham, P.T., Engelkemeir, A.G., Graczyk, D.G., Kumar, R. (1983) Oxygen-18 study of non-aqueous phase oxidation of sulfur dioxide. Atmospheric Environment, 17, 625-632.

Holt, B.D., Kumar, R., and Cunningham, P.T. (1982) Primary sulfates in atmospheric sulfates: estimation by oxygen isotope ratio measurements. Science, 217, 51-53.

Klemm, R.F., and Gray, J.M.L.,1982, Acidity and chemical composition of precipitation in central Alberta, 1977-1978. Procedings of the Acid Forming Emissions in Alberta and their Ecological Effects Workshop. Edmonton, Alberta March 9-12, 1982, 153-179.

Krouse, H.R. and Herbert, M.K., 1988, *Sulphur and carbon isotope studies of food webs. In: Diet and Subsistence: current archeological perspectives.*B.V. Kennedy and G.M. LeMoine (eds.), Proceedings of the Nineteenth Annual Conference of CHACMOOL, Arch. Assoc. of the University of Calgary. 315-322, 1988.

Myrick, R.H. and Hunt, K.M. (1998) Air quality monitoring in Alberta: 1996 data report. Air Issues Monitoring Branch, Alberta Environmental Protection, I494-A9803, 197 pp.

Rowe, R.D., Mohtadi, M//.F., Havlena, J.J., Exall, D.I., Boyle, J.M., Banjamin, S.F. (1977) The short range chemistry of sour gas plant plumes. In: Proceedings of Alberta S Gas Workshop III. Sandhu, H.S., and Nyborg, M. (eds.) University of Alberta, Edmonton, Alberta, November 17-18. 27-38.

Saltzman, E. S., G. W. Brass, and D. A. Price, 1983, The mechanism of sulfate aerosol formation: chemical and sulphur isotopic evidence. *Geophys. Res. Lett.* **10**, 513-516.

AIR POLLUTION DISPERSION MODELLING IN SURROUNDING OF INDUSTRIAL ZONE OF CITY PANCEVO

Zoran Grsic, Predrag Milutinovic, Milena Jovasevic-Stojanovic, Dragan Dramlic, Marko Popovic[*]

1. INTRODUCTION

Information about air pollution at the territory of the municipal Pancevo (~30x30 km), recently was attainable only from two air monitoring stations located in the city, in spite of suburbs of the city are in touch with faces of the industrial plants which routinely emit more than million tons of air pollutants in the atmosphere yearly.

In the industrial zone of Pancevo there are three factories, which emit air pollutants to the atmosphere from chimneys and vents of the heights up to 150 meters. Numbers of undeclared small sources are not accounted in analysis of air pollution in the zone of influences of these factories.

Only mentioning of the industrial complex, which is consisted of factories like oil refinery, fertilizer factory, and petrochemical complex makes association of huge quantities of air pollutants emitted to the atmosphere from such one place. In the case of Pancevo, from the official information sources, it means 9600 tons of SO_2 per year and 6400 tons of NOx. Quantification of the other air pollution constituents like CO_2, NH_3, Cl ......, are not included in this analysis. Information sources of air pollution emission, used in this paper are material balances from technological process made in the factories project phase. Industrial complex of Pancevo was built about 40 years ago, mentioned

---

[*] Zoran Grsic, Milena Stojanovic-Jovasevic, Institute of Nuclear Sciences Vinca, POB 522, SCG-11001 Belgrade
Predrag Milutinovic, Dragan Dramlic, Marko Popovic, Institute of Physics, Pregrevica 118, SCG-11000 Belgrade

emission balances are still in use, but it is well known that situation about air pollution emission is much worst, because in factories installations are old and small amount of funds was directed to the environmental protection purposes.

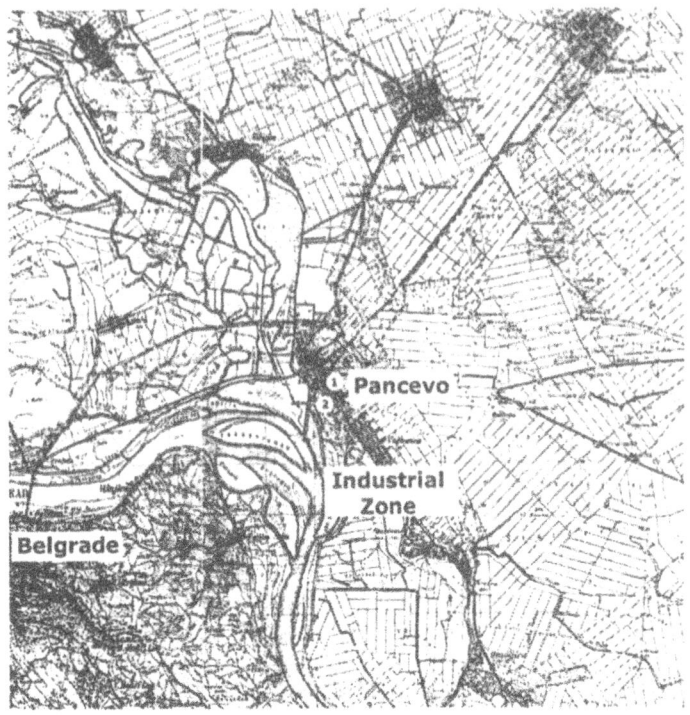

Figure 1. Map of Pancevo territory. Scale 1:300000.
1,2 air quality monitoring sites

No one chimney in the industrial zone is covered with continual monitoring of air pollutants emission. Data about air pollution emission originate not only from the technological balances but from the random controls, which are doing once or twice per year, but in the most case they are doing rarer , only once in few years.

Only two air quality monitoring stations existed at the territory of the range 900 square kilometers and they are located very close to each other. Some episodes with raised air pollution in the populated part of Pancevo, citizens signalized, but this air monitoring system did not recorded them due to it's location and meteorological conditions.

Main reason that even so big industrial area has not appropriate air quality monitoring network is at first bad Serbian and federal legislative relating to the environmental protection, then very bad economic situation in the country especially in the past ten to fifteen years and very low level of ecological education of the employees and especially of the factories leaders.

Experiences from the period of bombing, with every day events of the worst chemical accidental scenarios, "instantly" educated employees in the factories, their leaders and citizens of Pancevo, which resulted in defining the project of automation existed meteorological station established during bombing and realization of a fully automated system for continual air pollution control in routinely and accidentally situations.

2. AUTOMATED AIR QUALITY CONTROL SYSTEM

Main goal of the project is establishing of one automated system for air quality assessment at the whole municipal Pancevo territory.

Central part of the system is PC located in control room in City Hall to which are connected automated meteorological station and three the other PCs from factories in industrial zone.

Central PC operates like server which processes meteorological data, saves and checks emission inventory of every factory and on dependence of information about type of air pollutants emission runs accidental or routine atmospheric dispersion models.

Task of local PCs is to inform central PC about changing in air pollutants emission and to transfer all relevant data in the case of an accident. Operators may have the same figure on it's PC screens, or some operator can on it's own demand starts air pollution assessment from arbitrary chosen industrial sources. An operator can change air pollutants emission base only from the factory where he is authorized and all data base changing are documented and kept in archive in local and central PCs.

2.1. Hardware

At the XXVth NATO/CCMS meeting in Belgium, semi automated meteorological station mounted at the skyscraper of City Hall Pancevo was presented[1]. Through realization of the mentioned project this station has been fully automated, so that data from the sensors for wind speed and direction, one wind speed sensor, two air temperature sensors, global and net radiation sensor, air pressure and rain gage are getting in real time. The acquisition system of the meteorological station, capable of receiving sixteen analogue signals simultaneously, is controlled by a micro controller which packs received signals into a message and sends them using a direct connection to the central PC.

2.2. Software

Software is consisted of two parts, software for processing data coming from sensors and exchange information between central and local PCs (system software) and atmospheric dispersion software.

2.2.1. System software

The software developed for the System enables user to:
- receive data from data acquisition system which acquires data from meteorological sensors
- memorize and display this data in numerical and/or graphical form

- transfer this data to the Oil Refinery, Fertilizer Factory and Petrochemical Complex
- update the data related to the pollution sources
- receive data about pollution sources acquired in the Oil Refinery, Fertilizer Factory and Petrochemical Complex
- display spreading of the pollution superimposed on the map of Pancevo city or alternatively on the map of the entire Pancevo municipal territory together with the numerical value of the corresponding maximum air pollution concentration at the computing domain
- graphically display the degree of air pollution with different colors; additionally clicking at a specified point user obtains numerical value of the pollution in that point
- control the performance of the entire automated system, and especially the state of the interconnections
- display at the monitor and/or to print out all air pollution concentration data for a specified date together with the relevant meteorological situation

Software uses the Windows operating system enabling simple and easy communication with the System.

Entire automated system behaves like a LAN and is based on the TCP/IP protocol.

2.2.2. Atmospheric dispersion software

In designing atmospheric dispersion software starting point was limitation which originates from the characteristics of automated meteorological station and of quality of the data about air pollution sources. Important moment in making decision about atmospheric dispersion model for industrial zone in Pancevo was experience with automated meteorological station in the Institute of nuclear sciences Vinca, especially during bombing. Addition requirement related to the number of 999 air pollution sources from whole territory of municipal Pancevo , which have to be included in the source bases, caused choice of the straight line Guassian dispersion model for routine air pollutants release and puff dispersion model of the Gaussian type for instantaneous releases.

Diffusion equation for routine releases:

$$
C(x,y,z) = \frac{Q}{2\pi\sigma_y\sigma_z u} \exp\left(-\frac{1}{2}\frac{y^2}{\sigma_y^2}\right)
$$
$$
\left\{\exp\left[-\frac{1}{2}\frac{(z-H)^2}{\sigma_z^2}\right] + \exp\left[-\frac{1}{2}\frac{(z+H)^2}{\sigma_z^2}\right]\right\}
\tag{1}
$$

where:

$C(x,y,z)$ concentration at x,y,z

Q uniform emission rate of pollutants

130

H effective emission height

σ_y, σ_z standard deviations of plume concentration distribution in the horizontal and vertical

u mean wind speed affecting the plume

Diffusion equation for instantaneous releases:

$$C(x,y,z,t)=\frac{q}{\sqrt{(2\pi)^3}\,\sigma_x\sigma_y\sigma_z}\cdot\exp\left\{-\left[\frac{(x-ut)^2}{2\sigma_x^2}+\frac{y^2}{2\sigma_y^2}\right]\right\}$$
$$\times\left\{\exp\left[-\frac{(z-H)^2}{2\sigma_z^2}\right]+\exp\left[\frac{(z+H)^2}{2\sigma_z^2}\right]\right\} \quad (2)$$

Where ut is the downwind travel distance covered by the cloud through the time t, standard deviations σ_i are functions of ut and H is the effective point source height, or height of the cloud center. Other variables in the formula have usual meanings with exception that q represents the total mass released into individual puff.

From wind speed data, solar radiation-day and net radiation-night or temperature vertical gradients stability is derived, which lead to the standard deviations of concentration distribution in accordance with some empirical set of σ curves.[2]

Both models are constitutive part of the automated meteorological station which supplies them with appropriate meteorological data. Input source parameters are: time of integration, coordinates and height of the sources, its diameters, source strength (Q,q), temperature of pollutants and level above ground for concentration calculation, effective source/puff heights and appropriate physical dimensions of the sources. They can operate in the prognostic mode or in the diagnostic mode. Further, plume/puff rise and effective stack height, are calculated using the Briggs model and wind power low, respectively. Models take into consideration topography, dry and wet deposition.

3. INPUT DATA

Automated meteorological station supplies models with appropriate meteorological input data. Situation is much different with input data concerning to air pollutants emission and its sources. No one chimney in Serbia has continual emission control. All data about air pollution sources and its emissions are from the factories data bases, which originate from the project documentation and from the temporary, very rare control of chimneys gas emissions.

Duty of every operator in factories is to enter any changing in air pollution sources conditions, all such changing are recorded and archived in central PC and in local PC where changing was made.

After establishing continual air pollution control in industrial zone Pancevo in accordance with new environmental protection act of Serbia which has to go into effect this year, data about air pollutants emission will directly enter the models.

Information we have used in analysis presented in this paper is that from 39 chimneys factories emit about 29 tons SO2 per day and from 31 chimneys about 19 tons NO2 per day.

4. MODELING AIR POLLUTION CONCENTRATION

Using mathematical models, space continual fields of SO2 and NO2 concentrations are presented at the whole area of the municipal Pancevo, covering and two mentioned air monitoring points. Input data for these calculations were source characteristics and its strength for every factory from industrial zone, together with meteorological data collected at the automated meteorological station.

Period of analysis is April 1st 2002., when automated meteorological station was put in experimental operation, to January 31st 2003.

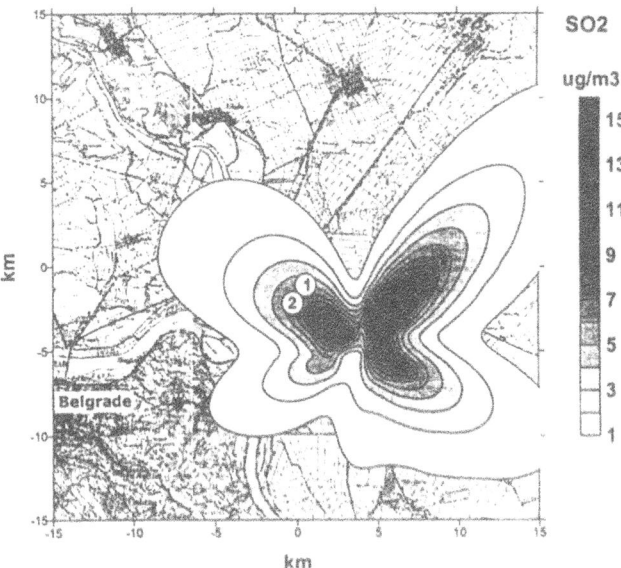

Figure 2. SO2 concentration averaged for the period April 2002.-January 2003.

Modeled values of SO2 concentrations are 5.8 and 6.4 µg/m3 for monitoring sites marked with 1 and 2 respectively. In the same time measured values were 11.8 and 13.2 µg/m3

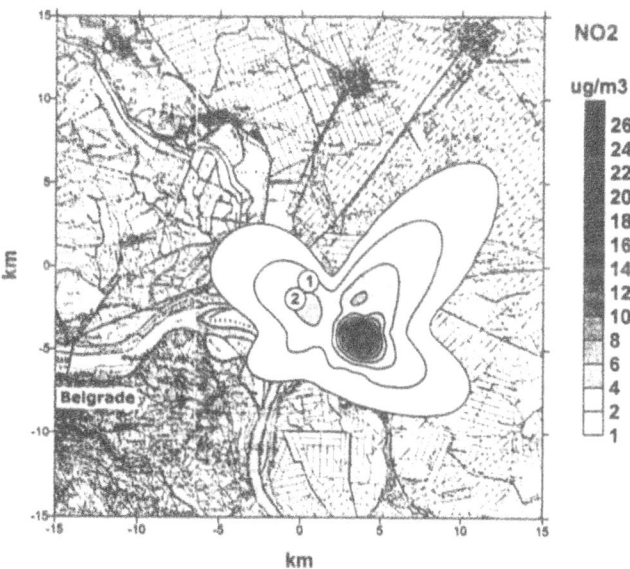

Figure 3. NO2 concentration averaged for the period April 2002.-January 2003.

Modeled concentrations of NO2 were 4.7 and 5.2 µg/m3 while measured values were 18.9 and 17.9 µg/m3 for monitoring sites 1 and 2 respectively..

Results of comparing measured and modeled concentrations of the air pollutants simultaneously emitted from the chimneys, vents and the other sources in the Oil Refinery, Petrochemical complex and Fertilizer factory show about two times lower modeled values for SO2 and three times lower modeled values for NO2.

Possible reason for this differences is that emission of SO2 and NO2, which factories have in their data bases are not correct and that they emit few times more air pollutants to atmosphere than they proclaimed.

5. CONCLUSION

Differences between modeled and measured values probably come from incorrect input data about emission SO2 and NO2, so first step will be to check them.

From modeled results it is obvious that air quality monitoring network with only two monitoring stations is not solution for solving air pollution problems in Pancevo and that appropriate air monitoring network must be established.

6. REFERENCES
1. Z.Grsic, P.Milutinovic, M.Jovasevic-Stojanovic, M.popovic,*Air Pollution Modeling and its Application XV*, Kluwer Academic/Plenum Publishers,2002, pp.509-511
2. IAEA Safety Guide No. 50-SG-S3, 1980, Atmospheric Dispersion in Nuclear Power Plant Sitting, Viena

DISCUSSION

S.A.- AKSOYOGLU   Without chemical reactions included in the mode, you cannot compare model results with measurements, because especially NO_x levels change due to photochemical reactions.

Z. GRSIC   I think that main reason for disagreement between measured and modeled NO2 concentrations lies in incorrect emission data base for NOx and their check is in progress. In the next phase we plan to include chemical module in the software.

MODELING THE IMPACT OF GLOBAL CLIMATE AND REGIONAL LAND USE CHANGE ON REGIONAL CLIMATE AND AIR QUALITY OVER THE NORTHEASTERN UNITED STATES

C. Hogrefe, J.-Y. Ku, K. Civerolo, B. Lynn, D. Werth, R. Avissar, C. Rosenzweig, R. Goldberg, C. Small, W.D. Solecki, S. Gaffin, T. Holloway, J. Rosenthal, K. Knowlton, and P.L. Kinney[*]

1. INTRODUCTION

In recent years, the focus of global climate change research has shifted towards the assessment of regional scale consequences. This paper describes the design and initial results of a modeling study aimed at simulating the effects of global climate change and regional land use change on climate and air quality over the northeastern United States in order to project the associated public health impacts in the region. To this end, modeling tools on a variety of scales are being linked. Specifically, regional climate models are linked to both a global climate model and a regional land-use change model. Outputs from regional climate simulations are subsequently used both to assess changes in public health due to heat stress and to simulate regional and urban air quality. Finally, results from these air quality simulations are coupled to health impact models. This paper focuses on the air quality modeling aspect of the project. Global and regional climate change

[*] C. Hogrefe, ASRC, University at Albany, Albany, NY. K. Civerolo and J.-Y. Ku, New York State Department of Environmental Conservation, Albany, NY. B. Lynn, C. Rosenzweig and R. Goldberg, NASA-Goddard Institute for Space Studies, New York, NY. D. Werth and R. Avissar, Duke University, Durham, NC. C. Small, S. Gaffin, T. Holloway, J. Rosenthal, K. Knowlton, and P.L. Kinney, Columbia University, New York, NY. W.D. Solecki, Montclair State University, Montclair, NJ.

Air Pollution Modeling and Its Application XVI, Edited by
Borrego and Incecik, Kluwer Academic/Plenum Publishers, New York, 2004

135

could conceivably alter regional and urban air quality in a variety of ways through both direct and indirect effects. Rising temperatures could have a direct effect on chemical reaction rates and mixed-layer heights, while changes in synoptic circulation patterns might influence the transport and mixing of pollutants as well as the occurrence of conditions conducive to high ozone concentrations. Additionally, anthropogenic emissions of the ozone precursors NO_x and VOC are also expected to change in future decades, and it is necessary to understand whether future ozone air quality is more sensitive to changes in emissions or changes in climate. In this paper, we present initial results from regional-scale simulations for 3-months summer seasons under current and future climate conditions.

2. MODELS AND DATABASE

2.1. Emissions processing

Emissions are the driving force for climate change (through an increase in greenhouse gas emissions) and are also a critical input to the air quality model. The IPCC SRES (IPCC, 2000) has generated various future emissions scenarios based on projections of population, technology change, economic growth, etc. In this paper, we use the SRES "A2" scenario. This scenario is characterized by a large increase of CO_2 emissions, relatively weak environmental concerns, and large population increases (15 billion by 2100). The greenhouse gas emissions from this scenario are used as inputs to the global and regional climate models described below. Because the SRES emission scenarios are global in nature, their spatial resolution is not adequate for regional air quality modeling. Therefore, we use the county-level EPA 1996 National Emissions Trends (NET96) inventory as basis for air quality modeling. This emission inventory is processed by the Sparse Matrix Operator Kernel Emissions Modeling System (SMOKE) (Carolina Environmental Programs, 2003) to obtain gridded, hourly, speciated emission inputs for the air quality model. Future year emissions are estimated by multiplying the base year pollutant inventory with the spatially-uniform growth factors for the SRES "A2" scenario. "A2" U.S. VOC emissions are estimated to increase by 8%, NO_x emissions by 29.5% and SO_2 emissions by 49% by the 2050s relative to the 1990s.

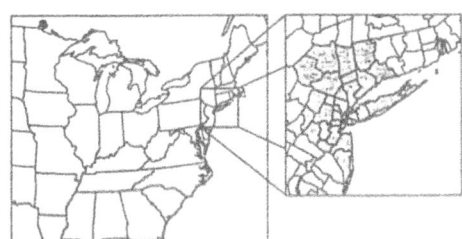

Figure 1: Map of the 36 km CMAQ modeling domain. The 31-county region around New York City is highlighted in the insert.

136

2.2. Global and regional climate modeling

Global climate simulations for the time period 1990 – 2100 were performed with the Goddard Institute for Space Studies (GISS) 4°x5° resolution Global Atmosphere-Ocean Model (Russell et al., 1995). Output from these simulations was then used as boundary conditions for regional climate models over the U.S. for individual summer seasons. Two regional climate models are used in this project, namely, ClimRAMS (Liston and Pielke, 2000) and MM5 (Grell, 1994). In this paper, only MM5 model results are presented. MM5 simulations driven by GISS-GCM boundary conditions were performed for the 10 summers from 1990-2000 and 2050-2060 at a horizontal resolution of 108 km over the continental U.S. The hottest summers in each decade were selected to perform additional simulations with two nested grids at 108 km and 36 km horizontal resolution; the 36 km grid covers the eastern U.S.

2.3. Air Quality modeling

Using the SMOKE-processed emissions and the 36 km MM5 regional climate simulation for the hottest simulated summer in the 1990s and 2050s, air quality simulations were performed using the Community Multiscale Air Quality (CMAQ) model (Byun and Ching, 1999). The modeling domain consists of 68x59 horizontal and 16 vertical grid cells and is depicted in Figure 1. The insert in this figure highlights the 31-county region around New York City that is the focus of the public health impact research of this project.

2.4. Observations

To evaluate the ability of the GISS-GCM and MM5 regional climate models to reproduce the present day climate, we obtained gridded monthly mean meteorological observations described in New et al. (2000). Additionally, we obtained 10 years of ozone observations at 21 monitors in the 31 county region in Figure 1 from the EPA's AIRS system.

3. RESULTS AND DISCUSSION

3.1. Regional climate modeling

MM5 simulations driven by GISS-GCM boundary conditions were performed for the 10 summers from 1990-2000 and 2050-2060 at a horizontal resolution of 108 km over the continental U.S. The hottest simulated summers in each decade (1994 and 2056, respectively) were selected to perform additional simulations with two nested grids at 108 km and 36 km horizontal resolution for subsequent air quality modeling. The summers from 1990 to 1998 simulated with the 108 km MM5 were compared both to outputs from the GCM and to gridded observations (New et al., 2000) in terms of monthly mean temperatures. The results of this comparison are presented in Table 1. The root mean square error is lower and the correlation with observations is higher for the MM5 simulation than the GCM simulation. While the GCM has a negative bias, the MM5 has a positive bias of smaller absolute magnitude. Both simulations underestimate the observed interannual variability measured by the standard devation of July montly mean temperatures over the 10 years at each grid. These results suggest that the MM5 simulation is indeed improving the representation of regional-scale climate compared to the GCM, and that the selection of the hottest summers in the 1990s and 2050s for the higher resolution 36 km simulation will yield a plausible realization of extreme summertime conditions both in the present and in a future climate.

3.2. Air quality modeling

3.2.1. Current climate and emissions

Before applying the photochemical model for simulations with projected emissions or future climatic conditions, it is important to assess the ability of the system to reproduce features evident in present-day ozone observations when driven by MM5 regional climate fields from the hottest summer in the 1990s and 1996 NET base year emissions. Because day-to-day fluctuations in the model and observations cannot be directly compared (the MM5 simulations are driven by GCM fields rather than actual observation-based reanalysis data such as NCEP), a comparison has to focus on different aspects of observed and predicted ozone distributions. Table 2 shows the number of days

Table 1: Evaluation of monthly mean July temperature 1990 – 1998 simulated by the GISS-GCM and MM5 at 108 km resolution. Model prediction are compared to the gridded observational data set described in New et al. (2000).

		RMSE	Bias	Correlation	Interannual variability (% of observed)
July Temperature	GCM	2.95	-1.33	0.83	66%
	MM5	2.22	0.92	0.90	66%

Table 2. Number of station-days exceeding daily maximum 1-hr ozone concentrations of 120 ppb at 21 monitors in the New York City region

Rank	Year	Observations 1991 - 2000	CMAQ driven by the hottest MM5 summer in the 1990s
1	1991	130	
2	1999	74	
3	1995	73	
4	1997	68	
5	1993	63	
6	1994	43	
7	1998	24	22
8	1996	21	
9	1992	20	
10	2000	18	

on which daily maximum 1-hr ozone concentrations exceeded 120 ppb at 21 ozone monitors in the 31-county area highlighted in Figure 1 for each summer from 1991 – 2000. The years are ranked from the year with the most exceedances (1991) to the year with the lowest exceedances (2000). Also shown in Table 2 is the number of exceedances at these monitoring stations predicted by the CMAQ simulation discussed above. It is evident that this simulated summer – although the hottest MM5 summer – does not rank among the highest ozone summers in the 31-county area during the 1990s. However, the number of exceedances falls within the observed range. Additional comparison with observed distributions of daily maximum ozone concentrations at these monitors (not shown here) indicate that the predictions replicate the overall shape of these distributions well.

An additional method of comparing observed and predicted variations is the use of spectral decomposition techniques as described in Hogrefe et al. (2000). To this end, we decomposed observed and predicted time series of ozone concentrations into fluctuations occurring on the intra-day (several hours), diurnal (12-48 hours), synoptic (2-21 days), and baseline (greater than 21 days) time scales and computed the variance for each of these components. For the observations, this analysis was performed for the years with lowest and highest total variance during the decade (2000 and 1991, respectively) as well as for the average over all 10 years. The results, presented in Figure 2, show that the CMAQ simulation (Figure 2d) overestimates the total variance compared to either the lowest variance observed summer (Figure 2a), highest variance observed summer (Figure 2b), or average observed summer (Figure 2c). However, the relative variance distribution among the different temporal components is quite similar between observations and predictions, although the predictions tend to underestimate the relative importance of the intra-day component while they tend to over-estimate the relative importance of the synoptic component. While it would be desirable to perform CMAQ simulations for

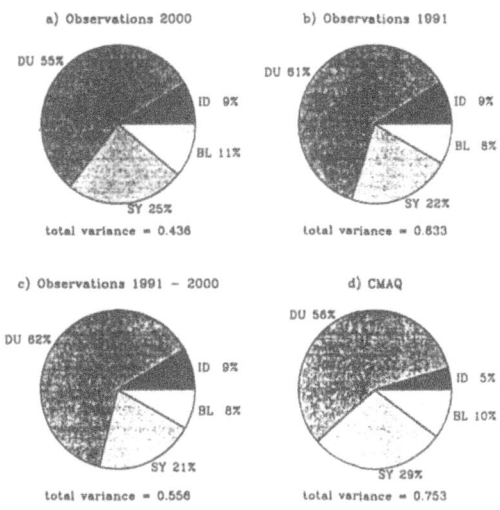

Figure 2: Relative contributions of the intra-day (ID), diurnal (DU), synoptic (SY) and baseline (BL) components to the total variance. Ozone observations for 2000 (a), 1991 (b), average observations 1991 – 2000 (c) and CMAQ simulation (d).

multiple summers to perform additional analyses, these initial results suggest that the GCM/MM5/CMAQ system – when applied to simulate a summer under current conditions – is able to replicate observed variability and, thus, can be applied to assess the relative impact of changes in future emissions and climate.

3.2.2. Effect of future emissions

The SRES "A2" scenario used in this study is characterized by an increase of VOC, NO_x and SO_2 emissions as described in Section 2.1. Distributions of daily maximum ozone concentrations from two simulations using the current-climate MM5 run with both present and future emissions are presented in Figure 3. The distributions show daily maximum ozone concentrations for the 31-county area (Figure 3a) as well as for all model grid cells (Figure 3b). The increase in NO_x emissions leads to an increase in the number of ozone titration events in some urban grid cell as evidenced by the slight downward shift of the distribution for the future emission simulation in Figure 3a. This effect is not visible in Figure 3b, indicating that the predicted increase in NO_x and VOC emissions will lead to an increase of the number of ozone exceedance days on a domain-wide basis. These results also indicate that – while the focus of the health impact studies is on the 31-county region – the simulation of future air quality needs to consider both urban and regional-scale effects.

3.2.3. Effects of changing climate and emissions

Analysis of the meteorological fields produced by the 36 km MM5 simulation for the hottest summer of the 2050s in the "A2" scenario reveals that temperatures in the 31-county area are expected to increase by about 2°C compared to the hottest summer in the 1990s. This increase is evident for all hours of the day. Likewise, future humidity levels are predicted to rise during all hours of the day. Additionally, distributions of the ventilation coefficient (calculated as the product of boundary layer height and wind speed and used as a measure of pollutant dilution) show that this quantity decreased for the future year simulation compared to the base year simulation. These changes in meteorological conditions are expected to have a profound impact on predicted ozone concentrations. To examine the isolated and combined effects of changing meteorology and changing emissions, four simulations were performed with all possible combinations of current and future climate and emissions. The results of these simulations are summarized in Tables 3a-b in terms of the number of ozone exceedance days for each simulation. Table 3a shows the results for the 21 ozone monitors in the 31-county area, while Table 3b shows the results for the entire modeling domain. It can be seen that for this set of simulations the number of 1-hr exceedance days is more sensitive to climate changes rather than emission changes both for the 31-county area and on a domain-wide basis. The highest number of exceedance days is predicted for the simulation using both future climate and emissions. This indicates that the predicted changes in the chemical composition of the atmosphere and the changes in physical parameters such as temperature, humidity, and vertical mixing both contribute to a deterioration of future air quality in these simulations.

4. FUTURE RESEARCH

While the results presented above indicate that the changes in regional climate have a bigger impact on air quality than predicted changes in emissions for the single summer seasons simulated here, this needs to be confirmed by performing simulations for additional summers in both present and future climate conditions. Specifically, the changes in predicted ozone exceedance days still are within the range of observed decadal variability (compare Tables 2 and 3a). Simulations for five summers during the 1990s and 2050s at 36 km resolution are currently underway and are expected to improve our understanding of the importance of climate change on predicted ozone concentrations. In a parallel effort, land-use projections for the 31-county region highlighted in Figure 1 have been generated as part of this project (http://www.csam.montclair.edu/luca/) To integrate these projections into the regional climate and air quality models, higher-resolution simulations will be performed for selected extreme heat and ozone events. These higher resolution simulations are also expected to better capture the complex flow patterns in New York City, and to provide higher-resolution information for the health impacts analysis. Data from the current runs will also be used to perform an initial analysis of health impacts of the changing regional climate and air quality (Knowlton et al., 2003). Moreover, simulations will also be performed for the decade of the 2080's and for the SRES "B2" scenario that is characterized by a slower growth of emissions.

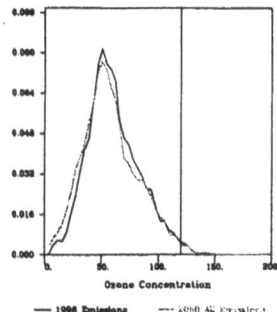 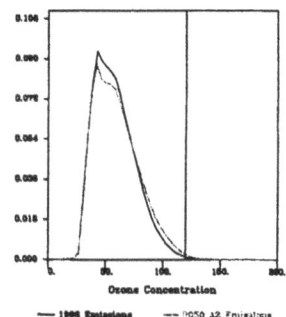

Figure 3: Distributions of daily maximum ozone concentrations using the hottest MM5 summer for the 1990s with current emissions (solid black curve) and future emissions (dotted grey curve) over the 21-county area (left) and the entire modeling domain (right).

Table 3a. Number of times daily maximum 1-hr ozone concentrations at 21 stations in the 31-county area exceed 120ppb during the time period from June 1 – August 31

	1990s Hottest Summer	2050s A2 Hottest Summer
1990s Emissions	22	69
2050s A2 Emissions	27	71

Table 3b. Number of times daily maximum 1-hr ozone concentrations at all 4,012 grids cells exceed 120ppb for the time period from June 1 – August 31

	1990s Hottest Summer	2050s A2 Hottest Summer
1990s Emissions	786	3,266
2050s A2 Emissions	1,745	7,489

5. SUMMARY

The design and initial results of a modeling study to simulate the effects of climate and land-use change on air quality and public health have been described in this paper. These initial results suggest that MM5 simulations forced by boundary conditions from the GISS-GCM yield plausible realizations of extreme summertime conditions both in the present and in a future climate. The results also indicate that the predicted changes in the chemical composition of the atmosphere and the changes in physical parameters such as temperature, humidity, and vertical mixing both contribute to a deterioration of future air quality. Finally, steps for future research in the NYCHP were discussed.

6. ACKNOWLEDGMENTS

This work is supported by the U.S. Environmental Projection Agency under STAR grant R-82873301.

7. DISCLAIMER

Although the research described in this article has been funded in part by the U.S. Environmental Protection Agency, it has not been subjected to the Agency's required peer and policy review and therefore does not necessarily reflect the views of the Agency and no official endorsement should be inferred.

8. REFERENCES

Byun, D.W. and Ching, J.K.S. (eds.), 1999. Science algorithms of the EPA Models-3 Community Multiscale Air Quality Model (CMAQ) modeling system. *EPA/600/R-99/030*, U. S. Environmental Protection Agency, Office of Research and Development, Washington, DC 20460.

Carolina Environmental Programs, 2003: Sparse Matrix Operator Kernel Emission (SMOKE) Modeling System, University of Carolina, Carolina Environmental Programs, Research Triangle Park, NC.

Grell, G. A., J. Dudhia, and D. Stauffer, 1994: A description of the fifth-generation Penn State/NCAR Mesoscale Model (MM5). *NCAR Technical Note*, 138 pp., TN-398 + STR, National Center for Atmospheric Research, Boulder, CO

Hogrefe, C., S. T. Rao , I. G. Zurbenko, and P. S. Porter, 2000: Interpreting the Information in Ozone Observations And Model Predictions Relevant to Regulatory Policies in the Eastern United States; Bull. Amer. Meteor. Soc., 81, 2083 - 2106.

IPCC, 2000. Special Report on Emissions Scenarios. Nacenovic, Nebojsa and Swart, Rob (eds.), Cambridge University Press, Cambridge, United Kingdom, 612 pp.

Knowlton et al., 2003: Modeling public health impacts of climate change in the New York metropolitan region, 5th International Conference on Urban Climate, September 1-5, 2003, Lodz, Poland (submitted)

Liston, G.E., and R.A. Pielke, 2000: A climate version of the Regional Atmospheric Modeling System, Theor. Appl. Climatol., 66, 29-47.

New, M., Hulme, M. and Jones, P.D., 2000: Representing twentieth century space-time climate variability. Part 2: development of 1901-96 monthly grids of terrestrial surface climate. *Journal of Climate* 13, 2217-2238

Russell, G.L., J.R. Miller, and D. Rind 1995: A coupled atmosphere-ocean model for transient climate change studies. *Atmos.-Ocean* 33, 683-730

DISCUSSION

T. HALENKA

Could you estimate the possible effect of ozone production increase backward on GHG radiative forcing?

C.HOGREFE

In our project, the regional air quality model does not provide feedback to a global chemistry-climate model. Thus, we cannot estimate the change in global radiative forcing due to increasing regional-scale ozone air pollution.

F. LEFEBRE

Which kind of LBC for you CTM did you use and do you plan to perform coupled climate-chemical GCM-MM5-CTM calculations?

C.HOGREFE

Because the regional air quality model CMAQ is not coupled to a global chemistry model in our project, we used climatological constant boundary conditions. At this point, we do not plan to perform fully coupled climate-chemistry simulations at both the global and regional scale.

D.STEYN

Before conducting MM5 modelling, you have to show that the GCM captures (for 1990s) the frequency of occurence of synoptic circulation types that favored the generation of ozone episodes. We tried this for the west coast and showed the GCM did not perform well in this sense. Have you done this?

C.HOGREFE

No, we have not done this because we do not think that the GCM is the right tool to capture these circulation pattern. Therefore, in our project we use MM5 to simulate synoptic scale circulation patterns. Our analysis of both simulated meteorology and ozone suggests that the GCM/MM5/CMAQ system is able to reproduce observed variability. However, we will further evaluate MM5's ability to capture the frequency of occurence of certain circulation patterns using the synoptic typing methodology you suggested.

N.MOUSIOPPOULOS Biogenic emissions will undoubtedly play a significant role in your analysis:
1. To what extent do you consider the impact of changed temperatures / irradiance on biogenic emissis
2. Can you estimate the importance of land use changes (typically desertification) in this context?

C.HOGREFE We use the BEIS2 model to estimate biogenic emissions. In BEIS2, emissions are sensitive to temperature and irradiance, and, therefore, we do take into account the effect of changing meteorology and climate on this aspect of biogenic emissions. At the current stage of our research project, we have not yet incorporated projected land use changes into the regional climate/air quality modeling, but we will add this linkage in the near future. By doing so, we will be able to estimate the impact of changes in biogenic emissions due to changes in land use on simulated ozone concentrations.

SIMULATION OF METEOROLOGICAL CONDITIONS AND SURFACE OZONE CONCENTRATIONS WITH MM5 AND CAMX IN ISTANBUL

Ümit Anteplioğlu, Selahattin İncecik, and Sema Topcu[*]

1. INTRODUCTION

Istanbul is located at about 41°N;29°E with a population of over 10 million people. According to 2000 figures, there are about 1.5 million motor vehicles in the city.

The city is divided by the Bosphorus as the European and Asian parts. The city separates the Sea of Marmara from the Black Sea. Photochemical air pollution in Istanbul is a new environmental issue since the first incidence was reported in 1999. The city of Istanbul has recently experienced high ozone days.

This study includes mesoscale meteorological model MM5 simulations to understand the meteorological influences in the form of high ozone days during 16-19 July 2001. MM5 had never been used before in this area. It was the first time the model was used in this area in connection with ozone studies. Furthermore, the CAMx photochemical model was selected as the chemical model to be used in the study.

2. EPISODE CHARACTERISTICS

The history of photochemical air pollution in Istanbul is not long. The city experienced serious SO_2 and TSP pollution until to mid-90s. Fuel switching caused disappearing the classical air pollution in the city (Topcu et al., 2003). Topcu and Incecik (2002) explained the preliminary results of the ozone levels in urban atmosphere of Istanbul. Mesoscale and regional factors that influenced to the 16-19 July 2001 episode are discussed in Topcu and Incecik (2003). Weak morning surface winds, early morning stable conditions and higher precursor concentrations are caused higher ozone concentrations. These

[*] Ümit Anteplioğlu, BÜ. Kandilli Observatoy &ERI , 34680, İstanbul, Turkey. Selahattin Incecik and Sema Topcu, Istanbul Technical University,Department of Meteorology, 34469 , Istanbul, Turkey.

Air Pollution Modeling and Its Application XVI, Edited by
Borrego and Incecik, Kluwer Academic/Plenum Publishers, New York, 2004

145

conditions are most favorable for ozone formation. On the other hand, strong NNE winds in the morning hours are associated with synoptic pressure systems in the region. Convective atmospheric conditions and increased instability may cause peak ozone concentrations. Surface ozone plays an important role in affecting the regional climate and causing harmful effects on health. In order to understand the impact of the wind on surface ozone in the region, it is necessary to determine the region's airflow features using mesoscale model.

3. METHOD

3.1. MM5

The PSU/NCAR MM5 mesoscale model is a limited-area, nonhydrostatic, terrain-following sigma-coordinate model designed to simulate or predict mesoscale atmospheric circulation. MM5 Version 3.5 31 half sigma level has 4 domains from surface up to 100 mb. This model was utilized in a one-way nested formation with four domains. Domain 1 has a horizontal resolution of 81 km, and 31x37 horizontal grid points. Domain 2 has a grid resolution of 27 km and dimensions of 34x40. Domain 3 has a resolution of 9 km and dimensions of 61x67 and the inner domain (Domain 4) has a grid resolution of 3 km and dimensions of 91x139. Analysis nudging is used above planetary boundary layer for only outermost domain.

3.2. CAMx

The Comprehensive Air quality Model with extensions (CAMx) is a Eulerian photochemical model that allows for integrated assessment of gaseous and particulate air pollution over many scales (Environ, 2000). The CAMx simulates the emission, dispersion, and chemical reactions by using the continuity equation for each chemical species on a system of nested 3-D grids. The chemical mechanisms supported by CAMx are based on the Carbon Bond-IV mechanism. 22 vertical layers up to 6000 meter are used. In this study we used the fourth domain as the inner grid resolution. The required input data include gridded wind, temperature, pressure, water vapor, cloud cover, rainfall fields, emission and air quality data in the form of initial and boundary conditions and land use data.

Unfortunately, it is not possible to obtain a full emission inventory in the city. For this purpose, vehicle traveled data provided as on road mobile source emissions, which are day specific, are adjusted from 1999 to 2001 (Anteplioglu et al., 2001). For this purpose we have considered only a limited inventory for the pollutants obtained from motor vehicles.

4. RESULTS

The results give of regional scale circulations over the city. Figure 1 and 2 gives MM5 winds at the surface on 16 and 17 July at 1100 UTC. Both figures show that the winds blowing different directions during the noon hours over Istanbul. The city has

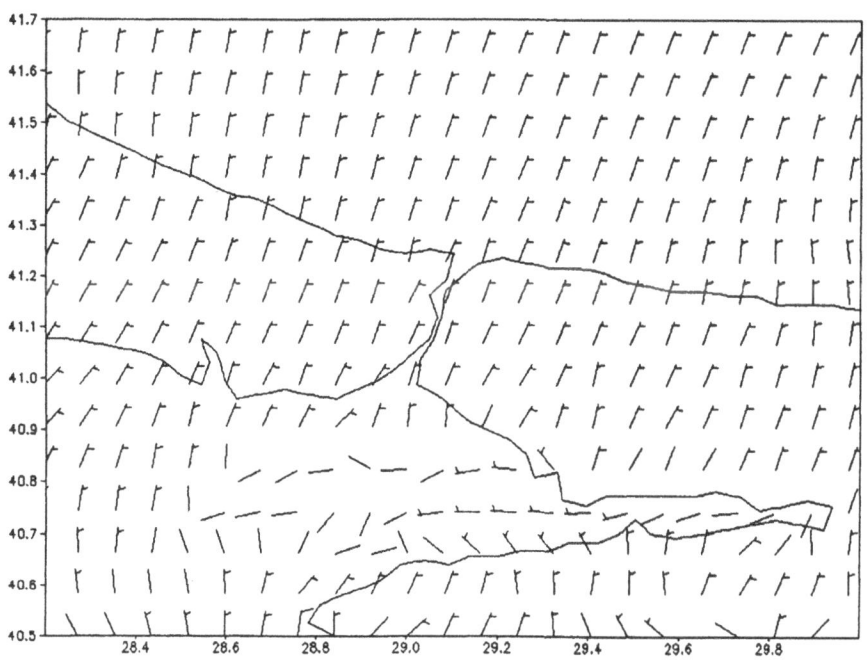

Figure 1. MM5 winds at surface on 16 July at 1100 UTC.

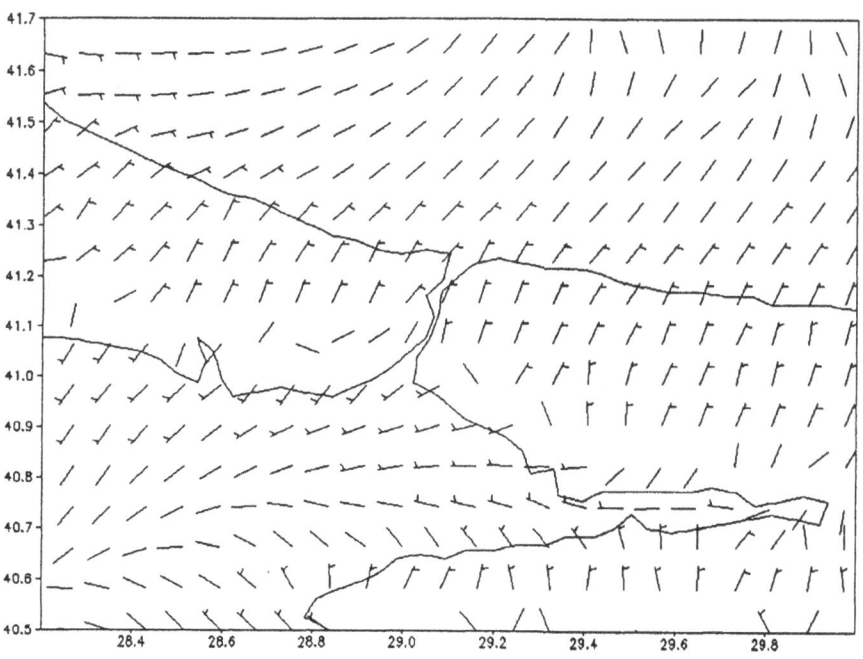

Figure 2. MM5 winds at surface on 17 July at 1100 UTC.

experienced a sea breeze circulation on July 17, 2001 during the noon hours. The winds are almost northerly on the coasts of Black Sea throughout the day depending on the synoptic pressure pattern. However, on that day, the regions on the coasts of Marmara, has experienced the winds blowing in the south from the morning hours. This circulation, which seems to be a typical seas breeze is gradually weaken by the mid afternoon hours. The wind components along the north-south cross section give a loop towards to the city center. This causes the precursor pollutants emitted in the city center are transported to the coastal area again.

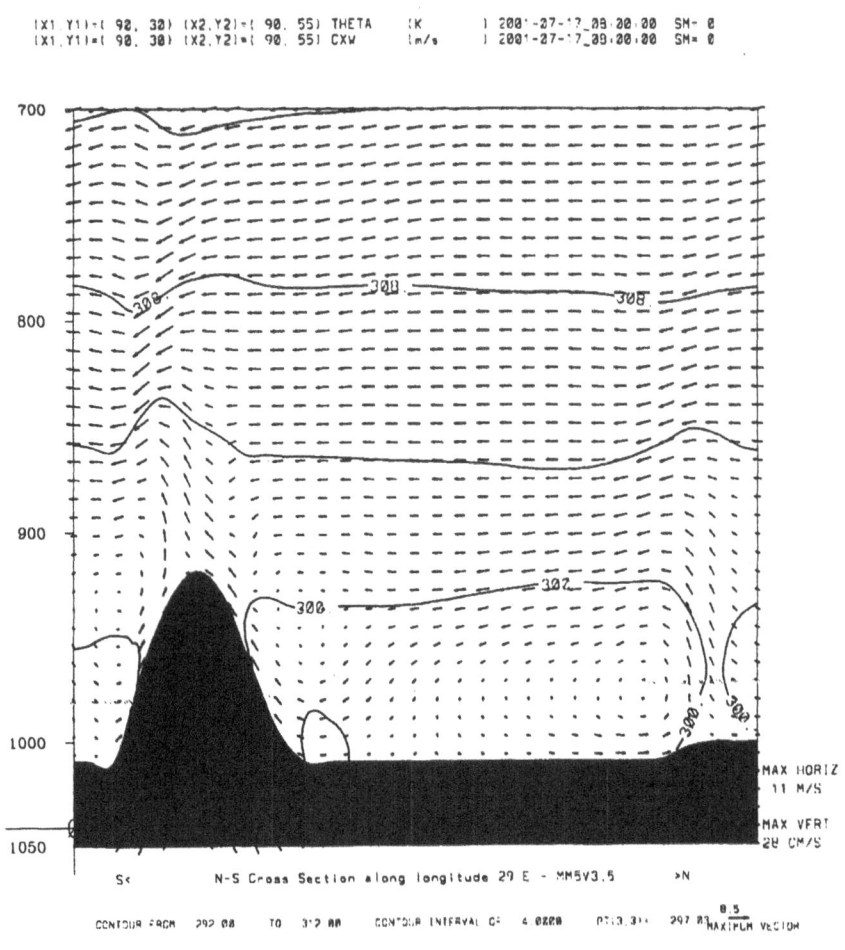

Figure 3. The wind vectors over longitude of the 29th on 17 July at 0800 UTC

In order to understand the regional wind circulation N-S Cross section long the longitude 29°E is shown (Fig. 3). Pollutant transport has been located in along the

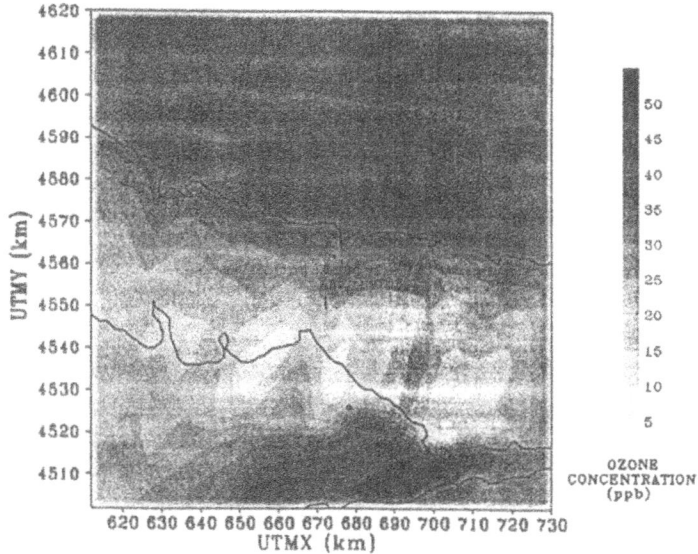

Figure 4. Simulated ozone concentrations by CAMx on 16 July at 1100 UTC.

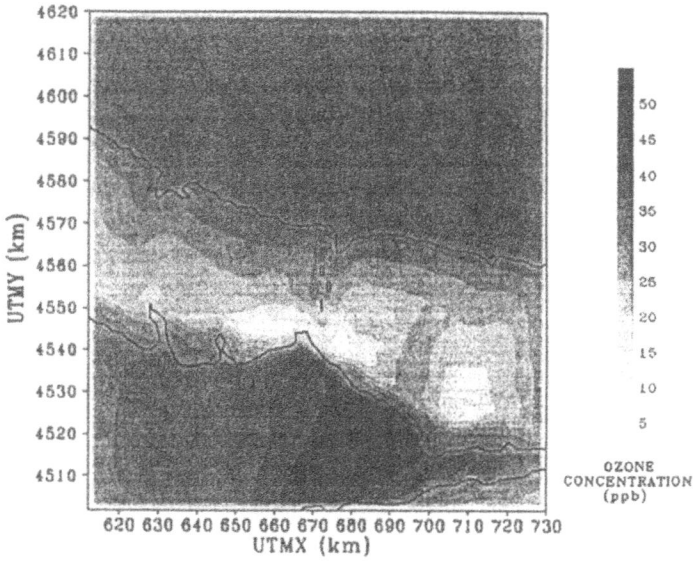

Figure 5. Simulated surface ozone concentrations by CAMx on 17 July at 1100 UTC.

149

circulation. Air pollution concentrations are higher in the afternoon hours especially in Izmit Bay, which is located in the south of the region. The most important chemical and paint industries and the biggest petroleum refinery are situated in Izmit Bay. According to the circulations, the topography of the region and the chemical industries may lead to higher concentration around the up slopes areas. The simulations of ozone concentrations by CAMx are shown in Figs. 4 and 5.

5. CONCLUSION

The objectives of the study include both MM5 simulations to understand meteorological factors in the formation of high ozone days and a CAMx Eulerian photochemical model results which allows us integrated assessment of ozone concentrations over urban and regional scales. Initial and boundary conditions, meteorological data and emission inventory need to be improved and observational nudging, verified land use data will be used. This study allows us an idea about the risks caused pollutants coming from Istanbul under significant meteorological conditions.

6. REFERENCES

Anteplioglu U., 2000, *Investigation of surface ozone by photochemical-dynamic model in Istanbul*, (in Turkish) PhD Thesis, Istanbul Technical University,pp 147.

Anteplioglu, U, Topcu, S., and Incecik,S., 2002, An application of a photochemical model for urban airshed in Istanbul, *Air Pollution Modeling and Its Application*, C.Borrego and G.Schayes, Ed., Kluwer Academic/Plenium Publishers, New York, pp 167-175.

Environ, 2000, User's guide to comprehensive air quality model with extensions (CAMx), Environ International Corporation.

Topcu, S, and Incecik, S., 2002, Surface ozone measurements and meteorological influences in the urban atmosphere of Istanbul, *Int. J Environ. and Pollution,* 17, 390-404.

Topcu, S, and Incecik, S., 2003, Characteristics of surface ozone concentrations in urban atmosphere of Istanbul: A case study, *Fresenius Environ. Bulletin.* 12.

Topcu, S, Incecik, S. and Unal, Y.S., 2003, The influence of meteorological conditions and stringent emission control on high TSP episodes in Istanbul, *Environ. Sci.. and Pollution Res.*, 8 , 24-32.

DISCUSSION

B.AINSLIE   Where on the map of your model domain are the ozone
                         measuring stations?

S.INCECIK   The model domain was not given on the map. The model was
                         utilized with the inner domain which has a grid resolution of 3
                         km and dimensions of 91×133.

AURAMS RUNS DURING THE PACIFIC2001 TIME PERIOD – A MODEL/MEASUREMENT COMPARISON

P.A. Makar[1,*], V. S.Bouchet[2], L-P. Crevier[2], A.P. Dastoor[1], S. Gong[1], W. Gong[1], S. Menard[2], M. D. Moran[1], B. Pabla[1], S. Venkatesh[1], L. Zhang[1]

1. INTRODUCTION

Research on the numerical prediction of chemically speciated gas and particle phase components of the atmosphere has been driven by public health studies linking both gases and particles to adverse health effects. Three dimensional air-quality models containing detailed chemical and physical processes for gas and particle formation (Meng et al., 1997; Dennis et al., 1996; Ackermann et al., 1998) provide a means of linking and describing the complex non-linear processes leading to these adverse air-quality health outcomes. While the physical and chemical processes leading to particle formation are domain-independent, their boundary conditions (such as local meteorology, topography, and emissions) may vary greatly between different domains. The comparison of a regional air quality model to measurements made in different locations and more than one model domain is therefore desirable for the evaluation of the model's prediction accuracy.

Measurement data that may be used for model comparison may be grouped into two broad classes, monitoring network and measurement intensive data. Data from monitoring networks usually is available from multiple locations across a wide spatial domain, and over long time periods, thus providing a means of evaluating model performance as an average over large spatial scales (Bouchet et al., 2003, these proceedings). The monitoring data at any given site may have limitations in terms of the level of chemical speciation, the time interval over which measurements are averaged (typically a day or longer) and the frequency of measurements (e.g. measurements may be made only one day in six, resulting in long time periods in which no data is available for analysis and comparison to regional air-quality model results). The data from measurement intensives, in contrast, are highly speciated and time-resolved, but are

[1] Makar, Dastoor, Gong, Gong, Moran, Pabla, Venkatesh, Zhang: Meteorological Service of Canada, 4905 Dufferin Street, Downsview, Ontario, M3H 5T4, Canada
[2] Bouchet, Crevier, Menard: Meteorological Service of Canada, CMC, 2121 Route TransCanadienne, Dorval, Québec, H9P 1J3, Canada

Air Pollution Modeling and Its Application XVI, Edited by
Borrego and Incecik, Kluwer Academic/Plenum Publishers, New York, 2004

usually made only at select sites for a limited total time. Both forms of data are therefore useful for model evaluation purposes; the monitoring data for evaluating the model's overall performance, and the intensive data for evaluating the details of the model processes in a given location.

This study describes the comparison of the Meteorological Service of Canada's AURAMS model to measurement data in a domain covering much of Western Canada and the North-Western United States of America, during the period August 25 – August 31, 2003. This time interval comprises the final week of a four-week intensive measurement campaign to study the processes leading to atmospheric particle formation that took place in the Greater Vancouver Region of Canada's west coast, "Pacific2001" (Li et al., 2002). Regular monitoring data from six networks were also collected during the same time period, over a much larger spatial domain than the measurement intensive.

Comparisons of AURAMS model results to measurements on other domains, and a detailed study of mass consistency and conservation effects in the model, may be found in two other studies described in this volume (Bouchet et al., 2003; Gong et al., 2003, respectively.

2. AURAMS: A Brief Description

AURAMS (A Unified Regional Air-quality Modelling System) is comprised of five independent codes that are run in sequence in order to produce an air-quality simulation.

The first of these is the Canadian Emissions Processing System (Moran et al., 1997; Makar et al., 2003), a set of codes that convert raw emissions data (criteria air contaminant annual or seasonal activity levels, VOC speciation emissions profiles, spatial and temporal disaggregation factors, etc.) into temporarily varying, spatially gridded mass fluxes for each of the major emissions categories (Major and Minor Point Sources, and Mobile, Non-Mobile, and Biogenic Area sources).

The second part of the system is one of two meteorological forecast codes (the Meteorological Service of Canada's MC2 (Benoit et al., 1997) or GEM (Côté et al., 1998) models), run in an "offline" mode to produce input meteorological input data for air-quality simulations.

The third part of the system, the AURAMS Preprocessor, extracts and converts input data files of geophysical constants, land-use categories, meteorological information, and the emissions fields created by CEPS to unformatted binary files for the grid used in a given study. The geophysical fields include the terrain height, vertical coordinate information in modified Gal-Chen coordinates (Benoit et al., 1997), vegetation type and surface roughness at or above each gridpoint in the model's horizontal domain. The fifteen land use categories are used in gas and particle deposition calculations (Zhang et al., 2001, 2002). Meteorological variables used by AURAMS include the components of the wind (see Gong et al., 2003, these proceedings), 3D fields for temperature, specific humidity, coefficients for the vertical diffusion of heat and momentum, cloud fractions (cumulus, stratiform and total), total cloud water content, liquid precipitation flux, solid precipitation flux, cloud water composition (water, ice/snow, rain), rain evaporation rate, cloud water to rain conversion tendency, and 2D fields for the friction velocity, anemometer height, boundary layer height, stability function (phi(z/L)), bulk Richardson number at the surface, cosine of solar zenith angle, surface pressure, solar irradiance at the ground, ground temperature, ice fraction over water, snow depth over land, and the grid cell locations in latitude and longitude.

The fourth component of the system is the Chemical Transport Model (CTM), the "core" of AURAMS that uses the above input data to predict the temporal and spatially varying concentrations of the gas and particle constituents of the atmosphere. The CTM of the AURAMS model is based on that of its predecessor, the CHRONOS model (Pudykiewicz et al 1997), with extensive modifications for detailed particle processes. The AURAMS CTM is both long and complex, and space consideration prevent a detailed description of all of its components; this and the two companion papers in these proceedings (Bouchet et al., Gong et al.) will give only an overview of the model's main features.

Control of gas and chemical speciation, input, output and the main CTM processing options is through a single "masterlist" file, read in at the start of processing, which may be modified by the user depending on the complexity of the information desired. The masterlist includes species-specific controls for the operators allowed to act on each chemical component, the particle distribution resolution (a bin approach is used), the emissions control (by source stream), emissions reduction scenario scaling factors, and options for output for comparison to measurement data (see Bouchet et al., 2003, these proceedings). The masterlist allows process and diagnostic level control of individual gas and particle bin variables throughout the CTM. The configuration used in these runs included 42 gas-phase species, and 7 aerosol species (sulphate, nitrate, ammonium, sea-salt, black carbon, organic carbon, and crustal material) distributed over 12 bins ranging in size from 0.01 to 40.96 μm diameter.

The model makes use of a single forward operator splitting on its main processes, with an operator-splitting step that is a multiple of the meteorological data output time-step and may be set by the user (a step of 900 seconds was used in the Pacific2001 runs that follow). The main operators, in the order in which they are called in the code once meteorological and emissions data are input, are as follows:

Wind correction: input meteorological wind fields are corrected for mass consistency and conservation (see Gong et al., 2003, these proceedings).

Dry deposition velocities: calculated for all species (Zhang et al., 2001, 2002).

Major point source emissions: added as a vertically distributed flux.

Advection: a semi-Lagrangian method with a second order interpolation (based on a Taylor series expansion) is used (Pudykiewicz et al, 1997; Gong et al., 2003, these proceedings).

Vertical diffusion: solved in the vertical column for each transported variable as a tridiagonal system of equations for the finite difference approximation of the diffusion equation, with area-source emission (or deposition) fluxes as the lower boundary condition, and a zero-flux upper boundary condition (Pudykiewicz et al, 1997).

Gas-phase chemistry: the speciation and mechanism used in these runs is the ADOM-II speciation (42 gas-phase species total, Lurmann et al. (1986)), solved using a vectorized version of the method of Young and Boris, and modified for the calculation of secondary organic aerosol formation using the method of Odum et al. (1996).

Particle dynamics (nucleation, coagulation, condensation, dry deposition, below-cloud scavenging, aerosol activitation): A 12-bin sectional approach was employed, using the numerically optimized Canadian Aerosol Module (Gong et al., 2002).

Cloud processes: Cloud processes in current AURAMS include aerosol activation, aqueous-phase chemistry (mass transfer and oxidation in droplets), and wet deposition (precipitation scavenging of gases and aerosols and evaporation). The aqueous-phase chemistry is adapted from the ADOM-II mechanism and solved with a vectorized Young-Boris predictor-corrector scheme (Gong et al, 2003).

Inorganic heterogeneous chemistry: two different options are available in the model – the ISORROPIA code of Nenes et al. (1998), or a more numerically efficient code based on the ISORROPIA algorithms, HETV (Makar et al., 2003).

The fifth and final component of AURAMS is the post-processing code. The output of AURAMS is available on a highly time-resolved individual species and size-bin basis, but measurement data are frequently available only on a coarser time and cut-size. The CTM described above includes options for summing of particle bin data by species, but further processing is required to extract site-specific data for model comparison. This module is described in more detail in Bouchet et al. (2003), these proceedings.

3. MODEL DOMAIN, NETWORK DATA, STUDY PERIOD CHARACTERISTICS

Model simulations were performed on a 148 by 124 grid with a horizontal resolution of 21 km and 28 terrain-following layers in the vertical from the ground to 30 km. The data from four monitoring networks (CAPMoN, NAPS, IMPROVE and AIRS), with a total of 1519 stations available for model–measurement intercomparison purposes. While the number of *potential* monitoring stations is large, it should be noted that the number of stations with useable data was smaller. All of these networks provide 24 hour averaged data on a sometimes intermittent basis (e.g. once a week) and not all of the species desired for model comparison are available across all stations. The number of observations available for comparison to model results thus varies in space, time and between chemical species. The size resolution of the monitoring network particulate data are usually only available over broad size ranges necessitating aggregation of the model species – here, model-measurement comparisons are made for $PM_{2.5}$ sulphate, nitrate, and organic carbon, and PM_{total} sulphate, ammonium, and nitrate. Only two $PM_{2.5}$ ammonium records were available during the model simulation period; insufficient for comparison purposes.

The model simulations encompassed the period August 25 – August 31, 2001 (with a two day a priori spin-up period). This period was chosen due to observations made during the measurement intensive (Li et al., 2002) indicating rapid particle formation in the Greater Vancouver/Georgia Basin region of Canada's west coast, subsequent to the passage of a frontal system on August 23rd. The highest particle levels were recorded between the 27th -28th of August, with total PM levels reaching 20 $\mu g/m^3$. Measurement analysis to date (Li et al., 2002) has showed that all components of particle mass increased during the time period, thus providing a good test for model simulations.

4. MODEL RESULTS AND ANALYSIS

A sequence of three AURAMS-generated aggregated PM$_{2.5}$ sulphate concentration contour plots are shown in Figure 1. The Figure illustrates the rapid build-up of particle mass at five hour intervals during the period August 27th – August 28th.

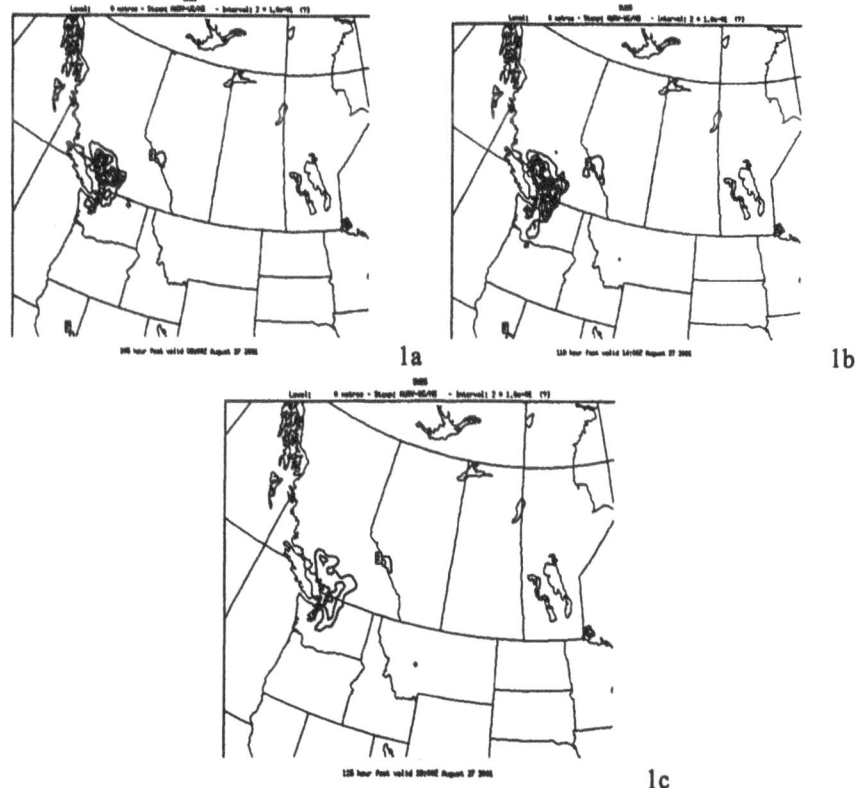

Figure 1. AURAMS aggregated PM$_{2.5}$ SO$_4$ contours (interval 2µg/m^3) at (a) 9:00 UT, (b) 14:00 UT, and (c) 19:00 UT on August 27, 2001.

The model clearly captures the increase in particulate matter levels along the west coast (left side of Figure) ; model-predicted PM2.5 SO$_4$ increases from levels of 2 µg/m^3 in the five hours prior to (1a), through 10 µg/m^3 in 1(a), 14 µg/m^3 in 1(b), dropping to 6 µg/m^3 in 1(c). A second, relatively minor sulphate high occurs at the same time left of centre – this is likely associated with natural gas /oil refining near the border between the Canadian provinces of British Columbia and Alberta.

While capturing the timing and total mass of the high particle concentration event reasonably well, the fractionation of that mass between individual chemical species is not as accurate. This is shown below in Figure 2, in which the 24-hour aggregated model values have been compared to observations from the 4 monitoring networks in the simulation domain.

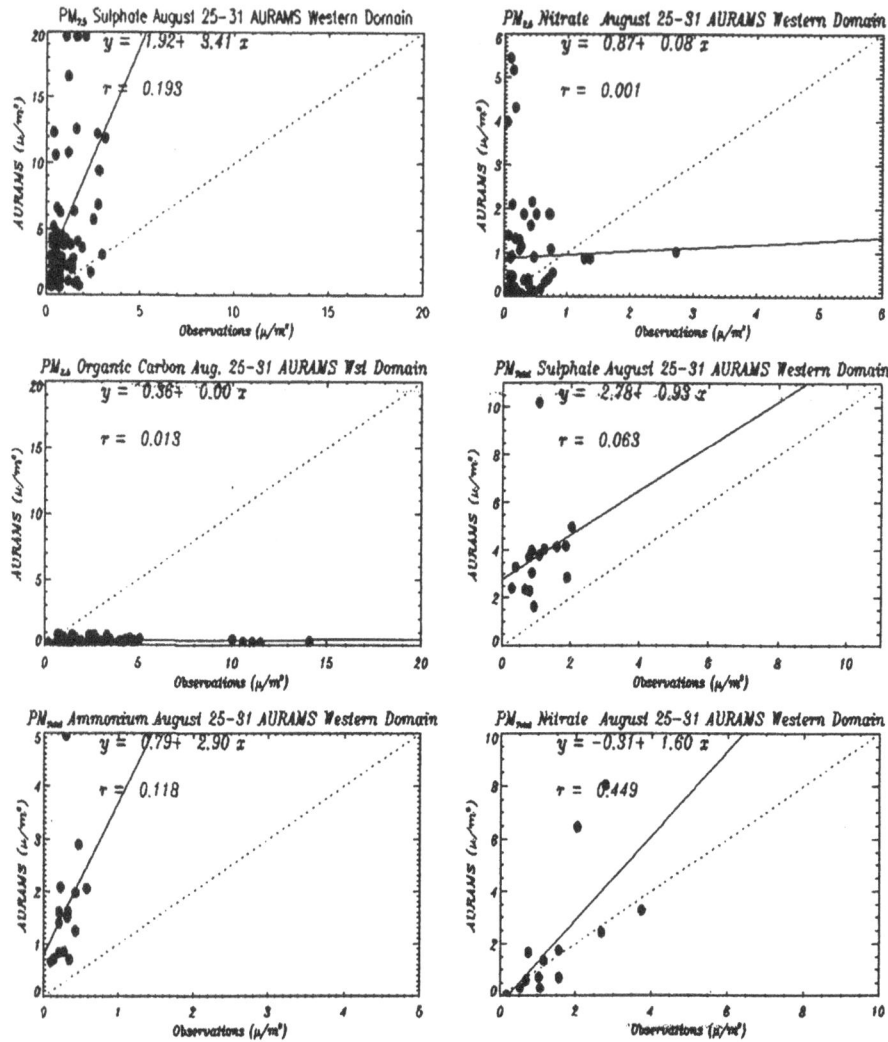

Figure 2. Model – Monitoring network comparisons for (a) PM2.5 SO$_4$, (b) PM2.5 NO$_3$, (c) PM$_{2.5}$ Organic Carbon, (d) Total PM SO$_4$, (e) Total PM NH$_4$, and (f) Total PM NO$_3$.

The likely reasons for the over-predictions and under-predictions vary depending on the chemical species. Considering each species in turn:

PM$_{2.5}$ Sulphate (2(a)): The model over-predicts fine-mode particle sulphate. This may be related to the details of the cloud processing module employed in AURAMS. Here, only stratus cloud processing of sulphate was considered in detail. Recent work by Gong et al. (2003) suggests that cumulus cloud processing may account for a larger share of sulphate processing into cloud water than stratus.

PM$_{2.5}$ Nitrate (2(b)): The model values are widely scattered, with not clear trend due to the two sets of outlying points near each axis. These will be studied in more detail to determine whether over- or under-predictions are associated with different regions. For example, under-predicted nitrate values are likely in regions of high sulphate due to

158

interactions between these components via inorganic heterogeneous chemistry. However, it is not clear from this initial analysis whether the low nitrate values are associated with high sulphate values.

PM$_{2.5}$ Organic Carbon (2(c)): The model values considerably under-predict relative to the measurements. Source apportionment work during the Pacific2001 study indicates that a large proportion of the organic carbon may be primary, and other work (Makar et al, 2003) has shown that the emission of primary OC particles may have a strong feedback effect on secondary organic aerosol formation. The primary particle emissions module of AURAMS is still under development – only secondary formation is considered at the moment. This is the likely cause of the observed under-predictions.

Total PM Sulphate (2(d)): While the model-measurement relationship is closer to unity than for fine particles, the model values have a vertical offset of 2.8 µg/m^3. Again, this probably relates to the cloud processing parameterizations used in this version of the model.

Total PM Ammonium (2(e)): These results differ from AURAMS runs in eastern domains (see Bouchet et al., these proceedings), where the tendency is towards ammonium under-prediction. The locations of the observation stations will be examined in the future, to determine the cause for these regional differences.

Total PM Nitrate (2(f)): The model values have a tendency towards over-prediction, in contrast to the fine mode values.

The above discussion illustrates one of the limitations of the use of monitoring network data in model inter-comparisons: lacking information as to which of the data points are co-located in time and space, analysis of the causes for model over- and under-predictions may be difficult. For this reason, future work will center on comparisons with the Pacific2001 measurement intensive data, with its considerably more detailed and co-located chemical speciation and time resolution.

5. CONCLUSIONS

The results shown above, along with the comparison on a different domain described elsewhere in these proceedings (Bouchet et al., 2003), are the first evaluations of the newly created AURAMS. While the model does a reasonable job of reproducing the large scale features of particle formation, biases in the model results noted here (fine mode sulphate over-predictions, organic carbon under-predictions) suggest the need for improvements/upgrades to some of the model processes. In addition to these upgrades, further intercomparisons with the more detailed Pacific2001 measurement intensive data are planned.

REFERENCES

Ackermann, I.J., H. Hass, M. Memmesheimer, A. Ebel, F.S. Binkowski and U. Shankar, 1998: Modeal aerosol dynamics for Europe: Development and first applications. Atmos. Environ., 32, 2981-2999.

Benoit, R.,M. Desgagné, P. Pellerin, S. Pellerin, Y. Chartier and S. Desjardins, 1997. The Canadian MC2: a semi-Lagrangian, semi-implicit wideband atmospheric model suited for finescale process studies and simulation. Mon. Wea. Rev., 125, 2382-2415.

Bouchet, V.S., M.D. Moran, L.-P. Crevier, A.P. Dastoor, S. Gong, W. Gong, P.A. Makar, S. Menard, B. Pabla, L. Zhang: 2003. Wintertime and summertime evaluation of the regional PM air quality model AURAMS (these proceedings).

Côté, j., J.-G. Desmarais, S. Gravel, A. Méthot, A. Patoine, M. Roch and A. Staniforth, 1998. The operational CMC/MRB Global Environmental Multiscale (GEM) model. Part I: Design considerations and formulation. Mon. Wea. Rev., 126, 1373-1395.

Dennis, R.L., D.W. Byun, J.H. Novak, K.J. Galluppi, C.J. Coats, and M.A. Vouk, 1996: The next generation of integrated air quality modeling: EPA's Models-3. Atmos. Environ.., 30, 1925-1938.

Gong et al., J.Geophys.Res., in press, 2002.

Gong, W., A.P. Dastoor, V.B. Bouchet, S. Gong, P.A. Makar, M.D. Moran and B. Pabla, 2003: Cloud processing of gases and aerosols in a regional air quality model (AURMAMS) and its evaluation against precipitation-chemistry data, Proceeding of the Fifth Conference on Atmospheric Chemistry: Gases, Aerosols, and Clouds, 2.3 (CD-ROM), AMS.

Gong, W., P.A. Makar and M.D. Moran: 2003. Mass-conservation issues in modelling regional aerosol (these proceedings).

Li, S.-M. and B. Thomson, 2002: PACIFIC2001 Air quality study – an overview. Proceedings, Fall AGU.

Lurmann, F.W., A.C. Lloyd, and R. Atkinson, 1986. A chemical mechanism for use in long-range transport/ acid deposition computer modelling, J. Geophys. Res., 91, 10905-10936.

Makar, P.A., M.D. Moran, M.T. Scholtz, A. Taylor, 2003: Speciation of volatile organic compound emissions for regional air quality modelling of particulate matter and ozone, J. Geophys. Res. (in press).

Makar, P.A., V.S. Bouchet, A. Nenes: 2003. Inorganic Chemistry Calculations using HETV – A Vectorized Solver for the SO_4^{2-}-NO_3^--NH_4^+ system based on the ISORROPIA Algorithms, Atmos. Environ. (in press).

Meng, Z., D. Dadub, and J.H. Seinfeld, 1998: Size-resolved and chemically resolved model of atmospheric aerosol dynamics, J. Geophys. Res., 103, 3410-3435.

Moran, M.D., M.T. Scholtz, C.F. Slama, A. Dorkalam, A.Taylor, N.S. Ting, P.A. Makar, and S. Venkatesh, 1997. An overview of CEPS1.0: Version 1.0 of the Canadian Emissions Processing System for regional air-quality models, Proc., 7th AWMA Emission Inventory Symp., 28-30 Oct. Research Triangle Park, N.C., Air and Waste Manage. Assoc., aPittsburgh.

Nenes, A., C. Pilinis, and S.N. Pandis, 1998. ISORROPIA: a new thermodynamic equilibrium model for multiphase multicomponent marine aerosols, Aquatic Geochem., 4, 123-152.

Odum, J.R., T. Hoffman, F. Bowman, D. Collins, R.C. Flagan, and J.H. Seinfeld, 1996. Gas/particle partitioning and secondary aerosol formation. Environ. Sci. Technol., 30, 2580-2585.

Pudykiewicz, J.A., A. Kallaur and P.K. Smolarkiewicz, 1997. Semi-Lagrangian modelling of tropospheric ozone, Tellus, 49B, 231-248.

Zhang, L., M.D. Moran, P.A. Makar, J.R. Brook, and S. Gong, 2001. Asize-segregated particle dry deposition scheme for an atmospheric aerosol module. Atmos. Environ., 35, 549-560.

Zhang, L., M.D. Moran, P.A. Makar, J.R. Brook, and S. Gong, 2002. Modelling gaseous dry deposition in AURAMS: a unified regional air-quality modelling system. Atmos. Environ., 36, 537-560.

160

DISCUSSION

N.CHAUMERLIAC You mentioned cloud processes and heterogeneous chemistry included in your model? Can you tell more about if this can influence some of the discrepancies between model results and observation.

W. GONG These are some of the areas that we are trying to evaluate our model representations. However the difficulty is the lack of suitable observations. In terms of cloud processing of aerosols the data is extremely lacking. We (MSC) are planning a field study next summer focussed on cloud-aerosol processes, and hopefully we will have some data to evaluate and improve our model representations. The heterogeneous chemistry is being evaluated against monitoring network data and measurement intensive data from the Pacific2001 campaign in ongoing work.

D.STEYN Your grid spacing is 21 km: at best you will resolve features at 40km, and can be fairly certain, of 80 km. The Fraser Valley has a horizontal dimension of 100km. Will you model at a finer resolution. If so, how fine ?

W. GONG This is our first step in attempting to simulate the Pacific 2001 experiment. This run, with 21 km resolution and reasonably large spatial domain, allows us to examine long-range transport of pollutants from Vancouver-Seattle to other parts of the Western Canada and the North-Western USA. In addition, it allows us to accurately characterize the inflow boundary conditions for the subsequent higher resolution runs we have planned. Our initial concern is to resolve the regional scale features accurately, then move to higher resolution runs. The next step will be to run the model (nested) on a more focussed domain with finer resolution. The grid spacing (in horizontal) that we are considering is 3 to 5 km depending on the available emission inventory.

P. BUILTJES 1. What are the reasons for the improvement in observed versus calculated SO4-2?

2. Do you have information about the accuracy of the BC observations?

W. GONG

1. Regarding the improvement in sulphate comparison, we can attribute this to mostly two things in this case. Firstly, there was some confusion in the definition of standard-temperature-pressure (STP) for some of the network sites in the earlier comparison (shown in Figure 2 in the manuscript), which had been corrected in this Pacific2001 run here. The other thing is the use of the sub-optimal blending for mass-conservation correction in the more recent simulation as opposed to the column scaling correction in the earlier model run. As shown in the presentation earlier (paper # 5.11) the column scaling correction for mass conservation tends to over correct at lower levels due to higher concentration normally found near the surface. This is not the case with the sub-optimal blending scheme.

2. Regarding the accuracy of EC/BC measurements, I am not too sure about that. Yes, it is important to know the accuracy of measurement as well when we compare our model results with the observations, particularly if there is a systematic error in observations.

ESTIMATES OF FUTURE PM$_{2.5}$ LEVELS IN SOUTHEASTERN UNITED STATES

M. Talat Odman, James W. Boylan and Armistead G. Russell[*]

1. INTRODUCTION

This modeling study was part of an integrated assessment conducted by the Southern Appalachian Mountains Initiative (SAMI) formed by eight southeastern states listed in Table 1, U.S. EPA, U.S. Forest Service, and the National Park Service. Its objective was to assess whether the mandated emissions reductions of the primary airborne pollutants, including sulfur oxides (SO$_x$), nitrogen oxides (NO$_x$) and volatile organic compounds (VOC) will be enough to protect and preserve the ecosystems and natural resources of the Southern Appalachian Mountains, especially in Class I areas. First, a comprehensive air quality modeling system that can address ozone, particulate matter (PM), and acid deposition was developed. Then, the system's ability to characterize the processes that affect air quality in the region was evaluated. This paper focuses on fine particulate matter (PM$_{2.5}$) that was studied for its impact on the visibility at the vistas of the region. Episodes that characterize the annual distribution of visibility were selected and modeled. The model results were compared with measurements of daily averaged PM$_{2.5}$ concentrations taken at Class I areas. The performance of the model was evaluated for good, moderate and bad visibility days to understand its predictive capability over the entire spectrum of visual air quality. Then, air quality simulations were conducted for the year 2010 using emission projections that take into account not only the growth in the region but also the controls that are expected to be implemented. Finally, the potential impacts of further controlling SO$_2$, NO$_x$, VOC, and NH$_3$ emissions were estimated with a new sensitivity analysis technique that facilitates the design of regional control strategies.

A detailed description of the study can be found in Odman et al. (2002). In this paper, after a brief review of model performance and simulation results for the year 2010, sensitivity of PM$_{2.5}$ levels to SO$_2$ emissions from different geographic sub-domains will be discussed. For various Class I areas, source regions with significant adverse effects will be identified.

---

[*] Talat Odman, Georgia Institute of Technology (GIT), Atlanta, Georgia, 30332-0512, talat.odman@ce.gatech.edu, Telephone: 404-894-2783, Fax: 404-894-8266. James Boylan, Georgia Department of Natural Resources, Environmental Protection Division, Air Protection Branch, Atlanta, Georgia, 30354. Armistead Russell, Georgia Institute of Technology, Atlanta, Georgia, 30332-0512.

2. METHODOLOGY

The Urban-to-Regional Multiscale (URM) model used in this study is a comprehensive air quality model that can calculate ambient levels of $PM_{2.5}$ along with gaseous pollutants (Boylan et al., 2002). URM uses the sectional approach where particles with diameters in a certain range are assumed to have the same composition. Here, four size bins or sections were used: 1) less than 0.156 μm, 2) 0.156-0.625 μm, 3) 0.625-2.5 μm, and 4) 2.5-10.0 μm. The first three size bins correspond to $PM_{2.5}$ while the combination of all four bins corresponds to PM_{10}. Inorganic aerosols are treated as an equilibrium system of sulfate, nitrate, ammonium, chloride, sodium and water (Nenes et al., 1998). Organic aerosols are lumped into a single species that condenses from oxidation of VOC (Pandis et al., 1992). Gravitational settling of particles is treated with a resistance approach similar to the dry deposition of gas-phase species.

2.1. Simulation of $PM_{2.5}$ Levels

Using URM, nine episodes (one winter, four spring and four summer episodes) were simulated between the years 1991-95. The modeling domain covered the eastern half of the United States. A multiscale grid placed the finest horizontal resolution of 12 km over the SAMI region and coarsened gradually towards the boundaries of the domain. Seven unequally spaced vertical layers with finer resolution near the surface extended to a height of approximately 13 km. Meteorological and emission inputs to URM were generated using the "Regional Atmospheric Modeling System" (Pielke et al., 1992) and the "Emission Modeling System" (Wilkinson et al., 1994), respectively. The initial and boundary conditions for $PM_{2.5}$ were set using data from the Interagency Monitoring of Protected Visual Environments (IMPROVE) network. In general, aerosol species have longer lifetimes than the gaseous species; therefore, the uncertainty in their initial and boundary conditions may have a larger impact on modeled concentrations.

Simulations of the year 2010 were conducted using the same meteorological inputs used for the basecase simulations but different emissions, which were projected from 1990 to 2010. The projections assumed reductions of VOC and NO_x as mandated by the Clean Air Act Amendments (CAAA) to comply with the 1-hour ozone standard, reductions of SO_2 and NO_x from utility sources under Title V of the CAAA, and reductions of NO_x and VOC from mobile sources under Tier I tailpipe standards and fuel rules. In addition, the projections assumed emission reductions from several recently promulgated regulations: regional NO_x reductions which will be included in "State Implementation Plans" to reduce ozone, NO_x and VOC reductions due to the implementation of Tier II and low sulfur rules, and VOC reductions resulting from "Maximum Achievable Control Technology" standards. They did not include the emissions reductions that might be required for the 8-hour ozone standard, the new $PM_{2.5}$ standard, or the regional haze rule. The annual emissions of SO_2, NO_x, VOC, $PM_{2.5}$ and NH_3 by source category from the eight SAMI states are illustrated in Figure 1. The levels of these emissions both in 1990 and in 2010 are shown for comparison. While the annual emissions of SO_2, NO_x, VOC in the 8 SAMI states decreased by 23, 24 and 12 percent, from 1990 to 2010, $PM_{2.5}$ and NH_3 emissions increased by 5 and 49 percent, respectively. The initial conditions of SO_2, sulfate and ammonium for the year 2010 were reduced in proportion to SO_2 emission reductions. Similarly, the initial conditions for NO_x were scaled by the reductions in NO_x emissions. However, all boundary conditions were kept

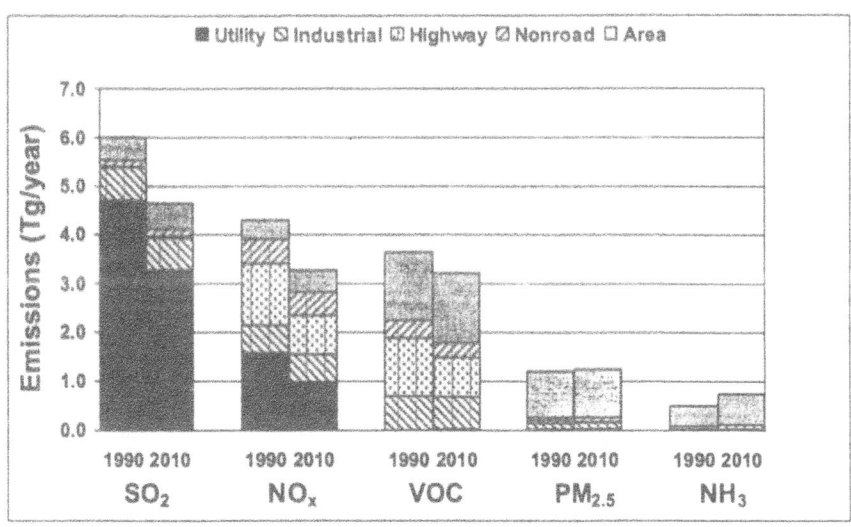

Figure 1. Annual emissions of SO_2, NO_x, VOC, $PM_{2.5}$ and NH_3 by source category in eight SAMI states: 1990 and 2010 projections.

the same as those used for basecase simulations because of the large uncertainty involved in determining future pollutant levels for Canada, Mexico and western U.S.

2.2. Analysis of Sensitivity to Emission Reductions

URM is equipped with the decoupled direct method (DDM) (Yang et al., 1997) for sensitivity analysis. The conventional sensitivity analysis involves running the model by perturbing one type of emission at a time and comparing the results to the original run with unperturbed emissions. With DDM, the sensitivity of pollutant concentrations to various emissions can be calculated simultaneously. Therefore, sensitivity analysis with DDM is much more efficient than with conventional methods. The first order sensitivity coefficient of the concentration of species i, C_i, to the emission of species j, E_j, is defined as:

$$S_{ij} = \frac{\partial C_i}{\partial E_j}.$$

(1)

For convenience the subscripts are dropped in what follows. By differentiating the original set of model equations for concentrations, one can obtain a second set of equations for first order sensitivity coefficients. The form of the two equation sets is very similar therefore solving both sets entails computational savings. The sensitivity coefficients in Eq. (1) can be used in calculating the change in concentration. For this, C can be expanded in a Taylor series for small changes ΔE in emission E around its original value E_0 as:

$$C(E_0 + \Delta E) = C(E_0) + \frac{\partial C}{\partial E}(E_0) \times \Delta E + \frac{\partial^2 C}{\partial E^2}(E_0) \times \Delta E^2 + \cdots \qquad (2)$$

The parentheses in Eq. (2) denote functional relationships. As a first order approximation, if the terms with second and higher order derivatives are ignored and the definition of the first order sensitivity coefficient in Eq. (1) is used then:

$$\Delta C = C(E_0 + \Delta E) - C(E_0) \approx S(E_0) \times \Delta E. \qquad (3)$$

The difference between the actual change and the change calculated from the DDM sensitivity coefficients have been compared for several species-emissions combinations. In general, the difference is small for small $\Delta E/E_0$ (e.g., up to 30 percent) unless there is a highly nonlinear relationship between C and E (Hakami et al., 2003). For those cases, adding the second order term produced a sufficiently accurate approximation. Differentiation of the equations for first-order sensitivities yields a set of equations for second-order sensitivity coefficients. This set can also be solved efficiently with DDM. While DDM is strongly recommended for the design of emission control strategies, the effectiveness of a selected strategy should always be verified by rerunning the model.

Using first order DDM in the simulation with 2010 emissions, the sensitivities of PM levels to SO_2, NO_x and NH_3 emissions were calculated. NO_x sources were further broken down as elevated and ground level sources. Each individual SAMI state and five surrounding regions (Central, Midwest, Northeast, Southeast, and the rest of the domain) were targeted as different geographic source areas for the emission reductions. A list of the states in each region can be found in Table 1.

3. RESULTS AND DISCUSSION

Using observations at Great Smokey Mountains (GRSM) and Shenandoah (SHEN), weights were assigned to each one of the nine episodes so that annual averages can be calculated from episodic modeling results.

Table 1. List of states in each source region used for sensitivity analysis.

Region	States
SAMI	Alabama (AL), Georgia (GA), South Carolina (SC), North Carolina (NC), Tennessee (TN), Kentucky (KY), Virginia (VA), West Virginia (WV)
Midwest (MW)	Ohio, Michigan, Indiana, Illinois, Wisconsin
Northeast (NE)	Maryland, Delaware, District of Columbia, Pennsylvania, New Jersey, New York, Connecticut, Massachusetts, Rhode Island, Vermont, New Hampshire, Maine
Central (CN)	Louisiana, Arkansas, Missouri, Iowa, Minnesota[*], Texas[*], Oklahoma[*], Kansas[*], Nebraska[*]
Southeast (SE)	Florida[a], Mississippi
All Others (AO)	South Dakota[a], North Dakota[a], Canada, Major Bodies of Water

[*] Entire state is not contained in the modeling domain

3.1. Comparison of Simulated PM$_{2.5}$ Levels to Observations

The levels and composition of PM$_{2.5}$ were evaluated by using IMPROVE measurements taken at Class I areas of the SAMI region. The agreement between modeled and observed PM$_{2.5}$ concentrations (Figure 2) is similar to or better than other modeling studies (Seigneur, 2001) although longer episodes covering a wider range of meteorological conditions were simulated in this study. The slight underestimation in PM$_{2.5}$ may be, in part, due to water, which is believed to be the primary unidentified component of IMPROVE measurements, not being included in the model estimates. The sulfate, ammonium, elemental and organic carbon components that form about 75 percent of PM$_{2.5}$ in southeastern U.S. all have normalized mean errors around 40 percent. Normalized errors are larger for less abundant components such as soils and nitrate. Modeled soils consist of more species than those measured by IMPROVE and their emissions are highly uncertain. Nitrate constitutes a very small fraction of PM$_{2.5}$ in the SAMI region and its concentration is sensitive to small errors in other constituents such as sulfate and ammonium.

3.2. PM$_{2.5}$ Levels Simulated for the year 2010

Annual average PM$_{2.5}$ levels simulated for the year 2010 were analyzed at selected receptors within the SAMI region. In general, the levels of PM$_{2.5}$ are expected to decrease at Class I areas but some components may increase. Sulfate is estimated to decrease by 6 percent or 0.3 µg/m^3 at GRSM and by 16 percent or 0.8 µg/m^3 at SHEN. Nitrate levels will increase at GRSM due to additional ammonia becoming available in response to

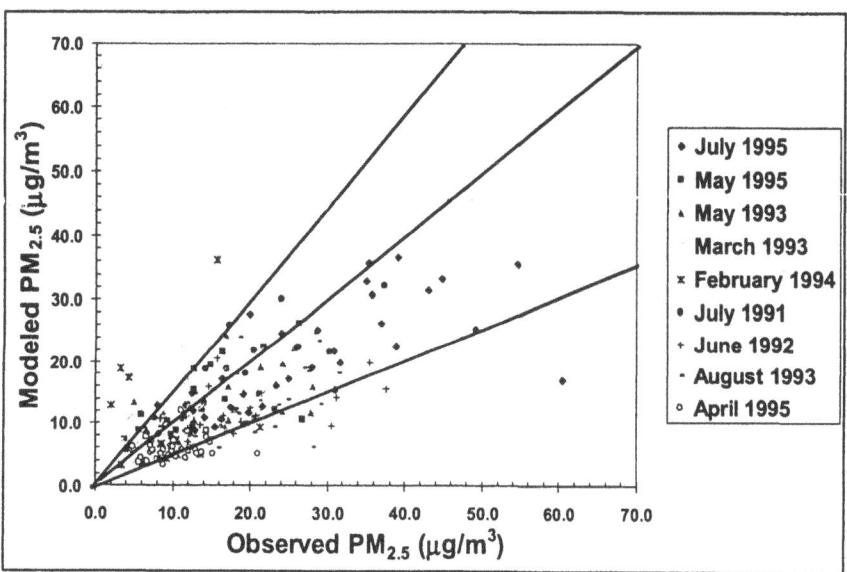

Figure 2. Modeled daily average PM$_{2.5}$ concentrations versus IMPROVE measurements. Also shown are the 1:1 and ±50% bias lines.

reductions in SO₂ and increases in NH₃ emissions. On the other hand, nitrate will decrease at SHEN due to the proximity to reduced NO$_x$ sources and relatively larger distance from the sources of increasing NH₃ emissions. Ammonium concentrations will increase slightly at GRSM but decrease at SHEN by 10 percent or about 0.2 μg/m³. Elemental carbon is expected to decrease while soils are expected to increase. Organic carbon is expected to remain the same at both sites.

3.3. Sensitivity of PM₂.₅ Levels to SO₂ Emission Reductions

Using the sensitivities calculated by DDM in Eq. (3) the change in PM₂.₅ levels were estimated for a 10 percent reduction in SO₂ emission. Figure 3 shows the response of annual average sulfate levels at several Class I areas to 10 percent SO₂ emission reductions in different geographic sub-domains. The 2010 levels of sulfate are also shown on the right axis so that the absolute change can be estimated from the percentage changes on the left axis. For example, a 10 percent reduction in SO₂ emissions uniformly throughout the domain would result in a 6 percent decrease in annual average sulfate levels at GRSM. This corresponds to a 0.3 μg/m³ decrease from the estimated 2010 level of 4.5 μg/m³. About half of this decrease is attributed to Tennessee (TN) and smaller fractions to the other 12 sub-domains assuming they all reduced their SO₂ emissions by 10 percent. The Class I areas are ordered from southwest to northeast along the x-axis of Figure 3. The largest reductions in sulfate levels in Class I areas of Alabama (SIPS) and Georgia (COHU) are due to SO₂ emission reductions in Alabama (AL). The sites in Tennessee (GRSM) and North Carolina (JOCK, SHRO, and LIGO) are most impacted by SO₂ emissions from TN; those in Virginia (JEFF and SHEN) and West Virginia (OTRC and DOSO) are most impacted by SO₂ emissions from the Midwest (MW). The sulfate levels in Shenandoah would decrease by 0.2 μg/m³ if all SO₂ emissions were reduced by 10 percent domain wide and about 30 percent of this decrease can be attributed to the

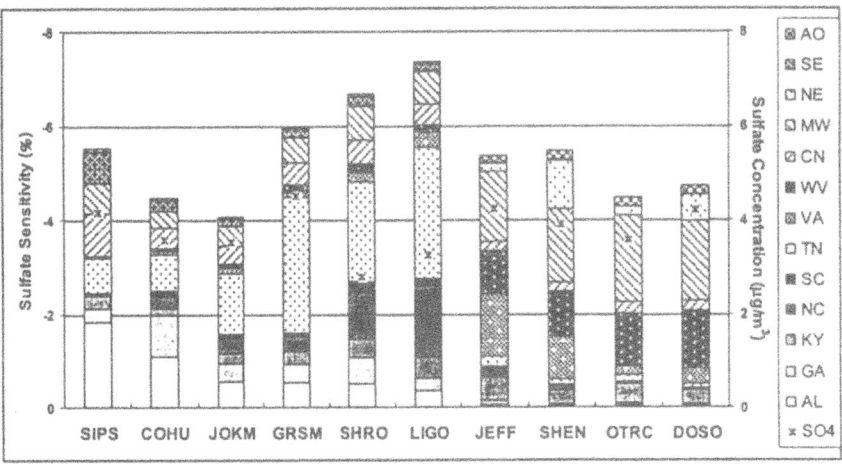

Figure 3. Annual average fine sulfate concentrations (·) and sensitivities for ten Class I areas to 10% reduction in SO₂ emissions from each geographic sub-domain.

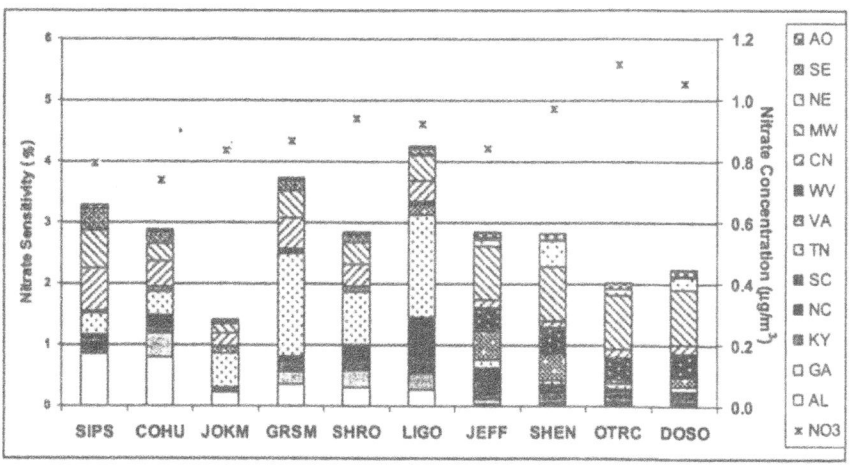

Figure 4. Annual average fine nitrate concentrations (•) and sensitivities for ten Class I areas to 10% reduction in SO₂ emissions from each geographic sub-domain.

Midwest region.

The response of annual average nitrate levels to SO₂ emission reductions in different geographic sub-domains are shown in Figure 4. For most Class I receptors, the fractional contributions of different sub-domains to nitrate increases are similar to the contributions to sulfate decreases. A 10 percent reduction in SO₂ emissions uniformly throughout the domain resulted in a 3.7 percent or 0.03 $\mu g/m^3$ increase in annual average nitrate levels at GRSM; 45 percent of this increase can be attributed to the SO₂ emissions in TN. The nitrate levels in Shenandoah would increase by 2.8 percent or 0.03 $\mu g/m^3$ in response to a 10 percent reduction of SO₂ emissions domain wide; about 30 percent of this increase can be attributed to the Midwest region.

4. CONCLUSIONS

URM, a comprehensive air quality model was applied to the study of regional haze, as well as ozone and acid deposition, in southeastern U.S. The model's performance in reproducing observed PM₂.₅ levels was comparable or better than the performance of other models in similar studies. URM was used to simulate how PM₂.₅ and other pollutant levels would change at Class I areas of the Southern Appalachian Mountains in the year 2010. It was found that the air quality of the region would generally improve due to regulations mandated under the Clean Air Act Amendments and other regulations. More stringent emission controls are likely to provide additional improvements. In response to SO₂ emission controls, annual average sulfate levels are expected to decrease significantly at Class I areas. However, SO₂ controls and increasing ammonia emissions may result in increases of nitrate levels due to more ammonia becoming available to react with nitric acid and form ammonium nitrate.

169

Using the direct sensitivity analysis feature of URM, sensitivities of ozone, PM, and acid deposition levels to the emissions of SO_2, NO_x, and NH_3 emissions from eight SAMI states and five surrounding regions were estimated. It was found that the 2010 sulfate levels would decrease between 4.0 and 7.3 percent if SO_2 emission were reduced by an additional 10 percent uniformly everywhere. On the other hand, nitrate levels would increase between 1.2 and 4.2 percent. In general, Class I receptors showed the greatest response to emission reductions in the nearest sub-domains: 1) Alabama and Georgia sites to emission reductions from Alabama, 2) Tennessee and North Carolina sites to emission reductions from Tennessee, and 3) Virginia and West Virginia sites to emission reductions from the Midwest. These results, especially specific sensitivities to emission reductions from geographic sub-domains are very useful for the design of emission control strategies. However, they should not be used as a substitute for full-scale modeling to determine the effectiveness of a control strategy.

5. ACKNOWLEDGEMENTS

This research was supported by the Southern Appalachian Mountains Initiative (SAMI) under contract number 012.

6. REFERENCES

Boylan, J. W., Odman, M. T., Wilkinson, J. G., Russell, A. G., Doty, K. G., Norris, W. B., and McNider, R. T., Development of a comprehensive multiscale one atmosphere modeling system: application to the southern Appalachian mountains, *Atmos. Environ.* 36, 3721-3734 (2002).

Nenes, A., Pilinis, C., and Pandis, S. N., ISORROPIA: A new thermodynamic equilibrium model for multiphase multicomponent inorganic aerosols, *Aquat. Geochem.* 4, 123-152 (1998).

Hakami A., Odman M. T. and Russell A. G., High-order, direct sensitivity analysis of multidimensional air quality models, *Environ. Sci. Technol.*, in press (2003).

Odman, M. T., Boylan, J. W., Wilkinson, J. G., Russell, A. G., Doty, K., Norris, W., McNider, R., Mueller, S. F., and Imhoff, R. E., SAMI Air Quality Modeling Report, Final Report, 205 pp., Southern Appalachian Mountains Initiative, Asheville, NC, 2002.

Pandis, S. N., Harley, R. A., Cass, G. R., and Seinfeld, J. H., Secondary organic aerosol formation and transport, *Atmos. Environ.* 26A, 2269-2282 (1992).

Pielke, R. A., Cotton, W. R., Walko, R. L., Tremback, C. J., Lyons, W. A., Grasso, L. D., Nicholls, M. E., Moran, M. D., Wesley, D. A., Lee, T. J., and Copeland, J. H., A comprehensive meteorological modeling system—RAMS, *Meteor. Atmos. Phys.* 49, 69-91(1992).

Seigneur, C., Current status of air quality models for particulate matter, *J. Air &Waste Manage. Assoc.* 51, 1508-1521 (2001).

Wilkinson, J. G., Loomis, C. F., McNally, D. E., Emigh, R. A., and Tesche, T.W. Technical Formulation Document: SARMAP/LMOS Emissions Modeling System (EMS-95), AG-90/TS26 & AG-90/TS27, Alpine Geophysics, Pittsburgh, PA, 1994.

Yang, Y.J., Wilkinson, J.W, and Russell, A.G., Fast direct sensitivity analysis of multidimensional photochemical models, *Environ. Sci. Technol.* 31, 2859-2868 (1997).

EVALUATING SEASONAL MODEL SIMULATIONS OF OZONE IN NORTHERN ITALY

Guido Pirovano, Cesare Pertot, Veronica Gabusi, and Marialuisa Volta[1]

1. INTRODUCTION

High concentrations of tropospheric ozone are quite usual in Mediterranean areas during summer seasons and cause adverse effects on human health and natural ecosystems. In order to capture ozone climatological behaviour and select effective reduction strategies, long-term simulations can be performed. Several authors (Hogrefe et al., 2001; Schmidt et al., 2001; Kasibihatla et al., 2000) present seasonal simulations on regional scale performed by photochemical systems. Usually models are run with a spatial resolution ranging from 36 km to 0.5° over domains extending up to 2-3000 km in each direction. In this paper an urban scale long-term simulation is evaluated. The simulation domain is located in Northern Italy, including Milan metropolitan area. It is characterised by complex terrain, high urban and industrial emissions and a close road network. Due to the presence of the Alps, winds are often weak and circulation shows very heterogeneous patterns. Besides, during anticyclonic stable conditions breeze can develop causing daily circulation of polluted air masses between urban and rural areas. The critical anthropogenic emissions, the frequent stagnating meteorological conditions and the solar radiation regularly cause high ozone level episodes. Due to the described meteorological features, pollutants can give raise to a complex spatial distribution, that is difficult to describe by means of mathematical models even using a detailed spatial resolution. Moreover, in several cases, spatial variability can induce a representativeness reduction of measurement stations, making critical the comparison with model results.

The simulations have been performed over 1996 summer season by two chemical and transport models, CALGRID (Yamartino et al., 1992) and STEM-FCM (Silibello et al., 2001), fed by emission and meteorological pre-processors, respectively POEM (Catenacci et al., 1999) and CALMET (Scire et al., 1999). The initial and boundary

[1]Guido Pirovano and Cesare Pertot, CESI, Via Rubattino, 54, 20134 Milano, Italy. Veronica Gabusi and Marialuisa Volta, D.E.A. Brescia University, Via Branze, 38, 25123 Brescia, Italy

Air Pollution Modeling and Its Application XVI, Edited by
Borrego and Incecik, Kluwer Academic/Plenum Publishers, New York, 2004

conditions have been obtained by means of a nesting procedure from the EMEP Lagrangian Photoxidant Model (EMEP, 1999).

Modelling system capability to actually reconstruct both temporal and spatial features of ozone concentrations has been assessed evaluating some statistical indexes by means of performance indicators (US EPA, 1991) and graphical tools. Statistical indexes have been calculated in correspondence of a selection of monitoring stations located on areas with different orographic and emissive features. The reference monitoring stations have been classified into two groups taking into account (1) the ozone concentration pattern and trend and (2) the precursor emissions estimated in the neighbors of the measurement sites.

Moreover, the long-term impact of ozone concentrations has been assessed, according to the WHO and European guidelines, estimating the AOT40 index for crops and forest protection and AOT60 index for health protection.

2. THE MODELLING SYSTEMS

The long-term simulations have been performed by two modelling systems, providing the description of the meteorology, emissions, transport and chemical transformation of photochemical pollutants in the atmosphere. Both of them include the same emission and meteorological pre-processors, respectively POEM and CALMET. The systems differ in the photochemical model: one implements CALGRID and the other one STEM-FCM.

The diagnostic meteorological pre-processor CALMET provides (1) 3D wind and temperature fields and (2) turbulence parameters, merging background fields provided by ECMWF model with measurements. The emission processor POEM has been designed to produce present and alternative emission fields; it can be applied to the CORINAIR data and it considers diffuse and main point sources coming from different activity sectors. Model outputs are the results of three step algorithm: (1) the spatial disaggregation, (2) the time modulation, (3) the NMVOC split and lumping procedures to the species required by the transport model.

CALGRID and STEM-FCM are Eulerian three-dimensional photochemical models. They implement an accurate advection-diffusion scheme in terrain-following coordinates, with vertical variable spacing. Both models use SAPRC-90 mechanism (Carter, 1990), including 54 chemical species and 129 reactions. As for the kinetic equations integration, CALGRID implement the QSSA (Quasi Steady State Approximations) solver (Mathur et al., 1998), while STEM-FCM implement the IEH solver (Sun et al., 1994).

3. THE SEASONAL SIMULATION

The selected simulation domain, 240x232 km^2 wide, includes the whole Lombardia Region with several cities as well as rural areas. It is a complex terrain region located in the Po Valley and it is one of the most industrialised and populated area of Northern Italy. Industries and a close road network are the most relevant pollution sources in the basin. The area has been subdivided according to a grid system having 60 per 58 horizontal cells, with 4 km step size and 11 vertical increasing layers (from 20 to 3900 m a.g.l.). The simulations have been performed over six months, from April 1 to September 30 1996.

Emission fields have been estimated by POEM model. Table 1 shows the diffuse annual emission in the simulation domain. The main NO_X sources are the road transport and the industrial processes, which contribute respectively to the 74% and the 13% of total emissions. Again, the prominent role of road transport is evident, contributing to almost 40% of total anthropogenic VOC emissions. The use of organic solvents for various applications is a source of primary interest for anthropogenic VOC emissions, followed by waste treatment and by industrial production processes. Analysing VOC emissions, it should be noted that biogenic emissions play an important role and should not be neglected.

Table 1. Total diffuse emissions over the domain $[Mg \cdot y^{-1}]$

Activity	NO_X	CO	SO_2	VOC
Industry	48,076	185,824	55,920	31,459
Solvents	-	-	-	165,005
Transport	273,850	1,176,955	38,544	171,825
Waste	10,469	495,712	5,551	25,120
Agriculture	85	4,540	-	209
Biogenic	36,370	-	-	28,844
Total	368,850	1,863,031	100,015	422,462

Meteorological fields have been provided starting from topography information and ground-based and upper-level meteorological data. Land-use information was used to define geophysical parameters needed as input to turbulence and deposition modules. As for meteorological data, hourly ground level data measured in 31 stations were available. Vertical wind structure and temperature fields were provided both by two radio-sounding stations and by ECMWF (European Centre for Medium range Weather Forecast, Reading, UK) model 6-hourly analyses. Each data set describes different aspects of the atmospheric circulation over the studied area: ECMWF fields outline synoptic features, while ground measurements account for local effects. Final wind fields have been obtained using ECMWF profiles as initial guess field and then merging measurements, by means of an objective analysis procedure.

Boundary conditions have been defined on the basis of EMEP model, a one-layer Lagrangian trajectory model covering the whole of Europe, with a 150x150 km² resolution, and providing 6-hourly boundary layer mean concentrations. Fields estimated by the EMEP model have been interpolated in time and extrapolated above PBL height, according to estimated mixing-height. Concentrations have been assumed to be well mixed in the PBL and exponentially decreasing with height for all species, except for ozone, for which a free atmosphere (3000 m a.s.l.) background level of 60 ppb has been assumed.

4. SYSTEM PERFORMANCE EVALUATION

Model evaluation is a necessary part in the iterative cycle of model development, testing and confidence building procedure. In fact, model evaluation supports the credibility of models and is an indispensable part of both the model development process and the model applications. In this study the model skill in simulating the observed concentrations along a seasonal period has been primarily investigated. In literature the model reliability is assessed (1) by graphical comparisons between measured and simulated concentrations at some specific locations and (2) by proper performance indicators.

Model results have been evaluated processing 1-hourly and 8-hourly daily maximum ozone concentrations, as they are considered among the most relevant parameters describing respectively the model capability to estimate peak values and mean daily trend (Sistla et al., 2001; Schmidt et al., 2001). Model performances have been also assessed with regard to NO_2 (not shown) comparing afternoon mean concentrations that represent a statistic with a greater spatial representativeness.

To compare model results versus observations a reduced set of monitoring stations has been selected for the domain of interest by means of cluster analysis techniques. Stations have been grouped in two sets: the former including 4 stations belonging to the Milan area (HEDA - High Emission Density Area) and the latter comprising 3 stations placed in areas with lower emission density (LEDA).

4.1. The Simulation Results: Graphical Comparison

Figure 1 shows time series of 1h daily maximum concentration of ground level ozone. Graphs are relative to the average of HEDA and LEDA groups and also to single stations. Models exhibit a similar behaviour and are able to follow quite well the seasonal evolution of measured concentrations. At HEDA stations some underestimations of highest peaks are observed. Inside metropolitan area observed peaks are generally due to an intense local photochemical production that models are not always able to reconstruct. Particularly, underestimations could derive from a too low VOC/NO_X ratio that prevents higher concentrations from developing. Figure 2 shows the comparison between 25th, 50th, 75th and 95th percentile of measured and computed time series of 1h daily max. Results are shown for each station and also as average and merge of the stations belonging to each group. HEDA stations are very homogenous confirming the underestimation of measured values that becomes more relevant in correspondence of the highest peaks. Differently HEDA are more heterogeneous, but they follow well measured distribution as a group. This could mean that ozone behaviour is correctly captured at the domain mean scale, while local transport effects need to be better described. Finally figure 3 shows the comparison between computed and measured AOT indexes estimated over 6 months. Also in this case, models exhibit similar results, with the exception of Ispra where STEM computes higher values than CALGRID. AOT60 estimates are more biased than AOT40, because the former is based only on high concentrations (more than 60 ppb) and, consequently, is more sensible to discrepancies between measured and computed values.

4.2. The Simulation Results: Statistical Assessment

The statistical model evaluation has been performed defining a minimum set of statistical, on the basis of US EPA recommendations. In particular, (1) the Normalised Bias (MNBE), (2) the Normalised Absolute Gross Error (MNGE) and (3) the Correlation

Coefficient have been estimated for the daily maximum 1-h ozone concentrations and for the daily maximum 8-h ozone concentrations (Table 2).

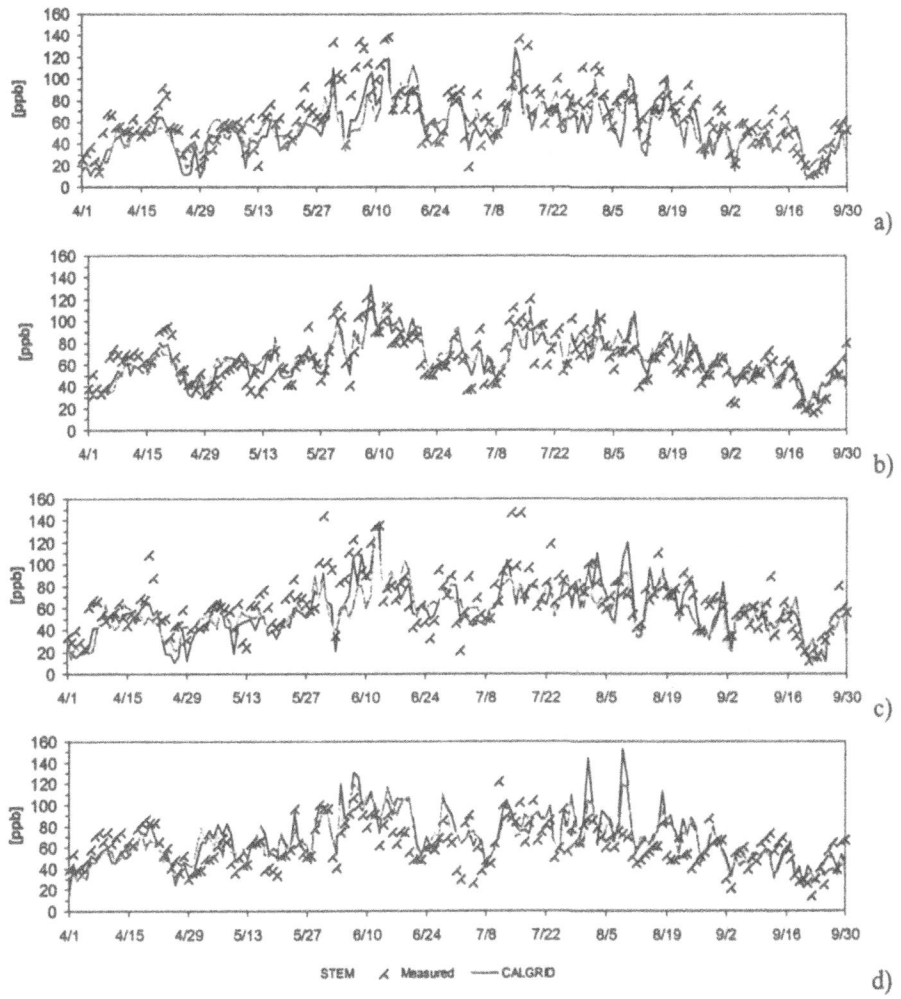

Figure 1. Computed and measured 1h daily maximum ozone concentrations: a) average of HEDA stations; b) average of LEDA stations; c) Vimercate; d) Ispra.

The EPA recommends a range of ±5-15% for the bias error and a range of 30-35% for the gross error. Ranges have been initially defined with regard to 1h daily maximum concentration for short applications, so, in this contest, they have been used as a qualitative reference target. The computed statistic screening reveals satisfying performances for both models. The index values agree the EPA suggestions and are comparable to recent regional simulation results performed by Hogrefe et al. (2001), Sistla et al. (2001), and Schmidt et al. (2001).

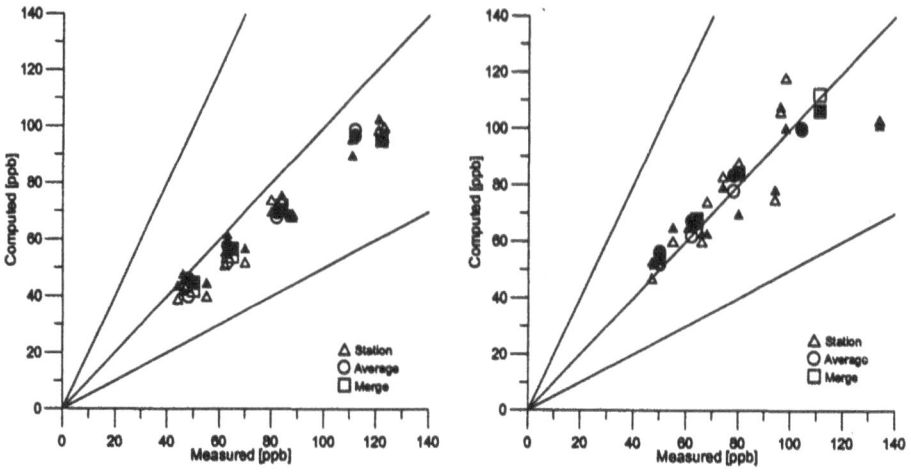

Figure 2. Computed and measured 1h daily maximum ozone percentiles for HEDA group (left) and LEDA group (right). STEM: empty symbols; CALGRID: full symbols.

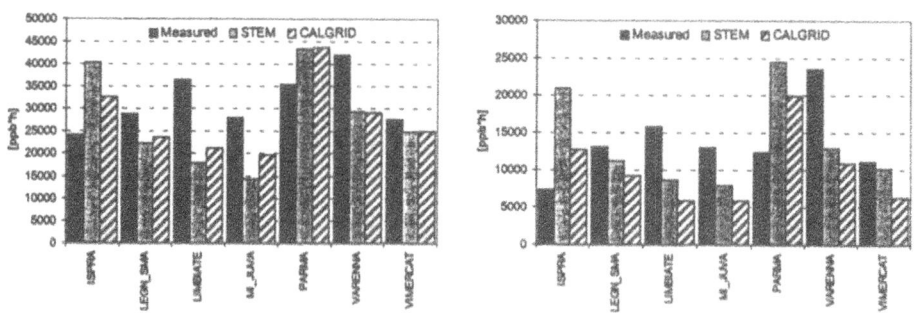

Figure 3. Computed and measured AOT40 forest (left) and AOT60 (right).

Performances are similar for both variables pointing out that models have been able to describe both peaks and daily pattern. Moreover, correlation coefficients, ranging from 0.6 to 0.8, highlight that simulations correctly reproduce the diurnal evolution of the mesoscale circulation features that influence ozone patterns at the domain scale. Biases are generally better at aggregated level than for each single station, particularly for HEDA group, confirming that local scale dispersion features are not yet well resolved.

5. OZONE EXPOSURE ASSESSMENT

Figure 4 reports AOT40 simulated fields for forests calculated from April to September by STEM and CALGRID. As can be noted, AOT patterns are quit similar, with highest values placed in the same areas. Besides, models have been able to account for complex terrain effect on ozone patterns.

176

Table 2. EPA Performance indicators. Indicators are computed for each station and for each group expressed both as average and as merge of the respective stations

	1-h daily max						8-h daily max					
	MNBE [%]		MNGE [%]		Correlation		MNBE [%]		MNGE [%]		Correlation	
	CG[a]	S-F	CG	S-F	CG	S-F	CG	S-F	CG	S-F	CG	S-F
Legnano	4.4	-6.3	27.6	25.3	0.69	0.77	14.0	-3.1	32.6	25.6	0.74	.80
Limbiate	-15.4	-21.2	27.5	29.0	0.60	0.73	-11.2	-22.4	26.2	31.1	0.66	.75
Juvara	1.4	-11.1	31.2	31.6	0.64	0.70	14.3	-5.1	36.6	33.5	0.70	.74
Vimercate	-5.6	-9.1	28.3	25.6	0.62	0.72	1.8	-3.3	28.5	25.6	0.68	.75
HEDA avg	-3.1	-14.2	24.0	25.2	0.70	0.80	1.6	-12.0	23.4	24.9	0.75	.82
HEDA mrg	-3.8	-12.0	28.8	27.9	0.69	0.72	5.8	-8.5	31.5	29.0	0.68	.75
Ispra	13.3	15.2	25.6	27.7	0.60	0.65	19.7	23.4	29.6	32.8	0.58	.66
Parma	21.6	18.5	30.4	27.1	0.58	0.67	25.5	20.0	32.2	28.1	0.64	.71
Varenna	-1.3	-6.6	25.2	26.2	0.67	0.70	0.6	-5.1	23.1	25.4	0.73	.74
LEDA avg	8.7	5.0	20.9	18.8	0.74	0.81	11.7	8.9	22.2	20.1	0.74	.81
LEDA mrg	12.1	11.0	28.7	27.2	0.60	0.60	16.2	15.2	30.8	29.5	0.60	.64

[a] CG = CALGRID; S-F = STEM-FCM

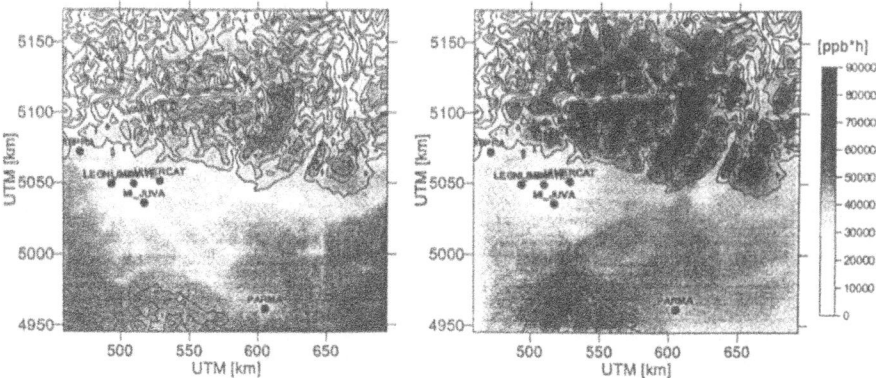

Figure 4. AOT40 forest estimated by STEM-FCM (left) and CALGRID (right).

Anyway, CALGRID and STEM exhibit different estimations in the southern part of the Alps. This is probably due to some differences in the vertical dispersion treatment. AOT40 is very high over the whole domain, greatly exceeding UE recommended target of 10000 [ppb*h] over the 6 months. Moreover, highest values are observed outside urban areas in correspondence of first relieves that favour the accumulation of secondary pollutants produced in the plain urbanised areas.

6. CONCLUSIONS

The results of a long-term simulation of photochemical pollution over Lombardia Region (Northern Italy), carried out with two different modelling systems, have been evaluated using graphical comparison and model performance indicators. Both CALGRID and STEM-FCM systems are able to reproduce the overall temporal and spatial behaviours of measured ozone concentrations, with similar levels of performance and meeting the EPA requirements in most cases. Anyway, the photochemical process description needs to be improved in urban area, as well as the transport algorithms in rural ones. Obtained results confirm that Lombardia region is exposed to high ozone concentrations for all summer season both in urban and rural areas. Particularly, rural areas bordering on main cities suffer for the highest concentrations due to the presence of complex terrain that induces ozone accumulation.

7. ACKNOWLEDGMENTS

CESI contribution of this paper has been supported by the MICA (Italian Ministry of Industry, Trade and Handicraft) in the frame of Energy Research Program for the Italian Electric System (MICA Decree of January 26, 2000), Project on "ESTERNA". This work has been partially supported by MIUR (Italian Ministry of University and Research) and AGIP-Petroli.

8. REFERENCES

Catenacci, G., Riva, M., Volta, M., and Finzi, G., 1999, A model for emission scenario processing in Northern Italy, in: *Proc. of EUROTRAC Symp'98*, Borrell, P. M. and Borrell, P. (Editors), WITPRESS, pp. 720-724.

Carter, W.P.L., 1990, A detailed mechanism for the gas-phase atmospheric reactions of organic compounds, *Atmospheric Environment*, 24, 481-518.

EMEP, 1999, Transboundary Photo-oxidant air pollution in Europe. EMEP/MSC-W *Status report 2/99*. DNMI, Norway.

Hogrefe, C., Rao, S. T., Kasibhatla, P., Hao, W., Sistla, G., Mathur, R., and McHenry, J., 2001, Evaluating the performances of regional-scale photochemical modeling systems: Part II – Ozone predictions, *Atmospheric Environment*, 35, 4175-2188.

Kasibhatla P., and Chameides, W. L., 2000, Seasonal modelling of regional ozone pollution in the eastern U.S., *Geophysical Research Letters*, 27(9), 1415-1418.

Mathur, R., Young, J. O., Schere, K. L., and Gipson, G. L., 1998, A Comparison of Numerical Techniques for Solution of Atmospheric Kinetic Equations. *Atmos. Environ.*, 32, 1535-1553.

Scire, J. S., Robe, F. R., Fernau, M. E., and Yamartino, R. J., 1999, A User's guide for the CALMET meteorological model – Version 5.0, *Internal report*, Earth Tech inc.

Schmidt, H., Derognat, C., Vautard, R. and Beekmann, M., 2001, A comparison of simulated and observed ozone mixing ratios for the summer of 1998 in Western Europe, *Atmospheric Environment*, 35, 6277-6297.

Silibello, C., Calori, G., Pirovano, G. and Carmichael, G.R., 2001, Development of STEM-FCM modelling system: Chemical mechanisms sensitivity evaluated on a photochemical episode, Proc. of *APMS'01- Int. Conference*, Paris, April 2001.

Sistla, G., Civerolo, K., Hao, W., and Rao, S. T., 2001, A comparison of measured and simulated ozone concentrations in the rural areas of eastern U.S. during summer 1995, *J. Air & Waste Manage. Assoc.*, 51, 374-386.

Sun, P., Chock, D. P. and Winkler, S. L., 1994, An implicit-explicit hybrid solver for a system of stiff kinetic equations, *Journal Of Comp. Physics*, 115, 515-523.

US Environmental Protection Agency, 1991, Guideline for regulatory application of the Urban Airshed Model, *EPA-450/4-91-013*, Research Triangle Park, NC 27711.

Yamartino, R.J., Scire, J.S., Carmichael, G.R. and Chang, Y.S., 1992, The CALGRID mesoscale photochemical grid model - I. Model formulation, *Atmospheric Environment*, 26A, 1493-1512.

AIR QUALITY OPERATIONAL PROCESS ANALYSIS BY USING MM5-CMAQ OVER THE IBERIAN PENINSULA: A WEB BASED APPLICATION

R. San José, J. L. Pérez[*] and R.M. González[**]

1. INTRODUCTION

A major function of air pollution models is to predict the spatial and temporal distributions of ambient air pollutants and other species. The three dimensional complex Eulerian models, output concentration fields of these species are determined by solving systems of partial differential equations. These equations define the time-rate of change in species concentrations due to a series of physical and chemical processes (e.g. emissions, chemical reaction, horizontal advection, etc.). Grid models can be configured to provide quantitative information on the effects of the chemical reactions and another atmospheric proceses that are being simulated. This process análisis has been implemented in the Models-3 Community Multiscale Air Quality Modelling System (CMAQ). In this contribution we have implemented the MM5-CMAQ modelling system in operational mode over the Iberian Peninsual with 27 km spatial resolution (nesting level 1) to allow an Internet user to access by using a Graphical User Interface and over a full spatial and temporal domain different chemical and physical atmospheric process in detail. The XADV (advection in E-W direction), YADV (advrection in N-S direction), ZADV (vertical advection, ADJC (mass adjustment for advection), DIC (horizontal difusión), VDIF (vertical difusión), EMIS (emissions), DDEP (dry deposition), CHEM (chemistry), AERO (aerosols) and CLDS (cloud proceses and aquous chemistry) have been implemented in the web interface (HTML/CGI). Each of above process can be extracted for every map location defined by the user. The system performs a bilinear

[*] Environmental Software and modelling Group, Computer Science School, Technical University of Madrid (UPM) (Spain), Campus de Montegancedo, Boadilla del Monte 28660 Madrid (Spain) http://artico.lma.fi.upm.es.
[**] Department of Geophysics and Meteorology, Faculty of Physics, Universidad Complutense de Madrid (Spain).

interpolation from the grid values to obtain the user value at any specified spatial location in the map reference by using GRADS. The results are presented over the Internet under operational daily basis.

1. INTRODUCTION

research in the last decades on air pollution modelling systems has grown extraordinary in parallel with the important advances on computer power capabilities and INTERNET development. During 90's the mesoscale meteorological modules and the air quality transport models developed over limited domains and with simulations of the basic chemical reactions in the atmosphere. During the second generations of atmospheric air pollution models (Venkatram et al. (1989); Carmichael et al. (1991); McRae and Seinfeld (1983); San José et al. (1994)) a considerable attention was given to the 3d Eulerian simulations and the operational basis. A clear limitation os these models was found in several areas: the limited information provided by the boundary conditions so that long range transport was almost not present. Simplified aqueous and cloud processes were present and poor attention was given to biogenic emissions and the impact on the chemical reactions and subsequent products. This contribution focuses on results from simulations over short – medium range by covering al, the spatial domains affected by the temporal simulation. The so-called Third Generation of Air Pollution Models (Peters L.K. et al. (1995)) is present by using the MM5-CMAQ modelling system and comparing the results at high resolution level (almost street level) with the OPANA model (San José et al. 1999) – representative of the Second Generation of Air Pollution Models but with full on-line linkage between meteorology and chemistry -. The results show an important variability which represent the air pollution concentrations from the air pollution monitoring stations in a much more realistic way. The preliminary results – due to the limited period of time of study – illustrate the importance of nesting the adequate number of levels to catch all the necessary information cming from the surrounding areas which will affect the local environment in an important manner.

2. THE MM5-CMAQ MODELLING SYSTEM

Investigation on atmospheric process to evaluate the impact on air quality concentrations of different atmospheric dynamical and chemical component process is an important area of research in the last years. The use of state-of-the-art air quality mesoscale Eulerian three dimensional models is almost a need to appreciate the impact on the determination of the air quality concentrations of the different physical and chemical process in the atmosphere. The MM5-CMAQ atmospheric modelling system is an state-of-the-art air quality modelling system developed by EPA (U.S.) and PSU/NCAR (Peters L.K. et al. (1995)). MM5 is a very robust tool with multiple applications world wide to provide reliable meteorological information at mesoscale level (although a global version of MM5 is already existing). The CMAQ module is a dispersion and chemical module which uses MM5 output data in a consistent way to evaluate the air quality concentrations at the case study model domain. In this contribution we have used MM5-CMAQ as a tool to analyze the impact on air concentrations of the atmospheric process such as chemistry, horizontal and vertical diffusion and advection, deposition and emissions. These models are a successful updating and adaptation of the so-called second generation of AQM's.

The MM5 ((Chen F. and J. Dudhia (2001); Warner T.T. and H.M. Hsu (2000); Stensrud D. J. et al. (2000); Ge et al. (1997))is built over a mother domain with 36 x 36 grid cells (81 km spatial resolution) and 23 vertical levels. This makes a domain of 2916 x 2916 km. The nesting MM5 level 1 model domain is built over a 69 x 66 grid cells (27 km spatial resolution) and 23 vertical levels, which makes a model domain of 1863 times 1782 km centered over the Iberian Peninsula. CMAQ model domains are 30 x 30 grid cells for mother domain and 63 x 60 over the nesting level 1 model domain. CMAQ mother domain lower left corner is located at (-1215000 m, -1215000 m) at the reference locations (-3.5W, 40N) and the first and second standard parallels (30N, 60N). The CMAQ nesting level 1 lower left corner is located at (-891000, -810000) with the same reference locations. The 9 km MM5 spatial resolution model domain has 54 x 54 grid cells, the 3 km MM5 spatial resolution model domain has 33 x 39 grid cells and finally the 1 km MM5 spatial resolution model domain has 30 x 30 grid cells. The corresponding CMAQ model domains are: 48 x 48 km, reference (-216000, -216000) in Lambert Conformal projection with 9 km spatial resolution; 27 x 33 grid cells, reference (-54000, -9000) with 3 km spatial resolution and finally, 24 x 24 grid cells, reference (-27000, 33000) with 1 km spatial resolution.

Figures 1 and 2 show an illustration of the MM5 nesting architecture over the Madrid domain.

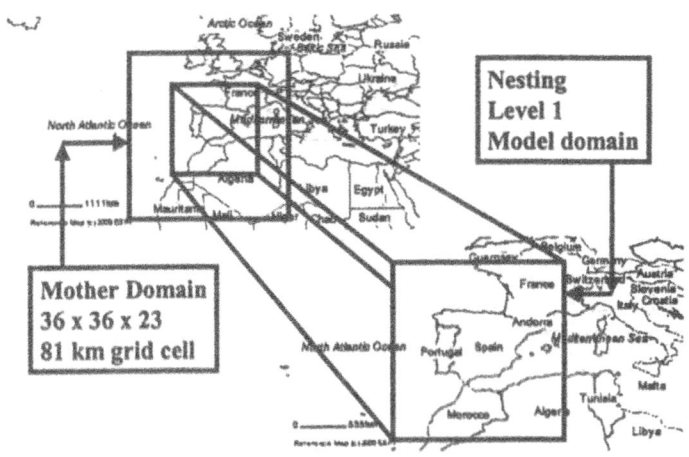

Figure 1.- Mother and nesting level domains for the MM5 architecture in this experiment.

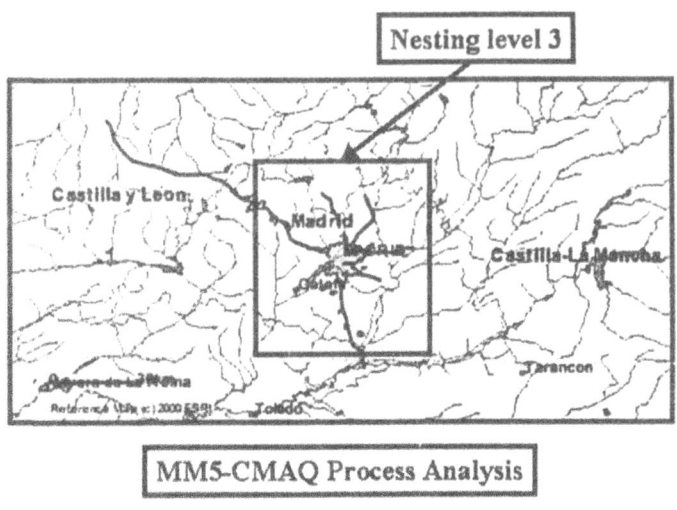

Figure 2.- Nesting levels 2 and 3 for the MM5 architecture in this experiment.

We have selected for this simulation the piece parabolic method (PPM) Colella and Woodward (1984). The updated version of the CBM-IV mechanism (36 species, 94 reactions including 11 photolytic reactions) with the isoprene chemistry mechanism and the Arrhenius type rate constant expressions. We have also used aerosol formation processes and aqueous chemistry. Finally we have used the Modified Euler Backward Iterative (MEBI) Solver which is mechanism dependent but it is at least as precise as the QSSA but the computing time required is much lower.

The emission inventory is obtained by using the Emission Database for Global Atmospheric Research – EDGAR from RIVM and TNO. This emission inventory is for 1° x 1° spatial resolution and for annual dataset. The emission inventory is used as it is for mother domain and for nesting domains we use information from Digital Chart of the World and USGS EROS Data Center for refining the spatial resolution up to 4 km. The time profiles for hourly emission data sets for all domains are obtained from examples in Europe from IER emission dataset (University of Stuttgart, Germany) at different European regions. These calculations are made with our emission module called EMIMO 2.0. The biogenic emissions are obtained by using our module called BIOEMI 2.0 which calculates the isoprene, monoterpene and natural NOx emissions for our expriment. The experiment is carried out over the February, 4-8, 2002 period of time.

3. RESULTS

The Eulerian model use the so-called operator splitting technique. This technique provides an easy method to obtain quantitative information about the contribution of individual processes to total concentrations. In this contribution we have activated the following process to be analysed: XYADV, horizontal direction; ZADV, vertical advection; ADJC, mass adjustment for advection; HDIF, horizontal diffusion; VDIF, vertical diffusion; EMIS, emissions; DDEP, dry deposition; CHEM, chemistry; CLDS, cloud process and aqueous chemistry. In Figure 3 we show the surface 120 hours averaged ozone concentrations for the model domain (3 km spatial resolution with 27 x 33 grid cells for CMAQ modelling domain). Figure 4 shows the chemical process analysis for 120 hour averaged ozone concentrations over the 3 km CMAQ-Madrid model domain. Figure 5 and 6 show different process analysis for a specific grid cell (17,12) with observed and simulated ozone concentrations. Finally Figure 7 shows the NO2 time series for different process (including emissions). The GRADS and FERRET software tools are used to provide surface interpolation of modelling data as shown in figures 3 and 4.

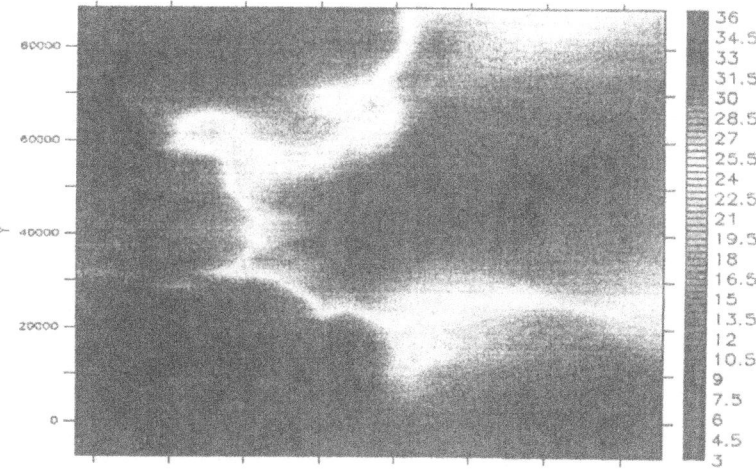

Figure 3.- 120 hour average ozone concentrations for the February, 4-8, 2002 period over the 3k spatial resolution domain centred over the Madrid area.

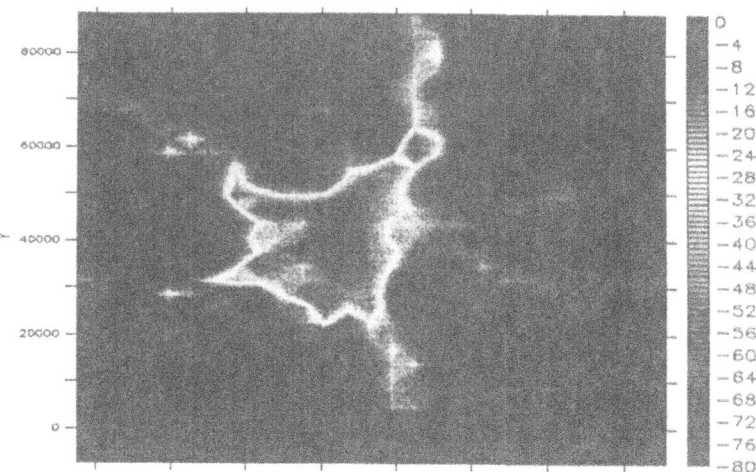

Figure 4.- Chemical process analysis averaged over the 3 km spatial resolution CMAQ modelling domain (Madrid area).

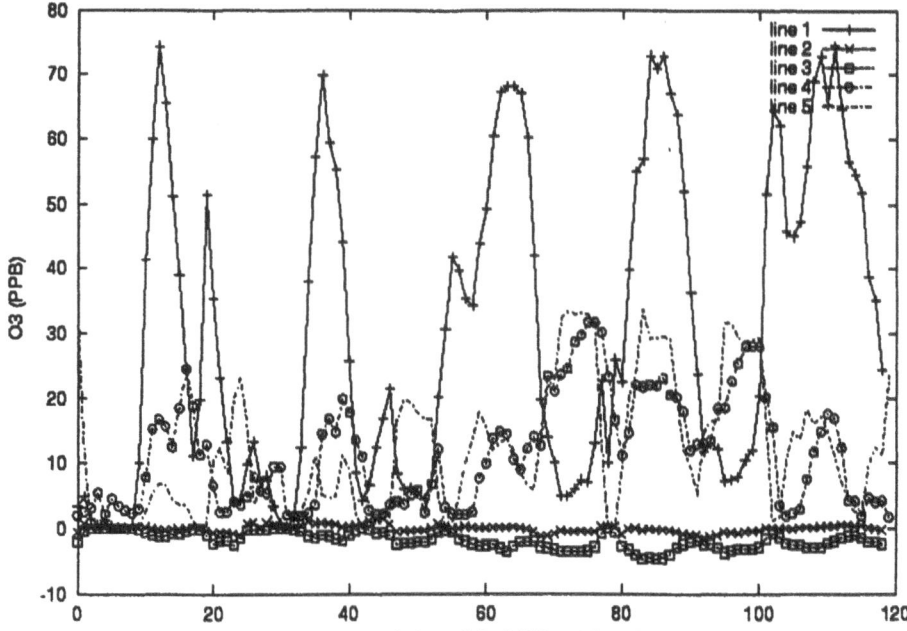

Figure 5.- Different process (vdif, hdif, ddep) analysis and observed and simulated ozone concentrations for the grid cell (17,12) for 120 hour time series.

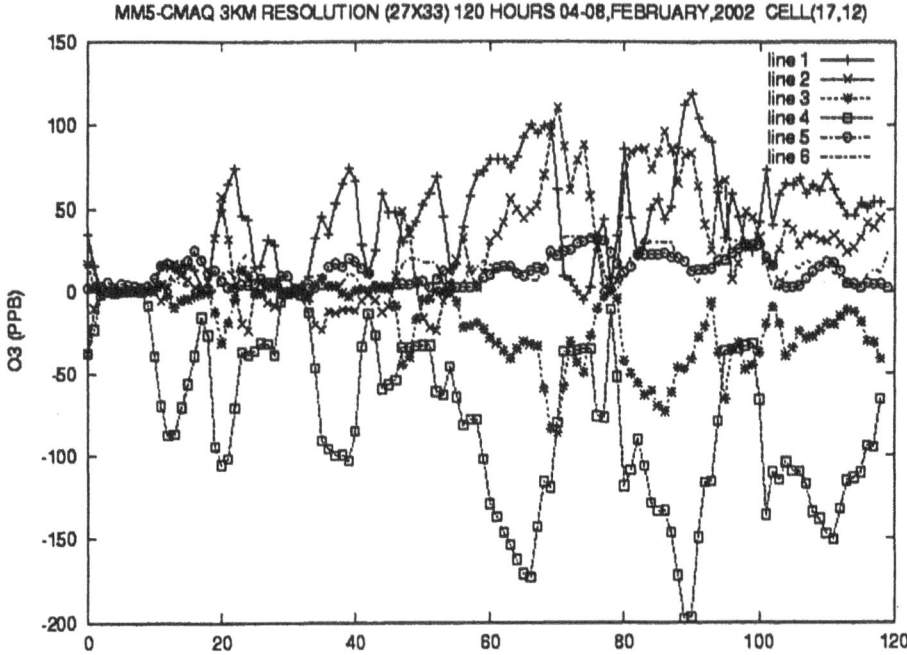

Figure 6.- Different process (xyadv, zadv, adjc, chem) analysis and observed and simulated ozone concentrations for the grid cell (17,12) for 120 hour time series.

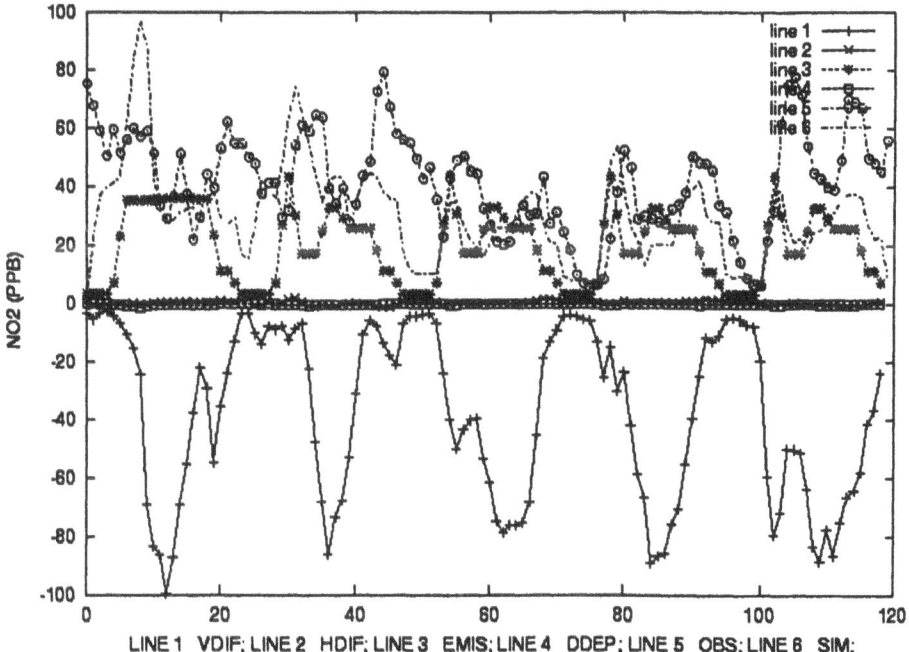

LINE 1 VDIF; LINE 2 HDIF; LINE 3 EMIS; LINE 4 DDEP; LINE 5 OBS; LINE 6 SIM;

Figure 7.- Different process (vdif.hdif,emis,ddep) analysis and observed and simulated NO2 concentrations for the grid cell (17,12) for 120 hour time series.

The results shows that different process are described in a physically and chemically consistent way. Also, the figures show that the comparison between observed and simulated values is quite significant, particularly for ozone concentrations. The vertical diffusion for the formation of NO2 plays an essential role at least in this station and for this simulation (fig. 7). From Figures 5 and 6 we see that chemical process plays the more important role on the ozone production and also – but with lower importance – the vertical and horizontal transportation. Both process – chemical production and vertical and horizontal transportation - play opposite roles (positive values assume that production is higher than consumption and negative values the opposite).

Further work is focused on developing and implementing a full integrated analysis software (post-processing of process analysis) which will give us detailed and integral information about where and when the maximum changes (production rates) for a specific process are produced. This tool will be essential for analysis the impact of industrial emissions or any other emission source in the surrounding areas with a full detail in real-time of the physical and chemical process expected (forecasting mode).

4. SUMMARY

We have implemented the CMAQ model (Community Multiscale Air Quality Modelling System) linked with the MM5 mesoscale meteorological model in a cluster of PC-Linux platforms managed by scripts. The GRADS and FERRET software tools are used in on-line mode to provide over the Internet the information required by the Internet user.

The use of the process analysis module in CMAQ allows the user to have a full detailed integral information of all the physical and chemical process involved in the air quality simulations. This includes detailed information at every grid cell and every time step. Simulations performed in this preliminary experiment show that the performance of this module is fully adequate to the expected information.

5. REFERENCES

Carmichael G.R., Peters L.K. and Saylor R.D. (1991). The STEM-II regional scale acid deposition and photochemical oxidant model –1. An overview of model development and applications. *Atmospheric Environment* **25A**, 2077-2090.

Chen, F., and J. Dudhia, 2001: Coupling an advanced land-surface/hydrology model with the Penn State/NCAR MM5 modeling system. Part I: Model implementation and sensitivity. Mon. Wea. Rev., 129, 569-585.

Colella P. and P.R. Woodward (1984): The piecewise parabolic method (PPM) for gas-dynamical simulations. *J. Comp. Phys* **54**, 174-201.

Flassak T. and Moussiopoulos N. (1989) Simulation of the sea breeze in Athens with an efficient non-hydrostatic mesoscale model. In: Man and his ecosystem., **Vol. 3**, Eds. L.J. Brasser and W.C. Mulder. Elsevier. Amsterdam. 189-195.

Ge, Xiaozhen, Li, Feng, and Ge, Ming, 1997: Numerical analysis and case experiment for forecasting capability by using high accuracy moisture advectional algorithm. Meteorology and Atmospheric Physics, Vienna, Austria, 63(3-4), 131-148.

Jacobson M.Z. and Turco R.P. (1994) SMVGEAR: A sparse-matrix vectorized gear code for atmospheric models. *Atmospheric Environment*, **28**, 273-284.

McRae G. J. And Seinfeld J.H. (1983) Development of a second generation mathematical model for urban air pollution. II. Evaluation of model performance. *Atmospheric Environment*, **17**, 501-522.

Peters L.K., Berkowitz C.M., Carmichael G.R., Easter R.C., Fairweather G., Ghan S.J., Hales J.M., Leung L.R., Pennell W.R., Potra F.A., Saylor R.D. and Tsang T.T. (1995) The current state and future direction of Eulerian models in simulating the tropospheric chemistry and transport of chemical species: A review. *Atmospheric Environment*, **29, 2**, 189-222.

San José R., Rodríguez L. And Moreno J. (1994). An application of the "Big Leaf" deposition approach to the mesoscale meteorological transport and chemical modelling in a three dimensional context. Proceedings of the EUROTRAC Symposium. Eds. Borrel P.M., Borrell P., Cvitas T. and Seiler W. 620.

San José R., Rodríguez M.A., Pelechano A. And González R.M. (1999) Sensitivity studies of dry deposition fluxes. In Measuring and Modelling Investigation of Environmental Processes. WITpress *Computational Mechanics Publications*. ISBN: 1-853125660. Ed. R. San José. 205-246.

Stensrud, D. J., J. -W. Bao, and T. T. Warner, 2000: Using initial condition and model physics perturbations in short-range ensemble simulations of mesoscale convective systems. Mon. Wea. Rev., 128, 2077-2107.

Venkatram A. and Karamachandani P. K. (1989). The ADOM II scavenging module: incorporation of improved cumulus cloud module. ENSR Doc. No. 0780-004-205.

Warner, T. T. and H. M. Hsu, 2000: Nested-model simulation of moist convection: The impact of coarse-grid parameterized convection on fine-grid resolved convection through lateral-boundary-condition effects. Mon. Wea. Rev., 128, 2211-2231.

DISCUSSION

N. MOUSSIPOULOS   Your approud is sound for episod applications and as an online tool, but do you really think that it will prove useful for long-term assessments that are required by the EU legislation?

R. SAN JOSE   My feeling is that the one year simulations required by the EU can be performed properly with these kind of applications although several weeks are needed. My experience is that the Cities and even companies asking for environmental assessments can wait one or two months for the complete task of such a kind of simulation. On the other hand more and more cities are installing real time forecasting systems which could store the data somewhow or run during the year the "historical" simulation to have it ready at the end of the year.

LOCAL AND URBAN EFFECTS OF
AIR POLLUTION IN SOUTHERN GERMANY

Peter Suppan[1]

1. INDRODUCTION

The political changes in Europe with the opening of the East European Countries and the Reunion of Germany has changed the traffic situation in Central Europe dramatically. Especially in the South East of Germany in the greater area of Munich, the traffic (air and ground) has increased significantly. The increasing vehicle movements in and around Munich (the Munich airport has to manage more than 25 million passengers a year) contributes strongly to the NO_x and VOC emissions and therefore also to the air quality of the region. High O_3 and NO_2 concentrations in the vicinity of Munich during the summer reflect very well the influence of the traffic to the air pollution situation. To assess the influence of the emissions to the air pollution, several modelling campaigns were performed in the greater Munich area.

In the vicinity of Augsburg in May 2001 a measuring campaign and a modelling case study took place to assess the contribution of the city to the surrounding region. The results show clearly the influence of the nearby highway to the air quality in that region.

Within the project MOBINET (Mobility in the conurbation of Munich) new forms of mobility services, innovative traffic technologies and multimodal traffic management (sustainable mobility) and their impact on air quality were studied. For this purposes air quality modelling for a episode in the year 2000 and a future trend scenario with an emission forecast for 2010 (without the MOBINET measures) were performed with the Multiscale Climate Chemistry Model (MCCM). In a second stage the expected small contributions (<10 %) of the MOBINET measures to the overall mobility the changes in the emission inventory and therefore to the air quality will be a part of future modelling calculations. By comparing these investigations, a classification of the different MOBINET measures can be carried out.

---

[1] Institute for Meteorology and Climate Research, Environmental Atmospheric Research, Forschungszentrum Karlsruhe GmbH, Kreuzeckbahnstr. 19, D-82467 Garmisch-Partenkirchen, Germany

Air Pollution Modeling and Its Application XVI, Edited by
Borrego and Incecik, Kluwer Academic/Plenum Publishers, New York, 2004

2. OVERVIEW

2.1 Working Areas within MOBINET

The program MOBINET contains 4 working areas (A-D), which are linked to the centralized working area (E). In the following part a overview about the working areas will be given (more detailed information can be obtained at http://www.mobinet.de)

2.1.1   Working Area A: Improving Multimodal Transport

The main task of improving multimodal transport are that almost half a million commuters take their car into Munich every work day, causing congestions on the approach roads and parking problems; therefore these commuters from the surrounding regions shall be motivated to use public transport more often. This objective can be achieved by improving the quality of public transport offers and by managing the scarce parking facilities in the inner city. Main objectives in working area A are:

- Reducing private car traffic through transferring non-essential car-trips on to reliable and attractive PT (public transport) offers
- Improving PT reliability and information
- Improving the adaptability of PT supply to the structure of actual traffic demand
- Optimising the transfer points between transport subsystems
- Creating incentives for more intensive PT usage

It is expected that within this working area a decrease and/or a redirection of emissions, will have the largest impact to the air quality.

2.1.2   Working Area B: Optimising Traffic on the Main Road Network

Modern technologies enable new qualities of traffic control when they are based on the prognoses of actual traffic flow. In the conurbation of Munich controlling procedures are used area wide to make more efficient use of the main road network. Traffic control systems are implemented on four levels typical for an agglomeration area (Figure 1). The main goals in this working area are:

- to relieve secondary streets and residential areas through a highly efficient and intelligent main road network
- to reduce waiting times and stop-and-go manoeuvres at signal installations
- to divert traffic away from congestions
- to disperse and level-out network load
- to utilize more efficient the traffic infrastructure
- to reduce **air pollution**
- to increase traffic safety
- to accelerate road-bound public transport services

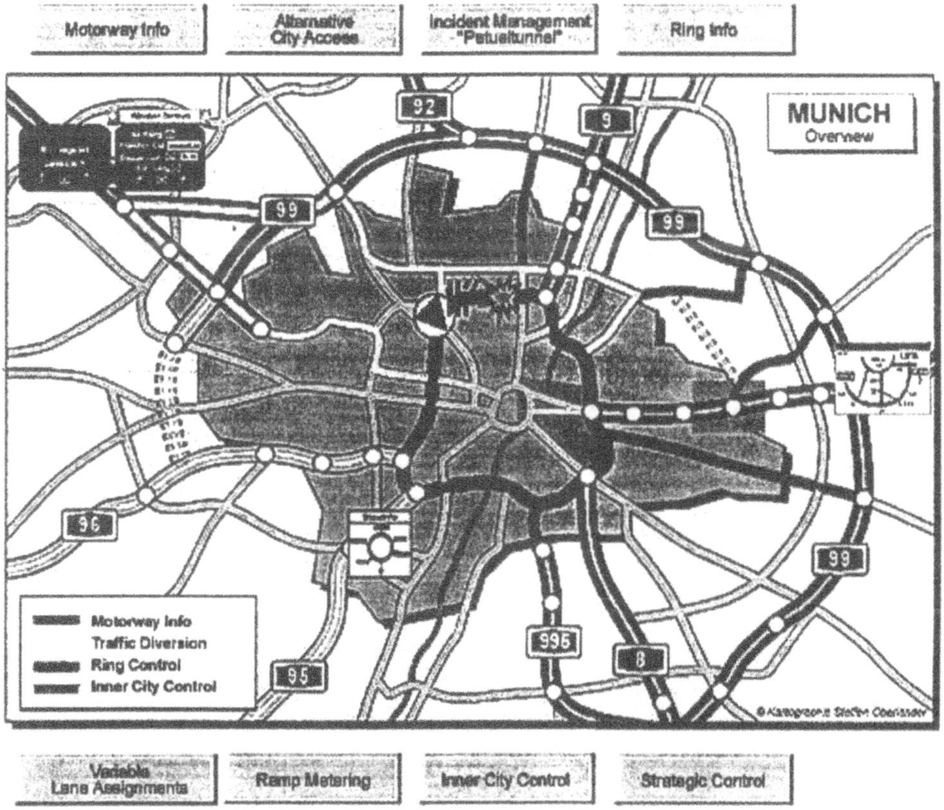

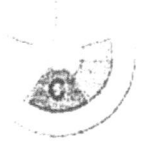

Figure 1. Optimised traffic information on the main road network.

Also within the working area B a redirection and/or reduction of emissions will be expected, which will be a subject of modelling calculations on the impact on air quality.

2.1.3 Working Area C: Multimedia-Information Services

Increasing mobility requirements generate more and more traffic, which is why new solutions are needed. Innovative mobility services can promote the efficiency in the overall traffic system through improved information and enhanced services. A key task in the area of "Multimedia-Information Services" is to provide data and link the mobility-relevant information from traffic, tourism and environment in the Munich conurbation and its surrounding recreational regions. In short: offering detailed travel information for anyone at any time. Summarized following goals are foreseen in this working area:

- Developing new technologies for pre-trip and on-trip information
- Allowing for a wide scope of activities and means of transport

- Providing individual and collective information systems
- Possibilities for choosing from alternative routes
- Optimising personal routes

As this working area is mainly addressed to transmit information, it will be expected that these tasks will not have a wide influence on emissions.

2.1.4   Working Area D: Innovative concepts for the mobile society

Within this working area new information and communication technologies, new forms of work organisation as well as the continued trend towards an event society is a subject of research. These factors have a reciprocally reinforcing effect and are particularly marked in conurbations. They promote new mobility patterns, which necessitate an expansion of the range of traffic-planning and traffic-policy strategies and measures. Through innovative concepts for the mobile society MOBINET seeks to utilise the potentials of new mobility patterns in order to shape and foster sustainable mobility. The main goals are:

- Promoting sustainable mobility
- Focussing on customer demands
- Developing new sustainable mobility services
- Improving traffic safety
- Securing the transferability of results
- Promoting further development of this approach after project-end
- Developing suitable evaluation instruments

Also this working area will not have a basic influence on the distribution and amount of the emission fields.

2.1.5   Working Area E: Management centre and data network

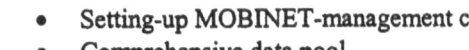

The management centre interlinks the various traffic systems of the Munich conurbation into a network. The projects are:

- Setting-up MOBINET-management centre
- Comprehensive data pool
- Interlinking the traffic systems via the data network
- Generating the present traffic situation
- Multimodal traffic strategies
- Long term development of business concept

With the goals which have to be achieved:

- Providing data
- Supporting traffic control
- Integrated traffic management
- **Air Quality Impact Assessment**

2.2 Emission Scenarios within MOBINET

After successfully introducing of all scenarios within 5 to 10 years (2000 to 2010) from working area A to D also a change in the emissions for specific grid cells can be expected. The most significant contribution will be given by working area A and B, e.g. due to the redirection of the traffic and/or reducing the numbers of congestions in and around the conurbation Munich. A first estimation of the influence of the introduction of the MOBINET scenarios within a time scale of 10 years to the emission reduction/redirection gave a reduction of less than 10 % (MOBINET Q2-Workshop, 2003). As the introduction of the so called "MOBINET demonstrators" is not yet finished a final approximation of the influence to the emission fields will be available end of 2003.

In order to classify the MOBINET measures, it will be necessary to discuss the trend of future emission scenarios from 2000 to 2010. This implies a basic model run for an episode within the year 2000 and a scenario calculation for the year 2010. After running these scenarios (without the MOBINET measures), model calculations including the MOBINET measures shall be performed.

3. CURRENT AND FUTURE EMISSION RATES

Based on the emission inventory of the year 2000, in Figure 2 the NO_x traffic emissions are shown in tonnes / (km^2 year). The emission fields clearly indicate the highways with emission contributions of more than 25 tonnes / (km^2 year). More detailed information can be seen in the MOBINET domain (conurbation of Munich), where the ring highway and the highway A9 and A92 to the airport in the north of the domain show very high contributions to the NO_x emissions. Already previous investigations have shown that this region will be a "hot spot" for air quality impact studies (e.g. Graf and Tremmel, 1996, Suppan and Graf, 1998). At the airport itself – in 2003 approximately 25 Mill. passengers were using the airport – the highest emission rates with close to 300 tonnes / (km^2 year) NO_x are given. In June 2003 a second terminal has opened which can handle another 25 million passengers. It is expected that in 2010/2015 approximately 50 million people will use this airport.

The future emission reduction for passenger cars on highways in Germany is shown in Figure 3. Based on the year 2000 a reduction of 50 % for the VOCs in 2005 is foreseen. The reduction forecast for the other pollutants will be much smaller with 30 % for CO and 40 % for NO_x and PM10. Until the year 2010 a target of another 6 to 10 % of a reduction is foreseen. By taking into account the sum of all road emissions in Bavaria, the highway reduction scenario will not differ significantly (not shown here).

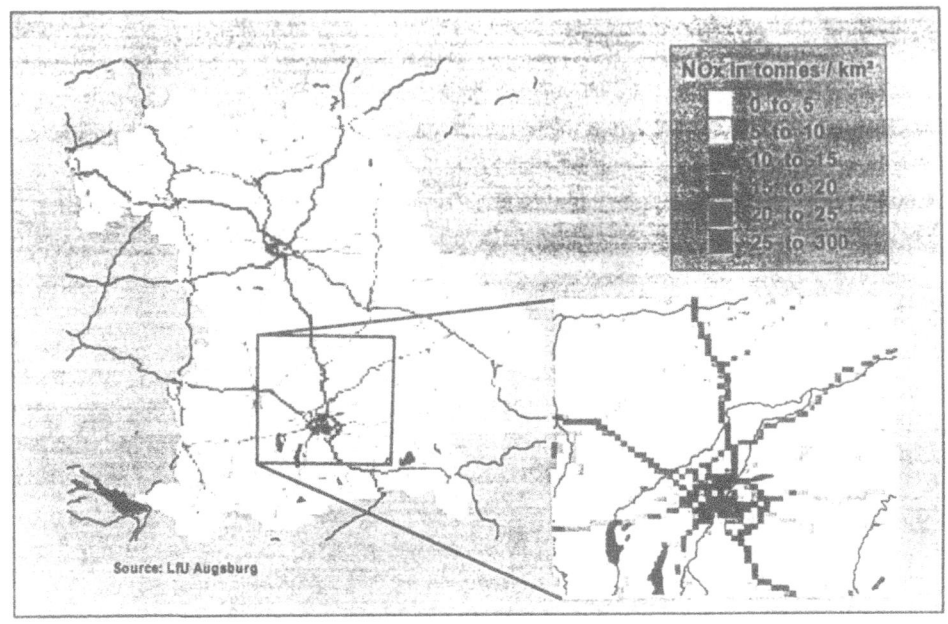

Figure 2. NO$_x$- traffic emission inventory for Bavaria for the year 2000 and for the MOBINET domain (right).

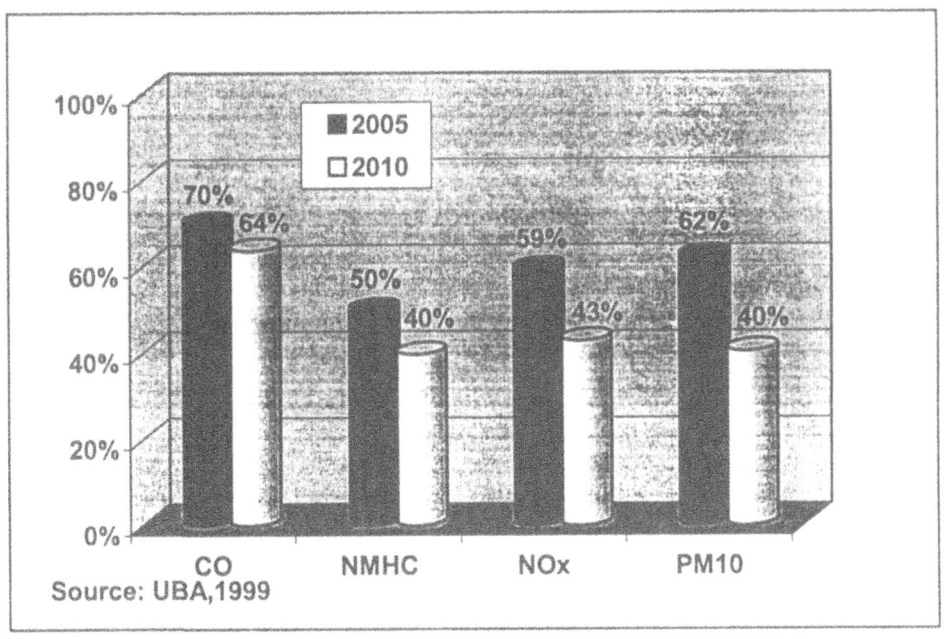

Figure 3. Emission reduction for passenger cars on highways based on the year 2000.

194

In detail the highway emissions can vary significantly e.g. shown in Figure 4. Here the change of the emissions for passenger cars at the highway A9/A92 to the FJS-Airport in the northern part of Munich was calculated until the year 2015. The traffic data for the calculation of the emissions were made available by BMW AG, the traffic emission factors by UBA. The graph is showing a smother reduction for VOCs, NO_x and PM10, but for CO the forecast will show an increase due to the amount of cars at this specific highway from 2000 to 2005. The prediction for 2015 shows an increase between 4 and 9 % for all emissions.

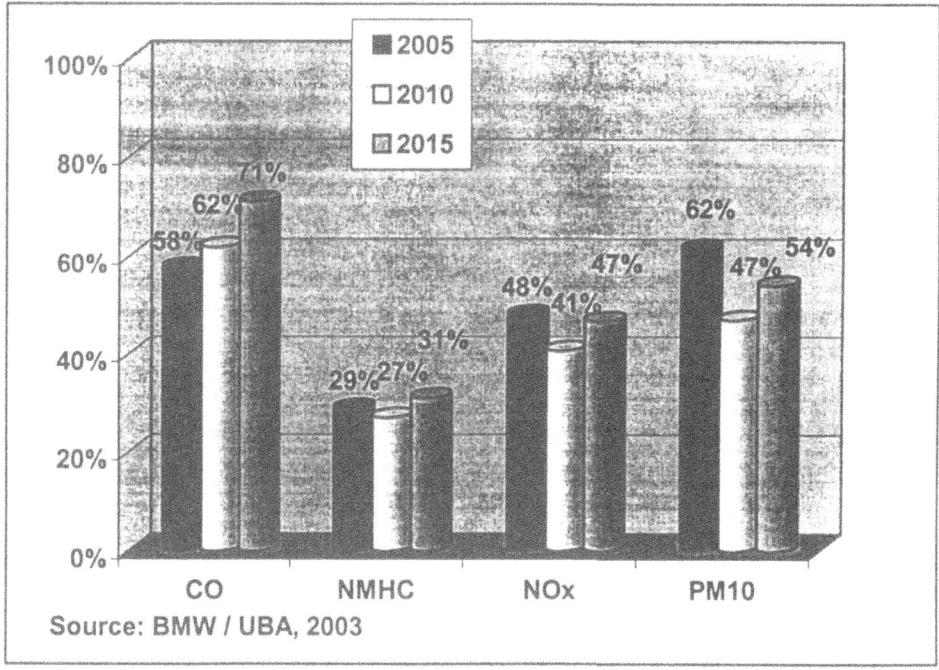

Figure 4. Change of emissions for passenger cars on the highway A9/A92 based on the year 2000.

The reason for this increase on emissions can be attributed to the increase of passengers on the FJS-Airport with about 50 Mill. people in the year 2010/2015. As not all passengers will use the public transport system, the traffic to the airport will also increase as it can be seen in Figure 5.

This development of traffic emissions is in contrast to the general assessment of the traffic emissions in Germany, therefore this will play a significant role in the assessment of the impact of the MOBINET measures to the air quality. To study this impact dispersion model calculations have to be performed with the current emission situation as well as with a future emission inventory. As it will be very difficult to extract the impact of MOBINET measures out of the long term trend of the air quality, it will also be important to select very carefully the time period for the calculations. Based on this a period with high ozone and nitrogen dioxide concentrations was selected.

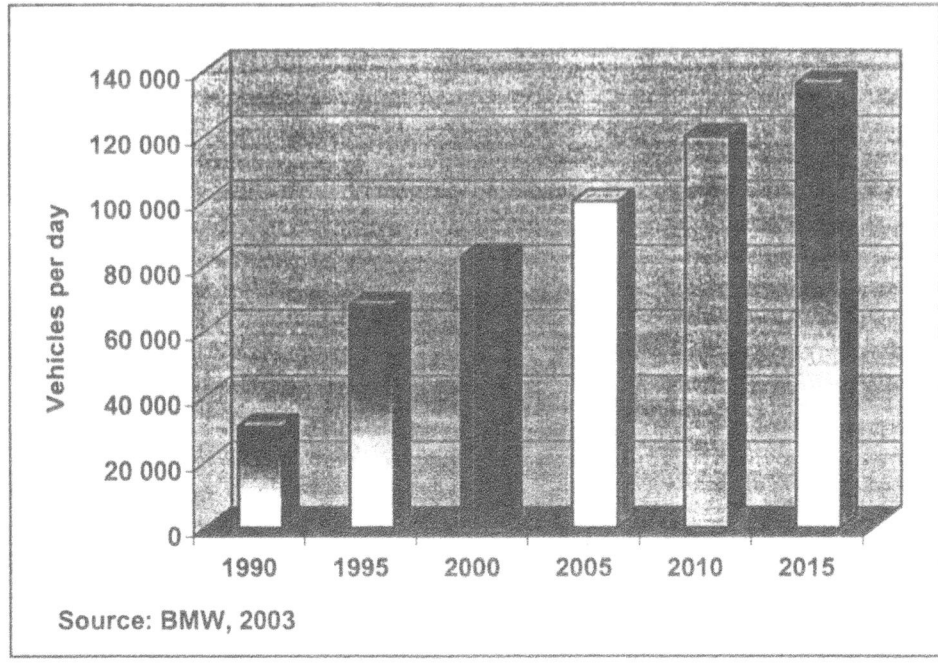

Figure 5. Number of passenger cars from 1990 to the year 2015 on the feeder roads A9/A92 to the FJS-Airport.

4. MODELLING CALCULATIONS

4.1 Dispersion Model

The modelling system that was used is the Multiscale Climate Chemistry Model (MCCM). As MCCM is based on MM5 (Fifth-Generation NCAR / Penn State Mesoscale Model) the online coupling of meteorology and chemistry provides fully consistent results with no interpolation of data in contrast to off-line coupled chemistry and transport models. The model includes a detailed description of the gas-phase chemistry (RADM) and the deposition of pollutants. The anthropogenic emissions can be treated as different sources and on an hourly basis. The biogenic emissions were calculated online during every time step. The multiscale nesting capability allows a minimisation of the boundary effects. A parallel computing architecture accommodates a fast performance of the model (Grell et al., 2000). The model also includes non-hydrostatic dynamics and four-dimensional data assimilation capability as well as many options for modelling microphysical processes. In association with the gas phase chemistry module 22 photolysis frequencies are computed according to cloud cover, ozone, temperature and pressure in the model atmosphere. In operation this model simultaneously calculates the meteorological and air chemistry changes over the model domain and provides time

196

dependent three-dimensional distributions of all the major inorganic and organic species which are relevant for the oxidant formation.

4.2 Model Setup

The period for the modelling calculations was set to the 18. - 22. June 2000. Within this period, a steady increase of the daily ozone maximum was measured until the 21st of June, were the highest concentrations of ozone in 2000 were measured. Afterwards a front passed the domain of interest and the ozone concentrations went back to typical medium levels.

The simulations were carried out with a nesting strategy of 4 nested domains, starting with a European wide domain (40 x 40 grid cells with a 54 km x 54 km grid resolution) down to the domain of interest with a grid resolution of 2 km x 2 km and a number of 51 .x 51 grid cells. In the vertical 25 non-equidistant levels were used with the lowest level (centre) at 15 m above ground.

4.3 Model Validation

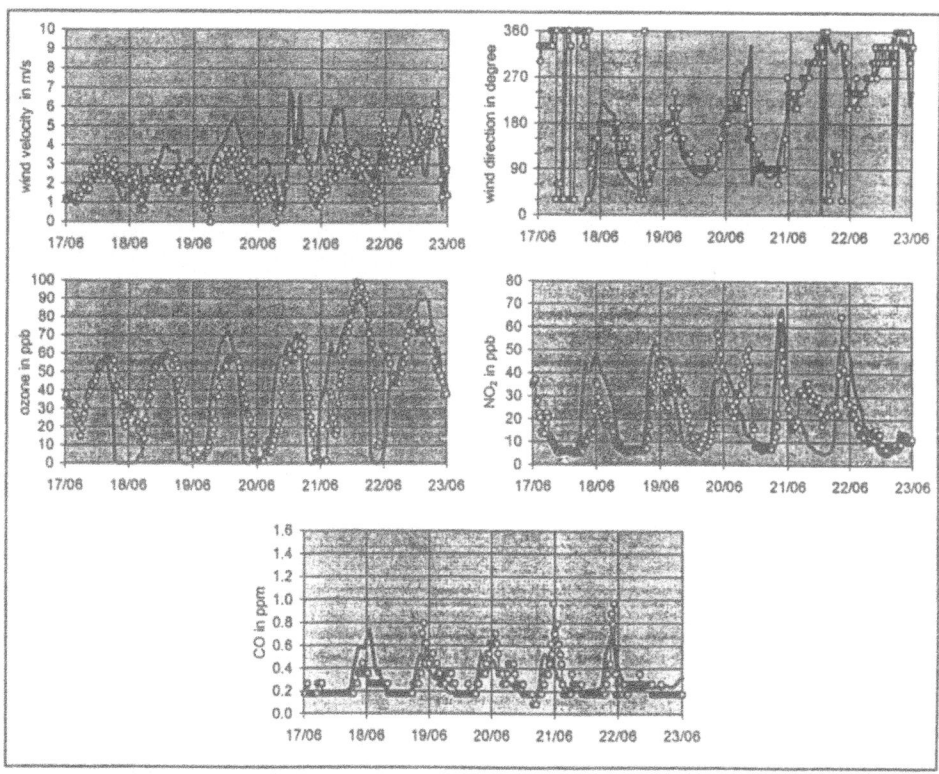

Figure 6. Example of comparisons between model calculations and measurements for some meteorological and chemical parameters.

The period of interest was characterised by prevailing south easterly wind directions at the beginning of the calculations. At the end and during the period with high ozone values the prevailing wind direction was from west (Figure 6).

The whole period was characterized by moderate wind velocity (3-5 m/s). Further validation examples Figure 6 show also very good agreement between the model results and the measurements for ozone, nitrogen dioxide and carbon monoxide. The model did very well reproduce the maximum value for ozone during the 21st of June.

4.4 Model Scenario Calculations

Based on these results a scenario calculation for the year 2010 was performed. Using the same meteorological conditions as in 2000, the impact to the air quality could be demonstrated in Figure 7. These figures show the relative change of O_3 and NO_2 from 2000 to 2010. On the main traffic networks around the city of Munich the increase of ozone, especially in the northern part of Munich was up to 15 %. In the same way the nitrogen dioxide concentrations did show a significant decrease with more than 40 % at grid cells where the O_3 concentrations did rise.

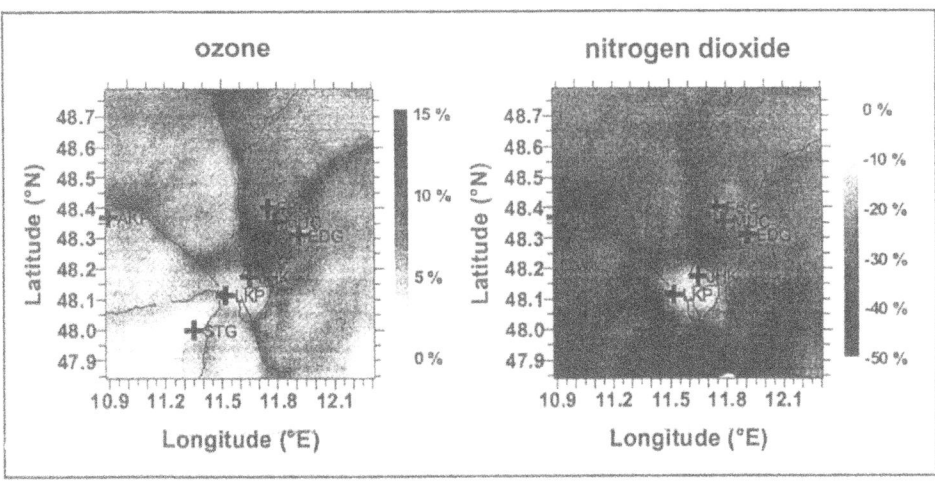

Figure 7. Relative change of O_3 (left) and NO_2 (right) concentrations (scenario 2010 – base case 2000) as 24 hour mean values from the 21st of June during a episode were the highest ozone concentrations in the year 2000 in southern Bavaria were measured. The solid lines reproduce the highways within the domain. Included are also some measuring stations (e.g. LKP & JHK, Munich measuring locations) and some points of interest (e.g. MUC- FJS-airport of Munich).

The results can be related to an earlier project where the influence of the city of Augsburg, to the local air quality was a task of interest (Schädler and Suppan, 2001). Within this project also the influence of highway emissions to the local air quality was a subject of interest (Suppan and Schädler, 2002). To assess the influence of the highway emissions to the ozone and nitrogen dioxide concentration fields, the line source (highway) between Munich and Augsburg (ca. 35 km) was switched off and the emissions were set to the surrounding levels. This could be attributed to a reduction of 80 % of the traffic emissions. As it can be seen in Figure 8, the increase on O_3 concentrations was up to 10 % whereas the NO_2 concentrations did show a decrease of close to 25 %. The red line marks the region of influence of this specific emission reduction. Taking into account the specific meteorological conditions with prevailing strong north easterly winds and temperatures less than 25°C, the results did fit very well to the estimation for the MOBINET domain. During the MOBINET time period the maximum temperatures were above 30°C and the expected reduction for traffic emissions was approximately 60 % for NO_x and the VOCs

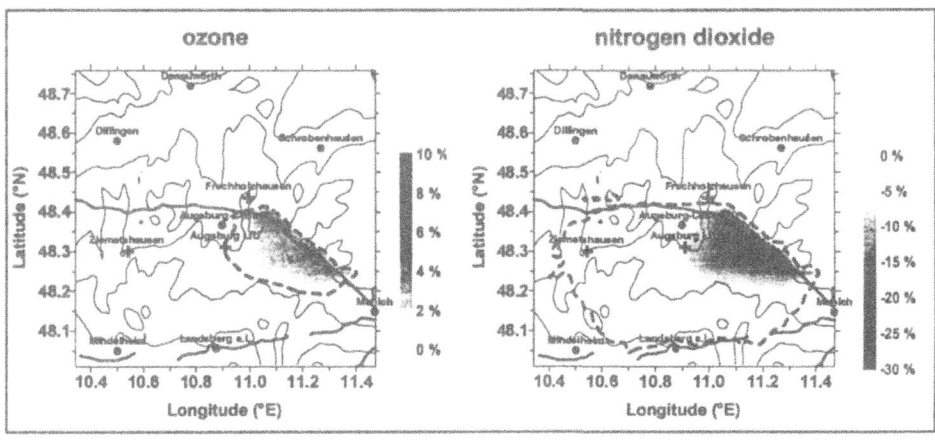

Figure 8. Relative change of O_3 (left) and NO_2 (right) concentrations as 72 hour mean values from the 21st to the 23rd of May 2001 (the dashed line indicates the reduction influenced area).

5. CONCLUSIONS AND FUTURE WORK

In general MCCM is given reliable results for the spatial and temporal variation of pollutant concentrations. The model results on O_3 and NO_2 show that the structure of the modelling domains accurately accounts the urban and regional representation of both pollutants over the modelling domains and therefore this infrastructure provides a broad integrated assessment of the photochemical situation in urban and regional areas. For future emission scenario calculations, these are fundamental facts, which are necessary to give correct answers on the assessment of future air quality impacts. Especially the question of the impact of the MOBINET measures to the air quality with an estimated influence of less than 10 % to the composition of the emission fields can be evaluated by such calculations.

It could be demonstrated that the northern part of Munich, due to its outstanding change of traffic emissions within the next 10 years, can be a field of promising MOBINET investigations. At specific traffic influenced locations the O_3 concentrations will rise 15 % and the NO_2 concentrations will decrease with more than 40 % within the next 10 years. These results go inline with former investigations in the vicinity of Augsburg where an increase of 6-8 % of O_3 and an decrease of 25 % of NO_2 has resulted from a 80% emission reduction.

After receiving the emission estimation of the different MOBINET investigations from the participating partners, the next step will be modelling calculations based on the meteorology and emission inventory (including the MOBINET measures) of the base case year 2000. Source/receptor analysis for single grid cells will give an answer of possible MOBINET influences to the impact of the air quality on the local scale.

6. REFERENCES

Graf, J. and Tremmel, H.G., 1996: *Simulation der mittleren räumlichen Ausbreitung von Flugzeugemissionen im Umfeld des Großflughafens München*, Abschlußbericht, Institut für Physik der Atmosphäre DLR, Oberpfaffenhofen, 1996

Grell, G., Emeis, S., Stockwell, W.R., Schoenemeyer, T., Forkel, R., Michalakes, J., Knoche, R. and Seidl, W., 2000: *Application of a a multiscale, coupled MM5/chemistry model to the complex terrain of the VOTALP valley campaign*. Atm. Environ. 34, 1435-1453

MOBINET Q2-Workshop, 2003, Munich, BMW Research Centre, 29.01.03

Suppan, P., Graf, J., 1998: *The Impact of an Airport on Regional Air Quality at Munich/Germany*, Int. J. Environment and Pollution

Schädler, G. and Suppan, P., 2001: *Untersuchungen zur Wechselwirkung Stadt-Umland im Raum Augsburg*, Forschungsbericht des Bayerischen Staatsministeriums für Landesentwicklung und Umweltfragen, Eigenverlag, Nov. 2001

Suppan, P. and Schädler, G., 2002: *Assessment of highway pollution during specific meteorological conditions in southern Germany*, The Science of the Total Environment, Special Edition: Proceedings of the Seventh International Symposium, Highway & Urban Pollution, 22-23 May, 2002, Barcelona

DISCUSSION

R.C. SILVIA   Did you account for the traffic induced turbulence and how in your model?

P. SUPPAN   Within this modeling study we did not consider traffic induced turbulence. In general this will not have an effect if you use a grid cell resolution of 2 km x 2 km, as we have done it in this case. Traffic induced turbulence will play a major role within street canyons and/or a grid cell resolutions less than 100 m.

GLOBAL AND LONG-RANGE TRANSPORT

Chairperson: G. Kallos

Rapporteur: M. Tayanc

CITYDELTA: A EUROPEAN MODELLING INTER-COMPARISON TO PREDICT AIR-QUALITY IN 2010

Philippe Thunis and Cornelis Cuvelier[*]

1 INTRODUCTION

As a contribution to the modelling activities in the Clean Air For Europe (CAFE) programme of the DG-Environment of the European Commission, an open model inter-comparison exercise has been launched by the JRC-IES in collaboration with EMEP[1], IIASA[2] and EUROTRAC[3] to explore the present and future (2010) urban air quality in some European cities as predicted by different atmospheric chemistry-transport dispersion models. The different emission-reduction scenarios investigated in this project as well as the range of responses resulting from this model inter-comparison is being used in the cost-effectiveness analysis of CAFE with the aim to balance Europe-wide emission controls against local measures. This exercise proposes therefore a better understanding of the benefits and limitations of regional emission control against local measures.

The model inter-comparison focuses on ambient levels of particulate matter and ozone in urban areas. It addresses health-relevant matrices of exposure (e.g., long-term concentrations) to fine and coarse particles as well as to ozone.

Comparisons are currently conducted for 8 European cities: Berlin, Copenhagen, Katowice, London, Marseille, Milan, Paris and Prague and involve the active participation of more than 15 modelling groups in Europe. Each of them is asked to perform a long-term validation simulation for a 300x300 km area around each city: a six months run for Ozone and a full year run for PM (1999 has been chosen as base year for emissions and meteorology), and a number of (NO_x, VOC, $PM_{2.5}$ and PM_{coarse}) emission-reduction scenario simulations for the year 2010. All input data (emission inventories, meteorology and monitoring data) have been made available to the participants through the Web.

In this paper, an overview of the methodology is presented.

[*] P. Thunis and C. Cuvelier, Institute for Environment and Sustainability (IES), Joint Research Centre (JRC), TP 280, 21020 Ispra, Italy

Air Pollution Modeling and Its Application XVI, Edited by
Borrego and Incecik, Kluwer Academic/Plenum Publishers, New York, 2004

205

2 OBJECTIVES

CityDelta aims at providing guidance on how urban air-quality could be implemented in a Europe-wide evaluation of the cost-effectiveness of emission control strategies. CAFE expects information from integrated assessment models on a cost-effective balance between emission control measures that should be taken at the European/Community level and measures that should be best left to the local city Authorities. For this integrated assessment task the central questions are:

- What is the influence of local versus regional emission control on health-relevant indicators for fine particles (PM_{10}, $PM_{2.5}$) and ozone in urban air?
- How regional model (e.g. with a typical spatial resolution of $50*50$ km^2) predictions differ from predictions obtained with finer resolved models?
- What is the range of agreement between different photochemical-dispersion models on the level of responses to emission changes?

These questions suggest for the first phase the primary focus on identifying the range of responses of models towards emission reductions (deltas in emissions) and providing recommendations on how to include urban air-quality into integrated assessment modelling.

In parallel with the inter-comparison, an evaluation of model performances against monitoring data is carried out for the 1999 base case, although it is not the main objective of CityDelta. Equally, it is not an objective to identify the best model.

3 A FOCUS ON LONG TERM SIMULATIONS

Assessments of impact on health and vegetation require information about the long-term exposure to various air pollutants. For ozone, studies revealed relations between the extent of vegetation damage and the AOT40, i.e. the hourly ozone concentrations in excess of 40 ppb cumulated over the vegetation growth period (three or six months). For health impacts, no-effect thresholds for short-term concentrations either are not identified (in the case of PM) or are believed to be low (for ozone). Thus, to capture the full impacts of PM and ozone over longer time periods (1 year), it will be necessary to include in the assessment all days on which critical values are exceeded. Single short 'peak episodes' (like a few extreme summer days) are not of primary interest. Such extreme situations are, by definition, rare cases and the public health importance is therefore much smaller than the problem due to the "general 'moderately high' exposure" situation.

For health-relevant ozone exposure, the analysis will be restricted to periods when ozone typically exceeds the discussed threshold values (i.e. the spring/summer half year), whereas vegetation damage can be estimated from calculations for three-month time slices of the vegetation periods. The CityDelta exercise recognizes this need and requests participating models to deliver the required information. Daily simulations over 12/6 months form the baseline of the model inter-comparison.

Ultimately, it will be necessary to address the meteorological conditions of several years to capture the inter-annual meteorological variability, but this appears to be unfeasible in the current phase of CityDelta which focuses on 1999 as reference year for validation.

4   INPUT DATA

The quality of the results obtained in the frame of CityDelta depends largely on the quality of the input data. Moreover, a correct interpretation of the results is only possible when model simulations are performed in comparable conditions, i.e. using similar input data. In this exercise, a main constraint to all participants has been to use identical emissions, meteorological year and boundary conditions. Emission inventories were provided at both the local and regional scale. The meteorological year 1999 has been chosen for the base case and the 2010 scenarios (although high resolution meteorological data have been provided to the participants, the only requirement was to simulate the year 1999). Finally the EMEP Eulerian model[4] has been used to provide boundary conditions for all city domains.

4.1   Boundary Conditions

The EMEP Eulerian model provided boundary conditions for the 1999 base case and the 7 emission scenarios in 2010 (see below) for 25 gas-phase species and SO_4, ammonium sulfate, ammonium nitrate, primary PM_{coarse} and $PM_{2.5}$. These boundary conditions are provided on 20 vertical levels ranging from surface to approximately 16 km. The lumping of the different species into each chemical mechanism as well as the interpolation procedures were left free to each modelling group.

4.2   Meteorology

The only constraint concerning meteorological data is that 1999 was chosen as the year to be modelled. Each modelling group has been left free to choose its own meteorological driver. Nevertheless, thanks to Meteo-France, meteorological fields with a 10 km spatial resolution from the ALADIN model[5] have been provided to the participants for the cities of Berlin, Milan, Paris and London.

4.3   Monitoring

Monitoring data for the different cities have been compiled with the help of local organisations and prepared for CityDelta. Stations directly on city streets have been eliminated whereas other stations have been classified into urban, suburban and rural. Monitoring data are available for the whole year 1999 on an hourly basis for the following pollutants: O_3, NO_2, NO, PM_{10} and eventually $PM_{2.5}$. The following table shows for each city and each pollutant the number and type of the measurement stations.

	O_3	NO_2	PM_{10}	$PM_{2.5}$	Urban	Suburb	Rural
Berlin (15 stations)	15	15	3	0	13	0	2
Copenhagen (3 stations)	3	3	0	0	1	0	2
Katowice (9 stations)	5	9	8	0	6	1	2
London (13 stations)	13	13	8	2	6	4	3
Marseille (14 stations)	12	14	7	0	12	1	1
Milan (13 stations)	13	13	4	0	7	3	3
Paris (15 stations)	14	15	4	0	11	3	1
Prague (8 stations)	4	8	8	0	2	3	3

Table 1: Available monitoring data for each of the eight CityDelta cities. Columns 2-5 indicate the number of stations available per pollutant whereas Columns 6-8 give information on the station classification (suburban, urban or rural)

4.4 Emission Inventories

One major point in the preparation of the CityDelta exercise has been to ensure that all models make use of a similar set of emissions. On the continental scale, emissions are based on EMEP/TNO[6] whereas at the urban scale, high spatial resolution emission inventories were prepared by local city groups.

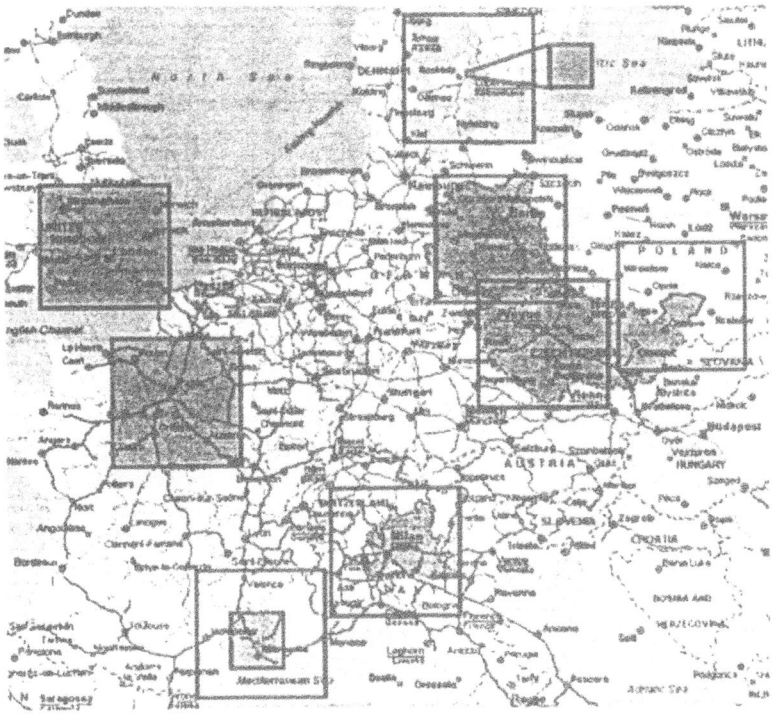

Figure 1. Spatial coverage of the high resolution emission inventory (grey shaded area) within the computational domain of 300x300 km2 indicated by the solid squares around each city.

For cities where the high resolution emission inventory did not cover the entire 300 x 300 km^2 area over which results are requested, EMEP emissions were used as complement.

Note that double counting of the emissions was avoided by subtracting city emissions from EMEP cell totals.

An overview table is provided below summarising the characteristics of each of the CityDelta cities. For most of the cities, NMVOC were split according to AEA-Technology[7] (AEAT). The lumping of these VOC species into each chemical mechanism has been left to the participants.

	Resolution	Extension	Sector speciation	Time speciation	Gas-phase species	PM species	VOC Speciation
Berlin	2 km	Berlin, Brandenburg, Sachsen, Sachsen-Anhalt	Into 5 SNAP + point sources	Month Day Hour	CH_4, CO, NH_3, NMVOC, NO_x, SO_2	PM10 PM2.5 based on PM10	Into 227 species based on AEAT at SNAP 1
Katowice	5 km	Upper Silesian Province	SNAP 1	Similar to Berlin	SO_2, NO_2, CO, NMVOC, NH_3, CH_4, CO_2, N_2O	PM10 TSP PM2.5 based on PM10	227 species based on AEAT at SNAP 1
London	2 km	300 x 300 km2	SNAP 1	Month Day Hour	CO, VOC, NO_x, SO_2	PM10	227 species based on AEAT at SNAP 1
Milan	5 km	Lombardy Region	SNAP 1	Season Day (3) Hour	CO, NO_x, SO_2, NH_3, NMVOC, CH_4	PM25, PM10 and TSP	426 species based on AEAT at SNAP 1
Paris	3 km	300 x 300 km2	7 activiy sectors.	Month, day (2) hour at SNAP 3 level	CO, CO_2 NO_x, SO_2, CH_4, NH_3 NMVOC	PM10 and PM2.5 scaled on EMEP NOx/PM correlations	185 species based on Genemis at SNAP 3 or 4
Prague	5 km	Czech Republic	SNAP 1	Similar to Berlin	CH_4, CO, NH_3, NMVOC, NO_x, SO_2,	PM10 PM2.5 based on PM10	227 sp based on AEAT (ref) SNAP 1

Table 2: Summary of the high resolution emission inventories available in CityDelta. More details are available on the CityDelta Web page[8]. The number of specific days within the week is indicated in brackets in the time split column. SNAP stands for Selected Nomenclature for Air Pollution.

5   EMISSION SCENARIOS

Emission scenarios for 2010 have been prepared by IIASA with emission reductions specific to each country, pollutant and activity sector. In this exercise, three of these activity sectors have been considered: traffic and stationary sources split into high (above 50 m) and low level emissions.

The emission scenarios include:

a)   the 2010 CLE (Current Legislation) scenario
b)   additional NO_x reductions corresponding to the MFR (Maximum Feasible Reduction) scenario
c)   additional NO_x reductions corresponding to the (MFR+CLE)/2 scenario
d)   additional VOC reductions (MFR)
e)   Combined MFR VOC and NO_x reductions
f) PM_{coarse} reduced to the MFR level
g) $PM_{2.5}$ reduced to the MFR level

For each of these scenarios, emission reductions have been imposed in a similar manner at the continental and local scale. Levels of emission reductions for NO_x, VOC, PM_{coarse} and PM_{fine} are shown in the figure below for the 7 countries of interest.

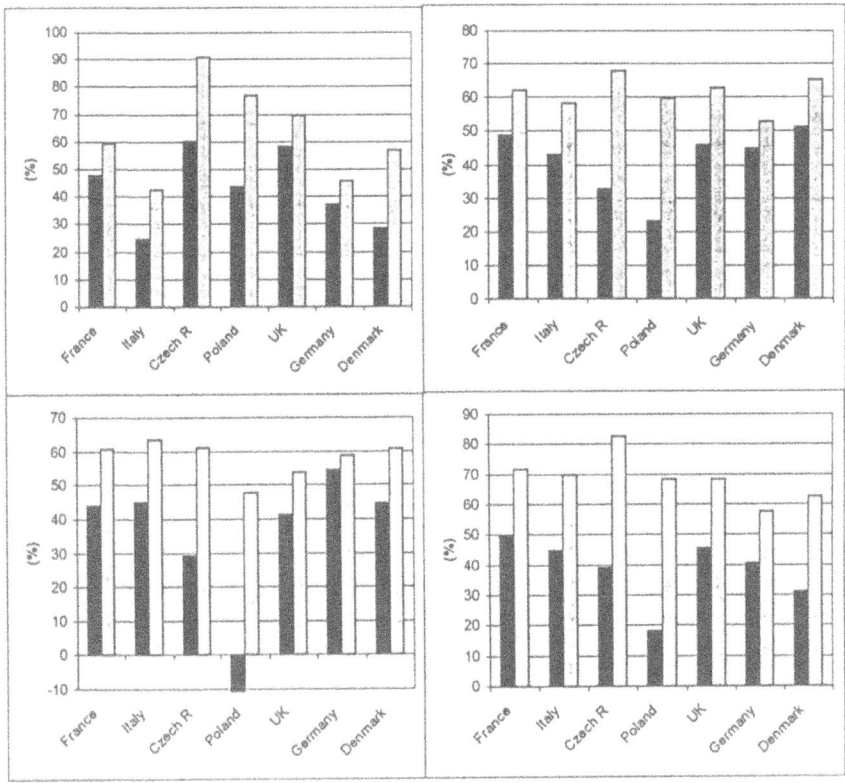

Figure 2. CLE (black) and MFR (grey) emission reduction (in percentage) compared to the 1999 base case emissions for PM_{coarse} (upper left), NO_x (upper right), VOC (lower left) and PM_{fine} (lower right)

210

In order to investigate the effectiveness of further reductions taken at the city scale, three additional scenarios have been designed which include emission reductions of NO_x, VOC and combined NO_x-VOC only into the urban part of the domain (typically over a 50x50 km area).

6 PARTICIPATING GROUPS

About 15 European modelling groups are participating in this exercise. The table below summarises the deliveries planned for each of these groups. Of course, each group does not perform simulations over the whole set of CityDelta cities and emission scenarios but many participate on multiple cities and with different model resolutions. Note that in this table, models differ not only by their spatial resolution but also by their formulation and approach. While the majority of the models perform continuous simulations throughout the whole year, a few approaches are based on selecting statistically representative episodes and aggregating them to obtain a full year picture. This set of results and its variety helps in the interpretation of the many deltas, city, scenario, spatial resolution, ...

Contact Person	Model	Paris	London	Prague	Katow	Milan	Berlin	5 km	10 km	50 km
M. Bedogni	CAMX					O		x		
R. Berkowicz	THOR		OP		OP		OP	x	x	x
F. Brocheton	MOCAGE	O	O	O	O	O	O			x
	MOCAGE	O				O			x	
L. Volta	CALGRID					O		x		
N. Moussiopoulos	OFIS	OP	OP	OP		OP	OP			
	MUSE	O	O	O		O	O		x	
G. Pirovano	STEM					O		x		
R. Stern	REM3						OP	x		
G. Schaedler	MCCM			OP			OP	x		x
C. Honore	CHIMERE	OP	OP	OP	OP	OP	OP			x
	CHIMERE	OP	OP	O		OP	OP	x		
L. Tarrason	EMEP	OP	OP	OP	OP	OP	OP			x
J. Brechler	SMOG			O				x	x	
B. Denby	EPISODE						O	x		
P. Builtjes	LOTOS	OP	.				OP	x	x	x
C. Philippe	TRANSCHIM	O		O				x	x	
O. Hellmuth	MUSCAT						OP	x		
F. De Leeuw	EUROS	O			O	OP		x		
C. Borrego	CAMX		OP					x		

Table 3: List of participating modelling groups in the CityDelta exercise. For each city and model, the letters OP indicate either Ozone (O) or PM (P) simulations. The three last columns indicate the spatial resolution at which models are run.

7   INTERPRETATION OF THE RESULTS

For each city and scenario, participants deliver results with either 5, 10 or 50 km spatial resolution. The choice of the procedure to interpolate the model results to a predefined latitude-longitude grid was left to the participants. According to the selected resolution, the domain over which results were submitted covered 500 x 500 km^2 for the 50 km resolution and 300x 300 km for the 5 and 10 km spatial resolutions. NO_2 and O_3 concentrations were required with an hourly frequency for a 6 months period from April to September whereas daily averaged PM_{10}, $PM_{2.5}$, sulfate, nitrate and ammonium were requested for the whole year.

The validation of the model results with the 1999 monitoring data as well as the interpretation of the deltas (differences among model results as function of the city, emission scenario, spatial resolution,…) follows a two-steps procedure:

a)   Each modelling group performs a first validation and interpretation of its results using some of the most relevant indicators.

b)   Modelling groups submit data to the JRC-IES for processing. A subset of the full dataset is then used to perform an intercomparison of the results. The JRC-IES developed a graphical interpretation tool which has been made freely available to the CityDelta community with which each participating group may evaluate its own results and compare them to others. The tool is flexible enough to guarentee a visualisation of various types of indicators. Usage of a common tool also facilitates the discussions between CityDelta participants.

This graphical tool includes the following four different applications:

1.   MONIT which allows users to visualize the characteristics (time series and statistics) of the monitoring stations for which measurements are available on a city by city basis.
2.   VALID which provides an evaluation of the model results against measurements for the 1999 base case at all available monitoring stations. Model results may be analysed using different types of indicators (bias, NMSE, Exceedance days, AOT, …) with flexible threshold values and averaging time and periods.
3.   DELTA which provides an intercomparaison of the deltas resulting from the different models at 9 pre-selected locations for each city. Indicators similar to those in VALID are available
4.   PLANE which provides a 2-D surface vision for a few monthly averaged indicators, namely exceedance days and AOT values.

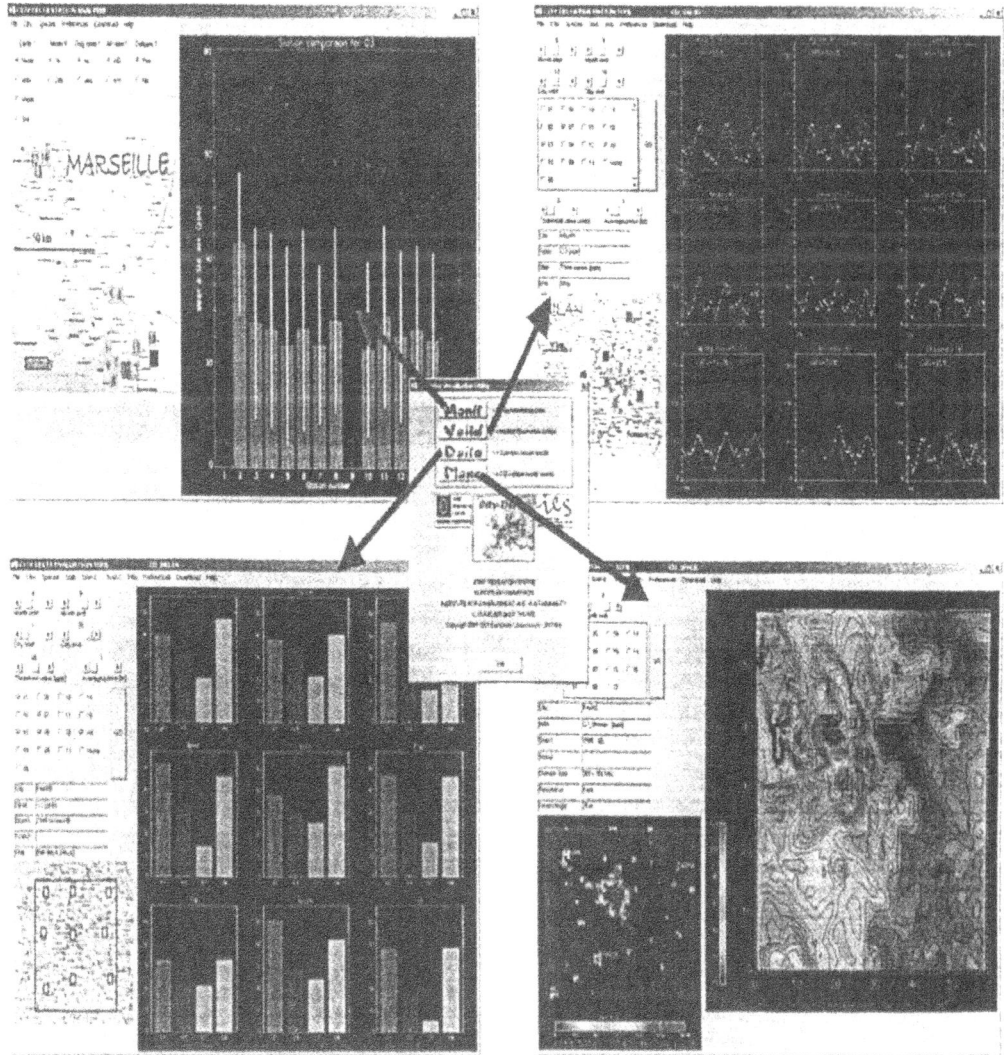

Figure 3. Graphical interpretation tool with its monitoring (upper left), validation (upper right), delta visualisation (lower left) and spatial indicators (lower right) components.

New results are processed and made available through the CityDelta Web page to the participants who can compare their own simulations with others. Note that anonymity is guaranteed through the use of model identification numbers.

8   PRELIMINARY RESULTS

The fourth City-Delta workshop took place in Valencia in April 2003 with a main objective being the interpretation of preliminary results. More than 220 6-months gas-

213

phase and 90 yearly aerosol-phase simulations were available at that date. Although preliminary, a few conclusions have been drawn regarding modelled long term ozone. In general, fine scale resolution model results (5 and 10 km) did not show a significant improvement compared to coarse ones (50 km) and the variability looked larger among different models with similar resolution than across scales with similar models. All models exhibited difficulty in reproducing the nighttime behaviour, showing a clear overestimation. On the other hand, NO_2 predictions versus measurements were significantly improved with finer resolutions.

Regarding PM, models generally showed a clear underestimation of long-term mean concentrations and some difficulties in reproducing the seasonal variations of the measurements.

9   CONCLUSIONS

In collaboration with EMEP, IIASA and EUROTRAC, the JRC-IES has launched a contribution to the CAFE programme (DG. ENV.) in the area of air quality modelling. The objective is to explore the present and future (2010) urban air quality predicted by different atmospheric chemistry-transport dispersion models. Both long-term exposure to O_3, fine and coarse PM fractions are investigated in this exercise.

A methodology has been set up to allow all participants to visualise and analyse not only their own results but also all available model results. The graphical interpretation tool now incorporates results from 6 cities, 15 models for different emission reduction scenarios and allows the interpretation of differences (deltas) among model results in terms of spatial resolution, model formulation, city, emission reductions, …

The fourth CityDelta worskhop in Valencia concluded in a lack of PM results and future efforts will focus in increasing the available results by selecting specific cities, models and scenarios to provide sufficient data for a meaningful interpretation of the results. The final CityDelta workshop is planned in fall 2003.

10   REFERENCES

1 Co-operative programme for monitoring and evaluation of the long-range transmission of air pollutants in Europe (EMEP). http://www.emep.int
2 International Institute for Applied Systems Analysis (IIASA). http://www.iiasa.ac.at
3 The EUREKA project on the transport and chemical transformation of trace constituents in the troposphere over Europe (EUROTRAC). http://www.gsf.de/eurotrac
4 EMEP Eulerian Photochemical model:
5 ALADIN homepage: http://www.cnrm.meteo.fr/aladin/index_common.html
6 EMEP-TNO. Co-operative programme for monitoring and evaluation of the long-range transmission of air pollutants in Europe (EMEP). Technisch Natuurwetenschappelijk Onderzoek (TNO)
7 Passant, N. R., 2002, Speciation of UK emissions of non-methane volatile organic compounds, AEAT/ENV/R/0545 Note.
8 CityDelta homepage: http://rea.ei.jrc.it/netshare/thunis/citydelta

A NESTED HEMISPHERIC MODEL FOR SIMULATIONS OF ATMOSPHERIC CO$_2$

Camilla Geels, Jesper H. Christensen, Jørgen Brandt, Lise M. Frohn and Kaj M. Hansen[*]

1. INTRODUCTION

The atmospheric distribution of CO$_2$ and the related sink and source processes, over the oceans as well as over the continents are still not well understood (Falkowski et al., 2000). The main focus within carbon cycle research has moved from the global scale down to more regional scales, in order to obtain both a better scientific understanding of the important processes and in order to meet the requirements by the international policy community (Tans and Wallace, 1999). Mesoscale transport models with a high spatiotemporal resolution are important tools for this purpose.

The Danish Eulerian Hemispheric Model (DEHM) has through new developments at the National Environmental Research Institute been extended to include multiple nests. Thereby high-resolution operations related to regional scale air pollution phenomena are made possible.

Currently, several versions of the DEHM model are developed with the following purposes in mind:

- Transport, transformation and deposition of reactive and elemental mercury (Christensen et al., 2002).
- Concentrations and depositions of various pollutants (Frohn et al., 2002b) through the inclusion of an extensive chemistry scheme.
- Transport and exchange of atmospheric CO$_2$ (Geels et al., 2001, 2002; Geels, 2002).
- Transport and exchange of persistent organic pollutants in the atmosphere.

In this paper the focus is on the CO$_2$ model with one nest, which is applied in order to study the spatial and temporal variations of atmospheric CO$_2$ over Europe. The observed variability in the CO$_2$ concentration is driven by a complex interplay between surface exchanges, atmospheric long-range as well as small-scale transport and mixing. The nested CO$_2$ version of the DEHM model has accordingly given a unique possibility for a

---

[*] National Environmental Research Institute, Department of Atmospheric Environment, Frederiksborgvej 399, P. O. Box 358, DK-4000, Roskilde, Denmark.

Air Pollution Modeling and Its Application XVI, Edited by
Borrego and Incecik, Kluwer Academic/Plenum Publishers, New York, 2004

detailed study of CO_2 and, for example, of the relative importance of the various drivers and sink/source types and areas. This study is a step towards a better understanding of the regional source pattern over Europe. A general description of the CO_2 version of the DEHM model system and the included sink and source parameterisations is given in the following. Thereafter a few examples from the model evaluation and simulations are discussed.

2. THE MODEL SYSTEM

The original version of the DEHM model was developed for studying the long-range transport of pollution (mainly SO_2 and SO_4) to the Arctic (Christensen, 1997). Therefore, the model domain is centred at the North Pole and covers the majority of the Northern Hemisphere with a horizontal resolution of 150 km at 60°N. Recently anthropogenic and natural surface fluxes for CO_2 have been implemented in the model and the evaluation against observations shows good performance. This version of the model has for example been used to study the transport and exchange of CO_2 within the North East Atlantic region (Geels et al., 2001). The model has been extended to include a nested domain over Europe with a horizontal resolution of 50 km at 60°N. Figure 1 (left) shows the hemispheric mother domain as well as the European sub-domain. The two domains are defined on a 96 × 96 grid and are coupled by a two-way nesting method. The input to the nested domain is introduced via the lateral boundaries, while the feedback to the mother domain occurs over the nest interior. The model is divided into 20 unevenly distributed vertical layers defined on terrain following σ-levels. The finest resolution is within the lowest few kilometres and the last level is approximately 16 km above the surface. The applied numerical schemes have all been carefully tested for the hemispheric model (Christensen, 1997 and references herein) and for the nested model version (Frohn et al., 2002a).

A schematic diagram of the nested model system with the data flow from the meteorological driver and from the three source-functions for atmospheric CO_2 is also given in Figure 1 (right). In order to obtain the required high-resolution meteorological data field the MM5 modelling subsystem (Grell et al., 1995), has been implemented as a meteorological driver between the data from ECMWF (2.5° × 2.5°) and the DEHM model. In the current study the domains and resolutions are the same in MM5 and DEHM and the meteorological data are archived at a 3 h interval.

2.1. Parameterisations of CO_2 fluxes

The net surface exchange of CO_2 is included as three source-functions representing the main natural and anthropogenic sources/sinks for atmospheric CO_2.

First, CO_2 emissions due to combustion of fossil fuels are based on a 1°×1° inventory for the year 1990 (Olivier et al., 1996). In an attempt to update the emission map, information on national emission levels (Marland et al., 2001) for the period 1990-1998 is applied. Furthermore, a weak seasonal cycle with higher emissions during wintertime north of 30°N is assumed (Rotty, 1987). Secondly, the exchange with the oceans is based on an air-sea gas exchange parameterisation (Wanninkhof, 1992) and a surface pCO_2 climatology (Takahashi et al., 1999). Finally, biospheric surface fluxes of CO_2 are based

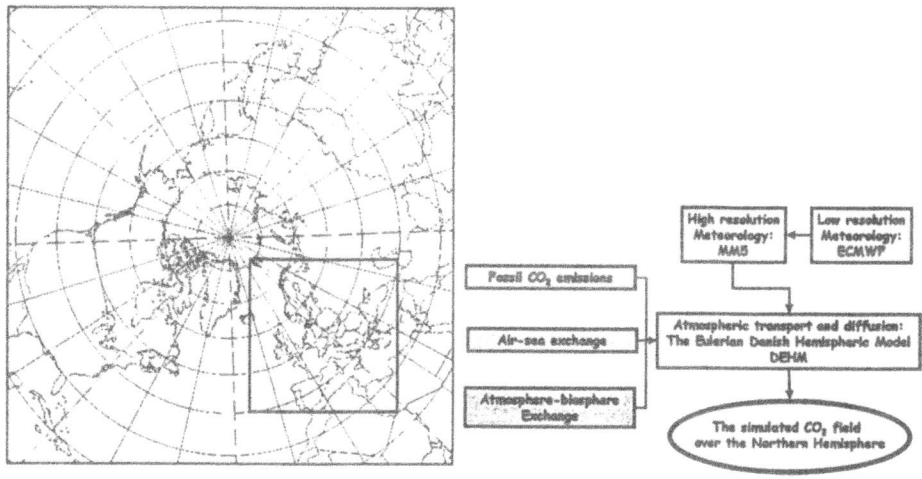

Figure 1. *Left:* the full hemispheric mother domain and the nested domain over Europe. *Right:* A schematic view of the CO_2 version of the DEHM system.

on a combination of daily fluxes estimated by the TURC model (Lafont et al., 2002) and a smooth seasonal cycle from the SDBM model (Knorr and Heimann, 1995). Both models are diagnostic type models forced by satellite data and information on various meteorological parameters. The diurnal cycle of the exchange is computed by scaling daily values of gross primary production proportional to radiation, while respiration is distributed with the diurnal temperature cycle. The resulting hourly flux fields have a horizontal resolution of $1° \times 1°$ and cover the year 1998.

Model results for 1998 will be discussed in section 3. A global background concentration of 364 ppm is applied as initial value in the model for this year. Likewise, the concentration at the lateral and upper boundaries of the model domain is assumed constant and equal to this value.

3. MODEL EVALUATION

The overall variability of atmospheric CO_2 in the Northern Hemisphere is dominated by a strong seasonal cycle controlled by the exchange with the terrestrial biosphere (e.g. Heimann et al., 1998). Therefore, accurate parameterisations of this source-function is crucial for valid simulations of the CO_2 variability and various parameterisations of complexity has been tested (Geels, 2002). The model has been run for different periods within the time interval 1991 - 1999 where CO_2 observations from several monitoring sites are available. The monitoring network for CO_2 consists of two types of observations. One type is the so-called flask observations where air is sampled in containers, typically with a weekly frequency. Traditionally these sites have been placed at remote locations and on mountain tops. Therefore, the observations are representative of background values of CO_2. Recently the monitoring network for continuous sampling has grown and includes today several sites within the continents. Thereby, the resulting time series will

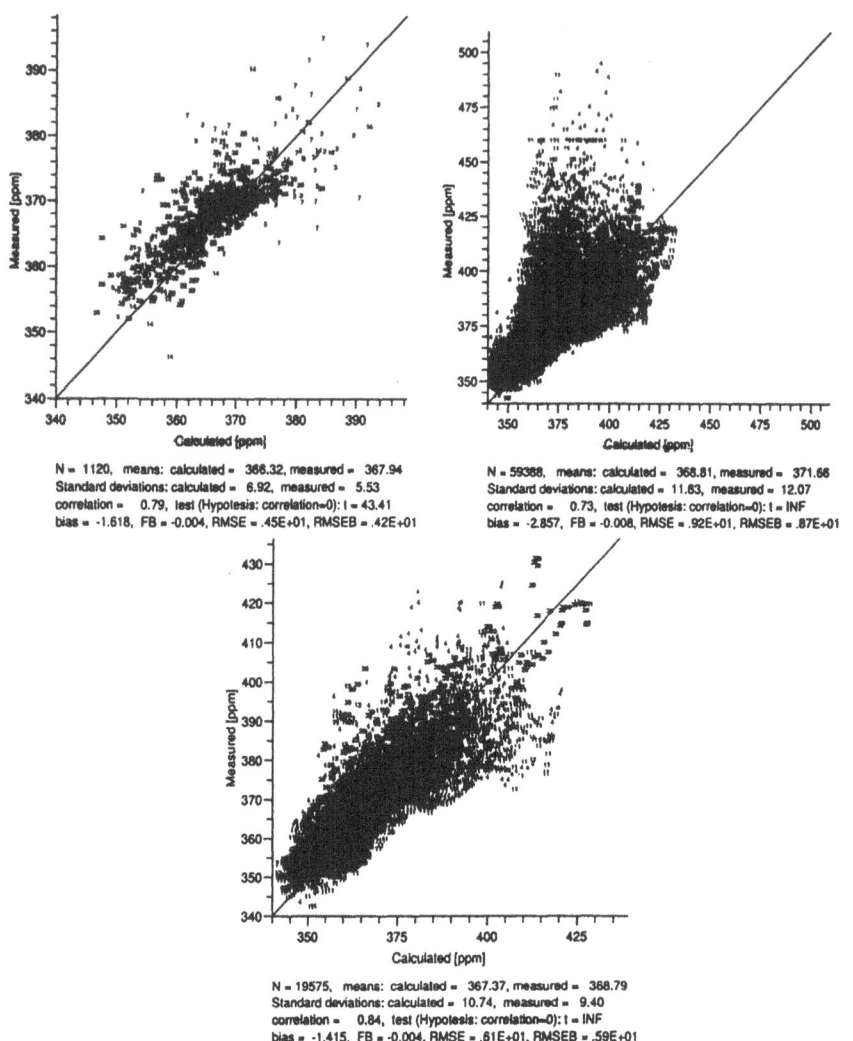

Figure 2. *Upper left:* comparison of weekly flask observations and simulations at 27 sites within the hemispheric model domain. *Upper right:* comparison of hourly mean observations and simulations at nine continuous sites within Europe. *Lower panel:* Comparison of the daytime selected (10:00 – 17:00 local time) observations and simulations at the nine European sites.

reflect the variability on annual and down to diurnal time scales and will be influenced by local source/sink fields for CO_2.

Comparing the various model runs to observations yields that the observed seasonal cycle with amplitudes ranging 10 – 15 ppm is generally captured, especially when the diagnostic biospheric models constrained by satellite data are applied. For the current model set up examples of the comparison between model results and observations are given in Figure 2. Various statistical parameters are given below the scatter plots. These includes the number of data samples, (N), mean values, standard deviation, correlation coefficients with test for significance, bias, fractional bias (FB), root mean square error (RMSE) and RMSE based on values where the bias has been subtracted (RMSEB). It should be noted that some of these parameters will be influenced by the chosen background mean (here 364 ppm), nevertheless they are helpful for comparing different model results. In 1998 more than 1100 flask observations are available from the 27 flask sites located in the Northern Hemisphere. Comparing the observations and model simulations for these locations results in a correlation coefficient of 0.79 (Figure 2, left). Hence, the overall seasonal variability is captured at these background locations.

Focusing on the nested domain the hourly mean model results are in Figure 2 (right) compared to hourly mean observations from nine European locations where continuous observations are obtained. On an hourly basis the more than 59,000 data point results in a somewhat lower correlation coefficient of 0.73 and as expected both observations and model results display more variability than at the background flask sites. The model underestimates especially the very high concentrations observed at some of these locations. Due to the vicinity of local source areas, variations in planetary boundary layer (PBL) depth etc., CO_2 observations from a given location can be more or less spatial representative. This is important to bear in mind, mainly when observations from low-elevation continental sites are compared to model simulations. Haszpra (1999) reports for two locations in Hungary that only data obtained during the early afternoon hours can be considered as regionally representative for the CO_2 concentration. In order to exclude the non-representative values, observations and model results from the daytime hours (here defined as 10:00 – 17:00 local time) are compared in Figure 2 (lowest plot). The correlation coefficient is seen to increase notably and also the other statistical parameters indicate that the agreement between the observed and simulated variability is higher during the daytime. The main reason for the underestimation of the highest night-time CO_2 values is the MM5 model's ability to resolve the PBL depth with the applied resolution. Generally the interplay between the mixing within the PBL and the direction of the biospheric flux (uptake/release) also known as the rectifier effect, is important to resolve in model studies like the present (Denning et al., 1996).

4. EXAMPELS OF TIME SERIES

In order to study the model performance in more detail two examples of monthly time series are discussed in this section. Tracer transport can be regarded as a linear problem and it is therefore possible to decompose the source field for CO_2. The model system has thereby also been run with the various source-functions separately and these model runs can be used to get a better understanding of the relative importance of the various source types. In Figure 3 the observed hourly time series for June and December (1998) are compared to the model simulations. The data are for a continental low-altitude European site, K-Puszta, in Hungary and a local mean has been subtracted in order to ease the comparison. The model run including all three source-functions as well as the separated components are given in Figure 3.

During a typical summer month the CO_2 field is seen to be modulated by a strong diurnal cycle with a maximum during night and a minimum in the early afternoon. The model is as expected from the statistical analysis seen to capture this variability both with regards to size and timing. An exception is again the very high values observed during night at a few occasions.

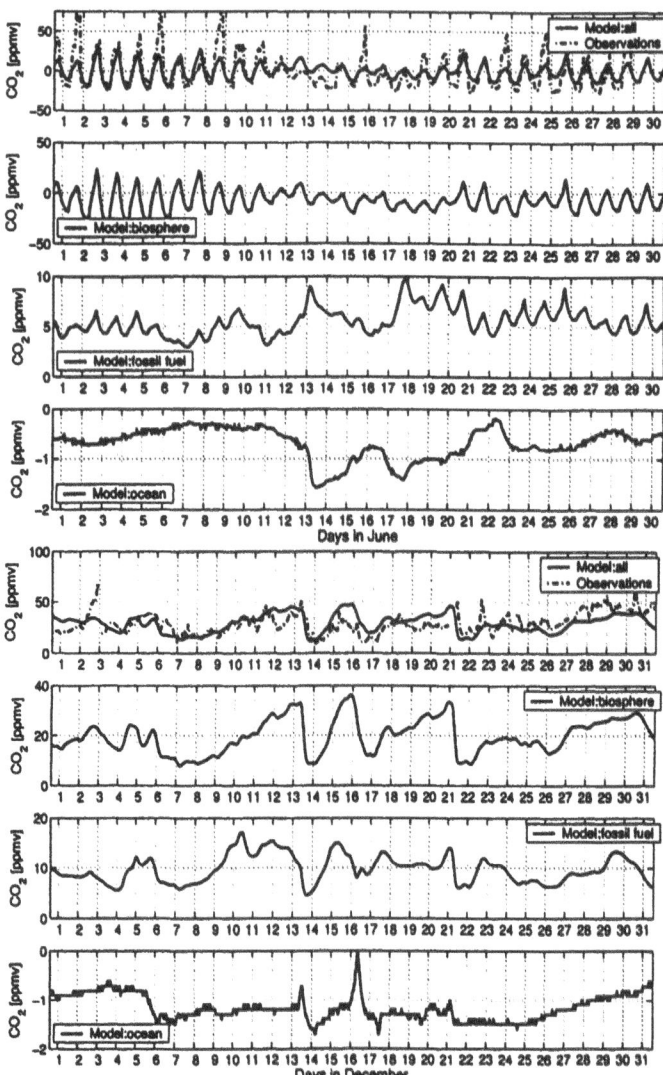

Figure 3. Hourly observed and simulated time series for atmospheric CO_2 at the Hungarian site K-Puszta. For the month of June (top) and December (below), 1998.

The separated model simulation show that the signal is dominated by the biospheric component with the respiratory/photosynthesis induced positive/negative CO_2 values. Over the continents also the diurnal variation in the boundary layer height is important for the surface concentration of CO_2 and the diurnal cycle in the fossil fuel component is caused by this variability in the mixing processes. The regular pattern is in all the time series seen to be damped for a period of several days in the middle of the month. An analysis of the meteorological conditions shows that a frontal system develops and passes the region during these days leading to e.g. an increased cloud cover and hence a decrease in the availability of incoming radiation. The latter will lead to a similar decrease in the photosynthetic activity and thereby to a damping of the biospheric component. In the same period a few events are seen in both the fossil fuel and oceanic components, which is caused by the advection of air from the North Atlantic and the western part of Europe.

During the month of December the model is again seen to capture the overall tendencies and the main events observed. The relative importance of the components has changed and the contributions from the biospheric and fossil fuel field are nearly equal and in phase during this winter month. From the time series it is evident that several of the events are caused by a gradual accumulation of biospheric and fossil CO_2 within shallow boundary layers that typically form during stable periods in this region. When the stable weather conditions break up variations in the larger scale concentration gradient and the vigorous wintertime atmospheric conditions lead to transport from e.g. the North Atlantic. During such episodes the oceanic component is seen to lower the CO_2 level by approximately 1.5 ppm.

5. CONCLUSIONS AND FUTURE WORK

A nested hemispheric model for simulations of atmospheric CO_2 has been developed. Here the model simulations for one year have been described and validated against observations. By comparing the model results to both weekly background observations as well as continuous observations from within Europe it has been shown that the model captures the overall variations on seasonal and down to diurnal time scales. The model can include the three main source-functions for atmospheric CO_2 separately. Thereby it can be used to analyse the relative importance of the various source types. Here the results for a summer and a winter month at a location in Central Europe have been presented. These examples illustrate how the interplay between the meteorological forcing and the flux fields can lead to day-to-day variations in the surface concentrations. Both parts should therefore be properly resolved when modelling CO_2 in continental regions.

The nested DEHM system is also part of the European AEROCARB project, where inverse modelling exercises currently are under development. The main objective is to quantify the European carbon fluxes more comprehensively than previous studies, which is done by combining high-resolution transport models and continuous observations. However, in order to do so it is important to understand the processes connected to the observed concentration variability. The representativenes error related to low-altitude continuous observations also needs to be quantified by for example detailed modelling studies like the present.

6. ACKNOWLEDGMENTS

The model is partly developed and funded within the framework of the EU cluster project CARBOEUROPE, sub-project AEROCARB. This project is also acknowledged for making the TURC and SDBM model data available. We thank the NOAA/CMDL Carbon Cycle Group (USA) and Global Atmospheric Watch for providing the various CO_2 data.

7. REFERENCES

Christensen, J. H., 1997, The Danish Eulerian Hemispheric Model - a three-dimensional air pollution model used for the Arctic, *Atmos. Environ.*, 31 (24), pp. 4169-4191.

Christensen, J. H., Brandt, J., Frohn, L. M., and Skov, H., Modelling of mercury with the Danish Eulerian Hemispheric Model. *Atmos. Chem. Phys.* Submitted December 19, 2002.

Denning, A. S., Randall, D. A., Collatz, G. J., and Sellers, P. J., 1996, Simulations of terrestrial carbon metabolism and atmospheric CO_2 in a general circulation model, Part 2: Simulated CO_2 concentrations, *Tellus*, **48B**, pp. 543-567.

Falkowski, P., Scholes, R. J., Boyle, E., Canadell, J., Canfield, D., Elser, J., Gruber, N., Hibbard, K., Högberg, P., Linder, S., Mackenzie, F. T., Moore, B., Pedersen, T., Rosenthal, Y., Seitzinger, S., Smetacek, V., and Steffen, W., 2000, The global carbon cycle: a test of our knowledge of Earth as a system, *Science*, **290**, pp. 292-296.

Frohn, L. M., Christensen, J. H., and Brandt, J., 2002a, Development of a high resolution nested air pollution model – the nummerical approach, *J. Comp. Phys.*, **179** (1), pp. 68-94.

Frohn, L. M., Christensen, J. H., and Brandt, J., 2002b, Development and testing of numerical methods for two-way nested air pollution modelling, *Phys. Chem. Earth*, **27** (35), pp. 1487-1494.

Geels, C., Christensen, J. H., Hansen, A. W., Kiilsholm, S., Larsen, N. W., Larsen, S. E., Pedersen, T., and Sørensen, L. L., 2001, Modelling concentrations and fluxes of atmospheric CO_2 in the North East Atlantic region, *Phys. Chem. Earth*, **106** (10), pp. 763-768.

Geels, C., Christensen, J. H., Frohn, L. M., and Brandt, J., 2002, Simulating spatiotemporal variations of atmospheric CO_2 using a nested hemispheric model, *Phys. Chem. Earth*, **27** (35), pp. 1495-1505.

Geels, C., 2002, Simulating the current CO_2 content of the atmosphere: Including surface fluxes and transport across the Northern Hemisphere, PhD. Thesis, National Environmental Research Institute, Frederiksborgvej 399, P. O. Box 358, DK-4000 Roskilde, Denmark, pp. 236.

Grell, G. A., Dudhia, J., and Stauffer, D. R., 1995, A description of the Fifth-Generation Penn State/NCAR Mesoscale Model (MM5), NCAR/TN-398+STR, NCAR Technical Note, June 1995, p. 122, Mesoscale and Microscale Meteorology Division, National Center for Atmospheric Research, Boulder, Colorado.

Haszpra, L., 1999, On the representativeness of carbon dioxide measurements, *J. Geophys. Res.*, **104** (D21), pp. 26953-26960.

Heimann, M., Esser, G., Haxeltine, A., Kaduk, J, Kicklighter, D. W., Knorr, W., Kohlmaier, G. H., McGuire, A. D., Melillo, J., Otto, R. D., Prentice, I. C., Sauf, W., Schloss, A., Sitch, S., Wittenberg, U., and Würth, G., 1998, Evaluation of terrestrial carbon cycle models through simulations of the seasonal cycle of atmospheric CO_2: First results of a model intercomparison study, *Global Biogeochem. Cycles*, **12** (1), pp. 1-24.

Knorr, W., and Heimann, M., 1995, Impact of drought stress and other factors on seasonal land biosphere CO_2 exchange studied through an atmospheric tracer transport model, *Tellus*, **47B**, pp. 471-489.

Lafont, S., Kergoat, L., Dedieu, G., Chevillard, A., Karstens, U., and Kolle, O., 2002, Spatial and temporal variability of land CO_2 fluxes estimated with remote sensing and analysis data over western Eurasia, *Tellus*, **54B**, pp 820-833.

Marland, G., Boden, T. A., and Andres, R. J., 2001, Global, Regional, and National fossil fuel CO_2 emissions, In: *Trends, A compendium of data on global change*, Carbon Dioxide Information Analysis Center, Oak Ridge National Laboratory, U.S. Department of Energy, Oak Ridge, Tenn. U.S.A.

Olivier, J. G. J., Bouwman, A. F., Van der Maas, C. W. M., Berdowski, J. J. M., Veldt, C., Bloos, J. P. J., Visschedijk, A. J. H., Zandveld, P. Y. J., and Haverlag, J. L., 1996, Description of EDGAR Version 2.0: A set of global emission inventories of greenhouse gases and ozone-depleting substances for all anthropogenic and most natural sources on a per country basis and on 1°×1° grid. National Institute of Public Health and the Environment (RIVM) report no. 771060 002 / TNO-MEP report No. R96/119.

Rotty, R. M., 1987, Estimates of seasonal variation in fossil fuel CO_2 emissions, *Tellus*, **39B**, pp. 184-202.

Takahashi, T., Wanninkhof, R. H., Feely, R. A., Weiss, R. F., Chipman, D. W., Bates, N., Olafsson, J., Sabine, C., and Sutherland, S. C., 1999, Net sea-air CO_2 flux over the global oceans: an improved estimate based on the sea-air CO_2 difference, Proceedings of the 2nd International Symposium, CO_2 in the Oceans, Center for Global Environmental Research, National Institute for Environmental Studies, pp. 9-14.

Tans, P. P., and Wallace, W. R., 1999, Carbon cycle research after Kyoto, *Tellus*, **51B**, pp. 562-571.

Wanninkhof, R., 1992, Relationships between wind speed and gas exchange over the ocean, *J. Geophys. Res.*, **97**, 5, pp. 7373-7382.

DISCUSSION

D.STEYN   I presume the negative concentrations are deviations around a background mean value? If this is so, did you use the observed or modelled mean value? If both, does this not affect the comparability of observed and modelled values?

C.GEELS   In the time series plots an observed/modelled local background mean has been subtracted from the observed/modelled record in order to ease the comparison. We do not see this as a problem for the following reason:

As DEHM is a regional model, it is necessary to prescribe the concentrations at the lateral/upper boundaries. In this experiment, the lateral boundaries are fixed at a global mean CO_2 value (same as for the initial conditions) throughout the simulations for 1998. The overall concentration level will hence be controlled by this boundary value. The European regional is, however, far from the boundaries and sensitivity studies show that the effect of the boundary condition is negligible on the synoptic variability analysed in the present study. On shorter time scales is therefore reasonably to assume that comparability of observed and modelled variability are unaffected by the subtraction of the specific mean.

THE ROLE OF CUMULUS PARAMETERISATION IN GLOBAL AND REGIONAL SULPHUR TRANSPORT

Trond Iversen and Øyvind Seland[*]

1. INTRODUCTION

In connection with regional acidification studies and investigation of Arctic haze and hemispheric-scale transport, limited-area chemistry-transport models (CTMs) on limited domains horizontally as well as vertically have been used. More recently, global models are being used for oxidized sulphur components for the purpose of calculating of possible impacts of sulphate particles on climate. To some surprise sulphur modeling has proven more difficult when integrated in global circulation models (GCMs) than experience from limited-area models, and to some extent by global CTMs. In particular the vertical distribution appears to be wrong. This also shows up as a considerable mismatch between ground-level measurements and calculations in source-regions. The inter-comparison exercise by Barrie et al. (2001) emphasized this problem, and it was confirmed in experiments with a new scheme by Iversen and Seland (2002). That paper presented results from using an extended version of the NCAR CCM3 atmospheric GCM to calculate sulphate and black carbon (BC). Also the result of the NCAR-group's own sulphur model produced these biases (Rasch et al., 2000).

Sensitivity tests by Iversen and Seland (2002) revealed that too efficient vertical exchange in deep cumulus clouds in combination with underestimated scavenging by convective precipitation, can explain large portions of these error biases. In that paper some rather arbitrary and radical assumptions were made to avoid the biases. Here we test some more physically sound methods to account for processes involving deep convection. These methods still leave considerable errors, but there are other physico-chemical processes that partly may account for those. Examples include clouds in the boundary layer influencing the efficient oxidation rate and deposition of sulphate. Here we focus on the vertical transport and scavenging in connection with deep convection only, and we use the NCAR CCM3 model as before.

[*] Trond Iversen and Øyvind Seland, Department of Geophysics, University of Oslo, P.O.Box 1022, Blindern, N-0315 Oslo, Norway. (trond.iversen@geofysikk.uio.no, oyvind.seland@geofysikk.uio.no)

Air Pollution Modeling and Its Application XVI, Edited by
Borrego and Incecik, Kluwer Academic/Plenum Publishers, New York, 2004

2. CUMULUS PROCESSING OF CONTAMINANTS

Processes inside and nearby deep cumulus clouds that may influence aerosols are precipitation scavenging, vertical transport in updrafts and downdrafts, aqueous phase oxidation of precursors, and coagulation processes between sub-saturated particles and cloud droplets. The efficiency of the two latter processes depends on availability of oxidizing agents, the relative motion of droplets and interstitial particles, and the effective fraction of time air parcels are in cloudy versus clear air. We do not see an immediate way to improve the handling of these processes in present climate models. For the two first mentioned processes we have some thoughts on how at least parts of the weaknesses can be accounted for.

2.1 Scavenging by cumulus precipitation

Precipitation scavenging by cumulonimbi clouds is generally more vigorous than for stratiform precipitation. Over typical a model time-step (15-20 min.), lower level air parcels may be sucked into cumulonimbi and detrained in the upper free troposphere. In this process contaminants will be exposed to precipitation inside and below the cloud. In-cloud scavenging is the generally much more efficient for aerosol particles. In-cloud processes involve saturation and growth of much of the hygroscopic particles. In particular for cumulus clouds, the realized super-saturation can be sufficient to turn most hygroscopic particle mass into cloud droplets. These cloud droplets may quickly add to the precipitation release by efficient coalescence in updrafts (auto-conversion). Below-cloud scavenging of particles is entirely determined by impaction of falling precipitation, which normally is inefficient for unsaturated particles in the accumulation mode (Seinfeld and Pandis, 1998, pp. 1020-1026).

In a model grid column where conditions for triggering sub-grid cumulus parameterization are present, an ensemble of cumulus towers of varying depth is assumed. In the NCAR model a version of the scheme developed by Zhang and McFarlane (1995) is used. This scheme produces a vertical profile of precipitation intensity depending on the influxes of water vapor and the strength of the vertical mass-fluxes. The vertical mass fluxes are estimated from the convective available potential energy in the height range of conditional stability, which is the closure assumption for the parameterization.

For cumulonimbi we propose to apply in-cloud scavenging efficiencies in all layers below the level of maximum rate of precipitation release. Air in those layers is prone to become cloudy over a significant fraction of the model time-step, and during those time-slots to be exposed to strong turbulent mixing. In the original version of the model, below-cloud scavenging rates are used below the cloud-base and the scavenging efficiency in any layer is multiplied by the minor fraction of cumulus cloudiness.

2.2 Mixing of air between updrafts and downdrafts

The overall effect of cumulus parameterization on the resolved scale is an effective vertical redistribution of quantities over a model time-step. The redistribution is a consequence of the atmosphere being conditionally stable over a layer of air. It takes place in saturated updrafts, in downdrafts caused by falling precipitation, and in the slowly subsiding air between the clouds. Clear air entrains into the up- and downdrafts by mixing and sub-grid convergence, and detrains after dehydration. The scheme of Zhang and McFarlane (1995) assumes an ensemble of clouds-plumes that only detrains air at the level where rising air become negatively saturated buoyant. This may well be an adequate assumption for the water-budget, but it will probably underestimate mixing that exchange other airborne constituents. So far we have no remedy for this.

Possibly more seriously is the lack of exchange between rising air in updrafts and sinking air in downdrafts. These turbulent air masses exist side by side with strong and opposite vertical winds. We assume that the mixing between updrafts and downdrafts are sufficiently efficient so that the updraft contaminant flux is counteracted by the downdraft mass fluxes. This may probably be too efficient, but has the benefit that we do not need to specify an exchange coefficient and solve a diffusion equation.

Consider a model layer no. k, where k=1 for the uppermost and k=K for the lowermost layers. Convective mass fluxes (transport of mass through a horizontal grid square per time unit) are given at interfaces between the layers, and these interface levels are numbered k+1/2 (below) and k-1/2 (above). Mass fluxes are zero for ½ and K+1/2. In a grid-column conditioned for deep moist convection, the parameterization assumes an ensemble of Cu clouds with strong mass-fluxes in the saturated fraction and a slower and exactly compensating subsiding motion in the unsaturated air. Air from the saturated updrafts (u) and downdrafts (d) may detrain into the ambient unsaturated air in layer k with rates d^u_k and d^d_k respectively. Entrainment rates for ambient air that becomes saturated are e^u_k and e^d_k. If $q^u_{k-1/2}$, $q^d_{k-1/2}$, and q_k are mixing ratios of a contaminant in the updraft, the downdraft and in the ambient air at the level given with the index, we can write the contaminant budget equation for layer k as follows:

$$(q_t)_k = (F^u_p + F^d_p + F^a_p)_k \quad ; \quad k = 1,...,K$$

where q is the mass mixing ratio for a contaminant, index t and p means derivative w.r.t. time and pressure, F are vertical contaminant fluxes due to the deep convective motion, and superscripts u, d and a signifies saturated updraft, saturated downdraft, and compensating vertical motion in ambient unsaturated air. Note that the fluxes are positive when directed upwards. Assuming that mass-fluxes satisfy a maximum Courant number of 1, mass consistency for positive definite quantities are secured (P. J. Rasch, NCAR, personal communication) by assuming zero fluxes at k=1/2 and k=K+1/2,

$$F^u_{k+1/2} = M^u_{k+1/2} q^u_{k+1/2} \quad \text{and} \quad F^d_{k+1/2} = M^d_{k+1/2} q^d_{k+1/2} \quad ; \quad k = 1,...,K-1$$

and for the compensating ambient-air currents for k=1,..., K-1:

$$F^a_{k+1/2} = -M^u_{k+1/2} \min\{q_{k+1/2}, q_k\} + M^d_{k+1/2} \min\{q_{k+1/2}, q_{k+1}\}$$

Contaminant mixing ratios for k=1,...,K in updrafts are determined by closing the flux budget over layer no. k of pressure thickness Δp (flux out + detrainment = flux in + entrainment):

$$\left[M^u_{k-1/2} + d^u_k \Delta p + a M^d_{k-1/2} \right] q^u_{k-1/2} = \left[M^u_{k+1/2} q^u_{k+1/2} + e^u_k \varphi_k q_k \Delta p \right]$$

and in downdrafts:

$$M^d_{k+1/2} q^d_{k+1/2} = \left[M^d_{k-1/2} q^d_{k-1/2} - e^d_k \varphi_k q_k \Delta p \right]$$

Here we have defined a tracer (a) which is zero in the original model version and 1 in the version which assumes mixing between updrafts and downdrafts.

2.3 Other model adjustments

The basic model version we run is thoroughly described in Iversen and Seland (2002), but with a few adjustments based on results in that paper. Here we only summarize the changes. We no longer keep separate track of accumulation mode SO_4 produced by coagulation of nucleation mode particles in dry air. This minor fraction of SO_4 particles is merged with the fraction produced by condensation onto pre-existing accumulation mode particles and is denoted $SO_4(a)$ (formerly a1 and a2), whilst the portion produced by oxidation in cloud droplets is denoted $SO_4(aw)$ (formerly a3). The prognostic components calculated by the model are then: DMS (di-methyl-sulphide); SO_2; externally mixed nucleation/Aitken-mode sulphate ($SO_4(n)$) produced in gas phase; internally mixed sulphate produced in gas phase condensed on background accumulation-mode particles or coagulation of nucleation-mode sulphate on the same ($SO_4(a)$); internally mixed sulphate produced by aqueous phase oxidation ($SO_4(aw)$); externally mixed nucleation/Aitken mode black carbon ($C(n)$); externally mixed and fractal black carbon ($Cx(a)$); internally mixed black carbon which is either emitted in that phase or produced by coagulation of the externally mixed black carbon with background particles ($C(a)$).

The below-cloud scavenging efficiencies given in Iversen and Seland (2002) are divided by 10 for accumulation mode particles in closer agreement with Seinfeld and Pandis (1998, pp. 1020-1026): $E_{bc}=0.01$ for $SO_4(aw)$ and 0.02 for $SO_4(a)$, $C(a)$, and $Cx(a)$. For cumulonimbi, the in-cloud scavenging efficiency for $SO_4(aw)$ is increased to 1 (from 0.8 in all clouds). Finally, black carbon (BC) emitted from biomass burning is distributed vertically over the three lowermost model-layers in the same way as sulphur from fossil fuel combustion. This is a consequence of deep vertical exchange in rising hot plumes from forest fires (wild or initiated by man).

EXPERIMENTS

We have run a set of 4 experiments to demonstrate different aspects of the properties of deep convection on the global distribution of sulphate and black carbon. **E1**

228

is the new basic run equivalent to the basic run in Iversen and Seland (2002), except that some parameter values are adjusted as mentioned in the previous paragraph. There is no vertical transport of the contaminants by the cumulus mass fluxes in this test. In E2, vertical transport by the mass fluxes is included with a=0, i.e. the original scheme. In E3 the wet scavenging of cumulonimbus precipitation is enhanced by assuming that the efficient scavenging ratios for soluble gases and particles below the level of maximum rate of convective precipitation generation are those valid for in-cloud conditions. The rational is that air in those levels under influence of deep convection, is saturated over a considerable fraction of the model time-step. In the final experiment, E4, an efficient exchange of air between updrafts and downdrafts is assumed, and in the formula for q^u we assume a=1, which reduces the updraft transport of contaminants by the downdraft mass fluxes.

Emissions are taken from IPCC (provided by Dr. J. Penner) and are supposed to be valid for the year 2000. Each experiments is run for 5 years repeating the annual cycle for sea surface temperature. Only results for the three latter years are used for evaluation.

SELECTED RESULTS AND DISCUSSION

We focus on the results for sulphur but also present some results for black carbon. The space only allows selected results. Table 1 showes global budgets for sulphate and BC calculated for the four experiments and compared to results from a selection of other models. The burden and turnover time for SO_2 are quite insensitive to the different ways of treating processes in convective clouds. The reason is that deposition competes with oxidation in determining the fate of SO_2. The numbers for sulphate is, on the other hand, very sensitive to the treatment of convective cloud processes. The treatment of vertical transport in deep cumulus clouds and the way wet scavenging by convective precipitation is parameterized, have decisive consequences for the abundance of sulphate. This is also evident by comparing the results from the two papers of Chin and co-workers. The 1996-paper has a statistical treatment of the convective cloud processes, whilst that from 2000 is more direct, as in Rasch et al (2000) and Koch et al (1999). There are other differences, but the influence on the burden and turnover of sulphate is striking.

In Table 2 similar budget numbers are given for black carbon. These budgets split

Table 1. Global budget parameters for the production of airborne particulate sulphate.

	SOx source (TgS/a)	SO2 dep. (%)	SO4-prod Aq. (%)	SO4-prod Gas. (%)	SO2 burden (TsS)	SO2 T (days)	SO4 source (TgS/a)	SO4 wetdep. (%)	SO4 burden (TgS)	SO4 T (days)
E1 = no conv. transp.	90.4	43.5	40.9	14.1	0.40	1.6	52.7	79	0.60	4.1
E2 = full conv. transp.	90.4	35.7	45.2	17.3	0.52	2.1	60.0	86	2.40	14.6
E3 = E2 +incr. wet dep.	90.4	43.7	41.4	13.5	0.42	1.7	52.7	93	0.63	4.4
E4 = E3 +decr. conv. transp.	90.4	45.5	40.6	12.4	0.39	1.6	51.0	92	0.44	3.1
Rasch et al.(2000)	81	32[a]	55[a]	12	0.4	1.9	55[a]	93[a]	0.60	4.0[a]
Koch et al. (1999)	82.3	43.4[a]	38.4[a]	15.9	0.56	2.6	46.6[a]	80.3[a]	0.73	5.7[a]
Chin et al.(2000)	92.5	56.6	26.5	15.1	0.43	1.8	40.7	85.3	0.63	5.8
Chin et al.(1996)	95.6	48.6	43.5	7.8	0.34	1.3	49.1	89	0.53	3.9

Table 2. Global budget numbers for production and turnover of black carbon.

	Hydrophobic BC			Total BC		
	source (TgC/a)	burden (TgC)	T (days)	source (TgC/a)	burden (TgC)	T (days)
E1 = no conv. transp.	10.5	0.06	2.0	12.4	0.21	6.1
E2 = full conv. transp.	10.5	0.14	4.9	12.4	0.57	16.8
E3 = E2 +incr. wet dep.	10.5	0.09	3.1	12.4	0.22	6.6
E4 = E3 +decr. conv. transp.	10.5	0.08	2.7	12.4	0.18	5.3
Koch (2001) (case S)	12.4	0.04	1.1	12.4	0.15	4.4
Cooke *et al.*(1999) (fossil fuel)	-	-	-	5.1	0.07	5.3

T is turnover times.

between hydrophobic and hydrophilic BC. In our model, all BC produced by fossil fuel combustion and half of that produced by biomass burning is assumed emitted in hydrophobic mode. The turnover to hydrophilic BC is brought about by coagulation with hydrophilic particles. We have neglected the effect of condensation of hydrophilic material in this connection. The effect of deep convection is profound, and it is crucial to treat scavenging in a consistent way.

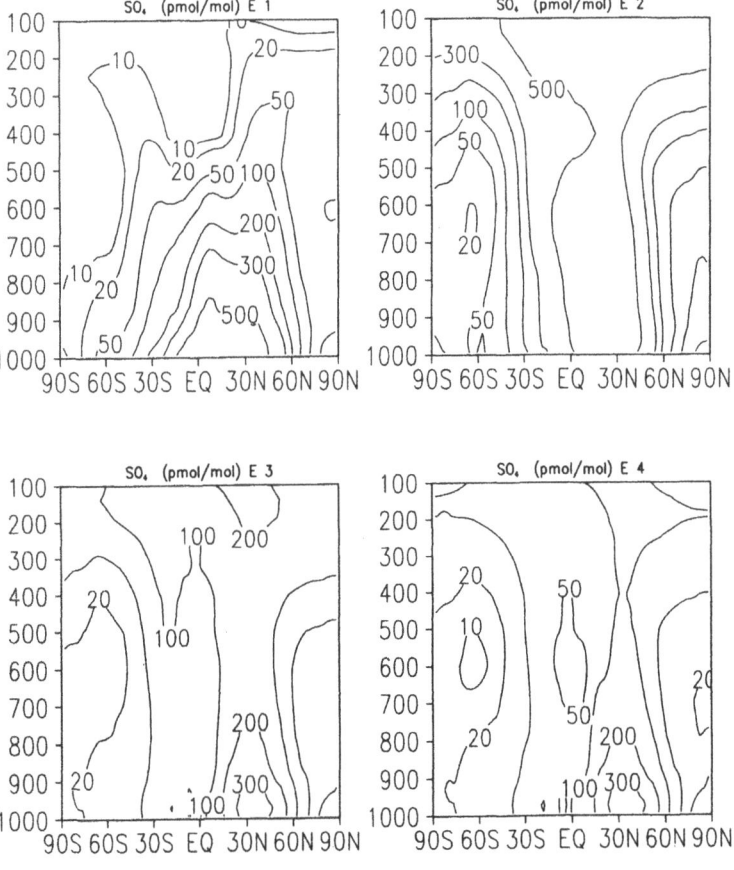

Figure 1. Calculated sonally averaged mixing ratios of airborne sulphate over three model years, for the four experiments (E1, E2, E3, and E4).

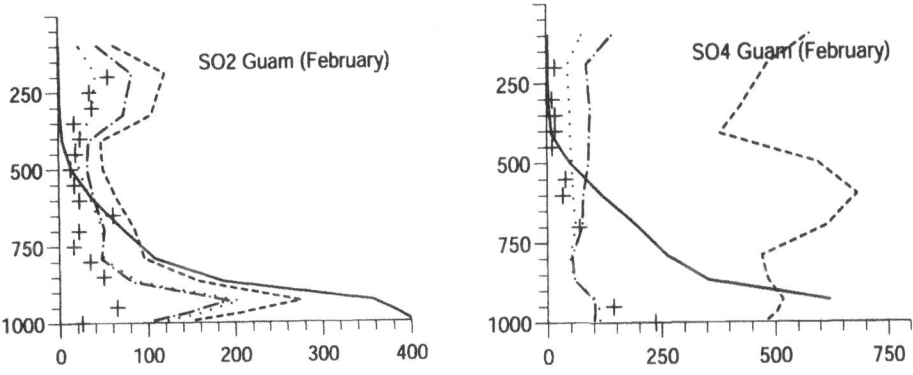

Figure 2. SO₂ (left) and Sulphate (right) measured in a flight campaign in February over Guam (crosses). data from Pacific Exploratory Mission (PEM) (Barth *et al.*, 2000). Calculations averaged over three February months are shown as lines: E1=solid; E2=dashed; E3=dashed-dotted; and E4=dotted line.

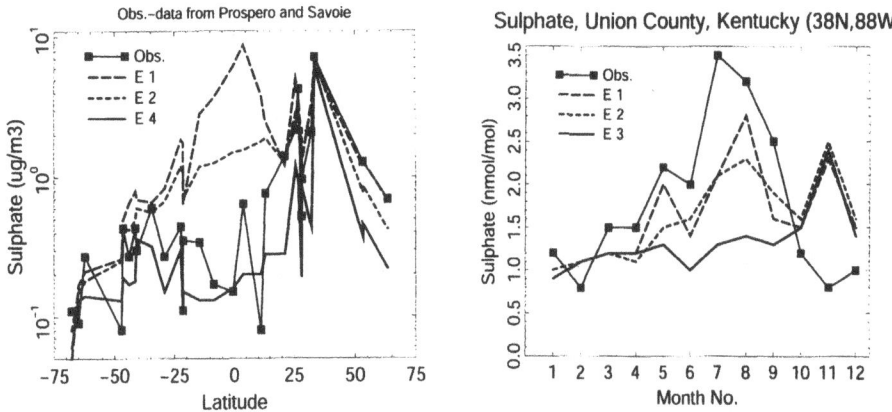

Figure 3. Ground-level concentrations calculated sulphate in the experiments E1, E2, and E4 and compared with measurements. Left: Annual sulphate at marine sites in the Atlantic Ocean; Middle: monthly sulphate at Union County, Kentucky, USA.

Figure 1 shows the profound effects of the modeled convective transport on the vertical sulphate distribution. Convective transport in plumes that only detrain at the level of negative buoyancy leads to higher concentrations in the upper troposphere than at the middle levels. Increased scavenging and exchange between updrafts and downdrafts improves the situation. The single comparison with aircraft data in Figure 2 confirms this.

Figure 3 shows that the exaggerated low level tropical concentrations are not fully improved by including convective transport. Increased scavenging is imperative. Results from the North American site indicate that scavenging during summer may then be too efficient at mid-latitudes. The treatment of the effect of convection in and around ITCZ should probably differ from that in local convective storms in the summer extra-tropics.

ACKNOWLEDGEMENTS

This work is part of the AerOzClim and RegClim projects financed by the Research Council of Norway. Computational costs are covered by a grant from the Research Council's Programme for Supercomputing. We are grateful to P. J. Rasch, J. E. Kristjansson, and A. Kirkevåg for co-operation and discussions around the use of the NCAR CCM3 model.

REFERENCES

Barth, M. C., Rasch, P. J., Kiehl, J. T., Benkowitz, C. M., and Schwartz, S. E. (2000) Sulfur chemistry in the National Center for Atmospheric Research Community Climate Model: Description, evaluation, features, and sensitivity to aqueous chemistry. *J. Geophys. Res.*, 105, 1387-1415.

Chin, M., Jacob, D. J., Gardner, G. M., Foreman-Fowler, M. S., and Spiro, P. A. (1996) A global three-dimensional model of tropospheric sulfate. *J Geophys. Res.* 101,18,667-18,690.

Chin, M., Rood, R. B., Lin, S.-J., Müller, J-F., and Thompson, A. M. (2000) Atmospheric sulfur cycle simulated in the global model GOCART: Model description and global properties. *J Geophys. Res.* 105, 24,671-24,687.

Cooke, W. F., Liousse, C., Cachier, H., and Feichter, J. (1999) Construction of a 1x1 fossil fuel emission data set for carbonaceous aerosols and implementation and radiative impact in the ECHAM4 model. *J Geophys. Res.* 104, 22,137-22,162.

Barrie, L.A., Yi, Y., Leaitch, W.R., Lohmann, U., Kasibhatla, P., Roelofs, G.-J., Wilson, J., McGovern, F., Benkovitz, C., Melieres, M.A., Law, K., Prospero, J., Kritz, M., Bergmann, D., Bridgeman, C., Chin, M., Christensen, J., Easter, D., Feichter, J., Land, C., Jeuken, A., Kjellstrom, E., Koch, D. and Rasch, P. (2001) A comparison of large-scale atmospheric sulphate aerosol models (COSAM): overview and highlights. *Tellus*, 53B, 615-645.

Iversen, T., and Seland, Ø., 2002: A scheme for process-tagged SO4 and BC aerosols in NCAR CCM3: Validation and sensitivity to cloud processes. *J. Geophys. Res.*, 107 (D24), 4751, 10.1029/2001JD000885.

Koch, D. (2001) Transport and direct radiative forcing of carbonaceous and sulphate aerosols in the GISS GCM. *J Geophys. Res.* 106, 20,311-20,332.

Koch, D., Jacob, D., Tegen, I., and Chin, D. (1999) Tropospheric sulfur simulation and sulfate direct radiative forcing in the Goddard Institute for Space Studies general circulation model. *J Geophys. Res.* 104, 23,799-23,822.

Rasch, P. J., Barth, M. C., Kiehl, J. T., Schwartz, S. E., and Benkovitz, C. M., (2000): A description of the global sulfur cycle and its controlling processes in the National Center for Atmospheric Research Community Climate Model, Version 3. *J. Geophys. Res.*, 105, 1367-1385.

Seinfeld, J.H. and Pandis, S.N. (1998) *Atmospheric Chemistry and Physics. From air pollution to climate change.* John Wiley and Sons, Inc., USA. 1326 pp.

Zhang, G.J. and McFarlane, N.A. (1995) Sensitivity of climate simulations to the parameterization of cumulus convection in the Canadian Climate Centre general circulation model. *Atmos. Ocean.*, **33**, 407-446.

DISCUSSION

R. SAN JOSE   How do you initialize the model? What period of time have you covered?

Ø. SELAND   The model is initialized using a given initial dataset from National Center for Atmospheric Research, NCAR. This datset is computed from a steady-state simulation using a fully coupled climate system model. The aerosols are initialized by using a spin-up time of two years before any aerosol statistics are calculated. The model is run in climate mode, thus the calulation is not done for any specific year. The model is however forced by climatological values for greenhouse gases and sea surface temperatures. The aerosol emissions are given by IPCC as valid for the year 2000, although the numbers are probaly more valid for the early 1990ies.

234

LONG-RANGE TRANSPORT OF A SEVERE ASIAN DUST (YELLOW SAND) OBSERVED IN KOREA DURING 21-23 MARCH IN 2002

Soon-Ung Park* and Hee-Jin In*

1. INTRODUCTION

Yellow Sand (Asian dust), which is a typical example of mineral aerosol frequently originates in the Sand desert, Gobi desert and Loess plateau in Northern China and Mongol during the spring season (In and Park, 2002). A severe dust storm was observed in Korea during 21-22 March in 2002. The maximum concentration of PM_{10} observed in the air pollution monitoring sites scattered over South Korea exceeded 1,000 $\mu g\ m^{-3}$ which is more than 10 times higher than that of the non-dust storm period.

In and Park (2002) have simulated a very long lasted Yellow Sand event that was observed over Korea for the period of 14-24 April in 1998 using a three-dimensional eulerian air quality model with the fifth-generation mesoscale model (MM5, Pennsylvania State University / National Center for Atmospheric Research) output of meteorological fields and the estimated threshold friction velocity based on the results of Westphal et al. (1987) and Gillette (1981). However, the simulated dust concentration was underestimated the observed total suspended particulate (TSP) concentration in South Korea. The purpose of this study is to improve the dust emission mechanism used in In and Park (2002) by using the statistically analyzed World Meteorological Organization (WMO) synoptic reporting data for seven years from 1996 to 2002 in the source regions and to simulate long-range transport of an intense Yellow Sand event observed during 21-22 March 2002 in Korea to understand important physical mechanisms that result in an extraordinarily intense Yellow Sand phenomenon in South Korea.

2. MODEL DISCRIPTION

The meteorological model used in this study is the operational meteorological model of the Regional Data Assimilation and Prediction System (RDAPS) in Korea Meteorological Administration (KMA). This model is developed on the basis of MM5 version 3 and defined in the x, y, and σ coordinate with a horizontal resolution of 30 km and 33 vertical layers up to the 50 hPa level.

The domain of the model is shown in Fig. 1. Four source regions of Yellow Sand: the Sand, Loess, Gobi (Chun, 1996; In and Park, 2002) and Mixed soil are indicated in Fig. 1.

---

* School of Earth and Environmental Sciences, Seoul National University, Seoul, 151-747 Korea

Air Pollution Modeling and Its Application XVI, Edited by
Borrego and Incecik, Kluwer Academic/Plenum Publishers, New York, 2004

The simulation covers the first dust-rise report in the source region on 19 March to the end of the Yellow Sand event observed in Korea on 23 March 2002.

The aerosol model is exactly the same as in In and Park (2002) that includes physical processes such as three-dimensional advection, diffusion, dry and wet depositions in the σ coordinate system (Westphal et al., 1987). In calculating dust concentration, particles of 0.2 – 74 μm in diameter are divided into 11 size bins with the same logarithm intervals.

To delineate dust-emission regions and occurrence frequencies of dust storms three hourly reporting present weather codes at 730 WMO synoptic stations are analyzed for the period of seven springs (March – May) from 1996 to 2002 and the results are given in Fig. 2. Dust storms and dust rises occur in the most parts of northern China and Mongol with the maximum occurrence frequencies in the central southern parts of Mongol, Badain Jaran and Tenggeri deserts in central northern China. Since the dust storms include not only the dust-rise but also the blowing dust, the soil type map in the analysis domain has been used to delineate the emission region of Yellow Sand. Fig. 1 also shows the spatial distribution of four different types of soil (Gobi, Sand, Loess and Mixed soil) in the dust emission regions.

The threshold wind speed is determined based on the 3 hourly observed wind speed, dust storm and dust rise reports, relative humidity, and precipitation at synoptic weather reporting stations (Fig. 2) located in the source regions of the Gobi, Sand, Loess and

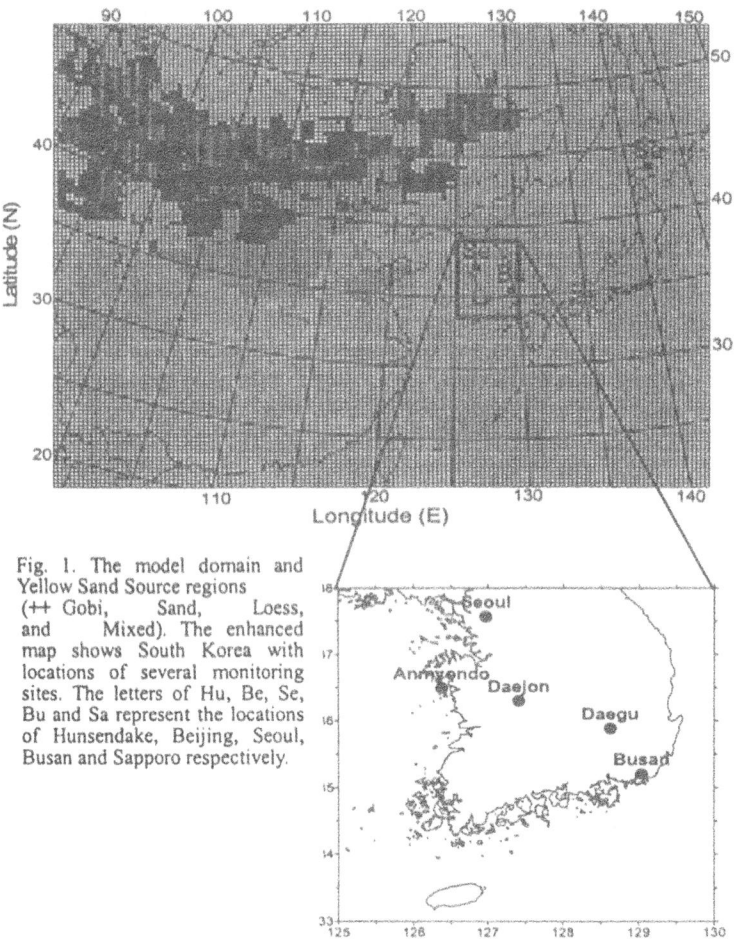

Fig. 1. The model domain and Yellow Sand Source regions (++ Gobi,   Sand,   Loess, and   Mixed). The enhanced map shows South Korea with locations of several monitoring sites. The letters of Hu, Be, Se, Bu and Sa represent the locations of Hunsendake, Beijing, Seoul, Busan and Sapporo respectively.

236

Mixed soil (Fig. 1). There are 162 WMO weather reporting stations in the source regions, among them 24 stations are in the Gobi region, 66 stations in the Sand region, 47 stations in the Loess region and 25 stations in the Mixed soil region. The data analysis indicates that there is no precipitation report at the station where dust storm or dust rise is reporting. The threshold wind speed is defined as the wind speed at the normalized cumulative ratio of the dust occurrence frequency being 3.5 % (Fig. 3a). On the other hand, the upper limit value of the relative humidity for the occurrence of the dust-rise is determined using the value of $e^{-1.5}$ of the exponential optimum regression equation of the dust occurrence percent versus relative humidity at each source region (Fig. 3b). These criteria are used as the conditions for the dust emission in the source regions (Table 1).

The dust flux F_a is estimated as

$$\frac{dF_a}{d \log r} \propto r^{1.5}$$

where r is the radius of a particle and the total emission amounts of uplifted dust whose radius in the range of 0.1 ~ 37 μm can be estimated as (e.g., Westphal et al, 1987)

$$F_a = 2.633 \times 10^{-13} u_*^4 \quad \text{if } u_* \geq u_{*t}$$
$$F_a = 0 \quad \text{if } u_* \langle u_{*t}$$

where u_{*t} is the threshold friction velocity. The estimated dust-emission flux is modified by the land-use types (Table 2) in such a way that

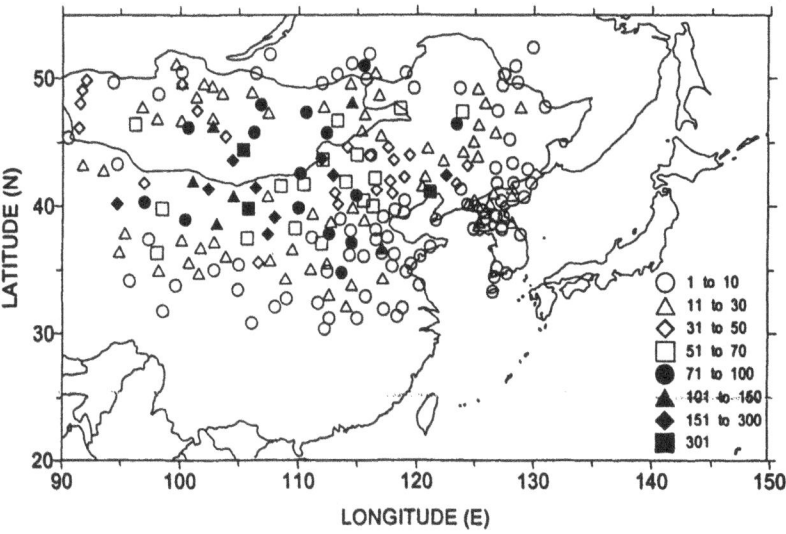

Fig. 2. The spatial distributions of occurrence frequencies of dust storms and dust rises obtained from 3-hourly WMO surface reporting data for 7 springs (March to May) from 1996 to 2002.

Table 1. Conditions for the dust-rise in the source region

Source region	Threshold wind speed (m s⁻¹)	Upper limit Relative Humidity (%)	Precipitation
Gobi	9.5	60	None
Sand	7.5	35	None
Loess	6.0	30	None
Mixed soil	9.2	45	None

237

(a) Wind speed   (b) Relative Humidity

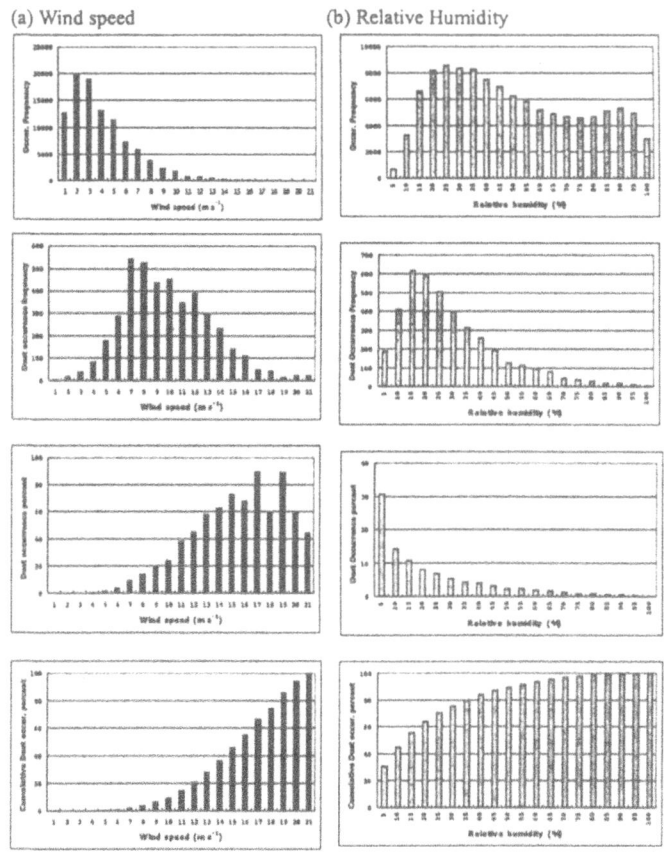

Fig.3. Occurrence frequencies of (a) the wind speed, dust storm and /or dust rise, percent ratio of the occurrence frequency of dust rise to that of the wind speed and the normalized cumulative dust occurrence percentage with respect to the wind speed at the Sand region. (b) The same as in (a) except for the relative humidity obtained from 3-hourly WMO surface reporting data for 7 spring seasons (March to May) from 1996 to 2002.

$$F_a = \sum_i (1 - f_i R_i) \times 2.633 \times 10^{-13} u_*^4 \qquad \text{if} \quad u_* \geq u_{*_t}$$

where f_i is the fractional coverage of i type of vegetation and R_i is the reduction factor of i type of vegetation in a grid. The U.S Geological Survey (USGS) satellite data with the horizontal resolution of 1 km are used for determination of the vegetation coverage in each emission grid.

3. RESULTS

The spatial distributions of the vertically integrated daily mean dust concentration with the wind vectors at the height of 1500 m ($\sigma = 0.82$) are shown in Fig. 4. A large amount of dusts emitted in southern Mongolia and central northern China on 19 March (Fig. 4a) moves southeastward and continuously supplied by dusts emitted in northeastern China and the Loess region on 20 March (Fig. 4b), yielding a wide area of dust laden air extending over Manchuria to central China.

Table 2. USGS vegetation categories with the surface roughness length (Zo) and the dust emission reduction factor (R).

Type	Description	Zo (m)	Reduction factor (R)
A	Urban and Built-up land	1.00	1.0
B	Dry Cropland and Pasture	0.02	0.4
C	Irrigated Cropland and Pasture	0.02	0.6
D	Mixed Dry / Irrigated Cropland and Pasture	0.02	0.5
E	Cropland / Grassland	0.02	0.5
F	Cropland / Woodland	0.02	0.7
G	Grassland	0.02	0.6
H	Shrubland	0.03	0.7
I	Mixed Shrub / Grassland	0.03	0.75
J	Savanna	0.02	0.8
K	Deciduous Broad leaf Forest	0.05	0.9
L	Deciduous Needle leaf Forest	0.05	0.9
M	Evergreen Broad leaf Forest	0.05	0.9
N	Evergreen Needle leaf Forest	0.05	0.9
O	Mixed Forest	0.05	0.9
P	Water	0.001	1.0
Q	Herbaceous Wet land	0.002	1.0
R	Wooded Wet land	0.003	1.0
S	Barren or sparsely vegetated land	0.01	0.1
T	Herbaceous Tundra	0.003	1.0
U	Wooded Tundra	0.003	1.0
V	Mixed Tundra	0.002	1.0
W	Bare ground Tundra	0.001	1.0
X	Snow or Ice	0.001	1.0

Thereafter, the dust laden air moves northeastward following the synoptic weather system, covering all over the Korean peninsula on 21 March (Fig. 4d). On 22 March (Fig. 4e), the northern parts of dust layer moves southeastward while the southern parts of dust layer moves east-northeast ward in association with wind fields. These dust-laden airs merge together over northern Japan. Thereafter, the dust-laden air passes over the Korean peninsula on 23 March 2002 (Fig. 4f). The evolutional patterns of dust-laden air simulated by the model quite resemble to those of the aerosol index obtained by TOMS (not shown here).

Fig. 5 shows the time series of modeled PM_{10} concentrations averaged for the layers of below 100 m, 100-1,500 m and above 1,500 m from the ground at five sites shown in Fig. 1. The estimated emission rates at the source site and Beijing while the observed PM_{10} concentrations at Seoul and Busan in Korea are shown in Fig. 5. Among sites, the source site and Beijing belong to the source regions (Figs. 5a and b) and are directly affected by dust emissions. Therefore, the PM_{10} concentration in the lowest level (below 100m from the ground surface) tends to be maximum and decreases with height. However, away from source regions including Seoul (Fig. 5c), Busan (Fig. 5d) and Sapporo (Fig. 5e), the maximum concentration tends to occur in the middle layer (100 – 1,500 m layer) first rather than in the lowest layer due to the effects of long-range transport and deposition.

The model simulates quite well the observed PM_{10} concentration at Seoul (Fig. 5c). Two peak values of the observed PM_{10} concentration at Seoul are well simulated by the modeled PM_{10} concentration in the lowest layer with slightly different times of occurrences. However, the starting time of the Yellow Sand event observed in Korea coincides with that of concentration increase in the upper layer whereas the ending time of the event coincides with the time of the concentration in the lowest layer being near zero.

The particle size is divided into three different classes for our convenience; Size 1 (the particle diameter, D 2.23   ), Size 2 (2.23 D 10   ) and Size 3 (10 D 74   ).

The temporal variations of model estimated dust concentration at different height with different size spectra in the source region (Hunsendake) and away from the source region (Busan) are given in Fig. 6. In the source region of Hunsendake (Fig. 6a), the larger size has higher concentration in all layers with higher concentration reduction rate with height. The maximum mean concentration of larger particle (Size 3) of about 11,600 m^{-3} occurs at 0200 UTC 20 March in Layer 1 (below 100 m above ground) whereas the PM_{10} concentration (sum of Size 1 and Size 2 concentrations) is about 2,700 m^{-3} at that time. This maximum concentration decreases greatly with height to be about 6,000 m^{-3} in Layer 2 (100 - 1,500 m layer) and less than 1,000 m^{-3} in Layer 3 (1,500 m layer) for Size 3 with less or no reduction for PM_{10} concentration up to Layer 2. Above Layer 2 the concentration of PM_{10} is very low, suggesting high diffusive character of small particles in the boundary layer.

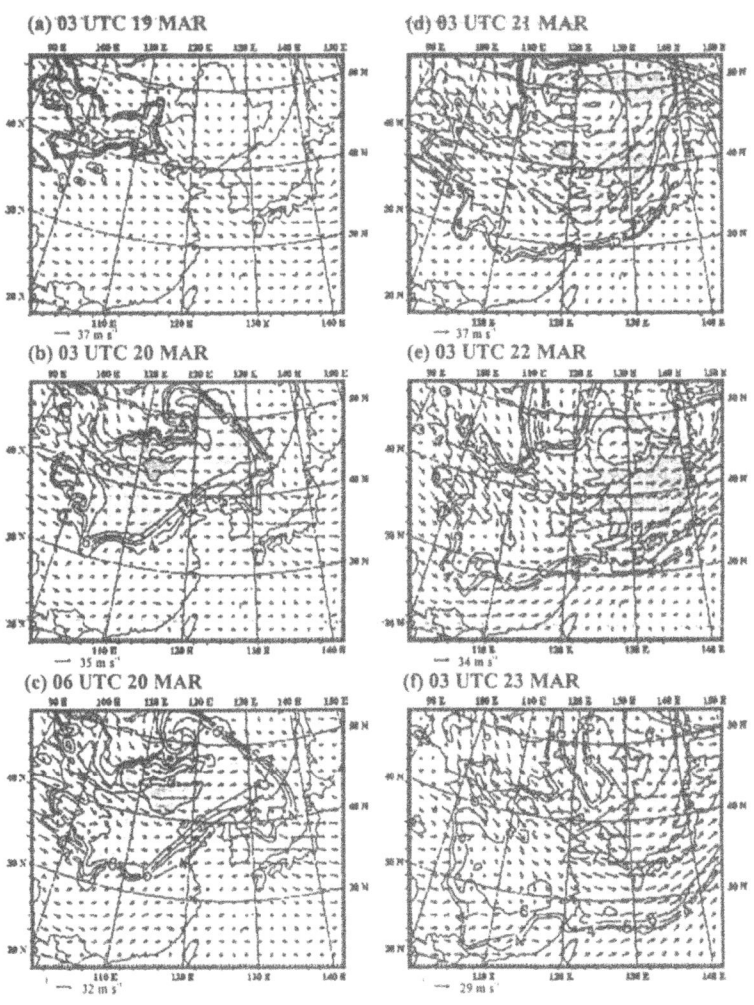

Fig. 4. Spatial distributions of the vertically integrated TSP concentration on 19 -23 March 2003 expressed in common logarithm ($\mu g\ m^{-2}$) with the wind vector at the height of 1500m.

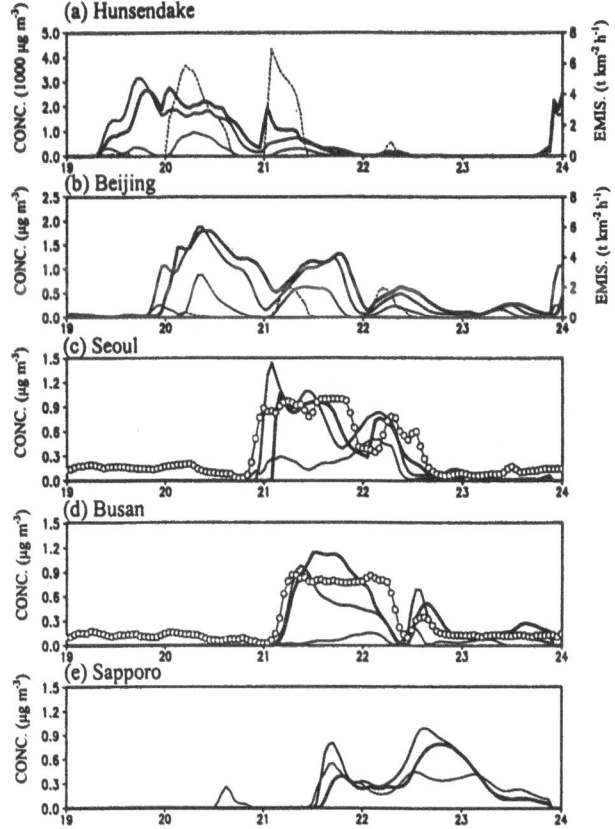

March 2002

Fig. 5. Time series of modeled mean PM$_{10}$ concentration ($\times$ 1000 µg m^{-3}) averaged for the layers of surface to 100 m (——), 100 – 1500m (——), above 1500 m (——) with emission rates (——, t km^{-2} h^{-1}) and observed surface PM$_{10}$ concentrations (- - - -) at (a) Hunsendake, (b) Beijing, (c) Seoul, (d) Busan and (e) Sapporo.

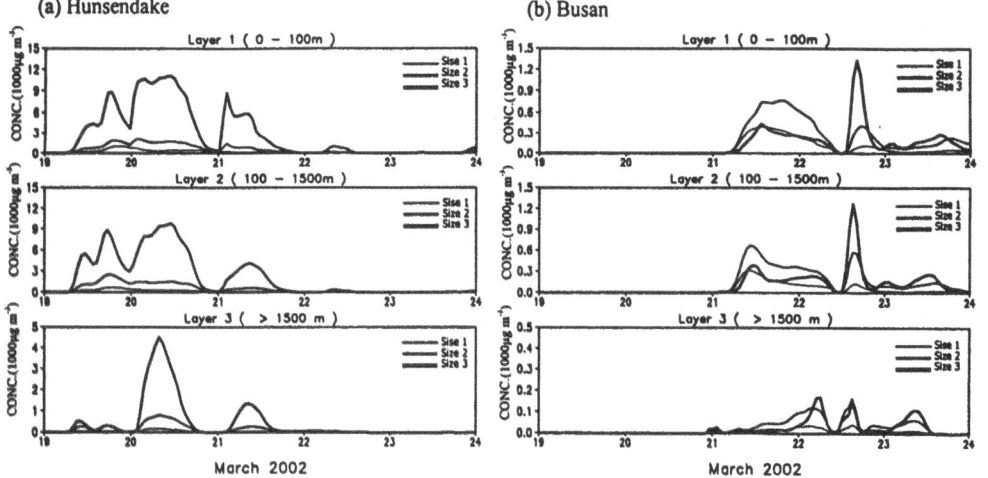

Fig. 6. Temporal variations of modeled mean concentrations ($\times$ 1000 µg m^{-3}) of Size 1 (——diameter $\leq$ 2.23 µm), Size 2 (——2.23 < diameter $\leq$ 10 µm) and Size 3 (——diameter > 10 µm) at three different layers (Layer 1 $\leq$ 100 m, 100 < Layer 2 $\leq$ 1,500 m, and Layer 3 > 1,500 m) in the (a) source region (Hunsendake) and (b) away from the source region of Busan in Korea.

241

It is worthwhile to note that the time series of the concentration of Size 3 has multi-peak values in the lower layer (Layer 1). However, it becomes more simplified with height. This is due to the fact that in the lower layer the concentration of Size 3 is controlled by both advection and local emission but the large particle (Size 3) has such a large terminal velocity that it cannot penetrates into higher level whereas the locally emitted large particles possibly penetrate into the upper layer resulting in high concentration in Layer 3 (Fig. 6a). In fact, there are strong dust emissions at around 0100 - 1900 UTC 20 and 0200 – 1200 UTC 21 March at Hunsendake (Fig. 5a) when the high concentrations of Size 3 occur in Layer 3 (Fig. 6a).

At Busan in Korea away from the source region (Fig. 6b) the concentration of PM_{10} (sum of Sizes 1 and 2) is much higher than that of Size 3 in all layers with a decreasing trend of concentration with height for the first dust storm observed from 0300 UTC 21 to 0900 UTC 22 March. The maximum concentration of PM_{10} and Size 3 is about 1,100 m^{-3} and 500 m^{-3} in Layer 1, respectively but they become to about 1,000 m^{-3} and 400 m^{-3} in Layer 2 with much reduced concentrations in Layer 3 (Fig. 6b). However, the second dust storm observed from 12 UTC to 21 UTC 22 March at Busan (Fig. 6b) has different concentration distribution patterns with size. The concentration of Size 3 is much higher than that of PM_{10}. The maximum concentration of Size 3 is about 1,400 m^{-3} but that of PM_{10} is about 550 m^{-3} with a decreasing trend with height. The different distribution pattern of the concentration with size between the first dust storm and the second one observed at Busan suggests the different pathway of dust storms. The first dust storm observed in Korea seems to be originated further away from Korea than the second one.

4. CONCLUSIONS

Long-range transport of a severe Yellow Sand event observed in Korea for the period of 21-23 March 2002 has been simulated using the three dimensional eulerian transport model with RDAP meteorological outputs together with dust emission parameterizations based on statistical analysis of WMO synoptic reporting data of dust storms and dust rises for seven springs from 1996 to 2002 in the Yellow Sand source regions.

The dust-rise conditions including the threshold wind speed, the upper limit of relative humidity and precipitation in the dust source region are obtained from the statistical analysis of WMO synoptic reporting data. The threshold wind speeds are found to be 9.5, 7.5, 6.0 and 9.2 m s^{-1} and the upper limit of relative humidity is found to be 60, 35, 30 and 40% in the Gobi, Sand, Loess and Mixed soil regions, respectively. These emission conditions yield much better results in estimating the starting and ending times of Yellow Sand and the dust concentration observed in Korea compared with the previous result of In and Park (2002). The proximity of the severe dust storms occurred at the Hunsendake desert area has caused the extraordinarily high dust concentration in Korea. The present model can be used effectively in forecasting the dust storm and the quantitative estimation of dust concentration.

Acknowledgements
    This research is partially supported by Climate Environment System Research Center at Seoul National University and Meteorological Research Institute of Korea Meteorological Administration.

References
Chun, Y.-S., 1996. Long-range transport of Yellow Sand with special emphasis on the dust rise conditions in the source regions. Ph. D. dissertation. Department of Atmospheric Sciences, Seoul National University, pp. 30-45 (Korean).
Gillette, D.A., 1981. Special Paper Geological Society of America 186, 11-26.
In, H.-J., Park, S.-U., 2002. A simulation of long-range transport of Yellow Sand observed in April 1998 in Korea. *Atmos. Environ.* 36, 4173-4187.
Westphal, D.L., Toon, O.B.,Carlson, T.N.,1988. A case study of mobilization and transport of Saharan dust. *J. Atmos. Sci.* 45, 2145-2175.

DISCUSSION

E. GENIKHOVICH   I used to think that the process of saltation is of major importance in dust storms why it is not the case in your study?

S.-U. PARK

G. KALLOS   The dust uptake depends on soil moisture and not that much necessarily on atmospheric relative humidity, why you decided to parameterize on atmospheric rlt? How about saltation processes?

S.-U. PARK

STUDYING POLLUTION LEVELS IN BULGARIA BY USING A FINE RESOLUTION DISPERSION MODEL

Zahari Zlatev and Dimiter Syrakov[*]

1. INTRODUCTION

The Danish Eulerian Model (DEM) is described in Zlatev (1995). A new version of the model is used in this study. The space domain of this and old versions contains the whole of Europe. However, this domain is now discretized on a 480×480 grid (10×10 km grid-squares). Comparisons with the original version (50×50 km grid-squares) runs are also made as to show the advantages of the refinement.

It was not possible to get input data on this high-resolution grid, neither emission data nor meteorological one. The emission data on a 96×96 grid was evenly distributed on the 25 small grid-squares obtained from a 50 km grid square during the transition to 10 km resolution. Simple linear interpolation (both in space and in time) is used to prepare meteorological data for the refined grid. Of course, this is a very crude approximation but the comparisons indicate that even by this rough approach some improvements can be seen when the refined grid is used.

Three emission scenarios with meteorology for 1997 were run.

- In the first scenario, we shall call it the **Basic Scenario,** the emissions for 1997 from the EMEP inventories (Vestreng and Støren, 2000) were used.
- In the second scenario, we shall call it the **Bulgarian Scenario**, again the emissions for 1997 from the EMEP inventories were used, but the Bulgarian emissions were set to zero.
- The third scenario, **Scenario 2010**, is obtained by modifying the EMEP emissions for 1990 by using the factors given in Amann et al. (1999).

Results obtained by applying these three scenarios will be discussed in the following sections.

---

[*] Zahari Zlatev, National Environmental Research Institute, Department of Atmospheric Environment, Frederiksborgvej 399, P. O. Box 358, DK-4000 Roskilde, Denmark, E-mail: zz@dmu.dk, Dimiter Syrakov, National Institute of Meteorology and Hydrology, 66, "Tzarigradsko chaussee" Bulvd., 1784 Sofia, Bulgaria, E-mail: dimiter.syrakov@meteo.bg

Air Pollution Modeling and Its Application XVI, Edited by
Borrego and Incecik, Kluwer Academic/Plenum Publishers, New York, 2004

2. VALIDATION OF THE MODEL RESULTS

Comparisons of calculated by the model concentrations with measurements taken at many EMEP stations located in different European countries have been used. Only data from "representative" stations are utilized. A station is declared as a representative one if (i) it measures at least 15 days in every month of 1997 and (ii) it is not located very high (not higher than 1300 m. over the sea level). Comparisons of measurements with model results obtained when the Basic Scenario is run are given in Fig. 1. It is immediately seen that the agreement between modeled and observed concentrations is quite good.

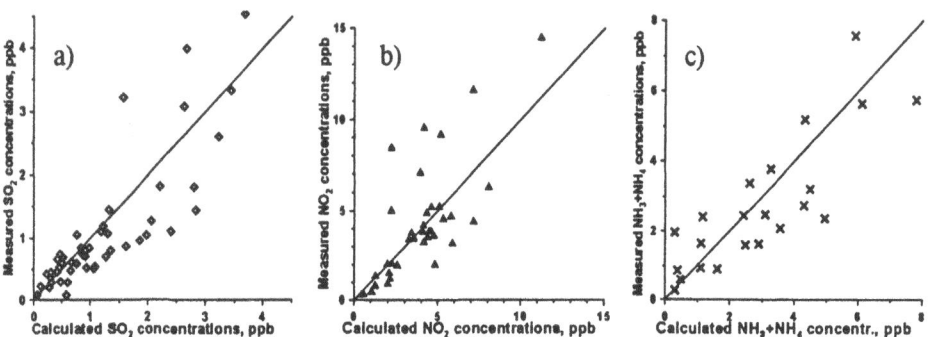

Figure 1. Comparison of measured and calculated on the 480×480 grid concentrations: a) SO_2 for 49 measurement stations from the EMEP network; b) NO_2 for 37 measurement stations from the EMEP network; c) NH_3+NH_4 for 22 measurement stations

The results from the comparisons (not only these illustrated in Fig. 1) are summarized in Table 1. It can be seen from Table 1 that the mean concentrations in air of almost all pollutants are overestimated by the model but the computed means over the finer grid are closer to the measured ones. The concentrations in precipitation are also overestimated but rough grid model estimations are closer to the measured values. The crude approach in refining meteorological data (including precipitation distribution) can play some role in this evidence. Some further advantages of using fine resolution grids will be pointed out in the next sections.

Table 1. Comparison of model results with measurements. All chemical species for which there are more than 10 measurement stations are presented in the table.

Pollutant	Number of stations	Measured means	Computed means 96×96 grid	Computed means 480×480 grid	Correlation 96×96 grid	Correlation 480×480 grid
SO_2	49	1.08	1.31	1.22	0.82	0.84
NO_2	37	4.20	4.01	3.92	0.73	0.74
NH_3+NH_4	22	2.70	3.22	2.93	0.85	0.83
SO_4	54	0.65	1.02	0.86	0.71	0.63
HNO_3+NO_3	25	0.95	1.27	1.11	0.82	0.81
O_3	69	29.86	30.31	29.48	0.66	0.66
NH_4	13	2.18	2.88	2.38	0.75	0.79
SO_4^P	18	0.60	0.77	0.88	0.89	0.90
NH_4^P	16	0.58	0.76	0.89	0.80	0.88
NO_3^P	17	0.46	0.38	0.32	0.71	0.76

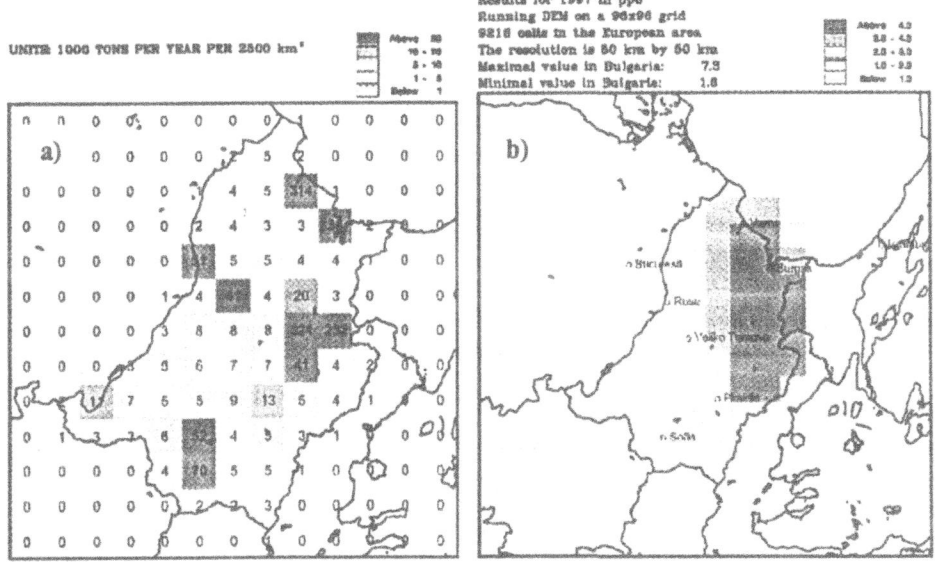

Figure 2. 1997 sulfur dioxide distribution in the region of Bulgaria: a) Bulgarian SO₂ emissions for 1997 estimated in 50 km resolution; b) Mean SO₂ concentrations in 96×96 grid (50 km resolution)

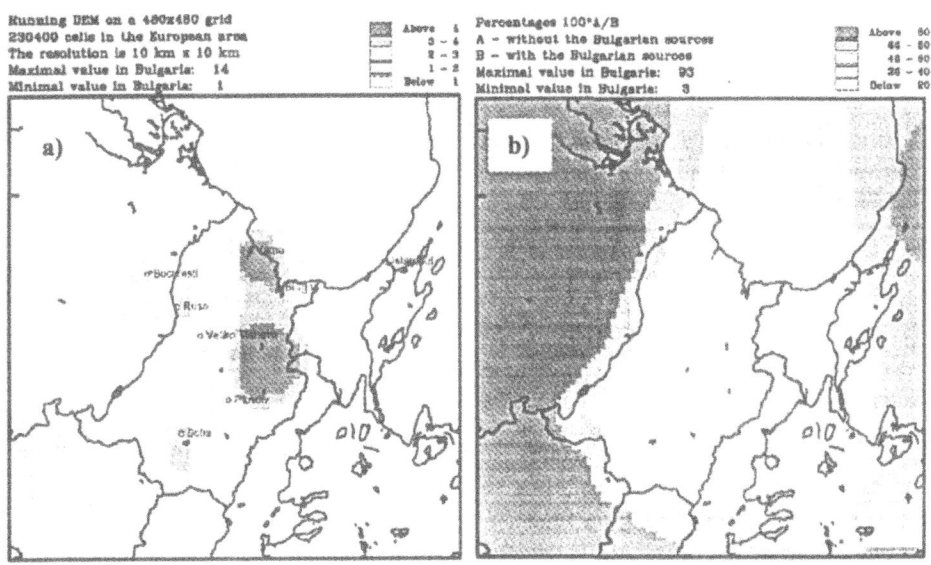

Figure 3. SO₂ concentration distribution for 1997 calculated in the 480×480 grid (10 km resolution): a) Mean concentrations for 1997 in ppb; b) Contribution from European sources to the Bulgarian SO₂ pollution levels.

3. CONTRIBUTION FROM EUROPEAN SOURCES TO BULGARIAN POLLUTION LEVELS

Two runs of DEM, the Basic Scenario and the Bulgarian Scenario, were performed in order to evaluate the contributions from the sources in the other European countries to the Bulgarian air pollution levels. The contributions are then calculated by dividing the results obtained in the second run at every grid-square with the corresponding results from the first run.

It should be emphasised that the above approach will not guarantee exact results even when the model is perfect (which is never the case). This is so because an implicit assumption is made that all the processes in the model are linear when we attempt to evaluate the contribution from the European sources to the Bulgarian pollution levels by setting the Bulgarian emissions to zero. There are, of course, many non-linear processes in any comprehensive air pollution model (for example, most of the chemical reactions are described by non-linear functions). Nevertheless, this approach is often used and it is believed that it gives good results (the results in many EMEP studies are based on similar assumptions; see, for example, EMEP, 1998).

The only pollutants, for which the chemical transformations have more or less linear character, are the sulfur dioxide and sulfate. In Fig. 2 the distribution of Bulgarian SO_2 emissions for 1997 are shown together with the annual SO_2 field for the same year as calculated on 96×96 grid. It can be seen that the annual SO_2 concentration field is quite smoothed in comparison with the emission distribution. The most polluted area is situated around the two most intensive SO_2 sources in the region – the thermal power plants (TPP) "Maritza-Iztok" and "Varna".

The calculations made on 480×480 grid are shown in Fig. 3a. Here the much more details of annual SO_2 concentration distribution can be found. The main sources are reflected much well than in Fig. 2b. Of course, one can not require perfect coincidence between the local source distribution and global concentration fields. The sources outside the country also contribute to the pollution level in Bulgaria. Their influence can be seen in Fig. 3b where the ratio "Bulgarian Scenario/Basic Scenario" for SO_2 annual concentration is plotted. The comparison with the emission field shows that the influence of outer sources is much higher in the clean areas of the country. The contribution of local sources is more than 90% in the areas around the most powerful polluters.

The improvement of the quality of the results when fine resolution computation is applied is much more suggestive in case of NO_2 pollution (NO_2 is one of the most important precursors in the process of ozone formation). In Fig. 4a,b the NO_2 emissions and calculated concentrations on the coarse 96x96 grid are presented. It can be seen that there are very small variations in the annual concentration field. The calculations on high-resolution grid give much more details as it can be seen in Fig. 4c. Fig. 4d presents the ratio "Bulgarian Scenario/Basic Scenario" for NO_2 annual concentration. The comparison with the emission field shows again that in the regions with intensive Bulgarian sources the influence outside is relatively small (30-40 % of the all pollution). It must be noticed that in case of SO_2 the influence of outer sources in these regions is much smaller. This is because the emissions in the most polluted by SO_2 areas are up to 100 times larger than in the remaining areas (Fig. 2a). For the NO_2 emissions the corresponding figure is much smaller. This explains why the contribution of the local NO_2 sources is only about 60 %.

The results concerning the influence of the European sources on the pollution levels in Bulgaria are summarized in the fourth column of Table 2. It is seen that the contribution from the European emission sources varies in a very wide range.

The final conclusion from this section is that it is desirable:

- to use fine resolution models which allow to see more details and to identify better the trends and
- to plot the results in the area (together with the respective maps) in order to see better the trends in the variations of the size of the contributions from different sources.

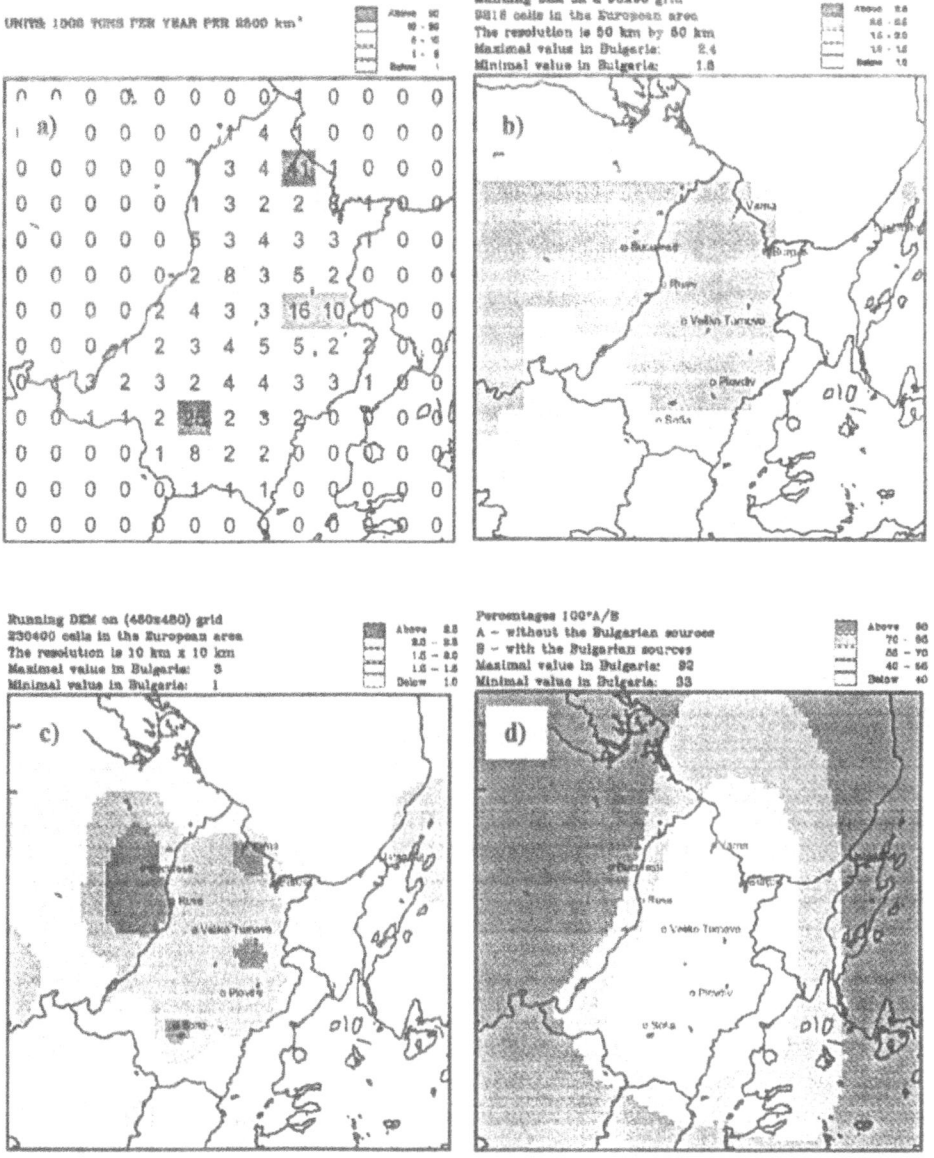

Figure 4. 1997 nitrogen dioxide distribution in Bulgaria and its surroundings: a) Bulgarian NO₂ emissions for 1997 estimated in 50 km ×50 km resolution; b) Mean NO₂ concentrations for 1997 estimated in the 96×96 grid (50 km resolution); c) Mean NO₂ concentrations for 1997 estimated in the 490×480 grid (10 km resolution); and d) European sources' contribution to Bulgarian NO₂ pollution for 1997.

4. EXPECTED POLLUTION LEVELS IN BULGARIA FOR YEAR 2010

Two runs of DEM, one with the Basic Scenario and the second one with Scenario 2010, were performed in order to evaluate the changes resulting by replacing the emissions for 1997 with emissions, which are expected to take place after year 2010. The changes are then calculated by dividing the results obtained in the second run at every grid-square with the corresponding results from the first run.

Table 2 Summary of the results obtained by the three scenarios. The numbers under "Max. value" and "Min. value" are giving the maximal and minimal annual means of the concentrations for 1997 in the Bulgarian area (the maximal and minimal means in the whole space domain of DEM are given in brackets). In the column under "From Europe" the intervals are given, in which the contributions (in percent) from foreign sources are varied in the Bulgarian area. The intervals in which the changes (in percent) vary in the Bulgarian area when Scenario 2010 is used are given in the column under "Changes in 2010".

Pollutant	Min. value	Max. value	From Europe	Changes in 2010
SO_2	1 (0) ppb	14 (60) ppb	[3% - 93%]	[60% - 74%]
NO_2	1 (0) ppb	3 (15) ppb	[33% - 92%]	[93% - 129%]

The emissions in Scenario 2010 are calculated, as mentioned in Section 1, by modifying the EMEP 1990 emissions by the factors given in Amann et al. (1999). The expected (in 2010) emissions in Europe are in general smaller than the European emissions in 1990. This means that one should also expect that the European pollution levels in 2010 (and after this year) would in general be less than the corresponding pollution levels in 1990. However, this is not necessarily true when we compare the pollution levels in Europe for 1997 with the expected pollution levels in 2010. The figures that are given in Table 3 explain why one should expect some pollution levels in Bulgaria and in some other countries in Eastern Europe to be increased in the transition from 1997 to 2010 when the factors from Amann et al. (1999) are used in Scenario 2010. The same effect has also been observed in Havasi and Zlatev, 2002).

Table 3. The total emissions in Bulgaria (in 1000 tonnes per year as SO_2, NO_2 and NH_3, respectively) for 1990, 1997 and 2010. The emissions for 1990 and 1997 are from Vestreng and Støren (2000). The emissions for 2010 are obtained from the emissions for 1990 by using the factors given in Amann et al. (1999). The changes (in percent) of the total Bulgarian emissions in the transition from 1997 to 2010 are given in brackets

Pollutant	1990	1997	2010
SO_2	2008	1365	884 (65%)
NO_x	361	225	303 (135%)
NH_3	144	77	128 (166%)

The situation in 1997 is much closer to the present situation than that one in 1990. Therefore, it is clear that the comparison of the pollution levels in 1997 with the pollution levels that are expected in 2010 gives a better evaluation of the relationship between the present pollution levels and the expected pollution levels in 2010.

It should be mentioned that the changes due to the use of the 2010 emissions do not vary in a very wide range. This is demonstrated in Fig. 5 where the changes in SO_2 and NO_2 concentrations due to 2010 source estimates are shown.

The following conclusions can be drawn from the results shown in Table 2 and Fig. 5.

- The calculations indicate that the SO_2 levels will be reduced over the whole Bulgarian area. This should be expected because the SO_2 emissions in Bulgaria will be reduced in 2010 comparing with 1997 (see Table 3);

- The calculations show that there are some places in Bulgaria where the NO_2 pollution levels will be reduced in 2010 in spite of the fact that the total Bulgarian NO_x emissions are expected to increase comparing again with the emissions in 1997. This indicates that the transport of NO_2 pollution to Bulgaria will be reduced because some of the countries, which contribute to the Bulgarian pollution levels, will reduce their NO_2 emissions.
- It is expected that the NH_3+NH_4 pollution levels in Bulgaria will be increased in practically the whole country in 2010 comparing with the pollution levels in 1997. This is not a surprise because the Bulgarian NH_3 emissions in 2010 are expected to be higher than the corresponding emissions in 1997 (see Table 3).

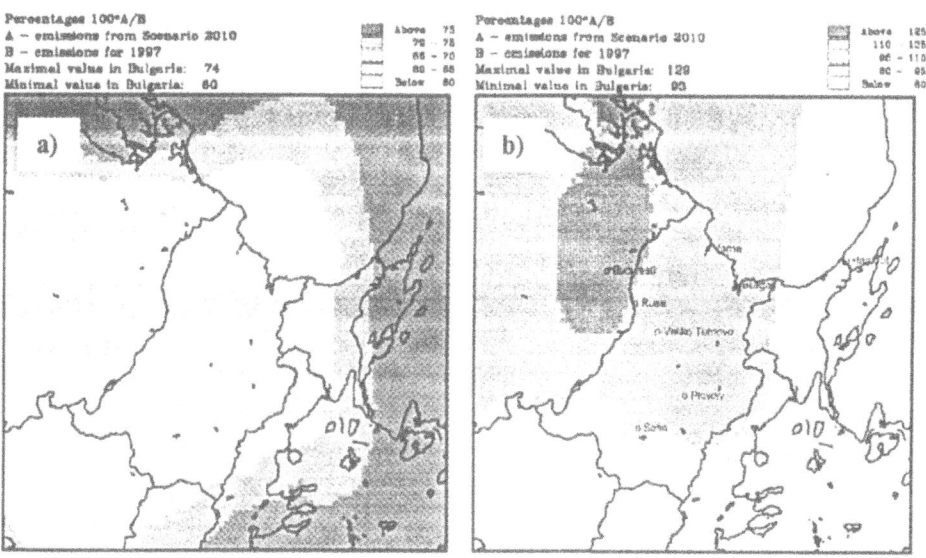

Figure 5. Changes in SO_2 (a) and NO_2 (b) annual concentrations when meteorology for 1997 and source estimates for 2010 (after Amann et al., 1999) are used.

5. CONCLUDING REMARKS AND PLANS FOR FUTURE RESEARCH

Some results concerning (i) the air pollution levels in Bulgaria, (ii) the contributions of foreign sources to these levels as well as (iii) the expected changes of the air pollution levels as a result of applying the predicted for 2010 European emissions have been presented and discussed.

The use of a high-resolution grid gives a better fit of calculated concentrations and depositions to the measurements. Moreover, much more details can be seen in the space distribution of pollution levels, in spite of the fact that the input data (both the meteorological data and the emission data) is approximated in a rather crude way. Comparisons of the pollution levels with the emission sources in the area show that the more detailed information obtained by using the fine resolution grid is also qualitatively more correct.

An interesting question arises: *Will the results be improved if the quality of the input data is improved?* We plan to carry out some research in this direction in the near future.

REFERENCES

Amann, M., Bertok, I., Cofala, J., Gyartis, F., Heyes, C., Kilmont, Z., Makowski, M. Schöp, W. Syri, S., 1999, Cost-effective control of acidification and ground-level ozone. *Seventh Interim Report*, IIASA A-2361 Laxenburg, Austria.

EMEP, 1998, Transboundary acidifying air pollution in Europe, *EMEP/MCS-W Report 1/1998*, Norwegian Meteorological Institute, P. O. Box 43 – Blindern, N-0313 Oslo, Norway.

Havasi, A. and Zlatev, Z., 2002, Trends of Hungarian air pollution levels on a long time-scale. Atmospheric Environment, to appear.

Vestreng, V. and Støren, E., 2000, Analysis of UNECE/EMEP emission data. *MSC-W Status Report 2000*, EMEP MSC-W Note 1/00., Norwegian Meteorological Institute, P. O. Box 43 – Blindern, N-0313 Oslo, Norway.

Zlatev, Z., 1995, *Computer Treatment of Large Air Pollution Models*. Kluwer Academic Publishers, Dordrecht-Boston-London.

AEROSOLS AS ATMOSPHERIC CONTAMINANTS

Chairperson: S. Incecik

Rapporteur: M. Z. Boznar

MODELLING OF ATMOSPHERIC MERCURY TRANSPORT, CHEMISTRY AND DEPOSITION: RECENT ACHIEVEMENTS AND CURRENT PROBLEMS

Alexey Ryaboshapko, Alexey Gusev, Ilia Ilyin, and Oleg Travnikov[*]

1. INTRODUCTION

Heavy metals (HMs) and first of all lead, cadmium and mercury are among the most important pollutants in the framework of the Convention on Long-Range Transboundary Air Pollution (CLRTAP). This seems to be natural in view of their long lifetime in the atmosphere, their high toxicity and capability to accumulate in trophic chains. Most European countries, Canada and the USA have prepared a special Protocol on control of HM emissions to the atmosphere, which is at the stage of ratification now. The Protocol implementation is supported by the activity of two Centres of CLRTAP – Chemical Co-ordinating Center (monitoring of HMs) and Meteorological Synthesizing Center "East" (modelling of HMs). The latter one is called upon to develop mathematical models of the long-range atmospheric transport of HMs, to use them for the assessment of transboundary pollution in Europe and to inform the participating countries of the current situation and long-term trends of HM pollution levels.

The major mass of HMs in the atmosphere is connected with aerosol particles, which have their maximum of mass-size distribution in sub-micron range. This circumstance promotes their atmospheric transport from emission sources over distances of thousands of kilometres and enriches HM background concentrations everywhere in the Earth atmosphere. On a regional level the ecological impact of HMs is caused by their long-term accumulation in soil and freshwater ecosystems.

Mercury holds a special place among HMs. This is connected with a number of reasons. First of all, mercury is toxic in most chemical forms. It is capable of generating organo-metallic compounds, which can be easily involved into biotic processes. From the viewpoint of atmospheric transport modelling the most important feature of mercury is its occurrence in the atmosphere in four very different physical-chemical forms – as gaseous elemental mercury (GEM), as reactive gaseous mercury (RGM), as organic compounds (e.g., dimethyl mercury - DMM) and as total particulate mercury (TPM) in the composition of solid aerosol particles. Moreover, physical-chemical processes result in transformations of one form to another in the atmosphere. Such processes take place in

[*] Alexey Ryaboshapko, Alexey Gusev, Ilia Ilyin, and Oleg Travnikov, EMEP Meteorological Synthesizing Center "East", Arhitektor Vlasov str., 51, Moscow 117393 Russia

the gaseous phase, in the aqueous phase of cloud droplets and on the surface of solid particles. An important point is that lifetime of elemental mercury in the atmosphere is so long that it should be considered as a global pollutant. Hence, models of the global level should be used in this case. Taking into account such diversity of mercury properties it is possible to state that modelling of atmospheric transport of any other metal can be considered as a specific case of mercury modelling.

2. MERCURY IN THE ATMOSPHERE – MODELING APPROACHES

When developing a conceptual scheme of any model it is necessary, first of all, to take into account main properties of a substance simulated. Such characteristics for mercury are presented in Table 1. The analysis of the data in the table shows that the most abundant form is gaseous elemental mercury. Organic mercury compounds are not typical for the atmosphere because of their high chemical instability. Oxidised forms (gaseous and aerosol) occur in the atmosphere in noticeable quantities and are capable of being transported over long distances.

Table 1. Properties of mercury forms in the atmosphere

Form and typical species	Atmospheric lifetime	Typical regional concentrations	Dry uptake by underlying surface	Solubility in water and washout	Chemical transformations
Elemental gaseous, Hg^0	0.5-1.5 yr	1.5-2.5 ng/m^3	Weak uptake by plants	Weak, practically no washout	Oxidation in gas and aquatic phases
Oxidised gaseous, $HgCl_2$	0.5-2 days	10-50 pg/m^3	As high as for HNO_3	High, similar to HNO_3	Reduction in aquatic phases
Organic, gaseous, $Hg(CH_3)_2$	Minutes - hours	Not measured	Practically no uptake	Weak, practically no washout	Very fast oxidation
Particulate, HgO	1-3 days	10-100 pg/m^3	Similar to fine particles	Partly, similar to fine particles	No

Knowledge of the behaviour of different mercury forms in the atmosphere and other media, of mechanisms of their transformations and scavenging allows to simulate mercury exchange between the media and atmospheric transport over different distances up to the global level. As usual, two main types of models are used – box models to describe the exchange between environmental compartments, and transport models as applied to the atmosphere. Below models of atmospheric transport of mercury species on the regional and global level are considered.

Modern models of atmospheric mercury transport are built upon the description of the following basic processes:

- natural and anthropogenic emissions of Hg in different physical-chemical forms;
- dispersion and transport within the atmosphere;
- physical-chemical transformations in the gas phase and in the aqueous phase of clouds;
- wet removal and dry uptake by the underlying surface;
- biochemical processes in ecosystems and Hg re-emission to the atmosphere.

To meet the HM Protocol requirements a task has been formulated to develop an operational model of mercury atmospheric transport for the evaluation of contamination of the entire Northern Hemisphere (hemispheric level) and for calculations of transboundary pollution of individual European countries (regional level). Such a task suggests a necessity of consideration of all the processes mentioned above. Currently, the model is under development in its two modifications – regional and hemispheric versions.

Both regional and hemispheric models are of Eulerian three-dimensional type and consider main processes governing transport and deposition of HM in the atmosphere - advection, diffusion, dry/wet removal and mercury chemical reactions (Travnikov and Ryaboshapko, 2002). The regional version operates within the EMEP region (including Europe, the northern part of Africa, part of Middle East, the North Atlantic and part of the Arctic) and has spatial resolution 50*50 km at 60°N (http://www.emep.int/grid/). The hemispheric version covers the whole Northern Hemisphere with resolution 2.5°×2.5°. In the vertical the model domain consists of eight irregular terrain-following layers up to the tropopause. The three-dimensional atmospheric transport of the models is based on the flux-form Bott advection scheme (Bott, 1989).

Each form of mercury can be scavenged from the atmosphere both by precipitation and uptake by the underlying surface. However, removal rates can differ by orders of magnitude. Wet removal is assumed as the first order process. Elemental mercury and gaseous organic mercury compounds are weakly soluble in water. On the contrary, gaseous inorganic compounds (for example, mercury chloride) can be very easily scavenged by precipitation. Properties of mercury chloride, which determine wet removal intensity, are close to such properties of nitric acid. It is assumed that the washout ratio for both compounds is equal to $1.4*10^6$ (Jonsen and Berge, 1995; Petersen et al., 1998). Aerosol particles which contain the main mass of mercury have their distribution maximum around 0.7 μm (Milford and Davidson, 1985). Presence or absence of mercury within a particle does not change its aerodynamical properties because mercury makes extremely small contribution to particle mass. Taking this into account it is reasonable to consider the wet removal process of TPM as that for sub-micron particles. In the model the value of washout ratio is equal to $5 \cdot 10^5$. Usually this value is applied to sub-micron particles (Iversen et al., 1989).

Until now the question on ability of elemental mercury to be taken up by the underlying surfaces remains open. Flux chamber experiments show that weak uptake can take place, especially in the case of vegetation-covered surfaces. To take into account dry deposition of GEM we adopted the following assumptions: (a) dry uptake occurs only over vegetation during daytime; (b) it is proportional to cosine of the solar zenith angle with the coefficient 0.03 cm/s for forests and 0.01 cm/s for other vegetation types; (c) it linearly drops with the surface temperature from 20°C to 0°C and is absent under negative temperatures. Dry deposition velocity for mercury organic compounds is assumed to be extremely low (taking into account their physical-chemical properties). On the contrary, for very soluble RGM it should be limited only by aerodynamic resistance of the atmosphere. In this work (as well as in (Petersen et al., 1998)) it is accepted that dry deposition velocity for RGM (by analogy with nitric acid) is equal to 0.5 cm/s. Dry deposition velocity of aerosol particles is calculated at each modelling step as a function of surface type, its roughness and atmospheric stability in accordance with ideas suggested by Wesely et al. (1985) and Ruijgrok et al. (1997).

The most important characteristic of mercury, which determine its rate of removal from the atmosphere, is its capability to be involved into different chemical reactions within the atmosphere. These reactions include both oxidation of elemental mercury and reduction of mercury compounds. The basic pathways of chemical transformations considered in the model are showed in Figure 1. One can see that the chemical processes can occur both in the gaseous phase (air) and in the aqueous phase (droplets of clouds or fog). After cloud drop evaporation mercury remains in a solid particle (TPM form). In its turn mercury-carrying particle can enter the liquid phase due to rainout or washout. It is also should be mentioned that mercury chemistry is strongly linked to chemistry of sulphur, ozone, chlorine, radical OH in the atmosphere.

Chemical reaction rates, chemical equilibrium parameters for the "liquid-solid" system and Henry's law constants for "gas-liquid" partitioning used in the modelling scheme are showed in Table 2. The values presented in the table are taken from (Hall, 1995; Tokos et al., 1998; Ryaboshapko and Korolev, 1997; Ryaboshapko et al., 1998; Munthe, 1992; Lin and Pehkonen, 1998; Petersen et al., 1998; Sander, 1997).

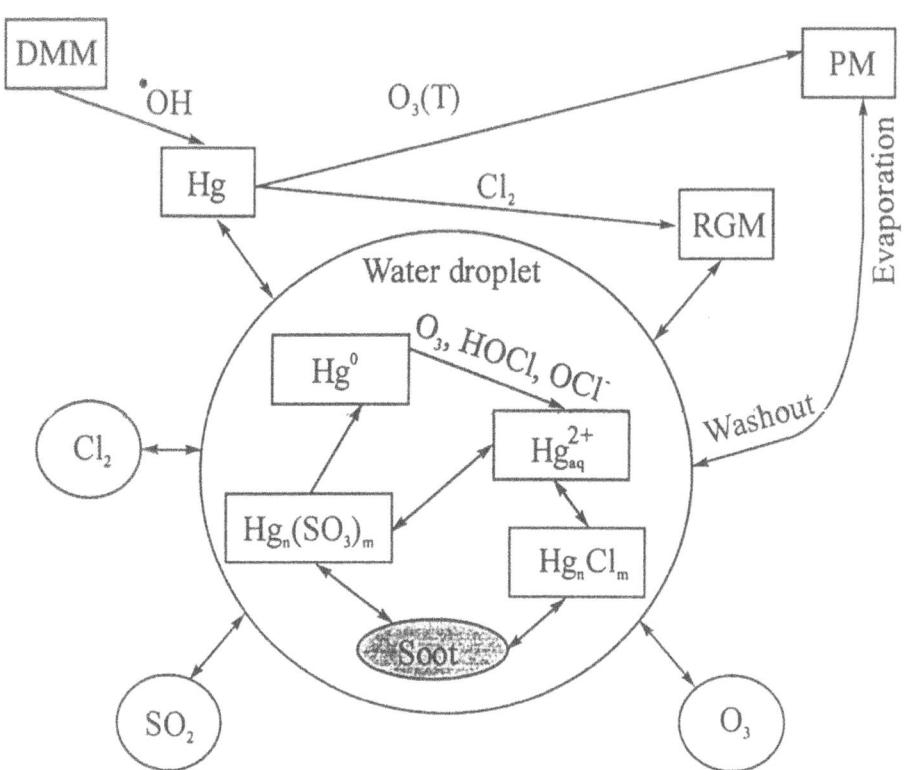

Figure 1. The scheme of physical-chemical transformations of mercury in the atmosphere.

Table 2. Reaction rates, chemical equilibrium parameters and Henry's constants

Reaction / Equilibrium	Constant	Comment
$Hg^0_{gas} + O_3 \rightarrow TPM$	3.2E-20 cm^3/molec/s	At 298K, f(T)
$Hg^0_{gas} + Cl_2 \rightarrow RGM$	4E-16 cm^3/molec/s	
$DMM \rightarrow (+ \ ^{\bullet}OH) \rightarrow Hg^0_{gas}$	2.3E-5 s^{-1}	Rough estimate
$Hg^0_{aq} + O_3 \rightarrow Hg^{2+}_{aq}$	7.8E-14 cm^3/molec/s	
$Hg^0_{aq} + HOCl_{aq} \rightarrow Hg^{2+}_{aq}$	2.09E6 M^{-1} s^{-1}	
$Hg^0_{aq} + OCl^-_{aq} \rightarrow Hg^{2+}_{aq}$	1.99E6 M^{-1} s^{-1}	
$Hg_n(SO_3)_m \rightarrow Hg^0_{aq}$	4.4E-4 s^{-1}	
$Hg^0_{gas} \Leftrightarrow Hg^0_{aq}$	2.88 at 298K	Dimensionless, f(T)
RGM ($HgCl_{2 \ gas}$) $\Leftrightarrow HgCl_{2 \ aq}$	2.93E7 at 298K	Dimensionless, f(T)
$Hg^{2+}_{aq} \rightarrow Hg_n(SO_3)_m$	1E-3	At SO_2=1 ppb; pH=4.5
$Hg^{2+}_{aq} \Leftrightarrow Hg_nCl_m$	1 : 167000	At [Cl$^-$]=1E-4 M
$O_{3 \ gas} \Leftrightarrow O_{3 \ aq}$	0.28 at 298K	Dimensionless, f(T)
$Cl_{2 \ gas} \Leftrightarrow Cl_{2 \ aq} \Leftrightarrow HOCl_{aq} + OCl^-_{aq}$	2.7E5	At [Cl$^-$]=1E-4; pH=4.5
$(Hg_n(SO_3)_m$ or $Hg_nCl_m)_{aq} \Leftrightarrow$ soot	1 : 6	

Many reactions of mercury in the atmosphere are temperature-dependent. Besides, changes in temperature shift conditions of equilibrium in the "gas-water" system. Usually such dependence is not taken into account by many modellers, although reaction rates and equilibrium constants can vary as much as several times over a real range of temperatures in the atmosphere. The most important temperature-dependent processes are gas-phase reaction of oxidation by ozone, and the exchange of elemental mercury, mercury chloride and ozone in the "air-water" system. In the model these dependences are described by the following equations:

$$k (Hg^0 + O_3) = 2.1E-18 \bullet exp (-1246/T) \qquad \text{(derived from (Hall, 1995));}$$
$$H_{Hg} (T) = 9.84E-3 \bullet T \bullet exp [2800 * (1/T-1/298)] \qquad \text{(Ryaboshapko and Korolev, 1997);}$$
$$H_{O3} (T) = 9.51E-4 \bullet T \bullet exp [2325 * (1/T-1/298)] \qquad \text{(Sander, 1997);}$$
$$H_{HgCl2} (T) = 1.054E5 \bullet T \bullet exp [5590 * (1/T-1/298)] \qquad \text{(Ilyin et al., 2002),}$$

where k is the constant of corresponding reaction rate, cm^3/molec/s; H is Henry's law constant for a given compound (dimensionless value); T is the temperature, °K.

In principle, mercury is a chemically stable element. However, in the gaseous phase elemental mercury can slowly be oxidised by such reactants as ozone and chlorine. The oxidation proceeds more intensively in the aqueous phase of cloud droplets. The oxidation leads to a disturbance of equilibrium in the "air-water" system and to solution in water additional portions of mercury from air. As a result, mercury can be accumulated in cloud water. However, this process is limited by the existence of a negative feedback in the system – oxidised mercury can be reduced to the elemental state due to the formation of mercury-sulphite complexes ($Hg_n(SO_3)_m$) and their subsequent decomposition followed by elemental mercury release (Pleijel and Munte, 1995).

This sulphite mechanism of reduction results in some non-trivial relationships. First, the more polluted is the atmosphere by sulphur dioxide, the lower is mercury concentration in cloud water. Second, the higher is acidity of cloud water, the lower is the fraction of sulphite-ion and, consequently, the higher is the intensity of mercury removal. Third, the higher is chloride-ion concentration in cloud water, the lower is sulphite-ion concentration and, consequently, the slower is the reduction process.

Properties of mercury compounds in solid particle composition play a very important role in mercury chemical transformations. First of all, it concerns their solubility in water – the higher is the fraction of soluble compounds, the higher is the fraction reduced to the elemental state. This rule is also applied to the processes of sorption-desorption. For example, a part of water-soluble mercury can be strongly adsorbed by soot particles within the water phase. This adsorption can prevent reactions from mercury reduction.

On the whole, it should be acknowledged that the level of our knowledge on atmospheric mercury chemistry is still remain to be insufficient for reliable modelling parameterisation of mercury atmospheric transport. It is exemplified by recently discovered phenomenon of mercury depletion events (MDE) in the Arctic (Schroeder et al., 1998). The essence of the phenomenon is the fact that during spring months usually chemically stable elemental mercury is rapidly oxidised. During a very short period of time its concentration drops practically to zero. About in a month the phenomenon can come to its end as quickly as it has begun. After MDE very high concentrations of mercury can be found in snow pack, as high as it is typically measured in highly industrialised areas. Rough estimates show that MDE can explain elevated levels of Arctic ecosystem pollution by mercury.

By the moment the first attempts were made to describe MDE in terms of atmospheric chemistry (Lu et al., 2001; Lindberg et al., 2002). In springtime when open water appears, different halogenous radicals and other chemically active compounds enter the atmosphere. Particles of sea-salt are also emitted to the atmosphere from the oceanic water surface. When air temperature is lower than 0°C some quick photo-chemical heterogeneous reactions of elemental mercury oxidation can take place on the surface of sea-salt crystals. Halogen-containing species like BrO radical can serve as oxidants in these reactions. In any case, the rate of oxidation is correlated with BrO concentration in air. Such explanation suggests that synchronous combination of some conditions is required to launch the MDE – sunlight, appearance of open oceanic water, low air temperature (below 0°C). Rising of the air temperature to positive values by the middle of June prevents from occurrence of heterogeneous reactions.

In the model described an attempt was made to estimate the importance of MDE for Arctic mercury pollution. Since there is an obvious lack of quantitative information on MDE mechanisms, modelling description is chosen by such a manner that changes of calculated concentrations are similar qualitatively to observed concentration variations. The air temperature near the surface was chosen as a trigger parameter to launch and stop MDE.

It is very important for models of regional and hemispheric level to determine correctly concentrations on the modelling borders. For the northern and western borders of the EMEP domain there are data on long-term observations of elemental mercury concentrations (Berg et al., 1999; Ebinghaus et al., 2002). TPM concentration at the domain boundaries is taken as 0.75% of GEM concentration. These data are used in the model as mercury concentrations in air masses coming into the domain through the borders. Concentrations of RGM in the incoming air masses are taken as zero. Initial concentrations of all mercury forms are accepted to be equal to zero, and the spin-up time for the regional model version makes up 1 month. For the hemispheric version the initial elemental mercury concentration and constant concentration in the Southern Hemisphere is accepted to be 1.5 ng/m^3. In this case the spin-up time is taken equal to 2 years.

3. MERCURY EMISSIONS TO THE ATMOSPHERE

Mercury always occurs in the atmosphere, however, human activity disturbed considerably its natural cycle. It is believed that global natural mercury flux to the atmosphere from the World Ocean was 2000 t/y, and the same from land (Seigneur et al., 2001). Natural content of mercury in the atmosphere as a whole was 1800 t and it rose due to human activity at least 2 times during the last 150 years (Lamborg et al., 2002). It is very important that natural sources practically do not emit oxidised mercury forms, which are very toxic and accessible for biota. For anthropogenic emission sources the contribution of these forms can make up 80%.

The most important input information for any model of atmospheric transport of any pollutant is the emission and its distribution in space and time. It is known now that the natural emission on land depends on mercury content in soils, soil properties, surface temperature and solar radiation (Carpi and Lindberg, 1998; Poissant and Casimir, 1998; Gustin et al., 1999). The natural emission intensity is especially high in regions where underlying rocks are enriched by mercury. In this work the land surface was divided into 4 types with different mercury emission intensity: (1) permanent glaciers; (2) background soils; (3) soils of mercury geochemical belts, and (4) soils of mercury deposit regions. The following temperature dependence was chosen for land natural emission:

$$F_{Hg} = \begin{cases} A_s \exp(-10^4/T_s), & T_s > T_0 \\ 0, & T_s \leq T_0 \end{cases}$$

where the flux intensity F_{Hg} is expressed in ng/m^2/hr; A_s is equal to 6.4· 10^{14} for background soils, 3.2· 10^{15} for mercury belts, and 6.4· 10^{15} for areas of Hg deposits; T_s is the surface temperature (K); $T_0 = 273$K. This dependence allowed to estimate the field of land natural emission.

It was also calculated that the total natural emission in European region makes up 220 t/y. The emission field with 50*50 km resolution was constructed on the base of spatial distribution suggested by Axenfeld et al. (1991).

Mercury emission from sea surface is observed everywhere, however its intensity is very variable. It is assumed that the formation of elemental mercury in seawater takes place involving biota (Kim and Fitzgerald, 1986). For modelling purposes the total oceanic emission was distributed in space in proportion to values of primary biological production of organic carbon. It is important to mention that natural emission both from land and oceanic surface is represented by mercury elemental form.

The main contributions to anthropogenic mercury emission are made by such processes as solid fuel combustion, waste incineration, and chlorine-alkali production with the use of mercury electrodes. For 1995 the total mercury anthropogenic emission in the Northern Hemisphere is about 1900 t/y (Pacyna and Pacyna, 2002; Travnikov and Ryaboshapko, 2002). The emission data for Europe including emission spatial distribution, emission heights and mercury speciation have become available recently

(Pacyna et al., 2001). A very important peculiarity of the anthropogenic emission is that it is represented by all physical-chemical mercury forms.

Another type of mercury emission is exhalation from previously contaminated soils and waters. This secondary anthropogenic emission or re-emission was estimated for Europe (Ryaboshapko et al., 1998) on the base of consideration of some parameters: mercury lifetime in soils, proportionality of re-emission intensity to the rate of accumulation in soils during XX century, similarity of re-emission field to the accumulated deposition field, and so on. It was found that re-emission in Europe by the end of the century made up about 50 t/y. In some heavy contaminated areas the re-emission can make a significant contribution to the total mercury emission.

4. MODEL VALIDATION

In accordance with the ideas suggested by Borrego et al. (2000) there are no "good" or "bad" models, they can be appropriate or not for solving of a given problem. Quality of a model is defined as a "fitness to purpose". An EMEP/WMO/UNEP workshop on heavy metal modelling came to a conclusion that the accuracy of model prediction should be currently on the level of the factor of 2 (Ilyin et al., 2000). A model, which can provide such an accuracy, can be considered as an appropriate tool for the purposes of the HM Protocol.

There are two approaches to model validation. The first one is based on a direct comparison of modelling results with the available data of observations. It is *a priory* clear that the more observational data and the higher density of monitoring networks, the more reliable the results of the comparison. The second approach can be considered as an indirect one. In this case the results of a given model are compared with the results of other models which accuracy is considered to be *a priory* higher. Since EMEP mercury model should be operational, i.e. to work in the regime of year-by-year calculations, it cannot take into account all possible details of mechanisms involved. Hence, it makes sense to compare its results with the results of the most advanced comprehensive scientific models.

In Europe there are 5 permanent monitoring stations to measure elemental mercury in air and 9 such stations to measure mercury concentrations in precipitation. All of them are located in the north-western part of Europe, hence the results of measurements are close to each other, and for comparison purposes it makes sense to consider them as a single totality. Mean observed values and mean values calculated by regional model for 1999, as well as ranges of their variations are given in Table 3. One can mention very good agreement for GEM. This is not surprising because the variability of the concentrations is low. For mercury in precipitation the agreement of mean measured and calculated values is on the level of 30%, however, for one of the stations the difference exceeds the factor of 2.

Unfortunately, no one of monitoring station measures RGM and TPM forms on the routine basis. Hence, for the comparison purposes the results of two-week measurement experiment carried out simultaneously at four European stations were used (Ebinghaus et al., 2002). In all, 21 samples of RGM and 55 samples of TPM were collected during November 1999. The results of the comparison of mean concentration values for all samples as well as ranges of variations are also given in Table 3. For RGM the correlation between observations and calculations is not high (the correlation coefficient is only 0.30), while for TPM it is much higher (CC = 0.69).

Table 3. Mean values and variation ranges of the observed and modelled concentrations of different mercury forms in air and precipitation

Parameter	GEM concentration, ng/m^3	Hg in precipitation, ng/L	RGM, pg/m^3	TPM, pg/m^3
Observations	1.68 (1.40-1.96)	8.4 (4.2-11.1)	7.3 (2.0-20.9)	24.0 (0.5-109.8)
Model calculations	1.95 (1.63-2.05)	11.4 (6.8-19.1)	9.6 (0.2-52.6)	22.5(0.7-86.1)
Observation/Model	0.86	0.74	0.76	1.07

The hemispheric version of the model was verified on the base of observational data on mercury wet deposition (16 monitoring stations) and GEM in air (5 stations) in Europe and North America (Travnikov end Ryaboshapko, 2002). The model quite well reproduces annual concentrations of the elemental form (mean deviation is less than 25%). Wet deposition values for a separate station can differ as much as 2 times but on the average the discrepancy is on the level of 40% with the correlation coefficient equal to 0.54.

To estimate capabilities of the EMEP operational model the comparison with the most comprehensive scientific models was organised (Ryaboshapko et al., 2002). In this modelling experiment 6 scientific models (developed by experts from Bulgaria, Canada, Denmark, Germany and the USA) participated. All the models calculated concentrations of different mercury forms in different parts of Europe for a two-week period. The results of this comparison are presented in Table 4. The table demonstrates that all parameters calculated by the EMEP model are within the limits obtained by other models. This fact confirms that the quality of the operational model is at least on the level of the most advanced scientific models. One can mention rather wide scattering of the modelling results, especially for RGM.

Table 4. Comparison of concentration values obtained by different mercury models

Country, model	GEM, ng/m^3	RGM, pg/m^3	TPM, pg/m^3
Bulgaria, EMAP	1.26	7.5	20.6
Canada, GRAHM	1.99	14.5	48.9
Denmark, DEHM	1.73	1.6	27.7
Germany, ADOM	1.14	9.0	19.0
USA, CMAQ	2.23	16.4	56.0
USA, HYSPLIT	1.88	9.3	31.7
EMEP, MSCE-Hg	2.13	9.6	22.5

5. MODELLING RESULTS

Satisfactory validation of the model gave a possibility to map concentration and deposition patterns over Europe (50*50 km) and the entire Northern Hemisphere (2.5*2.5 degree). Figure 2 presents the field of total mercury deposition over the EMEP modelling domain in 2000. Oxidized mercury forms (RGM and TPM) make the greatest contribution to the total deposition. Since a considerable fraction of these forms has basically anthropogenic origin, the deposition maxima are usually strongly associated with the anthropogenic sources. Areas of elevated depositions are located in Poland, the east of Germany, the north-east of Spain, in Greece, and the north of France (up to 250 $g/km^2/y$). Relatively low depositions can be indicated in the central part of the Scandinavian Peninsula. The computed deposition values there do not exceed 8 $g/km^2/y$.

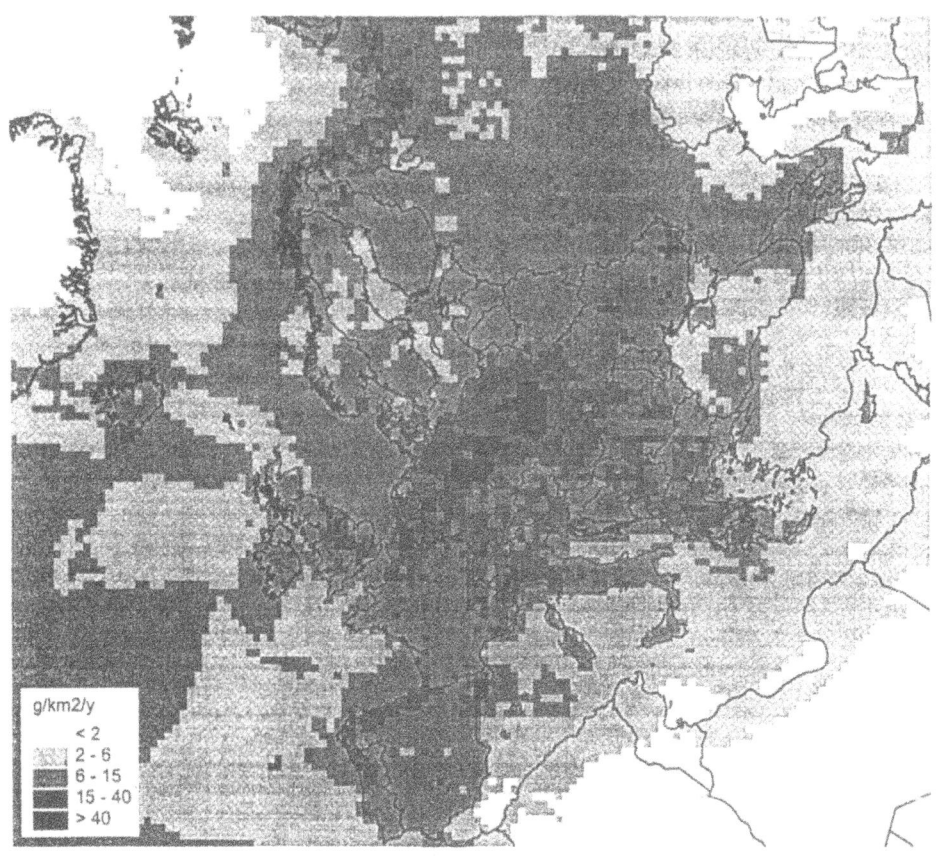

g/km2/y
< 2
2 - 6
6 - 15
15 - 40
> 40

Figure 2. Model predicted deposition of mercury over Europe in 2000.

The budget of emission, transport and deposition of mercury was calculated for each Party of the Convention. The main sources of mercury pollution were identified for each individual country. Table 5 shows the data on two most important countries-sources of mercury for some European countries-receptors in 2000 (the matrix for all European countries can be found in Ilyin at al. (2002)). Besides, the table presents percentage contributions of these countries-sources to the total deposition on the territory of the countries-receptors as well as contributions from their own emission sources. For countries of Central Europe contributions from transboundary air pollution and from own national sources dominate. At the same time, for northern countries like Sweden natural and inter-continental sources give the main input into mercury air pollution.

264

Nevertheless, in the southern parts of the Scandinavian countries the contribution of transboundary pollution is noticeable.

The consideration of the budget items for the region as a whole shows that anthropogenic emissions of the GEM form to the atmosphere do not practically define its amount in the model reservoir and in depositions onto the underlying surface. Only a small part of elemental mercury emitted to the atmosphere from European sources deposits within the region and the bulk of this amount is transported outside the region and enters the global mercury cycle. Relatively high background concentrations of elemental mercury in the atmosphere result in large advective fluxes across the lateral limits of the domain.

Table 5. Main countries-sources of transboundary pollution and their contributions (in %) to depositions of mercury on countries-receptors in 2000.

Countries-receptors	Major countries-sources and their contribution in %		Own sources, %	IND[*]
Belgium	France – 38	Germany – 4	30	20
Bulgaria	Romania – 9	Greece – 7	36	40
Czech Republic	Germany – 18	Poland – 14	31	26
France	Spain – 7	Switzerland – 2	43	43
Germany	France – 4	Switzerland – 2	61	23
Greece	Bulgaria – 3	Romania – 1	69	24
Netherlands	France – 20	Belgium – 13	16	30
Poland	Germany – 10	Czech Rep. – 4	61	18
Portugal	Spain – 6	France - <1	51	43
Romania	Hungary – 4	Poland – 3	41	39
Russia	Ukraine – 3	Poland – 3	13	76
Spain	Portugal - 3	France - 1	55	41
Sweden	Germany - 7	Poland - 7	1	73
Turkey	Greece - 5	Bulgaria - 2	<1	86
Ukraine	Poland - 7	Romania - 3	32	46

[*] IND – Indeterminate natural and inter-continental anthropogenic sources

Anthropogenic emissions of oxidised gaseous compounds and particulate mercury to the atmosphere are most significant for the environment of Europe. Emissions of these very forms cause high deposition levels in the central part of the continent. The main scavenging mechanism of these forms is washout by precipitation. However, one should bear in mind that oxidation of elemental mercury in the liquid phase also makes its contribution to washout. Only about 10% of oxidised mercury directly emitted from anthropogenic sources is transported outside the region. The main part is deposited within the region.

The hemispheric version of the model allows estimating the intercontinental mercury transport from major industrial regions and pollution level of such remote ecologically sensitive areas as the Arctic. Particularly it was applied to the assessment of the influence of recently discovered phenomenon of mercury depletion events on the Arctic contamination. Figure 3 presents the field of mercury deposition over the Arctic region. As seen annual deposition fluxes can exceed 20 g/km^2/y in some areas, which is comparable with those over polluted industrial regions. MDE provides additional annual deposition of mercury to the vulnerable Arctic ecosystems equal to 50 t/y or about 20% of the total deposition.

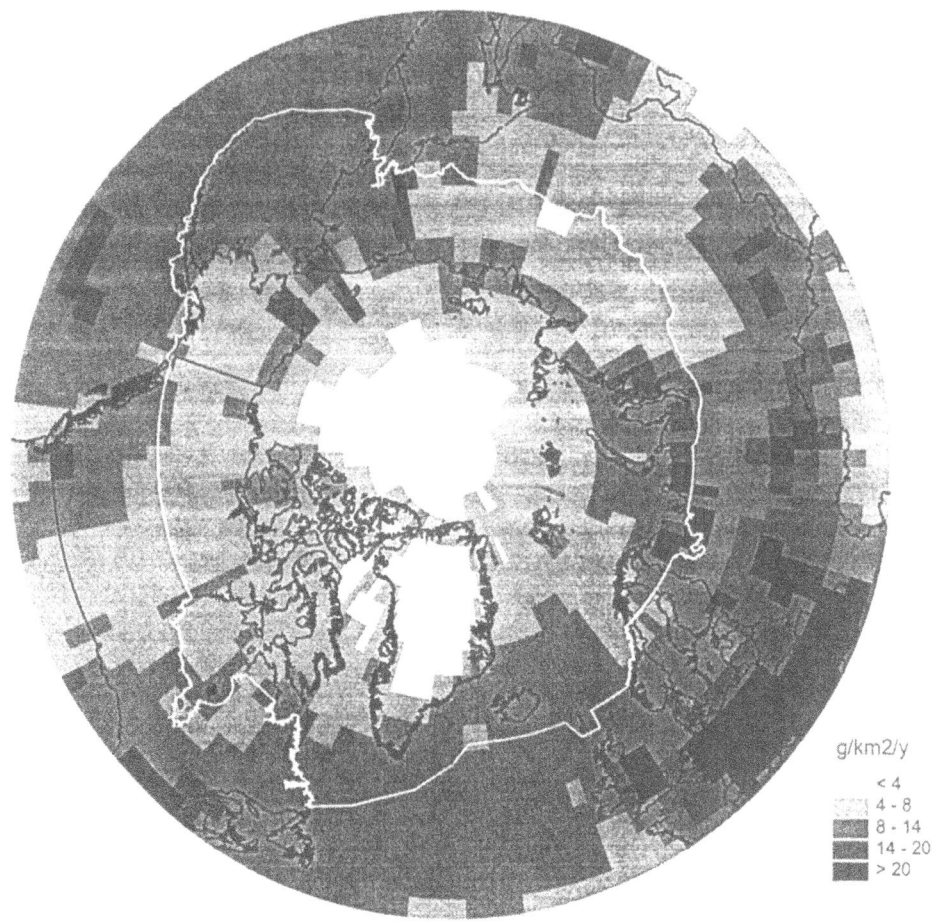

Figure 3. Spatial distribution of total annual deposition of mercury in middle and high latitudes of the Northern Hemisphere.

6. CONCLUSIONS

The level of mercury model development is currently high enough to consider the models as essential instruments for reliable assessment of pollution levels over Europe, for revealing the "source-receptor" relationships, and for evaluation of global mercury contamination. On the base of recent achievements obtained in atmospheric chemistry of mercury an operational model for calculations of mercury transboundary deposition in Europe and inter-continental transport in the Northern Hemisphere has been elaborated. The model was verified using the available monitoring data for Europe and North

America. Besides, a special numerical experiment was organised to compare the developed operational model with the most advanced scientific models of mercury atmospheric transport and deposition.

As a part of activity under HM Protocol implementation the model simulations were used for the evaluation of country-by-country pollution. Mercury concentration and deposition over European territory were mapped with 50*50 km spatial resolution. Atmospheric mercury balances for all European countries were assessed and the most important sources of transboundary contamination were revealed. Such a balance for the entire European region showed contributions of different mercury forms to mercury deposition. The hemispheric version of the model allowed calculating atmospheric mercury transport between the continents. An attempt was made to assess the role of newly discovered mercury depletion phenomenon on Arctic pollution.

Although the modelling scheme incorporates the significant progress in the understanding of mercury atmospheric pathways, our knowledge is still very restricted. The model should be developed further as additional information on emission speciation, redox processes in the gas and aqueous phases as far as scavenging parameters become available. Any long-term assessments of mercury pollution are impossible without consideration of processes of exchange between the atmosphere and other geophysical spheres. Hence, the next evident step of the mercury model development is a quantitative description of mercury pathways in soils and natural waters and a construction of a model of multi-compartment type.

REFERENCES

Axenfeld, F., Münch, J., and Pacyna, J.M., 1991, Belastung von Nord- und Ostsee durch ökologisch gefährliche Stoffe am Beispiel atmosphärischer Quecksilberkomponenten. Teilprojekt: Europäische Test-Emissionensdatenbasis von Quecksilber-Komponenten für Modellrechnungen". Dornier, Report 104 02 726, 99.

Berg, T., Bartnicki, J., Munthe. J., Lattila H., Hrehoruk, J., and Mazur, A., 1999, Atmospheric mercury in the European Arctic: Measurements and Modelling, *Atmos. Environ.*, **35**: 2569.

Borrego., Tchepel, O., and Carvalho, A.C., 2000, Model quality assurance, in *Transport and Chemical Transphormation in the Troposphere,* eds. P.Midgley, M.Reuther, and M.Williams, Springer, Berlin, pp. 21-26.

Bott, A., 1989, Reply to comment on "A positive definite advection scheme obtained by nonlinear renormalization of the advective fluxes", *Monthly Weather Review,* **117**: 2633.

Carpi, A., and Lindberg, S.E., 1998, Application of a teflon[TM] dynamic flux chamber for quantifying soil mercury flux: tests and results over background soil. *Atmos. Environ.*, **32**: 873.

Ebinghaus, R., Kock, H.H., Coggins, A.M., Spain, T.G., Jennings, S.G., Temme, Ch., 2002, Long-term measurements of atmospheric mercury at Mace Head, Irish west coast between 1995 and 2001, *Atmos. Environ.*, **36**: 5267.

Gustin, M.S., Rasmussen, P., Edwards, G., Schroeder, W., and Kemp, J., 1999, Application of a laboratory gas exchange chamber for assessment of in situ mercury emissions, *J. of Geophysical Research*, **104D**: 21873.

Hall, B., 1995, The gas phase oxidation of mercury by ozone, *Water, Air and Soil Pollution*, **80**: 301.

Ilyin, I., Munthe, J., Petersen, G., and Ryaboshapko, A., 2000, Numerical Models of Long-range Atmospheric Transport of Heavy Metals: Current State, and Direction of Further Development, In: WMO report No. 136 "WMO/EMEP/UNEP Workshop on Modelling of Atmospheric Transport and Deposition of Persistent Organic Pollutants and Heavy Metals", Vol. I. Geneva, Switzerland, 92.

Ilyin, I., Ryaboshapko, A., Afinogenova, O., Berg, T., Hjellbrekke, A.-G., Lee, D., 2002, Lead, cadmium and mercury transboundary pollution in 2000, EMEP/MSC-E Report 5/2002, 131 p.; http://www.msceast.org.

Iversen, T., Saltbones, J., Sandnes, H., Eliassen, A., and Hov, O., 1989, Airborne transboundary transport of sulphur and nitrogen over Europe – model descriptions and calculations, EMEP/MSC-W Report 2/89, Meteorological Synthesizing Centre – West, Oslo, Norway, 92 p.

Jonsen, J., and Berge, E., 1995, Some preliminary results on transport and deposition of nitrogen components by use of Multilayer Eulerian Model, EMEP/MSC-W Report 4/95, Meteorological Synthesizing Centre – West, Oslo, Norway, 25 p.

Kim, J.P., and Fitzgerald, W.F., 1986, Sea-air partitioning of mercury in the Equatorial Pacific Ocean, *Science*, 231: 1131.

Lamborg, C.H., Fitzgerald, W.F., O'Donnell, J., and Torgersen, T., 2002, A non-steady-state compartmental model of global-scale mercury biogeochemistry with interhemispheric atmospheric gradients. *Geochimica et Cosmochimica Acta*, 66: 1105.

Lin, C.-J., and Pehkonen S., 1998, Two-phase model of mercury chemistry in the atmosphere, *Atmos. Environ.*, 32: 2543.

Lindberg, S.E., Brooks, S., Lin, C.-J., Scott, K.J., Landis, M.S., Stevens, R.R., Goodsite, M., and Richter, A., 2002, Dynamic oxidation of gaseous mercury in the Arctic troposphere at polar sunrise, *Environ. Sci. Technol.*, 36: 1245.

Lu, J.Y., Schroeder, W.H., Barrie L.A., Steffen, A., Welch H.E., Martin, K., Lockhart W.L., Hunt R.V., Boila, G., Richter, A., 2001, Magnification of atmospheric mercury deposition to polar regions in springtime: the link to tropospheric ozone depletion chemistry. *Geophys. Res. Letters*, 28: 3219.

Milford, J., and Davidson, C., 1985, The size of particulate trace elements in the atmosphere – a review, *JAPCA*, 35: 1249.

Munthe, J., 1992, The aqueous oxidation of elemental mercury by ozone, *Atmos. Environ.*, 26A: 1461.

Pacyna, E.G., and Pacyna, J.M., 2002, Global emission of mercury from anthropogenic sources in 1995. *WASP*, 137: 149.

Pacyna, E.G., Pacyna, J.M., and Pirrone, N., 2001, Atmospheric mercury emissions in Europe from anthropogenic sources, *Atmos. Environ.*, 35: 2987.

Petersen, G., Munthe, J., Pleijel, K., Bloxam, R., and Kumar, A., 1998, A comprehensive Eulerian modeling framework for airborne mercury species: Development and testing of the tropospheric chemistry module (TCM), *Atmos. Environ.*, 32: 829.

Pleijel, K., and Munte, L., 1995, Modeling the atmospheric mercury cycle – chemistry in fog droplets, *Atmos. Environ.*, 29: 1441.

Poissant, L., and Casimir, A., 1998, Water-air and soil-air exchange rate of total gaseous mercury measured at background sites, *Atmos. Environ.*, 32: 883.

Ruijgrok, W., Davidson, C.I., and Nicholson, K.W., 1997, Dry deposition of particles. Implications and recommendations for mapping of deposition over Europe, *Tellus*, 47B: 587.

Ryaboshapko, A., Ilyin, I., Artz, R., Bullock, R., Christensen, J., Cohen, M., Dastoor, A., Davignon, D., Draxler, R., Ebinghaus, R., Munthe, J., Petersen, G., and Syrakov, D., 2002, Intercomparison study of numerical models for long-range atmospheric transport of mercury, EMEP/MSC-E Technical Note 10/2002, 19 p.; http://www.msceast.org.

Ryaboshapko, A., Ilyin, I., Gusev, A., and Afinogenova, O., 1998, Mercury in the Atmosphere of Europe: Concentrations, deposition patterns, transboundary fluxes, Meteorological Synthesizing Centre - East, EMEP/MSC-E Report 7/98, Moscow, 55 p.; http://www.msceast.org.

Ryaboshapko, A., and Korolev, V., 1997, Mercury in the atmosphere: estimates of model parameters. EMEP/MSC-E Report 7/97, Moscow, 60 p., http://www.msceast.org.

Sander, R., 1997, Henry's law constants available on the Web, *EUROTRAC Newsletter*, 18: 24; http://www.science.yorku.ca/cac/people/sander/res/henry.html.

Schroeder, W., Anlauf, K., Barrie, L., Lu, J., Steffen, A., Schneeberger, D., and Berg, T., 1998, Arctic springtime depletion of mercury, *Nature*, 394: 331.

Seigneur, C., Karamchandani, P., Lohman, K., Vijayaraghavan, K., and Shia R.-L., 2001, Multiscale modeling of the atmospheric fate and transport of mercury, *J. of Geophysical Research*, 106-D: 27795.

Tokos, J., Hall, B., Calhoun, J., amd Pretbo E., 1998, Homogeneous gas-phase reaction of Hg0 with H2O2, O3, CH3I, and (CH3)2S: implications for atmospheric Hg cycling, *Atmos. Environ.*, 32: 823.

Travnikov, O., and Ryaboshapko, A., 2002, Modelling of Mercury Hemispheric Transport and Depositions, EMEP/MSC-E Technical Report 6/2002, 67 p.; http://www.msceast.org.

Wesely, M.L., Cook, D.R., Hart R.L., and Speer, R.E., 1985, Measurements and parameterization of particulate sulfur deposition over grass. *J. of Geophysical Research*, 90-D: 2131.

DISCUSSION

G. KALLOS
1. Do you have the Hg model coupled directly on an atmospheric model or you just use the met model outputs?
2. How do you treat O_2, Cl and PM?

A. RYABOSHAPKO
1. We use outputs of a meteorological model developed and used in Russian Meteorological Center.
2. Molecular oxygen and PM are not taken into account in the chemical scheme. We consider ozone, black carbon and particulate mercury. Ozone and black carbon were calculated by our Norwegian colleagues. Particulate mercury is calculated by our model. Chlorine species are set as constant values.

E. GENIKHOVICH
As far as I understood the Arctic mercury depletion coinside with the occurence of the Arctic haze. I would suggest to analyze, if there is a possibility of capturing mercury or the aerosols forming this haze.

A.RYABOSHAPKO
Indeed, aerosol particles can absorb elemental mercury, however, only a small part of elemental mercury can be removed by this process. Most likely, we deal with chemical oxidation of elemental mercury. Probably, on the surface of aerosol particles.

SOME PRELIMINARY RESULTS CONCERNING THE HG BUDGET ESTIMATES FOR THE STATE OF NEW YORK

A. Voudouri[1], I. Pytharoulis[1], G. Kallos[1] and C. Walcek[2*]

1. INTRODUCTION

In this paper an attempt was made to identify the in/out of state contributions of mercury sources to the total deposited mercury over the State of New York. The transport, transformation and deposition of mercury in New York State are also discussed.

Two atmospheric models namely Regional Atmospheric Modelling System, RAMS (Pielke et al. 1992) and SKIRON/Eta (Kallos et al. 1997 and references therein) were used in the present study. The use of two atmospheric models allowed the inter-comparison of the results and, therefore improved understanding on the mercury modelling. Both models have been run for an individual episode, during which the highest quality and comprehensive emissions, observations of mercury concentration and deposition were available. Accurate assessment of our current understanding of atmospheric mercury cycle, requires both high-quality measurements of concentrations and deposition together with accurate estimates of Hg emissions. Model results and measurements have been compared during the 14 to 26 August 1997 simulation period for two different scenarios:

(a) when robust Hg measurements were available in and around NY State and
(b) when Hg emissions in NY State were not considered.

[*]
[1]University of Athens, School of Physics, AM&WF Group, University Campus, Bldg PHYS-5, 15784 Athens, Greece, voudouri@mg.uoa.gr
[2]State University of New York at Albany, Atmospheric Sciences Research Center, 251 Fuller Rd, Albany, New York 12203, USA

The wet deposition pattern calculated using both models has been compared with measurements provided by the Mercury Deposition Network (MDN). The model/observations inter-comparison showed a satisfactory agreement. Critical gaps in the current understanding of regional-scale transport deposition and fate of mercury in New York State still exist, despite the improved modelling capabilities and understanding of mercury as an air pollutant. However, the developed modelling systems can be useful tools for policy makers and be used in mitigating the impacts of mercury pollution.

2. MODEL SET-UP

The simulation performed with both models started at 0000 UTC on 14 August 1997 and ended at 0006 UTC 26 August 1997. The domain of both simulations (with and without the NY sources) covers the area of US East of the Rocky Mountains. The grid for RAMS has been selected with 90x90x30 points and 36 km horizontal grid increment. The coordinates of the center of the domain were at 36.926 °N and 85.037 °W.

For the SKIRON/Eta model the selected area extends from 21.5 °N to 48.8 °N and from 63.8 °W to 107.3 °W, centered at 36.9 °N and 85 °W, and covers the USA. This area covers the same area with the one used for RAMS with minor differences attributed to the SKIRON/Eta horizontal projection.

3. RESULTS AND DISCUSSION

The relative contributions of in-state mercury sources and out of state sources to the mercury deposition are an important issue for policy makers in New York State. In this study two simulations were performed, one using all available sources of NE USA and another without the New York State sources. The location of the sources for the first scenario with New York State sources is illustrated in Fig 1, while for the second scenario the New York State sources have been excluded. The mercury concentration calculated in both cases (with and without the New York State sources) was then compared. The concentration of mercury species is affected by several factors that influence the chemical and physical processes such as atmospheric reactions and deposition. It also depends strongly on flow conditions and source locations (Davies and Notcutt 1996).

Hg^0 is known to be a long-range transport pollutant. In both simulations, with all available sources and without the New York State sources, the results showed that the Hg^0 concentrations were high even over the Atlantic Ocean, where no sources exist, indicating a long-range transport. Moreover, for the second scenario, even though there were no anthropogenic sources over the State of New York the concentration of Hg^0 was increased up to 20% relative to the background value. For the first scenario the concentration of the pollutant over the New York State was usually more than 0.1 ng/m^3 greater than the concentration of Hg^0 for the second scenario indicating the strong influence of New York State sources on the local concentrations. Major differences of the Hg^0 concentrations appeared mainly downwind the State of New York, reaching up to 0.3 ng/m^3 in both scenarios.

Hg^2 can be removed in the vicinity of a few tens to a few hundreds of kilometres. For Hg^2, the differences in the concentration patterns for both scenarios are more evident over the State of New York (see Fig 2a when there are sources over the State of New York compared with the Fig 2b). The New York State concentrations of Hg^2 were more than 25 pg/m^3 higher for the first scenario, compared to the corresponding concentrations of the second scenario, especially in the areas near the anthropogenic sources. Finally Hg^p is known to be an intermediate distance deposited pollutant. The Hg^p concentrations were different within the State of New York and at intermediate from sources distances. Large differences between the Hg^p concentrations in both scenarios (reaching up to 25%) were calculated within the New York State. Moreover, the absence of emissions is likely to exert strong influence, not only on the concentration pattern of all species but also on their deposition pattern.

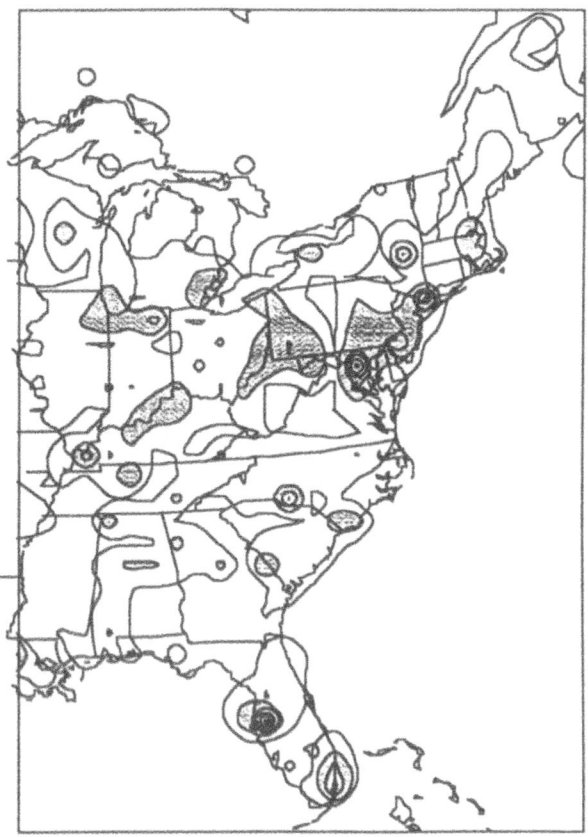

Figure 1: Spatial distribution of total gaseous and particulate emission rates in eastern North America. Emissions are summed over Lambert-conformal 80x80 km^2 areas, and contours are drawn at 90%, 70%, 50%, 30%, 10% and 1% of the maximum emission rate (west of Washington D. C.) of 23215 moles Hg per year per 80x80 km^2 area. Total emissions in domain = 5.26×10^5 moles year^{-1}.

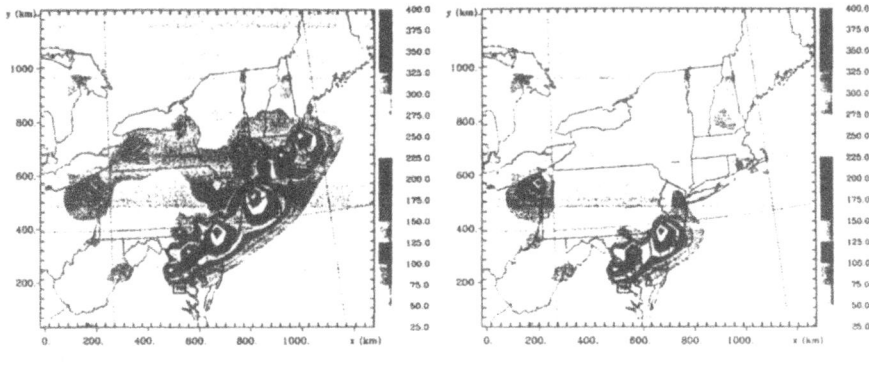

(a)   **(b)**

Figure 2: Hg^2 concentration at the first model level (~ 69 m) at 1200UTC on 14 August 1997, from RAMS model, (a) with and (b) without NY sources. The color scale corresponds to concentration (in pg/m^3).

Although the wet deposition is the predominant mechanism for the deposition processes the amount of mercury species deposited through different atmospheric processes was calculated for the simulation period. More specifically, the wet and dry deposition patterns of Hg^P, Hg^2 and Hg^0-adsorbed were estimated using both models RAMS and SKIRON/Eta. The dry deposition patterns of all three mercury species vary over sea and over land. The transport of mercury species is dependent upon the advective transport by the mean wind and transport by turbulent dispersion. The spatial and temporal variations on the dry deposition patterns can be determined through the similarities with the conventional pollutants. The dry deposition patterns of Hg^P and Hg^0-adsorbed depend on the pollutant concentration and the deposition velocity. The deposition velocity of Hg^P used in these simulations is a weighted average of 15 deposition velocities, corresponding to the 15 size intervals at which particles are distributed. Over regions with high humidity (e.g. over sea surface) greater deposition velocities are observed due to the dependence of the deposition velocity on the size of the particles. Larger particles are observed over these regions, as particles growth is relatively high under these conditions.

Wet deposition patterns of Hg^2 and Hg^P for the first scenario with the New York State sources, are illustrated in Figs 3a, 4a. The highest amounts of the Hg^2 are deposited near the sources. This is also consistent with the literature (Schroeder and Munthe 1998). Hg^2 is also highly soluble so it dominates the wet deposition pattern of gaseous mercury. The wet deposition pattern of Hg^2 has several similarities with the wet deposition pattern of Hg^0-adsorbed, but the Hg^2 deposited amounts are higher. The amount of Hg^P, wet deposited is higher over the mountainous areas, following the total amount of precipitation that is higher over the specific areas (see Fig 4a).

When sources of mercury are not considered, the dry and wet deposited amount of all species over the selected area of New York is lower. The sources of mercury increase the concentration of all species in the State of New York. Therefore, as sources of the pollutant located over the State of New York are excluded for the second simulation scenario, the mercury deposited amount is lower. Major differences between the two simulations are evident for Hg^2 and Hg^P since these species transported in short and intermediate distances respectively (see Figs 3a,b, 4a,b).

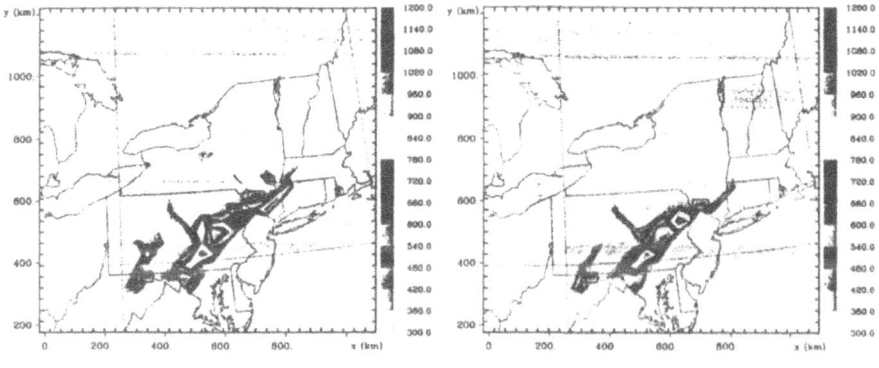

(a)   **(b)**

Figure 3: Wet deposition of Hg^P (in ng/m^2) at 0000 UTC on 26 August 1997 after 14 days of simulation, estimated from RAMS model (a) with and (b) without NY sources.

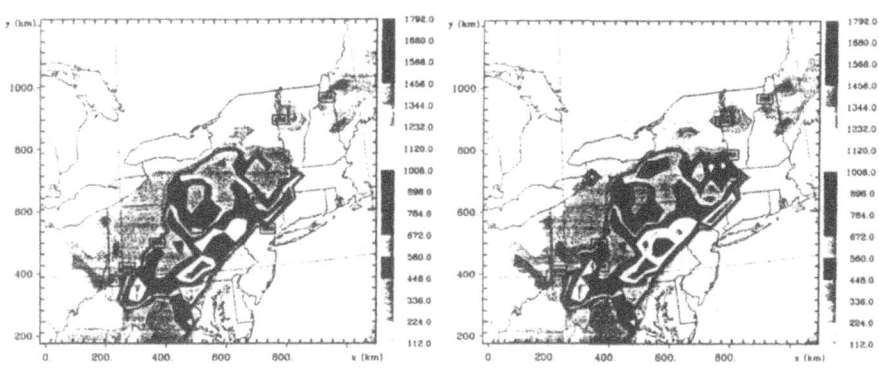

(a)   **(b)**

Figure 4: Wet deposition of Hg^2 (in ng/m^2) at 0000 UTC on 26 August 1997 after 14 days of simulation, estimated from RAMS model (a) with NY sources and (b) without NY sources.

The total deposited mercury (wet and dry) for the simulation performed using all available mercury sources was also averaged over the New York State. More specifically, the wet and dry deposition of Hg^P, Hg^2 and Hg^0-adsorbed were calculated using both modelling systems, namely RAMS and SKIRON/Eta. Dry and wet deposition of all mercury species, accumulated for the simulation period and averaged over the entire domain of the State of New York, is shown in Fig. 5. The model results are in good agreement, since the calculated amounts of deposited mercury do not exhibit major differences. However, SKIRON/Eta system calculated consistently higher amounts of wet and dry deposited mercury. The differences in the total (wet and dry) deposited amounts of mercury between the two systems are attributed mainly to the larger amounts of wet deposited mercury calculated from SKIRON/Eta. It is known that the wet deposition pattern of mercury is strongly dependent on the precipitation pattern. The Betts-Miller-Janjic (Betts and Miller, 1986, Janjic, 1994) convection scheme that SKIRON/Eta uses tends to overestimate the precipitation areas, without overestimating the precipitated water. On the contrary, the RAMS microphysical scheme calculates higher amounts of precipitation in more restricted areas, leading to local peaks.

In general, the agreement between the calculated total deposited amounts of both models is quite good (Fig. 5). This indicates that they are reliable tools in the study of mercury and that these modelling systems can be used for longer period simulations and estimations of mercury depositions.

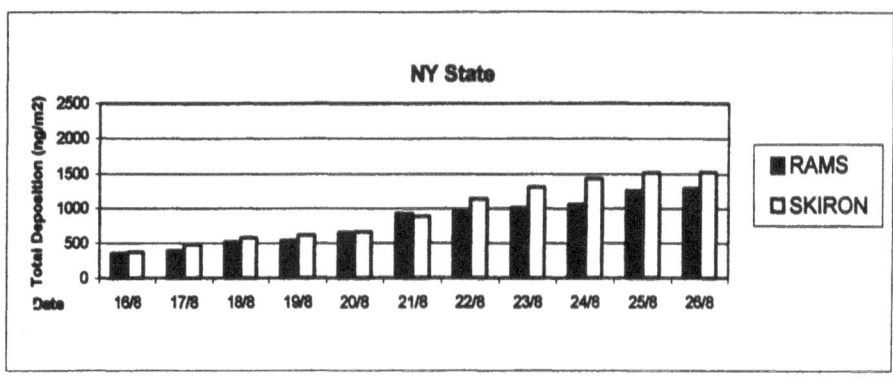

Figure 5: Total deposition of mercury (ng/m²) calculated from RAMS and SKIRON/Eta systems (with NY State emissions) averaged over NY State from 14 to 26 August 1997.

4. OBSERVATIONS-MODEL CALCULATION INTERCOMPARISON

Deposition measurements were available from several locations of the NE part of the US. More specifically, the Mercury Deposition Network (MDN) provided wet deposition measurements at sites upwind and downwind of NY State only. MDN deposition observations at selected sites within the MDN, namely Allegheny Portage at Pennsylvania, Bridgton and Greenville at Maine have been compared with the accumulated wet deposition of mercury from both models in both scenarios and the results are illustrated in Figs 6-8.

The available observations for these stations represent the weekly measured wet deposition of all mercury species, for the periods 12 to 19 August 1997 and 19 to 26 August 1997. Only two deposition observations are available for the model simulation period. However, an attempt was made to inter-compare model outputs and observations. Since no information for the starting hour during the sampling period is available, the observations have been compared with the 0000UTC model outputs. Accumulated wet deposition of all mercury species have been calculated for all 12 days of simulation. The wet deposition values of all three mercury species (Hg^P, Hg^2 and Hg^0-adsorbed) have been accumulated from the initial time of the simulation, for both cases (with and without NY sources) for the entire simulation period. A similar accumulation has also been made for the observations, in order to achieve greater consistency between the observations and model calculations.

The inter-comparison between model calculations and observations was made with both models, for both scenarios. Both models tend to overestimate the deposited amounts of mercury. The observations seem to be higher for the 12 to 19 August 1997 period, at Bridgton and Greenville compared to the model calculated values. On the contrary, when observations for the periods 12 to19 August and 19 to 26 August are accumulated, they are lower than the model-calculated total deposited amount. The overestimated deposited quantities of mercury from both models are within the acceptable limits, taking into account the observation errors, the uncertainties of the observation network and the fact that weekly measurements are used. It is worth mentioning that even a small shift (temporal or spatial) in the model estimated rain pattern during the observation period, can strongly influence the model deposition values.

When the NY State local emissions were not used during the simulation period, the accumulated wet deposition is (as expected) lower, indicating the strong influence of the emissions at the sites in and around the NY State (see Figs 6-8).

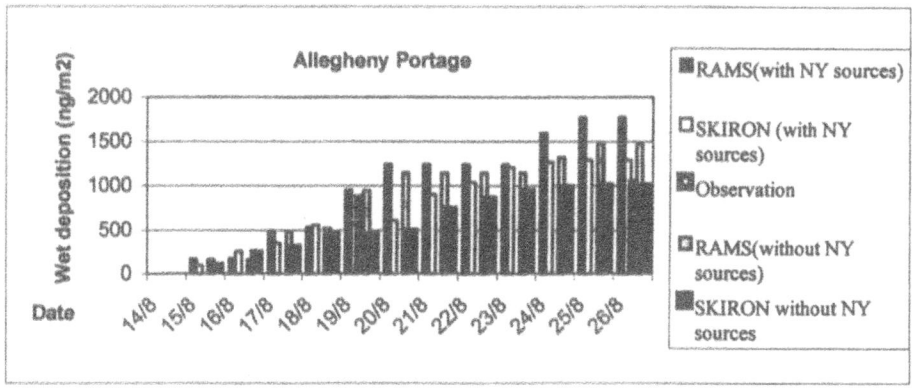

Figure 6: Comparison of observations of wet deposition of mercury (ng/m²), RAMS and SKIRON/Eta outputs (with and without NY State emissions) at Allegheny from 14 to 26 August 1997. RAMS and SKIRON/Eta outputs are accumulated since the initial time of the simulation, while the observations correspond to weekly-deposited mercury.

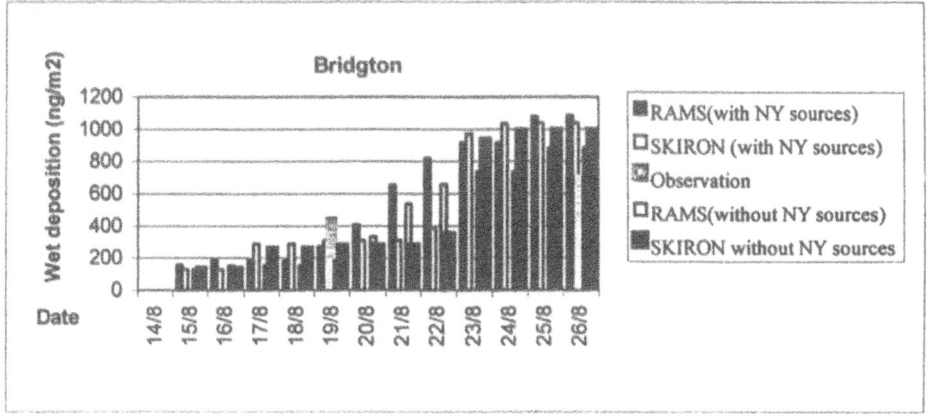

Figure 7: Comparison of observations of wet deposition of mercury (ng/m²), RAMS and SKIRON/Eta outputs (with and without NY State emissions) at Bridgton from 14 to 26 August 1997. RAMS and SKIRON/Eta outputs are accumulated since the initial time of the simulation, while the observations correspond to weekly-deposited mercury.

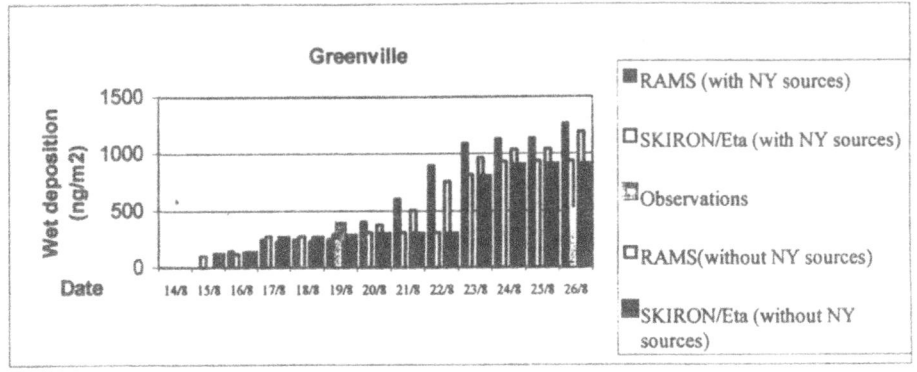

Figure 8: Comparison of observations of wet deposition of mercury (ng/m²), RAMS and SKIRON/Eta outputs (with and without NY State emissions) at Greenville from 14 to 26 August 1997. RAMS and SKIRON/Eta outputs are accumulated since the initial time of the simulation, while the observations correspond to weekly-deposited mercury.

5. CONCLUSIONS

This study focused on the regional and synoptic transport of mercury and the in/out of state contributions of mercury sources to the total deposited mercury over the State of New York. When the NY State local emissions were not used (second scenario), the accumulated wet deposition was (as expected) lower. The wet deposition of all three mercury species in the second scenario was reduced up to 15%.

In both models, a consistency is evident at most stations. However a longer simulation period combined with a large number of observations is absolutely necessary for more reliable conclusions. Wet deposition observations should be available on a daily basis to compare against modelled values. The deposition patterns estimated by the models can be used to overcome the difficulties in measuring the wet and dry deposition of mercury. The developed models should be considered as practical tools for studying the mercury processes and therefore can be useful to policy makers in assessing various emission control strategies.

6. REFERENCES

Betts, A. K. and M. J. Miller, 1986: A new convective adjustment scheme. Part II: Single column tests using GATE wave, ATEX and Arctic Air mass data sets, *Quart. J. R. Met. Soc.*, **112**, 693-709.

Davies, F., and Notcutt G., 1996: Biomonitoring of atmospheric mercury in the vicinity of Kilauea, Hawaii. *Water Air Soil Pollut.* 86, 275-281.

Janjic, Z. I., 1994: The step-mountain Eta coordinate: Further developments of the convection, viscous sublayer and turbulence closure schemes, *J. Atmos. Sci.*, **122**, 927-945.

Kallos, G., S. Nickovic, A. Papadopoulos, D. Jovic, O. Kakaliagou, N. Misirlis, L. Boukas, N. Mimikou, G. Sakellaridis, J. Papageorgiou, E. Anadranistakis and M. Manousakis 1997: The Regional weather forecasting system SKIRON. *Proceedings of the Symposium on Regional Weather Prediction on Parallel Computer Environments*, 15-17 October 1997, Athens, Greece. 109-122.

Pielke, R. A., Cotton W. R., Walko R. L., Tremback C. J., Lyons W. A., Grasso L. D., Nicholls M E., Moran M. D., Wesley D. A., Lee T. J., and J. H. Copeland, 1992: A comprehensive meteorological modelling system - RAMS. *Meteorol. Atmos. Phys.*, **49**, 69-91.

Schroeder, W., and J. Munthe, 1998 Atmospheric Mercury - An Overview. *Atm. Env.*, **32**, pp 809-822

CONTRIBUTION OF DESERT DUST TRANSPORT TO AIR QUALITY DEGRADATION OF URBAN ENVIRONMENTS RECENT MODEL DEVELOPMENTS

Papadopoulos, A., P. Katsafados, G. Kallos, S. Nickovic, S. Rodriguez, and X. Querol[*]

1. INTRODUCTION

Mineral dust, produced by wind erosion over arid and semi-arid areas of North Africa, may transport away to the Middle East, Mediterranean, Europe, even into and across the Atlantic Ocean (Kallos et al., 2002). This material transported away from its origin is considered as an important climate and environment modifier. Dust particles by absorbing and backscattering both the incoming solar radiation and the infrared outgoing radiation modify the Earth's radiation budget (Andreae, 1996). In addition, they alter the cloud microphysics processes acting as cloud condensation nuclei and having pH > 7.0 play a role in neutralization of the acid rains (Hedin and Likens, 1996). Also with the long-range dust transport, important nutrients are transported from their sources to other regions and may significantly modify the biogeochemistry of these marine and terrestrial ecosystems (Swap et al., 1996). For example, the deposition of the North African dust material on the Mediterranean Sea provides important nutrients, such as nitrogen species, phosphorus and iron, which may enhance the marine productivity. Some summer algal blooms in the Mediterranean Sea may be explained by such Saharan dust deposition (Dulac et al., 1996). Guerzoni et al. (1999) have estimated the amount of the atmospheric dust mass deposited on Mediterranean region to be ~40x10^6 tons. Even though, they turned out to this magnitude by measured atmospheric mass flux at 9 coastal sites (which is considered as a small number of sites for such a work), this is considered as a valuable estimation since it is the only one found so far.

As it was found from satellite observations and ground-based measurements, there is a large seasonal variability of the dust mobilization that depends on the source characteristics as well as the global atmospheric circulation (Ozsoy et al., 2001). During winter and spring the Mediterranean region is affected by two upper air jet streams: the polar front jet stream, originally located over Europe, and the subtropical jet stream which is typically located over northern Africa. The combined effects of these westerly jets in winter and spring support the propagation of extratropical cyclones towards East and Southeast, resulting

---
[*] A. Papadopoulos, National Centre for Marine Research, Athens, Greece 19013. P. Katsafados and G. Kallos, University of Athens, 15784 Athens, Greece. S. Nickovic, University of Malta, Valletta, Malta. S. Rodriguez, and X. Querol, Institute of Earth Sciences, Barcelona, Spain

Air Pollution Modeling and Its Application XVI, Edited by
Borrego and Incecik, Kluwer Academic/Plenum Publishers, New York, 2004

279

in dust plume intrusion in the Mediterranean. During summer the transported aerosols are almost twice as large as in winter (Husar et al., 1997) but the highest amount of dust transport is within the tropical easterly jet from Africa toward the tropical Atlantic, reaching the Caribbean Sea and North America (Perry et al., 1997; Kallos et al., 2002). Most of the Saharan dust events that transport significant amounts of dust towards the Mediterranean Sea and Europe occur during the low index circulation period of the year (cold and transient seasons as is described in Kallos et al. 1997, Rodriguez et al., 2001).

The dust amounts deposited on the surface are in proportion to the seasonal variability of the dust cycle in the atmosphere. To have a feeling of the magnitude and the geographical distribution of the dust deposition on ground surfaces and on coastal and open seas, the use of a credible numerical model is considered as essential. In this study, we illustrate and briefly describe a database of model-derived seasonal amounts of dust deposited on Mediterranean waters and European land. The SKIRON/Eta weather forecasting system coupled with a dust cycle model is used (Nickovic et al., 1997; Papadopoulos et al., 2002). The dust modules incorporate the state of the art parameterizations of all the major phases of the atmospheric dust life such as production, diffusion, advection and removal related also to particle size distribution (Nickovic et al., 2001). The SKIRON/Eta system is in operational use since 1998 providing 72-hour forecasts for the Mediterranean region (http://forecast.uoa.gr). The last modifications concerning the definition of the dust sources and the dust production mechanism enhance the forecast skill of the system to predict with a satisfactory accuracy the dust cycle in the atmosphere. Since dust is removed from the atmosphere due to mechanism such as gravitational settling and turbulent mixing and/or precipitation rates, we estimated separately dry and wet dust deposition. In this presentation the spatiotemporal distribution of the dust deposited over the Mediterranean waters and Europe by utilizing the database of the SKIRON/Eta outputs was made for the three-year period of 2000-2002.

2. DERIVATION OF SEASONAL AMOUNTS OF DUST DEPOSITED ON MEDITERRANEAN BASIN AND EUROPE

In order to derive the seasonal amounts of North African dust mass deposited on Mediterranean basin and Europe, the SKIRON/Eta system was integrated over the domain covering North Africa, Mediterranean and a big part of Europe and Middle East as it is illustrated in Figure 1. In the vertical, 32 levels were used stretching from the ground to the model top (15800 m). In the horizontal, the grid increment of 0.24 of degree was applied. At each model grid point, the dry and wet dust deposition was estimated, assuming them as an average of the sub-area around each one. The 24-hour dust concentration fields from the previous-day simulation cycle are used for initial conditions. Utilizing the first 24-hours model output of each daily simulation, we calculated the geographical distribution of the total dust mass deposited on surface for each month from January 2000 to December 2002 (36-month period). The dry and wet deposition of dust was calculated separately.

During winter, the dust plumes do not expand towards Europe because of the strong washout mechanisms occurring due to the cyclonic activity in the Mediterranean Region. In addition, during the winter, the dust extraction mechanisms are suppressed due to the crust formation over the soil due to rain that occurs from time to time.

During the transition seasons (spring and autumn) the dust plume shifts at more northerly latitude to Europe resulting significant amounts of dust deposited on Mediterranean Sea and also on European continent. For instance, during winter in the Central Italy the dust mass deposited is negligible but it reaches 2,5 gr/m^2 at the mean April. Besides, the most important dust events occurred in the area of interest during spring and in the beginning of the summer. In summer due to the fact that dust is also transport towards Atlantic, less amount of dust mass deposited on the Mediterranean Sea and the Europe.

The higher amounts of dust deposited over Europe are during the spring months. This is due to the enhanced "productivity" of dust source areas and the synoptic conditions in the Mediterranean Region. The dust production is enhanced during this period because of the breaking of soil crusts due to strong heating (Zilitinkevich, personal communication) and the quick passage of synoptic disturbances. The transport towards Europe is favored by the fact that the Mediterranean waters are relatively cool during spring months and the convective activities associated with rain are not significant to washout the dust.

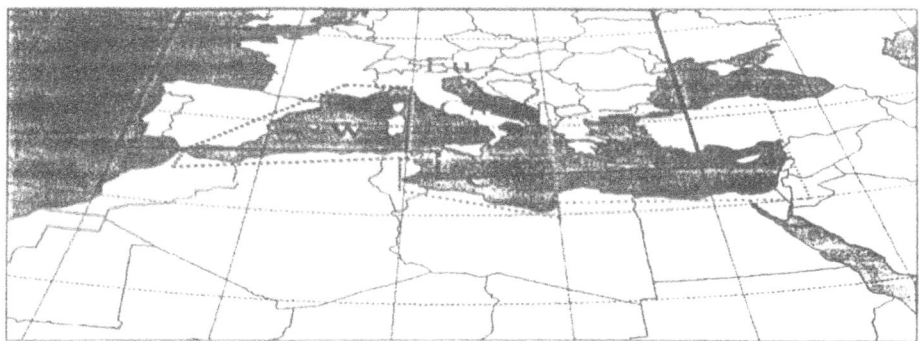

Figure 1. Model domain. Letters W, Cs, Cn, E and Eu denote the West, Central-Southern Part, Central-Northern Part, East Mediterranean sub-basins and the European part, respectively, over where the model-derived dust deposition are calculated.

During summer, dust transport events usually occur at the western part of the Mediterranean while in the East are rear because of the dominance of the trade wind systems from North to South.

The annual geographical distribution of the dust mass deposited on surface (in units of gr/m^2) for the years 2000, 2001 and 2002 is illustrated in Figures 2, 3 and 4, respectively. From this distribution we note that during 2000 much more dust mass has been deposited over Europe, with its total annual peak located in Italy, while in 2001 the dust mass deposited over Europe was less with its peaks shifted towards east. During the next year 2002 the amount dust mass deposited over Europe is much more than the previous years and its highest values are more spread over Italy and Balkan Peninsula and mainly at its westerly parts. This high variability in dust deposition from year to year suggests us to be careful when we estimate average annual depositions in specific locations and also the need for long time series (for several years) when we derive average amounts of deposited dust. Just one strong dust event can change significantly the local deposited quantities on monthly and annual base.

In order to check the amount of North African dust mass deposited over the Mediterranean Sea and Europe the model domain was divided in sub-regions. Then, the deposited (dry, wet and total) quantities over these regions were estimated. These sub-regions are called Europe (Eu), West Mediterranean (W), East Mediterranean (E), Central Mediterranean North (Cn) and Central Mediterranean South (Cs) and are indicated in Figure 1.

In Figures 5a, b and c, the monthly amount of dust deposited on the Mediterranean Sea and Europe is displayed for the years 2000, 2001 and 2002, respectively. As it is seen, the higher deposition occurs during the spring months with the autumn to follow. The dry deposition is higher over the Mediterranean Sea while the wet dominates over the European continent. Obviously, this is due to the amounts of rain over Europe and Mediterranean Sea. In these calculations we considered the dust amounts deposited only on the land grid points for the Europe while only the sea grid points are counted in the case of the Mediterranean Sea and its four sub-basins. The number of grid points used for each sub-area is shown in

281

Table 1. As we see, there are different numbers of grid points at each sub-region and this must be taken into the account when we try to interpret these amounts and figures.

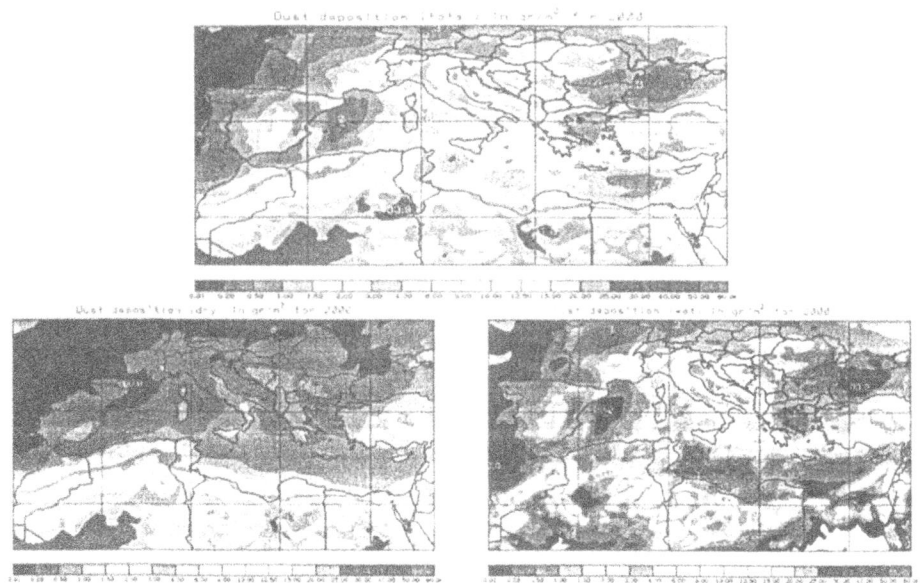

Figure 2. Annual dust deposition (total, dry, wet) for 2000.

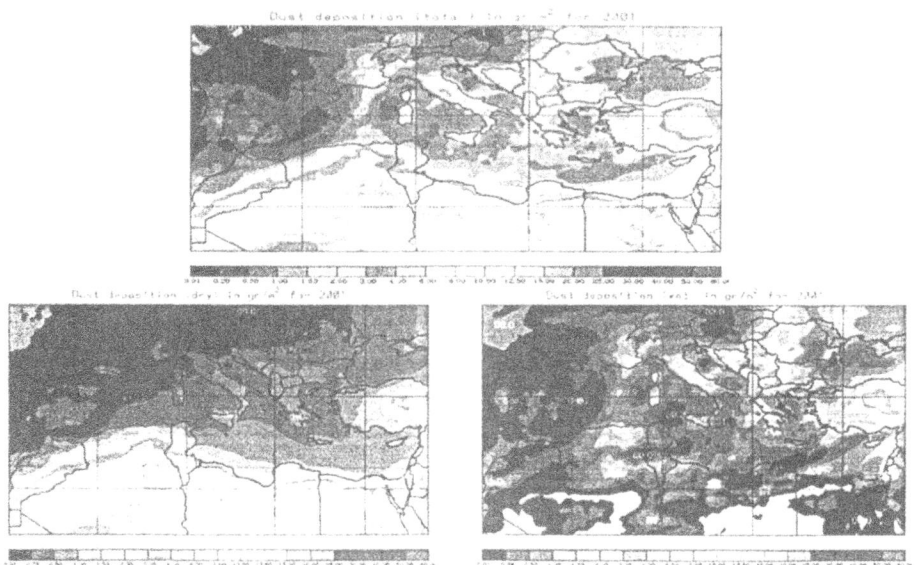

Figure 3. Annual dust deposition (total, dry, wet) for 2001.

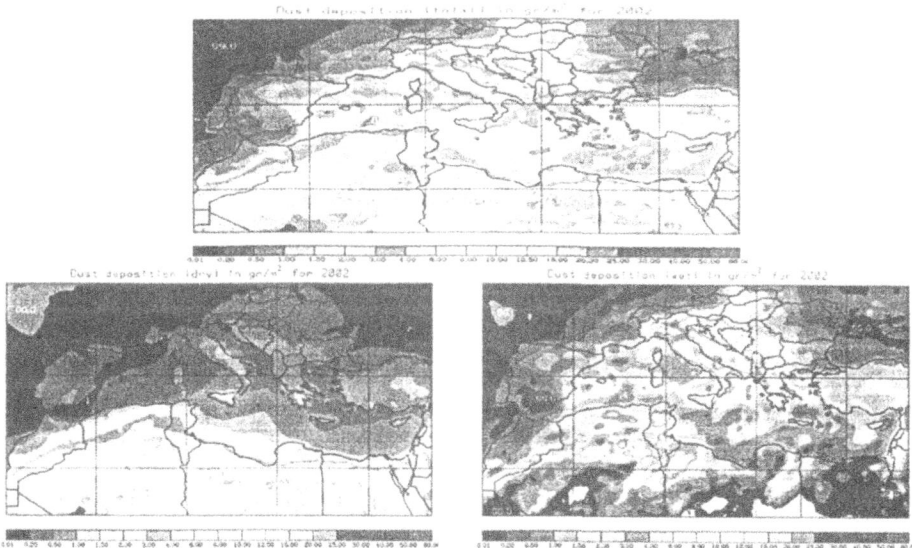

Figure 4. Annual dust deposition (total, dry, wet) for 2002.

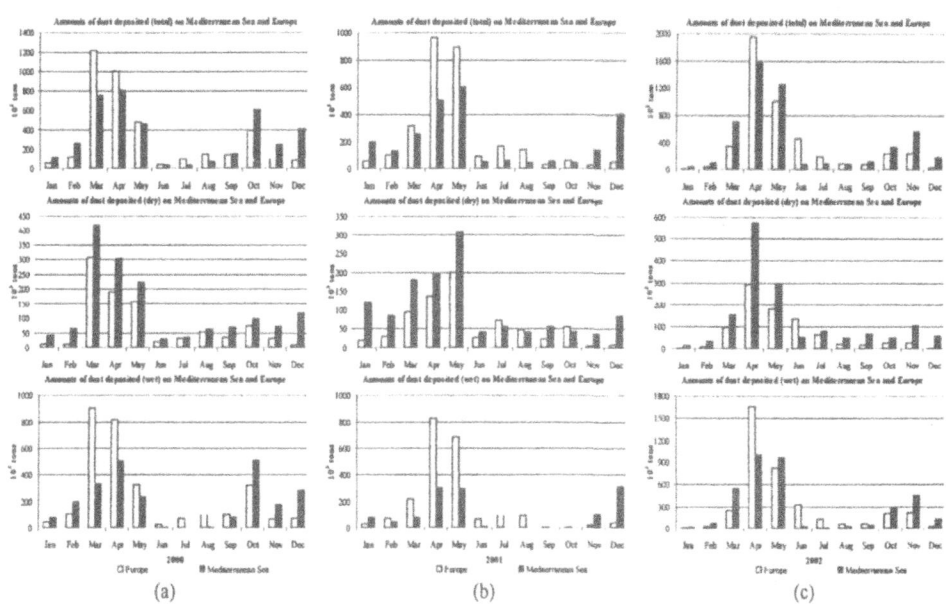

Figure 5. Monthly amounts of dust deposited (total, dry and wet) on Mediterranean Sea and Europe for (a) 2000, (b) 2001 and (c) 2002.

Table 1. Number of grid points used for each sub-area in the calculations of dust amounts deposited on the Mediterranean Sea and Europe

	EUROPE	MEDITERRANEAN				
		Whole	West	Central South	Central North	East
Land grid points	5783	-	-	-	-	-
Sea grid points	-	4344	1105	694	1000	1545

In order to elucidate the effects of dust inputs to the Mediterranean waters in various regions, the dust deposition was also calculated over the four sub-basins shown in Figure 1, denoting West, Central southern part, Central northern part and East Mediterranean Sea, for the two-year period. Central Mediterranean has been divided in southern and northern parts because the deposition patterns are different. The deposited amounts on each sub-region are illustrated in Figures 6a, b and c for the years 2000, 2001 and 2002, respectively. As it is shown, the Eastern part of Mediterranean receives a high amount of dust through wet deposition. This is due to the fact that the Middle East region receives most of its rain from the lows in the Cyprus and secondary from the lows moving along the coast of North Africa. These systems favor the dust production over East Saharan and then the transport towards East Mediterranean and Middle East.

Utilizing the estimated monthly dust deposition, the total annual amount of dust deposited on the Mediterranean Sea and Europe has been calculated. These quantities are tabulated in Table 2. As we see, the annual deposition of dust over the Mediterranean waters and European land is much higher for 2002 than the previous years 2000 and 2001.

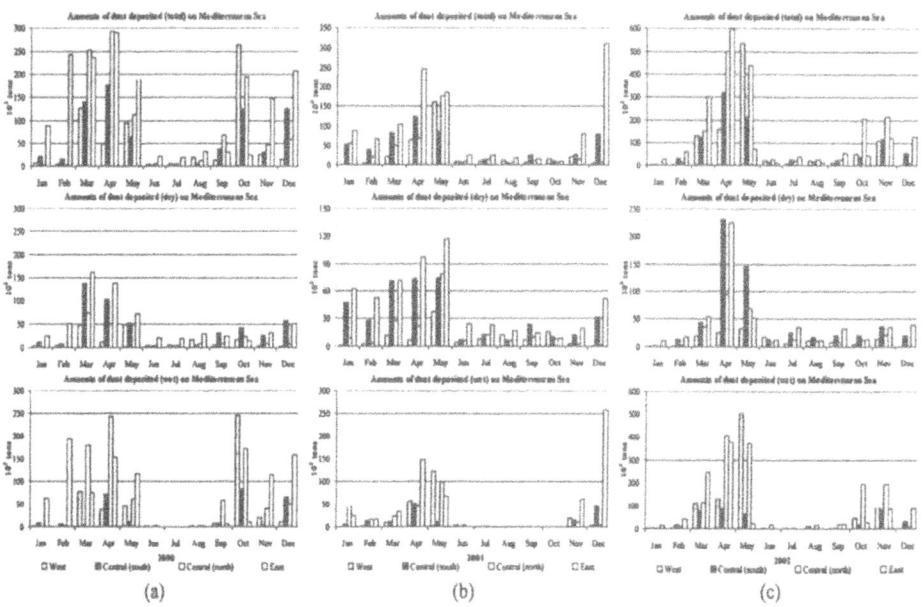

Figure 6. Monthly amounts of dust deposited (total, dry and wet) on four basins of the Mediterranean Sea for (a) 2000, (b) 2001 and (c) 2002.

Table 2. Total annual dust deposition (in 10^3 tons) on the Mediterranean Sea and Europe

year	Europe			Mediterranean Sea		
	total	dry	wet	total	dry	wet
2000	3914	936	2978	3962	1541	2421
2001	2909	725	2184	2500	1255	1245
2002	4723	883	3840	5133	1546	3586

According to the Guerzoni et al. (1999) the atmospheric dust mass deposited on Mediterranean region, is approximately 40×10^6 tons. The model-derived values presented above seem to be an order of magnitude less. This difference can be attributed to the following reasons:

1. The methodology followed by Guerzoni is characterized by low spatial representation. Guerzoni et al. turned out to their figures by calculating mean values for 3 sub-basins in the Mediterranean region from the measured atmospheric mass flux at 9 coastal sites and then estimating the total annual atmospheric dust flux for the whole Mediterranean basin. In addition, their methodology cannot separate the long-range transported Saharan dust from locally produced dust due to soil erosion and human activities.

2. By utilizing the numerical model there is the opportunity to estimate the dust fluxes at its grid point, assuming them as the representative values of an area of 0.24×0.24 degree^2. The version of the SKIRON/Eta system used for these calculations utilizes only one average dust particle diameter of 2 μm (i.e $PM_{2.5}$). With this assumption more dust particles can be transported further away but each of them carries less mass. This may lead to an underestimation of the total dust mass deposited near the North African coast because larger particles can be deposited there. This does not seem to be a problem for the northern coast of Mediterranean and Europe since this is the particle size that travels large distances. Regarding to measurements during some dust events, the model seems to simulate quite satisfactory the dust transport and deposition. This has been verified in the past during the MEDUSE project.

Further development of the SKIRON/Eta system has been performed during the last year in the framework of the ADIOS project. This new development includes a new scheme for dust production is applied based on a more sophisticated parameterization of dust mobilization process. Currently, four size bins are used in its operational application. An extensive testing of the new development is still under way. This is performed at the framework of ADIOS project and the cooperation with George Mason University and NASA.

4. CONCLUSIONS

The dust cycle in the atmosphere is considered as important due to several implications such as in climate, urban air quality, ecosystems, regional/mesoscale weather and rain. With the aid of the SKIRON/Eta modeling system, which is able to accurately simulate weather, conditions and the desert dust uptake-transport-deposition cycle, a database of model-derived seasonal amounts of dust deposited on Mediterranean Sea and Europe has been created. Due to the fact that dust is removed from the atmosphere by dry and wet deposition, in the model deposition scheme the dust is deposited on surfaces according to gravitational settling and turbulent mixing (dry) and to precipitation rates (wet).

In this study, the monthly and annual amounts of dust deposition (dry and wet) have been compiled. In addition, an attempt was made to analyze the dust deposition spatially. From this analysis we can say that the annual, seasonal and event variability is significant. This must be taken into the account when model-observations inter-comparison will be organized. There are cases where through a single episode the deposited dust in a certain location is equivalent to several-month deposition. The lack of available observations related to dust concentration in the atmosphere as well as deposition measurements did not permit us further model evaluation. The only possible comparison is between the simulated dust load fields and the Aerosol Index (AI)–the parameter that describes the level of UV absorption due to the existence of dust and smoke in the atmosphere (Herman et al., 1996). Despite the fact that there is no straightforward way to convert dust load to AI, the comparison of the position and shapes of the two fields may be used for a quality control of the dust model performance. This comparison showed a very good agreement between the satellite data and model outputs. Through the ADIOS project and the exchange of data with other partners, there will be an excellent opportunity to evaluate the dust transport and deposition against surface dust concentrations and more quantitative satellite images.

5. ACKNOWLEDGEMENTS

This study was performed within the framework of the ADIOS project (EVK3-2000-00604) funded by the European Commission – Research Directorate.

6. REFERENCES

Andreae, M. O., 1996, Raising dust in the greenhouse, *Nature* **380**:389-340.

Dulac,F., C.Moulin,C.Lambert,F.Guillard,J.Poitou,W.Guelle,C.Quetel,X.Schneider, U.Ezat, 1996: Quantitative remote sensing of African dust transport to the Mediterranean, in: *The impact of desert dust across the Mediterranean*, S. Guerzoni and R. Chester, ed., Kluwer Academic/Plenum Publishers, NY, pp. 25-49.

Guerzoni, S., R. Chester, F. Dulac, B. Herut, M.-D. Loye-Pilot, C. Measures, C. Migon, E. Molinaroli, C. Moulin, P. Rossini, C. Saydam, A. Soudine, and P. Ziveri, 1999, The role of atmosphere deposition in the biogeochemistry of the Mediterranean Sea, *Prog. Oceanogr.* **44**:147-190.

Hedin, L. O. and G. E. Likens, 1996, Atmospheric dust and acid rain, *Sci. Am.*, **12**:56-60.

Herman, J. R., P. K. Bhartia, O. Torres, C. Hsu, C. Seftor, and E. Celarier, 1996, Global distribution of UV-absorbing aerosols from Nimbus-7/TOMS data, *J. Geophys. Res.* **102**:16911-16929.

Husar R. B., L. M. Prospero, and L. L. Stowe, 1997, Characterization of tropospheric aerosols over the oceans with the NOAA advanced very high resolution radiometer optical thickness operational product, *J. Geophys. Res.* **102**:16889-16909.

Kallos, G., S. Nickovic, A. Papadopoulos, D. Jovic, O. Kakaliagou, N. Misirlis, L. Boukas, N. Mimikou, G. Sakellaridis, J. Papageorgiou, E. Anadranistakis, and M. Manousakis, 1997, The regional weather forecasting system SKIRON: An overview, in: *Proceedings of the International Symposium on Regional Weather Prediction on Parallel Computer Environments*, G. Kallos, V. Kotroni, and K. Lagouvardos, ed., ISBN: 960-8468-22-1, University of Athens, Greece, pp. 109-122.

Kallos, G., S. Nickovic, A. Papadopoulos, and P. Katsafados, 2003, Transport of Saharan dust towards the USA: Model simulation, *J. Geophys.Res.* (submitted).

Nickovic, S., D. Jovic, O. Kakaliagou, and G. Kallos, 1997a, Production and long-range transport of desert dust in the Mediterranean region: Eta model simulations, in: *Proceedings of 22nd NATO/CCMS International Technical Meeting on Air Pollution Modeling and Its Applications*, Sven-Erik Gryning and Nadine Chaumerliac, ed., Plenum Press, New York, pp. 15-24.

Nickovic, S., G. Kallos, O. Kakaliagou, and D. Jovic, 1997b, Aerosol production/transport/deposition process in the Eta model: desert dust simulations, in: *Proceedings of the International Symposium on Regional Weather Prediction on Parallel Computer Environments*, G. Kallos, V. Kotroni, and K. Lagouvardos, ed., ISBN: 960-8468-22-1, University of Athens, Greece, pp. 137-145.

Nickovic, S., G. Kallos, A. Papadopoulos, and O. Kakaliagou, 2001, A model for prediction of desert dust cycle in the atmosphere, *J. Geophys. Res.* **106**:18113-18129.

Ozsoy, E., N. Kubilay, S. Nickovic, and C. Moulin, 2001, A hemisphere dust storm affecting the Atlantic and Mediterranean in April 1994: Analyses, modeling, ground-based measurements and satellite observations, *J. Geophys. Res.* **106**:18439-18460.

Papadopoulos A., G. Kallos, P. Katsafados, and S. Nickovic, 2002, The Poseidon weather forecasting system: An overview, *GAOS* **8**:219-237.

Perry, K.D., T.A. Cahill, R.A. Eldred, and D.D. Dutcher, 1997, Long-range transport of North African dust to the eastern Union States, *J. Geophys. Res.* **102**:11225-11238.

Rodriguez, S., X. Querol, A. Alastues, G. Kallos and O. Kakaliagou, 2001, Saharan dust contribution to PM10 and TSP levels in Southern and Eastern Spain, *Atmos. Environ.* **35**:2433-2447.

Tegen, I., and I. Fung, 1994, Modeling of mineral dust in the atmosphere: Sources, transport, and optical thickness, *J. Geophys. Res.* **99**:22,987-22,914.

DISCUSSION

D.KORACIN	Have you done any systematic evaluation to see how the models predict spatial (horizontal and vertical) distribution of clouds using satellite data? That definitely has strong impact on your predictions.
G. KALLOS	Yes we have. Other researchers have done similar work for the atmospheric models we used. I can refer the recent paper of Mavromatidis and Kallos on JGR (Issue of May 2003).
D.SYRAKOV	What can you say about the last dry period over Balkans. Is there any dust period that time?
G. KALLOS	Dust episodes exist all time at the frequency of 2-3 episodes per month and for different parts of the Mediterranean. Of course during summer the episodes are rare in the Easter part of it due to the trade winds (etesians).
R.SAN JOSE	In our experience initializing the model with Aerosol Optical Index from NASA (satellite) could be good option. What is your opinion on that?
G. KALLOS	Satellite data do not provide vertical distribution of dust so they cannot be used for model initialization. There are some experiments doing such initialization but the entire methodoloy is under question for its credibility.
M.SOFIEV	A comment: In Finland we have made a quick check of the dust events and found that in most cases the dust comes from Caspian area rather than Sahara. The difference in the number of episodes is about of a factor of 3.
G. KALLOS	I agree with you. There are source areas in the Caspian Region too. The transfer towards Fnland is quite possible as you say.

NEW DEVELOPMENTS

Chairperson: G. Schayes

                   C. Borrego

                   Ph. Thunis

                   D. Steyn

Rapporteur: J. Flemming

                   D. Steyn

                   E. Renner

                   N. Moussiopoulos

ONCE MORE ON THE ADVECTION SCHEMES: DESCRIPTION OF *TRAP*-SCHEMES

Dimiter Syrakov[*]

1 Introduction

One of the key problems in air pollution modelling is the accuracy and the speed of the numerical schemes. The description of the advection processes still keeps being a real challenge for tracer dispersion modellers. Even the best chemical scheme or boundary layer parameterization is useless while the advection scheme produces considerable errors. In the same time, a tendency is observed comprehensive models to be used in long-term integration. This urges the modellers to search for compromise between the accuracy and the speed of computations. The same are the requirements of the models designed for performing on small computational platforms. A lot of methods and schemes were proposed in the literature (WMO-TCSU, 1979, Rood, 1987, Peters et al, 1995), but only a small number of them are suitable and are being in practical use. So far, none of the schemes possesses all properties of the exact solution of the advection equation.

One of the most widely used schemes is the one, elaborated by Bott (1989) and further improved by him (Bott, 1992, 1993). In the Bott scheme, the advective fluxes are computed utilising the integrated flux concept of Tremback et al. (1987). The fluxes are normalised and then limited by upper and lower values. The produced scheme is conservative and positively definite with small numerical diffusion. These properties make the Bott scheme very attractive for further improvements and optimisations. TRAP (Syrakov 1995, 1996, Syrakov and Galperin 1997, 2000) is such a daughter scheme.

[*] National Institute of Meteorology and Hydrology, 66 Tzarigradsko chaussee Bulvd., Sofia 1784, Bulgaria, elm: dimiter.syrakov@meteo.bg.

In the study, comparative tests of the Bott and two other schemes versus some variants of TRAP scheme are provided. Some useful properties and improvements of TRAP scheme are presented as well.

2 DESCRIPTION OF ADVECTION SCHEMES

As splitting is applied in the multi-dimensional case, the one-dimensional advection equation in non-divergent and in flux forms is considered here:

$$\partial C/\partial t + u\partial C/\partial x = 0, \qquad \partial C/\partial t + \partial(uC)/\partial x = 0, \tag{1}$$

$C(x,t)$ being the tracer concentration, $u(x,t)$ - the advection velocity, x - the space co-ordinate and t - the time. The simplest approach to numerical solving of Eq. (1) is the direct replacement of the derivatives by some difference approximations. Let homogeneous grid is introduced in space and time: $x \rightarrow x_i = i\Delta x$, $i=1,N_x$; $t \rightarrow t_n = n\Delta t$, $n=0,N_t$., Δx and Δt being the space and time steps. The corresponding grid values of velocity and concentration are $u(x_i, t_n)=u_i^n$ and $C(x_i, t_n)=C_i^n$. Keeping the mass balance in cell i, the flux form of Eq. (1) can be discretized as

$$C_i^{n+1} = C_i^n - \frac{\Delta t}{\Delta x}(F_{i+1/2} - F_{i-1/2}) \quad \text{or} \quad C_i^{n+1} = C_i^n - (Fr_i - Fl_i), \tag{2}$$

where $F_{i\pm1/2} = F(u,C)$ are the mass fluxes through the right and the left edges of the cell, Fr and Fl being masses transported through the edges for one time step. Fr and Fl can be positive or negative in dependence on the transport direction and $Fm=\text{sign}(u_m)Am$, $(m=r,l)$, where Am is the so called **flux area** (the shadowed area in Fig.1). Usually, the problem can be normalised by introducing the Courant number $U_i^n = u_i^n\Delta t/\Delta x$ and setting $\Delta x=\Delta t=1$. The schemes are **explicit** when the fluxes are calculated for the moment t_n. This is the case for the schemes discussed here, so the upper index n will be omitted further on. Practically, all possible schemes can be reformulated to the forms of Eq. (2); they differ in the way of determining of fluxes. A scheme is **mass conserving** if it is constructed in such a way that always $Fl_i = Fr_{i-1}$. As this condition is fulfilled for all schemes discussed here, mainly the flux at the right cell edge is considered further on.

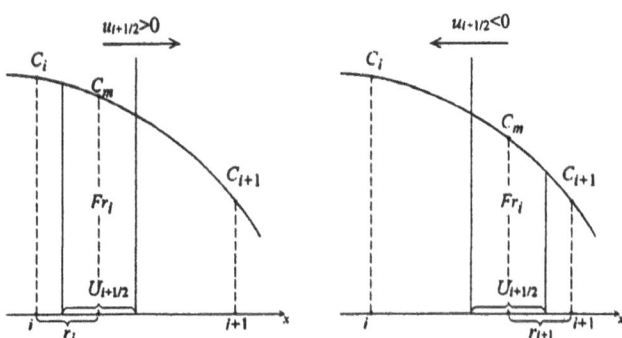

Figure 1. Right edge flux for positive and negative wind velocities

2.1 Holmgren's variant of MacCormack scheme - McC.

Holmgren (1994) developed some improved variants of the MacCormack scheme. After testing he used one of them in a meso-scale meteorological model. According to this scheme

$$C_i^{n+1} = 0.5\left\{C_i^n + C_i^* - U_i\left|(1+a)(C_i^* - C_{i-1}^*) - (a/2)(C_i^* - C_{i-2}^*)\right|\right\}$$
$$C_i^* = C_i^n - U_i\left[(1+a)(C_{i+1}^n - C_i^n) - (a/2)(C_{i+2}^n - C_i^n)\right] \qquad a=1/3. \qquad (3)$$

It can be noticed that this scheme is realised on a pattern of 5 grid points, so it is 4th order in space.

2.2 Roussel&Lerner's slope scheme - R&L.

Roussel and Lerner (1981) proposed a scheme, in which the flux area of cell i (Fig.1) is approximated by trapezium. The upper boundary of this trapezium is a straight line passing through the point (i,C_i) in case of positive velocity or through the point $(i+1,C_{i+1})$ when $U_{i+1/2}<0$ (the upstream concept). The authors constructed difference equations for the x- and y-slopes of this line and its' future values are predicted together with the concentration's ones. The height of trapezium is the distance passed by the mass particle which leaves the cell first and this distance is equal to the Courant number $U_{i+1/2}$ (for the right edge flux). The flux area is determined by multiplying this height by the linear estimate for the concentration in the middle of the passed distance. The starting grid concentration and the distance argument depend upon the direction of transport. They are C_i and $r_i=(1-U_{i+1/2})/2$ when $U_{i+1/2}>0$ and C_{i+1} and $r_{i+1}=-(1+U_{i+1/2})2$ in the opposite case.

2.3 Bott's scheme - BOT

Bott (1989) determines the flux area after the integrated flux concept (Tremback et al., 1987):

$$Ar_i = \int_{1/2-U_{i+1/2}}^{1/2} C_i^p(x)dx \text{ when } U_{i+1/2}>0, \qquad Ar_i = \int_{1/2}^{1/2-U_{i+1/2}} C_{i+1}^p(x)dx \text{ when } U_{i+1/2}<0, \qquad (4)$$

where $C_i^p(x)$ is a polynomial of order p that fits concentration line around reference point i best. After testing polynomials of different orders, Bott (1098) recommends the 4th order Lagrange polynomial

$$C^4(x)=a_0+a_1x+a_2x^2+a_3x^3+a_4x^4 \qquad (5)$$

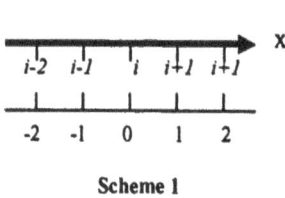

Scheme 1

If a 5-point pattern [-2,-1,0,1,2] has its origin in point i (Scheme 1), the coefficients in Eq.(5) are determined by solving a system of 5 simultaneous algebraic equations. The solution of this system is (Bott, 1989):

$$a_0 = C_i$$
$$a_1 = (-C_{i+2} + 8C_{i+1} - 8C_{i-1} + C_{i-2})/12$$
$$a_2 = (-C_{i+2} + 16C_{i+1} - 30C_i + 16C_{i-1} - C_{i-2})/24, \tag{6}$$
$$a_3 = (C_{i+2} - 2C_{i+1} + 2C_{i-1} - C_{i-2})/12$$
$$a_4 = (C_{i+2} - 4C_{i+1} + 6C_i - 4C_{i-1} + C_{i-2})/24$$

This local approach of polynomial fitting (separate coefficients for every grid point) leads to low numerical diffusion, but in case of steep gradients in concentration field the integrals in Eq.(4) can receive negative or unrealistic high values. Bott introduces limiters for the flux area

$$0 < Ar_i < C_i \text{ when } U_{i+1/2} > 0 \quad \text{and} \quad 0 < Ar_i < C_{i+1} \text{ when } U_{i+1/2} < 0. \tag{7}$$

As to assure the positive definiteness of the scheme, Bott introduces a normalisation of the fluxes, multiplying the flux area by the factor C_i/A_i (case of $U_{i+1/2} > 0$) or C_{i+1}/A_{i+1} (case of $U_{i+1/2} < 0$), where A_i is the area of the integral over the whole cell

$$A_i = \int_{-1/2}^{1/2} C_i^4(x)dx = a_0 + a_2/12 + a_4/80. \tag{8}$$

2.4 4th order TRAP scheme - TR4 (Syrakov 1995, 1996)

The TRAP concept is an alternative of the Integrated Flux one. Approximating the flux area with trapezium (as **R&L** does) the area is calculated by multiplying the Courant number (the trapezium height) by an estimate for the concentration referring the middle of the passed distance. In TRAP this estimate is obtained simply on the base of the Bott's polynomial fitting - Eqs.(5) - (6). The arguments of the polynomial are

$$r_i = (1 - U_{i+1/2})/2 \text{ at } U_{i+1/2} > 0 \quad \text{and} \quad r_{i+1} = -(1 + U_{i+1/2})/2 \text{ at } U_{i+1/2} < 0 \tag{9}$$

Bott's limiters, Eq.(7), and normalisation, Eq.(8), are also used. TRAP scheme appears to be a combination between the schemes of Bott (1989) and Roussel and Lerner (1981).

2.5 3rd order TRAP scheme using Bessel polynomial - TRB

The initial setting of Δt-value for any dispersion model utilising explicit advection scheme must keep the Courant stability condition $u\Delta t/\Delta x < 1$ fulfilled in the whole domain during all period of integration. Usually, some value of u greater than the strongest winds is fixed leading to small Δt. During the run, the predominant part of flux area calculations is made at small Courant numbers. This means that very often the flux areas are placed close to the cell edges, around the point $i+1/2$ in the case of i cell right edge. High precision of the approximation in this point is necessary for improving the accuracy of the schemes. It is well known that the Lagrange polynomial gives the best interpolating quality for $|r| \leq 0.25\Delta x$, i.e. close around the central point i. This is the reason for using different polynomials in the cases of positive and negative velocity. The right edge of the cell is placed at $r=0.5$ and the flux area (respectively, the middle of the flux trapezium) is placed near this point. In Syrakov and Galperin (1997) the Bessel interpolation polynomial is applied that gives best fitting quality for the region $0.25\Delta x \leq r \leq 0.75\Delta x$. As Bessel polynomials are always of an even degree, a third order interpolation is proposed there:

$$C^b(r) = b_0 + b_1 r + b_2 r^2 + b_3 r^3, \tag{10}$$

where the coefficients $b_k, k = \overline{0,3}$ are determined by (the i cell is referred)

$$
\begin{aligned}
b_0 &= C_i \\
b_1 &= (-2C_{i-1} - 3C_i + 6C_{i+1} - C_{i+2})/6 \\
b_2 &= (C_{i-1} - 2C_i + C_{i+1})/2 \\
b_3 &= (-C_{i-1} + 3C_i - 3C_{i+1} + C_{i+2})/6
\end{aligned}
\tag{11}
$$

The right edge flux area is calculated by multiplying the Courant number by the value of Eq. (10) estimated for $r_i = (1-U_{i+1/2})/2$. In this case, one and the same polynomial at positive and negative advection velocities is used. The limiting and the normalisation must be performed taking into account the upstream concept. For the normalization procedure, the area of the whole cell is necessary. It is determined by integrating the Bessel polynomial leading to

$$
\begin{aligned}
A_i &= (C_{i-1} + 22C_i + C_{i+1})/24 & \text{at } U_{i+1/2} > 0 \\
A_{i+1} &= (C_i + 22C_{i+1} + C_{i+2})/24 & \text{at } U_{i+1/2} < 0.
\end{aligned}
\tag{12}
$$

2.6 3rd order TRAP scheme using Lagrange polynomial - TR3.

As the rightmost edge of cell i is just in the middle of the set of 4 neighbouring points (i-1, i, i+1, i+2) it is natural to choose a polynomial of 3rd order as· to approximate concentration profile in this region:

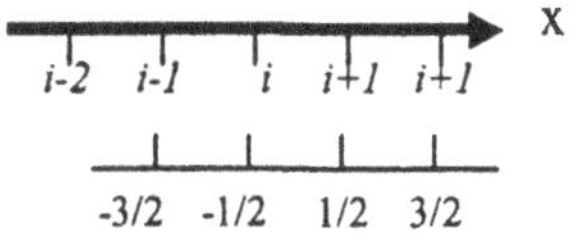

$$C^3(x) = c_0 + c_1 x + c_2 x^2 + c_3 x^3. \tag{13}$$

The 4-point pattern [-3/2,-1/2,1/2,3/2] has its origin at point i+1/2 (shifted pattern at **Scheme 2**) determining the coefficients:

Scheme 2

$$
\begin{aligned}
c_0 &= \left[9(C_{i+1} + C_i) - (C_{i+2} + C_{i-1})\right]/24 \\
c_1 &= \left[27(C_{i+1} - C_i) - (C_{i+2} - C_{i-1})\right]/24 \\
c_2 &= \left[-(C_{i+1} + C_i) + (C_{i+2} + C_{i-1})\right]/4 \\
c_3 &= \left[-3(C_{i+1} - C_i) + (C_{i+2} + C_{i-1})\right]/6
\end{aligned}
\tag{14}
$$

The flux area is calculated by multiplying the Courant number by the value of Eq.(13) at $r_i = U_{i+1/2}/2$. Bott's limiters Eq.(6) and normalisation Eq.(7) are applied, i cell area determined as

$$A_i = c_0 - c_1/2 + c_2/3 - c_3/4. \tag{15}$$

2.7 2nd order TRAP scheme using Lagrange polynomial - TR2.

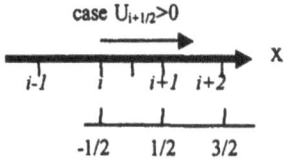

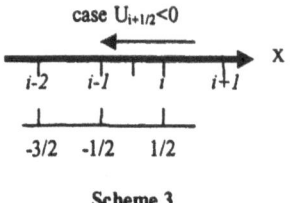

Scheme 3

It will be shown later that the decrease of the order of approximation from 4 to 3 leads to acceleration of computations without considerable change for the worse. Further decrease of this order and applying the shifted pattern from **TR3** is worth to be checked:

$$C^2(x)=d_0+d_1x+d_2x^2. \tag{16}$$

The 3-point grid pattern has again its origin in point $i+1/2$, but its orientation will depend on the transport direction as shown on **Scheme 3**. It is natural to use 2 points in upstream direction and one - in downwind one. The polynomial coefficients are:

at $U_{i+1/2}>0$ at $U_{i+1/2}<0$

$$
\begin{aligned}
d_0 &= (-C_{i-1}+6C_i+3C_{i+1})/8 \\
d_1 &= -C_i+C_{i+1} \\
d_2 &= (C_{i-1}-2C_i+C_{i+1})/2
\end{aligned}
\quad \text{and} \quad
\begin{aligned}
d_0 &= (3C_i+6C_{i+1}-C_{i+2})/8 \\
d_1 &= -C_i+C_{i+1} \\
d_2 &= (C_i-2C_{i+1}+C_{i+2})/2
\end{aligned}
\tag{17}
$$

The flux area is calculated by multiplying the Courant number by the value of Eq.(16) at $r_i=-U_{i+1/2}/2$. Bott's limiters Eq.(6) and normalisation Eq.(7) are applied, cell areas determined as

$$A_i=d_0-d_1/2+d_2/3 \text{ at } U_{i+1/2}>0, \qquad A_{i+1}=d_0+d_1/2+d_2/3 \text{ at } U_{i+1/2}<0. \tag{18}$$

2.7 Integrated Flux schemes analogous to TRB, TR3 and TR2

These schemes go with the above described TRAP schemes but the flux area is calculated according to the Integrated Flux concept. The right edge flux is calculated by integrating Eq.(10), Eq.(13) and Eq.(16) in the limits $[0.5-U_{i+1/2}, 0.5]$ for $U_{i+1/2}>0$ and $[0.5, 0.5-U_{i+1/2}]$ in the opposite case. Bott's limiters and normalisation are applied where the normalizing areas are calculated after Eq.(12), Eq.(15) and Eq.(18), respectively. The numerical experiments show that the simulation quality of these schemes is almost the same as its TRAP variants but they are slower and will not be discussed any more.

3 NUMERICAL EXPERIMENTS

The seven schemes from section 2 have passed different tests. Only part of the results - these ones from the two-dimensional rotational test (after Smolarkiewiecz, 1982) – will be presented in this section. Instantaneous releases with different initial profile are rotated with constant angular velocity. Keeping in mind that the initial concentration field is the exact solution of the advection equation after one or some full rotations, a number of criteria are established for estimating of the schemes properties.

Denoting the initial field as C^o and the final one as C, a number of estimates are introduced, part of them shown on Table 1.

Table 1 Estimates for simulation quality of numerical advection schemes (rotational test)

Estimate	Meaning
$\mathbf{Cmax} = \max(C_{ij})/\max(C_{ij}^o)$	Cmax<1 indicates presence of numerical diffusion
$\mathbf{Cmin} = \min(C_{ij})/\max(C_{ij}^o)$	Cmin<0 indicates absence of positive definiteness
$\mathbf{SM} = (\sum_{ij} C_{ij}^o - \sum_{ij} C_{ij})/\sum_{ij} C_{ij}^o$	SM - normalized difference of sum of masses. SM≠0 indicates absence of conservativeness
$\mathbf{SM2} = (\sum_{ij} C_{ij}^{o\,2} - \sum_{ij} C_{ij}^{2})/\sum_{ij} C_{ij}^{o\,2}$	SM2 reflects the second moment conservativeness of the scheme. The ideal advection scheme has SM2=0.
$\mathbf{DXc} = \sum_{ij} iC_{ij}/\sum_{ij} C_{ij} - \sum_{ij} iC_{ij}^o/\sum_{ij} C_{ij}^o$ $\mathbf{DYc} = \sum_{ij} jC_{ij}/\sum_{ij} C_{ij} - \sum_{ij} jC_{ij}^o/\sum_{ij} C_{ij}^o$	DXc and DYc estimate the displacement of the mass centre due to numerical effects. DXc = DYc = 0 after a number of full rotations indicate an ideal transport ability of the advection scheme
$\mathbf{DD} = (D - D^0)/D^0$, where $D = \sum_{ij} C_{ij}\left[(x - \mathbf{Xc})^2 + (y - \mathbf{Yc})^2\right]/\sum_{ij} C_{ij}$	D - mass dispersion around the mass centre DD indicates the degree of de-concentration of masses due to the numerical diffusion
$\mathbf{RC} = \sum_{ij}\left|C_{ij} - C_{ij}^o\right|/\sum_{ij} C_{ij}^o$	RC, RC2 - restoration capabilities (absolute and squared). In case of ideal advection scheme, they
$\mathbf{RC2} = \sum_{ij}(C_{ij}^2 - C_{ij}^{o\,2})/\sum_{ij} C_{ij}^{o\,2}$	are to be equal to zero after a number of full rotations
$\mathbf{T} = \Delta T_{calc}/\Delta T^{ref}$,	relative speed of performance

A grid field of 101×101 points with $\Delta x = \Delta y = 1$ is the test domain. After Smolarkiewiecz (1982) a rotational wind field with constant angular velocity of $\omega \approx 0.1$ (628 time steps per 1 rotation) and centre in point (51,51) is imposed on this area. Two types of sources - point-shaped one (a limited δ-function) and cone-shaped one - with maximum concentration of $C^o_{max}=3.87$ at point (76,51) are supposed. The cone base radius is $15\Delta x$. The quantitative results of all schemes for 1 and 6 rotations are presented on Table 2 and Table 3 for the point and the cone, respectively. The normalisation of the time of integration is made by the time of **TRB** scheme (ΔT^{ref}).

It can be noticed from Table 2 that none of the tested schemes describes satisfactorily the advection of a single disturbance in the concentration field (point source). There is no difference scheme, approximating the advection equation, capable to describe adequately this severe discontinuity in the concentration field. This can be seen in **Cmax** values, which fall down to some percents of the initial maximum. The steep gradients are squashed down immediately after the start of movement to a nearly-Gauss-shaped distribution that is advected further. The cone experiments show much better results as can be seen from Table 3.

Table 2. Estimates [%] of different schemes' performance. Instantaneous **point** source.

	point	McC	R&L	BOT	TR4	TRB	TR3	TR2
1 rotation	Cmax	6.40	5.79	3.78	3.81	3.53	3.53	2.40
	Cmin	-1.88	-.34	.00	.00	.00	.00	.00
	CM	.00	.00	.00	.00	.00	.00	.00
	CM2	-94.34	-96.04	-97.99	-97.99	-98.16	-98.16	-98.68
	DXc	.0	-.2	23.4	22.4	39.3	39.3	21.4
	DYc	.1	.1	-30.8	-26.6	-78.3	-78.3	-25.6
	DD	-	-	-	-	-	-	-
	RC	357.1	231.9	192.4	192.7	193.6	193.6	195.2
	RC2	93.7	92.4	94.4	94.4	95.4	95.4	96.5
	T	155	2276	766	110	100	109	89
6 rotations	Cmax	2.97	2.32	1.66	1.68	1.56	1.56	.98
	Cmin	-.78	-.14	.00	.00	.00	.00	.00
	CM	-.021	-.00	.00	.00	-.00	-.00	.00
	CM2	-97.5	-98.4	-99.1	-99.1	-99.2	-99.2	-99.6
	DXc	-.6	-.6	23.6	21.8	40.4	40.4	19.4
	DYc	.6	.4	-37.9	-27.6	-101.5	-101.5	-28.2
	DD	-	-	-	-	-	-	-
	RC	330.9	242.5	196.7	197.0	197.1	197.1	198.0
	RC2	96.5	97.0	97.5	97.5	97.9	98.0	98.6
	T	158	637	515	108	100	108	88

Table 3. Estimates [%] of different schemes' performance. Instantaneous **cone** source.

	cone	McC	R&L	BOT	TR4	TRB	TR3	TR2
1 rotation	Cmax	91.56	90.67	91.16	91.19	91.09	91.09	87.75
	Cmin	-1.45	-.61	.00	.00	.00	.00	.00
	CM	.000	.001	.000	.000	.000	.000	.000
	CM2	-.32	-.97	-.90	-.89	-.89	-.89	-1.95
	DXc	.0	-.1	.0	.0	.0	.0	﹁1
	DYc	.1	.1	-.2	.0	-.4	-.4	-.1
	DD	-.002	.552	1.793	1.819	1.964	1.964	2.952
	RC	3.681	2.935	3.444	3.443	3.698	3.698	4.582
	RC2	.079	.082	.107	.107	.114	.114	.182
	T	114	1361	659	120	100	108	90
6 rotations	Cmax	87.75	85.45	86.43	86.49	85.93	85.93	81.98
	Cmin	-2.11	-.89	.00	.00	.00	.00	.00
	CM	-.001	-.040	-.001	-.003	-.003	-.003	.000
	CM2	-1.31	-2.84	-3.43	-3.40	-3.41	-3.41	-8.19
	DXc	-.1	-1.4	-.1	-.3	-.3	-.3	-.6
	DYc	.4	.2	-.3	.5	-.1	-.1	.8
	DD	.00	.48	6.63	6.65	7.05	7.05	11.61
	RC	8.886	7.326	8.100	8.126	8.466	8.466	9.425
	RC2	.362	.330	.415	.420	.441	.441	.643
	T	168	329	500	109	100	109	89

These experiments (and other ones with other shapes of the initial concentration) show that **McC** and **R&L** schemes are not recommendable for using in tracer dispersion models because they are not monotone (lost of mass). The Bott scheme and the TRAP scheme of 4th order possess practically equal simulation properties, **TR4** being some times faster. The 3rd orders schemes, built with maximal approximation accuracy between the grid points (Bessel polynomial and Lagrange polynomial with shifted pattern) display almost the same properties as **BOT** and **TR4**, being in the same time faster. The fastest one is the scheme **TRB** and its time is used for normalisation of performance times of all other schemes. The 2nd order scheme is naturally faster than **TRB** but its simulation properties deteriorated. It still can be used in some applications requiring a high speed of performance.

The version **TRB** of the TRAP scheme shows almost the same characteristics as the original Bott one, but it is some times faster. Another important advantage of this low order scheme is the fact that it needs only two grid points at the border of the model domain for setting of boundary conditions. The 4th order Bott and TRAP schemes need three boundary points at every border.

As a final conclusion to this section the Bessel variant of TRAP scheme (**TRB**) can be recommend for practical use in air dispersion models as one with good simulation abilities possessing in the same time a good computational speed. This scheme is built in the EMAP model (Syrakov, 1995) and used successfully for different air pollution applications (Syrakov, 1996).

4. SIMULATION CAPABILITY OF TRAP SCHEME

In this section visualizations of some graphical experiments with TRB will be demonstrated. The first example, Fig.2, will be the simulation of one rotation of point-shaped and cone-shaped instantaneous sources, parameters given in Table 2 and 3.

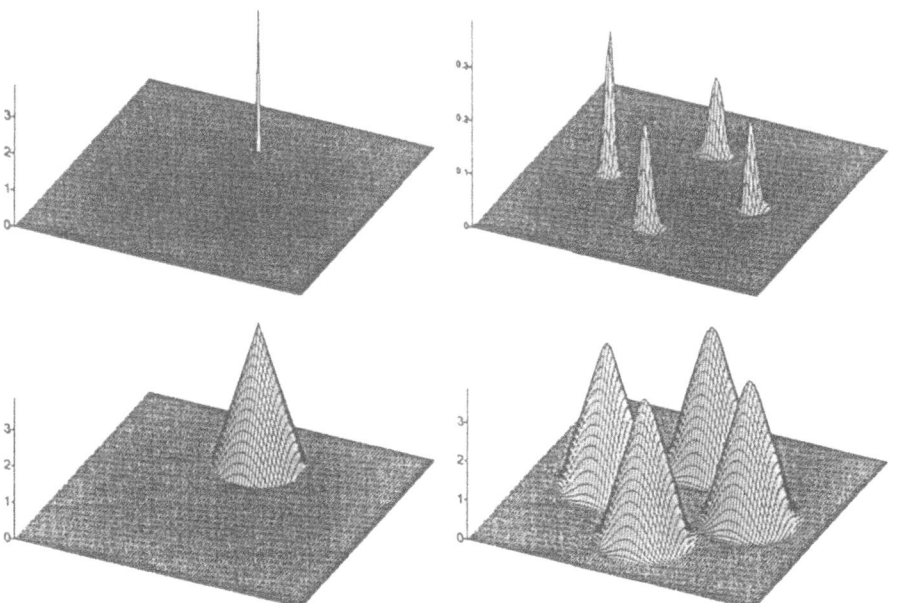

Figure 2. Rotation of point-shaped (up) and cone-shaped (down) instantaneous source by **TRB** scheme. The left plots present the initial condition. In the right plots the moved shape at every quarter of the rotation is shown. Note the difference in scales of upper plots.

It is clearly seen from Fig.1 that the description of a single disturbance advection is very bad; note that the dimension of the upper right z-axis is 10 times less than the upper left one. The cone is moved very well. Of course, the pick is little bit smoothed (so is the cone base) but this is normal reaction to the discontinuity there.

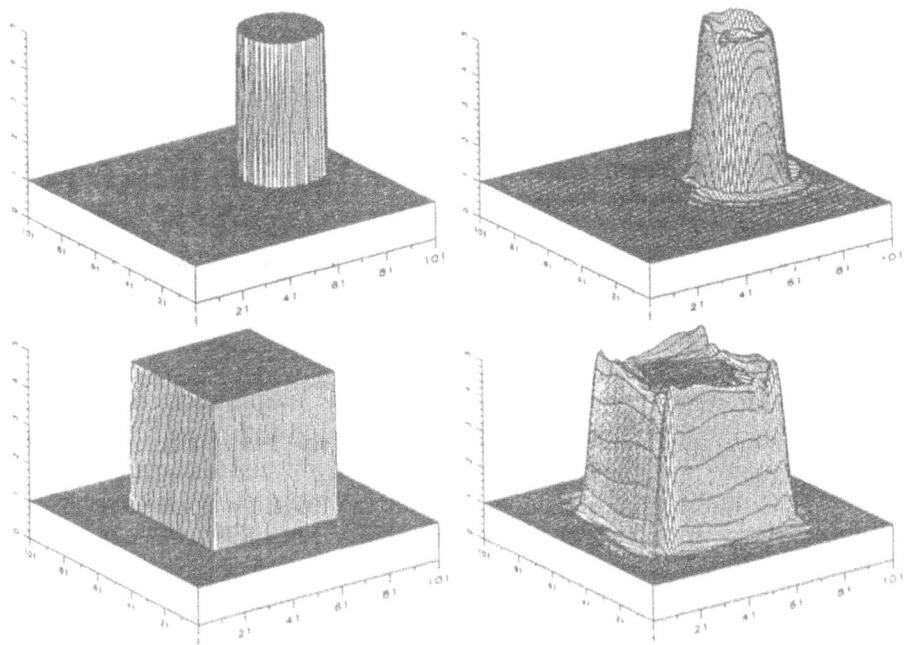

Figure 3. Rotation of cylinder (up) and cube (down) over background by **TRB**. The initial cube is placed in the centre of the field and is rotated around its centreline.

In Fig.3, another demonstration of the **TRB** capabilities is presented. In the first row, the initial and final (after 1 rotation) shape of a cylinder with height $3.87\Delta x$, radius $15\Delta x$ and initial centre at point (76,51). In the second row, a centred parallelepiped with base $50\Delta x \times 50\Delta y$ and height $3.87\Delta x$ is rotated over its axis. The figures are rotated over constant background field with height of $1\Delta x$. The one rotation appearance of these shapes is quite plausible keeping in mind the discontinuity in the initial distribution, but in common the description is very good. The tests over background are very important in proving the positive definiteness of a numerical scheme. When no background tests are performed an option for checking for and neglecting negative concentrations can keep the positive definiteness of the scheme. This limiter does not work when background tests are made. The undershooting and overshooting of the schemes usually dig "holes" in the background as to avoid the discontinuity. As shallow the holes as better the scheme. **TRB** demonstrates here a very good ability in this sense.

In Fig. 4, the rotation of an instantaneous line source is shown together with the rotation of continuous point source; both on background. The line source can present a moving front and has its initial position from point (60,51) to point (90,51). The shapes at every quarter of the rotation are presented. In the second graph, a continuous point source is modelled, a case simulating a real stack sources. It demonstrates the fast decrease of the initial concentrations and the transformation of the one-dimensional disturbances in three-dimensional Gauss-shaped objects continuously moving in the rotation field and overlapping one over another.

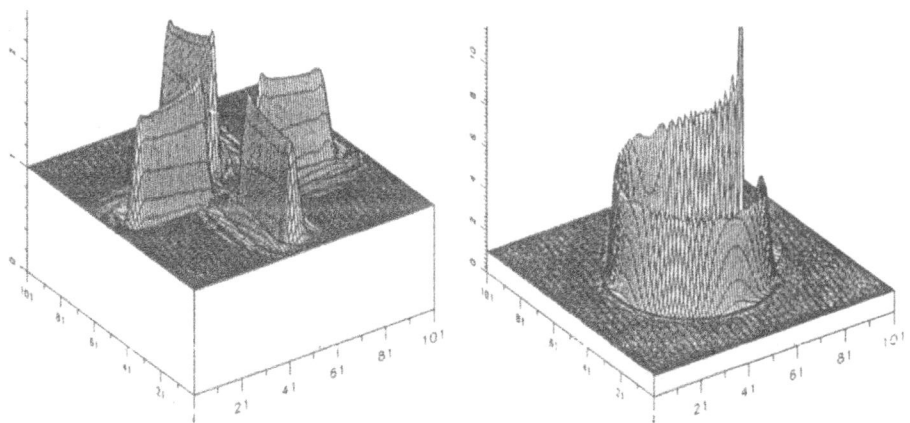

Figure 4. Rotation of instantaneous line source and continuous point source

The next demonstration of **TRB** abilities is the deformation test, proposed by Smolarkiewiecz (1982) and discussed in details in Staniforth et al. (1978). A cone with height 1 and radius 15 units is placed in the centre of the well known squared area of 101×points with $\Delta x=1$. The wind field for is defined by the streamfunction

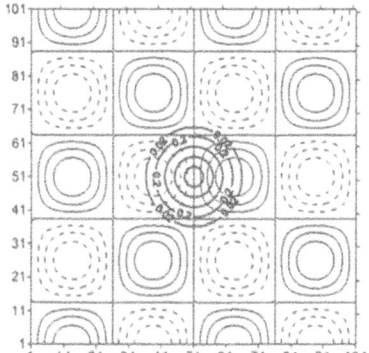

$$\psi(x,y) = A\sin(kx)\cos(ky),$$

where $A=8$, $k=4\pi/L$, $L=100$. The isolines of the streamfunction are shown in the figure: solid/dashed contours are positive/negative. The initial cone distribution is also shown, there.

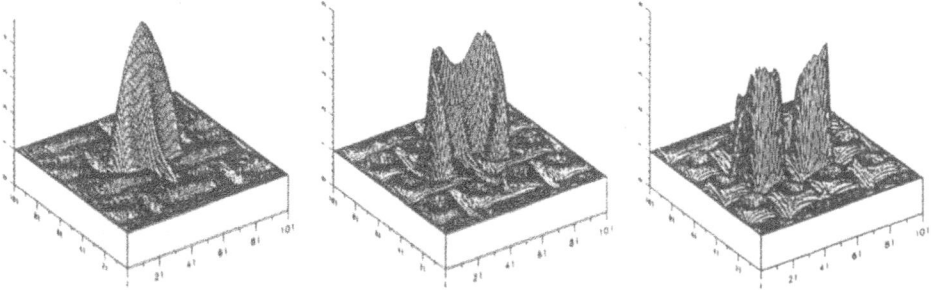

Figure 5. Cone in deformational flow after 19, 38 and 75 time steps ($\Delta t=0.7$)

Staniforth et al. (1978) presented the analytical solution of this advection problem. They pointed out that two regimes of simulation exist that have different evaluation criteria. For the first regime (short time period) numerical solution can be compared with the exact one in a quantitative manner, whereas for the second one (long time period) it should be evaluated on the basis of stability and qualitative similarity with an appropriate averaging of the exact solution. In Fig. 5, short time simulations of deformation test are shown. The numerical solutions after 19, 38 and 75 iterations nearly correspond to the analytical solutions presented in Fig. 3a-d of Staniforth et al. (1987) as well as those in Fig. 5a-d of Bott (1989). Long term calculations are not made because it is doubtful that such strong deformation can exist in the atmosphere for long time scales.

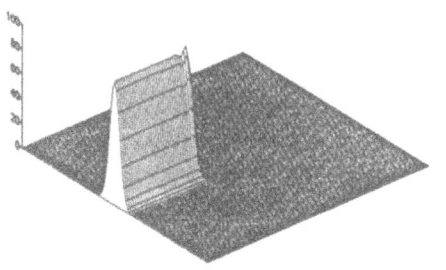

Figure 6. Movement of Gauss-shaped boundary profile into the computational area.

The last demonstration is the reaction of **TRB** to the boundary conditions. As the advective equation is first order in space, only one boundary condition can be set – some value at the incoming flow boundary. Usually values are prescribed to the first (closest) boundary points. As **TRB** is 3rd order in space it needs second boundary value at every boundary point. Here, the same values are prescribed to the second boundary point. In Fig. 6, the reaction of **TRB** to non-zero boundary values is demonstrated. A Gauss-shaped profile with height 100 units and dispersion 15 and centre at y=51 is set at points x=0,-1. A 40 steps at u=0.7, v=0 are made. Excluding the disturbance at the front of the shape, all other top points keep value of 100, the height of the boundary profile.

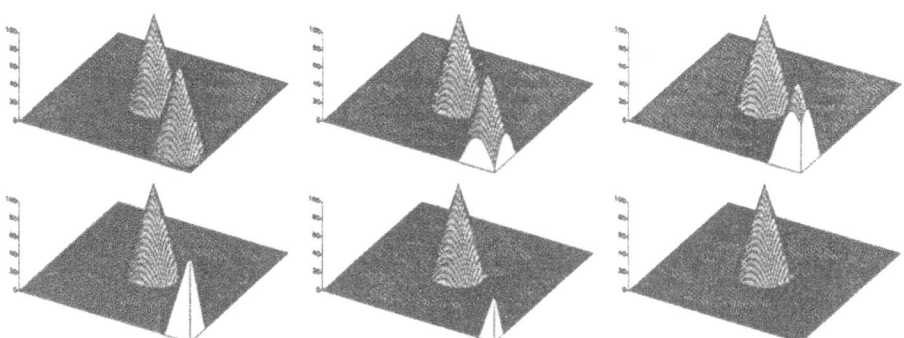

Figure 7. A cone shape is leaving the computational area as simulated by **TRB**.

Nevertheless the boundary condition at the outgoing flow boundary has not physical meaning it is not of less importance. The natural boundary condition there is the "open boundary" one. The simplest way is to set the concentration values at outgoing boundary points equal to zero. This boundary condition works but some delay of outgoing shapes is observed in this case. Here, an extrapolation over the two closest inner points is adopted

$$C_\Gamma = 2C_{\Gamma-1} - C_{\Gamma-2}, \qquad C_{\Gamma+1} - C_\Gamma, \tag{19}$$

where Γ denotes the closest boundary point. Some successive stages of a cone leaving the computational area are shown in Fig.7 demonstrating this useful property of **TRB**.

302

5. NON-HOMOGENEOUS VARIANTS OF TRAP-SCHEME. CONCLUSION

For many applications the troposphere and the atmospheric boundary layer are considered horizontally homogeneous and atmospheric models use constant horizontal grid step. Because of the specific vertical structure of the atmosphere (big gradients near to the ground and decreasing variability with height) the atmospheric models (pollution dispersion models, in particular) use non-homogeneous vertical discretization. Here, variants of **TRB** working on regular and non-regular non-homogeneous grid are presented (Syrakov, 2001).

The different grid steps Δz_k ($k=1, N_z$) make difficult the polynomial fitting of concentration profile necessary for **TRB** performance. Four main steps at every time step allow to overcome this problem and to use the above described **TRB**:

1. Conversion from concentration to mass-in-layer: $M_k = C_k \Delta z_k$;

2. Proper determination of the Courant number: $W_{k+1/2} = w_{k+1/2} \Delta t / \begin{cases} \Delta z_k, & w_{k+1/2} > 0 \\ \Delta z_{k+1}, & w_{k+1/2} < 0 \end{cases}$;

3. Performance of the homogeneous **TRB**, but using M_k^n instead of C_k^n;

4. Reverse conversion from mass-in-layer to concentration: $C_k^{n+1} = M_k^{n+1}/\Delta z_k$

Two grids in vertical are defined. The first one is 80-point regular grid with $\Delta z = 25$ m starting at 50 m, last point at 2000 m. The second grid is log-linear with parameters that allow covering the same distance by 24 grid points, $\Delta z_{min} = 11$ m, $\Delta z_{max} = 165$ m. Three initial concentration profiles - Gaussian shape, single disturbance (point source) and triangle – are moving upward with a velocity of 0.01 m/s.

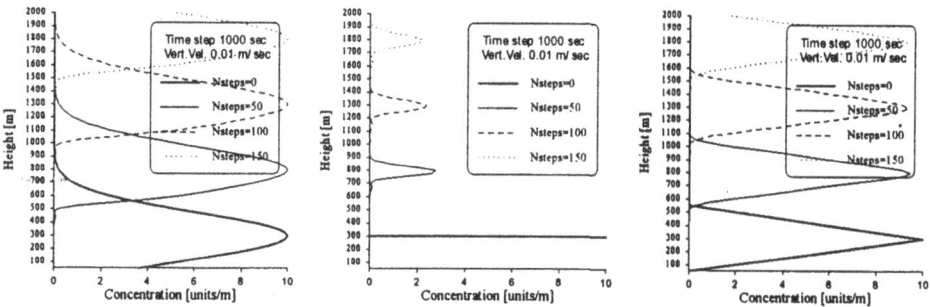

Figure 8. Performance of **TRB** advection scheme over homogeneous grid (upward motions).

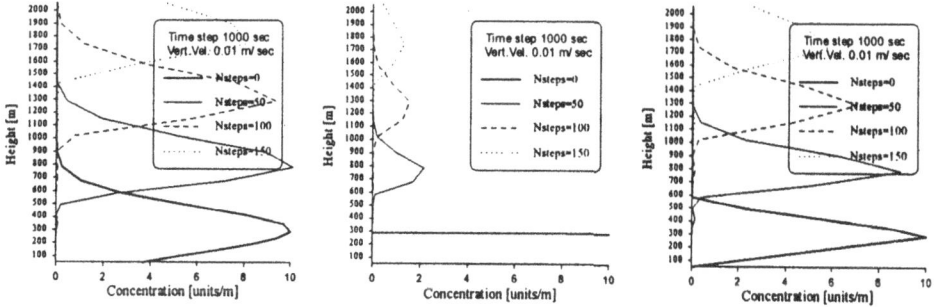

Figure 9. Performance of **TRB** advection scheme over non-homogeneous grid (upward motions).

In Fig.8 and 9 the evolution of the 3 different initial concentration profiles is shown as described by TRAP scheme over homogeneous and non-homogeneous grids. The Gaussian form is moving best, almost without deformations. The movement of the instantaneous point source is described worst. The triangle is a case between the above two ones. It can be noticed that the quality of description on the non-homogeneous grid gets worse. This is due not only to the disadvantages of the numerical scheme but mainly to the fact, that during its movement the object goes to areas with worse and worse resolution. The Figures and the quantitative estimates show that **TRB** scheme describes the advection of pollution forms without steep gradients piety well. The transition to non-homogeneous grid makes the description worse but, in spite this decrease of description quality, **TRB** in both cases of regular and non-regular grids demonstrates a rather good and fast performance and can be recommended for dispersion modelling applications.

ACKNOWLEDGEMENTS

Deep gratitude is due to Michail Galperin for its valuable comments and discussions.

REFERENCES

Bott, A., 1989, A positive definite advection scheme obtained by nonlinear renormalization of the advective fluxes, *Mon.Wea.Rev.*, **117**, pp. 1006-1015.

Bott, A., 1992, Monotone flux limitation in the area preserving flux form advection algorithm, *Mon.Wea.Rev.*, **120**, pp. 2592-2602.

Bott, A., 1993, The Monotone Area-preserving Flux-Form Advection Algorithm: Reducing the Time-splitting Error in Two-Dimensional Flow Fields, *Mon.Wea.Rev.*, **121**, pp. 2637-2641.

Holmgren, P., 1994, An Advection Algorithm and an Atmospheric Airflow Application, *J. Comp. Phys.*, **115**, pp. 27-42.

Russell, G.L., J.A. Lerner, 1981, A New Finite-Differencing Scheme for the Tracer Transport Equation, *J. Appl. Meteor.*, **20**, pp. 1483-1498.

Peters, L.K., C.M. Berkowitz, G.R. Carmichael, R.C. Easter, G. Fairweather, S.J. Ghan, J.M. Hales, L.R. Leung, W.R. Pennell, F.A. Potra, R.D. Saylor and T.T. Tsang, (1995), The current state and future direction of Eulerian models in simulation the tropospheric chemistry and transport of trace species: a review, *Atmospheric environment*, **29**, No 2, pp. 189-222.

Rood R.B., 1987, Numerical advection algorithms and their role in atmospheric transport and chemistry models, *Rev. Geophys.*, **25**, pp. 71-100.

Smolarkiewiecz, P.K., 1982, The multidimensional Crowley advection scheme, *Mon.Wea.Rev.*, **113**, pp. 1109-1130.

Staniforth, A., Côté, J., and Pudikiewicz, J., 1978, Comments on "Smolarkiewicz's deformational flow", *Mon.Wea.Rev.* **115**, 894-900.

Syrakov, D., 1995, On a PC-oriented Eulerian Multi-Level Model for Long-Term Calculations of the Regional Sulphur Deposition, in *Air Pollution Modelling and its Application XI, NATO/CMSS*, **21**, Gryning S.E. and Schiermeier F.A., ed., Plenum Press, N.Y. and London, pp. 645-646.

Syrakov, D., 1996, On the TRAP advection scheme - Description, tests and applications, in *Regional Modelling of Air Pollution in Europ*, Geernaert G., A.Walloe-Hansen and Z.Zlatev, eds, National Environmental Research Institute, Denmark, pp. 141-152.

Syrakov, D., M. Galperin, 1997 On a new Bott-type advection scheme and its further improvement, in *Proceedings of the first GLOREAM Workshop, Aachen, Germany, September 1997*, H. Hass and I.J. Ackermann, ed., Ford Forschungszentrum Aachen, pp. 103-109.

Syrakov, D., M. Galperin, 2000, On some explicit advection schemes for dispersion modelling applications, *Intrnational Journal of Environment and Pollution*, **14**, pp. 267-277.

Syrakov, D., 2001, On the Use of a Flux-type Advection Scheme on Non-homogeneous Grid, in *Proceedings from the EUROTRAC-2 Symposium 2000*, P.M. Midgley, M. Reuther, M. Williams, eds., Springer-Verlag Berlin, Heidelberg (CD Rom).

Tremback, C.J., J. Powell, W.R. Cotton and R.A. Pielke, 1987, The forward-in-time upstream advection scheme: Extension to higher orders, *Mon.Wea.Rev.*, **115**, pp. 540-555.

WMO-TCSU, 1979, Numerical methods used in atmospheric models, Vol. I and II, GARP Publication series, No 17, Sept. 1979.

DISCUSSION

D. G. STEYN	I am not convinced that the fine resolutions model has improved performance over the coarse resolution. Surely the only way to improve performance is to improve input data - especially emissions?
D. SYRAKOV	You are formally right. By this rough approach, using only linear interpolation for both meteorology and emission data, in fact we do not input additional information to the model. But the increase of resolution give as results, averaged for a cell 10x10 km and it is quite possible this result to be closer to the single point measurement, than the averaged for a cel 50x50 km. This explains the slight improvement of the comparison.
T. ODMAN	In terms of accuracy, how does the 4th order TRAP scheme compare to both scheme?
D. SYRAKOV	It produce the same values of all estimators as the Bott scheme being in the same time faster. The Bessel variant of the TRAP scheme has the same simulation characteristics being faster of 4th order TRAP scheme and decreasing the number of boundary grid points from 3 to 2.
N. MOUSSIOPOULOS	Could you comment on the multi-dimensionality aspects related to the advection parameterisation? How would the polynomial approaches you are using have to be modified in order to account for such aspects?
D. SYRAKOV	Here, the splitting on directions is used. First, advection in x-direction is performed for one time step over all domain. The obtained concentration values are used as initial field for advection only in y-direction. No special investigation of the splitting errors is made here, i.e. the presented results contain these errors.
E. GENIKHOVICH	Did you try the symmetrical splittings like product of aperators along x, y, y and x directions?
D. SYRAKOV	Yes, the order of applience of x- and y-advection is alternated at each time step.

MODELLING DIFFUSE SOLAR RADIATION IN THE CITY OF SÃO PAULO USING NEURAL-NETWORK TECHNIQUE

Marija Zlata Božnar*, Primož Mlakar*, Amauri P. Oliveira[#] and Jacyra Soares[#]

ABSTRACT

In this work a neural network technique is applied to estimate hourly values of diffuse solar radiation at the surface in São Paulo City, Brazil, using as input global solar radiation and other meteorological parameters measured from 1998 to 2002.

The diffuse solar radiation at the surface was directly estimated using a shadow-ring device. All available measurements are used to train a Perceptron neural network based model.

Feature determination techniques are used to select the optimal subset of meteorological parameters. Pattern selection techniques are employed to determine the most suitable subset of historical measurements for effective neural network based model training.

The results obtained here indicated the possibility of effective use of neural network based models for several tasks of pre-processing meteorological parameters when explicit formulas are not known. The models are especially useful for problems involving parameters in highly non linear relationship.

1. INTRODUCTION

The knowledge of the evolution of solar radiation components, at the surface, is important for climate studies, solar collector efficiency estimation, agricultural and several meteorological applications (Iziomon and Aro, 1999). Unfortunately, these quantities are not available, with the required spatial and temporal resolution, in Brazil. With most of its territorial area (8.547 millions of km^2) located in equatorial and tropical latitudes, Brazil has a very incipient solarimetric network. For instance, Pereira et al.

---

[*] Jozef Stefan Institute and AMES d.o.o., Jamova 39, SI-1000 Ljubljana, Slovenia
[#] Institute of Astronomy, Geophysics and Atmospheric Sciences, University of São Paulo, 05508-900 - São Paulo, Brazil

Air Pollution Modeling and Its Application XVI, Edited by
Borrego and Incecik, Kluwer Academic/Plenum Publishers, New York, 2004

(1996) reported only 22 ground stations with global solar radiation measured directly or estimated from sunshine hours. The importance and difficulties to estimate hourly distribution of solar radiation have been emphasised by Aguiar and Collares-Pereira (1992) and Oliveira et al. (2002a).

The major reason for the lack of direct measurements of diffuse and direct components of the solar radiation, at surface, is that these parameters require special set-ups like sun-tracking pyrheliometers or occulted pyranometers with shadow-bands or disks. These apparatus are expensive, difficult to operate in regular basis and require the use of correction factors in order to compensate the unwilling effects caused by the blocking device (Le Baron et al., 1990).

To investigate the inherent difficulties associated to operation of solar radiation measurements in a regular basis and to identify the effects caused by urban environment on the spatial representative of these measurements, an experimental site was set up in the City of São Paulo (Oliveira, et al., 2002a). In this site, measurements of several meteorological parameters, including global and diffuse solar radiation have been carried out regularly since April 1994. These measurements follow WMO recommendations (WMO, 1971) and have been carried out using a data acquisition system fully automatic. The diffuse component of solar radiation, at surface, has been measured by a shadow-band device (Oliveira et al., 2002b).

The City of São Paulo, with 10 millions habitants, together with 38 other smaller cities, form the Metropolitan Region of São Paulo (MRSP). This region is occupied by 16.5 millions habitants distributed over an area of 8051 km^2 and it is the largest urban area in South America and one of the 10 largest in the world (Fig. 1). Therefore, it is an ideal place to investigate the role played by urban environment on solar radiation.

The MRSP, with more than 4 million of motor vehicles (CETESB, 1999), is characterised as having moderate degree of contamination by particulate matter and other pollutants (Kretzschmar, 1994). Oliveira et al. (1996) detected a reduction of up to 18 % in the direct beam associated to a progressive increase in the concentration of particulate matter during a period of five cloudless days in the City of São Paulo. According to the authors, depletion in direct beam was partially compensated by an increase in diffuse solar radiation at surface. The long-term effects caused by particulate matter on the solar radiation field in São Paulo are still unknown.

The energy consumption in the MRSP is around 40 TWh per year (Eletropaulo, 1999). This is a significant fraction of all energy produced in Brazil and it is generated basically by hydroelectric power plants. To increase production of energy, Brazil is looking for other sources of energy because the available hydroelectric powers have been practically used up. Moreover, considering the environmental impact associated to the others available sources – nuclear, coal and gas burning power plants - solar radiation has to be prioritised by the local policymaker.

Oliveira et al. (2002c) used measurements of global and diffuse solar radiation at the surface in the City of São Paulo, between April 1994 and June 1999, to derive models to estimate diffuse solar radiation from values of global solar radiation at the surface. These models are based on the correlation between the diffuse fraction and clearness index for hourly, daily and monthly values. However, in the case of the hourly values, all expressions derived for São Paulo performed poorly.

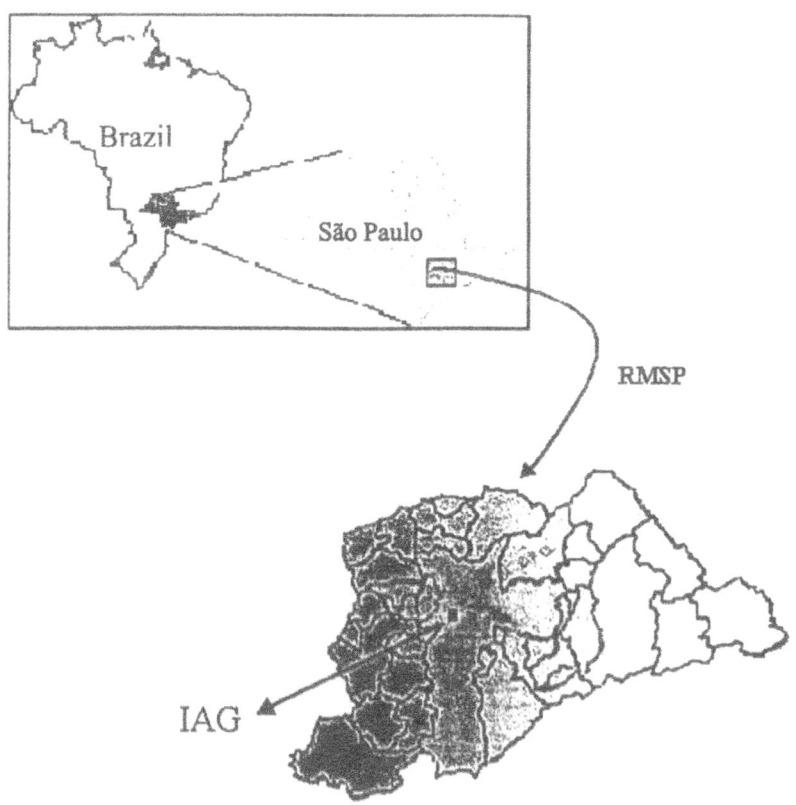

Figure 1. Schematic representation of the measurement site. The metropolitan region of São Paulo (RMSP) is located in the State of São Paulo, southeast of Brazil (Top). The measurements were carried out in IAG, located in the west portion of RMSP (bottom).

2. SITE, INSTRUMENTATION AND DATA SET

Several meteorological parameters, including global and diffuse components of solar radiation, at surface, have been regularly measured in the City of São Paulo, Brazil, since April of 1994. Diffuse component has been measured directly using a shadow-band device. Comparatively to other devices commercially available, this new device has a low cost and is much easier to operate (Oliveira et al., 2002b). The long wave downward atmospheric emission has been also measured in São Paulo, at the surface, using a pyrgeometer from Eppley, since 1997. The work described here it was used 2 years of data corresponding to measurements carried out in 2000 and 2001.

309

The measurements were taken on a platform located at the top of the building of the IAG-USP ("Instituto Astronômico e Geofísico da Universidade de São Paulo") at the "Cidade Universitária Campus", in the west side of the City of São Paulo, at 744 m amsl ($23^0$33'35"S, $46^0$43'55"W). Hereafter this measurement site will be called *IAG site* (Figure 1).

The City of São Paulo is located in the State of São Paulo, Brazil, at approximately 770-m amsl and 60-km westward from the Atlantic Ocean (Figure 1). Its climate - typical of subtropical regions of Brazil - is characterised by a dry winter during June-August and a wet summer during December-March. According to Oliveira et al. (2002a), the minimum values of daily monthly-averaged temperature and relative humidity occur in July and August (16°C and 74%, respectively), and the minimum monthly-accumulated precipitation occurs in August (35 mm). The maximum value of daily monthly-averaged temperature occurs in February (22.5°C) and the maximum value of daily monthly-averaged relative humidity occurs from December through January and from March through April (80%). The maximum value of monthly-accumulated precipitation occurs in February (255 mm). The shortest and the longest day light duration is 10.6 hours (June) and 13.4 hours (December) when the sun reaches the maximum elevation of 54° and 89°, respectively. The maximum value of monthly accumulated period of sunshine occurs in July (183 hours) and the minimum in September (149 hours). The maximum daily monthly-averaged cloudiness occurs in December (8.2 tenths) and the minimum in July (6.1 tenths).

3. THE CHOICE OF MULTI-LAYER PERCEPTRON NEURAL NETWORK (MLPNN)

There are several types of artificial neural networks and they can be divided into groups according to their possible use or according to their topology.

The choice of neural network type depends on the specific modelling problem. The aim here is to reconstruct the diffuse solar radiation as a function of other meteorological parameters. It is important that the used function should be non-linear:

$$y = f_{nonlinear}(x_{in_i})$$

The most suitable topology for modelling such system is the topology of Multilayer Perceptron neural network (Figure 2) with non-linear transfer function (Rumelhart, 1986, Lawrence, 1991). It can be mathematically verified (Hornik, 1991) that this topology, when having enough elements (neurons), is capable of modelling (with arbitrary small error) any "smooth" non-linear multivariable function. Practical consequence of this theorem is that most of the modelling problems can be solved using Perceptron Neural Network with one or two hidden layers and enough of neurons with non-linear transfer function. The theorem shows that the solution of the investigated problem exists but it is the "art of using neural networks" that gives the guidelines to achieve the solution. To clarify the neural network based model, as shown on the Figure 2, we can represent it with the following formula:

$$y = f_{nonlinear}\left(x_{in_i}, \omega_{jl}^n\right)$$

where:

- $x_{in\ i}$ is i^{th} input and
- ω_{jl}^n is the weight from j^{th} neuron in n^{th} layer to l^{th} neuron in n+1 layer

The structure can also be represented by a system of highly non-linear equations where x_1 to x_N are independent variables and ω_{11}^1 to ω_{mk}^2 are parameters that determine how several non-linear sigmoid functions are combined together to obtain the final modelled function.

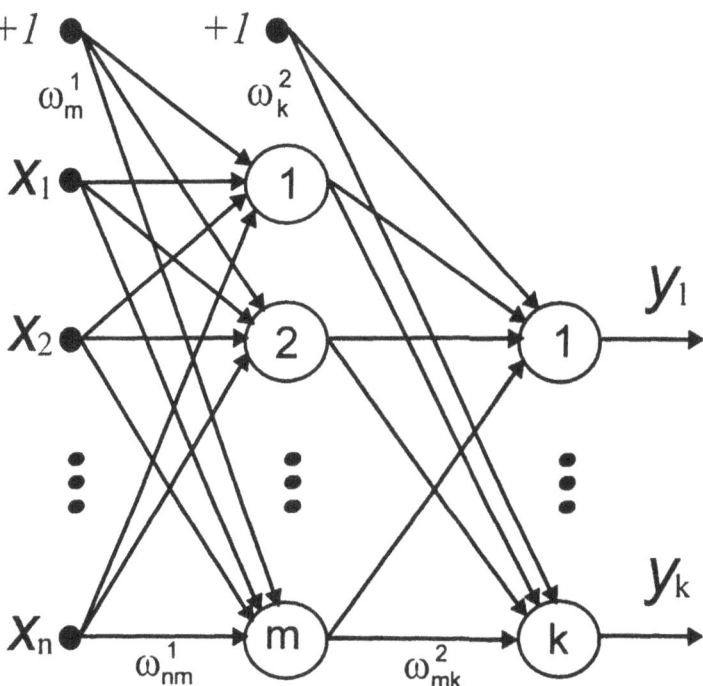

Figure 2. Topography of a typical Perceptron neural network with one hidden layer showing ω as interconnection weight

From this perspective the problem of establishing MLPNN based model is a problem of determination of ω parameters. We can only determine this parameters from the known (measured) points that our function consists of. This points are the vectors that consist of

311

values of independent variables and a value of dependent output variable for a particular realisation. In the language of neural networks these vectors are called *patterns* and the variables – components of the vectors are called *features*.

The process of training MLPNN is a process of ω-parameter determination using the known patterns. The parameters should be determined so that the cost function - average error - be minimal. The process of training is therefore a minimisation problem that can be solved using several numerical techniques. When dealing with MLPNN the most common algorithm used is the gradient descent method called "back-propagation". In back-propagation algorithm the learning patterns (vectors) are presented to the structure repeatedly and the parameters of the MLPNN are changed to finally achieve smaller value of the cost function. In short, we are looking for the "best fit function" over the chosen pattern set. Details of this process can be found in Božnar and Mlakar (2002).

4. THE ART OF USING NEURAL NETWORKS

In the previous section the main idea of establishing MLPNN based model was explained, but when modelling actual problems, the modeller should also overcome several steps that form "the art of using neural networks". This steps are:
- problem formulation,
- determination of all available inputs,
- determination of all available data sets (patterns),
- selection of modelling tool (particular implementation of neural network),
- first construction of simple model with all inputs and all available data,

- improvement of the model:
  - determination of input features,
  - pattern selection,
  - determination of MLPNN number of hidden neurons,
  - determination of training algorithm parameters,
  - MLPNN training with optimisation,
  - verification of the trained model.

The output feature (or features) is the natural phenomenon that we want to model using other available measurements.

Determination of the input features (Mlakar, 1997) is usually the first task. All available environmental parameter measurements and sometimes calculated parameters that may influence the modelled output feature are potential input features. The process of input feature determination can be the process of feature selection or the process of feature extraction. In the process of feature selection the modeller evaluates the relative importance of a particular input feature to the output. The most elegant way is using the trained neural network. One possible procedure is the determination of the contribution factors. They are determined from a neural network that was trained with all possible input parameters (simple model). The contribution factor of particular input feature is the sum of absolute values of the weights that lead from this input to all the neurons in the second layer. The higher this score is, more important the parameter is. More sophisticated and usually more accurate procedure is determination of Saliency metrics that is also based on a trained simple neural network, but it takes into account interconnection weights of all the layers (Mlakar, 1997; Boznar, Mlakar, 2001).

Once that the input and output features were determined it is necessary to decide the number of hidden layers, the quantity of neurons in this layers and the transfer function. The transfer function should be non-linear if we want to model non-linear processes. Usually a sigmoid function is suitable. The number of hidden neurons frequently depends on the number of features and the number of the training patterns.

After the feature selection it is necessary to determine carefully which patterns should be used (Božnar, 1997, Božnar, Mlakar, 2001). The patterns are usually historical measurements of the features. Each pattern represents values for all parameters – features (input and output ones) for a particular measuring interval. From the set of all patterns available, firstly the production set should be removed for the final verification of the trained model. The production set should not be used during the training process if we want to fairly judge the established model performance. Then the learning set should be chosen. The learning set must represent all the possible phenomenon of our modelled parameters that may occur. These are the patterns that will be used for neural network training and they are usually divided into two groups, *training set*, used with back-propagation algorithm and a smaller *testing set* used for optimisation during the learning process. It is known that when the MLPNN learns its training patterns in too much details it looses its generalising capabilities. The generalising capability allows the trained neural network to produce proper output also for the input pattern that differs a little from any of the pattern it has seen during the learning process. Therefore the testing set is used periodically during the learning process to test the performance of the neural network. The final model should be the neural network that gives the smallest error on the testing set and not the one that gives the smallest error on the training set. If the testing set is a good representation of the modelled phenomenon than the final model performance will be good.

Training algorithm parameters determine how quickly the model will be trained and if we will obtain global or relatively good local minimum of the cost function.After the completed learning process the model should be verified using the production set of patterns. It should be stressed that the cost function used for evaluation should clearly reflect the aim of the modelling (Mlakar 1997, Božnar, Mlakar 2001).

5. DIFFUSE SOLAR RADIATION MODEL

The data base of measurements from 1998 to 2002 was firstly divided into two parts. The latest year – 2002 was used as a production set for final verification of the models. From several available parameters input and output features were determined.

After several test it was noticed that the fraction of diffuse over global solar radiations was more suitable for modelling than the diffuse solar radiation itself. Therefore, the fraction became model output feature.

The input features were determined using selection based on the contribution factors and selection based on Saliency metrics. The initial models for feature selection were trained using all available patterns in the period 1998 to 2001 and most of the available parameters. The process of training and determination of the contribution factors and Saliency metrics was repeated several times so that a 95% confidence interval for a particular contribution factor or Saliency metrics was computed.

After several evaluations the following parameters were chosen for input features:

- hourly accumulated global solar radiation at top of atmosphere
- hourly accumulated global solar radiation at surface
- fraction global / top
- zenith angle of sun
- elevation angle of sun
- azimuth angle of sun
- hourly average air relative humidity
- hourly average vapour pressure
- hourly average longwave atmospheric downward radiation at surface

The parameters became input features if they were important at least in one of the two adopted criteria.

One of the important conclusions was that the long wave solar radiation is highly related to diffuse solar radiation.

The pattern selection process has two steps.

After the production set determination, all the good patterns, from the period 1998 to 2001, were used to train the model. From this learning set 10% of data was chosen randomly to test set for optimisation during the learning process. The rest of patterns were used for training set. The results of final model verified on production set were good.

In the second step we want to enhance the performance of the model in the range of high values of diffuse solar radiation. This patterns are especially important and difficult to reconstruct. In the total learning set we reduce the number of patterns that have low value of diffuse solar radiation simply by removing 40% of such samples randomly.

Standard backpropagation algorithm was used for training. The learning rate parameter was set to 0.5 and the momentum was set to 0.9.

6. RESULTS

On the Figure 3 a typical time series prediction is shown for the learning set. This proves the ability of the used MLPNN to memorise training patterns. If we properly select the learning patterns, so that they are representative for the whole modelling domain (Božnar, Mlakar, 2002), also the results on production set of patterns will probably be good.

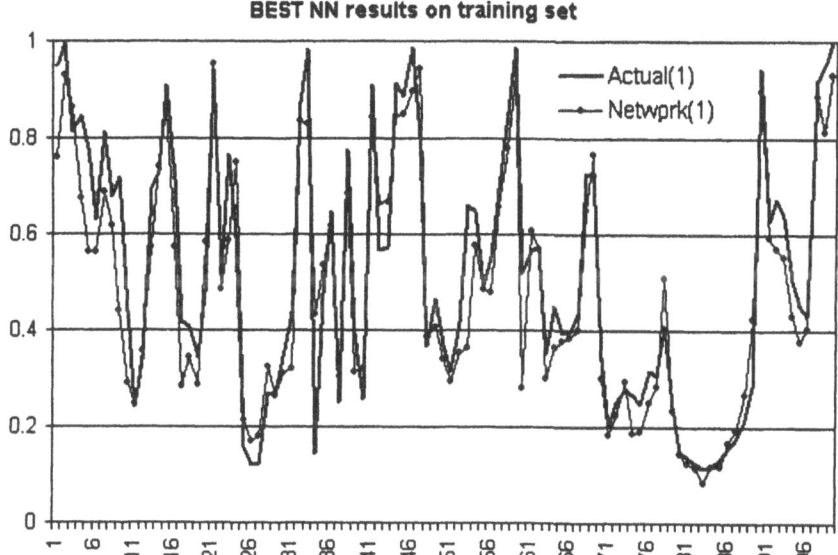

Figure 3. Time series prediction for the learning set shows the memorisation and reconstruction capabilities

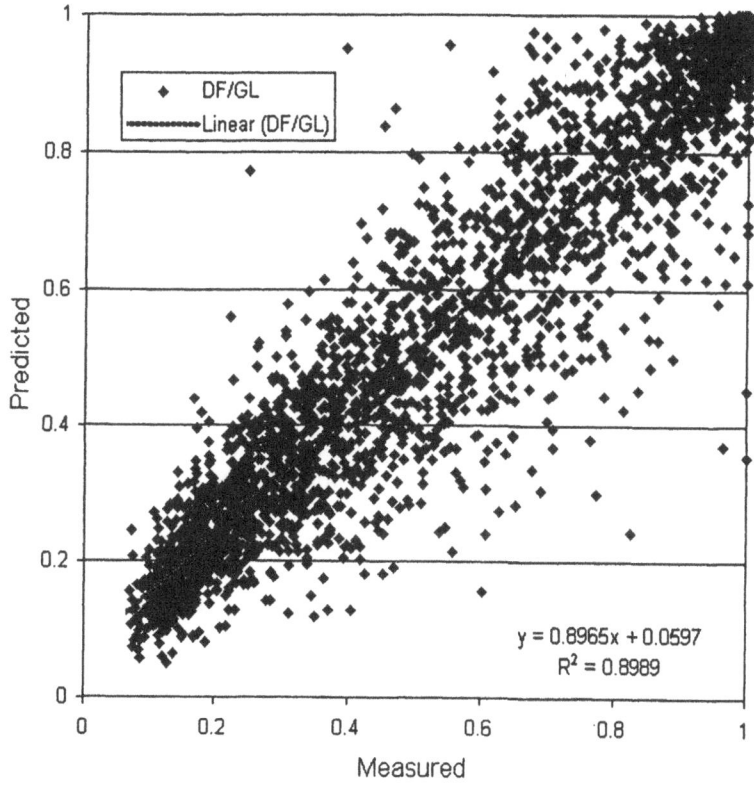

Figure 4. Scatter plot (diffuse/global actual) versus (predicted)

Figure 4 shows the scatter plot of actual versus predicted values of the model over the independent production set that was not used during the learning process.

Figure 5 shows a time series of actual and predicted values over the production set for a well reconstructed period.

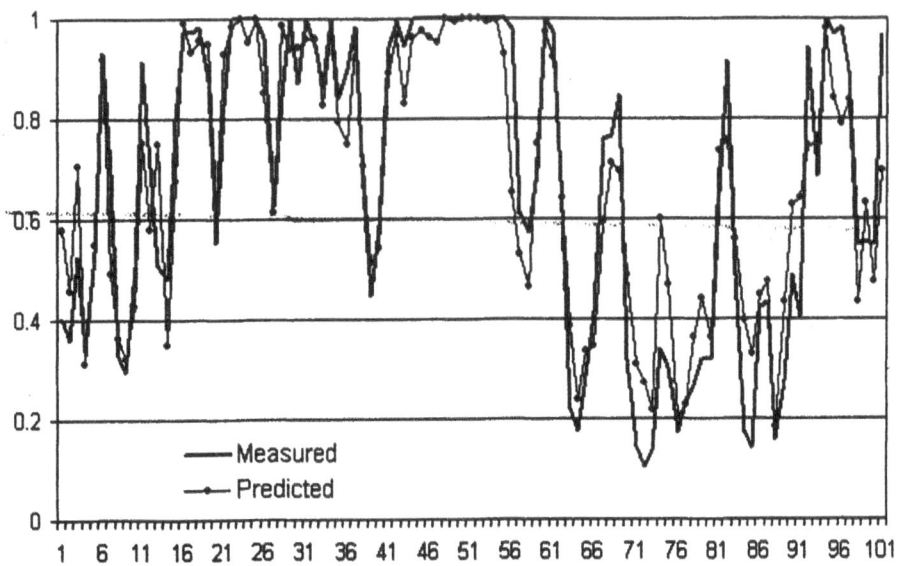

Figure 5. Example of good prediction over independent production set

Figure 6 shows a time series of actual and predicted values over the production set for a period that does not give so good results.

7. CONCLUSIONS

Compared with the correlation models (Oliveira et all, 2002c) this new Perceptron neural network based model has a better performance for hourly averaged values of diffuse solar radiation.

8. OTHER APPLICATIONS OF ARTIFICIAL NEURAL NETWORKS IN THE AIR POLLUTION SCIENCE

We started our work using artificial neural networks on a problem of SO_2 air pollution around Slovenian Thermal Power Plant Sostanj (Božnar, Mlakar, Lesjak, 1993). The method was based on Multilayer Perceptron neural network. We developed it further specially in by our opinion most important tasks of feature determination and pattern selection techniques.

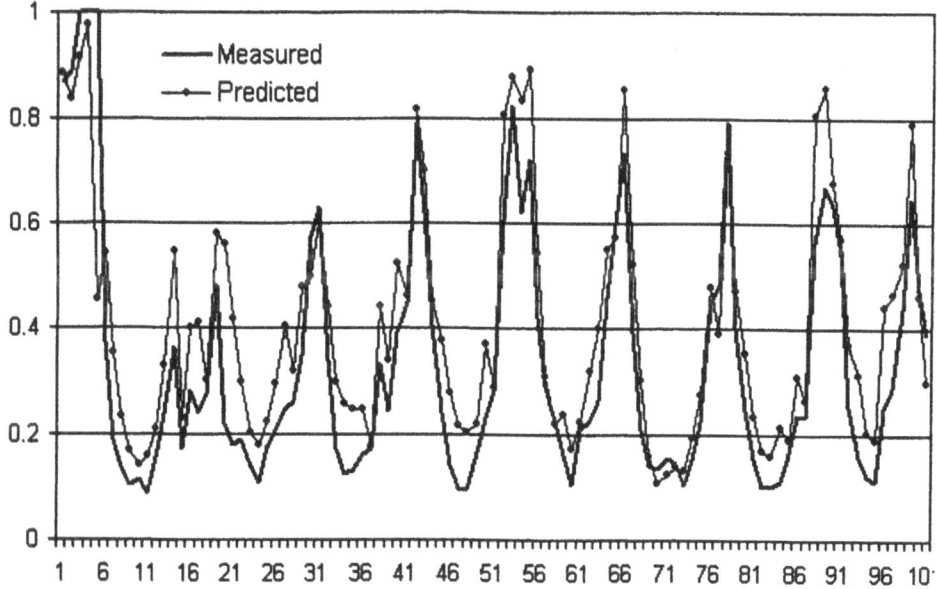

Figure 6. Example of poor prediction over independent production set

In the pattern selection techniques we also introduced the possible use of self organising Kohonen neural network.

We tested both neural networks also in the field of winds reconstruction and classification (Božnar, Mlakar 2002, Mlakar, Božnar 1996). And finally we applied the method to the problematic of diffuse solar radiation.

In last years also other authors used artificial neural networks in the field of air pollution modelling. An overview of applications is given in review article by Gardner, Dorling 1998.

9. ACKNOWLEDGEMENT

This research was sponsored by CNPq. (Conselho Nacional de Desenvolvimento Científico e Tecnológico) during the collaboration program Brazil-Slovenia carried out in 2002-2003 (Proc. CNPq No. 490017/02-9).On Slovenian side the research was sponsored by the AMES company and Slovenian Ministry for school, science and sport during the same colaboration program.

10. REFERENCES

Božnar, M., 1997, Pattern selection strategies for a neural network - based short term air pollution prediction model. V: ADELI, H. , ed., Intelligent Information Systems IIS'97, Grand Bahama Island, Bahamas, December 8-10, 1997. *Proceedings*. Los Alamitos, California: IEEE Computer Society, 340-344

Božnar, Marija Zlata, MLAKAR, Primož. Artificial neural network-based environmental models. V: GRYNING, Sven-Erik (ur.), SCHIERMEIER, Francis A. (ur.). *Air pollution modeling and its application XIV*. New York [etc.]: Kluwer Academic: Plenum Publishers, cop. 2001, 483-490.

Božnar, Marija Zlata, MLAKAR, Primož. Use of neural networks in the field of air pollution modelling. V: 25th NATO/CCMS International Technical Meeting on Air Pollution Modelling and its Application, 15-19 October 2001, Aula Magna, Louvain-la-Neuve, Belgium. *Preprints*. Louvain: Université Catholique de Leuvain, 2002, 265-272 .

Božnar, M., Lesjak, M., Mlakar, P., 1993, A neural network-based method for short-term predictions of ambient SO2 concentrations in highly polluted industrial areas of complex terrain. *Atmos. environ., B Urban atmos.*, vol. 27, 221-230.

CETESB, 1999: Technical Report (*in Portuguese*).

Collares-Pereira, M. and Rabl, A. (1979) The average distribution of solar radiation – correlation between diffuse and hemispherical and between daily and hourly insolation values, *Solar Energy*, 22, 155-164.

Eletropaulo, 1999: Technical Report (*in Portuguese*).

Hornik, K., 1991, Approximation capabilities of multilayer feedforward networks, Neural Networks, vol.4, 251-257

Gardner, M.W. and Dorling, S.R., *1998*, Artificial Neural Networks (The Multilayer Perceptron)- A Review of Applications in the Atmospheric Sciences. *Atmospheric Environment*, 32 (14/15), 2627-2636

Iziomon, M.G. and Aro, T.O., 1999: On the Annual and Monthly Mean Diurnal Variation of Diffuse Solar Radiation at a Meteorological Station in West Africa. *Meteorol. Atmos. Phys.*, 69, 223-230.

Kretzschmar, J. G., 1994: Particulate matter levels and trends in Mexico City, São Paulo, Buenos Aires and Rio de Janeiro. *Atmos. Environ.*, 28, 3181-3191.

Lawrence, J., 1991, Introduction to Neural Networks, California Scientific Software, Grass Valley

LeBaron, B. A., Michalsky, J.J. and Perez, R., 1990: A simple procedure for correcting shadowband data for all sky conditions. *Solar Energy*, 44, 249-256.

Mlakar, P., 1997, Determination of features for air pollution forecasting models, V: ADELI, H. , ed., Intelligent Information Systems IIS'97, Grand Bahama Island, Bahamas, December 8-10, *Proceedings*. Los Alamitos, California: IEEE Computer Society, 350-354.

Mlakar, P., Boznar, M., 1996, Analysis of winds and SO2 concentrations in complex terrain. V: Caussade, B. ed., Power, H. ed., Brebbia, C. A. , ed., *Air pollution IV : monitoring, simulation and control*. Southampton; Boston: Computational Mechanics Publications, 455-464.

Oliveira, A. P., Escobedo, J.F, Plana-Fattori, A., Soares, J. and Santos, P. M., 1996: Medidas de radiação solar na Cidade de São Paulo: Calibração de piranômetros e aplicações meteorológicas. *Revista Brasileira de Geofísica*, 14, 203-216. (*in Portuguese*)

Oliveira, A. P., A.J.Machado, J.F.Escobedo and J.Soares, 2002a: Diurnal evolution of solar radiation at the surface in the city of São Paulo: seasonal variation and modeling. *Theoretical and Applied Climatology*. 71 (3-4). 231 – 249.

Oliveira, A. P. Machado, A J. and J. F. Escobedo 2002b: Diurnal evolution of the solar radiation in the City of São Paulo, Brazil: Seasonal Variation and Modeling, *Meteorology and Atmospheric Physics*.

Oliveira, A.P., A.J.Machado, J.F.Escobedo, J.Soares, 2002c: Correlation models of diffuse solar radiation applied to the city of São Paulo (Brazil). *Applied Energy*. 71 (1), 59 – 73

Pereira, E.B., Abreu, S. L., Stuhlmann, R., Reiland, M. and Colle, S. (1996) Survey of the incident solar radiation in Brazil by use of Meteosat satellite data, *Solar Energy*, 2, 125-132.

Rumelhart, D. E., McClelland, J. L., 1986, Parallel distributed processing 1,2, MIT Press, Cambridge, MA

WMO, 1971: Guide to meteorological instruments and observing practices. Fourth Edition, Geneva, Secretariat of the World Meteorological Organization, *publication WMO N° 8*, 325 pp.

DISCUSSION

J. KELLER   Which are the assumptions for the non-linear transformations? Are they based on physics?

M. Z. BOZNAR   The non-linear transfer function that is a part of a neuron structure should be monotonusly increasing ($\partial f(x)/\partial x > 0$) and should be without singularities. There are several posibilities. We have choosen sigmoid transfer function. The selection of the function is not based on physics.

SCALING ANALYSIS OF OZONE PRECURSOR RELATIONSHIPS

Bruce Ainslie and D.G. Steyn[*]

1. INTRODUCTION

The importance of ozone production from mixtures of NOx and volatile organic compounds (VOCs) in the presence of sunlight was first recognized in the Los Angeles basin in the 1940s (Haagen-Smit, 1952). Since that time, a great deal of effort has been directed at understanding tropospheric photochemistry. Complicating this effort is the multitude of VOC species found in a polluted environment, and the extremely large number of ways these species can react – it is estimated that an explicit treatment of ozone formation would contain more than 20,000 reactions involving several thousand reactants and products (Dodge, 2000). To limit the number of species and reactions, different means of simplifying the chemistry have been employed which typically yield mechanisms with 30 to 50 species and between 80 and 200 reactions (Dodge, 2000). While this represents a significant reduction, the resulting mechanisms remain complex; solving the resulting chemical transformation require the largest fraction of computing time in a photochemical simulation.

In response to this difficulty, considerable research has been devoted to understanding photochemical systems in terms of a small set of variables. One way to restrict the problem involves using smog chamber data to empirically fit ozone concentrations to initial NOx and VOC concentrations (Akimoto et al. (1979), Sakamaki et al. (1982) and Chang and Rudy (1993)). Another approach has been to develop a highly parameterized set of chemical reactions involving only a few species (Azzi et al., 1992 and Venkatram et al., 1994). A third approach uses ambient ozone and NOx concentrations to infer the extent to which NOx has reacted (Chang et al., 1997 and Blanchard, 2000).

This paper presents a new method of understanding ozone production from NOx-VOC systems, in terms of a small set of variables, through the use of a scaling analysis.

[*] Atmospheric Science Programme, University of British Columbia, Vancouver, B.C., Canada

Air Pollution Modeling and Its Application XVI, Edited by
Borrego and Incecik, Kluwer Academic/Plenum Publishers, New York, 2004

321

2. SCALING ANALYSIS

A central (but generally unstated) belief in the physical sciences is that physical laws hold at all scales. However, for complex systems, the relative importance of different influences may change with scale. As a result, it often appears (to first order) that different laws hold at different scales i.e. fluid flow at high and low Reynolds numbers. Whenever a complex system does not show changing behaviour across different scales, we say a similarity relationship holds. In such instances, the complex system can be described using a simple statistical or geometrical framework where highly specific physical laws are not necessary. In this paper we describe ground-level ozone concentration in terms of similarity relationships without the need for specific chemical or atmospheric processes.

2.1 Scaling analysis and photochemical modeling

Before proceeding with the scaling analysis, it should be noted that scaling methods are already extensively used in photochemical modeling. In these models, there are three separate branches of investigation: emissions, meteorology and photochemistry. The emissions branch models the flux of pollutants for all major sources across the domain of interest. Except for large industrial point sources, actual emissions rates are seldom measured. Instead it is assumed that they scale with more easily measured surrogates. For instances, mobile source emissions are generally scaled to: vehicle kilometers traveled (VKT), fleet characteristics (age and vehicle type) and driving patterns (vehicle speed and engine load).

For meteorological modeling, the local dispersion of pollutants by turbulent mixing is often not explicitly modeled but calculated using similarity relationships. These relationships show how turbulent mixing scales with: height, wind speed, surface heating and temperature structure of the atmospheric boundary layer.

Finally, scaling relationships are also used in the development of chemical mechanisms. As mentioned before, modern mechanism contain only a fraction of the number of reactions and a smaller number of reactants and products. To achieve this condensation, organic compounds are grouped into manageable sets of VOC. This implies that across a restricted set scales (in this case molecular structure, OH-reactivity and emission levels), the complex photochemical behaviour of VOCs is unchanging.

2.2 Buckingham Pi Analysis

Our investigation makes use of the Buckingham Pi scaling analysis. This is a systematic method of dimensional analysis in which all relevant physical variables are placed into dimensionless groups. Since the number of relevant dimensionless groups must equal the number of original variables less the number of fundamental dimensional units, this method reduces the order of the problem (Bluman and Anco, 2002). The method is used to provide a description (similarity relationship) of the system in terms of dimensionless groups. Finally, an empirical curve must be found to quantitatively describe the resulting similarity relationship.

3. PROPENE-NOx SYSTEM

3.1 Dimensionless Groups

For a smog chamber experiment, the relevant physical variables which govern ozone photochemistry are: the initial precursor concentration (both NOx and VOC), the energy added to the system in terms of actinic flux (intensity and total) and temperature. In addition, the rate at which nitric oxide titrates ozone and the rate at which the VOC reacts with the hydroxyl radical characterize the fast photostationary state (PSS) cycle and the slower radical driven cycle. Thus, maximum ozone concentration could depend on as many as the 7 different factors given in Table 1.

Table 1. Physical Quantities which affect ozone concentrations in a smog chamber

Factor	Variable	Units
Initial VOC concentration	[VOC]	molecules cm^{-3}
Initial NOx concentration	[NOx]	molecules cm^{-3}
Chamber temperature	T	K
NO + O_3 titration rate	k_{NO}	molecules cm^{-3} s^{-1}
Average actinic flux	J_{av}	s^{-1}
Total irradiation time	Δt	s
VOC + $OH^\bullet$ reaction rate	k_{OH}	molecules cm^{-3} s^{-1}

From Table 1, we see that our problem has 3 fundamental dimensional units: *length, time* and *temperature*. In order to non-dimensionalize all physical quantities, three key variables must be chosen which contain the three dimensions. The variables: k_{no}, j_{av} and activation energy of the OH-reactivity (E/R) have been chosen which yields the following dimensionless groups:

$$\Pi = \frac{[O_3]_{max}}{j_{av}/k_{NO}}$$ Dimensionless Ozone concentration

$$\Pi_1' = \frac{[VOC]_o}{j_{av}/k_{NO}}$$ Dimensionless VOC concentration

$$\Pi_2 = \frac{[NOx]_o}{j_{av}/k_{NO}}$$ Dimensionless NOx concentration

$$\Pi_3 = j_{av}\Delta t$$ Total actinic flux

$$\Pi_4' = -\frac{E}{R}\left(\frac{1}{T}\right)$$ Dimensionless Activation Energy

It is more convenient to re-define Π_1 and Π_4 as follows:

$$\Pi_1 = \frac{\Pi_1'}{\Pi_2} = \frac{[VOC]}{[NOx]}$$ Ratio of Initial Precursor Concentrations

$$\Pi_4 = \exp\left\{\Pi_4'\right\}$$ Dimensionless Temperature Scale

The theory of dimensional analysis states that an ozone response surface can now be described in terms of these four dimensionless variables; a reduction in the order of the problem by three. Guided by the work of Johnson (1984), Blanchard (2000) and Chang and Rudy (1993), we assume a simple power law dependence of ozone on NOx i.e.:

$$\frac{\Pi}{\Pi_2^{\,a}} = \phi\left(\Pi_1, \Pi_3, \Pi_4\right) \qquad (1)$$

where the unknown similarity function (ϕ) now depends on only three dimensionless groups. To keep the analysis brief, we focus on the precursor dependence (Π_1), postponing a qualitatively description of temperature (Π_4) and actinic flux (Π_3) dependence to a later section.

3.3 OZIPR Simulations

To generate model output for our scaling analysis, the numerical solver OZIPR (Gery and Crouse, 1990) was used with the RADM2 chemical mechanism (Stockwell et al., 1990). Simulations were run with a constant temperature of 25°C and total actinic flux corresponding to Vancouver B.C. (latitude of 49.25°N) on June 21. Simulations started at 7:00 a.m. and ended at 6:00 p.m. (local daylight savings time). There was no dilution in the simulation. Propene was used as the sole VOC due to its relatively simple chemistry, high reactivity and because of its use as a reactivity standard (NRC, 1991). In the RADM2 mechanism, propene photochemistry is modeled using the OLT surrogate. In our simulations, initial OLT concentrations varying from 0 to 0.6 ppm and initial NOx from 0 to 0.15 ppm. The ratio of initial NO_2 to NO was fixed at 0.25.

Simulations were run for an 11 by 11 matrix by varying initial OLT and NOx concentrations in 10% increments. Model output consisted of maximum ozone concentration during the 11 hour simulation along with the initial OLT and NOx concentrations. However, before the model output was analyzed, some runs were excluded. Whenever the initial NOx concentration was zero, RADM2 produced no ozone and these trivial runs were removed. In addition, with no initial OLT, RADM2 produces ozone in low concentrations. These values, independent of OLT, represent ozone concentrations at the PSS. Since the scaling analysis is concerned with NOx-VOC chemistry, these were also excluded. After removal, each simulation consisted of 100 runs which were used with Eq. (1) to determine a similarity relationship.

3.4 Similarity Relationship

Various values of the exponent (a) in Eq. (1) were chosen in an attempt to get the dimensionless ozone data to collapse onto a single common curve. An exponent of $a = 0.60$ was found to produce the best collapse and the resulting curve, which represents the unknown similarity relationship (ϕ), is shown in Figure 1. From the Figure, we see very little scatter in the model output (hereafter referred to as "data") which presents a great improvement over earlier work by Chang and Rudy (1993). The value of a lies between the 0.5 suggested by Chang and Rudy (1993) and 0.66 proposed by Blanchard (2000).

Following the ideas of Ainslie and Steyn (2001) and Chang and Rudy (1993), the Weibull model was used to parameterize this relationship. However, to first check the appropriateness of this parameterization, the "data" were normalized and then Weibull transformed. In this way, plotting $ln(ln(1/(1-\phi/\gamma)))$ against $ln(R)$, should find the "data" falling on a straight line (where γ represents the normalizing factor)

Figure 2 shows the results of the 'Weibullized-data' plotted as a function of $ln(R)$. From this figure, it is evident that the "data" collapses onto not 1 but 2 straight lines; suggesting a change in chemical process occurring at $ln(R) = 1.4$. Included in this Figure are two boundaries separating the figure into 3 regions. Region I, lies to the right of the break and regions IIa & IIb to the left. It appears that for small $ln(R)$, that there is more scatter in the "data"; indicative of a transition to a different scaling regime. This transition, marked by the boundary between IIa and IIb, is difficult to objectively place and requires further investigation. We also see a slight hook in the "data" at $ln(R) = 3.5$. This represents ozone loss through $OLT+O_3$ reactions, which occurs for large excesses of OLT, and marks another change in the governing chemical process.

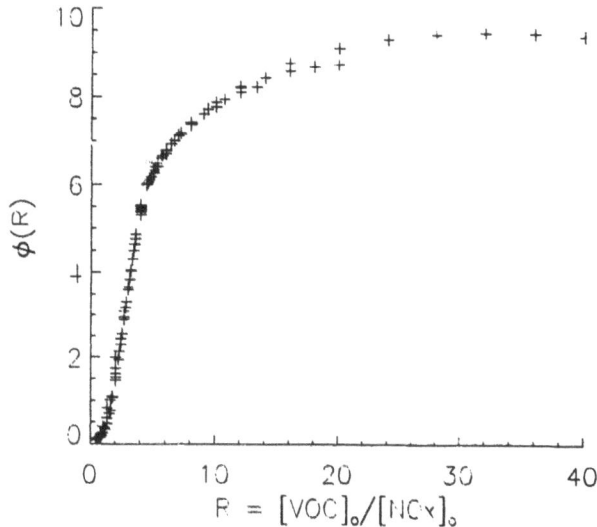

Figure 1. Dimensionless maximum ozone divided by dimensionless initial NOx raised to the 0.6 plotted against the ratio of initial OLT to NOx concentration.

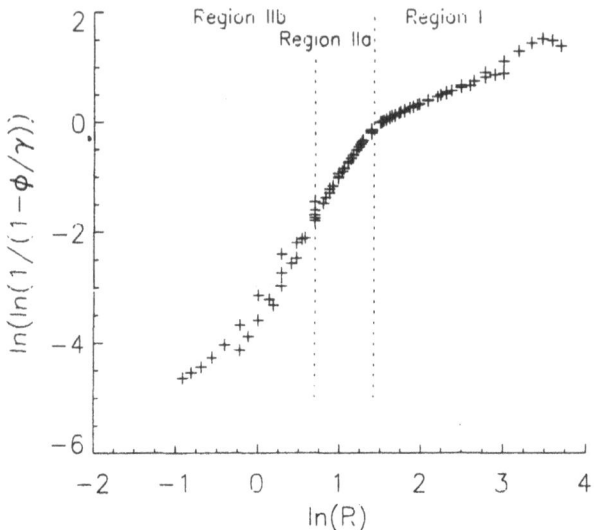

Figure 2. Weibull transformed OLT data plotted against the ln(R). Three regions of interest are also shown: regions I and IIa are separated by a change in slope while region IIb suggests the beginning of a new scaling regime.

4. WEX Model

A single curve for the similarity relationship, which captures the scaling break, is both continuous and differentiable, has been developed. The WEX model is a modified Weibull with shape parameter (α) that switches between two values at the scaling break:

$$\phi(R) = 1 - \exp\left\{-\lambda(R/\beta)^{\alpha(R)}\right\}$$

$$\alpha(R) = \left(\frac{\alpha_2 - \alpha_1}{2}\right)\tanh(R - \beta) + \left(\frac{\alpha_1 + \alpha_2}{2}\right)$$

$$= WEX(R; \alpha_1, \alpha_2, \beta, \lambda)$$

In this model α_1 and α_2 are the slopes of the 'Weibullized data', β is the R-value at the scaling break and λ is related to the value of the "Weibullized data' at $R = \beta$. When $\alpha_1 = \alpha_2$ and $\lambda = 1$ the Weibull model is recovered.

In Figure 3, an isopleth plot of the "data" that was used in Figures 1&2 is shown along with the WEX model used to fit the "data". A line of constant R, associated with the scaling break, has also been plotted. The ridgeline, as determined by the location where the sensitivity of ozone to changing NOx concentrations is zero has also been drawn. It has a value of $R = 4.8$, and appears to lie below the scaling break. From the Figure, we see that the similarity relationship fits the ozone isopleths quite well in both Region I and IIa. In region IIb the curve starts to diverge from the OZIPR output.

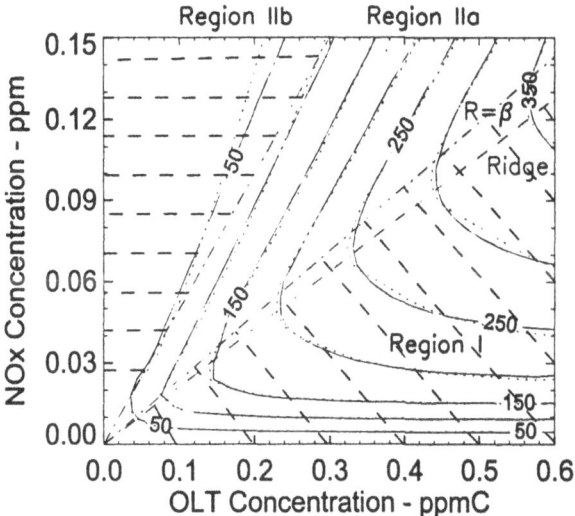

Figure 3. Ozone isopleths (in ppb) for OZIPR "data" (solid lines) and WEX model (dotted lines). Also shown is the ridgeline, scaling break (β),and boundaries for regions I, IIa and IIb in dashed lines.

5. UNIVERSALITY OF SCALING ANALYSIS

Further scaling analysis has been carried out for other VOC species and VOC mixtures. With the exception of unreactive species, all show a characteristic similarity function embodied by a Weibull plot having "data" cluster onto two distinct line segments separated by a scaling break. In general, unreactive VOCs (principally the alkanes) do not show scaling. Aromatics show more scatter than alkenes but are still captured by the WEX model. Carbonyl compounds show the best scaling. In addition, two other chemical mechanisms, the CBM-IV (Gery et al., 1989) and the SAPRC (Carter, 1990), have also been used to generate model output. Their output produce the same trends. Further analysis shows the four WEX parameters are related to overall effects of competing sub-processes within the model. For instance, the WEX slope α_1 is a measure of the VOC's NOx-inhibition.

The scaling dependence on environmental conditions (temperature and actinic flux) has also been examined. We find changing environmental conditions do not change the nature of the similarity function but rather alter the slopes (α_1 and α_2), scaling breaks (β) and λ-values of the data in the Weibull plane. The most sensitive parameter to these changes is the scaling break (β); increasing with increasing temperature and with decreasing actinic flux. The other WEX parameters are not strongly sensitive to changing environmental conditions.

By including the precursor, temperature and actinic flux dependencies together, it is possible to develop a single "universal" similarity function which describes the ozone response surface as a function of all 7 dependant variables.

6. CONCLUSIONS

This paper outlines a scaling analysis of ozone production in the presence of NOx and VOCs. The analysis allows us to characterize an ozone response surface (isopleth plot) with a 2-D similarity function. Results are independent of chemical mechanism and show consistent behaviour over a wide range of temperature and actinic fluxes. Furthermore, a Weibull transformation of the resulting similarity relationship reveals a fundamental switch in the underlying chemical processes occurring above the ridgeline.

A model has been developed to quantify the ozone response surface which holds in the NOx-limited region, around the ridgeline and for parts of the VOC-limited region. This represents a great improvement over parameterizations by Johnson (1984), Chang and Rudy (1993) and Blanchard (2000). Furthermore, the model parameters describe process in the mechanism and can be used to understand how chemical mechanisms produce ozone. In future work, the WEX model will be incorporated into an Eularian grid model using the concept of species age (Venkatram et al., 1994). There it is hoped that our photochemical parameterization will capture ozone sensitivity to changing emission levels and serve as an effective screening tool.

7. REFERENCES

Ainslie, B and Steyn, D.G., 2002. Revisiting ozone-precursor relationships. *In Air Pollution Modeling and Its Application XV*, Borrego and Schayes (eds.), Kluwer Academic Press, New York, pp.347-355.

Akimoto, H., Sakamaki, F., Hoshino, M., Inoue, G. and Okuda, M., 1979. Photochemical ozone formation in propylene-nitrogen oxide-dry air system. *Environmental Science & Technology* 13, 53-58.

Azzi, M., Johnson, G.M. and Cope, M., 1992. An introduction to the generic reaction set photochemical smog mechanism. *Proceedings of the 11th International Clean Air Conference 4th Regional IUAPPA Conference*, Brisbane, Australia, July 1992.

Blanchard, C.L., 2000. Ozone process insights from field experiments – Part III: extent of reaction and ozone formation. *Atmospheric Environment* 34, 2035-2043.

Bluman, G.W. and Anco, S.C., 2002. Symmetry and Integration Methods for Differential Equations. Applied Mathematical Sciences 154, Springer-Verlag, New York, 419 pp.

Carter, W.P.L., 1990. A Detailed Mechanism for the Gas-Phase Atmospheric Reactions of Organic Compounds *Atmospheric Environment*, 24A, 481-518.

Chang, T.Y. and Rudy, S.J., 1993. Ozone-precursor relationships: a modeling study of semi-empirical relationships. *Environmental Science and Technology* 27, 2213-2219.

Dodge, M., 2000. Chemical oxidant mechanisms for air quality modeling: critical review. *Atmospheric Environment* 34, 2103-2130.

Johnson, G.M., 1984. A simple model for predicting ozone concentration of ambient air. *Proceedings Of The Eighth International Clean Air Conference*, Melbourne, Australia, pp.715-731.

Gery M.W., Whitten G.Z., Killus J.P. and Dodge M.C., 1989. A photochemical kinetics mechanism for urban and regional scale computer modeling. Journal of Geophysical Research 94 (D10), 12 925- 12 956.

Gery, M. and Crouse, R., 1990. *User's Guide for Executing OZIPR*. Order No. 9D2196NASA, U.S. Environmental Protection Agency, Research Triangle Park, NC, 27711.

Haagen-Smit, A.J., 1952. Chemistry and physiology of Los Angeles smog. *Industrial and Engineering Chemistry* 44, 1342-1346.

Sakamaki, F., Okuda, M. and Akimoto, H., 1982. Computer modeling study of photochemical ozone formation in the propene-nitrogen oxides-dry air system. Generalized maximum ozone isopleth. *Environmental Science & Technology* 16, 45-52.

Stockwell, W.R., Middleton, Chang, J.S. and Tang, X., 1990. The second generation regional acid deposition model chemical mechanism for regional air quality modeling. *Journal of Geophysical Research* 95, 16343-16367.

Venkatram, A., Karamchandani, P., Pai, P. and Goldstein, R., 1994. The development and application of a simplified ozone modeling system (SOMS). *Atmospheric Environment* 28: 3665-3678.

DISCUSSION

E. GENIKHOVICH   Did you compare your computational results with data of field measurements?

B.AINSLIE   I have not compared my results with field data. I intend to compare it with smog chamber though.

V.P.ANEJA   Can this technique be used for other photochemical oxidants, such as hydrogen per oxide nitnic and nitrous acid?

B.AINSLIE   In principle, I believe it would work for these species as well. I have obtained similiar results when modelling odd-oxygen (Ox). However, my goal is to couple this parameterization with a simple model for emissions and meteorology to create a fast running photochemical model, so my insterest is primarily with ozone.

W.GONG   Do you have explanations on the fact that your break point (or transition) is above the "ridge line"?

B.AINSLIE   The ridgeline is somewhat arbitrary – there are several ways to define one. A better way to look at an isopleth diagram is with 3 regions: VOC-limited, ridge or transition region and Nox limited region. My break point identifies the beginning of the ridge or transition region. Physically, the break point signals at change in feedback loops that controls ozone production (P(O3)).

G.SCHAYES   Does this scaling approach allow to simplify the chemical part of the models?

B.AINSLIE   Yes it does. I have reduced a mechansim that contains over 60 coupled differential equations to an analytic expression with 6 parameters. It should make my screening model orders of magnitude quicker than the full photochemical model.

SOME ASPECTS OF TURBULENCE AND DISPERSION IN LOW WIND SPEED CONDITIONS

Domenico Anfossi, Dietmar Oettl and Gervásio A. Degrazia[*]

1. INTRODUCTION

In many places in the world low wind speed conditions (LWS, $\bar{u} \leq 1 - 2 \ m \, s^{-1}$) occur for a substantial percentage of time. Dispersion in these conditions is governed by meandering (low frequency horizontal wind oscillations), weak, layered and intermittent turbulence and air stagnation. These characteristics give rise to highly non-stationary and inhomogeneous diffusion conditions. Even when the stability (during night time) reduces the vertical dispersion, meandering disperses the plume over rather wide angular sectors. Thus, the resulting ground level concentration is generally much lower than that predicted by standard Gaussian plume models. As a consequence, different types of models should be used (see, for instance: Brusasca et al., 1992; Oettl et al., 2001). Oettl et al. (2001), in particular, put in evidence the presence of a negative lobe in the horizontal wind velocity Eulerian autocorrelation functions (hereafter EAF), attributed to the meander. An example of a large negative lobe in the longitudinal low wind EAF was also reported by Hanna (1983)

This paper is an attempt to improve our understanding of LWS turbulence characteristics. It is based on the analysis of hourly time series of sonic anemometer wind speeds.

Two sonic anemometer data sets, one containing wind observations in complex terrain (in a suburban area of the city of Graz, Austria) and the second one in a rather flat area (at Tisby site, about 45 Km west of Uppsala, Sweden), were analysed. LWS were defined as $\bar{u} \leq 1.5 \, m \, s^{-1}$. The coordinate system was rotated so that the cross wind component $\bar{v}$ and the vertical wind component $\bar{w}$ were zero.

This paper is divided into two parts: firstly some general characteristics of wind turbulence in LWS conditions were studied (consequently, we used the Graz data set, which contains a complete year of data, thus being able to capture all the diurnal and seasonal conditions), then we analysed in detail the shape of autocorrelation functions

---

[*]Domenico Anfossi, C.N.R., Istituto di Scienze dell'Atmosfera e del Clima, Torino, Italy, D. Oettl, Institute of Internal Combustion Engines and Thermodynamics, Graz University of Technology, Graz, Austria, Gervásio A. Degrazia, Universidade Federal de Santa Maria, Departamento de Física, 97105-900, Santa Maria, Brazil.

and looked for analytical forms fitting them (thus, we used the Tisby data set, since this area is rather flat and so any possible influence of terrain interactions with the flow can be excluded).

2. GENERAL TURBULENCE CHARACTERISTICS OF LOW WIND (GRAZ DATA SET)

Figure 1 shows the dependency of the observed σ_w on stability. As can be seen, there is almost no change in σ_w with varying stability $z L^{-1}$. Hence, it can be concluded u_* to be the main parameter that determines σ_w. The Graz observations support the

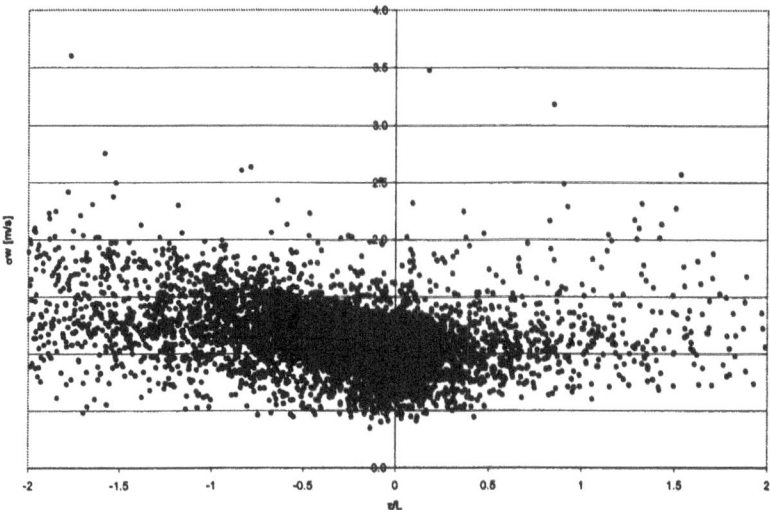

following relationship in between these two quantities (see Figure 2):

$$\sigma_w = 1.1 \cdot u_*$$
(1)

The coefficient in eq. (1) is slightly higher than the one in Oettl et al. (2001), where a similar relationship was given for $z L^{-1} > 0$. The total number of data points used in Figures 1 and 2 was n = 7989. Panofsky and Dutton (1984) suggested an expression for σ_w with a slightly higher coefficient namely 1.25 for $z L^{-1} > 0$. Stull (1989) gives a value of 1.58, and 1.0 to 1.58 for stable and neutral stability respectively. For convective conditions a relationship can be found in Stull (1989), where there is also a

Figure 1 – *Dependency of the observed σ_w on stability*

dependency of σ_w on L. As shown previously in Figure 1, such a dependency was not confirmed by the observations presented in this study.
Clearly, meandering of the flow, which occurs at LWS does not affect the vertical dispersion process.

As to the standard deviations of the cross wind and along wind component σ_v and σ_u, no satisfying expression was found for both quantities. However, some correlation exists between the friction velocity and the square root of the vector sum of σ_v and σ_u (see Figure 3):

$$\sqrt{\sigma_u^2 + \sigma_v^2} = 2.53u_* + 0.27 \qquad (2)$$

Equation (2) was found to be independent on stability. Stull (1989) lists an almost similar relationship for neutral and stable conditions, but without a shift in the y-axis and a higher coefficient, namely 2.92.

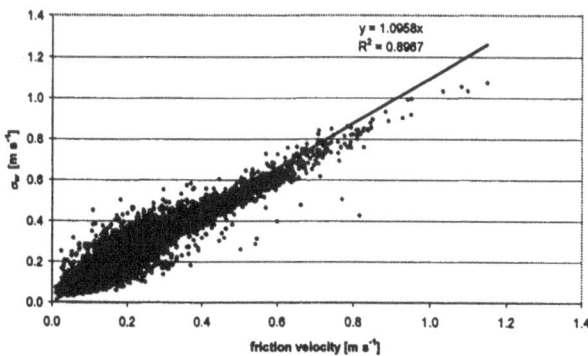

Figure 2 - *Scatter plot of observed σ_w and friction velocity*

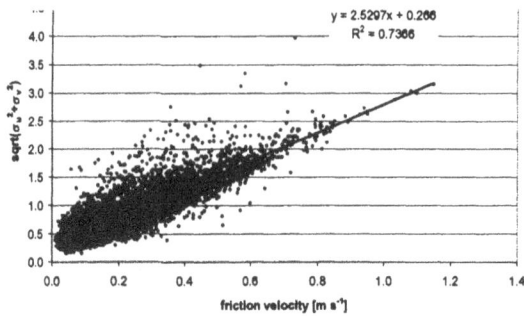

Figure 3 – *Scatter plot of observed* $\sqrt{\sigma_u^2 + \sigma_v^2}$ *and friction velocity*

3. BEST FIT OF LOW WIND EAFs

Let us remember the following classical results (Taylor, 1921):

$$\sigma_y^2(t) = 2\overline{v'^2} \int_0^t (t - \tau) R_I(\tau) d\tau \qquad (3)$$

$$T = \int_0^T R_I(\tau) d\tau \qquad (4)$$

$$R_I(\tau) = e^{-\frac{\tau}{T}} \qquad (5)$$

where: $\sigma_y^2(t)$ is the variance of the displacement of particles, $R_I(\tau)$ is the autocorrelation function of the wind velocity, T is the integral time scale. Since $R_I(\tau)$ is generally assumed to have the exponential form, namely

$$R_I(\tau) = e^{-\frac{\tau}{T}} \qquad (6)$$

eq. (3) becomes:

$$\sigma_y^2(t) = 2\overline{v'^2} \left[tT - T^2 \left(1 - e^{-\frac{t}{T}} \right) \right] \qquad (7).$$

However, eq. (6) does not reproduce EAFs in LWS conditions since EAFs show an oscillating behaviour with the presence of significant negative lobes. Consequently, we considered the following two $R(\tau)$ forms:

$$R_2(\tau) = e^{-\frac{\tau}{T_m}} \left(\cos\frac{\tau}{T_n} - \frac{T_n}{T_m} \sin\frac{\tau}{T_n} \right) \qquad (8)$$

proposed by Csanady (1973), and

$$R_3(\tau) = e^{-\frac{\tau}{(m^2+1)T_3}} \cos\frac{m\tau}{(m^2+1)T_3} \qquad (9)$$

suggested by Frenkiel (1953). Eqs. (8) and (9) contain two parameters (T_m and T_n, T_3 and m, respectively)., the first one can be associated to the classical integral time scale and the second one to the meandering characteristic time.

Csanady's form yields:

$$\int_0^\infty R_2(\tau)d\tau = 0 \qquad (10)$$

and $\qquad \sigma_y^2(t) = 2\sigma_v^2 \frac{T_m^2 T_n^2}{T_m^2 + T_n^2}\left[1 - e^{-\frac{t}{T_m}}\left(\cos\frac{t}{T_n} + \frac{T_n}{T_m}\sin\frac{t}{T_n}\right)\right] \qquad (11).$

Thus, the integral time scale is zero and $\sigma_y^2(t)$ reaches an asymptotic constant value for time approaching infinite (Csanady, 1973).

Frenkiel's forms gives:

$$\int_0^\infty R_3(\tau)d\tau = T_3 \qquad (12)$$

and

$$\sigma_y^2(t) = 2\sigma_v^2 T_3\left\{t + (m^2-1)T_3 - T_3 e^{-\frac{t}{(m^2-1)T_3}}\left[(m^2-1)\cos\left(\frac{mt}{(m^2+1)T_3}\right) + 2m\sin\left(\frac{mt}{(m^2+1)T_3}\right)\right]\right\} \qquad (13)$$

We point out that only in the case of Frenkiel's forms the statistics (autocorrelation function, time scale and $\sigma_y^2(t)$) computed for the LWS case reduces to the classical statistics (setting $m=0$ and $T_3 = T$ into eq. 9, 12 and 13).

Figures (4 – 6) show the typical results of the comparison among the EAF's computed on the Tisby data set (full line) and the best fits estimated from Csanady's form (eq. 8 – full line and crosses) and Frenkielis form (eq. 9 – full line and open circles); in all graphs the exponential EAF (eq. 5 - full line and pluses), is also shown. The first two plots refer to nocturnal conditions whereas the third one to daytime conditions.

From the subjective analysis of all graphs (u and v wind components relative to the whole Tisby datat set) emerges that eqs. (8) and (9) give a good or reasonable good fit in about 3/4 of cases. In about 10% of cases both equations give a bad or null agreement and this happens when the experimental EAFs do not exhibit a negative lobe (no meandering). The best fit of the time series having $\bar{u} > 2$ m/s whith eqs. (8) and (9) came out completely wrong, thus suggesting that LWS conditions are completely different from the more general case of moderate and strong wind and need a peculiar treatment.

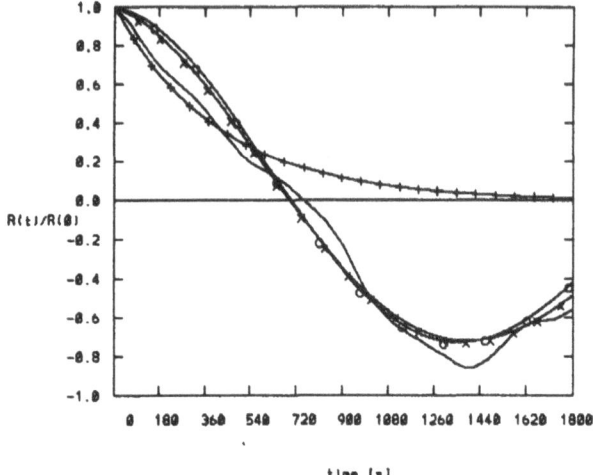

Figure 4 -- Comparison among EAF's; April 25, 02.00 L.T., $5\bar{u}$ = 0.3 ms^{-1}. Full line refers to the Tisby experimental EAF (u component), + indicate exponential EAF (eq. 6), x and o correspond to Csanady (eq. 8) and Frenkiel (eq. 9) EAF's, respectively.

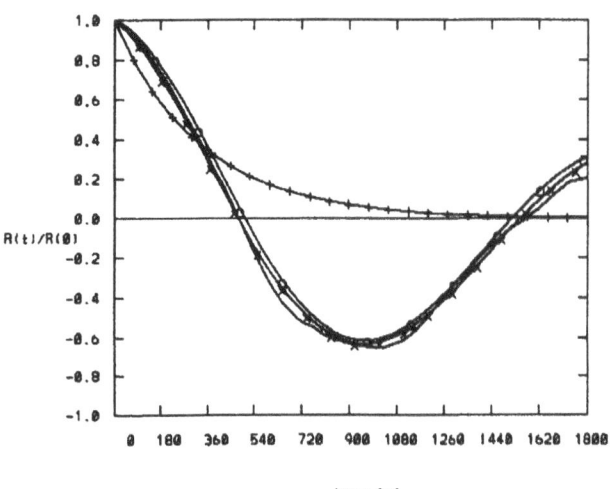

Figure 5 - As in Fig. 4 except for April 27, 01.00 L.T., $\bar{u}$ = 0.7 ms^{-1} (v component).

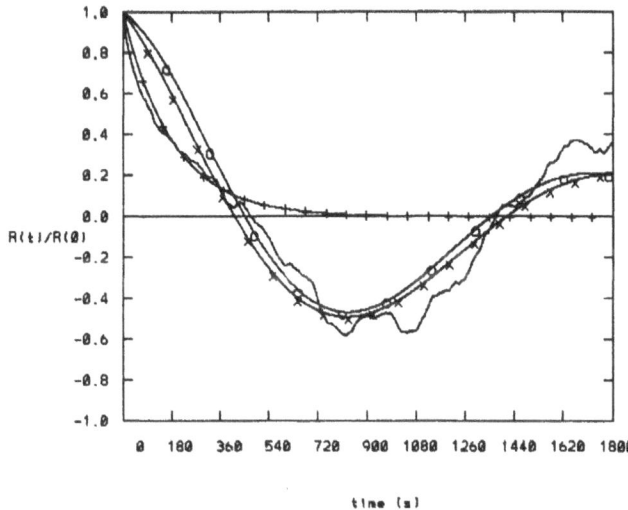

Figure 6 - As in Fig. 4 except for April 29, 11.00 L.T., $\bar{u}$ = 1.4 ms^{-1} (v component).

Meander is a general phenomenon (in the sense that it is found in any part of the earth in low wind conditions). As a consequence, it cannot be simply due to local effects, such as the presence of orography and/or terrain inhomogeneities, even if these last certainly may play a role in the process. In particular, no relationship between wind provenance and wind speed was found for the Tisby datat set, thus confirming that no preferential directions associated to LWS were present.

4. CONCLUSIONS

In this paper, we considered LWS conditions. We found that observed σ_w depend on u_*, but almost no change in σ_w with varying stability $z\,L^{-1}$ is observed. This suggests that u_* is the main parameter determining σ_w. No satisfying regression curve was found for σ_v and σ_u, but a correlation exists between u_* and the square root of the vector sum of σ_v and σ_u that is independent on stability. Autocorrelation functions $R\left(\tau\right)$ show an oscillating behaviour with the presence of significant negative lobes. Two relationships, due to Csanady (1973) and Frenkiel (1953), were fitted to the data. They contain two parameters, the associated, rrespectively, to the classical integral time scale and to the meandering characteristic time. It was found that both relationships fit the experimental $R\left(\tau\right)$ better than the classical exponential form (as expected) and that the Frenkiel's equation is to be preferred because it has the property of giving a non zero integral time scale and of collapsing of the statistics on the classical ones when the meandering effect is eliminated (i.e. by setting $m = 0$).

337

5. REFERENCES

Brusasca G., G. Tinarelli and D. Anfossi (1989). Comparison between the results of a Monte Carlo atmospheric diffusion model and tracer experiments. *Atmospheric Environment*, **23**, 1263 – 1280

Csanady G.T.: 1973, *'Turbulent diffusion in the environment'*, Geophysics and Astrophysics Monograps, Reidel, Boston 248 pp

Frenkiel F.N.: 1953. 'Turbulent diffusion: mean concentration distribution in a flow field of homogeneous turbulence'. *Adv. Appl. Mech,*. **3**, 61-107

Hanna, S. R.: 1983. 'Lateral turbulence intensity and plume meandering during stable conditions'. *J. of Appl. Meteor.*, **22**, 1424-1430

Oettl, D., R. A. Almbauer, and P. J. Sturm, (2001). A new method to estimate diffusion in stable, low wind conditions. *J. of Appl. Meteor.*, 40, 259-268

Panofsky, H. A., and J. A. Dutton, 1984: *Atmospheric Turbulence*. Wiley, 397 pp.

Stull, R. B., 1989: An Introduction to Boundary Layer Meteorology. *Kluwer Academic Publishers*, 666 pp.

COUPLING AN AEROSOL MODULE TO A DETERMINISTIC LONG TERM OZONE EXPOSURE MODEL

A. Arvanitis, N. Moussiopoulos[*]

1. ABSTRACT

OFIS is a model for describing pollutant transport and photochemical transformation and for simulating ozone exceedances over longer time periods. Simulations with OFIS are based on large-scale meteorological data, long range transport information and emission data over a longer period, typically one year.

In this study, the incorporation of an aerosol dynamics and chemistry module to OFIS is described. This module calculates particle number, mass and composition in two log-normally distributed size modes of a population of primary and secondary particles. Elemental carbon and dust are treated as primary particulate matter while organic carbon, sulphate, nitrate, ammonium and water formation from gases and vapours is also considered. The inorganic and organic particle mass is assumed to be in thermodynamic equilibrium with the surrounding gases. Coagulation of fine particles is also taken into account. Advection, vertical mixing and integration of all processes are treated accordingly by the host model.

Preliminary simulations were performed and the results are presented and discussed, in comparison to available observed data. Overall, the model appears to be capable of addressing the issues raised by the EU Air Quality Framework Directive.

2. INTRODUCTION

The EU Air Quality Framework Directive (96/62/EC) defines the legislative basis for assessment and management of air quality in European Union Member States. The

---

[*] Laboratory of Heat Transfer and Environmental Engineering, Aristotle University Thessaloniki, Box 483, GR-54124, Thessaloniki, Greece. E-mail thanos@aix.meng.auth.gr

directive stresses out the need of model application as a supplementary assessment method to reporting of monitoring data. Both this directive and the following Daughter Directives adopted rise a twofold challenge for the modelling research community; (i) estimating spatial distributions of pollutant concentrations and (ii) doing so for at least one year.

A territory covering information is essential as air quality management, in particular the intensity of any assessment requirements, is focused at zones of the territory featuring different air quality conditions. At the same time, the use of annually averaged concentrations and the annual frequency of exceedance of a limit value as an air quality index reflects the observed association of health effects to long term exposure to air pollutants.

In principle, territory covering concentration distributions could be calculated for longer time periods with state-of-the-art Eulerian Chemical Transport Models (Moussiopoulos, 2003). These models are, however, computationally expensive and applying those to selected zones of a territory for a whole year, although feasible, is rather impractical considering CPU-time and disk-space requirements.

In order to circumvent this problem, the OFIS (Ozone Fine Structure) model was developed (Sahm and Moussiopoulos, 1999; Moussiopoulos and Sahm, 2001). Originating from the photochemical dispersion model MARS (Moussiopoulos, 1995), OFIS simulates concentration changes due to chemical reactions and advection of species in each cell of the computational domain. The latter was limited to what appears to suffice for investigating the effect of an urban agglomeration on air quality in its surroundings: a gridded strip with a length of the order of 100 km and the width of the city, with the city in the centre, oriented to the prevailing wind direction for each day of the period considered. The height of each cell coincides with the mixing height thus varying with time. The restriction to this computational domain results in a high computational speed and a low output file size.

For covering the needs of the first Daughter Directive (1999/30/EC) we developed further and extended the OFIS model to simulate pollution levels of primary and secondary particles. This was achieved by incorporating an aerosol module comprising of an inorganic equilibrium solver for sulphate, nitrate, ammonium and water, a simple parameterisation of organics gas-particle exchange and an algorithm estimating the change in the shape of the particles size distribution over time.

The present paper describes how the implementation of the aerosol module to the host model was realised and presents the model results for a case study.

3. THE MODELS

The OFIS Model

OFIS is a robust and efficient model for simulating pollutant transformation in an urban plume. The model was developed to serve a twofold aim; (i) allowing authorities to assess urban air quality by means of a fast, simple and still reliable model and (ii) refining a regional model simulation by estimating the urban subgrid effect on pollution levels. The model uses an 1D Eulerian framework calculating numerically pollutant concentrations advected past a city assuming constant wind features and varying mixing height for each day of simulation. The computational domain of the model for a specific day with a prevailing wind from NE is illustrated in Figure 1. Depending on the spatial information of emissions available for the area surrounding the city, either geographically distributed gridded emission data or emission data integrated to urban, suburban and rural totals can be used. Model calculations are performed using a lumped atmospheric chemistry mechanism such as EMEP MSC-W or CBM-IV.

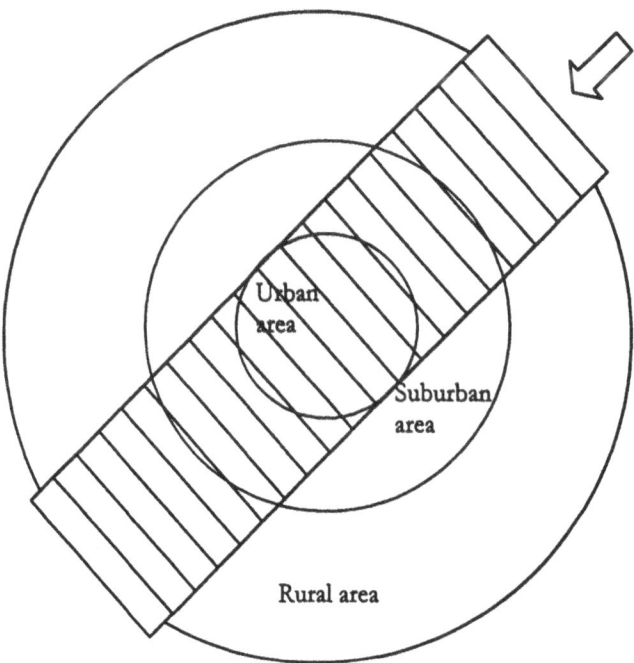

Figure 1. OFIS computational domain (gridded strip) and geographic distribution of emissions to three areas of different emission density: urban, suburban and rural area. The arrow indicates the prevailing wind direction.

For each species of the chemical mechanism in use the following equation is solved simultaneously in each grid cell of the computational domain:

$$\Delta c_i / \Delta t =$$
$$q_i / H_t + R_i(c_1, ..., c_n) + u(c_i^u - c_i) / \Delta x +$$
$$(c_i^{bc} - c_i) \max[0, \Delta H_t / (\Delta t \cdot H_t)]$$
(1)

where

c_i is the concentration of species i of the n species of the mechanism,

q_i is the emission rate of i,

H_t is the mixing height in the current time step,

R_i is the chemical production or destruction of i,

u is the daily average wind speed,

Δx is the length of a grid cell in the wind direction,

c_i^u is the upwind cell concentration of i and

c_i^{bc} is the concentration above the mixing height.

R_i in Eq. (1) is defined by the chemical mechanism in use. The third term on the right part of the equation refers to the advection discretised using a simple upwind scheme. c_i^{bc} and inflow boundary conditions can be derived from available monitoring data or, preferably, taken from results of a regional scale model. Meteorological input may be derived from either available measurements or the output of a mesoscale model. The mixing height can alternatively be calculated within OFIS: it can be estimated with an embedded subroutine as a function of friction velocity, Monin-Obukhov length and the Coriolis parameter utilising a vertical column atmosphere/soil radiation budget model. The numerical solution of equation system (1) is based on a variable step, second order BDF formula and a Gauss-Seidel iterative technique (Verwer 1994).

The model simulates separately each day of, typically, one year. Model results correspond to the diurnal variation of air pollutant concentrations on any specific day in the gridded strip illustrated in Figure 1, whereas for the part of the considered domain outside this strip pollution levels are assumed to coincide with the regional background. From the overall concept of OFIS it is apparent that the model aim is not to accurately reproduce concentration patterns in any location over the whole year, but rather to correctly report when and to what extent high concentration levels or a regulated threshold exceedance occur at an urban location or a spot downwind the city.

The Aerosol Module

The aerosol module used is a simplified version of a generic module described in Arvanitis *et al.* (2001). The module distinguishes two size modes, a fine and a coarse one, described by two log-normal distributions. Each distribution is defined by calculating total number of particles, geometric mean diameter and standard deviation of the distribution from the sixth distribution moment and total particle mass (Whitby, 1990). Total particle mass of the fine mode comprises of particle phase sulphate, nitrate, ammonium, organic carbon, elemental carbon and other chemically undefined primary particles, while total particle mass of the coarse mode comprises of sea-salt, dust and other primary particles.

So as to estimate what part of the inorganic species lies in the particle phase, an inorganic equilibrium model was used. Although the model takes into account species of crustal origin (Ca, K, Mg) and sea-salt (Na, Cl), limiting chemically reactive species to fine particles –for simplicity and low computational cost– excludes such species from this simplified model application. The error introduced from this omission is expected to be small when studying airsheds away from a coast and at not very dry conditions. For other cases, this assumption has to be evaluated.

For the SO_4-NH_4-NO_3 system the model takes into account the gas, solid and aqueous species of Table 1.

Table 1. Species considered in the inorganic equilibrium module.

Gas phase	Aqueous phase	Solid phase
HNO_3, NH_3, H_2O	NH_3, HSO_4^-, SO_4^{-2}, NO_3^-, NH_4^+, H_2O, H^+, OH^-	$(NH_4)_2SO_4$, NH_4HSO_4, $(NH_4)_3H(SO_4)_2$, NH_4NO_3

The equilibrium chemical reactions modelled and the equilibrium constants are presented in Table 2 (Kim *et al.* 1993). The reactions in the parentheses are additional, redundant reactions which, based on the suggestion of Villars (1959) for the numerical method used, speed up the solution time considerably. They result as the linear combination of the other reactions and their equilibrium constants are calculated the same way.

The equilibrium constant of each reaction is defined on the basis of species activities, which, for ionic species in the aqueous phase, are defined as the product of the activity coefficient and the molal concentration of the species (Seinfeld and Pandis 1998). The activity coefficient of a multi-component solution is a function mainly of the concentrations of the ions in the solution, and the binary activity coefficient of each ion pair. The latter is a function of the ionic strength of the solution and is estimated in this module with the Kussik and Meissner method (Kim *et al.* 1993; Kim and Seinfeld 1995). The multi-component activity coefficients are calculated following the Bromley method (Jacobson 1999a). The activities of solids equal unity.

Table 2. Reactions considered in the inorganic equilibrium module.

$HSO_4^- \leftrightarrow H^+ + SO_4^{-2}$
$NH_{3g} \leftrightarrow NH_{3aq}$
$NH_{3aq} + H_2O \leftrightarrow NH_4^+ + OH^-$
$HNO_3 \leftrightarrow H^+ + NO_3^-$
$H_2O \leftrightarrow H^+ + OH^-$
$(NH_4)_2SO_4 \leftrightarrow 2NH_4^+ + SO_4^{-2}$
$NH_4NO_3 \leftrightarrow NH_{3g} + HNO_3$
$NH_4HSO_4 \leftrightarrow NH_4^+ + HSO_4^-$
$(NH_4)_3H(SO_4)_2 \leftrightarrow 3NH_4^+ + HSO_4^- + SO_4^{-2}$
$(HSO_4^- + NH_{3g} \leftrightarrow NH_4^+ + SO_4^{-2})$
$(HNO_3 + NH_{3g} \leftrightarrow NH_4^+ + NO_3^-)$
$(HCl + NH_{3g} \leftrightarrow NH_4^+ + Cl^-)$

The system of equations of molalities of ions, partial pressures of gases, and concentrations of solids is rewritten following Jacobson (1999b) so that the same units, $\mu mol/m^3_{air}$, can be used for all species in any physical state. For any given initial concentrations of the species of Table 1, their equilibrium concentrations can be calculated with the mass flux iteration (MFI) method (Jacobson *et al.* 1996; Villars 1959). The MFI method requires the solution of one equilibrium equation at a time. After this is achieved – iteratively following Jacobson (1999a) – the updated concentration values are used to calculate the error in the other equilibrium equations. The equation of the maximum error is, then, solved (Villars 1959) and the procedure is repeated until the maximum error is reduced below a predefined threshold. Finally, the activity coefficients – held constant during the iterative procedure – are recalculated, and the water content of the particle is estimated with the ZSR method (Stokes and Robinson 1966; Jacobson 1999a). A few iterations of the final step are enough for the system to reach convergence.

When considering particles >1μm, inorganic species in different physical states do not always equilibrate chemically under common atmospheric conditions (Wexler and Seinfeld 1990). Since coarse mode particles are not treated as chemically reactive in the model, the equilibrium assumption introduces acceptable error and attractively low computational effort.

Organic partitioning between gas and aerosol phase is calculated according to the mass-based stoichiometric coefficients and absorption equilibrium partitioning coefficients of the products of xylene oxidation as calculated by Odum et al. (1997). A more comprehensive treatment of secondary organic aerosols requires modifications in the organic part of the gas-phase chemical mechanism which are beyond the scope of this first modelling approach.

In each of the aforementioned modes, the size distribution is a function of three parameters; the number concentration, the geometric mean diameter and the standard deviation:

$$n_i(\ln d_p) = N_i / (\sqrt{2\pi} \ln \sigma_{pgi}) \exp\left[-(\ln d_p - \ln d_{pgi})^2 / (2 \ln^2 \sigma_{pgi})\right] \qquad (2)$$

where N_i is the number concentration of the ith mode,

d_p is the particle diameter,

d_{pgi} is the geometric mean diameter of particle population of the ith mode and

σ_{pgi} is the standard deviation of the ith mode distribution.

So as to describe the changes of the shape of the distribution due to physical or chemical processes without constructing differential equations for the variation of geometric mean particle diameter or standard deviation (which are not physical but statistical quantities), the moments method is followed (Whitby and McMurry 1997; Williams and Loyalka 1991); For a lognormal distribution the kth integral moment, M_k, is a function of the 0th moment which equals the total number of particles, N, the geometric mean diameter and the standard deviation of the distribution (Whitby and McMurry 1997):

$$M_k = Nd_{pg}^k \exp(\frac{k^2}{2} \ln^2 \sigma_g) \qquad (3)$$

Following Whitby (1990), in addition to the 0th moment, N, which equals the total number of particles, we chose to model the third moment, M_3, and the sixth moment, M_6, of the distribution. The former was chosen because it is proportional to the total volume of the particles (Friedlander 1977) and the latter for simplifying the coagulation calculations (Whitby and McMurry 1997). This approach was also used by Binkowski and Shankar (1995) in the RPM model and by Ackermann *et al.* (1998) in MADE.

Coupling the Processes and Implementing to the Host Model

Concentrations of particle sulphate, nitrate, ammonium, other fine primary PM, organic carbon and elemental carbon, other coarse primary PM, dust and chemically inactive sea-salt are calculated as follows:

- Emissions of each species are interpolated for the current time step. Third moment emissions are calculated by converting mass emissions to total volume emissions for each mode.
- Sixth moment and number emissions are calculated according to the third moment and fixed particle diameter and standard deviation of the distributions.
- Advection is calculated for particle number, sixth moment and all species; equilibrium species (sulphate, ammonium, nitrate) are not yet phase-distinguished but are treated as total concentrations of all phases.
- Change in concentrations due to mixing height variation is taken into account.
- Equilibrium is calculated for the inorganics.
- Organic particulate matter is calculated from the mass of xylene oxidised during the time step.
- Total particle mass concentration is calculated by summing up the concentrations of individual species of the last time step and using the resulted values of the equilibrium model and the organic parameterisation.
- Total particle volume is calculated from total mass
- Intra-modal coagulation of the fine mode is calculated and the new geometric mean diameter and standard deviation are calculated for the new total volume of each mode.
- Production and loss terms of ammonia, sulphuric acid and nitric acid are included in the derivatives referring to overall ammonium, sulphate and nitrate concentrations.
- All derivatives are summed up and introduced in the second order BDF formula of the host model; the system of equations is solved subsequently to the gas phase system by the same Gauss-Seidel iterative technique (Verwer 1994).
- The error of the solution is checked and the whole procedure is repeated, if needed, with a reduced time step.

4. MODEL APPLICATION

The new version of OFIS, incorporating the above presented aerosol module, was applied and validated in the framework of CITY-DELTA. Main aim of this model intercomparison exercise is to explore the changes in urban air quality predicted by different dispersion models in response to changes in urban emissions (CITY-DELTA, 2001). More specifically, OFIS was applied to study PM pollution levels for the city of Milan.

The emission inventory used was compiled by Regione Lombardia (CITY-DELTA, 2001) and comprises CH_4, CO, NH_3, Total VOC (split into more than 400 individual species), NO_x, SO_2, PM_{10}, $PM_{2.5}$ and TSP (split into more than 35 individual species) emission rates for each hour of a whole year in a circa 300×300 km^2 grid of 5×5 km^2

resolution surrounding Milan. This data was classified as urban, suburban and rural emissions and was allocated to the three areas shown in Figure 1. The emission inventory species were assigned to the corresponding species of the EMEP MSC-W chemical mechanism and those considered in the aerosol module.

Boundary conditions for a number of species, including PM_{10}, $PM_{2.5}$ and inorganic particulate matter, originated from calculations of the Eulerian EMEP acid deposition model (Bartnicki et al., 1998) were used as upwind concentrations or mass entrained during the mixing height deepening.

Both boundary condition and emission data were delivered to us in specified formats by the CITY-DELTA coordinators (Thunis 2003).

Meteorological data needed by the model, such as prevailing wind features and temperature for each day, were calculated with the mesoscale model MEMO (Kunz and Moussiopoulos 1995).

All data refer to year 1999.

5. RESULTS AND DISCUSSION

In the following model simulation results are compared with PM_{10} measurements taken at the four sites shown in Figure 2. All values used for the charts shown in Figures 3, 4, 5 and 6 are 24-h averages. The model quite accurately reproduces the observed data with the exception of the site in Vimercate, where the model systematically underestimates PM_{10} concentrations. For the other three sites, the model follows observed tendencies, peaks and lows, resulting in well balanced scatter plots.

Although all sites are located in the urban area, they exhibit varying levels of PM pollution as it is evident from the exceedances charts. The OFIS model seems to be able to capture these differences; even if in Magenta VF OFIS overestimates the number of days of average values exceeding the limit value of 50 $\mu g/m^3$ of the first Daughter Directive (1999/30/EC), it still reflects the relative geographical distribution of exposure.

In conclusion, the OFIS model approach in assessing long term pollutants exposure in urban areas gives encouragingly accurate estimates even for the complicated system of primary and secondary particles. However, a full validation of the aerosol module would require more observed data and application of the model to urban airsheds of varying characteristics. A sensitivity analysis of the input parameters is in the near future planning of the authors in parallel to a comparison with a more complex aerosol module incorporated in the 3D Chemical Transport Model MARS (Moussiopoulos, 1995).

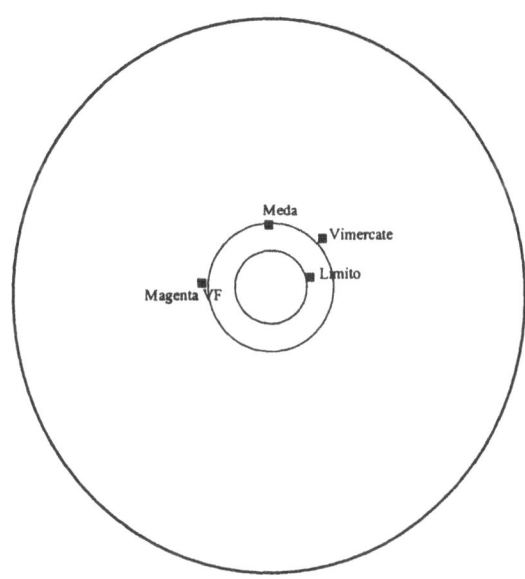

Figure 2. Location of stations where PM$_{10}$ measurements are taken in the Milan metropolitan area.

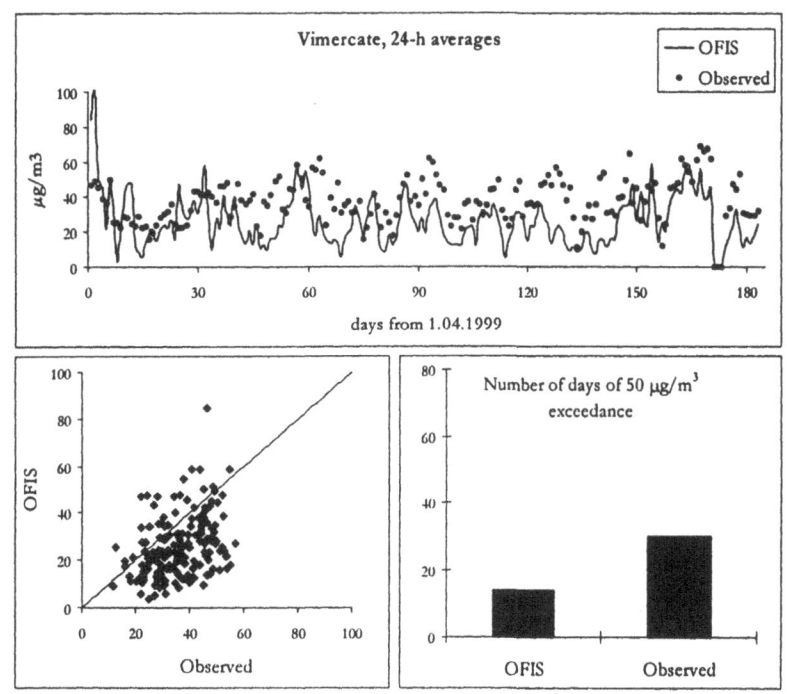

Figure 3. OFIS results vs observed data as 24-h averages for 1999 summer semester at Vimercate, Milan.

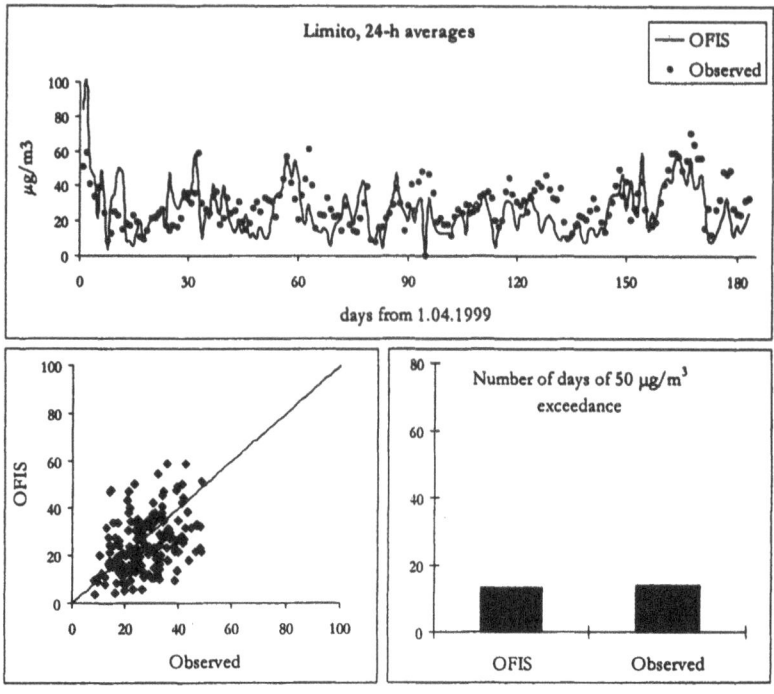

Figure 4. OFIS results vs observed data as 24-h averages for 1999 summer semester at Limito, Milan.

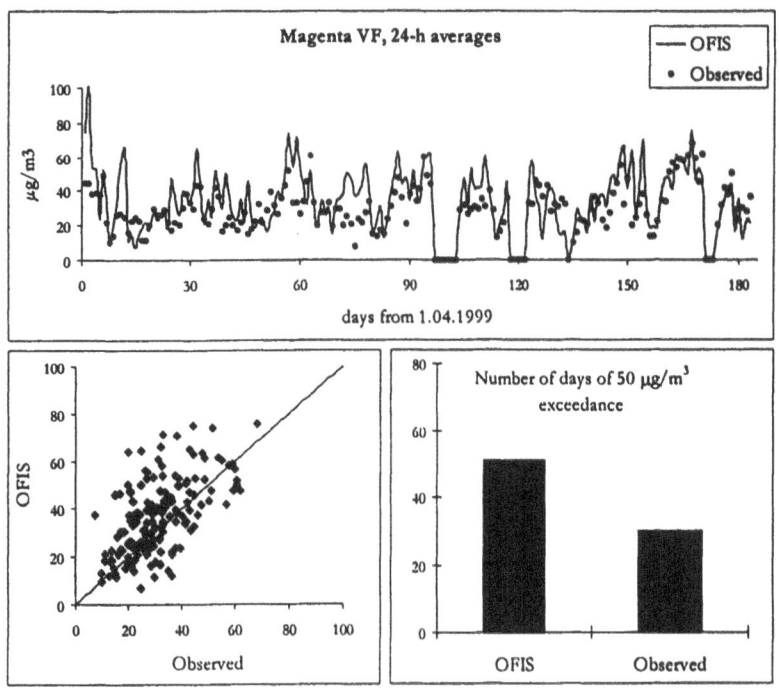

Figure 5. OFIS results vs observed data as 24-h averages for 1999 summer semester at Magenta VF, Milan.

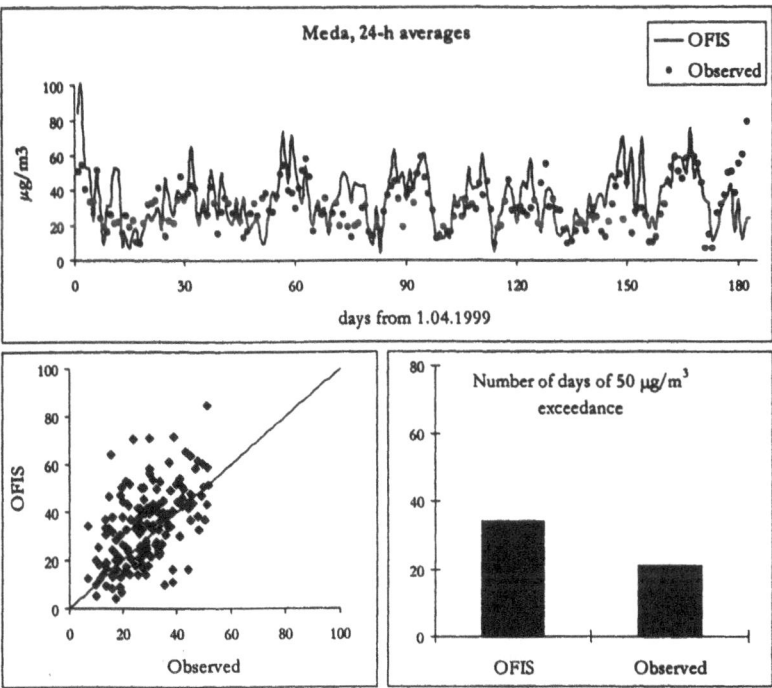

Figure 6. OFIS results vs observed data as 24-h averages for 1999 summer semester at Meda, Milan.

6. ACKNOWLEDGEMENTS

The authors are grateful to CITY-DELTA partners for kindly providing the data for the model application. The assistance of Chris Karavakas in preparing the model runs and the contribution of Apostolis Papathanasiou in finalising mixing height estimations of the model are also appreciated.

7. REFERENCES

Ackermann IJ, Hass H, Memmesheimer M, Ebel A, Binkowski FS, Shankar U (1998) Modal aerosol dynamics model for Europe: Development and first applications. *Atmos. Environ.* 32:2981-2999.

Arvanitis A, Moussiopoulos N and Kephalopoulos S (2001) Development and testing of an aerosol module for regional/urban scales, *Proceedings of the 2nd Conference on Air Pollution Modelling and Simulation, APMS' 01*, Champs-sur-Marne, 9-12 April, 277-288.

Bartnicki J, Olendrzynski K, Jonson J E, and Unger S (1998) Description of the Eulerian acid deposition model. In: *Transboundary acidifying air pollution in Europe. MSC-W Status Report 1998.* Part 2. Oslo (EMEP/MSC-W Report 1/98).

Binkowski FS, Shankar U (1995) The regional particulate matter model 1. Model description and preliminary results. *J. Geophys. Res.* 100:26191-26209.

CITY-DELTA (2001) http://rea.ei.jrc.it/netshare/thunis/citydelta/main.htm.

Frielander SK (1977) Smoke, dust and haze. John Wiley and Sons, New York.

Jacobson MZ (1999a) Fundamentals of atmospheric modeling. Cambridge University Press.

Jacobson MZ (1999b) Studying the effects of calcium and magnesium on size-distributed nitrate and ammonium with EQUISOLV II. *Atmos. Environ.* 33:3635-3649.

Jacobson MZ, Tabazadeh A, Turco RP (1996) Simulating equilibrium within aerosols and nonequilibrium between gases and aerosols. *J. Geophys. Res.* **101**:9079-9091.

Kim YP, Seinfeld JH (1995) Atmospheric gas-aerosol equilibrium: III. Thermodynamics of crustal elements Ca^{2+}, K^+, and Mg^{2+}. *Aerosol Sci. Technol.* **22**:93-110.

Kim YP, Seinfeld JH, Saxena P (1993) Atmospheric gas-aerosol equilibrium: I. Thermodynamic model. Aerosol Sci. Technol. **19**:157-181.

Kunz R, Moussiopoulos N (1995) Simulation of the wind field in Athens using refined boundary conditions. *Atmos. Environ.* **29**: 3575-3591.

Moussiopoulos N. (1995) The EUMAC Zooming model, a tool for local-to-regional air quality studies. *Meteorol. Atmos.Phys.* **57**: 115-133.

Moussiopoulos N, ed. (2003), Air Quality in Cities (SATURN Final Report), Springer, Heidelberg, 308 pp., in press.

Moussiopoulos N and Sahm P (2001), The OFIS model: an efficient tool for assessing ozone exposure and evaluating air pollution abatement strategies, *International Journal of Environment and Pollution* **14**: 597-606.

Odum JR, Hoffmann T, Bowman F, Collins T, Flagan RC and Seinfeld JH (1996) Gas-particle partitioning and secondary organic aerosol yields, *Environ. Sci. Technol.*, **30**: 2580-2585.

Sahm P and Moussiopoulos N (1999) Urban ozone levels in Europe: Simulations with the OFIS model, *Proceedings of the 6th Conference on Harmonisation within Atmospheric Dispersion Modelling for Regulatory Purposes*, Rouen, France, 11-14 October, 111-112.

Seinfeld JH, Pandis SN (1998) Atmospheric chemistry and physics: From air pollution to climate change. John Wiley and Sons.

Stokes RH, Robinson RA (1966) Interactions in aqueous nonelectrolyte solutions I. Solute-solvent equilibria. *J. Phys. Chem.* **70**: 2126-2130.

Thunis Ph (2003), JRC/IES, European Commission (private communication).

Verwer JG (1994) Gauss-Seidel iteration for stiff ODEs from chemical kinetics. *SIAM J. Sci. Comput.* **15**:1243-1250

Villars DS (1959) A method of successive approximations for computing combustion equilibria on a high speed digital computer. *J. Phys. Chem.* **63**:521-525.

Wexler AS, Seinfeld JH (1990) The distribution of ammonium salts among a size and composition dispersed aerosol. *Atmos. Environ.* **24A**:1231-1246.

Whitby ER (1990) Modal aerosol dynamics modeling. Ph.D. thesis, University of Minnesota.

Whitby ER, McMurry PH (1997) Modal aerosol dynamics modeling. *Aerosol Sci. Technol.* **27**: 673-688.

350

DISCUSSION

J. SLOAN   I assume your model uses equilibrium between gas phase and dissolved concentrations for soluble species (answer-yes). Since the time frame for transit of the airmass over a city centre (~5km) is relatively short, is this a valid assumption?

N. MOUSSIOPOULOS   For a typical wind speed over a city of 5m/s the airmass stays over the city for ~17 min. Submicron particles equilibrate in a few minutes; coarser particles equilibrate within hours. Obviously the equilibrium assumption holds for fine particels and not for coarse, thus in OFIS only fine mode particles are considered to be chemically interactive and in equilibrium while coarse are affected only by emission flux and advective and turbulent transport.

D. SYRAKOV   Can be the OFIS-model used for rough (without any space details) short-time forecast?

N. MOUSSIOPOULOS   There are two reasons why this is not possible: First of all, a forecast for the regional background situation would be needed, and to my knowledge such information is currently not available. Besides, the OFIS model concept does not allow describing accurately the pollution situation for specific receptors at individual times: The model was devised to yield long-term exceedance statistics rather than air pollutant concentrations during episodes.

GLOREAM-GLOBAL AND REGIONAL ATMOSPHERIC MODELLING

Peter Builtjes, Adolf Ebel, Hans Feichter and Annette Muenzenberg[*]

1. INTRODUCTION

The central purpose and scientific focus of GLOREAM, which stands for GLObal and REgional Atmospheric Modelling, was the investigation - by means of advanced and integrated modelling - of the processes and phenomena which determine the chemical composition of the troposphere over Europe and on a global scale. An essential part of GLOREAM has been the development of state-of-the-art models and the determination of the model capabilities and model performance. The interaction between the different modelling groups and modellers in Europe to discuss problems, to exchange information and to share knowledge formed the heart of GLOREAM.

GLOREAM was part of EUROTRAC-2, and was carried out over a period of 6 years, from 1997-2002.

The modelling subproject GLOREAM was the successor of the EUROTRAC, first phase, subprojects EUMAC - European Modelling of Atmospheric Constituents - and GLOMAC - GLObal Modelling of Atmospheric Chemistry. In EUMAC the focus had been on the development of an advanced model hierarchy for air pollution dispersion simulations. The central model was the three dimensional (3-D) long-range transport model EURAD covering Europe, which focused on episodic simulations. The aim of GLOMAC was the development and application of 3-D models of the global troposphere and the lower stratosphere focusing on ozone and acidification. The central models were MOGUNTIA, a relatively simple global model, and the on-line model ECHAM and the off-line model TM (Ebel *et al.*, 1997). GLOREAM was built on the experience gained in both projects and continued working in further model development and application (Builtjes *et al.*, 1999).

[*] Peter Builtjes, TNO-Apeldoorn, The Netherlands and Free Univ. Berlin, Germany. Adolf Ebel, Univ. of Cologne-EURAD, Germany. Hans Feichter, Max Planck Inst. for Meteorology, Hamburg, Germany. Annette Muenzenberg, DLR-BMBF, Bonn, Germany.

Air Pollution Modeling and Its Application XVI, Edited by
Borrego and Incecik, Kluwer Academic/Plenum Publishers, New York, 2004

353

GLOREAM was structured according to the following five, closely related working groups:

- model investigation and improvement on a European scale,
- model investigation and improvement on a global scale,
- computational aspects,
- model evaluation and validation,
- model application and assessment.

In GLOREAM a workshop was held every year, so that including the kick-off meeting a total of seven workshops have taken place. These workshops were in general attended by around 40 GLOREAM principal investigators and several guests. The number of peer-reviewed papers which have been written in the framework of GLOREAM was about 160 and about 25 theses. GLOREAM has also worked on capacity building. Quite a number of young scientists gave their first talks at GLOREAM workshops and many fruitful co-operations between institutes across Europe have been established.

The general aims as formulated at the start of GLOREAM at the beginning of 1997 were:

- to develop and improve three-dimensional regional and global scale atmospheric transport chemistry models,
- to investigate, with the aid of models, processes and their interaction that control the chemical composition of the troposphere,
- to apply complex and simplified models for specific environmental policy issues, and to assist other EUROTRAC-2 subprojects.

Looking to all the results obtained in GLOREAM, the double refereed papers, the theses and the reports describing the more policy-oriented results, it can be concluded that GLOREAM has done what it promised to do.

Major achievements are the following:

- Continental/regional scale and global models have been improved considerable and have now reached a stage of maturity with respect to tropospheric ozone. Although progress has been made with respect to aerosols, many aspects still require further research.
- Long-term (over years and more) model calculations are nowadays possible for most models on an hour-by-hour basis. Also ozone forecasting has become possible. This has been made possible by the increase of computer power and the improvements in the efficiency of numerical methods.
- Models of different complexity, state-of-the-art and models of intermediate complexity are now available, enabling the calculations of many scenarios. Model inter-comparison studies have been carried out as well as model evaluation and validation. In this way, model errors and flaws could be detected, and a beginning has been made to formulate criteria which models have to pass before they can be used in further studies.
- New numerical methods for nesting/scale interaction and data assimilation have been developed which have greatly improved the capabilities of the models to address new science and policy issues.

In the following some highlights of Gloream are presented. More information can be found at www.dmu.dk/AtmosphericEnvironment/gloream/ and in Builtjes et al. (2003).

2. AEROSOL MODELLING AND MODEL INTERCOMPARISON STUDIES

GLOREAM models are able to do long-term calculations with an hourly output of atmospheric composition. Most of them include photochemistry and aerosols. A model intercomparison study aiming on the aerosol composition and including six 3D-CTMs of different complexity have been undertaken for the European scale for the growing season in the year 1995 (April – September). The results presented in Figure 1 show a considerable scatter in the modelled versus observed aerosol SO_4. The same holds for total NO_3, with a tendency to overprediction.

These results also indicate that although clear progress has been made in aerosol modelling, there is also much more research and development needed before these models can be used to analyse abatement strategies with the same reliability achieved for photochemical models. The further research in aerosol modelling should also include improvements in the emission data both for primary and secondary precursor emissions, and a critical review of the observations.

See for more information: Hass et al., 2003.

GLOREAM models have also been used in a model intercomparison study carried out within the TOR-2 sub-project for the year 1997 aiming on the analysis of measured ozone trends (Roemer *et al.*, 2003). The response of the different models to emission scenarios for 1997 and 1987 has been calculated.

Although all models show a similar behaviour, there is a difference in the size of the response due to the emission changes. It could be shown that the ozone maxima are reduced considerably due to emission reduction whereas the average ozone concentrations show a reduction in the Mediterranean region but not in Central and Western Europe.

The increase in computer power and speed has made it possible to perform air quality model calculations very fast. As a consequence, several groups are currently capable of operational air quality forecasts. At the moment six groups have forecast systems available on the Web, namely:

- *http://artico.lma.fi.upm.es/*, developed by the Environmental Software and Modelling Group, at the Technical University of Madrid, Spain,
- *http://www.eurad.uni-koeln.de*, developed and implemented by EURAD, at the University of Cologne, Germany,
- *http://www.dmi.dk/vejr/index.html*, developed and implemented by the Danish Meteorological Institute,
- *http://luft.dmu.dk*, developed and implemented by NERI, Denmark,
- *http://trumf.fu-berlin.de*, developed and implemented by the Free University of Berlin, Germany,
- *http://www.smhi.se/sgn0102/n0205/baltichome_real/luftkvalitet.htm*, developed and implemented by SMHI, Sweden.

GLOREAM AEROSOL INTERCOMPARISON STUDY

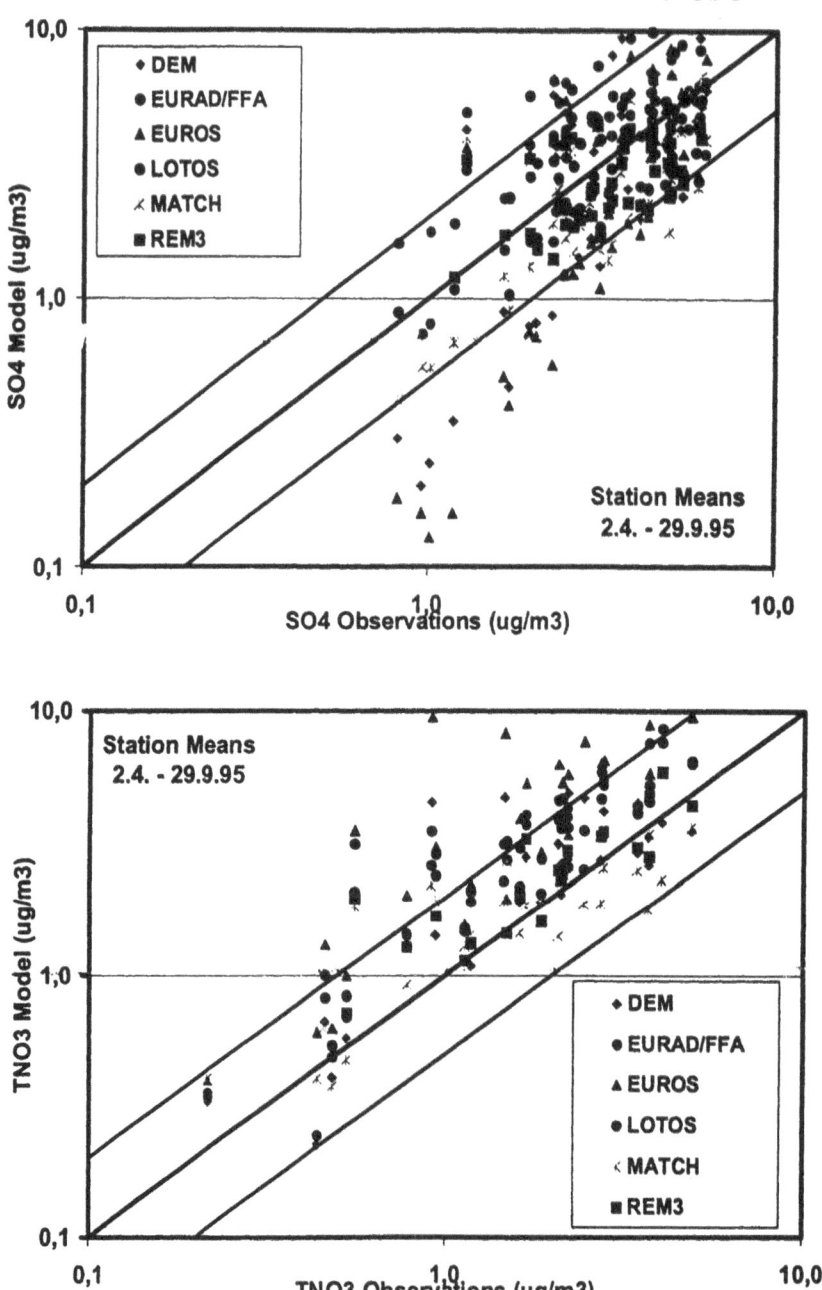

Figure 1. Comparison of observations from EMEP stations and model results from the aerosol model intercomparison study for the growing season in 1995. The numerical simulations include photochemistry and treatment of particulate matter in all of the participating 3D chemistry transport models.

The first system uses the MM5 (PennState/NCAR Mesoscale Model Version 5) and the CMAQ (Community Multiscale Air Quality Modelling System US EPA). The MM5 is driven by the initial conditions available on the NOAA web site and assures the CMAQ proper meteorological fields, which produces concentration fields of air pollutants over the Iberian Peninsula.

The EURAD forecast system consists of the mesoscale meteorological model MM5, the emission processor EEM (EURAD Emission Model) and the EURAD-CTM. The initial and boundary data for MM5 are obtained from the global AVN forecast (NCEP) at the start of the forecast cycle (00 UTC).

The Danish Meteorological Institute (DMI) makes a prognosis of the surface ozone concentrations produced by the system of models composed by the meteorological model HIRLAM (HIgh Resolution Limited Area Model), the numerical weather prediction model from the DMI and the DACFOS model (Danish Atmospheric Chemistry FOrecasting System). DACFOS consists of two components: a 3-D Lagrangian chemical-transport, receptor-point model (DACFOS_L) and a statistical after-treatment of the ozone forecasts from the chemical transport model (DACFOS_S), when real-time ozone measurements are available.

The model system at NERI, the so-called Thor-system, uses a nested model hierarchy from the hemispherical scale down to, in principle, the street-level. The meteorological driver used is the Eta-model.

At the Free University of Berlin the version of REM-3 of the year 2000 is used, in connection with real time weather forecast.

At SMHI in Sweden the model MATCH is used to perform real time ozone forecast.

Results of an intercomparison of ozone forecasting for the summer of 1999 can be found in Tilmes et al. (2002).

3. AIR QUALITY AND CLIMATE INTERACTIONS

The Danish Eulerian Model DEM has been used to calculate the impact of a climate change scenario for the last 30 years of the 21 th century on peak ozone levels. The results show an increase of up to 60 % in the number of days in which over Europe the 60 ppb level is exceeded and an increase of up to 15 % in the ozone daily maxima.

Similar studies have been performed by SMHI, Sweden. They used the meteorological data from dynamical downscaling experiments carried out at the Rossby Centre. The simulations indicate substantial impact of climate change on both deposition of sulfur and nitrogen and concentrations of surface ozone. The increase of ozone was simulated over large parts of central and northwestern Europe, in according with the Danish results.

The impact of climate change on the air quality over the Great Lisbon Area, and on the impact of winter weather patterns over Northern Portugal was studied by the Univ. of Aveiro. One of the results indicated less rain fall, and consequently a potential increase in forest fires.

These studies clearly show the needed to integrate climate change and air quality, both in research, modelling and policy.

4. GLOBAL MODELLING

Many attempts have been made in recent years to understand the physical, chemical and biological processes that constitute the earth-atmosphere system using global numerical models. Global climate change is the product of both, natural processes and a growing suite of relatively recent anthropogenic perturbations. Chemical processes in the troposphere and stratosphere are important to evaluate the extent of possible climate changes over the coming decades. The uptake and the release of various trace gases by the marine and the terrestrial biosphere determines to a large degree the chemical composition of the atmosphere. Trace gases such as CO_2, CH_4, N_2O and tropospheric O_3 (greenhouse gases 'GHG') trap the terrestrial infrared radiation and warm the troposphere and cool the stratosphere. Relatively small amounts of trace gases such as CO, NO_x and volatile organic compounds play a key role in atmospheric chemistry by affecting the tropospheric and stratospheric concentrations of ozone. It is now evident that agricultural and more recently industrial activities as well as land use practices have changed the atmospheric composition. As a consequence, the atmospheric volume mixing ratio of CO_2 has increased from 280 to 350 ppm, that of CH_4 from 0.7 to 1.7 ppm and that of N_2O from 0.28 to 0.31 between the years 1860 and 1990. Ozone, also a GHG, has nearly doubled in the troposphere whereas in the lower stratosphere it has decreased due to anthropogenic halocarbon emissions. Since ozone absorbs solar ultraviolet radiation, which heats the stratosphere, the stratospheric ozone depletion exerts a cooling in the stratosphere and the upper troposphere. Loadings of tropospheric aerosols (e.g. sulfate, nitrate and carbonaceous aerosols) have increased substantially over the past 150 years as a consequence of industrial activities. These aerosols enhance reflection of solar radiation both directly, by scattering light in clear air, and indirectly, by increasing the reflectivity and life-time of clouds. Aerosols act as cloud condensation nuclei and control the formation and the optical and physical properties of clouds.

In turn, atmospheric dynamics and cloud processes control the concentration and distribution of atmospheric constituents. Winds transport gaseous and particulate matter and loft dust and sea-salt aerosols into the atmosphere. The intensity of the solar radiation and the temperature determine the chemical reaction rates. Cloud droplets are chemical reactors and contribute to the formation of aerosol particles and the precipitation cleans the atmosphere from gases and particles. Vegetation and biogenic emissions are governed by meteorological parameters. In turn, vegetation plays a major role in regulating the hydrological cycle.

Global general circulation models (GCM) of the atmosphere used for climate studies have been recently explored to evaluate the atmospheric transport and the interactions of gaseous and particulate constituents.

MPI Hamburg (Bauer and Langmann, 2002) developed a model hierarchy in order to interprete results of a global model on regional scales (Figure 2). They nested in a global model a regional model (REMO) with a horizontal resolution of 50x50 km and 20x20 km. Further to analyse ozone measurements taken during the BERLIOZ measurement campaign, a non-hydrostatic model (GESIMA) with a resolution of 4x4 km was nested in the regional model.

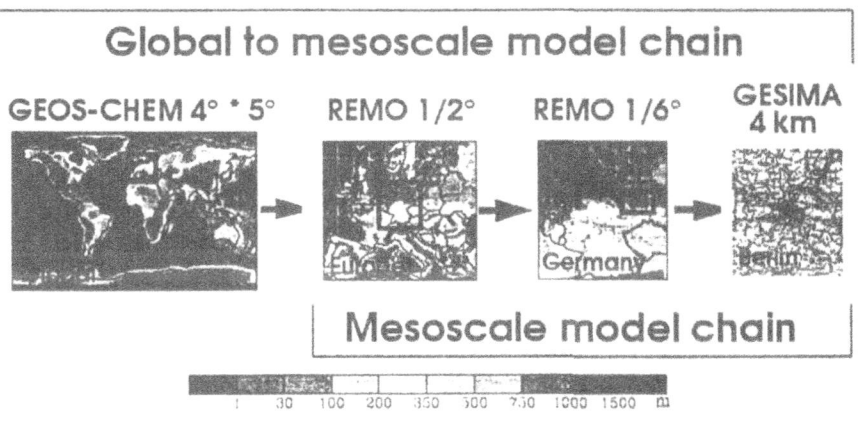

Figure 2. Model hierarchy from global to regional scales.

A new field for application of chemistry transport models is the chemical weather forecast. The goal is to provide the measurement community during campaigns with predicted distributions of some chemical key species to support the planning of flight tracks. The Max Planck Institute for Meteorology, Hamburg, developed a preliminary version of the new ECHAM5 general circulation model to provide forecasts of the chemical weather. This model version was equipped with several carbon monoxide (CO) tracer species using a technique called "tagging" to distinguish the influence of individual source regions and emission types on the simulated CO concentrations. The model was run in nudging mode which means relaxation of the model dynamics towards up-to-date forecast products from the European Centre for Medium Range Weather Forecast (ECMWF). The output is saved every two hours for up to five days in advance. The model version has been developed to support experimentalists in planning flight tracks. So far it has been applied in two field campaigns, GTE/TRACE-P and MINOS.

5. MODEL EVALUATION AND VALIDATION

Model evaluation includes three different elements. At first it is based on a strategy protocol. Then decision criteria for the success or failure of the model need to be defined prior to the simulations. Finally a number of model runs is carried out as the core activity.

In the evaluation strategy protocol the details of the performance tests are summarized including the time period and the model domain of the study. Agreement is needed in respect to the trace substance concentrations and/or the meteorological quantities that serve as target parameters. Furthermore, qualifying criteria for models like e.g. the existence of a documentation, may be introduced. It is essential that model quality objectives (MQOs) in respect to accuracy, precision, representativeness and completeness are defined and accepted by the modellers before they do their simulations. These MQOs depend strongly on the measurement techniques and the data sets which are available for comparison to the model results. They will be more stringent if quality controlled and assured data with good four dimensional coverage exists. On the other

359

hand, larger differences have to be accepted between measurements originating from a routine network and modelled data. It is desirable to select a collection of suitable data sets both showing a wide range of values for the target parameters and representing special atmospheric situations, like e.g. summer time high ozone smog episodes.

The criteria for the decision whether a model is able to successfully reproduce a significant portion of the measured data set or not are usually based upon predefined scores. Scores are quantitative measures. One option for such a score is to use the percentage of simulated data, which differs less than the pre-selected MQOs for accuracy and precision from the observations. Finally, one should always keep in mind that a score may be sufficient for one application, whereas it is not acceptable in another context.

The following procedure may be applied if more than one model is evaluated: The data used for comparison are open and as many test simulations as necessary are allowed for each model. The only constraint is that a fixed deadline exists for the delivery of the 'best' simulation results to an 'independent' group of scientists, who have no own model in the intercomparison. This group prepares the intercomparison along the lines that have been agreed upon in the strategy protocol using the selected scores.

Within GLOREAM a joint effort of up to eight modelling groups teamed in the German Tropospheric Research Program (TFS) has been carried out in order to estimate quantitative performance measures for Eulerian chemistry and transport models (CTMs) under summer and fall time meteorological conditions. Potential temperature, specific humidity and the concentrations of both ozone and nitrogen dioxide have been selected as target parameters for the following reasons: When three-dimensional data exist the potential temperature provides information about the atmospheric stratification (with consequences for the atmospheric turbulence and mixing) and the daily variations in the reaction rate coefficients. The specific humidity on the one hand behaves like a passive tracer in a cloud-free or fog-free atmosphere and thus gives indications about the net effect of the various transport processes in the boundary layer. On the other hand it reveals whether the photochemistry in the model runs in a dry or wet environment. Ozone is the prime indicator species for photochemical smog and nitrogen dioxide has been chosen because it gives an indication about existing precursor concentration levels. In addition, its rather small atmospheric concentrations are an additional challenge for the models. Table 1 summarizes the MQOs in respect to accuracy and precision for the four target values.

Table 1. Model quality objectives (MQOs) for the four target values used in the evaluation exercise carried out in the framework of the German Tropospheric Research Program (TFS, 1996-2000). A value of 10 percent of the median of the observations is used as MQO for specific humidity and ozone concentration, respectively. For the nitrogen dioxide concentration the corresponding value is 50 percent.

Model quality objective (MQO) for	FluMoB 26 and 27 July 1994	BERLIOZ 20 and 21 July 1998	
potential temperature	± 1.5	± 1.5	K
specific humidity	± 1.01	± 0.85	g/kg
ozone concentration	± 10	± 7	ppb(v)
nitrogen dioxide concentration	± 1.25	± 0.75	ppb(v)

For further information, see Builtjes et al. (2003).

6. OUTLOOK FOR THE FUTURE

Although substantial progress has been made in global and regional modelling, still gaps in knowledge exists which need further attention:

- Clear progress has been made in aerosol modelling, but many unknowns still exist. This is to a large extent due to the lack of sufficient and detailed aerosol observations which makes model evaluation in part not possible.
- Although model inter-comparison studies and model evaluation have been carried out, a complete and generally accepted model validation system is not yet in place, nor have all the models undertaken a full model evaluation.
- Numerical improvements are still necessary, especially to increase the computational speed/efficiency to enable the use of the full capabilities of data assimilation and (two-way) nesting.
- One of the weakest points in modelling is the proper treatment of clouds (also in relation to aerosols) and the treatment of vertical exchange.
- Air quality and climate are interconnected processes. This should be taken into account in future model improvements and model application and scenario studies.

7. REFERENCES

Bauer and Langmann (2002) An atmosphere-chemistry model on the meso-gamma scale: model description and evaluation, Atm Env. 36, 2187-2199

Builtjes et al (1999) Gloream Subproject Description EUROTRAC-2, ISS, Munich

Builtjes et al (2003) Gloream Final Report EUROTRAC-2, To be published, ISS, Munich

Ebel et al (1997) Tropospheric Modelling and Emission Estimation, EUROTRAC Report 7, Springer Verlag, Berlin, ISBN 3-540-63169-0

Hass et al (2003) Aerosol Modelling: Results and Intercomparison from European Regional Scale Modelling Systems, To be published, ISS, Munich

Roemer et al (2003) Ozone trends according to ten dispersion models, To be published , ISS, Munich

Tilmes et al (2002) Comparison of five Eulerian air pollution forecasting systems for the summer 1999 using the Geran ozone monitoring data, J. Atmos Chem. 42, 91-121

DISCUSSION

D.G. STEYN   I would like to applaud your model evaluation approach of
                establishing model acceptance / rejection criteria before
                performing the modelling. To do otherwise exposes us to great
                philosophical and technical difficulties.

P.BUILTJES   Thank you for your support in this. The experience obtained by
                Eberhard Schaller-Univ. Cottbus, Germany in the German TFS-
                program has shown the adequacy of this approach.

THE COMBINED EFFECT OF MECHANICAL AND THERMAL FORCING ON THE DISPERSION OF A PLUME: FINE-SCALE MODELING AND PARAMETERIZATION.

Alessandro Dosio[1], Jordi Vilà-Guerau de Arellano[1], Albert A.M.Holtslag[1] and Peter J.H Builtjes[2]

1. INTRODUCTION

Dispersion in the so-called Convective Boundary Layer (CBL) is driven by the combined effect of thermal forcing (buoyancy) and mechanical forcing (wind shear).

In conditions of strong convection, dispersion is influenced mainly by buoyancy. This situation has been largely studied in the past by menas of laboratory experiments (Willis and Deardorff 1976, Willis and Deardorff 1981, Deardorff and Willis 1985, Weil et al., 2002), field campaigns (Briggs 1983) and numerical simulations (Lamb 1982, Nieuwstadt and De Valk 1987, Henn and Sykes 1992, Liu and Eung 2001). These works showed that the turbulent field is characterized by large subsidence motion of cold air (downdraft) surrounded by strong narrow updraft of warm air.

In condition of weaker stability, however, the flow is forced by the combined effect of buoyancy and shear: numerical simulations by Moeng and Sullivan (1994) and Sykes and Henn (1989) showed that due to the increasing wind shear, the turbulent structure was modified and two dimensional roll structures appeared aligned with the mean wind. Since the behavior of a dispersing plume is related to the turbulence characteristics, the change in the relative importance between thermal and mechanical forcing is expected to influence the dispersion characteristics. In the past, however, only few studies (Mason 1992, Gopalakrishnan and Avissar 2000) investigated the behavior of a dispersing plume in conditions of weak and moderate convection.

In this study, we used a fine-scale numerical model (Large-Eddy Simulation, hereafter LES) to investigate the behavior of a plume emitted at two different heights in boundary layers with different combinations of surface heat flux (thermal forcing) and geostrophic wind speed (mechanical forcing). The simulated flows ranged from pure

[1] Meteorology and Air Quality Section, Wageningen University, 6701 AP Wageningen, The Netherlands.

[2] TNO_MEP, Apeldoorn, The Netherlands.

convective to near-neutral and they were classified according to scaling parameters such as u_*/w_* (where u_* is the friction velocity and w_* is the convection velocity scale) and $-z_i/L$ (where z_i is the height of the Boundary Layer and L is the Monin-Obukhov length). The influence of the increasing wind shear was studied on the following dispersion characteristics: mean plume height, vertical and horizontal spread and ground concentrations. Finally, parameterizations for the dispersion parameters in boundary layers driven by buoyancy and shear are proposed and compared to the LES results.

2. NUMERICAL SETUP AND SIMULATED CASES

LES is a numerical code that solves the most relevant scales of motion of a turbulent flow. A set of coupled non-linear differential equations of conservation for the main variables of a turbulent field (wind velocity, temperature, humidity, turbulent kinetic energy) is solved on a staggered numerical grid. An equation is added for the evolution of a passive scalar. The simulated domain (10km x 10km x 2km) is solved by a numerical grid with a resolution of about 60 m in the horizontal and 30 in the vertical direction.

Four different values of the geostrophic wind (0.5 m/s, 5 m/s, 10 m/s, 15 m/s) and three different surface heath fluxes (0.156 Km/s, 0.1 Km/s, 0.052 Km/s) were imposed as initial forcing condition to generate different turbulent flows. Four classes of boundary layers were defined according to the value of the characteristic velocity scale for flows driven by buoyancy and shear (u_*/w_* called shear-buoyancy ratio) and the dimensionless height $-z_i/L$: pure buoyancy (cases B1-B5, $-z_i/L > 40$, $u_*/w_* < 0.2$), shear-buoyancy-driven boundary layers (cases SB1-SB2, u_*/w_* about 0.3), shear-driven boundary layer (cases S1-S2, u_*/w_* about 0.46) and near-neutral boundary layer (NN, $-z_i/L = 1.9$, $u_*/w_* = 0.59$). The simulated cases and the scaling parameters are listed in Table 1.

Table 1. Simulated cases, forcing initial conditions and classification according to scaling parameters

Case	Geosptrophic wind (m/s)	u_*/w_*	$-z_i/L$
B1 – B5	0.5 – 5	< 0.21	> 40
SB1 – SB2	5 – 10	0.27 – 0.34	7.8 – 18
S1 – S2	10 – 15	0.46	3.5 – 4.0
NN	15	0.59	1.9

An Instantaneous Line Source of passive tracer was released at two different heights ($z_s/z_i = 0.078$ and 0.48 respectively, where z_s is the source height) in the simulated turbulent flows and the dispersion characteristics were analyzed. The line source can be equivalently interpreted as a Continuous Point Source via the time-space transformation $x = U{:}t$. For each simulated cases, three different realizations were performed; the results were finally averaged over the different realizations and the above-mentioned classification.

The dispersion parameters have been calculated following Nieuwstadt and de Valk (1987) as follows:

$$\sigma_z^2 = \frac{\int c(z-z_s)^2 dV}{\int cdV} \quad , \quad \sigma_y^2 = \frac{\int c(y-\bar{y})^2 dV}{\int cdV} \tag{1}$$

where c is the space-dependent concentration, $dV=dxdydz$ and $\bar{y}$ is the mean plume horizontal position, defined according to:

$$\bar{y} = \frac{\int cydV}{\int cdV}. \tag{2}$$

The dimensionless cross-integrated concentration C_y is calculated as follows:

$$C_y = \frac{z_i \int cdxdy}{\int cdV}. \tag{3}$$

All the results were scaled by the velocity scale:

$$w_m^3 = w_*^3 + 5u_*^3 \tag{4}$$

which takes into account both the characteristic velocity scales for convective- and shear-driven flows, as well as providing a suitable scaling for the second-order moments of turbulence (Moeng and Sullivan 1994).

Finally, a dimensionless space (time) was defined as follows:

$$X_m = \frac{w_m}{z_i} \frac{x}{U} = \frac{w_m}{z_i} t. \tag{5}$$

3. RESULTS AND DISCUSSION

3.1. Plume behavior

Figure 1 (a-d) shows the cross-wind integrated concentrations (C_y, Eq. 3) for a near-ground release as calculated by the LES in all the simulated cases. Figure 1e shows the ground concentrations for the respective cases.

The influence of the relative importance of buoyancy and shear on dispersion is clearly visible in the figure. In the pure convective cases (B1-B5) only the thermal forcing is present; the plume is rapidly lifted off by the strong thermals (updrafts) and a surface minimum ($C_y < 0.7$) and an elevated maximum ($C_y > 1.3$) are present at $X_m=2$ (Figure 1a) The plume reaches the ground at $X_m=3.5$ and at farer distances from the source the tracer is uniformly mixed within the whole mixed layer.

When the wind speed increases and the buoyancy becomes less dominant (figure 1b-1d) the tracer is advected horizontally for a longer time before being lifted off. In the

365

shear-buoyancy cases (SB1-SB2, figure 1b) an elevated maximum is still present, but the relative surface minimum is shifted towards greater distances from the source. In the shear cases (S1-S2, figure 1c) the elevated maximum is not evident and the surface minimum is present at $X_m=3$. Finally in the near-neutral case (NN, figure 1d), the wind keeps he tracer close to the ground up to large distances from the source, and neither elevated maximum nor surface minimum are present.

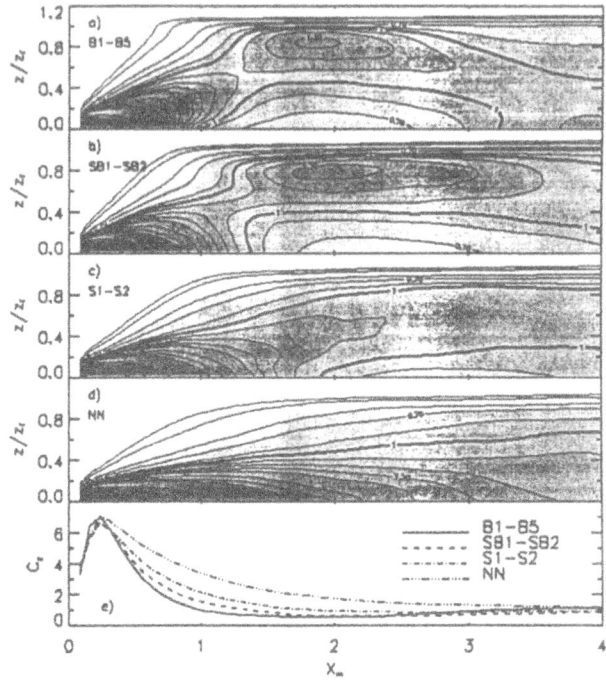

Figure 1. a-d: Normalized cross-wind integrated concentrations (Cy) as function of the dimensionless distance Xm for a near-ground release as calculated by the LES. e: Ground concentrations (Cy at z=0).

The different contribution of thermal and mechanical forcing influences strongly the dispersion characteristics of a plume emitted from an elevated source too, as shown by Dosio et al (2003).

3.2 Dispersion parameters and parameterization

The dispersion parameters σ_y and σ_z were computed explicitly from the concentration field calculated by the LES (Eq 1) as function of the dimensionless time. As shown by Dosio et al (2003) the main effect of the increasing wind shear is to reduce vertical dispersion and enhance horizontal spread due to the changing in the turbulent structure. Although many studies are available either in pure convective or in the neutral limit, attempts to parameterize the dispersion behaviors in boundary layers driven by both buoyancy and shear are scarcer.

We propose therefore parameterizations for the dispersion parameters, which include the combined effect of mechanical and thermal forcing and we evaluate them against the LES results.

4.1.1 Horizontal dispersion parameter

According to Venkatram (1988) he horizontal dispersion parameter can be divided in two components, the buoyancy-induced dispersion (σ_{yb}) and the shear-induced dispersion (σ_{ys}):

$$\sigma_y^2 = \sigma_{yb}^2 + \sigma_{ys}^2 . \tag{6}$$

From the LES results the buoyancy-generated dispersion can be parameterized as function of the turbulent parameters as follows:

$$\frac{\sigma_{yb}}{z_i} = \frac{\sigma_v / w_m X_m}{\left(1 + \dfrac{X_m}{2T_v} \dfrac{z_i}{w_m}\right)^{1/2}} , \tag{7}$$

where σ_v is the horizontal (fluctuation) velocity variance and the Lagrangian time scale T_v is parameterized as:

$$T_v = 1.7 \left(\frac{\sigma_v}{w_m}\right)^2 \frac{z_i}{w_m} . \tag{8}$$

The shear-produced dispersion is parameterized according to:

$$\sigma_{ys}^2 = \frac{a_0 S_*^2 \sigma_w^2 T_w X_m^3 z_i / w_m}{\left[1 + \left(\dfrac{X_m}{X_0}\right)^3\right]^{2/3}} , \tag{9}$$

where

$$X_0 = \left(\frac{z_i^4}{a_0 b_0 \sigma_w^4 T_w \tau_c}\right)^{1/2} \frac{w_m}{z_i} , \tag{10}$$

$a_0 = 0.09$, $b_0 = 60$, $\tau_c = 0.7 z_i / w_m$ and σ_w is the vertical velocity variance. The Lagrangian time scale for the vertical velocity T_w is parameterized with an expression similar to Eq. 8.

The dimensionless wind shear S_* is defined as follows:

$$S_* = \frac{V}{w_m} \theta_m , \tag{11}$$

where V is the total horizontal wind speed and θ_m is the difference between the geostrophic wind direction and the wind direction at the mean plume height.

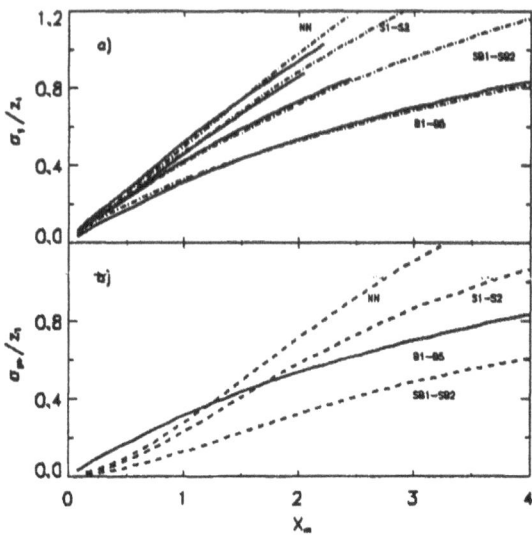

Figure 2. a): Comparison between the LES calculated horizontal dispersion (solid lines) and the parameterization (eq. 6, dashed-dotted lines) for all the simulated cases. b): Total dispersion in the pure buoyancy cases (solid line) and shear-produced dispersion (dashed lines) as computed by eq. 7 and 9 respectively.

Figure 2a shows the comparison between the horizontal dispersion calculated from the LES results (solid lines) and the parameterization (dashed lines) in all the simulated cases. The agreement is more than satisfactory. Figure 2b shows the shear contribution only (eq. 9) compared with the total dispersion for the pure buoyancy cases (in this case $\sigma_y = \sigma_{yb}$ as the shear is absent). The results show that for large values of the wind, the shear contribution is of the same order of magnitude or larger than the dispersion generated by buoyancy.

4.1.1 Vertical dispersion parameter

The vertical spread is strongly dependent on the source height (Weil, 1988). For a source above the surface layer ($z_s/z_i > 0.1$) the vertical dispersion parameter can be successfully represented by the formula (Taylor's law):

$$\frac{\sigma_z}{z_i} = 0.51 \frac{w_*}{w_m} X_m \qquad (12)$$

which fits the LES results accurately (see for more details Dosio et al, 2003).

For a near-ground release, at short distance from the source ($X_m < 1$) two opposite effects have to be taken into account: the reduction of the vertical spread due to the presence of the surface and the increasing contribution due to the shear. In the convective limit, the vertical spread satisfies the relationship:

$$\frac{\sigma_z}{z_i} = 0.52 X_m^{6/5}, \qquad (13)$$

whereas in the neutral limit it grows linearly according to:

$$\frac{\sigma_z}{z_i} = 0.6 \frac{u_*}{w_m} X_m. \qquad (14)$$

The following parameterization is proposed which combines the above-mentioned limits:

$$\frac{\sigma_z}{z_i} = \left[\left(0.52 \frac{w_*}{w_m} \right)^{12/5} X_m^{2/5} f + \left(0.6 \frac{u_*}{w_m} \right)^2 \right]^{1/2} X_m \qquad (15)$$

The function f accounts for the reduction of vertical spread discussed previously. It depends on the atmospheric surface layer stability according to:

$$f = \frac{\left(-0.2 \frac{z_i}{L} \right)^3}{\left[1 + \left(-0.2 \frac{z_i}{L} \right)^3 \right]} \qquad (16)$$

Expression (15) is in satisfying agreement with the LES results as shown in figure 3.

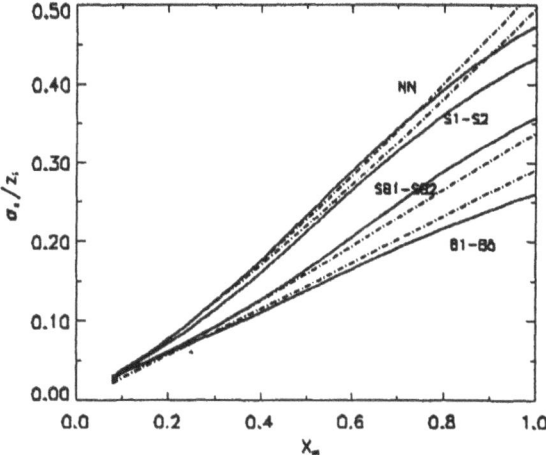

Figure 3. Comparison between the LES calculated vertical dispersion (solid lines) and the parameterization (eq. 15, dashed-dotted lines) for all the simulated cases.

5. CONCLUSIONS

The dispersion of a passive tracer in boundary layers driven by different combination of buoyancy and shear was studied by means of a fine-scale numerical model (LES). The turbulent characteristics of the boundary layer depend largely on the relative importance of thermal and mechanical forcing. In consequence, the dispersion proprieties of a tracer will depend on the ratio of these two forcings. The main effect of increasing wind shear is to reduce vertical spread and to enhance horizontal dispersion. The wind tends to advect the plume horizontally for a longer time, so that the ground concentrations are strongly affected by the increasing wind speed.

Parameterization for the vertical and horizontal dispersion parameters were proposed which take into account the influence on the combined effect of buoyancy and wind shear and they were compared with the LES results showing satisfactory agreement.

6. ACKNOWLEDGEMENTS

Alessandro Dosio is funded within the Centre of Expertise Emissions and Assessment – a cooperation between TNO and Wageningen University7 All computations were performed on TERAS (SGI Origin 3800) at the Academic Computing Center Amsterdam (SARA). The National Computer Facilities Foundation (NCF-341) sponsored use of these computer facilities.

7. REFERENCES

Briggs, G. A., 1993, Plume dispersion in the convective boundary layer. Part II: analysis of the CONDORS field experiment data, *J. Appl. Meteorol.* 32, 1388-1425.

Deardorff J. W. and Willis G. E., 1985, Further results from a laboratory model of the convective planetary boundary layer, *Bound-. Layer Meteo.*, 32, 205-236.

Dosio A. et al, 2003, Dispersion of a passive tracer in buoyancy- and shear-driven boundary layers, Accepted for publication on *J. Appl. Meteor.*, February 2003

Gopalakrishnan S. G. and Avissar R., 2000, A LES study of the impacts of land surface heterogeneity on dispersion in the convective boundary layer, *J. Atmos. Sci.*, 57, 352-371.

Lamb R. G. 1982, Diffusion in the convective boundary layer, in: *Atmospheric turbulence And Air Pollution Modeling*, Reidel Publishing Company, Dordrecht, pp. 159-229.

Liu C. H and Leung D. Y. C., 2001, Turbulence and dispersion study using a three-dimensional second-order closure eulerian model, *J. Appl. Meteorol.* 40, 92-113.

Luhar A. K., 2002, The influence of vertical wind shear on dispersion in the convective boundary layer and its incorporation in coatal fumigation models, *Bound-. Layer Meteo.*, 102, 1-38.

Mason P. J., 1992, Large-eddy simulation of dispersion in convective boundary layers with wind shear, *Atmos. Environ.*, 26(A), 1561-1571.

Moeng C. H. and P. P, Sullivan, 1994, A comparison of shear- and buoyancy-driven planetary boundary layer flows`, *J. Atmos. Sci*, 7, 999-1022.

Nieuwstadt F. T. M. and M. M. de valk, 1987, A large eddy simulation of buoyant and non buoyant plume dispersion in the atmosperic boundary layer, *Atmos. Environ.*, 21 2573-2587.

Schidt H. and U. Schumann, 1989, Coherent structure of the convective boundary layer derived from large-eddy simulation, *J. Fluid Mech.*, 200, 511-562.

Sykes R. I. and Henn D. S., 1989, Large eddy simulation of turbulent sheared convection, *J. Atmos. Sci.*, 46, 1106-1119.

Venkatram A., 1988, Dispersion in the stable boundary layer, , in *Lectures on Air Pollution Modeling*, American Meteorological Scoiety, pp. 228-265.

Weil J. C., 1988, Dispersion in the convective boundary layer, in *Lectures on Air Pollution Modeling*, American Meteorological Scoiety, pp. 167-227.

Willis G. E. and Deardorff J. W., 1974, A laboratory model for the unstable boundary layer, *J. Atmos. Sci.*, 31, 1297-1307.

Willis G. E. and Deardorff J. W., 1976, A laboratory model of diffusion into the convective planetary boundary layer, *Quart. J. Roy. Meteor. Soc.*, 102, 427-445.

Willis G. E. and Deardorff J. W., 1981, A laboratory model of diffusion from a source in the middle of the convectively mixed layer., *Atmos. Environ,*. 15, 109-117.

DISCUSSION

S.E. GRYNING   How is the effect of the source height taken into account is the parameterisation of σ_y and σ_z ?

P.BUILTJES   A distinction is made between low level ($Xm < 1$) and higher level sources in the parametrisation, and via the dimensionless wind shear

E.GENIKHOVICH   Your computational domain was 10x10km, i.e. you were able to resolve eddies with the time scale less than 1 hour. Does it mean that you were simulating only small-scale features and did not interested in low-frequency effects?

P.BUILTJES   The purpose was indeed to resolve the sub-grid scale features within a grid of 10 x 10 km2. The larger scale features are treated in the larger, regional scale 3-D Eulerian grid model LOTOS.

DATA ASSIMILATION FOR CTM BASED ON OPTIMUM INTERPOLATION AND KALMAN FILTER

Johannes Flemming[1], Maarten van Loon[2] and Rainer Stern[1]

1 INTRODUCTION

The aim of this paper is to compare the performance of two data assimilation schemes for the Eulerian chemistry transport model REM/CALGRID. Optimum Interpolation (OI) and Kalman Filtering (KF) have been applied to assimilate hourly O_3 and NO_2 observations in a model run for July 2001. The comparison comprises the structure of the obtained model error covariances and the analysed concentration fields. In addition, an example of an assessment of model parameters such as turbulent exchange coefficients by means of the Kalman filter is given.

Application of data assimilation techniques means combining two information sources: observation and modelling. The result of data assimilation is an analysis, i.e. a concentration field in a certain resolution. In active data assimilation the analysis replaces the fields in the state vector during the model run. In passive data assimilation the analysis, i.e. the blending of observations and model fields is done after the model run.

The analysis of consistent concentration fields is the main purpose for the use of data assimilation with CTMs. Data assimilation can therefore be considered as an interpretation and spatial and temporal interpolation of the observations.

Operational air quality networks are mainly designed to monitor high concentration levels in polluted areas. The spatial representativeness of these observations appears to be limited and is not well known. Therefore spatial information from Eulerian models is useful for interpolating and mapping of the observations (Flemming and Stern, 2002). The assimilation of satellite observations in CTMs helps to interpret the radiance information and to improve its vertical and temporal resolution (van Loon et al., 2002).

The techniques of OI and KF data assimilation are based on the theory of stochastic processes. They require knowledge about the statistical properties of the errors of the model and the observations. Model errors are treated as a spatial random field, which is determined by the mean, i.e. the model bias, and its covariance matrix. The observation error is considered as a spatially uncorrelated portion of the measurement value quantified by its variance. It is mainly a measure of the representativeness of the observations.

[1] flemming@zedat.fu-berlin.de, Institut für Meteorologie, Freie Universität Berlin, 12165 Berlin, Germany
[2] TNO-MEP Apeldorn, The Netherlands.

Air Pollution Modeling and Its Application XVI, Edited by
Borrego and Incecik, Kluwer Academic/Plenum Publishers, New York, 2004

The correct specification of all error statistics is of importance for the performance of the analysis scheme. The ratio between the variances of the observations and the model errors controls the impact of the observations in the analysis. The larger the observation error variance the less is adaptation of the model field towards the observed values.

The KF algorithm consists of several subsequent steps. It affects the whole model state vector whereas OI only applies to those fields for which observation are available. The core of the KF algorithm is the analysis step, which is equal to the OI. The fundamental difference between both schemes is the way in which the spatial covariance for the analysis step is obtained.

OI requires empirical covariance modelling which is limited either to spatial homogeneity or temporal stationarity. In contrast, the Kalman Filter uses the dynamical model to obtain the covariance structure of the model error. It reflects its time - space variability and reveals the relationship between model variables, which are not directly observed.

In the application presented in this paper ,the model error covariance is build up by an ensemble of different model states, which was obtained by introducing perturbations of model parameters as emissions or turbulent vertical exchange coefficients. However, this way of dynamical covariance modelling leads to the very high computational costs of the KF which 30-50 times higher than for OI.

2 DESCRIPTION OF THE CT-MODEL

REM/CALGRID (Stern et al., 2003) is a Eulerian grid model of medium complexity that can be used on the regional, as well as the urban, scale for short-term and long-term simulations of oxidant and aerosol formation. For the data assimilation study presented here, the model was applied at a modelled resolution of 0.25° latitude and 0.5° longitude for an area that covers Central Europe. The model was run with 4 dynamically changing layers and the CBM4 photochemistry scheme.

Meteorological input is derived from synoptic meteorological surface data, upper air data and climatological information (Reimer and Scherer, 1992).

3 AIR QUALITY DATA BASE AND OBSERVATION ERROR STATISTICS

The air quality database has been obtained from the networks of the German federal states. It includes about 250 measurement sites for O_3 and NO_2. The data quality assurance (EU-directive) demands 90% completeness for the annual time series.

By means of a hierarchical clustering the measurement sites were classified in air quality regimes (Flemming, 2001). For each species separately regimes such as "mountain", "rural", "suburban", "urban", "urban-traffic" and "traffic" were identified.

In the context of data assimilation the observation error is a consequence of the instrument error and more important of the limited spatial representativeness of the measurements for the CT-model resolution. It is defined as spatially uncorrelated part of the measurement with zero mean. Its correct specification is important because it controls to what extend the analysed field matches the measured value.

The estimation of the observation error variance is based on the observational method (Hollingsworth and Lönnberg, 1986). The observation error variance at a cer-

374

tain station is estimated by the extrapolation of the covariance field from surrounding stations by means of a spherical covariance model (Flemming, 2003).

Figure 1 shows the range of the estimated observation error standard deviation for NO_2 and O_3 for different air quality regimes. The typical observation error SD for ozone is about 5 ppb for all regimes indicating a smaller relative error for the regimes with higher mean ozone concentrations. For NO_2 the observation error increases with increasing concentrations from about 2 ppb for rural to about 8 ppb for the traffic sites. For the traffic sites the assumption of a zero mean observation error seems to be not correct. Hence, only stations of the rural and urban regimes are taken into account during the data assimilation both for NO_2 and O_3.

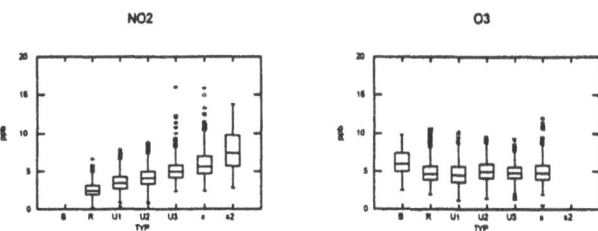

Figure 1 Estimated oObservation error standard deviation in ppb vs. air quality regime (TYP) for all stations in Germany (left NO_2 right O_3). B = mountain, R = rural, U1 = suburban, U2 = urban, U3 = urban, polluted, S = street, s2= Street extreme

4 THEORY OF OPTIMUM INTERPOLATION AND KALMAN FILTER

Kalman Filter (KF) and Optimum Interpolation (OI) are based on the theory of stochastic processes and control theory (Bouttier and Courtier, 1999). Their aim is to supply a best linear unbiased estimate (BLUE) of the concentration fields. Van Loon und Heemink (1997) describe rationale, approximations for linearity, and numerical aspects of the KF implementation used in this paper.

Active data assimilation with OI can be considered as a simplification of the KF-algorithm (eq. (1) - (5)). It relies on the same assumption as KF, such as no bias between observations and the model, uncorrelated and unbiased observation errors and Gaussianity of the errors.

The common principle of KF and OI-data assimilation is the sequence of the time forecast of the model state vector x by the CT model M (eq. (1)) and the analysis step (eq. (4)), which is the adaptation of the forecasted state vector x^{t+1} to the observations y. The observation operator H accounts for the transformation of the modelled variable to the observed one. Its imperfectness is expressed as part of the observation error.

The gain matrix **K** determines the influence of the observations in the analysis. The calculation of **K** reveals the statistical character of the analysis equation because **K** is obtained from the covariance matrix of the model error P_B and the observation error **R**. The analysed model state x_A replaces the model state for the next model integration step eq. (1). The new error covariance P_A of x_A is obtained by eq. (5)

The difference between the described KF-algorithm and OI is the determination of the model error covariance matrix **P**. In the KF **P** is propagated like the state vector by an approximation of the tangential linearised operator **M** of the dynamical model M. The imperfectness of the dynamical model is expressed in a model forecast error covariance matrix **Q**, which is added at every forecast of **P** eq. (2). For OI the covariance of the model error is estimated by empirical covariance modelling of the observation increments.

forecast of model state:

$$\mathbf{x}^{t+1} = M\mathbf{x}_A^t \qquad (1)$$

forecast of model error covariance matrix:

$$\mathbf{P}_B^{t+1} = M\mathbf{P}_A^t M^T + \mathbf{Q} \qquad (2)$$

computation of gain matrix:

$$\mathbf{K}^t = H\mathbf{P}_B^t (H\mathbf{P}_B^t H^T + \mathbf{R})^{-1} \qquad (3)$$

analysis of model state:

$$\mathbf{x}_A^t = \mathbf{x}_B^t + \mathbf{K}^t (\mathbf{y}^t - H\mathbf{x}_B^t) \qquad (4)$$

update of model error covariance matrix:

$$\mathbf{P}_A^t = (\mathbf{I} - \mathbf{K}^t H)\mathbf{P}_B^t \qquad (5)$$

5 COVARIANCE OF THE MODEL ERROR BY OI AND KF

5.1 Dynamical covariance modelling by the Kalman Filter

For complex CT models M the linear approximation M cannot be numerically handled since the covariance matrix **P** has a dimension of the square of the full state vector. Therefore an approximation of eq. (2) needs to be used which is in this case the Reduced Rank Square Root (RRSQRT) algorithm. The basic idea behind the RRSQRT algorithm is to avoid to work with the covariance matrix **P** but instead with its square root S (i.e. $SS^T = P$). The columns of S can be considered as different model states, which establish the variation, expressed in **P**. It can be shown that a numerical approximation of (2), i.e. the forecast of **P** can be achieved by forecasting each column of S by the numerical code of the dynamical model M.

By means of a principal component analysis of the covariance matrix **P** the q model states (modes), which contribute most to the variability of P, can be found. The RRSQRT algorithm reduces the amount of computational work considerably as the forecast of **P** is done by q model integrations of the columns of S.

The Kalman Filter assumes an imperfect model M by adding noise (matrix **Q**) at each time step of error covariance forecast P in (2). In the given implementation this is done by forecasting additional p model states where uncertain model parameters have been altered in the forecast step. Starting with an empty matrix S each time step adds p new perturbed model states to S. If the number of modes in S reaches a certain value, q the number of modes is reduced by the PCA.

In this application 3 model parameters have been selected to account for the model forecast error: The total NOx- and VOC emissions and the turbulent vertical exchange coefficient. The parameters are varied uniformly in the whole model domain by 25% percent by means of a "model noise factor". The question whether the injected noise leads to a realistic specification of **P** can be answered to some extend by comparing it with the results of empirical covariance modelling.

5.2 Empirical Covariance modelling for OI

The empirical covariance modelling of the model error covariance is based on the differences between observations and the modelled values. The ensemble for the estimation is therefore limited to the station locations. In order to obtain a full covariance matrix an empirical covariance model has to be established.

Inhomogeneous covariance modelling has been developed for passive data assimilation with REM/CALGRID assuming temporal stationarity of the errors (Flemming, 2003) . However, in active assimilation the current model state and its error is influenced by the analysis of the previous model time step. That's why a simple form of instantaneous covariance modelling has been applied, namely the fitting of a homogeneous and isotropic covariance function in exponential form. It expresses the covariance of the model error between two points merely as a function of their distance. Such a covariance function of the O_3 and the NO_2 field was estimated for every analysis step in the OI run.

Taking measurements into account for covariance modelling allows to determine and to correct the bias between model state and the observation. Here, the actual bias correction is based on the difference between the mean model state and the mean of the observations at the rural stations.

5.3 Comparison of Covariance structure of KF and OI

The KF produces a much more comprehensive and temporal and spatial distinguished covariance matrix then the simple covariance function of OI. However, the covariance of the KF depends strongly on the choice and the scope of the variation of the perturbed model parameters.

The empirical covariance modelling applied in OI uses model errors directly obtained by comparison with the measurements. The measurement seems to be a direct indicator what might be the real atmospheric state. However, the difference between observation and a CTM result does not mean in any case a model failure if a station is not representative for the model scale.

The comparison between KF and OI is limited because the OI covariance modelling is much simpler than that of KF. Therefore, only the model error, and a measure for the spatial structure of the covariance can be compared. This measure is the range which describes the decrease of covariance with increasing spatial distance. Figure 2 shows box whiskers – plots of the square root of model error variance, i.e. the covariance for zero distance, for every hour of the day. In the homogeneous OI approach this is one value for the whole field at each time step. For KF the variance is heterogeneous and in figure 2 all values at station locations are depicted. The variances have a similar cycle with highest values in the morning and the afternoon. The magnitudes of the KF and OI values are in the same range. KF values are smaller indicating that more model "noise" should be added. The same conclusion is valid for the variances of NO_2.

Empirical estimated covariances decrease with increasing distance in a range of about 100 – 200 km. This feature cannot be found in the KF covariances due to the uniform variation of the model noise parameters.

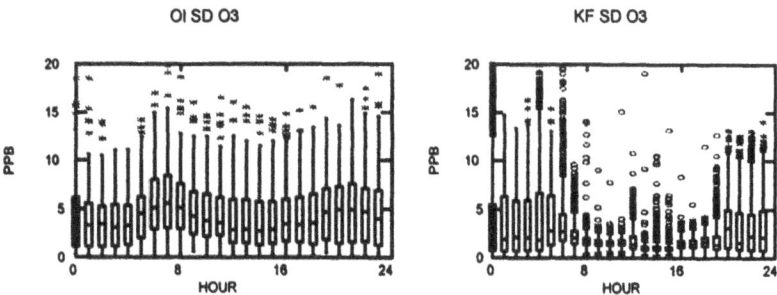

Figure 2 Daily cycle of the standard deviation of the model error variance of ground level O_3 obtained by empirical covariance modelling for OI (left) and by the Kalman filter (right).

6 ASSIMILATION PERFORMANCE

OI and KF have been applied for an assimilation run of REM/CALGRID for July 2001. Hourly Ozone and NO_2-Observations of the rural, suburban and urban air quality regimes have been assimilated. This selection has been made due to the lacking representativeness of the traffic stations for the model grid box size of about 25x25 km².

A proper assessment of the data assimilation performance should rely on a cross validation approach. This has been done for the OI in Flemming (2003) and for the KF in van Loon et al. (1999). The objective of this study is the direct comparison of the two schemes, based on the same observation error variances. Hence, the differences between the assimilation results are completely due to the different levels of complexity of the two schemes.

Figures 3 and 4 show the mean daily cycle of the measured, modelled and OI, respectively KF assimilated ground level concentrations at rural and urban stations for ozone (top) and NO_2. The assimilation of ozone leads to a better agreement with measurements whereas the assimilated NO_2 concentrations do not improve. It has to be investigated whether the good assimilation of O_3 has a negative cross effect on the assimilation of NO_2. The better OI performance for the rural stations is mainly as a consequence of the bias correction based on the rural stations. KF performance for ozone is better in the urban regimes due to the better description of the inhomogeneity.

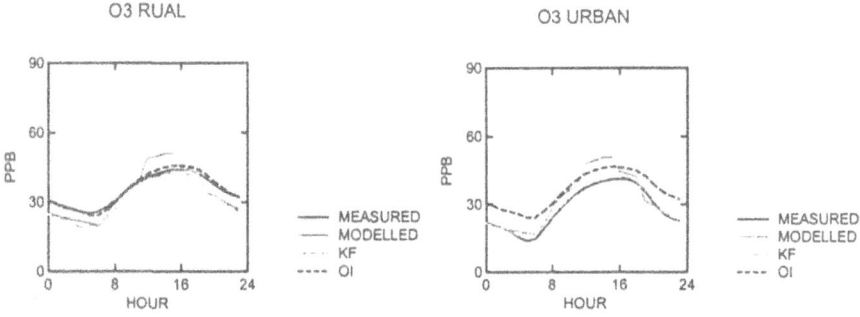

Figure 3 Mean daily cycles of the measured, modelled and assimilated (with OI and KF) ground level concentrations at rural (left) and urban (right) stations for Ozone.

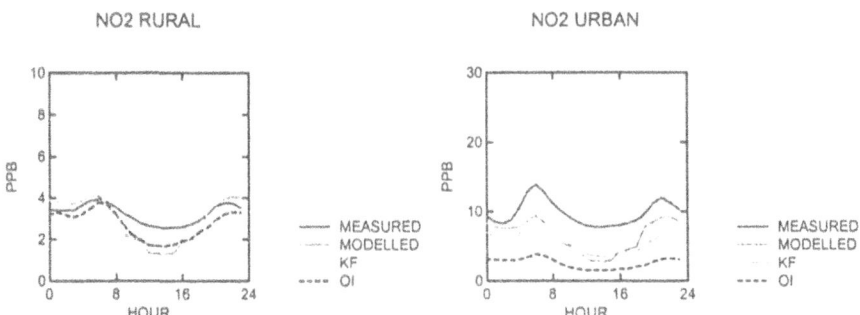

Figure 4 Mean daily cycles of the measured, modelled and assimilated (with OI and KF) ground level concentrations at rural (left) and urban (right) stations for NO_2.

7 ANALYSIS OF EMISSION AND PBL PARAMETERS BY KF

The noise factors have been made part of the model state vector and are therefore analysed by eq. (4) according to available observations. The covariance matrix developed by the KF triggers the extent of the change. Noise factor greater 0 zero indicate that an increase of the model parameter would yield a better agreement with the measurements. However, this relationship is of pure statistical character and assumes a linear responses between noise parameter and changes in the concentration fields.

The investigation of systematic features of the analysed model noise parameters can help to understand the reasons for weaknesses of the model and its input data. Due to dynamical vertical grid system, REM/CALGRID depends strongly on the daily variation of the PBL. Therefore, the daily cycle of the analysed model noise parameters are summarised in figureFigure 5. The most obvious daily pattern exhibits the vertical exchange coefficient. According to the analysis, increased vertical exchange during the night and less exchange during the day would lead to a better agreement with the measurements. The daily pattern of the NO_x–emissions is consistent with this conclusion. However, the statistical analysis cannot answer the question which of these parameters should be corrected in model. Further sensitivity studies are necessary to clarify this issue.

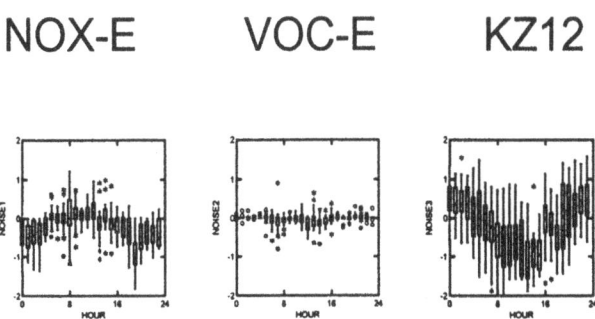

Figure 5 Daily variations of analysed model noise parameters for NOx- (left) , VOC-emission (center) and vertical turbulent exchange coefficient between model layer 1 and 2, KZ12 (right) . Box-Whiskers-Plots for all data in July 2001.

8 SUMMARY AND OUTLOOK

The paper compares sequential data assimilation with Kalman Filter (KF) and Optimum interpolation (OI) within the chemical transport model REM/CALGRID. Both schemes have been applied to assimilate ozone and NO_x in a model run for July 2001. The computational burden of KF is about 30-40 times higher than that of OI.

OI and KF have a similar theoretical basis but differ in their complexity and in the way the model error covariances are obtained. KF affects the whole model state vectors and determines an inhomogeneous covariance matrix by calculating several model states with different settings of internal model parameters (noise parameters) at each time step. In this application, the total NO_x and VOC- emissions and vertical turbulent exchange coefficient varied to account for the model error. In the OI data assimilation the covariance is estimated from the differences between observations and the modelled values . The covariance is expressed by a simple homogeneous covariance function.

Both methods work better for the assimilation of O_3 concentration than for that of the NO_2 – concentrations. In rural ozone regimes KF and OI performance is comparable, whereas the KF shows better results in urban regimes. This might be due to the more realistic description of the covariance by the KF.

The analysed KF-noise parameter have a systematic daily patterns, which suggest that larger turbulent exchange during the night and smaller during the day might improve model performance.

Further work will include KF runs with different and more spatially distinguished model noise parameters. The OI covariance modelling has to be improved by inhomogeneous approaches, which were developed for passive data assimilation (Flemming, 2003). Sensitivity studies will be carried out to see whether the analysis of the model noise parameters can lead to an improvement of the model performance.

9 ACKNOWLEDGEMENTS

This work has been initiated and funded by the German Federal Environmental Agency (Umweltbundesamt) within the project "F+E-Vorhaben 299 43 246".

10 REFERENCES

Bouttier, F. und Courtier, P., 1999, *Data assimilations concepts and methods*, in Data assimilations concepts and methods , eds. F. Bouttier und P. Courtier, Meteorological Training Course Lecture Series, ECMWF, Reading.

Daley, R., 1991, *Atmospheric Data Analysis*, Cambridge University Press, Cambridge.

Flemming, J., 2003, *Immissionsfelder aus Beobachtung, Modellierung und deren Kombination, PhD-thesis FU-Berlin, http://www.diss.fu-berlin.de/2003/71/.*

Flemming, J. , Stern, R. Reimer, E., 2002, *Data assimilation for CT-Modelling based on optimum Interpolation*, in ITM Air pollution modelling and its applications XV , eds. C. Borrego und G. Schayes, NATO CMS, Kluwer Academic / Plenum Publishers, New York.

Hollingsworth, A. and Lönnberg, P., 1986, *The statistical structure of short-range forecast errors as determined from radiosonde data, Part I: The wind field*, Tellus, 38:A, 111-136.

Reimer, E. and Scherer, B., 1992, *An operational meteorological diagnostic system for regional air pollution analysis and long-term modelling*, in: Air Pollution Modelling and its Applications IX., van Doop, H.., ed.., Plenum Press.

Stern, R., Yarmatino, R. und Graff, A., 2003, *Dispersion modeling within the European community's air quality directives: long term modeling of O3, PM10 and NO2*, in proceedings of 26th ITM on Air Pollution Modelling and its Application. May 26-30, 2003, Istanbul, Turkey.

Van Loon, M. and Heemink, 1997, A.W., *Kalman Filtering for non linear atmopsheric chemistry models: First experiences*, Technical Report MAS-R9711, CWI, Amsterdam.

Van Loon, M., Builtjes, P. H. J., Segers, A. und Roemer, M. , 1999, *Reactive Trace Gas Assimilation*,TNO-MEP report, Apeldorn.

Van Loon, M., Builtjes, P.and de Leeuw, G., 2002, *PM2.5 concentrations over Europe combining satellite data and modelling*, in ITM Air pollution modelling and its applications XV , eds. C. Borrego und G. Schayes, NATO CMS, Kluwer Academic / Plenum Publishers, New York.

DISCUSSION

E.GENIKHOVICH   At the 24th NATO/CCMS conference in the US I presented a paper with a version of the optimal interpolation technique for assimilation of data of the urban monitoring of air pollution. I think it is free from same deficiencies of OI mentioned in your talk. Did you try to check its applicability?

J.FLEMMING   No, I did not check it. If I remember your paper correctly the covariance estimation is based on modelled time series and is therefore available for every combinations of model grid points. It would allow omitting the assumption of spatial homogeneity and isotropy. However, using times series data for the estimation of the spatial covariance requires temporal filtering in order to assume the statistical independence of the data. A strong temporal correlation, e.g. due to the daily cycle, can lead to an overestimation of the spatial covariance.

R.SAN JOSE   Our experience with MM5 by assimilating surface and vertical met soundings indicate that by using both data sets the improvement is much better than using only surface met data. Have you tried to use vertical pollution data in your approach?

J.FLEMMING   Tropospheric concentration soundings would be very helpful for data assimilation with CTM. We know from sensitivity studies about initial conditions in ozone forecasting that changing only the ground layer concentration field does not lead to significant changes in a 24-hour forecast due to predominant sources (emissions) and sinks (deposition). Changes in the higher levels of the model do affect the forecast of the next day. Unfortunately, operational vertical concentrations sounding are not available. It might be helpful to assume an uniform ozone profile in the PBL during the day, which would allow to assimilate observation directly in the second an third model layer.

G.KALLOS   How many days do you use for the calculation of the covariance matrix and therefore the next "correlation cycle"? Do you use them for the entire season and few weeks or days? I mean in the operational use.

J.FLEMMING	In both OI and KF no time series data are used to calculate the spatial covariance matrix. The covariance matrix is calculated at every hour of the run. In KF assimilation the covariance calculation is based on an ensemble of different model states. In OI assimilation homogeneity and isotropy is assumed in order to estimate a covariance function, which depends only on the spatial separation. The ensemble is formed out of all station pairs in certain distance class.
R.BORNSTEIN	Isn't it true that this technique would work best for pollutants with smooth gradients, i.e., these whose concentrations are not dominated by point sources.
J. FLEMMING	I do agree but I would argue that the resolution of the model is main reason for this limitation. The concept of the observation error was introduced to account for the lack in the representativeness of the observations for the model grid box. I presented the success of the assimilation performance by means of the RMSE. If we trust in our estimations of the observation error variance, the assumed observation error is the lower limit for this RMSE, which would be obtained by the "perfect" model or assimilation run. It might be helpful to look at the ratio between RMSE and the observation error variance. The observation error variance is a climatological variable and we do not know the actual observation error. This is of course a further limitation, especially for observations, which are dominated by point sources.

MASS-CONSERVATION ISSUES IN MODELLING REGIONAL AEROSOLS

Wanmin Gong, Paul A. Makar, and Michael D. Moran[*]

1. INTRODUCTION

The problem of using mass-inconsistent winds in tracer advection is relatively well recognized in the air quality modelling community. As pointed out by Byun (1999), an inconsistency in the air-density and wind fields is equivalent to an additional source term in the tracer continuity equation, which will lead to mass-conservation violation in air quality models. There are a number of causes of mass inconsistency in the input air-density and wind fields for air quality models. For example, the continuity equation for air density is not always included as one of the prognostic equations in meteorological models, and air density is then a diagnostic variable; staggering of space and time grids in meteorological models for momentum and thermal variables can affect the accuracy of the solution of the continuity equation; furthermore, de-staggering in air quality (AQ) models can also introduce mass inconsistency. Various methods (both formal and ad hoc) have been used in practice to correct the error introduced by mass-inconsistent winds. This usually involves corrections to the winds to make them mass consistent (e.g., Odman and Russell, 1999) and adjustment to tracer fields through a correction to air density (e.g., Lu et al., 1997). Byun (1999) also suggested a number of ways of correcting the mass-inconsistency error in air quality models based on a formal analysis of his continuity equation formulated in a generalised coordinate system.

Besides mass-inconsistent input wind fields, advection schemes may also be a source of mass-conservation errors in AQ models. For example, semi-Lagrangian advection schemes, which have become quite popular in the past two decades, particularly in weather forecast models, for their good performance in stability and accuracy, do not formally conserve mass. However, the errors arising from this source are usually believed to be small and are not regarded by the meteorological community as a serious problem other than for climate studies, where long simulations are involved (Priestley, 1993; Gravel and Staniforth, 1994). Pudykiewicz et al. (1997) have shown that the semi-

[*] Wanmin Gong, Paul A. Makar, and Michael D. Moran, Meteorological Service of Canada, 4905 Dufferin Street, Downsview, Ontario, M3H 5T4, Canada

Air Pollution Modeling and Its Application XVI, Edited by
Borrego and Incecik, Kluwer Academic/Plenum Publishers, New York, 2004

383

Lagrangian scheme can performs well in chemical transport models due to its less diffusive nature compared to other Eulerian advection schemes.

In the present study, a simulation of regional aerosol over eastern Canada and the north-eastern US is carried out using a new air quality modelling system, AURAMS, which has been developed at the Meteorological Service of Canada. Spuriously high concentrations in modelled aerosol components were found to develop at isolated locations. Analysis shows that these "hot spots" arise from mass-conservation violations in the numerical advection scheme, i.e., the semi-Lagrangian advection scheme employed has difficulty conserving mass in the vicinity of steep gradients in tracer fields. In particular, a sustained feedback between cloud processes, which create sharp vertical gradients, and the mass-conservation error in the vertical advection leads to the development of these hot spots. It is then shown that this problem can be brought under control by ensuring mass conservation in the advection scheme.

In what follows, we will first describe the numerical simulation and the impact of mass-conservation errors in the advection scheme on the modelled aerosols; we will then discuss the ways to ensure mass conservation in advection, including a simple column-based mass constraint and the use of a blending scheme to achieve mass conservation in semi-Lagrangian advection; finally, we finish with a discussion and conclusions.

2. MODEL SIMULATION OF REGIONAL AEROSOLS AND MASS CONSERVATION PROBLEMS

The simulation of size- and composition-resolved tropospheric aerosols over eastern Canada and the northeastern U.S.A is carried out using AURAMS (A Unified Regional Air-quality Modelling System). AURAMS consists of these primary components: an emissions processor; a meteorological driver model; and a chemical transport model closely coupled with a size-resolved (sectional) and chemically speciated aerosol module. Processes represented in the chemical transport model include advection and diffusion,

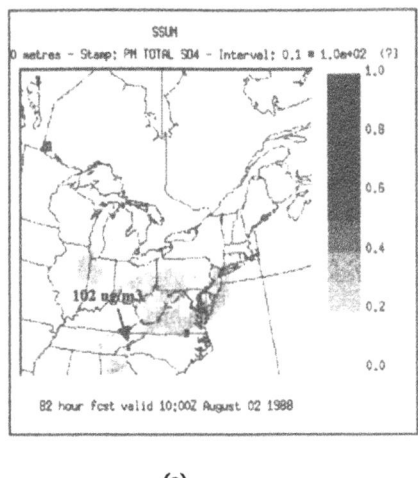
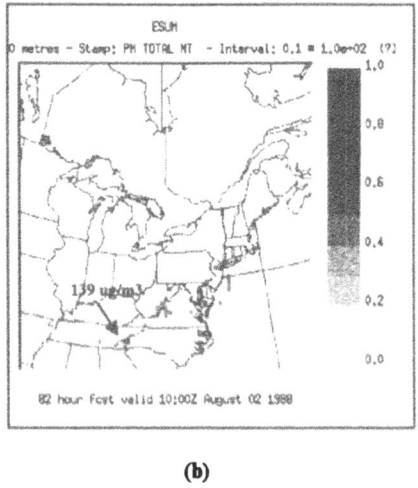

(a)   (b)

Figure 1. Modelled (a) particulate sulphate and (b) particulate nitrate concentrations (ug/m3) at the lowest model level 82 hour into the simulation, or at 10 Z, August 2, 1988.

gas-phase chemistry, gas-to-particle conversion (including secondary organic-matter aerosol formation and multi-phase thermodynamics of the sulphate-nitrate-ammonia system), aerosol microphysics, cloud processing of gases and aerosols (including activation, aqueous-phase chemistry, scavenging, and removal), dry deposition, and sedimentation. For a detailed description of the model see Moran et al. (1998) (also see Makar et al., 2003 and Bouchet et al., 2003 in these proceedings).

The AURAMS simulation was carried out for the period of August 1 to 6, 1988, with a horizontal resolution of 40 km (true at 60°N) and 28 levels unevenly spaced in the vertical between the surface and 30 km. The simulation starts from 00Z on July 30, 1988 to allow for a two-day spin-up. "Hot spots", i.e., isolated grid cells with concentrations much higher than (sometimes an order or two magnitudes of) any realistic peak values, were found in modelled aerosol fields 2-3 days into the simulation. Figure 1 shows two modelled aerosol components, sulphate and nitrate aerosols at hour 82 (10 Z, August 2, 1988). Hot spots can be seen in North Carolina near the border with Tennessee, at grid (27, 7). It should be pointed out that these hot spots did not develop when the precipitation-evaporation process was turned off in another simulation. Diagnosis of column-mass conservation (since evaporation acts to redistribute tracers in the vertical) and of total sulphur column budget after each cloud processes was carried out, and it was found that the evaporation process did not create erroneous mass in the vertical column and the sulphur budget was conserved after each of the cloud and aerosol processes. The column diagnosis was then extended to other operators in the chemical transport model, revealing that it was the advection operator that was responsible for the increase in column aerosol mass.

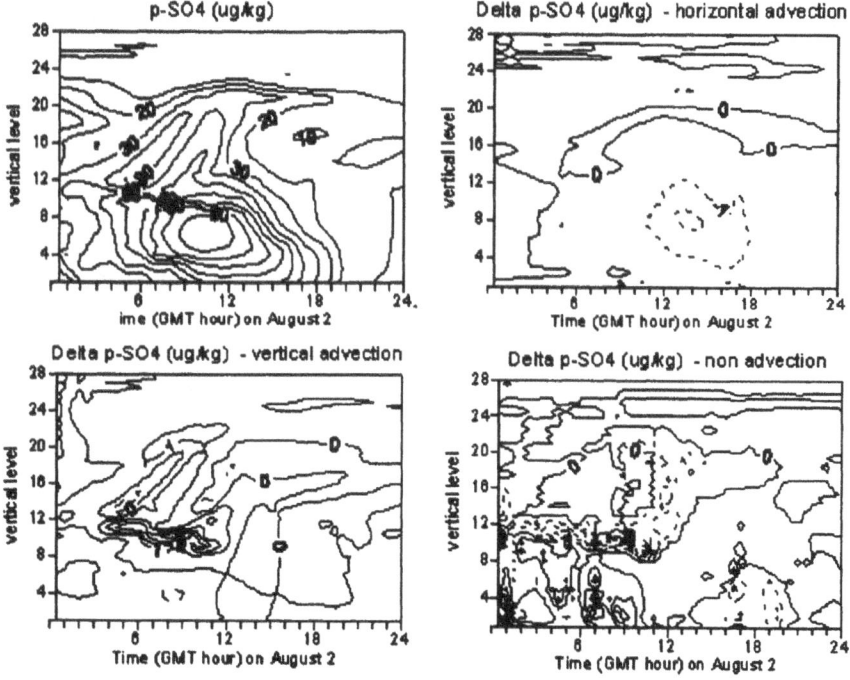

Figure 2. Top left : time-height plot of particulate sulphate (p-SO4) mass mixing ratio at grid cell (27, 7); top right: changes in p-SO4 due to horizontal advection (operator); bottom left: changes in p-SO4 due to vertical advection (operator); and bottom right: changes in p-SO4 due to other processes/operators.

Although a column mass change can be reasonably expected following advection, an increase in column mass due to advection at a location where the concentration is much greater than its surroundings throughout the vertical column is suspicious and an indication of erroneous mass creation. A first step to take to address this apparent mass-conservation problem was to make sure that the wind fields and the air density field used for the simulation are consistent. Following Byun (1999), mass consistency was ensured by first adjusting the vertical wind to satisfy the full continuity equation (for air density) in the model coordinate system and then applying a tracer concentration correction using advected air density. The enforcement of mass consistency in the wind fields resulted in some significant improvement to the hot-spot problem, primarily in the secondary organic aerosol component, with the impact mainly due to the adjustment of the vertical wind (Gong et al., 2002). However the hot-spot problem still remained after the corrections for mass-consistent wind, though to somewhat a lesser degree.

After making the mass-consistency corrections, some of the previous diagnostic procedures were repeated and advection was still found to be the only operator contributing to the increase in column mass at the hot-spot locations, which then led us to examine the advection scheme itself. AURAMS employs a semi-Lagrangian non-oscillatory advection scheme (Pudykiewicz et al., 1997; Smolarkiewicz and Pudykiewicz, 1992). The non-oscillatory property of the solution is achieved through an approach derived from the flux-corrected-transport idea (Smolarkiewicz and Grell, 1992), which controls under-shoot and over-shoot problems but does not ensure mass conservation in the solution. This can be particularly problematic in advection of a tracer field with steep gradients (discontinuity). Figure 2 shows time-height cross-sections of sulphate aerosol mass mixing ratio and incremental changes in the mixing ratio due to various operators at the hot-spot location, grid cell (27, 7). Here we have split the advection operator into two separate sub-operators, horizontal followed by vertical, in order to attribute the problem more clearly. This analysis revealed that the significant increase in mass mixing ratio in a region where sharp vertical gradients are found is contributed by the vertical advection, while the horizontal advection has very little impact on the change in the mass mixing ratio in the column. One of the benefits of splitting the advection operator is that one can expect the vertically-integrated mass to be conserved as far as the vertical advection operator (not component) is concerned (a case of 1-D advection), with consideration of fluxes through the lower and upper boundaries.

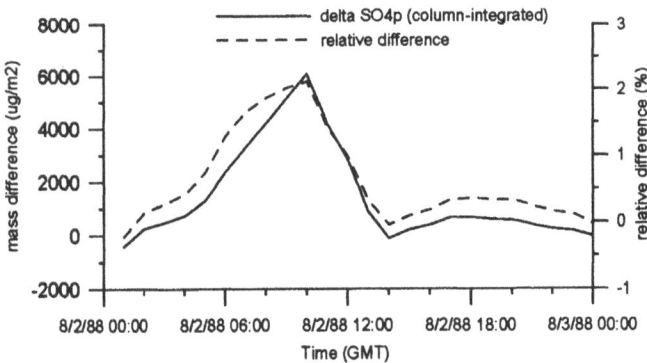

Figure 3. Change (absolute and relative) in column-integrated particulate sulphate mass at grid (27, 7) after vertical advection operator (at each 450-second time step).

Figure 3 shows the change in column-integrated mass (accounting for the boundary fluxes) following the vertical advection for particle sulphate. It can be seen that, corresponding to the period when significant increase in mass mixing ratio is seen in the region of steep gradients due to vertical advection (bottom-left panel in Figure 2), there is an excess of column-integrated mass generated by the vertical-advection operator (as much as 2 % of the column mass per time step). It can also be noted from Figure 2 that there is a sharp decease in the sulphate aerosol mixing ratio due to the 'non-advection" operators (bottom-right panel), corresponding to the increase due to vertical advection. It was determined that this decrease is largely due to the cloud-to-rain conversion process near the bottom of the cloud deck persisted throughout this period. The increase in mixing ratio in the region below the cloud base is due to evaporation. The picture emerging from the above diagnosis is that, for this case, the steep vertical gradient in particle mass mixing ratio is maintained by the cloud-to-rain conversion and below-cloud evaporation process, which leads to significant accumulation of mass-conservation errors in the numerical advection scheme in the vicinity of the steep vertical gradients. Without the mechanism to maintain the gradient the mass-conservation error would dissipate with the dampening of the gradient. Note that this explanation is consistent with the previous test in which the evaporation was turned off and no hot spot occurred. (More detailed analysis on the mass-conservation errors is documented in Makar et al., 2002).

3. METHODS TO CONTROL MASS-CONSERVATION ERRORS

3.1 Simple Column Mass Control (Scaling)

A simple and straightforward way of ensuring mass conservation for the (1-D) vertical advection operator is to perform a scaling on the final solution so that the adjusted solution yields the same column-integrated mass (accounting for the boundary fluxes) as that before the vertical advection operator. This approach was used as a quick way to test whether mass-conservation errors associated with the vertical-advection operator lead to the feedback observed in the model results and as a temporary correction to allow the continued use of the current interpolant and advection algorithm.

Specifically, for each tracer, the vertically-integrated mass in each grid column was calculated both before and after the vertical advection operator (corrected for boundary fluxes). The mass mixing ratios at each level were then multiplied by the ratio of the column mass prior to vertical advection to the column mass subsequent to vertical advection, for that column. An obvious weakness for this method is that, while mass conservation may be violated only locally at a few levels, the post-vertical-advection adjustment is applied at all levels in a grid column.

3.2 Application of a Sub-Optimal Blending Scheme for Achieving Mass Conservation in Semi-Lagrangian Advection

While the use of the above simple column mass control/scaling demonstrated (see next section) that by ensuring mass conservation the development of hot spots, due to the interaction between the mass-conservation error in the advection scheme and the discontinuous nature of the cloud processes, was prevented, it is a rather rudimentary solution to the problem as pointed out above.

Priestley (1993) attempted to address the lack of conservation in semi-Lagrangian scheme for advecting passive scalars. He proposed an approach that combines, in a sub-optimal fashion, linear interpolation and higher-order interpolation by imposing invariance of the first moment of the solution (i.e., mass in the case of passive tracer) as a constraint. The approach follows closely the method proposed by Bermejo and Staniforth (1992) for achieving monotonicity in semi-Lagrangian schemes. Gravel and Staniforth (1994) further extended Priestley's conservation scheme to the coupled shallow-water equations.

In the present work, we attempt to apply Priestley's scheme to the vertical advection in order to achieve mass conservation. If we denote U^L as the low-order solution and U^H as the higher-order solution (mixing ratio, in the present case), the blended solution $U^{M,C}$ would be

$$U^{M,C} = \alpha U^H + (1-\alpha)U^L,$$

where α is chosen so that

$$0 \le \alpha \le \alpha^{max},$$

while enforcing (in 1-D) the constraint

$$\int U^{M,C}(z)\rho(z)dz + F_{boundary} = \int U^0(z)\rho(z)dz = C,$$

where α^{max} ($0 \le \alpha^{max} \le 1$) is the optimal blending parameter for ensuring a monotonic solution U^M; $F_{boundary}$ accounts for the fluxes through the top and bottom boundaries due to advection. In the current model coordinate system the transformed vertical velocity is always zero at the lower boundary. Therefore, $F_{boundary}$ can be evaluated from

$$F_{boundary} = U(Z_T)W(Z_T)\rho(Z_T)J_T\Delta t,$$

where W denotes the vertical velocity in the transformed vertical coordinate (Z) and J is the transformation Jacobi. Following Priestley (1993), the condition for maximising α becomes

$$\sum_k \alpha_k (U_k^H - U_k^L)\rho_k\Delta Z_k + \alpha_T (U_T^H - U_T^L)\rho_T W_T J_T \Delta t$$

$$= C - \sum_k U_k^L \rho_k \Delta Z_k - U_T^L \rho_T W_T J_T \Delta t.$$

In our implementation, the second-order solution (based on a Taylor-series expansion), the linear solution, and the non-oscillatory solution from the original semi-Lagrangian scheme (Pudykiewicz et al., 1997) are used for U^H, U^L, and U^M respectively. The iteration for α_k, subject to the condition above, is carried out for every grid column following the same algorithm as described in Gravel and Staniforth (1994), which ensures that the sub-optimal blending only applies to the grid cells contributing to mass-conservation violations.

4. RESULT AND DISCUSSION

Figure 4 shows the time-height cross-sections of mass mixing ratio for sulphate aerosols and secondary organic aerosols at grid cell (27, 7) for August 2, from both the column-mass scaling and the sub-optimal blending runs compared to the original run. (Note that these runs were all conducted for the 4th day of the simulation only using restart fields). It can be seen that the mass conservation in vertical advection from both methods resulted in significant reductions in mixing ratio, up to 40% for sulphate and over 50% for secondary organic aerosols in peak values. The reduction in mixing ratio is seen to be greater from the column-mass scaling than from the sub-optimal blending particularly for sulphate aerosols. This is mostly due to the fact that the effective adjustment to the solution is spread out into the entire grid column and is proportional to the magnitude of mixing ratio at a given level in the case of the column scaling whereas the adjustment is more localised in the case of the blending method. This is illustrated in Figure 5 where, as an example, the adjustments through the vertical column at grid cell (27, 7) at time-step 72 (or 09Z on August 2, 1988) are compared for the two methods. It clearly shows that the adjustment from the blending approach is more focused at the levels where the largest mass-conservation errors are expected (in the vicinity of steep gradient). Makar et al. (2002) showed that by ensuring mass conservation in the vertical advection operator using column scaling, not only was the development of hot spots prevented but mass was also recovered at places where mass had been lost due to mass-conservation errors in the previous advection which led to better agreement with observations. Similar impacts are also seen from the sub-optimal blending scheme,

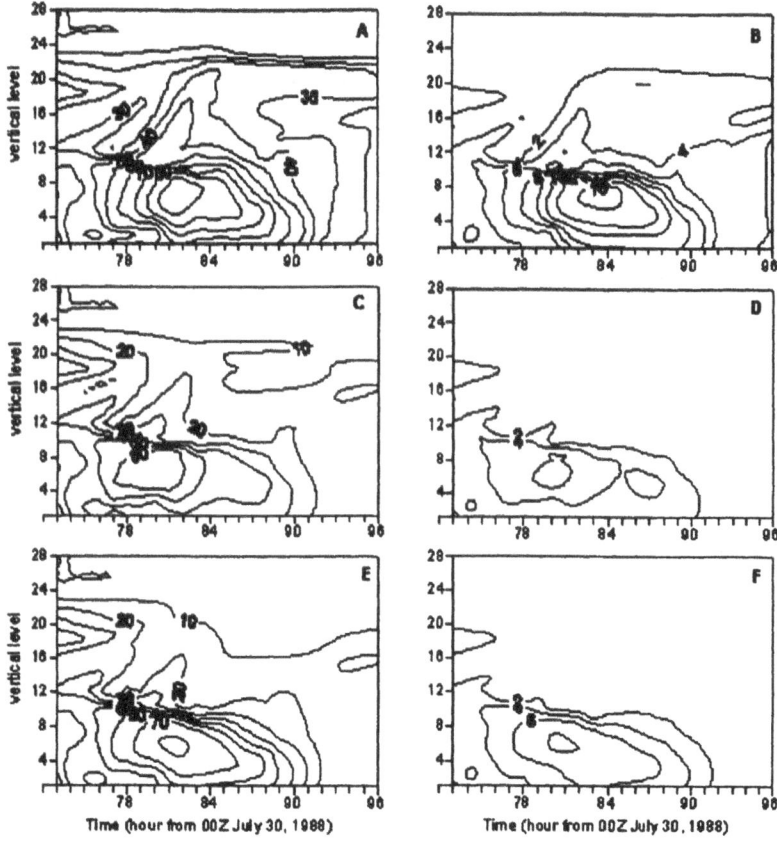

Figure 4. Time-height plot of mixing ratios (ug/kg) for p-SO4 (left column) and secondary organic particles (right colum) at grid cell (27, 7) for August 2, 1988. A and B – original advection with mass-conservation errors; C and D – column scaling; E and F – sub-optimal blending.

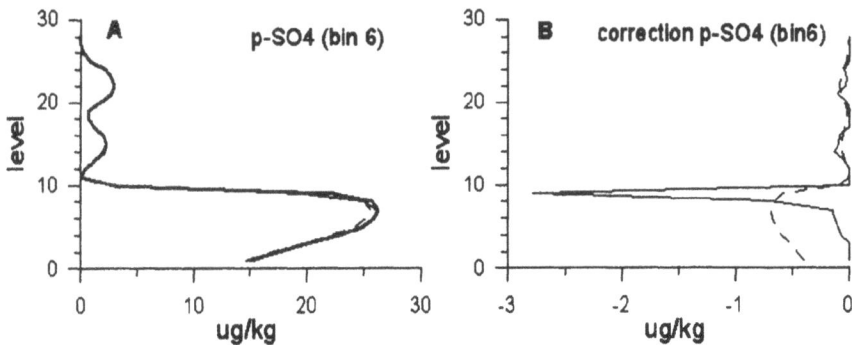

Figure 5. A. Vertical profile of p-SO4 mixing ratio (for size bin 6) at grid cell (27, 7) at 09Z August 2 : thick solid line – the original solution before correction; thin solid line – blended solution to ensure mass-conservation; dashed line – corrected with column scaling. B. The corresponding adjustment to the original solution using the blending scheme (solid) and using the column scaling method (dashed).

though with some differences in the magnitude. As indicated in Figure 5, the column scaling approach may be over-correcting closer to the surface, where the mixing ratios are greater, while under-correcting where the mass-conservation errors actually occur. The mass-conservation corrections also have a significant impact (not included here) on modelled tracers other than aerosols, e.g., surface ozone.

Statistics on the mass-conservation errors in the advection operators over the model domain were examined. The average mass-conservation error due to the vertical advection operator (without the mass-conservation correction) is less than 0.05% in magnitude (in a single column per time step) for most of the aerosol components over the model domain, whereas the magnitude (positive or negative) can be up to 4 – 7 % in particular locations. This suggests that although the mass-conservation errors in a semi-Lagrangian scheme may be small in a globally- or domain-wise-averaged sense, locally the errors can be significant in the presence of large spatial gradients. The present study shows that these errors can interact with other physical/chemical processes, leading to considerable deterioration in model results. In the present case the error can grow to as much as 20 - 100% over a few hours (or tens of time steps).

5. CONCLUSIONS

In this study, we have shown that mass-conservation errors due to numerical advection schemes (semi-Lagrangian in this case) can significantly deteriorate air-quality model results, especially through interactions with other processes that lead to steep gradients in the tracer fields. Methods of controlling such errors were explored, including a simple column-mass scaling and a sub-optimal blending scheme. It is shown that, by ensuring mass conservation by the advection operator, spurious mass increases that could lead to the development of tracer "hot spots" are prevented. The correction for mass-conservation errors works both in the case of mass excess (erroneous creation of mass) and in the case of mass deficit (erroneous loss of mass). In addition, the blending scheme seems preferable to the column-mass scaling correction for maintaining mass conservation since it adjusts the solution only locally where mass-conservation errors are found. Further work to implement and evaluate the full 3-D quasi-conservative semi-Lagrangian advection algorithm in the model appears warranted.

The study also suggests that the requirement for numerical advection schemes to be mass conserving may be even more important for modelling atmospheric aerosols than for other applications, since aerosol growth and removal processes can often lead to (and sustain) steep gradients in tracer fields. Better advection schemes that are mass-conserving as well as monotone and shape-preserving will be explored in the future.

ACKNOWLEDGEMENT

We thank the other members of the AURAMS development team, particularly Drs. V. Bouchet and S. Gong, for helpful discussions. Dr. S. Gravel of RPN/MSC kindly provided the code for the original sub-optimal blending algorithm for mass conservation.

REFERENCE

Bermejo, R. and A. Staniforth, 1992. The conversion of semi-Lagrangian advection schemes to quasi-monotone schemes. *Mon. Wea. Rev.*, 12, 2622-2632.

Bouchet, V.S., M. D. Moran, L-P. Crevier, A. P. Dastoor, S. Gong,, W. Gong, P.A. Makar, S. Menard, B. Pabla and L. Zhang, 2003. Wintertime and summertime evaluation of the regional PM air quality model AURAMS, these Proceedings.

Byun, D.W., 1999. Dynamically consistent formulations in meteorological and air quality models for multiscale atmospheric studies. Part II: Mass conservation issues. *J. Atmos. Sci.*, 56, 3789-3807.

Côte, J. and A. Staniforth, 1988. A two-time level semi-Lagrangian semi-implicit scheme for spectral models. *Mon. Wea. Rev.*, 116, 2003-2012.

Gong, W., P.A. Makar, and M.D. Moran, 2002. Mass-consistency errors and corrections for AURAMS. Unpublished report, available from Air Quality Research Branch, Meteorological Service of Canada, Downsview, Ontario, 30 pp.

Gravel, S. and A. Staniforth, 1994. A mass-conserving semi-Lagrangian scheme for the shallow-water equations, *Mon. Wea. Rev.*, 122, 243-248.

Lu, R., R.P. Turco and M.Z. Jacobson, 1997. An integrated air pollution modelling system for urban and regional scales: 1. Structure and performance, *J. Geophys. Res.*, 102, 6063-6079.

Makar, P.A., V. Bouchet, L.P. Crevier, S. Gong, W. Gong, S. Menard, M. Moran, B. Pabla, S. Venkatesh, 2003. AURAMS runs during the Pacific2001 time period – a model/measurement comparison., these Proceedings.

Makar, P.A., W. Gong and M.D. Moran, 2002. Mass-conservation corrections for AURAMS. Unpublished report, available from Air Quality Research Branch, Meteorological Service of Canada, Downsview, Ontario, 24 pp.

Moran, M. D., A. P. Dastoor, S.-L. Gong, W. Gong and P. A. Makar, 1998. Conceptual design for the AES unified regional air quality modelling system. Air Quality Research Branch, Meteorological Service of Canada, Downsview, Ontario M3H 5T4, Canada.

Odman, M.T. and A.G. Russell, 1999. Mass conservative coupling of non-hydrostatic meteorological models with air quality models, in *"Air Pollution Modelling and Its Applications XII"*, eds. S.-E. Gryning and E. Batchvarova, 651-660.

Priestley, A. 1993. A quasi-conservative version of the semi-Lagrangian advection scheme, *Mon. Wea. Rev.*, 121, 621-629.

Pudykiewicz, J.A., A. Kallaur and P.K. Smolarkiewicz, 1997. Semi-Lagrangian modelling of tropospheric ozone, *Tellus*, 49B, 231-248.

Smolarkiewicz, P.K. and G.A. Grell, 1992. A class of monotone interpolation schemes. *J. Comp. Phys.*, 101, 431-440.

Smolarkiewicz, P.K. and J.A. Pudykiewicz, 1992. A class of semi-Lagrangian approximations for fluids. *J. Atmos. Sci.*, 49, 2082-2096.

DISCUSSION

M.SOFIEV   1.If the advection scheme does not work- why not replace it with the correct one?

2.Practically any grid transformation ruins the vertical velocity as well as continuity equation balance. The only way to keep things under control is to compute the vertical velocity before the transformation corrections into the horizontal wind components.

W. GONG   1. Of course we could try with different advection schemes. For example the flux-based schemes are known to be conservative. However we may lose on efficiency and other desired properties. We inherited the semi-Larangian advection scheme from our meteorological model. For the sake of consistency we would prefer to use the same type of advection scheme for tracer transport as in the meteorological driver model. This would be particularly important when we eventually integrate our air quality model in-line with our meteorological model. Semi-Lagrangian advection schemes are know to be very efficient and less diffusive. However it does have deficiency in conservation, which, as we have shown here, can be a serious problem when it comes to air quality modelling. We have also shown that the use of a mass conservation correction to the existing semi-Lagrangian advection, based on a sub-optimal blending algorithm, is effective in achieving mass conservation. It would be desirable to have a more advanced advection scheme that is efficient, monotone, shape-preserving, and conservative for both our meteorological and air quality models.

2. Regarding the mass consistency issue, the problem arises, in our case, not so much from coordinate transformation (we use the same coordinate as in the meteorological model) but rather from de-staggering of the grids. To adjust wind fields to satisfy the continuity equation for air density, one could choose to adjust either the horizontal wind or the vertical wind. We have chosen to adjust the vertical wind as suggested in several existing work (Byun, 1999; Odman et al,. 1999).

E.GENIKHOVICH   A possible reason for the problem of non-conservation of mass couldbe in approximating separately different governing equations and different terms in these equations. Did you work with variational numerical schemes proposed by Marchuk, Penenko, Aloyau and their collaborators?

W. GONG

There is no doubt that errors may arise from solving the governing equation separately in an operator-splitting fashion. However we have shown that, in the case of our "hot-spot" problem, the mass conservation error arises from the advection scheme having difficulty conserving mass in the vicinity of sharp vertical gradient in concentration (the sharp gradient is, in this case, maintained by another process in the model). We have not worked with the variational schemes that you mentioned.

T.ODMAN

Did you try checking the mass conservation property of your model by using a buoyant (non-deposited) inert tracer: injecting a controlled amount of this tracer, blowing it around with the wind fields used in the model, until the tracer reaches the lateral or top boundaries and checking it the injected amount is preserved?
Getting rid of the hot spots does not guarantee that you have a mass conservative model. Conversely, the presence of hot spots does not necessarily mean there is a mass conservation problem. Hot spots can occur over lakes, rivers, or due to chemical formation if your indicator species is $SO4$ or SOA

W. GONG

We were considering a test on mass conservation by advecting inert tracer but never got round to doing it. However this type of test would easily indicate mass conservation problem due to mass-inconsistent wind but may not indicate the mass conservation problem that we have with the advection scheme. As shown, part of our "hot-spot" problem was due to the semi-Lagrangian advection scheme having difficulty conserving mass in the vicinity of large gradient. This type of mass-conservation problem would not show up if we were to advect a uniformly distributed tracer field. Regarding the "hot-spot" problem, it is true that it can be due to other causes. Nevertheless we have tracked down some mass conservation problems in advection that were contributing to the hot-spot problem we were having and we attempted to correct them as shown in this talk.

KALMAN FILTER ANALYSIS OF BOUNDARY LAYER OZONE IN EUROPE WITH A CHEMISTRY TRANSPORT MODEL

Remus G. Hanea,[1,2] Guus J.M. Velders[2] and Arnold W. Heemink[1]

1. INTRODUCTION

Elevated concentrations of ozone in the boundary layer can cause adverse affects to human and ecosystems. Ozone in the boundary layer is formed by chemical reactions of ozone precursors, nitrogen oxides (NO_x), volatile organic compounds (VOCs), carbon monoxide, and methane, under the influence of sunlight. The impact of ozone is not limited to the area close to where the ozone precursors are emitted. Boundary layer ozone is measured at various locations in Europe to obtain a regional image of its concentration.

Atmospheric chemistry-transport model have been developed to better understand the processes controlling the formation of ozone and to assess the potential effects of emissions reductions on the ozone concentration. Since the knowledge on the various physical phenomena is not complete and because any deterministic model is an approximation and simplification of the real dynamics and chemistry, models results may differ from observations. Using only observations of ozone a complete picture of the ozone concentrations in Europe can also not be obtained, due to measurements errors and the limited number of observations. Using data assimilation a better estimate of the real ozone concentrations can be obtained by combining available measurements with the information provided by a model. The method has already proved to be useful in some other fields of applications (see for example Verlaan and Heemink, 1996). In the recent years applications of data assimilation in the field of atmospheric chemistry and transport modeling have received a lot of attention (see for example, van Loon and Heemink, 1997; Segers et al., 2000; Elbern et al., 1997; Elbern and Schimdt, 2000). The basic idea of data assimilation is combining two sources of information: the model and the available measurements. It assumes that both of them are not perfect, and therefore are a source of uncertainties. The data assimilation technique we employ is the Kalman filter. Two algorithms of the Kalman filter are used here, the Ensemble Kalman filter (ENKF) and the Reduced Rank Square Root Kalman filter (RRSQRT). To apply the Kalman filter

[1] Delft University of Technology, PO Box 5031 GA Delft, Netherlands.
[2] Netherlands Environmental Assessment Agency, PO Box 3720 BA, Bilthoven, Netherlands.

Air Pollution Modeling and Its Application XVI, Edited by
Borrego and Incecik, Kluwer Academic/Plenum Publishers, New York, 2004

395

technique we need first to specify and to define the uncertainties in the measurements and in the model parameters in statistical terms.

In this work results are presented of the first simulations with the chemistry-transport model EUROS in combination with the Kalman filter for deriving assimilated ozone concentrations in Europe. The performance of the algorithms (ENKF and RRSQRT) is compared with respect to their ability to yield close agreement between assimilated and measured ozone concentrations. For the assimilation, the emissions of NO_x and VOC, the photolysis rates of NO_2 and ozone, and the deposition rate of ozone are considered as uncertain parameters and optimized by the Kalman filter. The influence of the parameters on the performance of the assimilation is studied. Ultimate goal of the data assimilation is to obtain the optimal estimation of the real state of atmosphere and the corresponding model parameters. In section 2 the data assimilation technique is discussed as well as the implementation of the ENKF and RRSQRT algorithms. A short description of the EUROS model is given in section 3, followed in section 4 by the definition of the statistical environment used in the simulations. The results of the data assimilation calculations are presented and discussed in section 5 and the conclusions in section 6.

2. DATA ASSIMILATION

The link between model and measurements is represented by data assimilation. This technique can be seen as an inverse modeling problem. We need to specify first the uncertainties in the model, and then use the observations that are available at every time to try to reduce these uncertainties and to get a good estimation of the true state of the model. This is achieved by extending the deterministic model to a stochastic model. Instead of a deterministic model given by

$$x(t_{k+1}) = f(x(t_k))$$ (1)

where $x(t_k)$ denotes the state of the model at time t_k and f is the model operator, we use the stochastic extension like

$$x(t_{k+1}) = f(x(t_k), w(t_k))$$ (2)

where $w(t_k)$ is the noise input vector. This vector represents the uncertainties in the model which need to be specified it. For practical purposes the noise parameters are assumed to be Gaussian and "white", which means that they can be described by a mean and covariance and they do not depend on the past, i.e. have the Markov property: $w(t_k) \sim N(0, Q(t_k))$. The observations also do not give complete information about the true state. Uncertainties in measurements need to be specified. The vector y^o that represents the set of measurements available at time t_k can be written as

$$y^o(t_k) = H(t_k)x(t_k) + v(t_k)$$ (3)

where the observations operator $H(t_k)$ assigns the ozone level in the surface layer of a grid cell to an observation and $v(t_k)$ is the measurement error. In this study it is assumed that the measurements errors are Gaussian and uncorrelated in time and space. For the true state there are two estimates: x^f the forecast of the model and y^o the measurements vector. In order to quantify the difference between the model and the observations we define the representation error. Let $d = H^T(t_k) - y^o(t_k)$ be the residue of the forecast at time t_k. The final goal of an assimilation procedure is to obtain a time series of assimilated or analyzed state $x^a(t_k)$ given the model and the measurements, with assimilated residues as small as possible.

2.1. Kalman filter

Among the most known methods to apply data assimilation are the variational methods and sequential methods. In Elbern et al. (1997) and Elbern and Schimdt (2000) a variational data assimilation is implemented using a 4D-var filter technique for a chemical transport model showing that this method can be applied successfully for nonlinear atmospheric models. Other examples of filter techniques and other ways to apply data assimilation can be found in the Proceedings of the SODA workshop on Chemical Data Assimilation (1999). Apart from variational methods, sequential methods have been developed from the beginning of 1980s (Gill et al., 1981). In our case we will use Kalman filter techniques for applying data assimilation. In the beginning the Kalman filter techniques were used especially in meteorological applications (Evensen, 1994). The good results of the implementation of this method led to the introduction of similar techniques in related fields, like hydrodynamics (Verlaan, 1998) and atmospheric chemistry (Segers, 2000). The first step in a Kalman filter algorithm is to specify the initial distribution for the true state. The true state is denoted with a superscript t:

$$x^t(t_0) \sim N(x^f(t_0), P^f(t_0)) \tag{4}$$

The second step, **Forecast step,** is to define the error between the true state $x^t(t_{k+1})$ and the model forecast $M(t_k) x^t(t_k)$, where $M(t_k)$ is the model operator:

$$x^t(t_{k+1}) = M(t_k)x^t(t_k) + w(t_k) \tag{5}$$

where $w(t_k) \sim N(0, Q(t_k))$ is the model error. The stochastic model completely defines the evolution of the distribution of the true state:

$$x^f(t_{k+1}) = E(x^t(t_{k+1})) = M(t_k)x^t(t_k) \tag{6}$$

$$P^f(t_{k+1}) = E[(x^t(t_{k+1}) - x^f(t_{k+1}))(x^t(t_{k+1}) - x^f(t_{k+1}))^T] = \\ = M(t_k)P^f(t_k)M(t_k)^T + Q(t_k) \tag{7}$$

The third step, **Analysis step,** in the filter algorithm is the analysis of the available data. If measurements are available at a certain time the mean and the covariance are replaced by analyzed equivalents given the new information. The observations at time t_k are stored in a vector:

$$y^o(t_k) = H(t_k)x(t_k) + v(t_k) \tag{8}$$

where $v(t_k) \sim N(0, R(t_k))$ is the representation error between observed data and true value. The analyzed values are calculated with a linear gain K

$$x^a(t_k) = x^f(t_k) + K(t_k)(y^o(t_k) - H(t_k)x^f(t_k))$$
$$P^a(t_k) = [I - K(t_k)H(t_k)]P^f(t_k)[I - K(t_k)H(t_k)]^T + K(t_k)R(t_k)K^T(t_k)$$
(9)

where the Kalman Gain is calculated by

$$K(t_k) = P^f(t_k)H(t_k)[H(t_k)P^f(t_k)H(t_k) + R(t_k)]^{-1}$$
(10)

The final result of the Kalman filter using the minimum variance gain is a time series of mean and covariance of the true state, equal to the conditional mean and covariance given all available data from the past:

$$x^a(t_k) = E[x^t(t_k) | y^o(t_k), y^o(t_{k-1}), y^o(t_{k-2}), \ldots]$$
$$P^a(t_k) = E[(x^t(t_k) - x^a(t_k))(x^t(t_k) - x^a(t_k))^T | y^o(t_k), y^o(t_{k-1}), y^o(t_{k-2}), \ldots]$$
(11)

These are the general equations for conventional Kalman filtering for linear systems. But in using this in for atmospheric chemistry there are some problems. $P^a(t_k)$ is very big (high computational burden) and is ill-conditioned, i.e. there is a large range for eigenvalues. To solve these problems some low-rank Kalman filter has to be used suitable for data assimilation in models with large state vectors. Two algorithms are used for this in combination with the EUROS model: Reduced Rank Square Root Kalman filter and Ensemble Kalman filter. Also, the relations in atmospheric chemistry are highly non-linear. To apply the theory described above a tangent linear model is needed.

2.2 Ensemble Kalman filter (ENKF)

This idea was introduced by Evensen (1994) and has been successfully used in many applications (Evensen and Van Leeuwen, 1996; Houtekamer and Mitchel, 1998). This Monte Carlo approach is based on representation of the probability density of the state estimate in an ensemble of possible states $\xi_1, \xi_2, \xi_3 \ldots \xi_N$. Each ensemble member is assumed to be a single sample out of distribution of the true state. Whenever necessary, statistical moments are approximated with sample statistics. Thus, the ensemble Kalman filter algorithm is based on representation of the probability density of the state estimate by a finite number of N randomly generated system states. An ensemble of N states $\xi_i^a(t_0)$ is generated to represent the uncertainty in $x(t_0)$. In the second step of the algorithm the stochastic model propagates the distribution of the true state:

$$\xi_i^f(t_k) = f(\xi_i^a(t_{k-1})) + w_i(t_k)$$
$$x^f(t_k) = \frac{1}{N} \sum_{i=1}^{N} \xi_i^f(t_k)$$
$$E^t(t_k) = [\xi_1^f(t_k) - x^f(t_k), \xi_2^f(t_k) - x^f(t_k), \ldots, \xi_N^f(t_k) - x^f(t_k)]$$
(12)

When the measurements become available the analysis step replaces the mean and the covariance with equivalent once calculated using the same classical algorithm this time in terms of $E^f(t_k)$. The advantages of this algorithm are that P^f is positive definite and that the linear tangent model is not required any more because the ensembles are propagated trough the model using the original operator like in Eq. (12). As a result the computational effort required for the ENKF is approximately N model simulations. The errors in the state are of statistical nature and decrease very slowly with the number of ensembles. This is a disadvantage of this Monte Carlo approach.

2.3. Reduced Rank Square-Root Kalman filter

Another approach for solving large scale Kalman filtering problems is to approximate the full covariance matrix of the state estimate by a matrix with a reduced rank. This approach was introduced by Cohn and Todling (1995, 1996) and Verlaan and Heemink (1995, 1997). Algorithms based on similar ideas have been applied by Pham (1998). The reduced rank approach can also be formulated as an ENKF where the q ensemble members are not chosen randomly, but in the direction of the q leading eigenvectors of the covariance matrix (Verlaan and Heemink, 1997). This algorithm can be applied and implemented because the matrix $P^a(t_k)$ is a low rank matrix. The algorithm approximates the errors covariance $P^{f,a}(t_k)$ for forecast and analyzed by

$$P^{f,a} = (V^{f,a})(D^{f,a})(V^{f,a})^T \tag{13}$$

where $V^{f,a}(t_k)$ contains the eigenvectors of the covariance matrix and $D^{f,a}(t_k)$ is the diagonal matrix with eigenvalues on the diagonal. The square-root of $P^{f,a}$ is denote by

$$L^{f,a} = (V^{f,a})(D^{f,a})^{1/2} \tag{14}$$

So, $L^{f,a}(t_k)$ will be a $n \times q$ matrix because it will take the first q eigenvalues of the matrix. Now the general Kalman filter algorithm can be rewritten in terms of $L^{f,a}$. In this case the advantage is that the original model operator is used for calculations and also the preserved characteristic of positive definite for P^f. This algorithm is more robust and need a large number of modes to be efficient.

3. EULERIAN CHEMISTRY-TRANSPORT MODEL EUROS

A number of models have been developed to describe the complex mechanism in the atmosphere in the formation of ozone. These models contain parameterizations of the various chemical, dynamical and radiative processes in the atmosphere, as well as information about the emissions of ozone precursors, nitrogen oxides, VOCs, carbon monoxide and methane. In this project we use the Eulerian atmospheric chemistry-transport model EUROS (EURopean Operational Smog, van Loon (1996), developed at RIVM (National Institute of Public Health and the Environment, Netherlands). The model is used for simulating emissions, transport processes, chemical transformation, and dry and wet deposition processes of various air-polluting components. The model can be used to examine the time and spatial behavior of SO_x, NO_x, O_3 and Volatile Organic

Compounds (VOCs) in the lower troposphere over Europe. The model area extends over a large part of Europe and uses a shifted pole coordinates system; i.e. the real North Pole is at 30^0 northern latitude and the equator has been shifted to 60^0 northern latitude in the new coordinate system. The grid consists of 52×55 grid cells with a $0.55^0 \times 0.55^0$ (about 60×60 km) longitude-latitude resolution. The vertical stratification of EUROS consists of four layers from the surface up to 3000 m. The height of the layers varies during the day due to meteorological processes.

In the model 15 species are taken into account: sulfur dioxide, sulfate, nitrogen oxide, nitrogen dioxide, ozone, hydroxyl radical, nitrate aerosol, ethane, butane, ethene, propene, xylene, isoprene, carbon monoxide, and nitric acid. Methane is also included, because it is important for ozone formation. Its value has been fixed to 1700 ppb. The EUROS model contains all the relevant physical and chemical processes: emissions, advection, vertical exchange due to fumigation processes, horizontal and vertical diffusion, dry and wet deposition, and chemistry. Several chemical schemes can be used in EUROS. For this first study we choose to use a simplified scheme with 15 chemical reactions, mainly because this limits the computation time for the simulations.

4. STOCHASTIC ENVIRONMENT

Knowledge of the model errors is crucial for a successful data assimilation simulation. Two types of model errors are identified; model input errors and modeling errors. The model input errors are due to errors in emissions, meteorological fields, and boundary conditions. The other type of errors is a consequence of the parameterization of the chemistry. In simulations with an atmospheric model the emissions are an important input parameter, but contain uncertainties. The emissions of NO_x and VOCs are taken here as uncertain parameters and put in a stochastic environment as

$$e_s(t_k) = \bar{e}_s^j(t_k)[1 + w_s(t_k)] \quad , s = \{NO_x, VOC\} \tag{15}$$

where $\bar{e}_s$ e is the deterministic value of the $s = \{NO_x, VOC\}$ emissions in a grid cell and w_s the noise parameter. Taking into account the complex chemistry of the EUROS model, the errors in chemical parameters will effect the computed concentrations. The photolysis rates of NO_2 and ozone are important for the formation and destruction of ozone and its diurnal cycle and therefore also considered as uncertain parameters. Another important parameter determining the ozone concentration is the deposition rate of ozone. The noise in the photolysis rates and in the deposition of ozone are implemented analogously as described in Eq. 15. For the emissions of NO_x and VOCs and for the deposition rate of ozone four different regions are defined for the noise parameters. This in shown in the equations with a superscript j attached to the deterministic values. The four different regions are 1) Netherlands, Belgium and Luxembourg, 2) Germany and Denmark, 3) France, and 4) United Kingdom and Ireland. These four regions are used to get a better explanation of the noise parameters and to obtain improved emissions for these four regions. The values for the noise parameters have been bounded between 0.2 and 5 (multiplication factors) in order to prevent unrealistic variations in the model input and instabilities in the model simulations. To improve the estimation of the assimilated concentrations of ozone and to reduce the uncertainties in emissions the model is

400

combined with available measurements. Time series of measurements at surface level are taken from the LML (Elzakker, 2001), EMEP (Hjellbrekke, 1998) and Airbase (EEA, 2002) databases. For the period studied (summer 1996) there are in total 135 measurement stations in Europe, marked as 'background" and "rural", with ozone surface data. This set of 135 stations is split in two: 65 stations are used for the assimilation process (assimilation stations) and the rest is used for diagnostic purposes (validation stations). The stations of these sets are chosen so that they are evenly distributed over Europe. For each ozone measurements it is assumed that the standard deviation of the error is equal to 10% of the measured concentration. This uncertainty accounts for both the uncertainty in the actual measured concentration at the specific station as well as for a representation error, reflecting the size of the grid cell.

5. RESULTS AND DISCUSSION

In this section several analysis are discussed to test, the combination between the Kalman filter and the EUROS model and to study the model sensitivity to several input parameters, i.e. number of modes, number of observations and specifications of the noise parameters. Several simulations were performed to see how the true state of the atmosphere improves using data assimilation with the Kalman filter algorithms. A simulation with the EUROS model and Kalman filter was performed for a period of one month, i.e. June 1996. For data assimilation, specification of the measurement and the model uncertainties are required. For this case, the model uncertainty is specified in terms of uncertain emissions for NO_x and VOCs, uncertain photolysis rates for NO_2 and O_3 and uncertain deposition rate for O_3. The standard deviation σ of w is set to 30% and the time correlation parameter to 24 hours for emissions of NO_x. For emissions of VOCs, which are in general believed to be more uncertain than that of the other components, the standard deviation was increased to 50% and the time correlation parameter remains the same. For both of the NO_2 and O_3 photolysis rates the standard deviation is set to 50% and the time correlation parameter is 6 hours. Finally, for deposition rate of O_3 σ is 30% and the time correlation parameter is 24 hours. The noise parameters were applied for all four regions for the emissions of NO_x and VOCs and for the deposition of O_3.

In the left panel of Fig.1 the measured, modeled (by EUROS) and assimilated ozone concentrations at measurement site Lagrano (Spain) are plotted. The assimilated time series for this station are in better agreement with the observations than the ones calculated by the model alone. This improvement can not only be observed at measurement sites of which the observations are used in the assimilation procedure (i.e. so called assimilation stations), it also holds for so-called validation stations. The right panel of Fig. 1 shows the concentration for the validation station Hellendoorn (Netherlands).

Table 1. Average absolute residuals ($\mu g\ m^{-3}$) averaged over all validation and assimilation stations in the Netherlands and in Europe for the 24 hour average, day time and night time average. ENKF 20 modes.

	24 hour		Day time		Night time	
	EUROS	KF mean	EUROS	KF mean	EUROS	KF mean
NL validation	26.1	17.7	26.7	17.4	25.6	17.9
NL assimilated	26.8	17.2	24.9	15.1	28.7	19.1
Europe validation	27.8	21.0	27.9	19.8	27.9	22.3
Europe assimilated	27.2	20.0	26.9	18.7	27.4	21.3

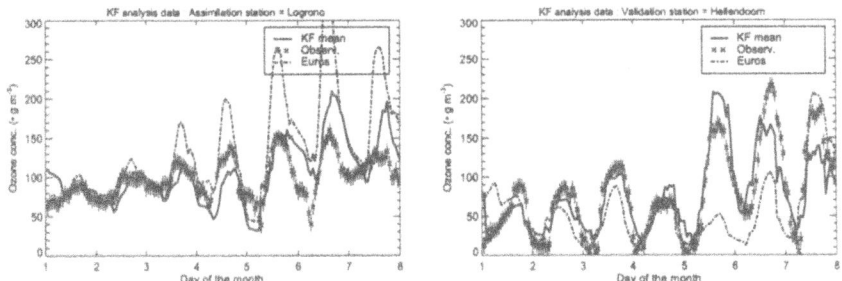

Figure 1. Ozone concentration (μg m^{-3}) at assimilation site Lagrano (left) and validation site Hellendoom (right) for the Kalman filter mean (solid line), the EUROS model (dash-dot), and the observations (crosses).

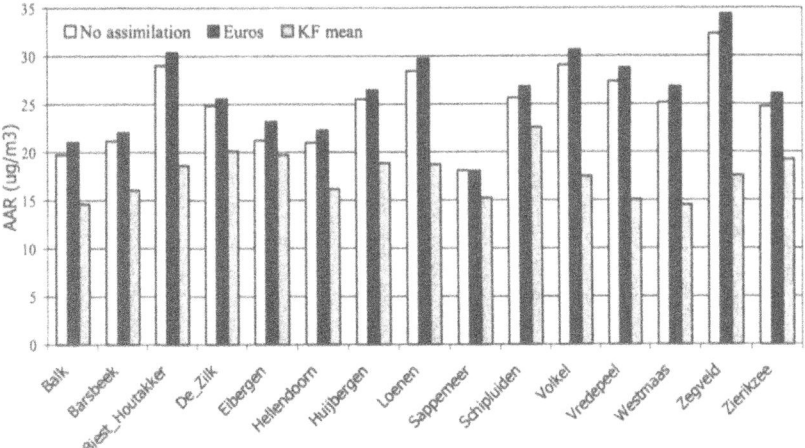

Figure 2. Diurnal average absolute residuals (μg m^{-3}) of the assimilation (ENKF 20 modes) for all the validation stations in the Netherlands for the model calculation, the Kalman filter mean and without assimilation.

To get a better idea about the improvement, the diurnal average residues of the ozone concentrations from simulations with and without assimilation are plotted in Fig. 2. For this case the measurements location are all the validation stations in the Netherlands. The residue is defined as the average of the absolute differences between the measured and modeled concentrations. The Figs 1 and 2 show that when data assimilation is applied, the residues in general decrease at both the assimilation and validation stations, indicating a proper functioning of the data assimilation. The average residues for validation and assimilation stations from the Netherlands and the whole Europe are presented in Table 1 for the 24 hours average, the daytime and night time average. The residues decrease by a significant amount for all times of the day.

Second part of the applications is about the impact of the input parameters on the performance of the assimilation. Several simulations were performed to test the sensitivity of the model; regarding the number of modes used in Kalman filter algorithms, the number of observations present in the model, and the specifications of the noise.

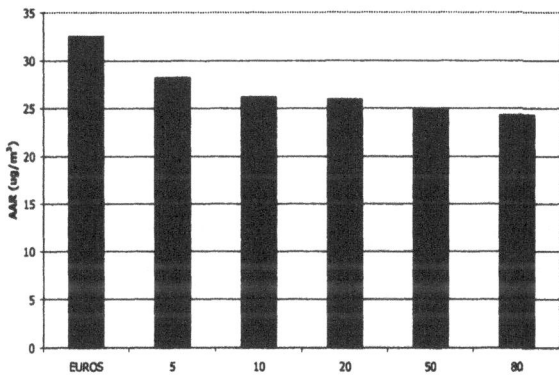

Figure 3. Average absolute residuals (µg m⁻³) for the ENKF algorithm with different numbers of modes for all the validation stations in Europe. The 'EUROS' bar is only a model run without assimilation.

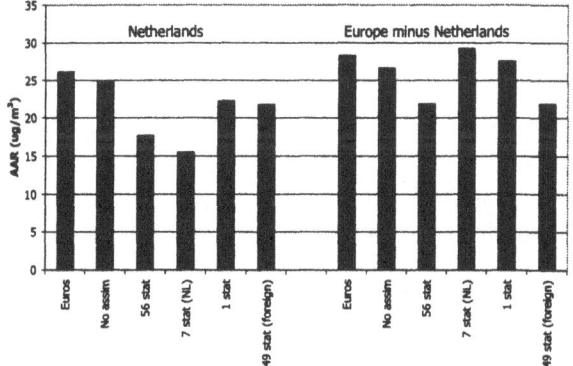

Figure 4. Average absolute residuals (µg m⁻³) for the assimilation (ENKF with 20 modes) with different numbers of observational sites for the average of the validation sites in the Netherlands (left) and in Europe without the Netherlands (right).

Computational efficiency is an important aspect of Kaman filter applications. The parameter, which is important for this, is the number of modes used in assimilation. The effect of the number of modes for the ENKF algorithm on the residuals is shown in Fig. 3. Shown here are the absolute residuals averaged over all the validation stations in Europe. The computational burden increases with the number of modes and the accuracy of the data assimilation procedure is also increases with the number of modes. Using 20 modes the largest improvements have been obtained, so more modes does not seem necessary for these assimilations.

Another parameter that has an important role in the data assimilation process is the number of observations used in the assimilation. One can ask, what is the impact of observations in Germany on the performance of the assimilation regarding the Netherlands? To answer this question a number of simulations were performed using different number of observations and looking at the performance of the assimilated concentrations in two different regions: the Netherlands and the rest of Europe.

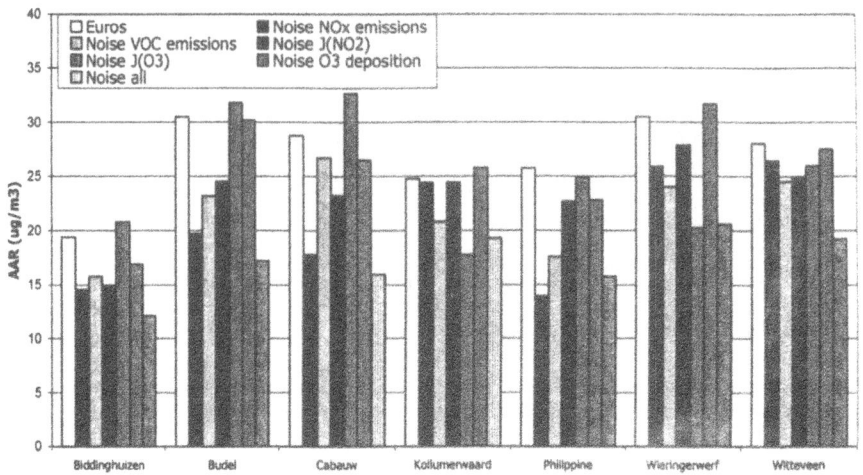

Figure 5. Average absolute residuals (μg m⁻³) with different noise parameters in the assimilation (ENKF with 20 modes). Shown are the averages for the assimilation stations in the Netherlands.

The performance of the assimilation was quantified by using the absolute average residuals. From Fig. 4 it can be observed that using 49 foreign (not in the Netherlands) stations for the assimilation there is an improvement of the information given by the model regarding the Netherlands site, but not so important than in the case of using only 7 stations in the Netherlands. It can also be observed that the quality of the information in the case of a simulation with 49 foreign stations is almost the same with in the case of a simulation with only one station in the Netherlands. In the right part of Fig. 4 the results of the same simulations are shown, but this time regarding to foreign site. The same pattern can be seen in the performance of the data assimilation. From this it can be concluded that the combination of the EUROS model and the data assimilation performs well locally, and the impact of the information coming from further away (i.e. foreign sites in this case) is limited for the Netherlands.

To know how much the uncertainties in noise parameters (emissions, photolysis and deposition) can be reduced the sensitivity of the assimilation to the noise parameters is studied. The value of the noise parameters gives an idea about the emissions that are used by the assimilations. In Fig. 5 the average values of the absolute residues for a uniform set of validation stations in the Netherlands are given using a different noise parameter in each simulation. The largest impact on the assimilation comes from the noise in the emissions of NO$_x$ and VOCs. The best estimation of the true state of the atmosphere is obtained in the case when the noise was uses in all parameters. Therefore, it can be concluded that even though the results using noise only in the emissions of NO$_x$ and VOCs are good, taking in account also noise in the other parameters yields significantly better results.

6. Summary

Large-scale numerical air pollution models can be used to predict ozone concentrations in the boundary layer from known emissions of ozone precursors. Ozone concentrations can also be obtained from observations at ground level. Both types of data contain uncertainties. We have shown that ozone concentrations can be improved using a Kalman filter in combination with an atmosphere model. The accuracy of the data assimilation increases with the number of modes. Using 20 modes the largest improvements are obtained, so more modes does not seem necessary for these assimilations. The combination of the EUROS model and the data assimilation performs well locally and the impact of the information coming from further away is limited. The uncertainties in emissions of NO_x and VOCs have the largest impact in the assimilation process.

7. REFERENCES

Cohn, S.E., and Todling, R., 1995, Approximate data assimilation schemes for stable and unstable dynamics, *J. Meteorological Soc. Of Japan,* **74** (1), 63-75.

EEA, 2002, Airbase from the European Environment Agency, Copenhagen, http://www.etcaq.rivm.nl.

Elbern, H., and Schmidt, H., 1997, Variational data assimilation for tropospheric chemistry modeling, *J. Geophys.Res.,* **102**(D13), 15967-15985.

Elbern, H., Schmidt, H., Talagrand, O., and Ebel, A., 2000, 4D-variational data assimilation with an adjoint air quality model for emission analysis, *Environmental Modelling and Software.,* **15** (6-7), 539-548.

Evensen, G., 1994, Sequential data assimilation with a nonlinear quasi-geostrophic model using Monte Carlo methods to forecast error statistics, *J. Geophys.Res.,* **99**(C5), 10143-10162.

Evensen, G., van Leeuwenl, P.J., 1996, Assimilation of Geosat altimeter data for the Agulhas Current using Ensemble Kalman filter with quasi-geostrophic model., *Mon. Weather Rev.,* **124**, 85-96.

Elzakker, B.G., 2001, Monitoring activities in the Dutch national air quality monitoring network, RIVM report no. 723101055, RIVM, Bilthoven, Netherlands.

Ghil, M., Cohn, S., Tavantzis, J., Bube, K., and Isaacson, E., 1981, Applications of estimation theory to numerical weather prediction, *Dynamic Meteorology: Data Assimilation Methods*, 139-224.

Heemink, A., Verlaan, M., and Segers, A., 2001, Variance reduced Ensemble Kalman filtering, *Mon. Weather Rev.,* **129**(7), 1718-1728.

Hjellbrekke, A.G., 1998, Ozone measurements 1996, EMEP/CCC 3/98, EMEP website http://www.emep.int.

Houtekamer, P., and Mitchell, H. L., 1998, Data assimilation using an Ensemble Kalman Filter technique, *Mon. Weather Rev.,* **126**, 796-811.

Pham, D.T., Verron, J., and Roubau, M.C., 1998, A Singular Evolutive Extended Kalman Filter for data assimilation in oceanography, *J.of Marine Systems*, 16(3-4), 323-240.

Segers, A., 2002, Data assimilation in atmospheric chemistry models using Kalman filtering, PhD thesis, Delft University of Technology, Delft, Netherlands.

Segers, A., Heemink, A., Verlaan, M., and van Loon, M., 2000a, A modified rrsqrt-filter for assimilating data in atmospheric chemistry models, *Environmental Modelling and Software*, **15**(6-7), 663-671.

van Loon, M., 1996, Numerical methods in smog prediction, PhD thesis, University of Amsterdam, Amsterdam.

van Loon, M., Builtjes, P., and Segers, A., 2000, Data assimilation applied to LOTUS: First Experience, *Environmental Modelling and Software*, **15**(6-7), 603-609.

Veders, G., Matthijsen, J., van Loon, M., van Oss, R., Sauter, F., and Segers, A., 2001, Smog forecasts with a chemistry transport model using data assimilation; possibilities of GOME tropospheric ozone observations, NRSP-2 Report 00-38, ISBN 90 54 11 336 7, Delft, Netherlands.

Verlaan, M., 1998, Efficient Kalman Filtering Algorithms for Hydrodynamical models, PhD thesis, Delft University of Technology, Delft, Netherlands.

405

AN EDDY DIFFUSIVITY DERIVATION FOR THE NON-ISOTROPIC CONVECTIVE DECAYING TURBULENCE IN THE RESIDUAL LAYER

Antonio G. Goulart, Gervásio A. Degrazia, Marco T. Vilhena, Domenico Anfossi, Davidson M. Moreira, Jonas C. Carvalho and Joecir Palandi[*]

1. INTRODUCTION

Eddy diffusivities are properties of turbulent flow that describe the magnitude of the dispersion in a Planetary Boundary Layer (PBL). Recently, a great number of works have been carried out to derive eddy diffusivities in a Convective and Stable Boundary Layer (CBL and SBL). However, little attention has been paid to derive vertical eddy diffusivities in a situation of decaying turbulence in the Residual Layer.

In a recent paper (Goulart et al., 2003) a theoretical model for describing the non-isotropic turbulent kinetic energy (TKE) decay in the Residual Layer was developed. The model, based on the Taylor statistical diffusion theory and the budget equation for turbulent kinetic energy in a decaying isotropic convective turbulence, was derived considering the buoyancy and transfer terms in the energy density spectrum equation. In that paper it was not possible to derive an analytical expression for the eddy diffusivity, due to the complexity of the obtained solution. A first approach to develop the eddy diffusivity in a decaying convective turbulence was proposed by Goulart et al. (2002). In that case, the above model was simplified by assuming that buoyant contribution could be disregarded and that isotropic turbulence was prevailing. The results of this model were compared with a decaying vertical eddy diffusivity obtained from LES data (Nieuwstadt and Brost, 1986). From this comparison, both vertical eddy diffusivities show a good agreement for small decaying times. For larger times, as a consequence of the -2 exponent obtained for the decaying vertical velocity variance calculated from LES, the LES vertical eddy diffusivity decays strongly faster.

---

[*] Antonio Goulart, Universidade Regional Integrada, Departamento de Ciências Exatas e da Terra, Santo Ângelo, Brazil. Gervásio A. Degrazia and Joecir Palandi, Universidade Federal de Santa Maria, Departamento de Física, 97105-900, Santa Maria, Brazil. Marco T. Vilhena, Universidade Federal do Rio Grande do Sul, Instituto de Matemática, Porto Alegre, Brazil. Domenico Anfossi, CNR, Istituto di Scienze dell'Atmosfera e Del Clima, 10133, Torino, Italy. Davidson M. Moreira and Jonas C. Carvalho, Universidade Luterana do Brasil - PPGEEAM, 92420-280, Canoas, Brazil.

Air Pollution Modeling and Its Application XVI, Edited by
Borrego and Incecik, Kluwer Academic/Plenum Publishers, New York, 2004

The novelty of the present work is that the obtained results demonstrate to describe for the whole decaying time the behavior of the decaying vertical eddy diffusivity derived from LES data. Our approach is as in the Goulart et al (2002) paper except that turbulence was assumed to be homogeneous but non-isotropic. In the budget equation for TKE,, the inertial energy transfer term is closed using a model suggested by Pao (1965) on the basis of dimensional considerations. Thus, we obtained a budget equation for TKE that was analytically solved for the three-dimensional (3-D) Energy Density Spectrum (EDS). This 3-D EDS is expressed in terms of an initial ($t = 0$) 3-D spectrum. To calculate this initial 3-D spectrum the mathematical formulation proposed by Kristensen et al. (1989) (non-isotropic case) was used, which allowed us to determine the 3-D spectrum of the homogeneous and non-isotropic turbulent flow from a known one-dimensional (1-D) spectrum.

To calculate the decaying vertical one-dimensional spectrum and the vertical eddy diffusivity a mathematical method using a weight function was utilized. This weight function informs about how this vertical component takes part in the formation of the 3-D EDS. Finally, the results of our theoretical model are compared with the results obtained from the above mentioned LES data for a decaying convective boundary layer.

2. THE SOLUTION OF ENERGY DENSITY SPECTRUM DYNAMICAL EQUATION IN DECAYING TURBULENCE

During the TKE decay in a CBL, as a first approximation, the buoyant and shear terms in the budget equation for 3-D EDS can be disregarded and turbulence can be assumed to be homogeneous. Consequently, the following 3-D EDS equation reads (Hinze, 1975):

$$\frac{\partial}{\partial t}E(k,t) = W(k,t) - 2\nu k^2 E(k,t) \tag{1}$$

where: t is time, k is the wave-number, $E(k,t)$ is the 3-D EDS, $W(k,t)$ is the inertial transport term and ν is the kinematic viscosity.

A turbulent flow contains eddies of different size or different wavelengths. The small eddies are subjected to the stress generated by large eddies. This field increases the vorticity of small eddies and, consequently, their kinetic energy. Thus, TKE is transferred from large eddies towards smaller and smaller eddies until the Kolmogorov micro-scale is reached, where the energy is dissipated as heat. This process is represented by the term $W(k,t)$ of equation (1). This term was parameterized by Pao (1965) for a turbulent isotropic flow on the basis of dimensional analysis, as follows:

$$W(k,t) = -\frac{\partial}{\partial k}\left(\alpha^{-1}\varepsilon^{1/3}k^{5/3}E(k,t)\right) \tag{2}$$

where α is the Kolmogorov constant and ε is the rate of molecular dissipation of TKE. By substituting equation (2) into equation (1) one obtains:

$$\frac{\partial E(k,t)}{\partial t} + \alpha^{-1}\varepsilon^{1/3}k^{5/3}\frac{\partial E(k,t)}{\partial k} + \frac{5}{3}\alpha^{-1}\varepsilon^{2/3}k^{2/3}E(k,t) - 2\nu k^2 E(k,t) = 0 \tag{3}$$

Considering the following dimensionless parameters, in which w_* is the convective velocity scale and z_i is the CBL height

$$t_* = \frac{w_* t}{z_i}, \qquad\qquad R_e = \frac{w_* z_i}{\nu}, \qquad\qquad \psi_\varepsilon = \frac{\varepsilon\, z_i}{w_*^3} \qquad (4)$$

equation (3) becomes

$$\frac{\partial E(k',t_*)}{\partial t_*} + \alpha^{-1}\psi_\varepsilon^{1/3}(k')^{5/3}\frac{\partial E(k',t_*)}{\partial k'} + \frac{5}{3}\alpha^{-1}\psi_\varepsilon^{1/3}(k')^{2/3}E(k',t_*) - \frac{2}{R_e}(k')^2 E(k',t_*) = 0$$

$$(5)$$

where $k' = kz_i$. A solution of (5) is:

$$E(k',t_*) = E(\xi,0)\left(\frac{k'}{\xi}\right)^{-5/3}\exp\left\{-\frac{3\alpha}{2R_e\psi_\varepsilon^{1/3}}\left((k')^{4/3} - \xi^{4/3}\right)\right\} \qquad (6)$$

where $\xi = \left\{(k')^{-2/3} + \frac{2}{3}\alpha^{-1}\psi_\varepsilon^{1/3}t_*\right\}^{-3/2}$, $\alpha \cong 1.5$ (Muschinski and Roth, 1993) and $E(\xi,0)$ is the initial ($t = 0$) 3-D spectrum.

3. APPLICATION TO THE PBL CONVECTIVE CASE

The dynamical equation describing the turbulent flow is valid just in 3-D space. Consequently, the spectrum $E(k,0)$ that represents the CBL initial condition in equation (6) is the CBL turbulent 3-D spectrum.

Differently from Goulart et al. (2002) we are here considering non-isotropic turbulence, and, as consequence, we will use the formulation proposed by Kristensen et al. (1989) to determine the initial 3-D spectrum. This formulation allows determining the 3-D spectrum of a homogeneous turbulent flow from known 1-D spectra, namely:

$$E_0(k,z) = k^3 \frac{d}{dk}\frac{1}{k}\frac{dF_u(k)}{dk} + 12 A_i m_i B_i^{-\frac{17}{6}} k^4 \sum_{n=0}^{3} C_n \int_{W_{1i}}^{\infty} \frac{Z_i^{3n-12}}{(Z_i^3 - 1)^5}\, dZ_i$$

$$ -\frac{84}{9} A_i m_i B_i^{-\frac{3}{2}} k^{\frac{4}{3}} \sum_{n=0}^{3} C_n \int_{1}^{W_{2i}} \frac{Z_i^{3n-12}}{(Z_i^3 - 1)^{n-5}}\, dZ_i \qquad (7)$$

with

$$W_{1i} = \left(1 + \frac{1}{\sqrt{B_i s}}\right)^{\frac{1}{3}}, \qquad W_{2i} = \left(1 + \sqrt{B_i s}\right)^{\frac{1}{3}}, \qquad A_i = a_i\left(\frac{1}{b_i}\right)^{\frac{5}{6}},$$

$$m_u = 2, \, m_v = m_w = -1,$$

$$C_0 = -\frac{55}{27}, \qquad C_1 = \frac{70}{9}, \qquad C_2 = -\frac{725}{72}, \qquad C_3 = \frac{935}{216}, \qquad A_i = a_i b_i^{-\frac{5}{6}},$$

$$B_i = b_i^{-2} \tag{8}$$

According to Degrazia and Anfossi (1998) the initial 1-D component of spectrum can be written as:

$$F_i(k,0) = \frac{a_i}{(1+b_i k)^{5/3}} \qquad i = u, v, w \tag{9}$$

where

$$a_i = \frac{0.98}{2\pi} c_i \left(\frac{z}{z_i}\right)^{5/3} z_i \psi_\varepsilon^{2/3} w_*^2 [(f_m^*)_i^c]^{-5/3} \quad \text{and} \quad b_i = \frac{1.5}{2\pi} \frac{z}{z_i} z_i \frac{1}{(f_m^*)_i^c},$$

with

$c_i = \alpha_i (0.5 \pm 0.05)(2\pi\kappa)^{-2/3}$, $\alpha_i = 1, 4/3, 4/3$ for u, v and w respectively (Champagne et al., 1977) $w_* = (u_*)_0 (-\frac{z_i}{\kappa L})^{1/3}$, $(f_m)_i^c = \frac{z}{G_i z_i}$, $G_u = 1.5$, $G_v = 1.5$, and

$$G_w = 1.8[1 - \exp\left(-\frac{4z}{z_i}\right) - 0.0003\exp\left(\frac{8z}{z_i}\right)].$$

To calculate the u, v e w 1-D components of the spectrum we consider that for a particular time instant t there is a relationship between the 1-D spectra and the 3-D spectrum given by the following expression:

$$F_i(k,t) = \alpha(k)\frac{\frac{1}{T}\int_0^T F_i(k,t)dt}{\frac{1}{T}\int_0^T E(k,t)dt} E(k,t) \tag{10}$$

where the ratio between the two integrals, that is between the time averages over the decay time T for each wave number, is a weight function that indicates as each 1-D component takes part in the construction of the 3-D spectrum and $\alpha(k)$ is an order unity parameter. The solution of equation (10) provides the 1-D components as a function of the 3-D spectrum.

410

$$F_i(k,t) = F_i(k,0)\exp[\int_0^t Q'(k,s)ds] \qquad (11)$$

In this case $F_i(k,0)$ is given by equation (9) where

$$Q'(k,s) = \alpha(k)Q(k,s) + \frac{1}{Q(k,s)}\frac{\partial Q(k,s)}{\partial s} \qquad (11a)$$

and $Q(k,s)$ given by

$$Q(k,s) = \frac{E(k,s)}{\int_0^t E(k,s)ds} \qquad (11b)$$

3.1. Comparison of velocity variance with LES results

Since, to our knowledge, there are no conclusive observations of the CBL turbulence decay process, we compare our theoretical expressions with LES data (Nieuwstadt, 1986). The velocity variance of the turbulent flow is calculated from the following equation:

$$\sigma_i^2(t;z) = \int_0^\infty F_i(k,t;z)dk \qquad (12)$$

Figure 1 shows the temporal evolution of the vertical velocity variance decay across the CBL ($0.2 < z/z_i < 0.8$), calculated from equation (9), (11) and (12) made dimensionless by w_*^2. Figure 1 shows that our theoretical model (solid line) agrees very well with the results simulated by LES (crosses; Nieuwstadt and Brost, 1986).

3.2. Derivation of a convective decaying vertical eddy diffusivity and comparison with LES data

The eddy diffusivities calculated by Batchelor (1949) can be identified with those from advection-diffusion equation when a homogeneous turbulent flow is considered. For large diffusion travel times the vertical eddy diffusivity has the form (Hanna, 1981)

$$K_\alpha(t;z) = \frac{\pi}{3}\frac{\beta_i}{U}\frac{\sigma_i^2(t;z)}{(k_m)_i} \qquad \alpha = x,y,z \qquad (13)$$

where $(k_m)_i$ is the wave-number of the spectral peak (associated to the energy-containing eddies), U is the mean wind speed, β_i is defined as the ratio of the Lagrangian to the Eulerian integral time scales.

Considering $\beta_i = 0.55\dfrac{U}{\sigma_i}$ (Hanna, 1981; Degrazia, G. and Anfossi, D., 1998) in equation (13) yields the following decaying vertical eddy diffusivity

$$K_\alpha(t;z) = \frac{0.55\pi}{3} \frac{\sigma_i(t;z)}{(k_m)_i} \tag{14}$$

where, for our model, $\sigma_i(t;z)$ is obtained from equations (11) and (12).

Nieuwstadt and Brost (1986) report that the vertical velocity spectrum (depicted in their Figure 14), computed for several dimensionless times, all have a maximum value for $(k_m)_w z_i \approx 4$. This allows calculating the vertical eddy diffusivity averaged across the boundary layer for different times t_*. In Figure 2 we show this decaying vertical eddy diffusivities. The crosses were calculated from Eq. (14) using LES data for σ_w. Solid line was calculated also from equation (14) using σ_w values derived from equations (11) and (12). We can see that the agreement is very good for the whole decaying time t_*. The following relationship represents a good fit to the decaying vertical eddy diffusivity calculated from equations (9), (11), (12) and (14):

$$\frac{K_z}{z_i w_*} = \frac{0.079}{\sqrt{1+2t_*^{1.7}}} \tag{15}$$

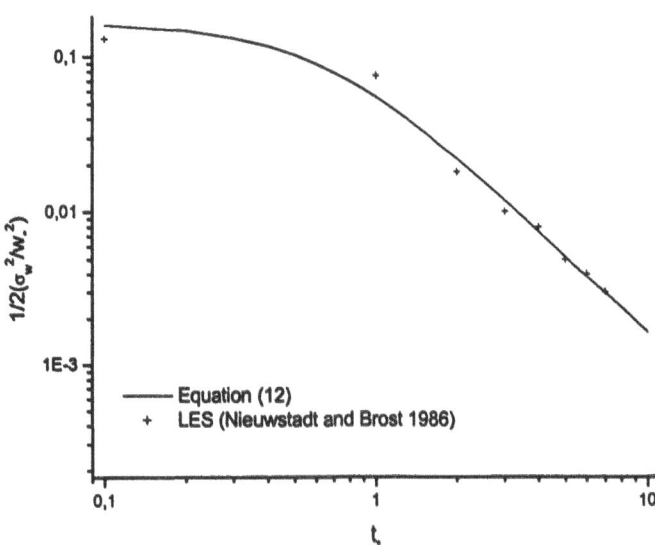

Figure 1 - Temporal evolution of the vertical velocity variance decay across the CBL ($0.2 < z/z_i < 0.8$), calculated from equation (9), (11) and (12) made dimensionless by w_*^2.

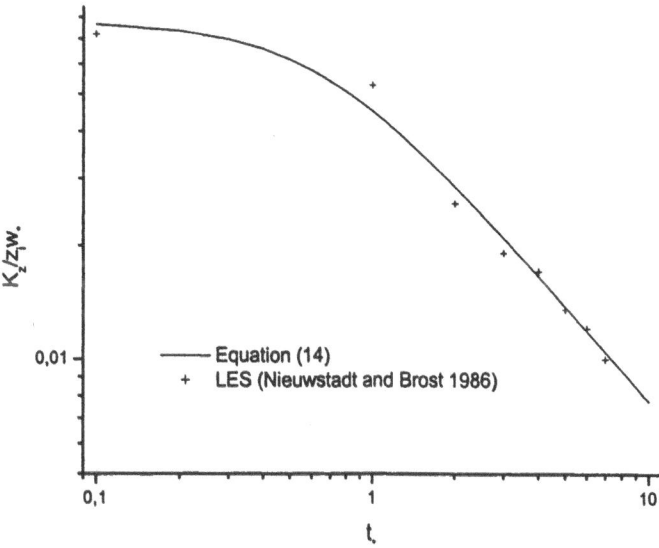

Figure 2 – Decaying vertical eddy diffusivity. The crosses were calculated from Eq. (14) using LES data for σ_w. Solid line was calculated also from equation (14) using values for σ_w determined from equations (11) and (12).

4. CONCLUSIONS

A general method to derive eddy diffusivities in a convective decaying turbulence in the Residual Layer is proposed. The method is based upon a model for the budget equation describing the 3-D energy density spectrum and the Taylor statistical diffusion theory. Firstly, on the basis of a dimensional analysis, the unknown inertial transport term present in the dynamical equation for the 3-D spectrum is parameterized. This allowed obtaining an analytical solution for the budget equation in terms of an initial 3-D spectrum that describes the observed turbulent spectrum in the daytime CBL. Due to the non- isotropic decaying convective turbulence, this initial 3-D spectrum is calculated from an 1-D spectrum by the use of a complex mathematical formulation developed by Kristensen (1989). Furthermore, to calculate the decaying vertical eddy diffusivity one needs the vertical one-dimensional spectrum. This vertical one-dimensional spectrum is derived from the decaying 3-D spectrum (solution of the budget equation for the 3-D energy spectrum), employing a weight function that allows to select the magnitude of this vertical spectral component in the formation of the decaying 3-D EDS.

Finally, the decaying vertical eddy diffusivity derived from the present theoretical model has been compared with that generated from LES data for a decaying CBL. From this comparison, we conclude that the theoretical approach developed in this study provides a very good physical description of the decaying process of the turbulence in a CBL. It is also relevant to mention that the proposed regression model for an eddy diffusivity, expressed by Eq. (15) (fitting of the theoretical model for K_z), for the non-

413

isotropic convective decaying turbulence in the residual layer, is found to be suitable for application in atmospheric diffusion models.

Acknowledgements

The work was partially supported by CNPq, FAPERGS and CAPES.

5. REFERENCES

Bachelor, G.K., 1949, Diffusion in a field of homogeneous turbulence. *Aust. J. Sci. Res.*, **2**: 437-450.

Champagne, F.H., Friehe, C.A., La Rue, J.C., Wyngaard, J.C., 1977, Flux measurements, flux estimation techniques, and fine-scale turbulence measurements in the unstable surface layer over land. *J. Atmos. Sci.*, **34**: 515-530.

Degrazia, G.A., Anfossi, D., 1998, Estimation of the Kolmogorov constat Co from classical statistical diffusion theory. *Atmos. Environment*, **32**: 3611-3614.

Goulart, A., Degrazia, G.A., Anfossi, D. and Acevedo, O., 2002, Modelling and eddy diffusivity for convective decaying turbulence in the residual layer. *15th Symposium on Boundary layers and turbulence*, American Meteorological Society, 267-268.

Goulart A., Degrazia G., Rizza U. and Anfossi D. (2003) "A theoretical model for the study of convective turbulence decay and comparison with Large-Eddy Simulation data". Boundary-Layer Meteorology, 107, 143-155

Hanna, S.R., 1981, Lagrangean and Eulerian time-scale in the daytime boundary layer. *J. Appl. Meteor.*, **20**: 242-249.

Hinze, J.O., 1975, Turbulence, Mc Graw Hill, 790p..

Kristensen, L., Lenschow, D., Kirkegaard, P. and Courtney, M., 1989, The spectral velocity tensor for homogeneous boundary layer. *Boundary Layer Meteorology*, **47**, 149-193.

Muschinski, A. and Roth, R., 1993, A local interpretation of Heisenberg's transfer theory. *Contr. Atmos. Phys.*, **66/4**, 335-346.

Nieuwstadt, F.T.M., Brost, R.A., 1986, The decay of convective turbulence. *J. Atmos. Sci.*, **43**: 532-546.

Pao, Y.H., 1965, Struture of turbulent velocity and Scalar Fields at Large Wavenumbers. *The Physics of Fluids*. **8**: 63-1075.

DISCUSSION

R. BORNSTEIN   Now, that you have modelled the decay phase, do you have any plans to include the effects from intermittent shear production?

D. ANFOSSI   Yes, one of our present aims is to study the influence of shear production on the decay process by considering the mechanical forcing term in the budgetequation for 3-D EDS.

G. SCHAYES   Your formulation is adequate if no sources of turbulence are present then only dissipation is active. This would imply that TKE should go to zero in a few hours time. However the few available observations show that TKE in the RL does not drop to zero. One should consider the effect of energy sources in the RL. Can you comment about this?

D. ANFOSSI   As I said in the Introduction of my talk, in a recently published paper (Goulart et al., 2003) we derived a theoretical model for describing the non-isotropic TKE decay in the R. L. that also included the buoyancy term. We can consider the solution here presented as a first approximation. From the dispersion point of view, it is to consider that the diffusion coefficient decreases with time less rapidly than TKE because the time scale TL increases with time. As I also answered to Prof. Bornstein, we are now trying to consider the wind shear term too. In this case the wind shear term will be regarded and a new parameterisation for this unknown turbulence production component will be proposed.Considering this new forcing in the budget eq. for 3-D EDS a new study of the balance between decaying and continuous developed turbulence will be simulated and better understood.

FORWARD AND INVERSE SIMULATIONS WITH FINNISH EMERGENCY MODEL SILAM

Mikhail Sofiev, Pilvi Siljamo[*]

1. INTRODUCTION

An evolution of the term "emergency modeling" to a large extent reflects the evolution of the other term: "public safety". The starting point was the Chernobyl catastrophe, when comparably safe plant suddenly turned to be one of the biggest threat to public health and life around European continent. Since then, it has become evident that the list of potentially dangerous installations can be extended further and further. The new threats of terrorists applying supposedly well-controlled agents like biological weapon or "non-standard" arms like dirty bomb, are forcing the scientists and decision-making authorities to review the methods and instruments used for the information support of the decision-making. In particular, it may appear that the source of the dangerous agent is not linked to any existing installation or simply unknown. Second, the net of potential sources is so dense that there can be no time to run the model to estimate the area of risk – it has to be known "in advance". Current paper presents some of possible responses to these challenges implemented in the modeling framework SILAM.

2. DESCRIPTION AND VERIFICATION RESULTS OF THE SILAM MODELING FRAMEWORK.

The Finnish emergency modeling framework SILAM (version 3.0), jointly developed by Finnish Meteorologcal Institute (FMI) and VTT Research Centre "Energy", consists of the Lagrangian dispersion model with the random-walk representation of the turbulent diffusion, meteorological and emission pre-processors, output post-processor and a set of internal interfaces and data handling and converting structures (Figure 1).

The core part of the framework – the Lagrangian dispersion model – follows the algorithm of (Eerola,1990), which includes iterations at each time step in order to get the best representation of the mean wind vector along the trajectory.

[*] Finnish Meteorological Institute, P.O.Box 503, Vuorikatu 24, Helsinki 00101, Finland

Air Pollution Modeling and Its Application XVI, Edited by
Borrego and Incecik, Kluwer Academic/Plenum Publishers, New York, 2004

417

The random-walk module contains three possible algorithms for calculations of the stochastic shift of the Lagrangian particle. The simplest one, mainly used for testing purposes, assumes uniform probability distribution of the particle random shift in a fixed-height layer, as well as in a fixed-size horizontal square drawn around the particle position. The second one takes into account the time-dependent height of the atmospheric boundary layer and dynamical relaxation of the turbulent motion. However, the boundary layer is still supposed to be well-mixed. Finally, the most sophisticated approach considers both Lagrangian time scale within and above the boundary layer, exponential relaxation of the movement as well as drift correction factor (the method is similar to that in SNAP model (Saltbones et al., 1996)).

Meteorological pre-processor, representing the largest part of the model code, prepares the input meteodata for the dispersion model, so that SILAM does not need any specific treatment of the input prior to calculations. Necessary quantities absent in the input files are produced "on-the-fly" during the simulations – prior to each time step.

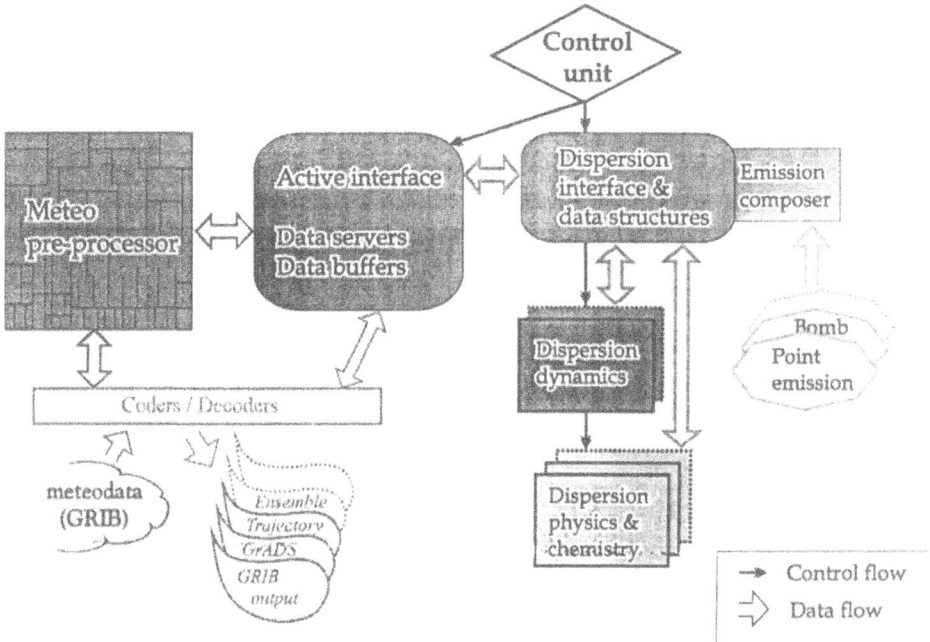

Figure 1. Parts of the SILAM framework version 3.0

Emission pre-processor is responsible for the distribution of the released pollutants between the lagrangian particles. An implemented algorithm ensures approximately the same mass attributed to each particle, regardless the number of sources, their strengths and time variability. Currently, there are two types of the sources. The first one is a point source characterized by its geographical location, elevation above the surface, and time-dependent release rate and speciation. The second one is a mushroom-shaped volumetric source create by a nuclear explosion.

A list of nuclides, which can be considered by SILAM includes 496 species with corresponding radioactive decay chains, from where a user is free to select sub-sets creating own radiological "cocktails".

Input of the meteorological data and output of the simulation results and processed meteo fields are supported by a set of IO coders and decoders, which currently cover 3 formats – GRIB for input and output, and GrADS and trajectory for output.

The model is written in FORTRAN-90 in an object-oriented style. This approach largely simplifies modifications and, in particular, introduction of new elements like new source types, parameterizations of physical, chemical or dynamic processes. The well-known problem of object-oriented programs – quite low efficiency – is mostly coped by the data buffers and data servers, which partially disclose the encapsulation and allow quick access to data via pointer structures.

Flexibility of the model allowed a universal approach to all meteorological data, regardless their origin, spatial or temporal characteristics. Prior to actual calculations, SILAM performs an analysis of both the input data and a simulation setup defined by user. Based on this analysis, the model builds a basic list of variables to be acquired from the input and a second list of quantities derived from the basic ones. It also computes the best-fit horizontal grid and vertical structure minimizing interpolation during computations and ensuring the coverage of output domain with maximum resolution.

2.1. Operational setup for the emergency response purposes

One of problems in the emergency response is a lack of time for the assessment of the situation and preparation of appropriate decisions, as well as incomplete information about the release characteristics. Consequently, it may be difficult to quickly formulate a realistic model setup, run the dispersion model and analyze the results within a short time period. Pre-compiled climatological estimates of affected areas do not help much because particular weather situation may be completely different from the long-term averaging.

Another solution implemented in FMI is based on routinely computed operational forecasts for areas of risk for a set of potential sources. These forecasts are based on actual weather predictions and assume some "standard" characteristics of the release. Current setup of such operational simulations covers five nuclear power plants in Finland, Sweden and Russia. For each source, a 54-hour forecast is produced automatically four times a day for an imaginary near-surface release of an inert agent. The release starts at standard meteorological times (00, 06, 12, 18 UTC) and lasts for 24 hours. Obtained forecast maps are automatically communicated to the authorities, providing necessary basis for quick decisions.

2.2. SILAM verification against ETEX – 1 experiment

Quality of the SILAM model was verified with the first ETEX experiment (Graziani et al., 1998). The model was run several times with different setup and input meteorological data. An example of the obtained pattern is presented in Figure 2.

Quantitative measures of the model accuracy can be expressed in many different ways. For the emergency situations, it is important that the overall position of the plume is reproduced correctly, while absolute concentration inside the clouds is less important. So, the key parameters are – probability for correct alert, missed alert, false alert and correct zero-report. Evidently, first two are situations when the observations have

419

reported presence of pollution and the last two are those when the stations have not observed anything. Taken for all ETEX stations, the SILAM results are the following:

p(correct zero-report) = 0.9   p(false alert) = 0.1
p(correct alert) = 0.6   p(missed alert) = 0.4

The last two numbers well correspond with the correlation coefficients between the SILAM results and the selected stations in Arch 1 and Arch 2 of ETEX-1 experiment. Arch 1 included 4 stations - NL5, B5, NL1, D44 - and Arch 2 contained 7 stations - DK5, DK2, D42, D5, CR3, PL3, H2. In both cases the time correlation was about 0.6, which places SILAM in the top-8 group of ETEX models out of 28 participants.

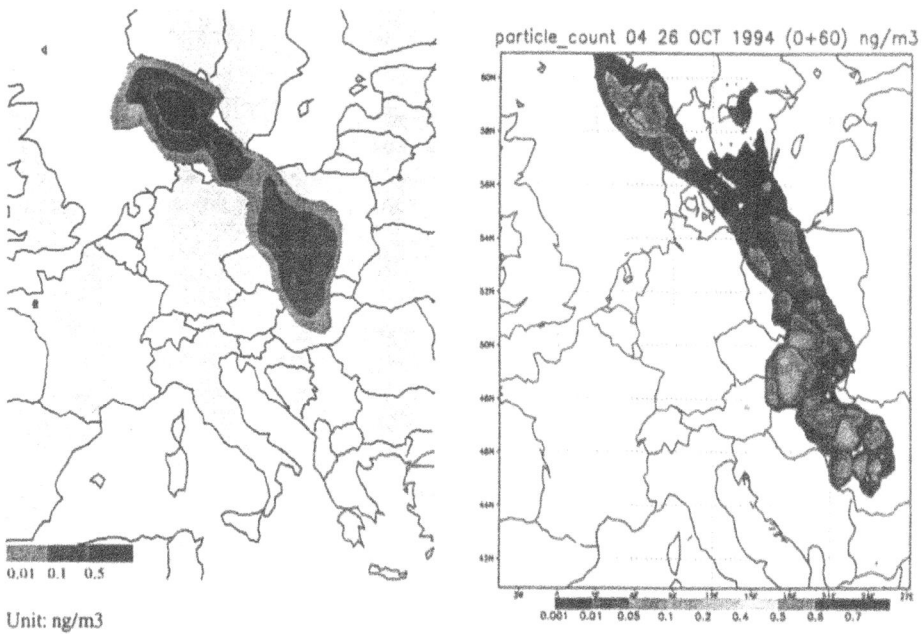

Figure 2. Observed (left) and simulated (right) positions of the pollution cloud after 60 hours since the release. The observed field is taken from ETEX Web site (http://rem.jrc.cec.eu.int/etex/)

Despite the very good performance in cloud positioning, SILAM under-estimates the near-surface concentrations, which may point to too intensive vertical mixing. Another reason is a limited representativeness of a single Lagrangian particle, which forces averaging within a layer of 200 metres depth. Otherwise non-monotonicity of the pattern became too high.

3. COUPLED FORWARD-AND-ADJOINT SIMULATIONS

3.1. Problem formulation

Until recently, the typical atmospheric dispersion problem dealt with known source and unknown pollution pattern created by this source – so-called forward problem. However, virtually in all cases the source characteristics are either uncertain or totally

420

unknown. In emergency response task it may happen that the source of pollution is not connected with stationary installations like nuclear power plants or was not timely reported (for example, this happened with Chernobyl). Then, the information comes from the measurement network only and identification of source becomes the key task leading to a full-size inverse problem.

Typical inverse problem is non-local and ill-posed, which means that it can not be formalized in differential equations and shows high sensitivity to variations of the input data, employed algorithms and their parameters. A straightforward way to solve an inverse problem is to apply the forward model many times trying to find the input configuration (in particular, emission source), which would lead to observed dispersion pattern. This is a heavy task, which, together with high sensitivity to input data, makes the inverse problems extremely difficult to approach.

A way to reduce the amount of computations and simplify the inverse problem solution is to use adjoint dispersion equations, which can be obtained from the forward one by changing sign of time and advection components (Marchuk, 1982) as follows.

Let denote $\varphi(x_i, t)$ as a concentration of some pollutant released from the emission field $f(x_i, t)$. Let dispersion conditions are described via wind $u_i(x_i, t)$ and diffusion coefficient $\mu(x_i, t)$. Let denote an intensity of all chemical and physical removal processes via $\sigma(x_i, t)$. Finally, let define the load functional M as a scalar product of the sensitivity p with the concentration φ. Then the forward dispersion problem will take the form:

$$L = \frac{\partial}{\partial t} + \frac{\partial}{\partial x_i}(u_i) - \frac{\partial}{\partial x_i}\mu_i\frac{\partial}{\partial x_i} + \sigma; \quad L\varphi = f; \quad M = (p, \varphi)$$

An adjoint problem with the adjoint differential operator L^* will take the form:

$$L^* = -\frac{\partial}{\partial t} - \frac{\partial}{\partial x_i}(u_i) - \frac{\partial}{\partial x_i}\mu_i\frac{\partial}{\partial x_i} + \sigma; \quad L^*\varphi^* = p; \quad M = (f, \varphi^*)$$

Here φ is a sensitivity distribution obtained as a solution of adjoint differential equation with the sensitivity p on the right-hand side.

The key feature of the above two problems is that the load functional M is the same in both of them. This means that user is free to choose between a solution of the forward dispersion equation with emission field on the right-hand side and with integration of the concentration with sensitivity OR a solution of the adjoint equation with sensitivity on the right-hand side and with integration of the sensitivity distribution with emission field.

3.2. Example of SILAM coupled forward-and-adjoint run

High flexibility of SILAM model enables to use it in both forward and adjoint simulations by changing a few lines in the input files. Below example illustrates the advantages and weaknesses of the adjoint problem solution and its relation to true inverse problem solution.

An experiment was created on the basis of an imaginary 1-hour release of a passive tracer near Sosnoviy Bor (Russia) at 6:00 UTC 30.10.2001, which, according to forward SILAM run, reaches Finnish coast line near Helsinki 24 hours later (see Figure 3, right panel). Time when concentration near Helsinki reaches its maximum (6:00 UTC 31.10.2001) was taken as a reference time for the adjoint model run.

For the adjoint run setup it was assumed that there was only one 1-hour long observation reporting elevated concentrations at the reference time. Duration of the run was 48 hours backward starting from the reference time. As seen from the Figure 3, the obtained sensitivity at Sosnoviy Bor points to the very moment of the release as the only possible one. This confirms that forward and adjoint problems can be solved numerically with similar efficiency and their results will give the same load assessment.

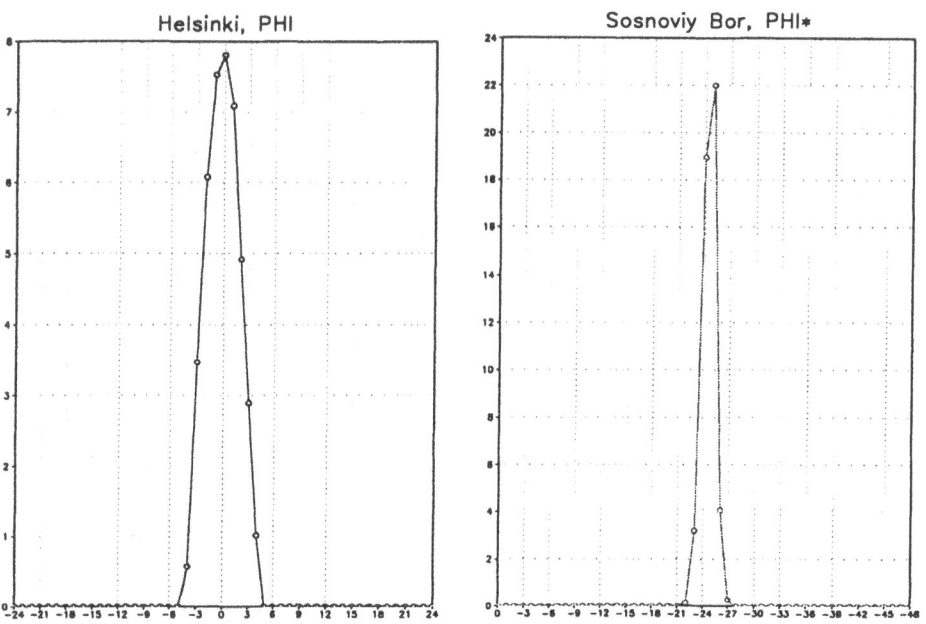

Figure 3. Time series of concentration near Helsinki (forward run, left) and sensitivity distribution near Sosnoviy Bor (inverse run, right). In both panes x-axis represent time with zero at 6:00UTC 31.10.2001

Above example showed that adjoint problem is indeed a dual formulation to the forward one. However, it is not an inverse problem solution. The sensitivity distribution just shows the areas where the sources can be, so that their plumes affect the receptor point at reference time. This, however, is not sufficient to localize the source and exclude the others from the consideration.

The second limitation is possibly high sensitivity of φ^* to input conditions. In the current case it can be illustrated by the following experiment. Charts in Figure 3 were computed by SILAM with 1-hour meteorological input from a single 48-hours long forecast data set. To simulate some minor disturbance of the meteorological fields, a subset with 6-hour time step was extracted from the main set. This sub-set was given to SILAM and intermediate times were reproduced inside the model via linear time interpolation. Resulting sensitivity distribution is shown in Figure 4, left panel. More significant disturbance of the meteorological data was introduced when the data set was created from several forecasts with 6-hour horizons. Each forecast was started from own analysis. The distortion appeared to be even more significant, as seen in Figure 4, right panel.

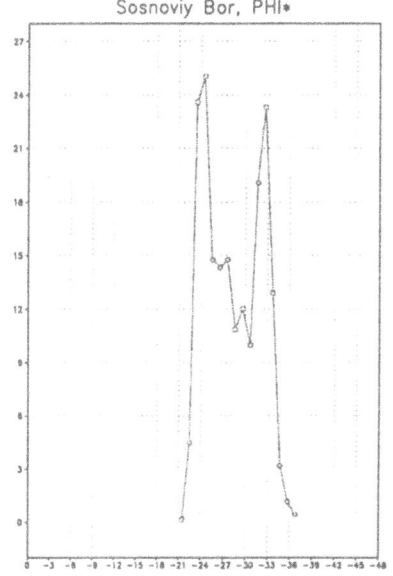 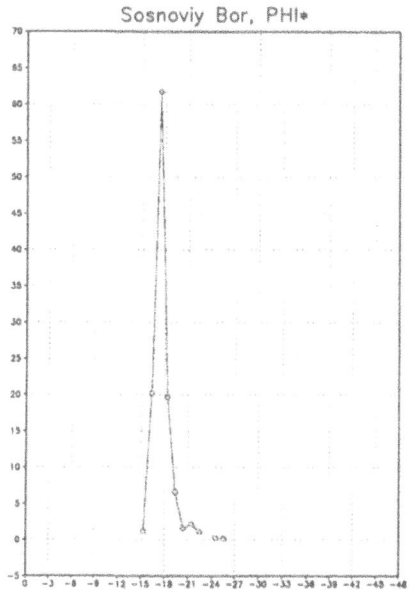

6-hour meteo time step, single analysis at $T_{ref} - 48$ hours

1-hour meteo time step, forecast is updated with analysis every 6 hours starting from $T_{ref} - 48$ hours.

Figure 4. Results of the sensitivity runs of the adjoint model.

Qualitatively, so high fluctuations can be explained by Figure 5 where the position of the sensitivity distribution φ^* is shown at the time of release ($T_{ref} - 24$ hours). As seen from the picture, the release point is located very close to the edge of the sensitivity distribution, so that a small distortion of that distribution may severely affect the time of arrival and departure of this distribution to and from the release point. This will not necessarily be the case in other situations but possibility of such unfortunate location of the source has to be taken into account.

Pattern Figure 5 illustrates one more important difference between the sensitivity distribution and a probability to have the source at some place (which is a solution of the inverse problem). In fact, any non-zero sensitivity means that the source can be located at corresponding place, while the absolute sensitivity value can not be directly transformed to the probability. More than that, there is an uncertainty area near the edge of the φ^* distribution where even zero φ^* does not mean zero probability to have the source. This is illustrated by the above experiment with three slightly different meteo data sets.

4. SUMMARY

Current paper outlines the construction and main features of the Finnish emergency modelling framework SILAM and presents its verification against ETEX-1 experiment. It is shown that application of the adjoint dispersion equation can simplify the solution of inverse dispersion problem but can not provide a complete solution. Strengths and weaknesses of such approach are illustrated with an artificial experiment with SILAM.

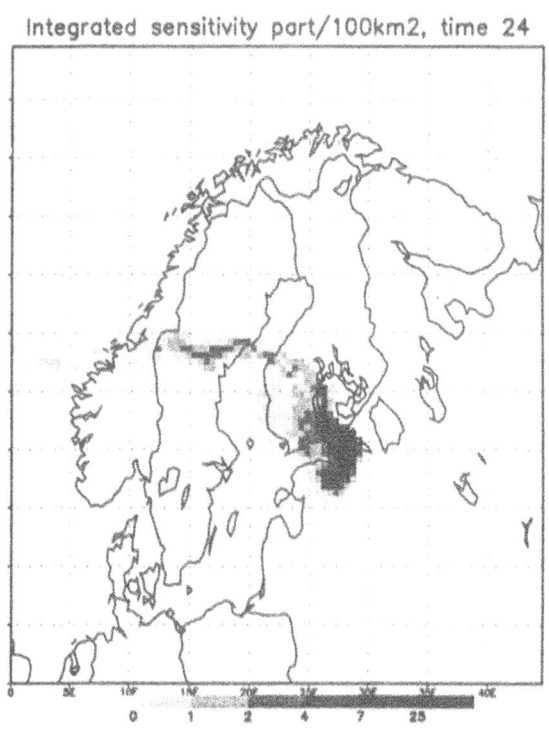

Figure 5. Position of the sensitivity distribution at reference time –24 hours

5. REFERENCES

Eerola, K., 1990, Experimentation with a three-dimensional trajectory model, *Meteorological Publications*, **15**, Finnish Meteorological Institute, Helsinki, 35pp.

Graziani, G., Klug, W., Mosca, S., 1998, Real-time long-range dispersion model evaluation of the ETEX first release. Office for Official Publications of the European Communities, Luxembourg, 214 pp.

Marchuk, G.I., 1982, Mathematical modeling in the environmental problems (in Russian). *"Nauka" publisher*, Moscow, 320 pp.

Saltbones, J., Foss, A., Bartnicki, J., 1996, Severe Nuclear Accident Program (SNAP). A real-time dispersion model.' *Air Pollution Modelling and Its Application XI.* Ed. Gryning, S.-E. & Schiermeier, F.A., Plenum Press, New York, pp.471-479.

424

DISCUSSION

D. KORACIN   How did you treat diffusion coefficient and actually turbulence since the diffusion and turbulence are not reversible?

M. SOFIEV   We have not reversed the turbulence or diffusion – in adjoint equations they are treated in exactly the same way as in forward equations (in our case, we used K–theory closure). As seen from the equation, only time and pure advection are reversed, while diffusion and sinks, which are irreversible, are not. This is the strength of the adjoint formalism, which makes it mathematically correct – it does not imply reversing of irreversible processes. From the other side, this very feature means that the adjoint formalism itself is not yet the inverse problem – there must be something after it, which transforms the sensitivity distribution to probability fields and then to inverse problem solution.

THE CHEMISTRY-TRANSPORT MODELING SYSTEM *LM–MUSCAT*: DESCRIPTION AND *CITYDELTA* APPLICATIONS

Ralf Wolke, Olaf Hellmuth, Oswald Knoth, Wolfram Schröder, Birgit Heinrich and Eberhard Renner *

1. INTRODUCTION

Air quality models base on mass balances described by systems of time-dependent, three-dimensional advection-diffusion-reaction equations. The solution of such systems is numerically expensive in terms of computing time. This requires the use of fast parallel computers. Multiblock grid techniques and implicit-explicit (IMEX) time integration schemes are suited to take benefit from the parallel architecture. A parallel version of the multiscale chemistry-transport code *MUSCAT* (MUiltiScale Chemistry Aerosol Transport) is presented which is based on these techniques (Wolke and Knoth, 2000).

The physical and chemical processes in the atmosphere occur simultaneously, coupled and in a wide range of temporal and spatial scales. These facts have to be taken into account in the numerical methods for the solution of the model equations. The multiblock techniques and IMEX schemes allow the use of different resolutions in space and also in time. The chemical reactions are described by large systems of extremely stiff ordinary differential equations (ODEs). In chemical terms, a stiff system occurs when the lifetime of some species are many orders of magnitude smaller than the lifetime of other species. To illustrate the magnitude of the problem, we note that the stiffness ratio for a typical tropospheric photochemistry system is usually greater than 10^{10}. Because explicit ODE solvers require numerous short time steps in order to maintain stability, most current techniques solve stiff ODEs implicitly or by IMEX schemes (Verwer et al., 1998; Knoth and Wolke, 1998a).

---

*Institute for Tropospheric Research, Permoserstr. 15, D–04303 Leipzig, Germany, *Email: wolke@tropos.de*

Air Pollution Modeling and Its Application XVI, Edited by
Borrego and Incecik, Kluwer Academic/Plenum Publishers, New York, 2004

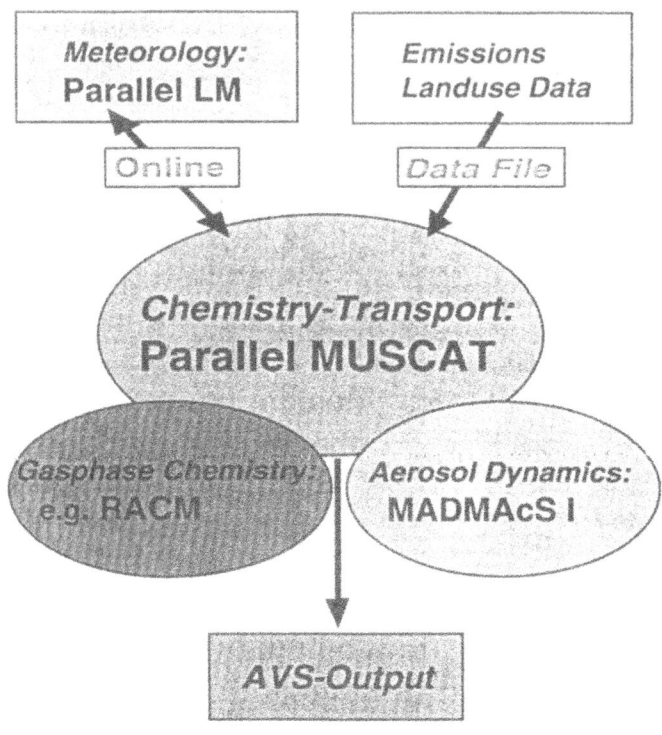

Figure 1: The coupled model system *LM–MUSCAT*.

The *MUSCAT* code is parallelized and is tested on several computer systems. Presently, *MUSCAT* has an online-coupling to the parallel, non-hydrostatic meteorological code *LM* (Doms and Schättler, 1999; Schättler and Doms, 1998) which is the operational regional forecast model of the German Weather Service. Both parallel codes work on their own predefined fraction of the available processors and have their own separate time step size control (Wolke et al., 2002). The coupling scheme simultaneously provides time-averaged wind fields and time-interpolated values of other meteorological fields (vertical exchange coefficient, temperature, humidity, density). Coupling between meteorology and chemistry-transport takes place at each horizontal advection time step only. In *MUSCAT*, the horizontal grid is subdivided into so-called "blocks". The code is parallelized by distributing these blocks on the available processors. This may lead to load imbalances, since each block has its own time step size control defined by the implicit time integrator. Therefore, well-suited dynamic load balancing is proposed and investigated.

The performance of the coupled system *LM-MUSCAT* (Figure 1) is discussed for long-term simulations of ozone and *PM* within the CITYDELTA[1] project. The model inter-comparison CITYDELTA, initiated by the Joint Research Center (JRC), focuses on health-relevant matrices of exposure (e.g., long-term concentrations) to fine particles and ozone in urban areas. The range of response resulting from the

---

[1] http://www.rea.ei.jrc.it/netshare/thunis/citydelta

model-intercomparison study will be used in the cost-effectiveness analysis of CAFE with the aim to balance Europe-wide emission controls against local measures. As our contribution to CITYDELTA, long-term simulations of ozone and PM for the area of Berlin are performed for the case szenario 1999 and seven emission scenarios for 2010 with modified NO_x, VOC and PM emissions. In the present paper, the model system and simulation approach are described as well as first project results are reported.

2. THE CHEMISTRY-TRANSPORT CODE *MUSCAT*

Chemistry-transport processes are described by systems of time-dependent, three-dimensional transport equations with reactive terms.

Multiblock Grid. In *MUSCAT* a static grid nesting technique (Wolke and Knoth, 2000; Knoth and Wolke, 1998b) is implemented. The horizontal grid is subdivided into so-called "blocks". Different resolutions can use for individual sub-domains in the multiblock approach. This allows fine resolution for the description of the dispersion in urban regions and around large point sources. This structure originates from dividing an equidistant horizontal grid (usually the meteorological grid) into rectangular blocks of different size. By means of doubling or halving the refinement level, each block can be coarsened or refined separately. This is done on condition that the refinements of neighbouring blocks differ by one level at the most. The maximum size of the already refined or coarsened blocks is limited by a given maximum number of columns. The vertical grid is the same as in the meteorological model.

The spatial discretization is performed by a finite-volume scheme on a staggered grid. Such schemes are known to be mass conservative because of the direct discretization of the integral form of the conservation laws. For the approximation of the surface integrals, point values of the mixing and its first derivative are needed on the cell surfaces. To approximate the mixing ratio at the surface we implemented both a first order upwind and a biased upwind third order procedure with additional limiting (Hundsdorfer et al., 1995). This scheme has to be applied to non-equidistant stencils which occur at the interface of blocks with different resolutions (Knoth and Wolke, 1998b).

Time Integration. For the integration in time of the spatially discretized equation we apply an IMEX scheme (Wolke and Knoth, 2000; Knoth and Wolke, 1998a). This scheme uses explicit second order Runge-Kutta methods for the integration of the horizontal advection and an implicit method for the rest. The fluxes resulting from the horizontal advection are defined as a linear combination of the fluxes from the current and previous stages of the Runge-Kutta method. These horizontal fluxes are treated as "artificial" sources within the implicit integration. A change of the solution values as in conventional operator splitting is thus avoided. Within the implicit integration, the stiff chemistry and all vertical transport processes (turbulent diffusion, advection, deposition) are integrated in a coupled manner by the second order BDF method. We apply a modification of the code LSODE (Hindmarsh, 1983) with a special linear system solver and a restriction of the BDF order to 2. The error control can lead to several implicit time steps per one explicit step.

Furthermore, different implicit step sizes may be generated in different blocks. The "large" explicit time step is chosen as a fraction of the CFL number. This value has to be determined for each Runge-Kutta method individually in order to guarantee stability and positivity. Higher order accuracy and stability conditions for this class of IMEX schemes are investigated in Knoth and Wolke (1998a).

Gas Phase Chemistry. The chemical reaction systems are given in ASCII data files in a notation that is easily understandable. For the task of reading and interpreting these chemical data we have developed a preprocessor. Contained in its output file are all data structures required for the computation of the chemical term and the corresponding Jacobian. Changes within the chemical mechanism or the replacement of the whole chemistry can be performed in a simple and comprehensive way. Several gas phase mechanisms (e.g. RACM, RADM2, CBM IV) are used successfully in 3D case studies. Time resolved anthropogenic emissions are treated in the model as point, area and line sources. It is distinguished between several emitting groups. Biogenic emissions are parameterized in terms of land use type, temperature and radiation. Dry and wet deposition processes are also included.

Aerosol Dynamics. For simulation of aerosol-dynamical processes the model *MADMAcS I* (Wilck and Stratmann, 1997) was included in *MUSCAT*. The particle size distribution and the aerosol-dynamical processes (condensation, coagulation, sedimentation and deposition) are described using the modal technique. The mass fractions of all particles within one mode are assumed to be identical. Particle size distribution changes owing to various mechanisms, which are divided into external processes like particle transport by convection and diffusion, deposition and sedimentation as well as internal processes like condensation and coagulation.

Parallelization. Our parallelization approach is based on the distribution of blocks among the processors. Inter-processor communication is realized by means of MPI. The exchange of boundary data is organized as follows. Since the implicit integration does not treat horizontal processes, it can be processed in each column separately, using its own time step size control. An exchange of data over block boundaries is necessary only once during each Runge-Kutta sub-step. Each block needs the concentration values in one or two cell rows of its neighbours, according to the order of the advection scheme. The implementation of the boundary exchange is not straightforward because of the different resolutions of the blocks. The possibilities of one cell being assigned to two neighbouring cells or of two cells receiving the same value must be taken into account. We apply the technique of "extended arrays": the blocks use additional boundary stripes on which incoming data of neighbouring blocks can be stored. Hence, each processor only needs memory for the data of blocks that are assigned to it.

Dynamical Load Balancing. Consider a static partition where the blocks are distributed between the processors only once at the beginning of the programs run time. Here, we use the number of horizontal cells (i.e., of columns) as measure of the work load of the respective block. Therefore, the total number of horizontal cells of each processor is to be balanced. This is achieved by the grid-partitioning tool PARMETIS (Karypis et al., 1998). It optimizes both the balance of columns and

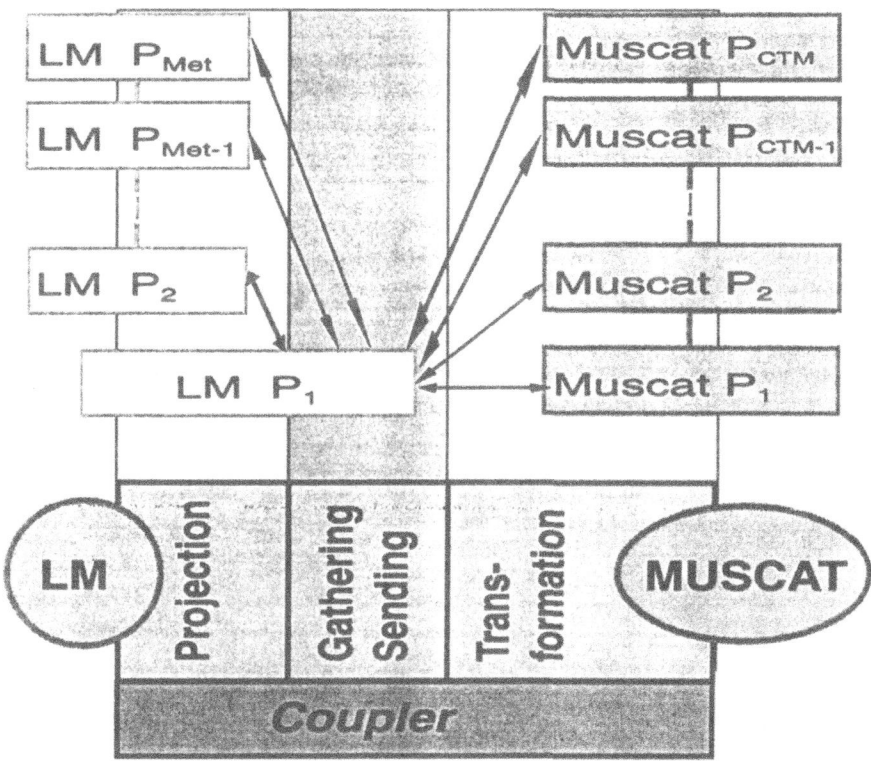

Figure 2: Communication structure in the model system *LM-MUSCAT*.

the "edge cut", i.e., it takes care of short inter-processor border lines. In order to improve the load balance, techniques for redistribution of blocks have been implemented (Wolke et al., 2001). A blocks work load is estimated using the numbers of Jacobian and function evaluations applied during a past time period. According to the work loads of the blocks, PARMETIS searches for a better distribution, besides minimising the movements of blocks. The communication required for the exchange of block data can be done by means of similar strategies as for the boundary exchange.

3. ONLINE-COUPLING TO THE PARALLEL METEORO-LOGICAL MODEL *LM*

The meteorological and the chemistry-transport algorithms have their own separate time step size control. The coupling procedure is adapted to the applied IMEX schemes in the chemistry-transport code. Each stage of the IMEX scheme requires a new calculation of the horizontal fluxes and an implicit integration cycle. All meteorological fields are given with respect to the equidistant horizontal meteorological grid. They have to be averaged or interpolated from the base grid into the block-structured chemistry-transport grid with different resolutions. The velocity field is supplied by its normal components on the faces of each grid cell, and their

431

corresponding contravariant mass flux components fulfill a discrete version of the continuity equation in each grid cell. Since *LM* solves a compressible version of the model equations an additional adjustment of the meteorological data is necessary. The velocity components are projected such that a discrete version of the continuity equation is satisfied. The main task of this projection is the solution of an elliptic equation by a preconditioned conjugate gradient method. This is also done in parallel on the *LM* processors. The projected wind fields and the other meteorological data are gathered by one of the *LM* processors. This processor communicates directly with each of the *MUSCAT* processor, see Figure 2.

4. LONG-TERM SIMULATIONS FOR CITYDELTA

In its first phase, CITYDELTA is focused on the following questions: (1) What is the influence of local versus regional emission reductions on health-relevant matrices for fine particle and ozone in urban air ? (2) What are the differences between predictions from regional models and those obtained with finer resolved models ? (3) What is the agreement between different scale dispersion models on the level of response to emission changes? To answer these questions, long-term simulations for selected European city areas will be performed by modeling groups from several European countries. Our group is focusing on the city area of Berlin. For long-term simulations it is of crucial importance to compromise between the degree of model sophistication and project requirements. In the present study, the model will be integrated over the whole year 1999 (reference year). Primary project intention is to determine so-called Deltas, i.e., differences between different models and their sensitivity against changes in emissions and grid resolution. Although model validation is of lower ranking in the CITYDELTA framework it is important for the assessment of the reliability of simulation results. Therefore, accompanying validation studies will be performed. CITYDELTA can best be described as "screening study".

To accommodate the project requirements, especially with regard to long-term integration, some model simplifications have been realized:

1. The chemical mechanism RACM (Stockwell et al., 1997) with 73 species and 237 reactions is used for this application. The *MADMAcS* aerosol model has been substituted by a more simplified aerosol model similar to that of the *EMEP* Eulerian model *MADE50* (Multi-level Acid Deposition model for Europe with 50 km resolution)(e.g., Berge, 1997; Jakobsen et al., 1997). In this approach, the formation of ammonium sulphat and ammonium nitrat aerosols is calculated from equilibrium reactions (mass-based approach to calculate $PM2.5$ and $PM10$).

2. To ensure comparability with other model results, in the framework of the multi-block grid structure an uniform grid resolution is used.

3. For consistency reasons, it is important that all models use similar boundary conditions. The boundary conditions for the chemical species have been provided by the *Regional EMEP* model. With regard to required CPU time and planned time schedule the option to provide boundary conditions from a preceding large scale *LM-MUSCAT* simulation has been dispensed (this issue will be pursued in parallel). The boundary conditions for chemistry include

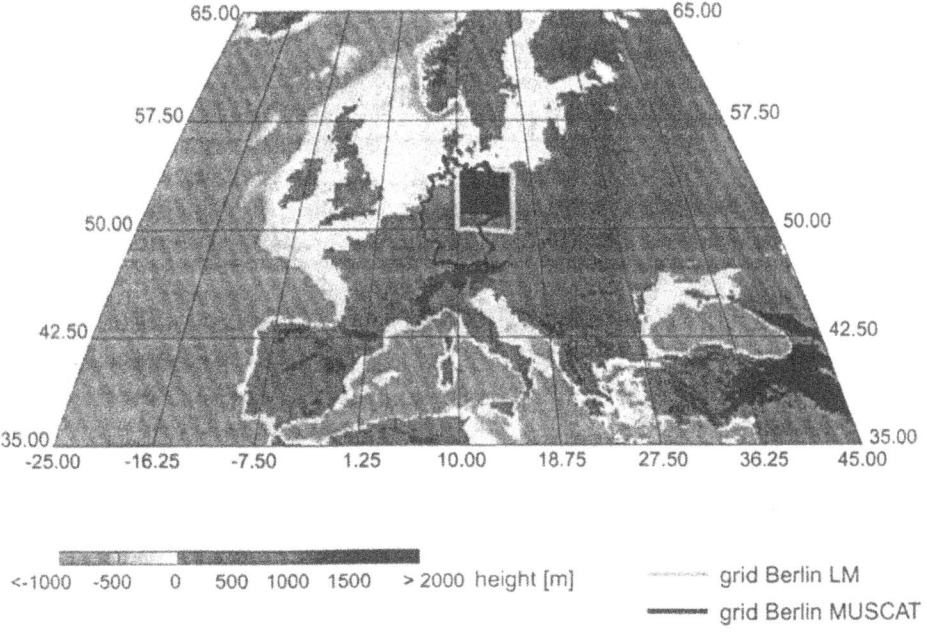

Figure 3: Location of *LM* and *MUSCAT* grid.

24 gas- and 5 aerosol-phase species on 20 *EMEP* vertical levels with a time resolution of 3 hours.

4. For emissions, the Berlin Emission Inventory provided by the Federal Environmental Agency and the FU Berlin has been taken (Stern, 2002). It contains area emissions with a spatial resolution of approximately 2 km and point sources of NO_x, $NMVOC$, SO_2, CO, $PM10$, $PM2.5$, and benzene. Source group categories of CORINAIR have been used. All Emissions are given as annual averages for the reference year 1999 for area and point sources. There is no information about NH_3 and biogenic VOC emissions. For point sources volumetric flow rate, stack height, exit temperature, stack area, monthly operation hours and emission mass flux are given. Monthly, daily and hourly time variations are also included in the emission inventory.

Six-hour analysis data of the global forecast model *GME* are available for initialization and large-scale forcing of *LM*. *GME* data from several databases have been "merged" to obtain an unique dataset with meteorological boundary conditions for the whole year 1999. The *GME* analysis data have been interpolated to the *LM* model domain (64 x 68 x 40 grid cells) with horizontal grid resolution of 0.0625° (approx. 7.5 km). The grid resolution of *MUSCAT* is identical to that of *LM*, but the *MUSCAT* grid has been imbedded into the *LM* grid (Figure 3).

To keep the long-term evolution of meteorological fields as close as possible

433

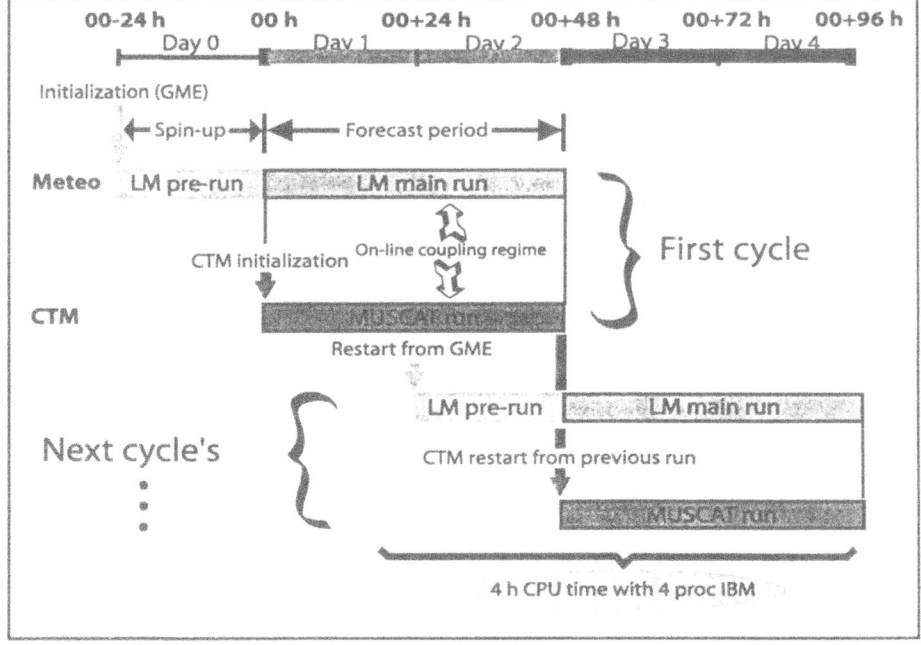

Figure 4: CityDelta simulation cycles.

nearby observations, the integration period has been subdivided into overlapping short-term forecast cycles. Each of these cycles consists of a one-day *LM* pre-run for spin-up of the meteorology followed a two-day coupled run of meteorology and chemistry-transport (Figure 4). The *LM* pre-run is initialized with *GME* data. For initializing *MUSCAT* final concentration fields of the previous cycle are taken. Thus, the CTM is actually restarted rather than initialized. While model initialization is an update of complete 3D fields, the external forcing via nudging is limited to the boundary grid cells of the model domain, only. The designed forecast cycle has been compared with other cycles and found to be suitable with respect to consistency of meteorological fields between two cycles.

Preliminary results from the pilot simulation May-June 1999 will be shown. Modelled wind and concentration fields reveals a rational and comprehensible overall evolution of meteorological fields as well as chemical concentrations and particulate matter in space and time. The development of meteorological fields corresponds well with synoptic weather situation. Also small scale convergences in the wind field variable in space and time obey common expectations of mesoscale modeling. Simulated wind, ozone and NO_2 concentrations during a ozone episode shown in Figure 5 as an example. In the last ten days of May 1999, large scale advection of ozone from southwest and southeast has lead to a clear enhancement of the background ozone concentration in the region of interest except for Berlin, where ozone remained at comparably low levels. Beginning in the late afternoon, enforcing in the evening and during the night ozone has been systematically reduced. This is pronounced along the motorways.

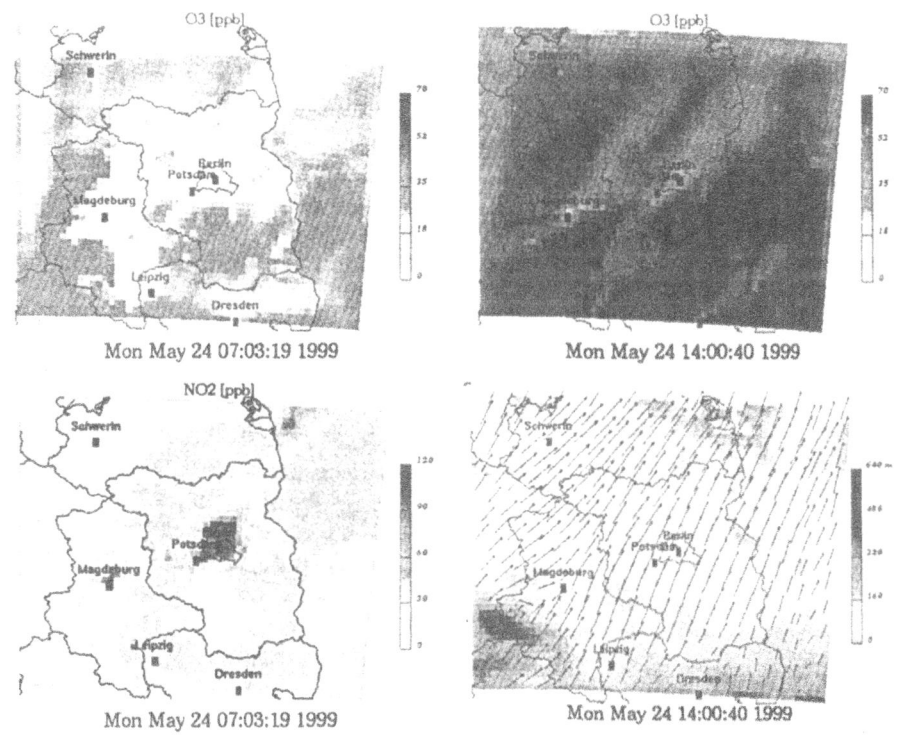

Figure 5: Ozone, NO_2 concentration fields and wind fields

Modelled data are compared with meteorological and air pollution measurements. For the rural station Burg, the agreement between observed and predicted time series of hourly ozone concentrations is encouraging (Figure 6). A closer look reveals, that ozone has been underestimated, especially at night, and NO_x is clearly over-predicted. One reason could be that the vertical exchange in *LM* is too low at night, leading to higher NO_x in the lowest layer than observed. Further investigations are necessary to trace back these discrepancies. Screening of *PM* simulations shows also a coherent evolution. For validation there are no $PM2.5$ observations available. $PM10$ has been measured at three stations, only.

5. CONCLUSIONS AND OUTLOOK

The model system *LM-MUSCAT* is a powerful instrument for environmental studies. The parallel efficiency is encouraging. The results from the CITYDELTA pilot study are encouraging too. The availability of a full coupled meteorology-chemistry-aerosol model with well-defined interfaces to external emission inventories and regional-scale forcing data for chemistry and aerosols, e.g., *EMEP*, is an important step forward in regular model applications on behalf of the European Community or other entities.

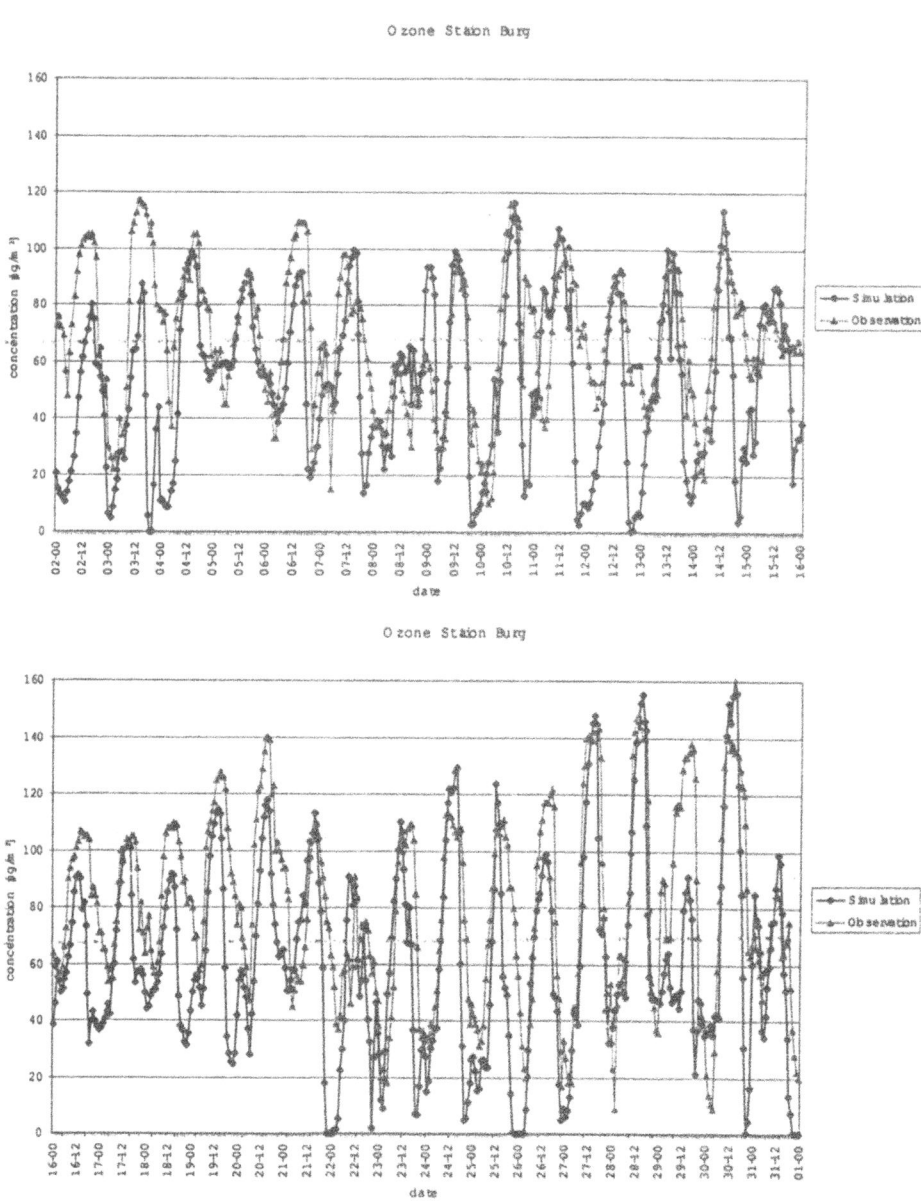

Figure 6: Modelled and observed ozone concentration, May 1999, Burg (rural)

Acknowledgements

The work was supported by the NIC Jülich, the DFG and the Ministry for Environment of Saxony. Furthermore, we thank the DWD Offenbach for good cooperation. Special thanks are due to CITYDELTA project steering group.

REFERENCES

E. Berge (1997), Transboundary air pollution in Europe. *MSC-W Status Report 1997, Part 1 and 2, EMEP/MSC-W Report 1/97*. The Norwegian Meteorological Institute, Oslo.

G. Doms and U. Schättler (1999), The Nonhydrostatic Limited-Area Model LM (Lokal Model) of DWD: I. Scientific Documentation (Version LM-F90 1.35). German Weather Service, Offenbach.

H.A. Jakobson, J.E. Jonson, and E. Berge (1997), The multi-layer Eulerian model: Model description and evaluation of transboundary fluxes of sulphur and nitrogen species for one year. *EMEP/MSC-W Note 2/97*. The Norwegian Meteorological Institute, Oslo.

A.C. Hindmarsh (1983), ODEPACK, A systematized collection of ODE solvers, in: R.S. Steplman et al., Eds., *Scientific Computing*, (North-Holland, Amsterdam), 55-74.

W. Hundsdorfer, B. Koren, M. van Loon, and J.G. Verwer (1995), A positive finite-difference advection scheme, *J. Comput. Phys.* **117**, 35-46.

G. Karypis, K. Schloegel, and V. Kumar (1998), *ParMETIS*. Parallel graph partitioning and sparse matrix ordering library. Version 2.0. University of Minnesota.

O. Knoth and R. Wolke (1998a), Implicit-explicit Runge-Kutta methods for computing atmospheric reactive flow, *Appl. Numer. Math.* **28**, 327-341.

O. Knoth and R. Wolke (1998b), An explicit-implicit numerical approach for atmospheric chemistry-transport modeling, *Atm. Environ.* **32**, 1785-1797.

U. Schättler and G. Doms (1998), The Nonhydrostatic Limited-Area Model LM (Lokal Model) of DWD: II. Implementation Documentation (Version LM-F90 1.35). German Weather Service, Offenbach.

R. Stern (2002), Emission data base Berlin-Brandenburg. Technical Note. Institute of Meteorology, Free University, Berlin.

W.R. Stockwell, F. Kirchner, M. Kuhn, and S. Seefeld (1997), A new mechanism for regional atmospheric chemistry modeling, *J. Geophys. Res.* **D22, 102**, 25847-25879.

M. Wilck and F. Stratmann (1997), A 2-D multicomponent aerosol model and ist application to laminar flow reactors, *J. Aerosol Sci.* **28**, 959-972.

R. Wolke and O. Knoth (2000), Implicit-explicit Runge-Kutta methods applied to atmospheric chemistry-transport modelling, *Environmental Modelling and Software* **15**, 711-719.

R. Wolke, O. Knoth, O. Hellmuth, W. Schröder, and J. Weickert (2001), Load-balancing in the parallel model system LM-MUSCAT for multiscale chemistry-transport simulations. Proceedings of the 5-th GLOREAM Workshop, Wengen (Switzerland, September 2001); http://www.dmu.dk/AtmosphericEnvironment/gloream.

R. Wolke, O. Knoth, E. Renner, W. Schröder, and J. Weickert (2002), in: H. Rollnik and D. Wolf, Eds., *NIC-Symposium 2001*, (John von Neumann Institute for Computing, Jülich), 453-462.

J.G. Verwer, W.H. Hundsdorfer, and J.G. Blom (1998), Numerical time integration for air pollution models. *Technical Report MAS-R9825*. CWI Amsterdam.

DISCUSSION

S.A.-AKSOYOGLU   Does the model take into account the effect of aerosols on the irradiance, i.e. photolysis rates?

R. WOLKE   The photolysis rates are given for clear-sky conditions. Cloud shadowing effects are taken into account. For this purpose the approach of the REMSAD model (http://remsad.saintl.com) is realized. In the current version, the simulated aerosol distribution has no feedback on the irradiance.

G. KALLOS   1. As far as understood what you described as "online coupling" is not exactly a dynamic coupling that means you solve the basic equality with the same methodology the same (or multiple) time steps. You described the treatment of pressure or LM and the chemistry model. This treatment by itself cannot be considered as "dynamical coupling"

2. Do you use the same projection in the horizontal and vertical as in LM?

3. You mentioned "dynamical balancing" for distribution the load between the CPUS: That means you redistribute the load between the processors after certain numbers of time steps. Is it correct? If yes how do you do it?

4. How about precipitation from LM and the usage on MUSCAT

R. WOLKE   Our "online coupling" is a dynamical coupling. The meteorological code LM and the chemistry-transport model MUSCAT work simultaneously and have their own step size control. The exchange of information between the models is performed by a coupler after each explicit time step (Wolke et al., 2002). The MUSCAT code solves the three-dimensional advection-diffusion equations with reactive terms. After the spatial discretisation using finite-volume techniques, the resulting large system of ODEs is integrated by an IMEX time integration scheme (Wolke and Knoth, 2000). All meteorological variables are considered as smooth functions in time. If meteorological fields are required during the integration then the coupler provides this information (as linear-interpolated or time-averaged field). The step size control of the IMEX scheme guarantees stability and a prescribed tolerance of the time integration error. Of course, the smoothness of the

438

solution and the meteorological fields is assumed in the order and stability analysis of the IMEX scheme (Knoth and Wolke, 1998a). The size of the explicit step is mainly determined by the CFL criterion. In the presented simulations explicit steps between 40 and 200 seconds are generated.

LM is a compressible model which uses the pressure as prognostic variable. Therefore the discrete continuity equation is not necessarily fulfilled. In our approach, all horizontal and vertical fluxes are corrected in such a way that the discrete continuity equation is satisfied and the distance between the corrected and the original fluxes of the meteorological code is minimized in a least squares sense.

In MUSCAT, a dynamical load balancing is used. During the simulation the blocks are redistributed in such a way that the work load is balanced. The redistribution is performed after a prescribed number of explicit time steps. The work load a_P of processor P is defined by

$$a_P = \Sigma_{B \in P} (N_F^B + 1.4\, N_J^B) * N_C^B$$

$B \in P$ stands for the blocks currently located on this processor, N_C^B is the number of columns of block B. For repartitioning, we use ParMETIS (Karypis et al., 1998) which is called if the ratio

$$\min_P (a_P) / \max_P (a_P)$$

gets below of a certain critical value. According to the work loads of the blocks, ParMETIS searches for a better distribution, besides minimising the movements of blocks. The communication required for the exchange of block data can be done by means of similar strategies as for the boundary exchange. In our tests a redistribution is performed after three explicit time steps. The LM runs on a fixed predefined number of processors. The balance of the work load is given by an equable decomposition of the LM grid. A load balancing between LM and MUSCAT is not implemented.

The grid scale precipitation as well as the parameterised convective precipitation are provided as three-dimensional arrays by the coupler directly from LM. The appropriate fluxes are averaged over an explicit time step interval. These three-dimensional time-averaged precipitation fluxes are used in MUSCAT. The description of rainout and washout processes is similar to the EMEP approach (http://www.emep.int).

SIMULATION AT NEIGHBORHOOD SCALE WITH CMAQ

Sylvain Dupont, Tanya L. Otte, and Jason Ching [*]

1. INTRODUCTION

We investigate air quality (**AQ**) modeling at very high spatial resolutions (on order of 1-km horizontal grid spacing), defined as neighborhood scales, as a means to perform human exposure and risk-based assessments (with emphasis on air toxic pollutants). AQ models perform better at this scale for those species which have inherently high spatial and temporal variability.

For this study, we use the U.S. EPA Models-3 Community Multiscale Air Quality (**CMAQ**) modeling system (Byun and Ching, 1999) using the Penn State/NCAR Mesoscale Model (**MM5**) (Grell et al. 1994) to provide meteorological input. CMAQ will be used to support modeling studies and air quality assessments of ozone, particulate matter, and air toxics at high resolution in Houston, Texas.

At neighborhood scales, the meteorological fields, the emissions, and subsequent air quality distributions are strongly influenced by the presence of the urban morphological structures of varying complexities. This application requires developing more detailed treatment of the influence of urban structures in the models and using additional urban morphological databases as input. Our first step is focused on the meteorology by improving the urban canopy representation inside MM5 because the meteorological fields have a large influence on the pollutant concentration fields. Indeed, they control the mixing height (**MH**), the effective volume in which pollutants are dispersed, the stability of the atmosphere which influences the vertical diffusion of pollutants, and the intensity of the turbulence which controls the intensity of pollutant dispersion and the interactions with other pollutants. A new set of advanced urban canopy parameterizations (**UCP**) developed inside MM5 is presented in section 2.

To prepare for the Houston study, the new MM5 versions are tested on Philadelphia (section 3), and their consequences on MM5 meteorological fields and CMAQ ozone fields are analyzed in section 4.

---

[*] U.S. EPA, Research Triangle Park, NC, USA (On assignment from NOAA Air Resources Laboratory)

Air Pollution Modeling and Its Application XVI, Edited by
Borrego and Incecik, Kluwer Academic/Plenum Publishers, New York, 2004

441

2. NEW MM5 VERSIONS FOR URBAN APPLICATIONS

The improvement of MM5 to represent the urban canopy has been made in three steps. The first step is characterized by an improvement of the dynamic fields parameterization inside the roughness sub-layer (**RSL**); the second step is focused on the improvement of the urban surface energy budget assessment; and, the third step corresponds to the integration of the previous steps. Each step is characterized by a new version of MM5: i) the DA-SLAB version which uses the drag-force approach (**DA**) to represent the dynamic effect of the buildings, ii) the RA-SM2-U version which uses the rural and urban soil model SM2-U (Soil Model for Sub-Meso scales, version Urbanized, Dupont, 2001) with the roughness approach (**RA**), and iii) the DA-SM2-U version which corresponds to an integration of previous versions where the DA is also extended to the vegetation canopy.

2.1. DA-SLAB version

As in many atmospheric models, MM5 represents the aerodynamic and thermodynamic characteristics of urban surfaces by applying the RA and by modifying the parameters of the rural soil model. Thus, the influence of surface obstacles is represented by gridded roughness length and a displacement height, and the surface exchange coefficients are calculated from Monin-Obukhov similarity theory. It is well known that the RA is unsatisfactory inside the RSL which is not a constant-flux layer. To simulate the urban climatology or the impact of a city on its environment, the RA can be assumed satisfactory if the vertical and horizontal sizes of the first layer cells are large enough, respectively, to include the RSL and to assume an horizontal surface homogeneity. To assess urban AQ at neighborhood scales, since primary atmospheric pollutants are emitted inside the RSL and thus the first chemical reactions and dispersion occur in this layer, it is necessary to generate detailed meteorological fields inside the RSL. To give MM5 the capability to simulate meteorological fields inside the RSL, a UCP has been developed by Lacser and Otte (2002) using the DA following the previous works of Brown (2000). With the DA, the lower level of the computational domain corresponds to the real level of the ground, and additional vertical layers are included within the urban canopy to allow more detailed meteorological fields in the RSL. Thus for the DA, the vertical distribution of buildings inside the canopy needs to be known. This UCP is implemented in MM5 via the TKE-based Gayno-Seaman planetary boundary layer (**GSPBL**) model.

In this UCP, the dynamic effects of buildings are represented by adding in the momentum equations a frictional force created by these buildings, and a source term in the turbulence kinetic energy (**TKE**) equation, both depending on the building frontal area density. The dynamic effects of buildings are represented with the DA whereas the vegetation and bare soil are represented with the RA. It is important to note that the RA is here applied with a thinner first vertical layer than in a typical application of MM5, and this reaches the limits of the Monin-Obukhov similarity theory used by the RA. In the UCP, it is assumed that the buildings affect the flow, virtually, because of their vertical surfaces but they do not take up any volume within the grid cell.

For this first step, the thermodynamics effects of the urban canopy have been implemented via the air temperature tendency equation following Chin et al. (2000), by considering the time-varying anthropogenic sources following Taha (1999) and the sensible heat flux from the roofs following Brown (2000). For the determination of this last flux, it is assumed that roofs have no heat storage capacity and that the rooftop longwaves radiate at the air temperature. The ground energy balance, deduced from the rural soil model "slab", includes the shadowing/trapping effect of the net radiation reaching the ground in the city canyons by considering the extinction of the radiation through the urban canopy using a simple exponential function (Brown 2000). The water budget of urban surfaces is not considered.

2.2. RA-SM2-U Version

When simulations in urban area are represented in MM5 by only one urban category, it is impossible to represent the complexity of the canopy and to differentiate urban districts (e,g., residential, downtown, industrial-commercial), which can have very different urban energy budget. To improve the representation of the thermodynamic effects of urban surfaces inside MM5, the soil model, SM2-U, which considers in detail both rural and urban surfaces, has been introduced with the RA.

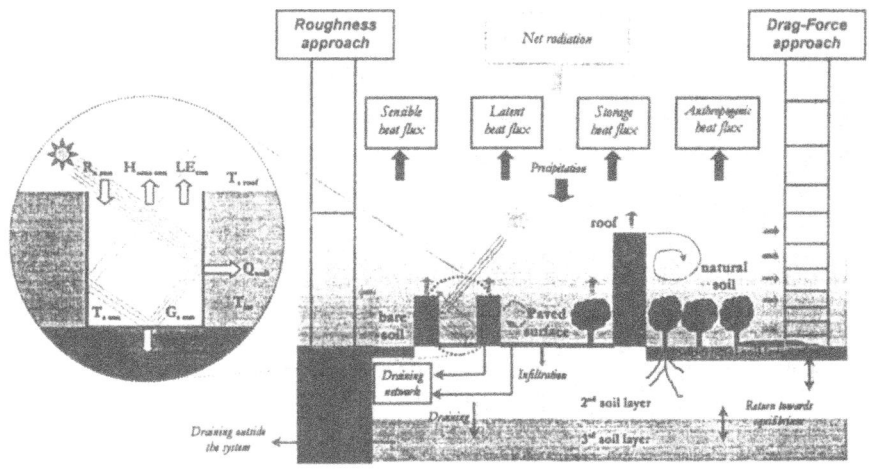

Figure 1. Scheme of the new urban canopy parameterizations inside MM5 using the soil model SM2-U with the roughness approach (left side) and with the drag-force approach (right side).

SM2-U corresponds to an extension to urban surfaces of the Force-Restore model ISBA (Noilhan and Planton, 1989). As indicated in Figure 1, SM2-U considers five surface types: natural soil, bare soil, paved surfaces, building roofs, and water. Each surface type is characterized in a cell by its surface density. Three soil layers are considered: a superficial soil layer for the natural surfaces, allowing the evaluation of the evaporation fluxes from the bare soils; a second soil layer representing the influence area of the vegetation roots; and a third soil layer used as a water reservoir to provide water to the second soil layer by diffusion in dry periods.

SM2-U determines the mean surface temperature and mean heat fluxes from the canopy from the average values of each surface type weighted by their area density. The building wall effects are considered by SM2-U through the equation of the paved surface temperature by introducing the heat stored by building walls, the wall resistance to atmospheric forcing, and by modeling the radiative trapping with an effective street canyon albedo parameterization deduced from Masson (2000). Thus, for dense urban districts, the paved surface temperature and heat fluxes correspond respectively to an effective mean temperature of street canyon surfaces and mean heat fluxes from street canyons. The anthropogenic heat fluxes are parameterized following Taha (1999)

2.3. DA-SM2-U Version

The DA-SM2-U version of MM5 corresponds to an integration of the DA with SM2-U by coupling the previous versions of MM5 and by using the work of Martilli et al. (2002) to improve the DA used in the DA-SLAB version (see Figure 1). Furthermore, in this version, the DA is extended to the vegetation to represent all urban and rural areas in the same way.

Thus, inside the canopy, the effects of buildings and vegetation are represented by adding i) in the dynamic equation a friction force induced by horizontal surfaces of buildings, and a pressure and viscous drag force induced by the presence of buildings and vegetation, ii) in the temperature equation the sensible heat fluxes due to buildings and vegetation, and the anthropogenic heat flux parameterized following Taha (1999), iii) in the specific humidity equation the humidity sources coming from the evapotranspiration of the vegetation and the evaporation of the water intercepted by buildings, and iv) in the TKE equation a shear production terms induced by horizontal surfaces of buildings, TKE sources induced by the presence of buildings and vegetation, and buoyant production terms from the sensible heat fluxes emitted by buildings and vegetation. The turbulence length scale (TLS) has been also modified inside the urban canopy, as proposed by Martilli et al. (2002), by adding a second length scale to consider the vortices induced by the presence of buildings; this modification has been also extended to the vegetation. All of these new terms are volumetric: the volume of buildings is considered in each cell whereas the volume of the vegetation is neglected. Note also that the turbulent transport in the vertical is also modified to consider the real volume of air in the cell.

With the DA, SM2-U determines the mean heat fluxes and mean surface temperature in each cell inside the canopy following the vertical distribution of buildings and vegetation. The net radiation flux inside the canopy is attenuated by the presence of the taller roughness elements, and the heat fluxes from the street canyon are distributed inside the street following the sky view factor of the walls and the flour.

3. APPLICATION FOR PHILADELPHIA

Our study focuses on simulating the meteorology and chemistry on Philadelphia, Pennsylvania (USA), on 14 July 1995, a day characterized by high pressure, abundant sunshine, abnormally high temperatures, and largely stagnant conditions on the eastern United States and southern Canada, which were favorable for the progressive production of ozone and transport of precursor pollutants. MM5 Version 3 Release 5 was run in a one-way nested configuration for several days in July including five nested MM5 computational domains 108-, 36-, 12-, 4-, and 1.33- km horizontal grid spacing. The new MM5 versions with UCPs are used only on the 1.33 km domain. Three simulations are made with the 1.33 km domain: one with the standard version of MM5, the second with the RA-SM2-U version, and the third with the DA-SLAB version. For DA-SLAB, ten layers were added in the lowest 100 m (lowest layer depth of 4 m) for both MM5 and CMAQ to resolve flow within the urbanized area. Simulations with the DA-SM2-U will be presented at the oral presentation.

These three cases use the GSPBL model and include the parameterization of the TLS of Bougeault and Lacarrère (1989) inside the PBL during unstable conditions, and extended to stable conditions for the RA-SM2-U version. For RA-SM2-U and DA-SLAB versions, seven urban subcategories have been roughly constructed inside the U.S. Geological Survey 24-category database to cover the urban area following Ellefsen (1990-91), with simplified vertical distributions of building for DA-SLAB.

The emissions inventory is the U.S. EPA National Emissions Trends Inventory, Version 3.11 for 1996. The CMAQ's version of the carbon-bond chemical mechanism, Version 4 (CB-4) is used where the eddy diffusivity field was modified to be based on the algorithms in the MM5's GSPBL scheme to ensure consistency with respect to influences of the in-canopy TKE.

4. RESULTS

In this section, a brief overview of the sensitivity of i) meteorological variables which affect the pollutant concentration and ii) urban ozone concentrations to the new MM5 UCPs is presented. More results will be shown during the oral presentation.

4.1. Meteorological Fields

Previous studies of the impact of our new MM5 UCPs (see Dupont et al., 2003) have shown that with the DA the simulated vertical profiles of thermodynamic and turbulent variables have a good behavior and are consistent with observed data in real urban area and wind tunnel studies: the shear stress and TKE are characterized by a maximum near the top of the canopy, followed by a decrease inside the canopy. The vertical profiles of the potential air temperature at an urban point (see Figure 2) show that simulations using the new UCPs reduce the tendency toward stable stratification. They even yield a neutral layer during the night because of the anthropogenic heat fluxes and also the heat released by urban surfaces for SM2-U which explains the warmer air with SM2-U. The higher nocturnal neutral layer, i.e. MH (Figure 3), with SM2-U is also explained by the use of the parameterization of Bougeault and Lacarrère (1989) which simulates larger values of the TLS than the GSPBL model. After sunrise (8 a.m. local time), the air temperature in the lower part of the atmosphere and the MH are higher for the DA-SLAB case. Indeed, for SM2-U the surface storage heat flux is larger than the sensible heat flux (not shown), which means that in the morning urban surfaces start to store energy before warming, whereas for DA-SLAB case the roof storage capacity is neglected. During the second part of the day, the three potential air temperature profiles have the same shape but the air temperature is always lower for SM2-U, since the surface storage heat flux is larger, and thus the sensible heat flux is lower. For the DA-SLAB case, the stability of the air inside the canopy is neutral during the night and slightly unstable during the day. Furthermore, the MH is always higher during the day, the mixing inside the RSL being larger with the DA because of the larger values of TKE at the top of the canopy that increase the TLS.

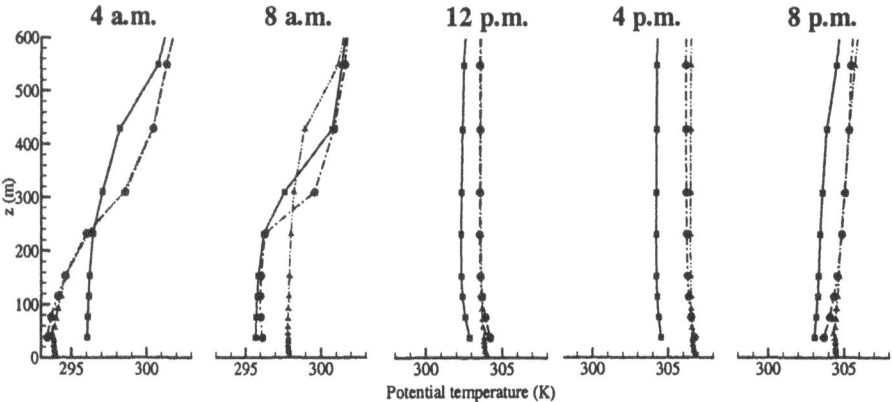

Figure 2. Vertical profiles of the potential air temperature for an urban point in Philadelphia center, simulated by the three following MM5 versions: standard (dashed dot line with circles), DA-SLAB (dashed dot dot line with triangles) and RA-SM2-U (solid line with squares). The indicated times are local times.

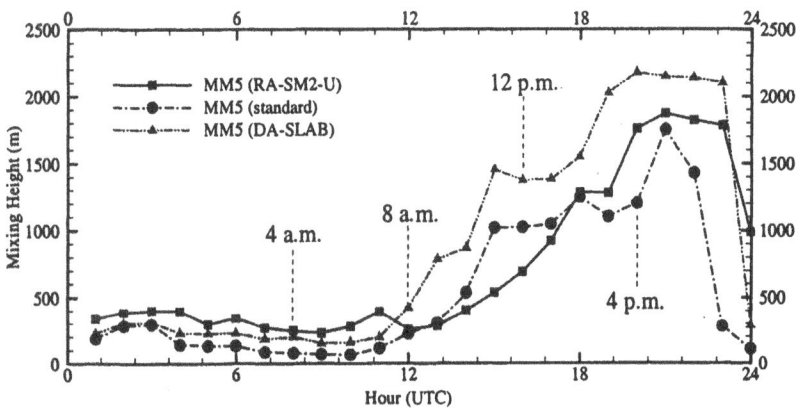

Figure 3. Time evolution of the mixing height for the same urban point as the Figure 2, simulated by the three MM5 versions.

4.2. Ozone Fields

For the same urban point as Figúres 2 and 3, the ozone profiles and the ozone concentration at the top of the canopy (20 m) are presented respectively in the Figures 4 and 5. As expected from the meteorological fields, the ozone concentration is larger during the night for the RA-SM2-U case, the mixing inside the RSL being larger and the concentration increasing with the height; thus the NO_x and its titrating effect on ozone are reduced. At 8 a.m. (local time), the ozone concentrations in the lower part of the atmosphere are very similar for the three cases because of the small values induced by the large concentration of NO_x at this time. During the daytime, the ozone concentrations near the surface are very similar for the DA-SLAB and standard cases, the differences in the upper part of the atmosphere are only due to the MH differences. During this period, the surface heat fluxes simulated from these two last cases are very similar. The ozone concentration simulated by the RA-SM2-U case is always lower; but from Figure 5, it seems that the ozone concentrations of RA-SM2-U are not in phase with concentrations of the two other cases. This interval of about 30 minutes can probably be explained by the hysteresis nature of the storage heat flux with the sensible heat flux which is stronger with the RA-SM2-U case, and thus delays the mixing layer development. After sunset (8 p.m. local time), the ozone concentration decreases quickly near the surface for the standard case because of the stable stratification of the atmosphere.

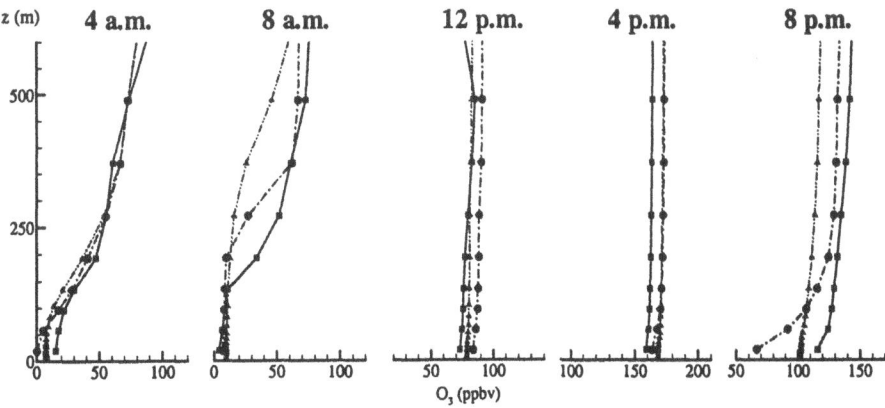

Figure 4. Vertical profiles of ozone for the same urban point as Figure 2, simulated by CMAQ using the three following MM5 versions: standard (dashed dot line with circles), DA-SLAB (dashed dot dot line with triangles) and RA-SM2-U (solid line with squares). The indicated times are local times.

5. SUMMARY

These first results show that the development of a more detailed urban canopy parameterization has an impact on boundary layer structure, which in turn influences the ozone fields especially during the night because of the heat released by the canopy. The increase of the number of layers and the improvement of the dynamics and turbulent profiles inside the canopy (Dupont et al., 2003) does not seem to have a large impact on the ozone fields. In these simulations, the CMAQ model and the emission inventory have not been changed to consider the more detailed canopy. Modification of the transport equations and more accurate treatment of the emission inventory inside the canopy will probably generate stronger modifications on ozone fields. The larger storage heat flux simulated by SM2-U after sunrise, larger than the sensible heat flux, seems to slightly delay the mixing height development which also delays the increase of ozone. SM2-U and the DA have given encouraging behavior on meteorological fields, and their coupling in the version described in section 2-3 of this paper is promising. These new versions of MM5 and CMAQ will be later applied on Houston where a large urban morphological database is in construction. CMAQ results will be evaluated against observations of the Texas 2000 Air Quality Study.

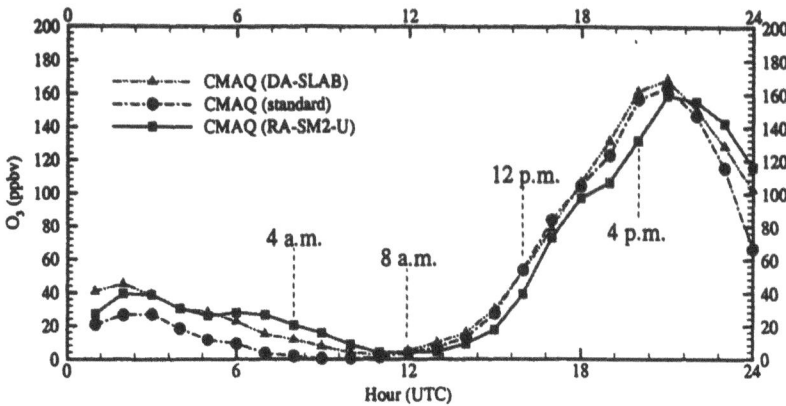

Figure 5. Time evolution of the ozone concentration at 20 m above the same urban point as Figure 2, simulated by CMAQ using the three MM5 versions.

Disclaimer. The information in this manuscript has been prepared under funding by the United States Environmental Protection Agency. It has been subjected to Agency review and approved for publication. Mention of trade names or commercial products does not constitute endorsement or recommendation for use.

6 REFERENCES

Bougeault, P., and Lacarrère, P., 1989 Parameterization of orography-induced turbulence in a mesobeta-scale model. *Mon. Wea. Rev.*, 117, 1872–1890.

Brown, M. J, 2000, Urban parameterizations for mesoscale meteorological models, *Mesoscale Atmospheric Dispersion.* Ed., Z Boybeyi, Wessex Press, 448 pp.

Byun, D. W., and Ching J. K. S., 1999, Science algorithms of the EPA Models-3 Community Multiscale Air Quality (CMAQ) Modeling System. EPA/600/R-99/030.

Chin, H. S., Leach, M. L., and Brown, M. J., 2000, A sensitivity study of the urban effect on a regional scale model: an idealized case. Preprints, *Third Sym. Urban Env.*, Davis, CA, Amer. Meteor. Soc., 76–77.

Dupont, S., 2001, Modélisation dynamique et thermodynamique de la canopée urbaine : réalisation du modèle de sols urbains pour SUBMESO, *Doctoral thesis*, Université de Nantes, France.

Dupont, S., Otte, T. L., Lacser, A., and Ching, J., 2003, Using MM5 to simulate the meteorological fields at neighborhood scale, The Fourth International Conference on Urban Air Quality-Measurement, Modeling and Management, 25-28 March 2003, Charles University, Prague, Czech Republic, 4 pp.

Ellefsen, R., 1990-91, Mapping and measuring buildings in the canopy boundary layer in ten U.S. cities, *Energy and Buildings*, 15–16, 1025–1049.

Grell, G., Dudhia, J., and Stauffer, D. R., 1994, A description of the Fifth-Generation Penn State/NCAR Mesoscale Model (MM5). NCAR/TN-398+STR, 138 pp.

Lacser, A., and Otte, T. L., 2002, Implementation of an urban canopy parameterization in MM5. Preprints, *Fourth Sym. Urban Env.*, Norfolk, VA, Amer. Meteor. Soc., 153–154.

Martilli, A., Clappier, A., and Rotach, M. W., 2002, An urban surface exchange parameterisation for mesoscale models, *Bound.-Layer Meteor.*, 104, 261–304.

Masson, V., 2000, A physically-based scheme for the urban energy budget in atmospheric models, *Boundary Layer Meteorology*, 98, 357-397.

Noilhan, J., and Planton, S., 1989, A simple parameterization of the land surface processes for meteorological models, *Mon. Wea. Rev.*, 117, 536-549.

Taha, H., 1999, Modifying a mesoscale model to better incorporate urban heat storage: A bulk parameterization approach, *J. Appl. Meteor.*, 38, 466–473.

DISCUSSION

D.KORACIN   How did you define and modify the vertical grid of
                   the MM5 model and how many layer you have within first 100
                   m?

S.DUPONT   The definition of the mesh and the interpolation of the
                   meteorological variables from the previous domain (4-km
                   horizontal grid spacing) to the smallest domain (1.3-km
                   horizontal grid spacing) for the boundary conditions are made
                   with the NESTDOWN program. 12 layers are within the first
                   100 m.

MODEL ASSESSMENT AND VERIFICATION

Chairperson:	P. Builtjes
	R. Bornstein
	D. Syrakov
	E. Genikhovich
Rapporteur:	R. San Jose
	A. I. Miranda
	J. Ferreira

REGIONAL CIRCULATIONS WITHIN THE IBERIAN PENINSULA EAST COAST

Oriol Jorba, Santiago Gassó and José M. Baldasano*

1. INTRODUCTION

The Iberian Peninsula is located at middle latitudes in the north hemisphere, in the most occidental part of Europe. It is limited at west by the Atlantic Ocean and by the Mediterranean Sea at east. Because these particularities and its complex orography, the development of intricate circulations within the Peninsula are common. The more usual synoptical situations affecting this region are westerly and northwesterly flows, and typical summertime barometric swamps (Martín, 1987; Jorba *et al.*, 2003). Photochemical air pollution episodes in the Iberian Peninsula are usually produced under summertime barometric swamps (MMAM, 2001).

The Mediterranean lands normally experience higher temperatures, greater amounts of sunshine and fewer rain days than western and central Europe during summer (Salvador *et al.*, 1997). The pressure system that dominates the region is the Azores anticyclone, contributing to the development of large thermally driven convective systems over the Peninsula and land-sea breezes over the coasts.

The complex orography of the Iberian Peninsula east coast, characterized by a first large range of mountains with heights between 500 to 3000 m at few kilometres inland, contributes to an intricate circulation development. A combination of sea breezes with upslope winds contribute to the advection and injection inland of the sea-coastal air masses. In this sense, the topography of the region plays an important role inducing the orographic forcing of the flows that are injected aloft. Furthermore, the location along the east coast of several urban regions contributes to the pollution of the zone, with the advection of primary pollutants inland that react photochemically, producing a complex pattern of air pollution within the east coast. High levels of O_3 along the Iberian Peninsula have been reported during summer (Martín *et al.*, 1991; Millán *et al.*, 1991; Toll and Baldasano, 2000; Barros *et al.*, 2003).

Millán *et al.* (1997) generally described the recirculations processes occurring along the Western Mediterranean Basin. During the day the sea breezes combine with upslope winds to transport coastal pollutants inland, while at the leading edge of the breeze front a large fraction of these pollutants are injected in their return flows aloft at heights ranging from 2 to 3 km. Once in those upper layers the pollutants move back toward the sea, and the compensatory subsidence creates "stratified reservoir layers" of aged pollutants,

---

*Oriol Jorba, Santiago Gassó and José M. Baldasano, Laboratory of Environmental Modeling, Department of Engineering Projects, Universitat Politècnica de Catalunya (UPC), SPAIN.

Air Pollution Modeling and Its Application XVI, Edited by
Borrego and Incecik, Kluwer Academic/Plenum Publishers, New York, 2004

stacked up to 2-3 km high, along the coast over the sea. These layers act as reservoirs and retain ozone from one day to the following days. The next morning the lowermost layers are drawn inland by the sea breeze, and the aged pollutants can react with new coastal emissions. With the aim to deep into the knowledge of these recirculation processes, a numerical weather prediction model (NWP) has been used. A synoptical situation affecting the Iberian Peninsula has been analysed and results are presented in this contribution.

2. MODEL CONFIGURATION

The NWP used in this work is MM5, the Pennsylvania State University (PSU)/National Center for Atmospheric Research (NCAR) Mesoscale Model 5, version 3, release 4 modeling system (Dudhia, 1993; MMD/NCAR, 2001), it is a non-hydrostatic mesoscale primitive equation model.

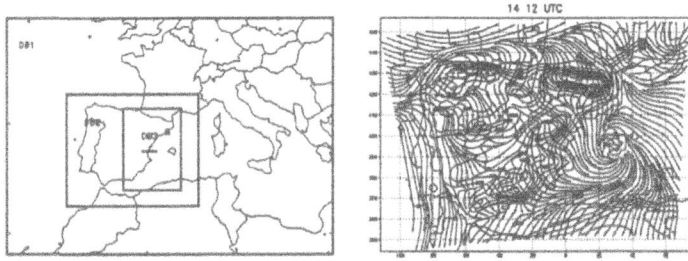

Figure 1. Left: Domains definition, radiosonde site (black square), cross section location (black line). Right: Surface streamlines for August 14th at 12 UTC.

The physics options used for the simulations were: the Mellor-Yamada scheme as used in the Eta model (Janjic, 1994) for the PBL parameterization, the Anthes-Kuo and Kain-Fritsch cumulus scheme (Kain and Fritsch, 1993), the Dudhia simple ice moisture scheme, the cloud-radiation scheme, and the five-layer soil model.

Three nested grids were defined (figure 1 left): a coarse grid having 72 km grid spacing with 35 x 49 cells covering the major part of Europe, a medium grid of 24 km with 49 x 61 cells covering the Iberian Peninsula, and finally a fine grid of the Iberian Peninsula east cost with 6 km of resolution and 141 x 105 cells. A one-way nesting approach was used.

Initialization and boundary conditions for the mesoscale model were introduced with analysis data of the ECMWF global model. Data were available at a 1-degree resolution (100-km approx. at the working latitude) at the standard pressure levels every 6 hours.

3. METEOROLOGICAL SITUATION

An intense episode of photochemical air pollution was produced during August 10th to 19th of 2000 in the north zone of the east Iberian coast. The highest ozone level of the year was reached in this region with a maximum of 273 µg/m³ in August 15th. The more intense part of the episode covered the days 14th to 16th.

The situation modelled for this contribution is the episode that covers August 13th to August 16th of 2000. This episode was characterized by a meteorological situation with weak synoptic forcing. In this case mesoscale phenomena, induced by the particular topography of the region, would be dominant. Thus, the description of the regional circulations occurring along the east Iberian coast will deep into the knowledge of the atmospheric phenomenas that contributes to develop the photochemical air pollution episode.

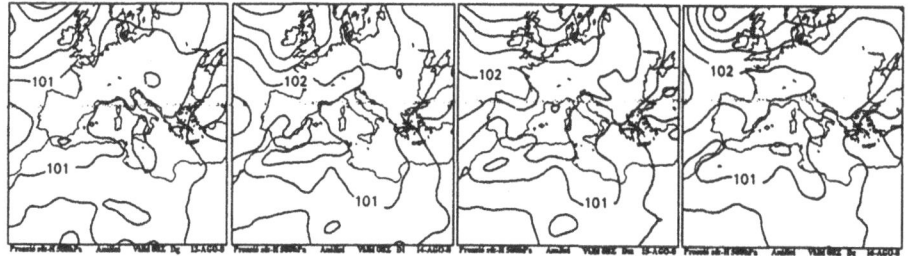

Figure 2. Synoptic situation of August 13th to 16th, 00UTC surface analysis.

Synoptic situation of the episode corresponds to a typical summertime barometric swamp over the Iberian Peninsula (figure 2). In height, a zonal flow blows aloft the Peninsula veering to the southeast having northwesterly winds affecting the northeast of the Peninsula. At surface, the high-pressure area is centred over south Atlantic Ocean, with the anticyclonic wedge affecting most part of the Iberian Peninsula, producing a typical barometric swamp along the easterly part of the region. Under this situation surface winds are low. This fact, and the strong daily solar heating, produced the development of mesoscale phenomena. These phenomena in the region are mainly sea-breezes, up-slope and down-slope winds and valley channelled winds. The heating during August 14th was so intense that a thermal low started to develop in the southeast of the Iberian Peninsula. The behaviour of the four days of the episode was similar, with low pressure gradient at surface and weak westerly northwesterly flows in height.

4. RESULTS: SUMMERTIME BAROMETRIC SWAMP - AUGUST 13th TO 16th OF 2000.

4.1. MM5 simulation

Results of the simulation with MM5 for the meteorological situation will be described. In order to validate the model behaviour, comparisons with surface measurements of wind and with the radiosonde launched at Barcelona site (see figure 1, left) have been done, and will be presented.

As explained before, under this situation the Iberian Peninsula was dominated by the Azores anticyclone, with very low pressure gradient. The sea-breeze regime, developed within all the coast of the Western Mediterranean Basin, induces an anticyclonic circulation over all the Western Mediterranean, with general subsidence over the region as noted by Millán et al. (1997). This anticyclonic circulation extends aloft reaching heights even of 4 km, the middle troposphere. In figure 1 (right) the surface streamlines are shown at 12 UTC for August 14th. The anticyclonic circulation is clearly developed. The formation of this circulation begins at August 13th, where the barometric swamp establishes over the Iberian Peninsula, and the centre of the circulation moves over the western Mediterranean in function of the sea-breeze strength. All the Western Mediterranean coasts present inland sea-breeze flows. Only the canalisation of the Pyrenees and the Central Massif blows into the Mediterranean, but the flow turns to the coasts by the thermal forcing. In low levels, this canalisation plays an important role, because it is the only pass to bring "new" air into the western Mediterranean basin.

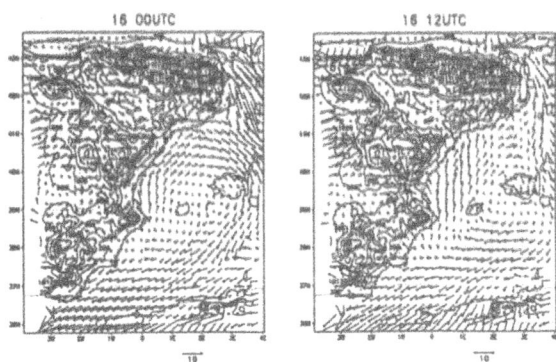

Figure 3. Domain 3 surface wind field for August 16th at 00 UTC (left) and 12 UTC (right).

A well developed sea-breeze regime is established along all the Iberian east coast for the analysed period, with breeze circulation cells up to 2 km height. This regime covers the central hours of the day, starting around 8-10 UTC and changing to a land-sea flow around 19 UTC. Figure 3 shows the surface wind field for August 16th at 00 and 12 UTC. It is remarkable the maintenance of the anticyclonic circulation over the sea at 00 UTC induced by the see-breeze regime of the previous days. At night, an offshore flow is established along the coastal zone. Inland, the downslope winds regime is dominant. At noon, the onshore winds are well developed along all the east coast. At 12 UTC, the breeze has reached the first mountainous chain producing important orographic injections. A common characteristic of this situation is the permanent entrance of air masses by the Pyrenees-Massif Central canalisation and the anticyclonic circulation during all day.

The strength of the sea-breeze and the complex orography of the east Iberian coast produce several injections of air due to orographic forcing. A cross section has been preformed in order to describe the recirculations in height (figure 4). A first plane coastal zone, a mountain chain, and a second higher mountainous chain characterize the orography of the cross section. Figure 1 shows the location of the cross section.

During August 13th, the development of the breeze produces three return flows around, 800, 1200 and 2300 m MSL. The major injection is produced when the breeze reaches the second mountainous chain. Figure 4 shows the major injections produced specifically at this point for August 13th, 14th and 16th. The convergence of the onshore flow and the inland westerly flow produces a chimney injecting air up to 3900 m MSL

456

for August 14th. This injection is more impressive for August 16th, where a chimney reaches heights over 5000 m MSL, and breaks the northwesterly circulation aloft. Important subsidence over the Mediterranean Sea is produced.

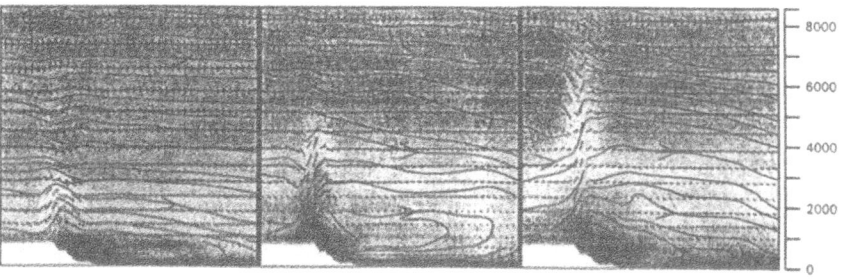

Figure 4. Cross section of wind field and the mixing ratio at the central part of the east Iberian coast (Left: August 13th at 15 UTC, centre: August 14th at 15 UTC, right: August 16th at 13 UTC).

Figure 5 shows two trajectories extracted from the MM5 simulation. It is noticeable the low speed of both trajectories, due to the low baric gradient dominant within the region. The subsidence over the sea and the anticyclonic circulation manage the evolution of both trajectories. The air at middle troposphere is forced to go down by the subsidence over the east coast, once in low levels the air masses recirculate over the sea with a possible return to the seaboard 3 days later.

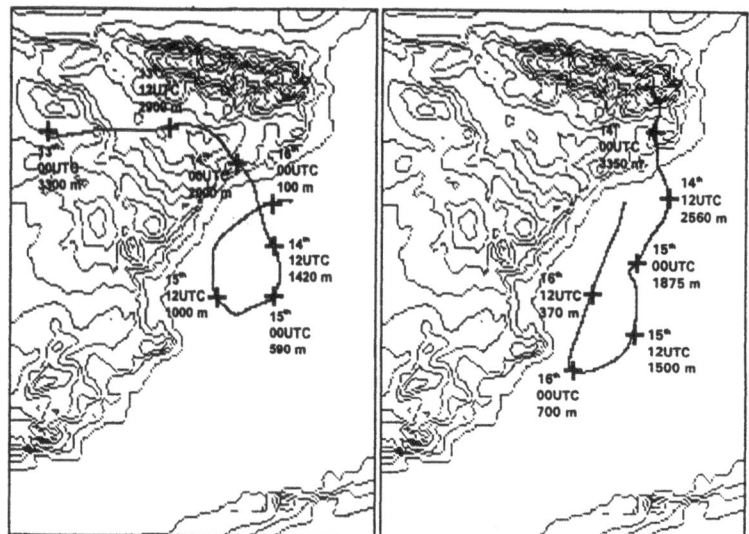

Figure 5. MM5 trajectories for August 13th to 16th (top view).

The combination of these circulations over the sea, and the orographic injections showed in figure 4 produces a recirculation of pollutants along the east coast that contributes to the development of episodes of photochemical air pollution in this region under barometric swamp conditions. Moreover, thermal injections of air in central parts of the Peninsula are common in summer (Salvador et al., 1997). These air masses are advected to the coast where the subsidence can introduce them to the recirculation processes described above. The accumulation of polluted air masses contributes to the degradation of the air quality of the east Iberian coast in summer.

4.2. Comparison with measurements

With the aim to validate the results and the explanation exposed above, comparisons with surface and aloft measurements have been done. 52 surface stations located at the northeast of the Iberian Peninsula have been used for the surface comparisons, and a radiosonde launched at the Barcelona city (North of the Iberian east coast) contributes to the verification of the model results.

The BIAS and the root mean square error (RMSE) of the wind velocity at 10 m AGL have been evaluated. In figure 6 the reader can see the evolution of these statistics for the period August 13th to 16th 2000.

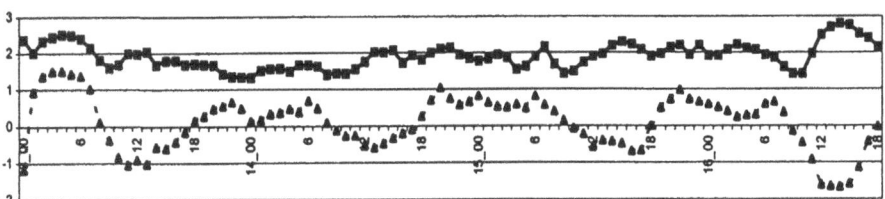

Figure 6. Evolution of the RMSE and BIAS of the wind velocity (ms⁻¹) at 10 m (Solid line: RMSE, dotted line: BIAS).

The absolute error remains below 2.5 ms⁻¹ for all the simulation with an increase at the evening of the last day. The error decreases after the cold start of the simulations and stabilizes around 2 ms⁻¹. The model tends to overestimate the nocturnal winds, and presents some problems to simulate the strength of the diurnal winds. The behaviour of the error has a daily cycle that tends to increase in magnitude as the simulation progress. Hence, the higher errors appear during august 16th.

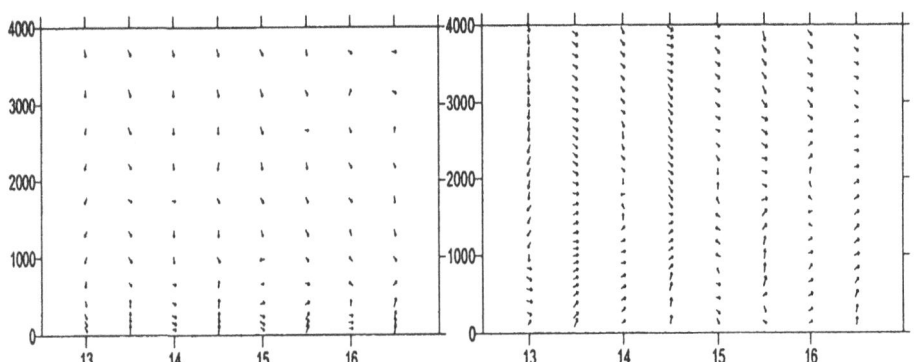

Figure 7. Vertical profile of the horizontal wind at Barcelona site for August 13th to 16th twice daily, 00 and 12 UTC (Left panel: MM5 results, right panel: Barcelona radiosonde data).

Data of the radiosonde routinely launched twice daily (00 UTC, 12 UTC) in Barcelona city has been used to validate the model results for the upper layers. Figure 7 shows the radiosonde and MM5 data of the period analysed for this site. The vectors represent the horizontal wind at a specific height.

The model results at noon present a marked sea-breeze cell structure. In most cases, the model simulates a narrow surface layer of see-breeze flow blowing southerly inland. The return of the flow is located between 500 and 1000 m AGL for the simulated days with a northwesterly direction. However, the structure is not so clear with the data measured by the radiosondes. In this sense, the inland flow blows more southwesterly than the model results and presents an extended layer reaching heights over 1500 m AGL, except for august 15th. This day has a more complex vertical structure with a first return at low levels, above 200 m AGL, and a second inland flow at 500m AGL with the return around 2000 m AGL. These profiles show the complexity of the circulations induced over eastern coast of the Iberian Peninsula due to its intricate orography.

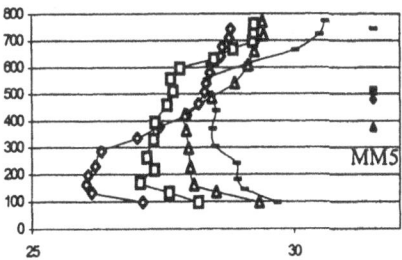

Figure 8. Potential temperature (°C) profiles of the radiosonde data launched at Barcelona site, and MM5 mixing layer height (m AGL) estimation (Diamonds: data of August 13th at 12 UTC, squares: data of 14th at 12 UTC, triangles: data of 15th at 12 UTC, dashes: data of 16th at 12 UTC).

Finally, in figure 8 we present the comparison of the mixing layer (ML) height between the model diagnostic at 12 UTC with the Eta PBL scheme, and the potential temperature profiles of the radiosonde data. The differences between the model prediction and the measurements are around 100 m, and for well developed ML the model tends to subestimate the height. The effect of the see-breeze front produces a limitation in the ML development at Barcelona site, with lower ML heights than it would be expected.

5. CONCLUSIONS

Results from a simulation of a summertime barometric swamp over the Iberian Peninsula have been shown in order to describe the circulations that are produced along the Iberian east coast. The complex orography of the seaboard and the influence of the warm Mediterranean Sea develop these circulations with the low synoptical forcing of the day.

The relevant phenomenas that drive all the circulations within the east Iberian coast and the western Mediterranean basin under a situation of barometric swamp have been described. The results of the simulation have shown the importance of the orographic features of the coastal region in the development of the recirculations observed.

Comparisons with measurements have been done in order to evaluate the reliability of the model results, and to validate the model in situations of low barometric forcings over the Iberian Peninsula. Reasonably agreement is produced between model results and observations.

Due to the characteristics of the high populated seaboard of the east Iberian Peninsula, and the particular circulations induced over the region several photochemical contamination episodes usually occur in summer under barometric swamp situations. This contribution helps to improve the knowledge of the processes responsibles of the high ozone levels in the region.

6. ACKNOWLEDGEMENTS

The authors wish to thank the Spanish Meteorological Institute (INM) for providing data from the ECMWF, and the Catalan Meteorological Service (SMC) for providing surface station and radiosonde data for validation. Simulations were run on an HP Exemplar V2500 belonging to the Supercomputing Center of Catalonia (CESCA). This work was developed under project IMMPACTE and REN2000-1020-C02-02/CLI.

7. REFERENCES

Barros N., Toll, I., Soriano, C., Jiménez, P., Borrego, C., and Baldasano, J. M., 2003, Urban photochemical pollution in the Iberian Peninsula: Lisbon and Barcelona airsheds, *J. Air&Waste Manag. Ass.* (in press).

Dudhia, J., 1993, A non-hydrostatic version of the Penn State-NCAR mesoscale model: Validation tests and simulation of an Atlantic cyclone and cold front, *Mon. Wea. Rev.* 121: 1493-1513.

Janjic, Z. I., 1994, The step-mountain eta coordinate model: Further development of the convection, viscous sublayer, and turbulent closure schemes, *Mon. Wea. Rev.* 122: 927-945.

Jorba, O., Pérez, C., Baldasano, J. M., Rocadenbosch, F., and López, M. A., 2003, Cluster analysis of back-trajectories arriving at Barcelona air basin, in *1ˢᵗ EARLINET Symposium on the Structure and Use of the Database derived form Systematic Lidar Observations*, Hamburg (Germany) 11-12 February.

Kain, J. S., and Fritsch, J. M., 1993, Convective parameterization for mesoscale models: The Kain-Fritsch scheme. The representation of cumulus convection in numerical models, K. A. Emanuel and D. J. Raymond, Eds., Amer. Meteor. Soc., p. 246.

Martín, J., 1987, Característiques climatològiques de la precipitació en la franja costera mediterrània de la Península Ibèrica, PhD. Thesis, Barcelona, edited by Institut Cartogràfic de Catalunya, pp. 111-129.

Martín, M., Plaza, J., Andrés, M. D., Bezares, J. C., and Millán, M. M., 1991, Comparative study of seasonal air pollutant behaviour in a Mediterranean coastal site: Castellón (Spain), *Atmos. Environ.* A 25: 1523-1532.

Mesoscale and Microscale Meteorology Division, National Center for Atmospheric Research, 2001, PSU/NCAR Mesoscale Modeling System Tutorial Class Notes and User's Guide: MM5 Modeling System Version 3, June 2001.

Millán, M. M., Artíñano, B., Alonso, L. A., Navazo, M., and Castro, M., 1991, The effect of meso-scale flows on the regional and long-range atmospheric transport in the western Mediterranean area, *Atmos. Environ.* A 25: 949-963.

Millán, M. M., Salvador, R., Mantilla, E., and Kallos, G., 1997, Photooxidant dynamics in the Western Mediterranean basin in summer: Results from European research projects, *J. G. R.*. 102(D7): 8811-8823.

Ministerio de Medio Ambiente, 2001, Medio Ambiente en España 2000, Madrid.

Salvador, R., Millán, M. M., Mantilla, E., and Baldasano, J. M., 1997, Mesoscale modelling of atmospheric processes over the western Mediterranean area during summer, *Int. J. Env. Pollution* 8: 513-529.

Toll, I., and Baldasano, J. M., 2000, Modeling of photochemical air pollution in the Barcelona area with highly disaggregated anthropogenic and biogenic emissions, *Atmos. Environ.* A 34: 3069-3084.

DISCUSSION

R.BORNSTEIN   Your return flow at 3 km is consistent with a general "rule of thumb" that it should be at an elevation of about 1.5 times the topographic elements.

J.M.BALDASANO   In a general way it follows this "rule of thumb", however, it depends in a great measure on the strength of the sea breeze and the characteristics of the mountainous chain.

S.E. GRYNING   In other presentations (at the conference) a grid size of 5 km has been advocated to be optimum for air pollution studies. Here you use 2 km to resolve the meteorological field. You comment as the grid size for a case in air pollution modelling.

J.M.BALDASANO   It is highly dependent on the complexity of the topography and the heterogeneity of the land-use of the studied region. I think that air pollution models have to take into account the particularities of the domain we are considering with its local characteristics (orography). In this sense, an accurate resolution is necessary in our region due to the complex terrain.

B.AINSLIE   Do you have any measurements to support your claim about the high level injections of moist boundary layer air aloft?

J.M.BALDASANO   We have observed this kind of phenomena with experimental aerosol measures that were made by lidar in several campaigns during 1992 and 2000-2002.

P.BUILTJES   Can you give already an estimate about how much ozone is produced locally in the Barcelona Area?

J.M.BALDASANO   The present estimation for Barcelona area is about 80%, nevertheless it depends logically on the daily hour due to an important recirculation factor of the air masses.

THE EVALUATION OF REGIONAL - SCALE AIR QUALITY MODELS

As part of NOAA's Air Quality Forecasting Pilot Program

Daiwen Kang, Brian K. Eder, and Kenneth L. Schere[*]

1. INTRODUCTION

A major component of the National Oceanic and Atmospheric Administration's (NOAA) Air Quality Forecasting Pilot Program has been the development and implementation of an evaluation protocol. This pilot program - which has as its eventual goal, development of an operational, National air quality forecasting system, enlisted three regional-scale air quality models to forecast O_3 concentrations across the eastern United States during the summer of 2002. The three models used in this demonstration include: a hybrid Lagrangian model based on NOAA's HYSPLIT (HYbrid Single-Particle Lagrangian Trajectory) model; an Eulerian model with coupled chemistry and meteorology developed at NOAA's Forecast System Laboratory, and MAQSIP (Multiscale Air Quality SImulation Platform), another Eurlerian model developed by Environmental Modeling Center of MCNC.

A suite of statistical metrics was identified as part of the evaluation protocol development to facilitate evaluation of both *discrete forecasts* (observed versus modeled concentrations) and *categorical forecasts* (observed versus modeled exceedances / non-exceedances) for both the maximum 1-hr (85 ppb) and 8-hr (125 ppb) forecasts produced by each of the models. Implementation of the evaluation protocol took place during a 25-day period (5-29 August 2002) utilizing hourly O_3 concentration data obtained from the U.S. EPA's Air Quality System (AQS) network.

---

[*] Daiwen Kang[#], Brian K. Eder[@], and Kenneth L. Schere[@], NERL, U.S. Environmental Protection Agency, RTP NC, 27711, USA. [#] On Assignment from University Corporation for Atmospheric Research, Boulder, CO 80301, USA. [@]On assignment from Air Resources Laboratory, National Oceanic and Atmospheric Administration, RTP, NC 27711, USA.

2. THE MODELING SYSTEMS

2.1. HYSPLIT-CheM

NOAA's Air Resources Laboratory (ARL)'s HYbrid Single-Particle Lagrangian Trajectory Chemistry Model (HYSPLIT CheM) is a hybrid Lagrangian-meteorological/Eulerian-chemical modeling system, that utilized meteorological output from the Mesoscale Meteorological Model (MM5) and emissions from the Sparse Matrix Operator Kernel Emissions (SMOKE) program [Stein et al., 2000]. The HYSPLIT CheM model assumes that the entire pollutant mass at each emission source is uniformly distributed among a number of "particles", each of which may be thought of as a capsule containing the various chemical species. These particles are advected, dispersed, and deposited throughout the simulation domain. The 24-hour ozone mixing ratios are forecasted for the eastern US based upon a 50 km resolution grid and 10 vertical layers. Ozone mixing ratios from the surface layer (75 m) are used in this study.

2.2. MM5-Chem

NOAA's Forecast Systems Laboratory (FSL)'s MM5-Chem modeling system [Grell et al., 2000] is a multiscale Eulerian air pollution prediction system based on MM5, which is coupled with the RADM2 chemical mechanism [Stockwell et al., 1990]. The biogenic emissions and chemical-transport-transformation calculations are performed as part of the MM5 simulation. Anthropogenic emissions data are from the EPA NET-96 data set. The model domain contains 27 km horizontal grid cells and 30 vertically stretched layers. Ozone mixing ratios from the surface layer (16 m) are used in this study.

2.3. MAQSIP

The Multiscale Air Quality Simulation Platform (MAQSIP) [McHenry, et al., 2003] is a comprehensive Eulerian grid model. The Real-Time version of MAQSIP (MAQSIP-RT) is a highly optimized version of MAQSIP that uses a modified version of the Carbon Bond IV chemical mechanism [Gery et al., 1989]. The embedded SMOKE provides emissions data and MM5 supplies the meteorological data. The model domain covers the eastern United States using 45 km horizontal grid spacing and 31 vertical layers. Results from the surface layer (38 m) are used for this evaluation.

3. O$_3$ DATA

The O$_3$ data employed in this evaluation were obtained from the Environmental Protection Agency's Air Quality System (AQS) (formerly the Aerometric Information Retrieval System (AIRS)). This database contains a multitude of hourly aerometric data, including O$_3$ concentrations (measured in parts per billion ppb), collected by state and local agencies at thousands of locations nationwide. For this study, depending on model domain, between 581 and 663 AQS monitors were used. For those model grid cells containing more than one monitor, the average concentration measured by the monitors was used.

4. STATISTICAL TECHNIQUES

4.1. Discrete Statistics

For the **discrete forecast** evaluation we calculated summary statistics along with two measures of **bias**: the Mean Bias (MB) and the Normalized Mean Bias (NMB); and two measures of **error**: the Root Mean Square Error (RMSE) and Normalized Mean Error (NME) as seen in Equations 1-4 below:

$$MB = \frac{1}{N}\sum_{1}^{N}(Model - OBS) \quad (1) \qquad NMB = \frac{\sum_{1}^{N}(Model - Obs)}{\sum_{1}^{N}(Obs)} \cdot 100\% \quad (2)$$

$$RMSE = \left(\frac{1}{N}\sum_{1}^{N}(Model - Obs)^2\right)^{0.5} \quad (3) \qquad NME = \frac{\sum_{1}^{N}|Model - Obs|}{\sum_{1}^{N}(Obs)} \cdot 100\% \quad (4)$$

4.2. Categorical Statistics

For the **categorical forecast** evaluation we calculated model Accuracy (A), Bias (B), Probability Of Detection (POD), False Alarm Rates (FAR) and Critical Success Index (CSI) based upon observed exceedances, non exceedances *versus* forecast exceedance, non exceedances for both the 1- and 8-hour standard. The equations for each metric are found below, where as seen in Figure 1, **a** would represent a forecast 1-hr exceedance (>125 ppb) that did not occur, **b**: a forecast 1-hr exceedance that did occur, **c**: a forecast 1-hr non-exceedance that did not occur and **d**: a non forecast 1-hr exceedance that did occur.

$$A = \left(\frac{b+c}{a+b+c+d}\right) \cdot 100\% \quad (5)$$

$$CSI = \left(\frac{b}{a+b+d}\right) \cdot 100\% \quad (6)$$

$$POD = \left(\frac{b}{b+d}\right) \cdot 100\% \quad (7)$$

$$B = \left(\frac{a+b}{b+d}\right) \quad (8)$$

$$FAR = \left(\frac{a}{a+b}\right) \cdot 100\% \quad (9)$$

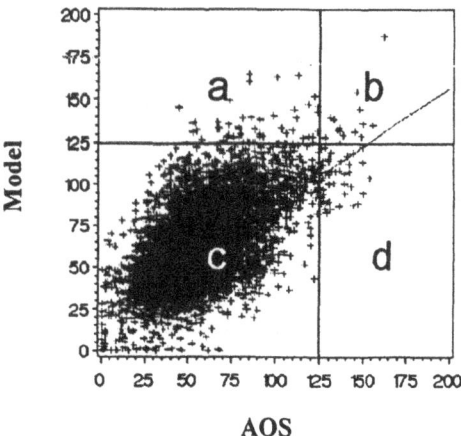

Figure 1. Example plot for categorical evaluation

5. RESULTS

5.1. Discrete Evaluations

As seen in Table 1, which provides results for the discrete forecasts, the three models varied in their ability to accurately predict the 1-hr and 8-hr maximum ozone concentrations. All three models over predicted concentrations as indicted by their positive MBs and NMBs. For the 8-hr predictions, HYSPLIT provided the best performance with a MB of only 0.53 ppb (NMB: 1.02%); while for the 1-hr prediction, MAQSIP performed best with a MB of 3.67 ppb (NMB 6.35%). In terms of error, MAQSIP out performed the other models for both the 1- and 8-hr maximum forecasts, producing the lowest RMSEs (14.85, 13.50 ppb) and NMEs (19.86, 21.03%) respectively. MAQSIP also provided better correlation coefficients for both the 8-hr (0.76) and 1-hr (0.73) as compared to MM5-Chem (0.69, 0.66) and HYSPLIT (0.60, 0.56). Note that for each model, the correlation coefficient associated with the 8-hr max was greater than the 1-hr max.

Table 1. Summary of discrete statistics for forecasts of maximum 1-hr and 8-hr O_3.

Models / Metrics	HYSPLIT		MM5-Chem		MAQSIP	
	Max1hr	Max8hr	Max1hr	Max8hr	Max1hr	Max8hr
MB (ppb)	4.81	0.53	7.42	6.71	3.67	5.04
NMB (%)	8.13	1.02	12.38	12.91	6.35	9.98
NME (%)	24.89	23.48	25.18	25.06	19.86	21.03
RMSE (ppb)	19.23	15.64	19.80	17.05	14.85	13.50
r	0.56	0.60	0.66	0.69	0.73	0.76

5.2. Categorical Evaluations

As shown in Table 2, the **Accuracy** (A) for each model prediction, which indicates the percent of forecasts that correctly predict an exceedance or non-exceedance, is generally greater than 90% (the exception being MM5-Chem's 8-hr exceedance forecast). The Accuracy of each model's 1-hr exceedance/non-exceedance prediction is considerably better than its 8-hr prediction, and in fact approaches 100% (perfection); however, care must be taken in interpretation of this metric, as it is greatly influenced by the overwhelming number of non-exceedances. To circumvent this inflation (which is common when evaluating the prediction of rare events like O_3 exceedances), the **Critical Success Index** (CSI) is often a better metric of model performance. The CSI provides a measure of how well the exceedances were predicted, without regard to the large occurrence of correctly predicted non-exceedances. For our evaluation, the CSIs for the 8-h exceedance ranged from 17.2 % for both MAQSIP and MM5-Chem to 5.7% for HYSPLIT. This indicates that MAQSIP and MM5-Chem were ~3 times better than HYSPLIT at accurately predicting 8-h exceedances. The model's ability to predict 1-hr exceedances were more similar, ranging from 11.1% for MAQSIP to 7.9% for HYSPLIT.

466

The **Probability Of Detection** (POD) metric is similar to the CSI, in that it measures the number of times a model predicted an exceedance when one actually occurred. In our evaluation, MM5-Chem had the largest PODs (35.6% for 8-hr, 29.2% for 1-hr), followed by MAQSIP (25.9%,17.0%) and HYSPLIT (7.02%, 18.2%).

Measures of **Bias** (B), which for a categorical forecast indicates if forecast exceedances (1- and 8-hr) are under predicted (B < 1) or over predicted (B > 1), vary across models and even within models (i.e. HYSPLIT). MAQSIP under predicts both 1- and 8 hour exceedances (B: 0.70 and 0.77, respectively), while MM5-Chem over predicts both, especially the 8-hr (1.43 2.29, respectively). HYSPLIT greatly underpredicts the 8-hr (0.31), yet overpredicts the 1-hr (1.48).

A fifth categorical metric, the **False Alarm Rate** (FAR), indicates the number of times that the model predicted an exceedance that did not occur. The FARs were high for the 1-hr forecast (ranging from 66.3% for MAQSIP to 77.3% for HYSPLIT) and even higher for the 8-hr forecast (ranging from 75.8% for MAQSIP to 87.8% for HYSPLIT).

Table 2. Summary of categorical statistics for forecasts of maximum 1-hr and 8-hr O_3

Models / Metrics	HYSPLIT		MM5-Chem		MAQSIP	
	Max1hr	Max8hr	Max1hr	Max8hr	Max1hr	Max8hr
A (%)	99.16	91.58	97.76	81.89	99.42	90.71
CSI (%)	7.89	5.67	9.75	17.15	11.11	17.16
POD (%)	18.18	7.02	29.25	35.60	17.02	25.91
B	1.48	0.31	2.29	1.43	0.70	0.77
FAR (%)	87.76	77.30	87.24	75.14	75.76	66.30

Scatter plots of the model forecasts versus AQS observations (for both the maximum 1- and 8-hr ozone concentrations) are provided in Figure 2. In addition to illustrating the exceedance threshold areas (which were used in calculation of the categorical statistics), the plots also provide least squares regression lines associated with each evaluation. As evident from the regression lines (which generally have intercepts ∃ 25 ppb) most of the over prediction common to each model occurs at the lower concentrations. (Note: HYSPLIT doesn't forecast concentrations lower than 30 ppb). The models all underpredict the higher ozone concentrations.

Evaluation of model performance over time was also performed as seen in Figure 3, where boxplots (denoting 75th, 50th, 25th percentiles, max., min. and mean) of simple bias (Model-AQS) are provided for each of the 25 days. Of the three models, MAQSIP generally exhibits the smallest bias variability across time and HYSPLIT the largest. It is interesting to note that the timing of the fluctuations of the bias above and below the zero bias line are generally "in-phase". This may be attributable to the fact that all three models used the same meteorological model (MM5) and that errors attributable to the meteorology may be perturbating through the forecasts – however further study is needed.

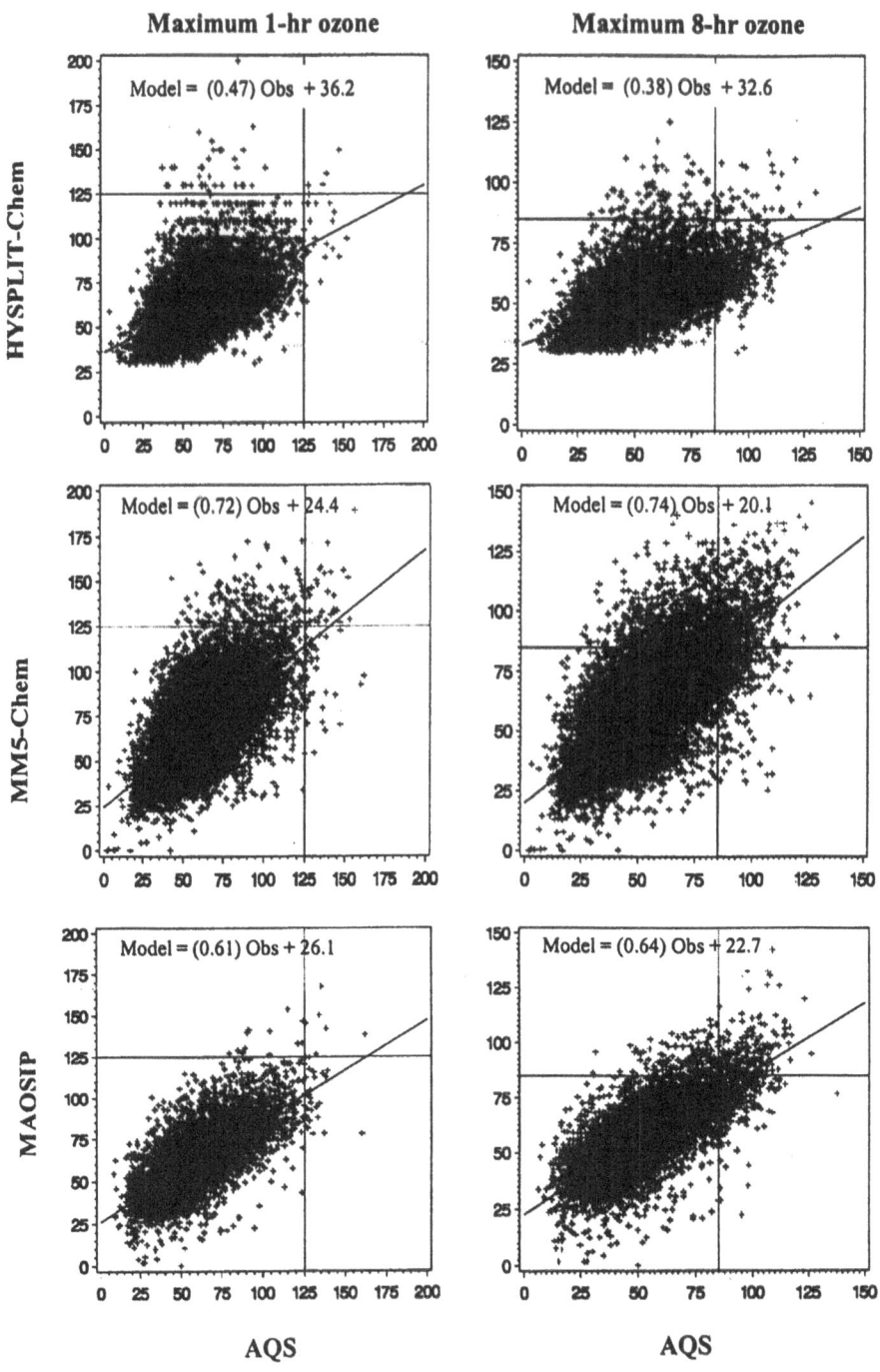

Figure 2. Scatter plots of the model versus AQS for both 1- and 8-hour maximum ozone concentrations (ppb) with exceedance thresholds and least squares regression indicated.

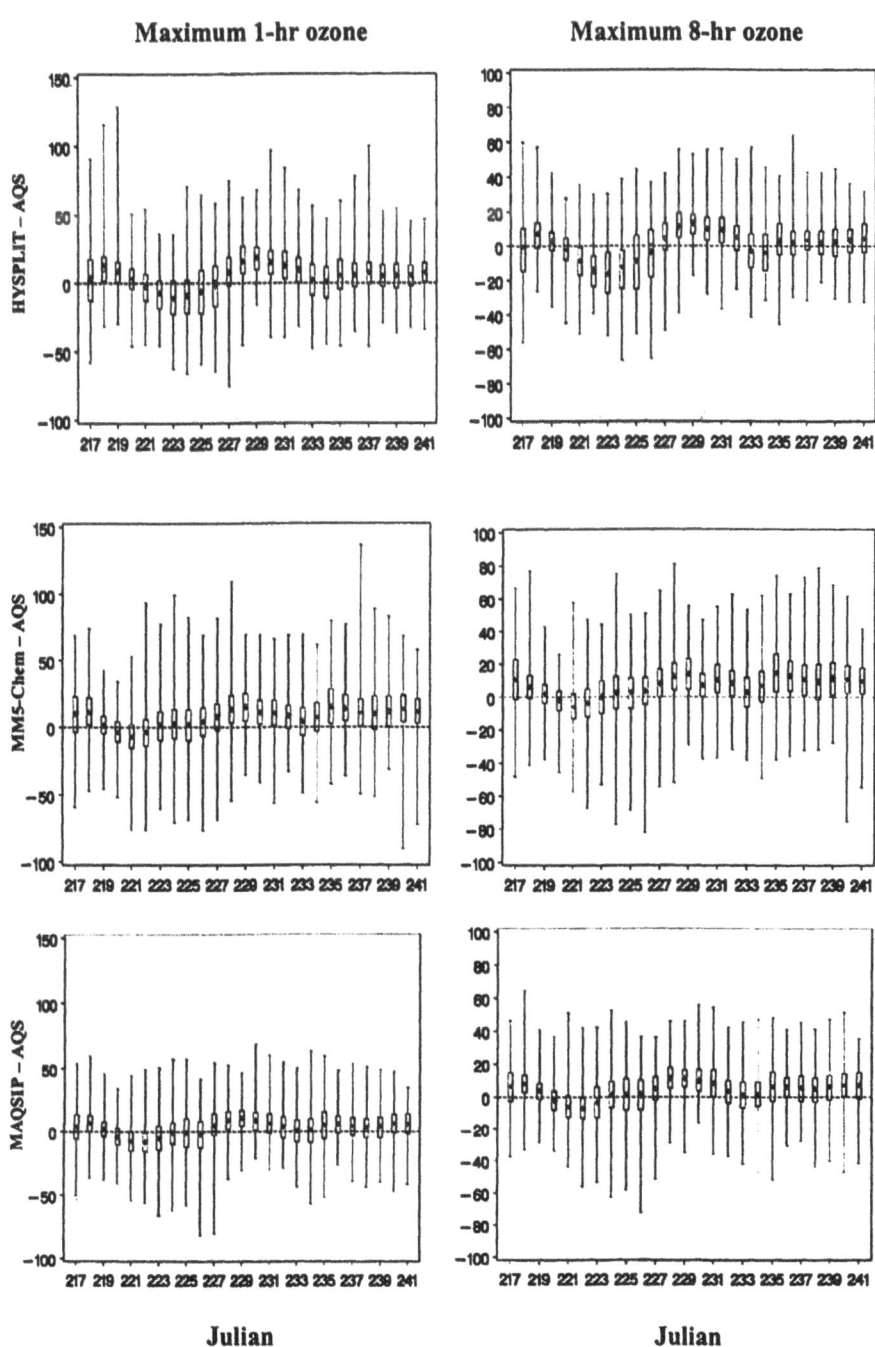

Figure 3. Boxplots of the variation across time of the (Model – AQS) for both 1- and 8-hour maximum ozone concentrations (ppb).

5. SUMMARY

The purpose of this research has been to develop and implement an evaluation protocol - as part of NOAA's Air Quality Forecasting Pilot Program, using three regional-scale air quality models, O_3 observations obtained from AQS and a suite of statistical metrics evaluating both discrete and categorical forecasts. This research has revealed that no single evaluative measure is sufficient, but rather a suite of measures are required to adequately characterize a model's performance.

Results from the discrete evaluation revealed that all three models tend to over predict O_3 concentrations (MB ranges from 0.53 ppb for HYSPLIT's max. 8-hr forecast to 7.42 ppb for MM5-Chem's max. 1-hr forecast) and that each produces substantial errors (RMSEs range from 13.50 ppb for MAQSIP's max. 8-hr forecast to 19.80 ppb for MM5-Chem's max. 1-hr forecast). The categorical evaluation revealed MAQSIP and MM5-Chem were much better at predicting 8-hr exceedances (CSIs: 17.2%) than was HYSPLIT (CSI: 5.7%), but that the models were more similar in their ability to predict 1-hr exceedances (CSIs ranged from 11.1% for MAQSIP to 7.9% for HYSPLIT). Measures of categorical Bias revealed that MAQSIP under predicted both 1- and 8 hour exceedances, MM5-Chem over predicts both, especially the 8-hr and that HYSPLIT greatly underpredicts the 8-hr, yet overpredicts the 1-hr. The False Alarm Rates or number of times that the model predicted an exceedance when none occurred were high for all models for the 1-hr forecast (ranging from 66.3% for MAQSIP to 77.3% for HYSPLIT) and even higher for the 8-hr forecast (ranging from 75.8% for MAQSIP to 87.8% for HYSPLIT). Evaluation of model performance over time revealed that MAQSIP exhibits the smallest bias variability across time and HYSPLIT the largest.

6. REFERENCES

Gery, M. W., G. Z. Whitten, J. P. Killus, and M. C. Dodge, A photochemical kinetics mechanism for urban and regional scale computer modeling, *J. Geophys. Res.*, 94, 12,925-12,956, 1989.

Grell, G.A., S. Emeis, W.R. Stockwell, T. Schoenemeyer, R. Forkel, J. Michalakes, R. Knoche, W. Seidl, Application of a multiscale, coupled MM5/chemistry model to the complex terrain of the VOTALP valley campaign, *Atmos. Environ.*, 34, 1435-1453, 2000.

McHenry, J.N., W. F. Ryan, N. L. Seaman, C. J. Coats, J. Pudykeiwicz, S. Arunachalum and J.M. Vukovich, A real-time Eulerian photochemical model forecast system: Overview and initial ozone forecast performance in the ne US corridor, In review, *Bull. AMS*, 2003.

Stein, A.F., D. Lamb, R.R. Draxler, Incorporation of detailed chemistry into a three-dimensional Lagrangian-Eulerian hybrid model: application to regional tropospheric ozone, *Atmos. Environ.*, 34, 4361-4372, 2000.

Stockwell, W.R., P. Middleton, J.S. Chang, X. Tang, The second generation regional acid deposition model chemical mechanism for regional air quality modeling, *J. Geophys. Res.*, 95, 16343-16367, 1990.

Acknowledgements The authors would like to thank the model developers for providing results of their simulations. Specifically we thank Dr. Ariel Stein of NOAA/ARL for providing results from the HYSPLIT model; Dr. Georg Grell of NOAA/FSL for providing results of the MM5-Chem; and John McHenry of Baron Advanced Meteorological Systems, LLC, for providing the MAQSIP results.

Disclaimer This document has been reviewed and approved by the U.S. Environmental Protection Agency for publication. Mention of trade names or commercial products does not constitute endorsement or recommendation for use.

ROOF LEVEL URBAN TRACER EXPERIMENT: MEASUREMENTS AND MODELLING

Sven-Erik Gryning[1], Ekaterina Batchvarova[1,2], Mathias W. Rotach[3], Andreas Christen[4] and Roland Vogt[4]

1. INTRODUCTION

In this study first results from a low-level source urban tracer SF_6 experiment are reported. The experiment was performed in the framework of the Basel UrBan Boundary Layer Experiment – BUBBLE - in an area of the city of Basel (Switzerland) named Kleinbasel. Extensive micrometeorological information on the vertical structure of the atmospheric turbulence within the street canyons and the overlying urban roughness sublayer as well as the flow field over the city was available. In traditional applied dispersion modelling the roughness sublayer is considered sufficiently shallow not to affect the atmospheric dispersion process and fluxes are considered to be constant with height near the surface. This is not the case in the roughness sublayer that exists above an urban area; here fluxes vary considerably with height.

The SF_6 tracer experiments were performed with near roof-level releases. The samplers were distributed close to roof level in a down-wind area stretching out to about 2.4 km. The tracer thus was released and sampled in the roughness sublayer. The part of Basel where the experiments were carried out is fairly homogenous in its city structure. The mean building height in the area is 15.5 m with a mean plan area density of 49%. During the campaign 4 successful tracer experiments were carried out, all in the afternoon

2. TRACER EXPERIMENT

The tracer SF_6 was released from the roof of a parking house about 1.25 times the building height. Only in one occasion the tracer release had to be made from a mobile crane at a different position. Samplers were located in a downwind sector of about 90° opening angle and located at 1.5 m above roof level. For most of the tracer releases

[1] Risø National Laboratory, DK-4000 Roskilde, Denmark; [2] National Institute of Meteorology and Hydrology, BG-1784 Sofia, Bulgaria; [3] Swiss Federal Institute of Technology. CH-8057 Zürich, Switzerland; [4] University of Basel, CH-4056 Basel, Switzerland

Air Pollution Modeling and Its Application XVI, Edited by
Borrego and Incecik, Kluwer Academic/Plenum Publishers, New York, 2004

samplers were located approximately on two arcs at 500 and 1000 m distance from the source. Additionally a profile along the center line of the expected plume extended up to about 2.4 km. Typically 12 sampler sites were operated in that way. The release of tracer typically started 30 min prior to the sampling and was kept constant. Tracer sampling was performed in bags. For most of the experiments, 6 bags were filled in sequence at each sampling location with a filling duration of 30 min for each. Thus a three-hour time series of near-roof concentrations is available at each of the sampling sites. Figure 1 illustrates the tracer measurements on June 26.

Bags were subsequently analyzed in the laboratory and a background concentration, that was measured for each release separately, was finally subtracted from the analyzed concentrations. Reproducibility of the observed concentrations was excellent. More detail about this tracer experiment will be published elsewhere

In order to make an estimate of σ_y, the lateral spread of the plume, as function of distance the concentration field over the area was estimated by interpolation among the measurements. Figure 2 shows the interpolated concentration field for June 26. Crosswind lines were laid out about 0.75 1.0 and 1.25 km from the source and cases that were well covered by the field measurements were selected for further analysis. In the case of June 26 the 0.75 and 1.0 km crosswind profiles were selected and the profile at 1.25 km disregarded. The interpolated concentrations were digitised along the selected crosswind lines and σ_y estimated from a best fit to a Gaussian distribution, Table 1.

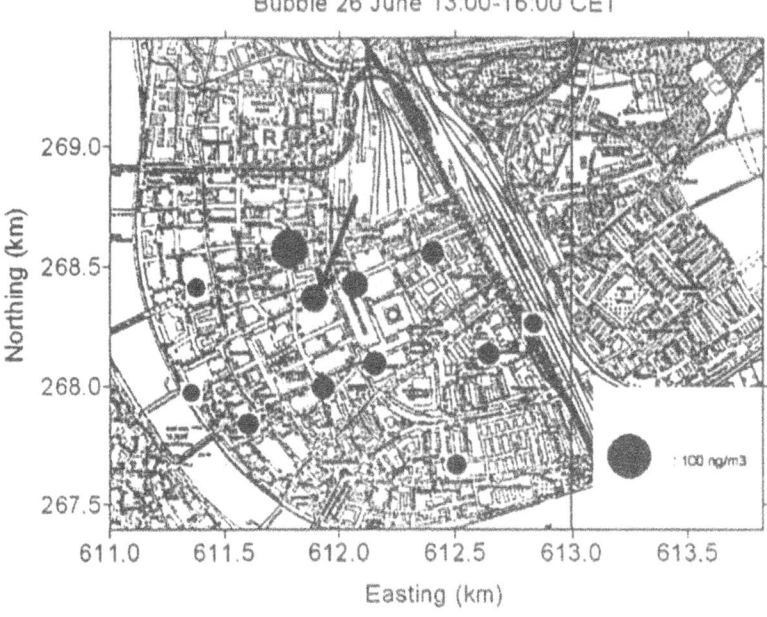

Figure 1. The tracer concentrations on 26 June, 13-16 CET. The release point is marked with R; at the tracer sampling positions the measured concentration is indicated with the area of the filled circle. For comparison a filled circle representing 100 ng m^{-3} is shown in the white box. The arrow shows the position of Sperrstrasse. Base map (c) copyright GVA BS, 25.10.2002.

472

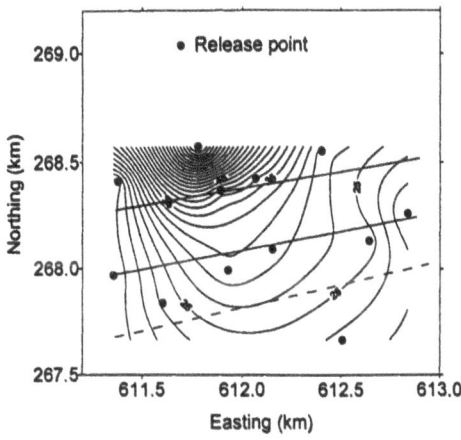

Figure 2. Isolines of interpolated tracer concentrations [ng m⁻³] averaged over the period for the experiment on 26 June, 13-16 CET. The filled circles designate the tracer release and sampling positions. The full lines represent crosswind profiles that formed the basis for the estimation of σ_y, see Table 1. The profile along the dashed line was not used due to insufficient data coverage. The co-ordinate system as in Figure 1.

Table 1. The lateral spread of the plume σ_y for the 4 experiments.

Experiment:	Downwind distance (m)	σ_y (m)
26 June (13:00-16:00 CET)	750 1000	537 672
4 July (15:00-18:00 CET)	750 1000	296 311
7 July (14:00-17:00 CET)	750 1000	393 424
8 July (15:00-18:00 CET)	1000 1250	425 437

3. METEOROLOGICAL MEASUREMENTS

From the BUBBLE network we use the observations from Sperrstrasse centrally situated in the tracer experimental area, Figure 1. There, Reynolds stress was measured at 6 levels on a tower, namely at 3.6, 11.3, 14.7, 17.9, 22.4 and 31.7 metres above ground. The local building height amounts to 14.6 m. An aerosol Lidar located within Basel 5 km from the experimental area gave information on the height of the mixing layer.

2.1. Meteorological parameters

In traditional applied models of atmospheric dispersion from low-level sources, the input consists of basic meteorological parameters such as the Obukhov length, the height of the mixing layer and a characteristic wind velocity. The Obukhov length is formed from parameters in the surface boundary layer where both the sensible heat flux $\overline{(w'T')}$ and friction velocity are constant as function of height. The friction velocity is:

$$u_* = \sqrt{-\overline{(u'w')}_o}$$

where the usually small lateral component $\overline{(v'w')}_o$ has been neglected. Here o denotes near surface values that are representative for the surface boundary layer. But how do we determine and apply appropriate scaling parameters over a rough surface like in urban environments? There observations (e.g., Rotach 2001) show that $-\overline{u'w'}$ exhibits a distinct profile with a maximum somewhere above roof level, Figure 3. By fitting a curve through the profile of observed measurements both the maximum value and the height z_m where it occurs can be found. In short Rotach (2001) argues that the Reynolds stress component at z_m actually reflects the drag that the flow aloft 'sees' from the bulk of the surface, and hence is a candidate for a scaling velocity with the usual definition in terms of momentum transfer to the surface

$$u_*^r = \sqrt{-\overline{(u'w')}_{max}} \quad .$$

This scaling velocity is called u_*^r in order to avoid confusion with the traditional definition of the friction velocity.

For the near-surface wind speed information from the sonic anemometer profile at Sperrstrasse (levels 17.9 m and 31.7 m) have been selected and interpolated to the height z_m. Concerning the sensible heat fluxes over urban areas often a maximum is observed slightly above roof level. Higher up the heat flux remains approximately constant while there is large variability inside the canyon. In this study the surface heat flux is obtained from averaging the observations of the two uppermost levels at Sperrstrasse (22.4 m and 31.7 m). The mixing layer height z_i was deduced from profiles of Lidar measurements, taken as the height of a major change in the backscatter of the signal.

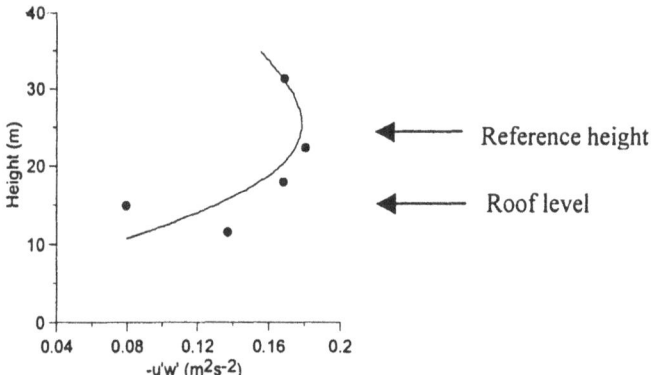

Figure 3. Profile of observed (filled circles) and fitted (full line) Reynolds stress component $-\overline{u'w'}$ at Sperrstrasse during the tracer experiment on 4 July.

474

Table 2. Meteorological conditions during the 4 tracer experiments; averaging times as in Table 1. The measurements represent the conditions at the height z_m .

Experiment	z_m (m)	u (m/s)	u_*^r (m/s)	w_* (m/s)	z_i (m)
26 June	21.7	1.28	0.41	2.31	1500
4 July	24.0	2.49	0.60	1.77	1200
7 July	19.2	1.44	0.31	2.22	1800
8 July	20.4	1.78	0.41	2.27	(2000)

2.2. Parameterisation of σ_v and of σ_w

The standard deviation of the crosswind fluctuations of the wind velocity, σ_v , is an important parameter for the lateral dispersion process of plumes. Similarly is σ_w , the standard deviation of the vertical fluctuations of the wind velocity, important in the description of the vertical spread of plumes. Here we compare commonly used parameterisations of σ_w and σ_v to the data from BUBBLE. At our disposition we have 6 half-hourly values of σ_w and σ_v from each of the four experiments. We apply the parameterisations recommended by Gryning et al. (1987). In the report of the COST Action 710 (Cenedese et al, 1998), these parameterisations were validated on a large number of data sets and found to perform well. They read:

$$\sigma_w^2 = u_*^2 \left[1.5 \left(\frac{z}{z_i} \right)^{2/3} \left(\frac{w_*}{u_*} \right)^2 \exp\left(-2\left(\frac{z}{z_i} \right) \right) + \left(1.7 - \left(\frac{z}{z_i} \right) \right) \right]$$

where the convective velocity scale is $w_* = \left((g/T)\overline{w'T'}\, z_i \right)^{1/3}$, with g for the acceleration due to gravity and T for temperature and

$$\sigma_v^2 = 0.35\, w_*^2 + (2 - z/z_i)\, u_*^2 \; .$$

Here we apply the parameterizations at the level of maximum shear stress, z_m inside the roughness sublayer over the BUBBLE urban area, using u_*^r for the friction velocity. In Figure 4 it can be seen that in general the parameterisations perform well over the urban area. The agreement is better for σ_w than for σ_v in accordance with the general experience from similar investigations over flat terrain. It can also be seen that the parameterised values have both σ_w and σ_v about 30% larger (dashed lines) than measurements. We do not attempt any immediate explanation for these systematic overestimations for σ_w and σ_v in the urban environment, but take them as an empirical fact and cope with them by simply reducing both parameterisations with 30%.

475

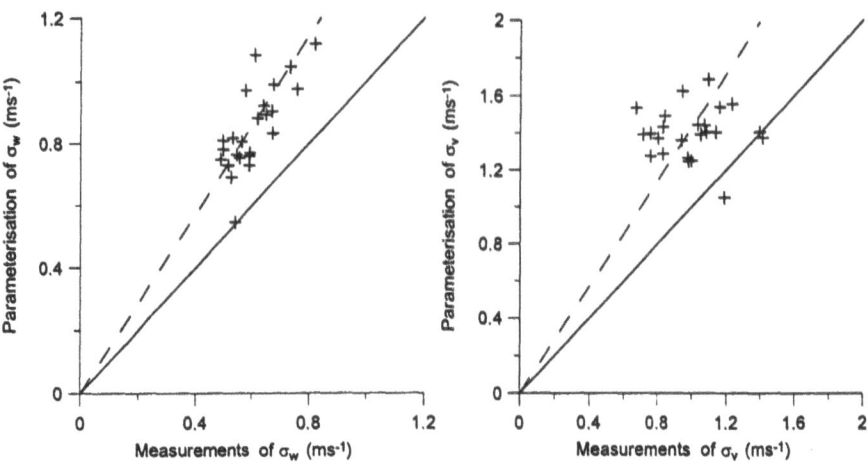

Figure 4. Observed half-hourly averaged versus parameterisations of σ_w (left panel) and σ_v (right panel).

4. EVALUATION OF TRACER CONCENTRATION DATA

4.1. Parameterisations of σ_y

Considering only the effect of atmospheric turbulence a simplified version of Taylor's famous formula for plume dispersion reads:

$$\sigma_y = \sigma_v t \, f_y(t/T_y)$$

where t is travel time of the plume and f_y is a function of the dimensionless travel time t/T_y where T_y is the Lagrangian time scale for the lateral dispersion process. The approximation

$$f_y = \left(1 + \sqrt{t/2T_y}\right)^{-1}$$

is often recommended for applied dispersion modelling (Gryning et al, 1987). For the unstable atmosphere it comes natural to connect the Lagrangian time scale to the time of transport between the surface and the mixing height:

$$T_y = z_i/\sigma_v$$

where σ_v is used as a characteristic velocity for the lateral spread of plumes. For atmospheric neutral conditions the mixing height in the usual sense for the convective atmosphere might not be present, in this case the vertical scaling height can be taken as $z_i = 0.2 u_*/f$ where f is the Coriolis parameter. By use of equations above and taking $\sigma_v = 1.7 u_*$ we have for neutral conditions $T_y \approx 1000\,s$. However in the above considerations the height dependence of T_y is neglected. This let Gryning et al (1987) distinguish between ground-level and elevated sources with $T_y = 200$ s recommended for ground-level sources and $T_y = 600$ s for elevated sources.

The above expressions for the lateral spread of the plume have been developed and validated mainly against data from low level sources over a rural area or from elevated sources over both rural and urban areas at high wind velocities. Such circumstances are very different from the conditions during the four BUBBLE tracer experiments where tracer release and concentration measurements were performed near roof level in an urban area during convective conditions and very low wind speeds.

The simulation of the lateral spread was performed in two steps. In the first one we use the observed σ_v values. Figure 5 shows the measurements and model simulations of σ_y using $T_y=200$ s, which is the recommended value for ground level sources, and $T_y=600$ s as suggested for elevated sources. Both assumed values of T_y in combination with the observed σ_v are seen to underestimate the lateral spread. The simulation was also performed with by use of $T_y = z_i/\sigma_v$ which is within a factor of 2 of the value of 600 s recommended for elevated sources. Use of this formulation improves the comparison with the measurements, which suggest that the high values of both the friction velocity and convective velocity typically for urban areas makes the plume behave more like an elevated source than as a ground level source.

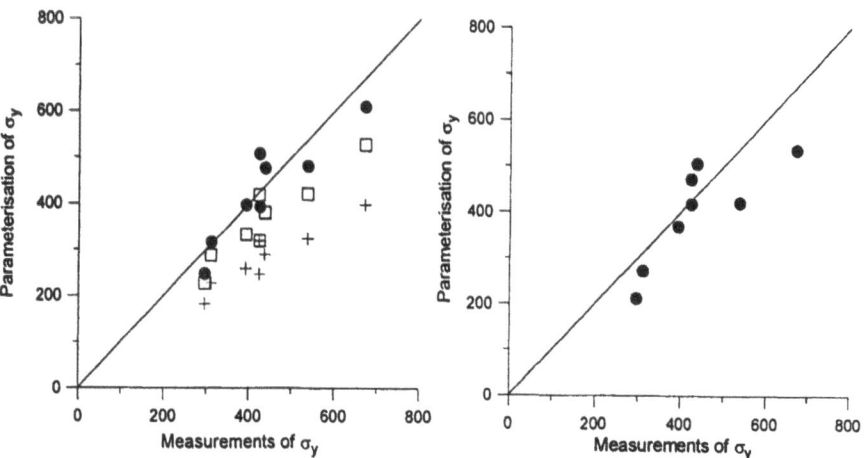

Figure 5. Measured and modelled values of σ_y . In the left panel the simulations are performed using observed values of σ_v and for $T_y =200$ s (+); $T_y =600$ s (□) and $T_y = z_i/\sigma_v$ (●). In the right panel the simulations performed by use of parameterised values of σ_v and $T_y = z_i/\sigma_v$ (●).

The comparison was also carried out with parameterised values of σ_v , Figure 5. The agreement is fair but not as good as the use of measured values of σ_v .

4.2. Numerical simulations

The numerical simulations will only be touched upon here as the work is ongoing. A Lagrangian particle dispersion model (LPDM) is used that can be run with parameterised turbulence profiles, Rotach (2001). Close to the surface the turbulence characteristics are parameterized specifically to match urban roughness sublayer observations. The simulation for the 26 June is shown in Figure 6. It can be seen that the present model does a reasonable job in reproducing the dispersion process. It also shows that the plume was caught by the samplers reasonably well and was neither drained into a street canyon nor lifted away from the surface by a large eddy due to highly convective conditions. It can be noted that the simulations reproduce the high concentrations well, while the low concentrations are somewhat underpredicted. Some splashing around in the wind field was observed during the experiment which might prevent the tracer plume from being completely advected out of the area; an effect that is not included in the simulation.

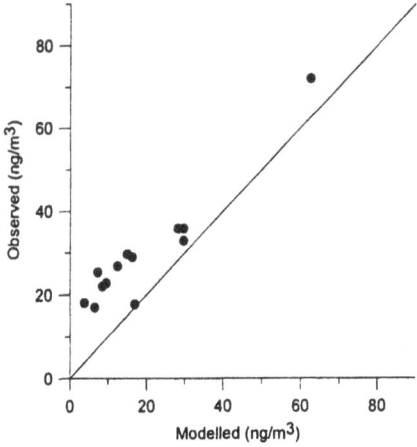

Figure 6. Observed and modeled near-roof concentrations at the 13 tracer sampling sites for the experiment of June 26 2002. Meteorological input data to the model are three-hourly averages and so are the observed concentrations. Parameterizations for the turbulence profiles are employed.

5. ACKNOWLEDGMENTS

The tracer study was funded by the Swiss Federal Institute of Technology (TH-35/02-1) and Risø National Laboratory. The involvement of EB was supported through an Institute Partnership financed by the Swiss National Science Foundation (7IP 065650.01). The modelling study is also supported by a NATO Linkage Grant (EST-CLG-979863).

6. REFERENCES

Briggs, G. A., 1985, Analytical Parameterizations of Diffusion: The Convective Boundary Layer. J. Clim. Appl. Met. **24**, 1167-1186.

Cenedese, A.; Cosemans, G., Erbrink, H.; Stubi, R., 1998, Vertical profiles of wind, temperature and turbulence. In: Harmonisation of the pre-processing of meteorological data for atmospheric dispersion models. COST action 710 Final report. Office for Official Publications of the European Communities.

Gryning, S.-E.; Holtslag, A.A.M.; Irwin, J.S.; Sivertsen, B., 1987, Applied Dispersion Modelling Based on Meteorological Scaling Parameters. Atmos. Environ. **21**, 79-89

Rotach, M. W., 2001, Simulations of urban-scale dispersion using a Lagrangian stochastic dispersion model. Boundary-Layer Meteorology, **99**, 379-410.

DISCUSSION

D. STEYN   On your profile of stress, there are few (no) points to determine the upper (decreasing) part of the profile. It seems therefore that the reference height z_m in rather imprecisely determined. What is the uncertainty in z_m, and what influence will this imprecision have on your subsequent analyses?

S.E.GRYNING   The momentum flux profile has a rather distinct behaviour. The momentum flux is rather low inside the street canyon, is has a very sharp gradient above the roof level, and reaches a maximum at z_m and then remains near constant as function of height. The uncertainty in the determination of z_m is likely not small, considering the sparseness of measurements of the momentum flux along the mast, but this is not critical for the analysis because of the small variation of parameters above z_m - in analogy with the conditions in the surface boundary layer.

B.TERLIUC   Can you describe the synoptic conditions during the experiment?

S.E.GRYNING   The lay-out for the tracer experiment was done according to the most predictable wind direction for the city of Basel during the afternoon in the experimental period. Due to the topographical features around the city a thermally driven wind system, so-called Clarawind, usually develops during cloud-free summer days with weak synoptic conditions. In the afternoon the Clarawind is characterised by a persistent north-westerly flow over the city. The meteorological conditions are very convective due to the large insolation. The meteorological conditions that favour the development of the north-westerly Clarawind in the afternoon in Basel, create in the morning drainage flows from the nearby mountains in the opposite direction.

M. KAASIK   In your data set available for other groups for model validation? Under which conditions?

S.E.GRYNING   A data-report on the tracer experiments with a thorough meteorological documentation is being prepared. The preliminary reference is: Roof-level SF_6 tracer experiments in the city of Basel (S.E. Gryning, E. Batchvarova, M.W. Rotach, A. Christen and R. Vogt). It is the intention to publish the report in the institute series of IAC-ETH.

COMPARISON OF THE SPACE-TIME SIGNATURES OF AIR QUALITY DATA FROM DIFFERENT MONITORING NETWORKS

Edith L. Gego[1], Christian Hogrefe[2], P. Steven Porter[3], John S. Irwin[1] and S. Trivikrama Rao[1]

ABSTRACT

Ambient air quality in the United States is measured by several regional air quality monitoring networks. Yet, differences in sampling protocol between the networks may not allow joint use of the data reported by different networks. In this study, we compare the space-time signatures of sulfate and nitrate fine particle mass concentrations reported by the Clean Air Status and Trend Network (CASTNet) and the Interagency Monitoring of PROtected Visual Environment Network (IMPROVE). First, a spectral decomposition technique was used to separate the low and high frequency variations in time series of pollutant concentrations at collocated IMPROVE and CASTNet sites. Through Principal Component Analysis (PCA) and Varimax orthogonal rotation, we determined the number of significant sulfate and nitrate modes of variation identifiable with both networks, and identify the mode of variation characterizing each monitoring site. In the case of sulfate, both networks allow identification of seven distinct modes of variation, each of which corresponds to a well-defined geographic area. PCA also suggests the existence of seven modes of variation for nitrate but, in contrast to sulfate, these modes of variations could not be linked to any unified geographic area. A combination of spectral decomposition and PCA reveals that the long-term fluctuations in sulfate at both networks are virtually identical - when they are averaged in homogeneous regions defined by PCA - between both networks.

---

[1] NOAA Atmospheric Sciences Modeling Division, On Assignment to the U.S. Environmental Protection Agency, Research Triangle Park, NC 27711, U.S.A.
[2] Atmospheric Sciences Research Center, University at Albany, Albany, NY 12222. U.S.A.
[3] Department of Civil Engineering, University of Idaho, Idaho Falls, ID 83401, U.S.A.
Corresponding author: S.T. Rao, 109 T.W. Alexander Drive, U.S. Environmental Protection Agency, Room E-240D, Research Triangle Park, NC 27711,Tel: (919) 541-4542, E-Mail: <Rao.ST@epa.gov >

Air Pollution Modeling and Its Application XVI, Edited by
Borrego and Incecik, Kluwer Academic/Plenum Publishers, New York, 2004

1. INTRODUCTION

The revisions to the fine particulate matter ($PM_{2.5}$) concentration standards required the monitoring of $PM_{2.5}$ concentrations in the United States. $PM_{2.5}$ is measured in several regional air quality monitoring networks. Yet, differences in sampling protocols, essentially linked to the objective of each network, may not allow combined use of the data reported by different networks, even if pertaining to the same species. To examine this issue, we use a spectral decomposition technique and rotated principal component analysis to compare the spatial and temporal characteristics of nitrate and sulfate concentrations measured by two of the oldest monitoring networks, the Clean Air Status and Trend Network (CASTNet) and the Interagency Monitoring of PROtected Visual Environment (IMPROVE). The techniques presented here for the purpose of network-to-network comparison can also be used to compare model estimates with monitoring data.

2. DESCRIPTION OF CASTNet AND IMPROVE SAMPLING PROTOCOLS

Created in 1990 for the purpose of measuring dry deposition flux, CASTNet comprises of over 70 monitoring sites in the United States (U.S.) that are mostly located in rural areas. The sampling device of CASTNet is a non size-selective three-stage filter pack located 10 m above ground level. In the eastern U.S., this filter samples air at a constant 1.5 l/m rate. Particulate nitrate and sulfate are collected on the first filter, composed of Teflon, and are quantified by ion chromatography. Filters are changed every week. Thus, the measured values represent 7-day average concentrations. All CASTNet data are adjusted to a temperature of $25^{0}C$ and a pressure of 1013 mbar before being reported.

Initiated in 1985, the IMPROVE network essentially aims at monitoring visibility conditions in Class I areas that include most National Parks and Wilderness areas. The IMPROVE network first developed in the western U.S. where most national parks are located, then expanded eastwards. By the end of 2001, 28 IMPROVE sites were operating in the eastern U.S. IMPROVE sampling devices consist of 4 independent sampling modules, located 3 m above ground level, and equipped with a sizing device designed to exclude particles larger than 2.6 µm. In most cases, sulfate concentration is calculated by stoichiometry, based on sulfur mass measurement. Sulfur is extracted from a Teflon filter and analyzed by X-Ray fluorescence. Sulfates are also measured by ion chromatography after extraction from a nylon filter; the nylon-based estimates being used a quality control of the Teflon based estimates. Nitrate particles are extracted from a nylon filter, preceded by an acidic vapor diffusion denuder to eliminate nitric acid vapor. Nitrates are determined by ion chromatography. Two 24-h integrated samples are collected each week; measured concentrations are reported under ambient temperature and pressure conditions, in contrast to the CASTNet method.

Further details about sampling protocols adopted by both networks are available at 'http://vista.cira.colostate.edu/improve/Publications/OtherDoc/IMPRO\VEDataGuide/ IMPROVEDataGuide.htlm' and at 'http://www.epa.gov/CASTNet'.

3. METHODS

3.1. Spectral decomposition of time series at collocated sites

Variations observed in pollutant time series data result from forcings operating on different time scales (Hogrefe *et al.*, 2001). The extent to which CASTNet and IMPROVE display these forcings was compared using filtered time series. Because of its simplicity and ability to deal with missing values, the Kolmogorov-Zurbenko filter (KZ filter) was used as the decomposition tool (Rao *et a.l*,1997). The KZ filter is a low-pass iterative moving average where the window size and number of iterations determine the threshold frequency (frequency at which 50% of variation is passed).

Sampling frequency and duration constrain filter design. Since IMPROVE data pertain to non-consecutive (1 sample every three days) 24-h air samples and CASTNet data are weekly averages, high frequency oscillations caused by the intra-day, diurnal or other high-frequency forcings are not identifiable. In addition, the limited duration of some series (1 year) limits observable low-frequency variation to a similar time frame. Therefore, two temporal components were made from IMPROVE data using a KZ filter with a window size of 15 days and 5 iterations. The threshold frequency for this filter is about 2.5 months^{-1}, with the smoothed time series having variations at frequencies less than 2.5 months^{-1} (the threshold), while the residuals (original observations – smoothed time series) have variations at frequencies greater than 2.5 months^{-1}. Similarly, temporal components were created for the CASTNet data (weekly averaged values) with a KZ filter having a window size of 3 weeks and 3 iterations. This filter was designed to also correspond to a threshold frequency of about 2.5 months^{-1}, so as to allow comparison of the low- and high-frequency signals of both networks.

Since the sampling rhythm and duration are different for the two networks, the length of CASTNet and IMPROVE time series, whether raw or decomposed time series, are not identical, even if pertaining to the same period of time. Therefore, the computation of correlation coefficients between these two time series is complicated. Rather, comparison of networks is realized by comparison of the low- and high-frequency signals of each network.

3.2. Principal component analysis and orthogonal rotation

Principal Component Analysis (PCA) is a statistical tool that allows determination of the number of distinct 'modes of variations' in a data set (Eder, 1989). PCA starts by computation of the covariance or correlation matrix between variables. In this case, a variable corresponds to the observed time series at a given location. For this study, all variables were standardized to a zero mean and a unit variance before evaluation of the correlation matrix, a procedure thought to limit the impact of heteroskedasticity (inequality of variances) and facilitate interpretation of results.

The first principal component (PC) is obtained by multiplying the original set of variables by the first eigenvector of the correlation matrix. The variance of that component is the first eigenvalue of the correlation matrix; it is also the part of the variance of the original data that can be explained by the first PC. Similarly, the second PC is obtained by multiplying the original variables by the second eigenvector of the correlation matrix. The variance of the second component is the second eigenvalue of the correlation matrix, and so on. Since higher order eigenvalues are progressively smaller, so are the portions of the variability in the original data that the higher order PCs explain. One may, therefore, consider that the information included in the original data set can be reasonably described by a limited number of PCs. The number of PCs retained represents what is referred to as the number of 'distinct modes of variations' or the number of clusters one may wish to differentiate in the data set.

Various techniques with varying degrees of empiricism are presented in the scientific literature to determine the number of PC to retain. No one approach is thought superior to the others. In this study, the number of clusters retained for each air pollutant and network is the number of eigenvalues greater than 1.

Rotating the principal components retained so as to increase their correlation with the original data, a procedure often referred to as Varimax (Kaiser, 1958), has been shown to facilitate interpretation of the principal components as well as identification of sites responding to the same mode of variation. Monitoring stations where nitrate or sulfate concentrations fluctuate in a similar manner are those that are more correlated with a given rotated principal component than with the others.

4. DATA

CASTNet and IMPROVE monitoring data are available on the internet at 'http://www.epa.gov/CASTNet/'and http://vista.cira.colostate.edu/improve/', respectively. This study utilizes nitrate and sulfate concentrations reported at sites located east of 100^0 longitude west (eastern U.S.). Spectral decomposition was applied to 5 collocated sites of both networks with a common monitoring period of at least one year. Since the KZ filter functions with incomplete data, missing values were not imputed.

Rotated principal component analysis was applied to data collected during 2001. Before 2001, the number of IMPROVE sites operating in the eastern U.S. was too limited to justify the use of PCA. Sites with more than 25% missing values were removed from the data set, limiting the number of sites incorporated in this study at 25 for the IMPROVE network and 41 for CASTNet. Missing values at a given location were replaced with the mean value at that location.

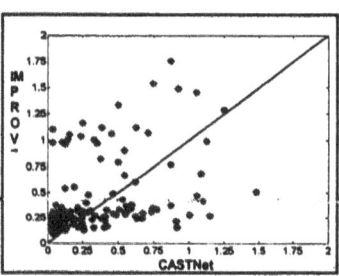

Figure 1. Weekly nitrate concentrations at GSMP from Jan. 1999 to Dec 2001.- CASTNet and IMPROVE estimates

5. RESULTS

5.1. Spectral decomposition

The results pertaining to one site only, Great Smoky Mountains National Park (IMPROVE site GRSM1, CASTNet site GRS420) are presented in details. The location of this site is marked by an arrow in Figure 4. The results characterizing Great Smoky Mountains National Park (GSMP) are typical of what was observed at other sites.

484

Justifying our approach, Figure 1 shows a scatter plot of weekly nitrate concentrations from CASTNet and IMPROVE between January 1999 and December 2001. Quasi-weekly MPROVE estimates were obtained by averaging the 24-h observations that fall within a given CASTNet week. The significant dispersion observable on Figure 1 proves that, at the time-scale of a week, both networks provide a different information. Caution should be exercised when using weekly estimates for model evaluation purposes. Spectral decomposition allowed us to compare the signals at other time-scales.

Figure 2 shows the results of the spectral decomposition of sulfate time series from 1999 to 2001. As shown in the panels A (raw data), IMPROVE data show more variability than their CASTNet counterparts. Yet, the low-frequency signals of both networks are very similar (panels B), even if, at GSMP, the IMPROVE signal is slightly less than that of CASTNet (about 12 % on average). The bias between the two networks is not identical at all sites. Recent installation of IMPROVE sites in the vicinity of CASTNet ones will allow further investigation of this issue. As seen in panels C, high-frequency CASTNet signals have a smaller variance than those of IMPROVE. The differences stem from variations in the sampling frequency (IMPROVE data are 24-h averages while CASTNet data are weekly averages and are therefore 'smoother'). CASTNet weekly sampling may compromise its ability to resolve high-frequency temporal variability.

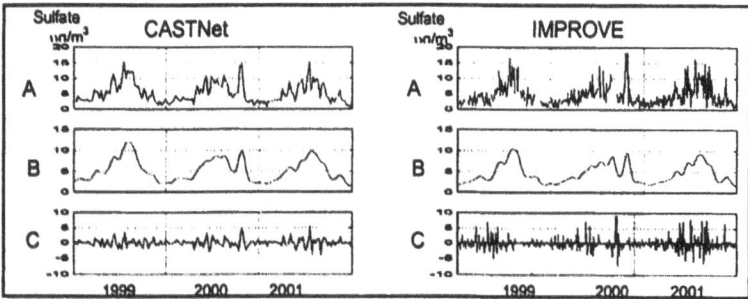

Figure 2. Decomposition of sulfate concentrations recorded from 1999 to 2001 at GSMP into low- and high-frequency signals (panel A: raw data, panel B: low-frequency signal, panel C: high-frequency signal)

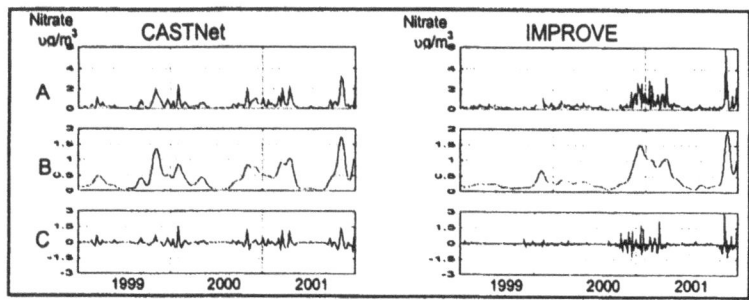

Figure 3. Decomposition of nitrate concentrations recorded from 1999 to 2001 at GSMP into low- and high-frequency signals (panel A: raw data, panel B: low-frequency signal, panel C: high-frequency signal)

Figure 3 displays the results of the spectral decomposition of nitrate time series also recorded at Great Smoky Mountains National Park. The low-frequency signals of the two networks show some resemblance (panels B) but that resemblance is not as close as that observed with the sulfate signals, suggesting prudence if one is combining the results of both networks. Differences between the two low-frequency nitrate signals are most pronounced during the winter season that is typically the high concentration season for nitrate (even if nitrate levels observed in Great Smoky Mountains National Park are fairly low).

Distinct sampling protocols may explain these dissimilarities. Nitrate is extracted from a Teflon filter in the CASTNet network while it is extracted from a nylon filter preceded by a denuder at IMPROVE sites. Advocates of IMPROVE protocols argue that nitrate concentrations based on extraction of a Teflon filter may be considerably underestimated because of volatilization of ammonium nitrate. In defense of CASTNet protocol, some say that nitrate estimates resulting from extraction of a nylon filter, even preceded by an HNO_3 vapor remover, will only be accurate if the denuder is efficient.

Similarly to what was observed with sulfate, the high-frequency components of CASTNet nitrate time series show significantly less variability than those of IMPROVE, differences once again linked to distinct sampling schedules.

5.2. Rotated principal component analysis

PCA of the sulfate concentrations measured at CASTNet sites included in this study indicates the presence of 7 distinguishable modes of variations (seven eigenvalues > 1). Performed on IMPROVE sites, PCA suggests that 6 groups would be sufficient . Yet, by analogy to CASTNet results, we also differentiated seven groups within the IMPROVE data set. Figure 4 shows the clusters formed from sulfate fluctuations in the CASTNet and IMPROVE networks. Each cluster corresponds to a distinct geographical region, clearly identifiable within both networks. Although the number of IMPROVE sites is too limited to accurately delineate the geographical boundaries of each mode of variation, it appears that the limits drawn for CASTNet zones I, II and III may be compatible with IMPROVE observations.

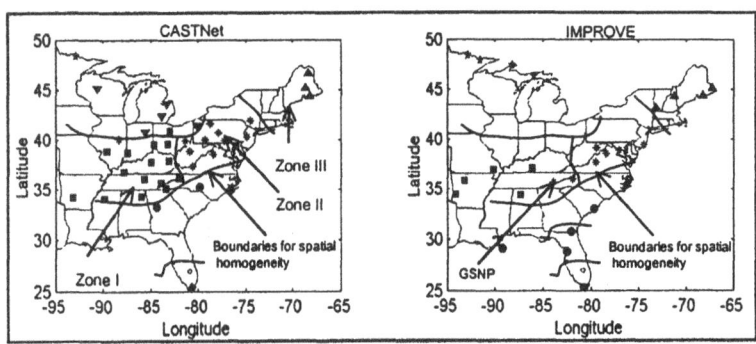

Figure 4. Identification of sites presenting similar modes of variation. Each mode of variation correspond to a different type of marker – case of sulfate

In the case of nitrate, the seven first eigenvalues of the correlation matrix calculated for CASTNet are greater than 1, seven homogeneous were therefore discerned in the original data set. Figure 5 shows the clusters identified for CASTNet and IMPROVE. Contrary to sulfate, nitrate modes of variations do not correspond to clearly distinct and unified geographical areas but are more sporadically encountered throughout the domain, an observation verified for both networks.

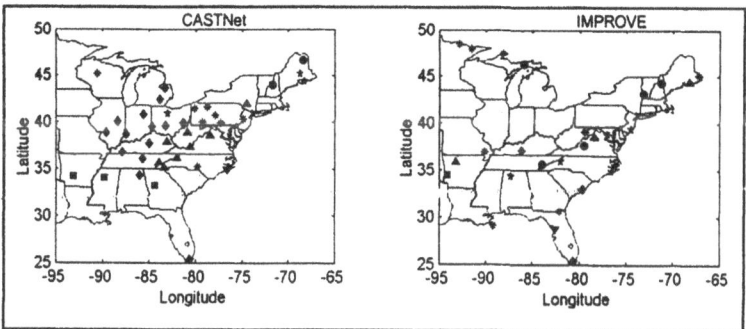

Figure 5. Identification of sites presenting similar modes of variation. Each mode of variation correspond to a different type of marker – case of nitrate

Differences between sulfate and nitrate behavior may be explained by the emission characteristics of their precursors and their associated chemistry. Sulfate results from the oxidation of SO_2 gas principally emitted in the Midwest by coal-fired power plants. The sources of SO_2 are, therefore, stationary. The extent of SO_2 drift from the source area and oxidation into sulfate is fixed by weather conditions and patterns that are quite 'geographically organized'. Nitrate is formed by oxidation of nitrous oxides. Some of these oxides are emitted from fixed combustion sources but the majority of them result from the combustion of motor vehicle fuel. In contrast to sulfate, the sources of nitrate precursors are less predictable.

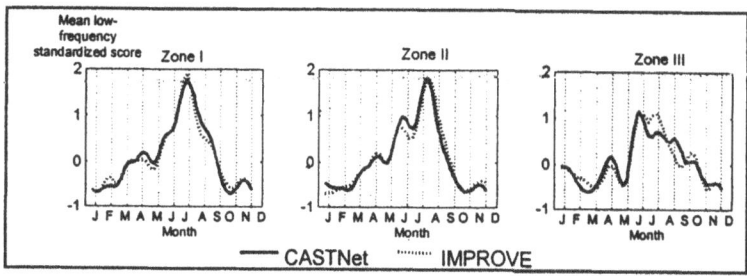

Figure 6. Low-frequency standardized signal of sulfate concentrations measured by CASTNet and IMPROVE in three homogenous areas – average of all stations in a each area.

5.3. Joint use of rotated principal component analysis and spectral decomposition

Figures 6 and 7 illustrate the power of combined use of spectral decomposition and PCA. Using sulfate data from previously defined homogeneous areas (see figure 4 – zones I, II and III), Figure 6 shows the average of the long-term standardized signals of all CASTNet or IMPROVE stations within the zone noted (see section 4.2 for standardization, section 4.1. for estimation of long-term signals). The average long-term standardized signals of the two networks are nearly indistinguishable. Comparison of the mean signals of zones I, II and III also indicates that each cluster identified by rotated PCA corresponds to a unique mode of variation (number of concentration peaks and durations different in zone I, zone II and zone III).

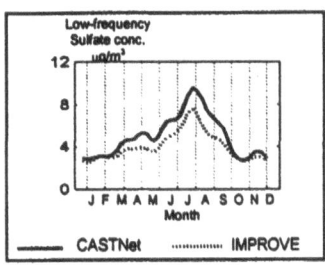

Figure 7. Average low frequency signal of sulfate concentrations measured by CASTNet and IMPROVE in zone I

Figure 7 shows the time series of the average low-frequency signal of both networks in zones I, in terms of real concentrations (rather than standardized scores). It appears that within these three zones, IMPROVE and CASTNet mean concentrations are different, probably because the sites monitored are not at identical locations. Yet, the concentration variations are very similar. This illustrates that, with rotated principal component analysis, the phasing of concentration oscillations is more important than their amplitude for the identification of the so-called modes of variations.

6. SUMMARY

The objective of this study was to compare the spatio-temporal characteristics of some environmental data (particulate nitrate and sulfate concentrations) reported by the CASTNet and IMPROVE networks. We first use a spectral decomposition technique to differentiate the low- and high-frequency signals in observed time series of nitrate and sulfate at IMPROVE and CASTNet collocated sites. This analysis shows that, while the low-frequency signals (50% cutoff period of 2.5 months) are fairly similar (especially in the case of sulfate, less so for nitrate), the high-frequency signals are very distinct, due to differences in the sampling frequency. PCA and Varimax rotation criterion were used to determine the number of distinct modes of variation identifiable for each air pollutant and network. In the case of sulfate, both networks allow identification of seven distinct modes of variations, each of which corresponds to a well-defined geographic area. For nitrate, PCA also suggests the existence of seven modes of variation, which are not linked to any unified geographic area, but are found more sporadically through the eastern U.S.

REFERENCES

Eder, B. K., 1989. A principal component analysis of SO_4^{2-} precipitation concentrations over the eastern United States, Atmospheric environment, 23 (12), 2739-1750.
Hogrefe C., S.T. Rao, P. Kasibhatla, G. Kallos, C. J.Tremback, W. Hao, D. Olerud, A. Xiu, J. Mc Henry and K. Alapaty, 2001. Evaluating the performance of regional-scale, photochemical modeling systems: Part I – meteorological predictions, Atmospheric environment, 35, 4159-4174
Kaiser H. F., 1958. The Varimax criterion for analytical rotation in factor analysis, Psychometrika, 23, 187-201
Rao S.T., I.G. Zurbenko, R. Neagu, P.S. Porter, J. Y. Ku and R. F. Henry, 1997. Space and time scales in ambient ozone data. Bulletin of the American Meteorological Society 78, 2153-2166.

Disclaimer: This document has been reviewed and approved by the U.S. Environmental Agency for publication. Mention of trade names or commercial products does not constitute endorsement or recommendation for use.

DISCUSSION

S.A.-AKSOYOGLU   Are there any NH_4 measurements in the network? They could help to evaluate NO_3.

C. HOGREFE   Thank you for your suggestion. Yes, there are NH_4 measurements avaiable, and we will compare these across both networks. The work we have done so far focused on developing a methodology to compare measurements from different networks, but we will expand our analysis to identify possible reasons for the observed differences in our future work.

T.ODMAN   When you paired the stations from the two networks, did you take into consideration the difference in the location of the monitors? For example, The Great Smoky monitor of IMPROVE may be sitting on a different ridge top than the CASTNET monitor, and may be at a different elevation. If there are such differences, did you make any corrections to account for them?

C. HOGREFE   For our analysis of measurements from co-located monitors, all but two of them were within one kilometer of each other and at almost the same elevation. Thus, we did not perform any corrections to the raw data. However, it is important to keep in mind that differences in the monitoring locations are just one possible reason for differences in the observations, along with different sampling intervals, sampling techniques or data reporting procedures. Note, our current analysis focused on determining features that are common between the two networks despite all these differences.

A.GERTLER SO_4 is regional and NO_3^- is land. The high regional background due to SO_4^2 how can the states meet the new annual $PM_{2.5}$ standard?

C. HOGREFE   I am not in a position to give a definite answer to this question, but the larger-scale features that our analysis detected in sulfate concentrations suggest that regional-scale controls of SO_2 would have a big impact on $PM_{2.5}$ concentrations. Further, one could expect that an additional reduction of $PM_{2.5}$ conentrations would occur from reductions in NO_x emissions contemplated to meet the new 8-hr ozone standard.

O.D.YAY — This question is in a way related with Andreani's question. Different from sulfate, some of the nitrate may be associated with the coarse particles depending on the formation mechanism. Could, for example, the IMPROVE data set be missing the coarse mode nitrate since it analyses only the fine fraction?

C. HOGREFE — This difference in sampling technique might indeed be responsible for some of the differences in nitrate features we detected from both networks. Again, our analysis was focused on finding robust features that are common between the two networks despite all these differences so that such features could be used in model evaluation.

DRY DEPOSITION OF COARSE SOLID PARTICLES IN PATCHY SUB-BOREAL LANDSCAPE

Marko Kaasik, Tiiu Alliksaar[*]

1. INTRODUCTION

The underlying surface effects to the dry deposition of airborne ingredients are well known at least in qualitative sense: rougher surface initiates stronger wind-induced turbulence and therefore, as a rule, larger deposition velocities. Nevertheless, the scatter in both cases – model versus model and model versus experiment – is often large. Obviously, the theory of turbulent exchange is not developed enough to satisfy high demands of surface-layer air pollution modelling.

Numerous earlier measurements (McMahon & Denison, 1979) show that at other similar conditions the dry deposition fluxes are larger on the areas with high plant cover (e.g. forest) than on the areas with lower one (e.g. grassland). The detail model calculations (Ruijgrok et al., 1995) show that deposition fluxes of particles with diameter about 10 µm can be about ten times larger in the forest than in the open landscape.

An operational dispersion-deposition model is intended to produce realistic deposition fluxes with moderate demands to the computer resources and limited (routinely available) meteorological data. Thus, a relatively simple surface-layer parameterisation is needed.

This paper is deals with the application of Monin-Obukhov surface layer theory to the airborne deposition fluxes. Recent updates for long-lasting inversion are taken into account. Theoretical results are compared with field studies in the north-eastern Estonia, the area which is found to be an appropriate test site due to relatively flat natural landscape and highly dominating well-identified point sources of particles (Kaasik et al., 2000).

2. BASIC CONCEPTS OF THE SURFACE-LAYER SCALING

Dry deposition can be included into an air pollution model, applying the bulk resistance concept. For airborne particles the total resistance (inverse value of deposition

[*] Marko Kaasik, Institute of Environmental Physics, University of Tartu, Tartu, Estonia, 50090. Tiiu Alliksaar, Institute of Geology, Tallinn Technical University, Tallinn, Estonia, 10143.

Air Pollution Modeling and Its Application XVI, Edited by
Borrego and Incecik, Kluwer Academic/Plenum Publishers, New York, 2004

velocity v_d) includes the aerodynamic resistance r_a and the resistance of viscous sub-layer r_b in sequence and the inverse value of gravitational settling velocity v_s in parallel with these:

$$1/r \equiv v_d = \frac{1}{r_a + r_b + r_a r_b v_s} + v_s. \tag{1}$$

The resistance of viscous sub-layer depends on the Brownian diffusivity of particles and is usually parameterised by means of the Schmidt number Sc and the Stokes number St:

$$r_b = \frac{1}{u_*(Sc^{-2/3} + 10^{-3/St})}, \tag{2}$$

where u_* is the friction velocity (see below). Both Sc and St can be easily evaluated using their definitions.

The general form (with slightly varying coefficients by different authors) of the aerodynamic resistance yielding from the classical Monin-Obukhov similarity theory is

$$r_a = \begin{cases} \frac{1}{ku_*}\left[\ln\left(\frac{z}{z_0}\right) + 5\left(\frac{z}{L} - \frac{z_0}{L}\right)\right] & \text{(stable)} \\[3ex] \frac{1}{ku_*}\left[\ln\left(\frac{z}{z_0}\right) + 5\ln\left(\frac{(\eta_0^{\,2}+1)(\eta_0+1)^2}{(\eta_r^{\,2}+1)(\eta_r+1)^2}\right) + 2\arctan\eta_r - 2\arctan\eta_0\right] & \text{(unstable)} \end{cases} \tag{3}$$

where $\eta_r = (1-15z/L)^{1/4}$ and $\eta_0 = (1-15z_0/L)^{1/4}$ with z as reference height. k is the von Karman constant ($k=0.4$ assumed here). The roughness length z_0 is in most practical cases evaluated applying the type of landscape, e.g. $z_0=1.4$ for boreal forest and 0.01 for snow-covered open peatland (EEA, 1992). The Monin-Obukhov length can be evaluated from its definition

$$L = \frac{u_*^3 c_p \rho \theta}{kgH}, \tag{4}$$

but we need u_* in both Eq. (3) and (4). Assuming that wind speed u at reference height is known from on-site measurements or derived from the meteorological analysis, we can express u_* from the scaling law for wind speed:

$$u_* = ku / [\ln(z/z_0) - \Psi(z/L)], \tag{5}$$

where classical Businger-Dyer form for universal function Ψ is

$$\Psi\left(\frac{z}{L}\right) = \begin{cases} -5\dfrac{z}{L} & \text{(stable)} \\[3ex] \ln\left[\left(\dfrac{\eta_r^{\,2}+1}{2}\right)\left(\dfrac{\eta_r+1}{2}\right)^2\right] + 2\arctan\eta_r - \dfrac{\pi}{2} & \text{(unstable)} \end{cases} \tag{6}$$

Coefficient 16 is commonly used in Eq. (6) instead of 15 as in Eq. (3), when calculating η. As shown by Zilitinkevich, equation (5) is not valid for long-lasting thermal inversion due to vanishing of residual layer and related influence of gravitational waves to the surface layer regime. For such a case Zilitinkevich (1998) proposed a theory, leading to a new formulation instead of Eq. (5):

$$u_* = (ku - C_{u1}C_{u2}Nz)/[\ln(z/z_0) + C_{u1}(z/L)], \qquad (7)$$

where the Brunt-Väisällä frequency $N = [(g/T)\partial\theta/\partial z]^{1/2}$, with the potential temperature gradient evaluated above the mixing height. The numerical values of empirical constants C_{u1} and C_{u2} are highly disputable due to large scatter of experimental data. Below two alternative sets are applied: indirectly calculated $C_{u1}=4.5$, $C_{u2}=0.2$ (Zilitinkevich et al, 1998) and directly evaluated $C_{u1}=2.1$, $C_{u2}=0.4$ (Zilitinkevich & Galanca, 2000).

Equations (4) and (5) or (7) with supplementary formula (6) constitute an algebraic system for u_*, which has exact solution (through the formulae of Cardano) only for stable stratification. In this paper the iterative numerical algorithm is used for all cases. It appears that if the (positive) Monin-Obukhov length, surface roughness length and reference wind speed are small enough, then the friction velocity approaches the zero value, i.e. the stably stratified surface layer is "screening" all the surface friction effects, resulting both aerodynamic and pseudo-laminar sub-layer resistances infinite. Such behaviour, although physically doubtful, is not a problem when dealing with coarse particles. Simply, we can assume, that gravitational settling is the only substantial deposition factor in very stably stratified surface layer. If the surface layer is stable enough, further increase of stability does not affect the resulting deposition velocity, which is solely determined by gravitational settling. Naturally, the gravitational settling velocity is the lower limit of deposition velocity for particles (Fig. 1).

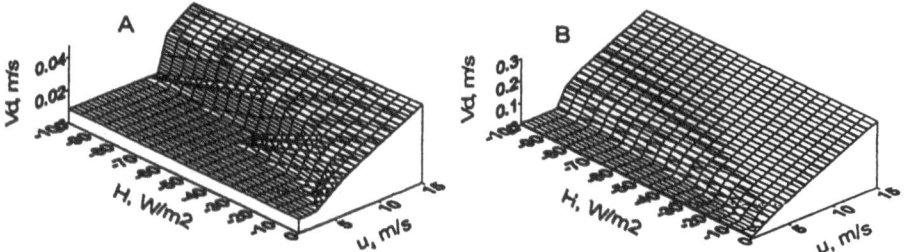

Figure 1. Total deposition velocities v_d depending on surface heat flux H and 10 m wind velocity u, calculated for underlying surface with roughness lengths 0.01 m (left) and 1.4 m (right) according to the Monin-Obukhov surface-layer theory. Low smooth surface (far left corner) is determined solely by gravitational settling according to Stokes law (parameters for particles see Subsec. 3.1).

3. EXPERIMENT

3.1. Field observations

Field data are based on sampling of atmospheric precipitation. Two measurement series are applied. For wintertime series (December 2 – 14, 2002) snow was sampled from a well-identified snow layer on the natural surface. Summer series (August 2 – 12, 2002) was collected using slightly modified EMEP (EMEP, 1996) precipitation sampling method. In order to collect the dryly deposited matter into the water reservoir, the plastic funnel of a sampler was washed with super-pure water at the end of sampling period.

The sampling area was located in the north-eastern part of Estonia, roughly at 59° N and 27,5° E. The sampling points were situated 15 – 45 km away from oil-shale-fired Narva power plants. These two power plants together are emitting nearly 50 000 tonnes of fly ash (limestone-like composition, about 30% CaO, Pets et al, 1985) annually. The average density of fly ash particles is assumed to be 2800 kg/m^3 and aerodynamic diameter 10 μm (estimation based on results by Aunela et al., 1995). The test area is sparsely inhabited, dominating landscapes are sub-boreal forest and bog. To distinguish the effects of regional emissions, some background samples were taken from similar landscape in the south-eastern part of Estonia (140 km away from Narva power plants).

Airborne particulate matter was identified following the deposition fluxes of calcium and sulphate ions and spheroidal fly ash particles. Due to its mineral nature airborne calcium is definitely an indicator of particulate matter. This assumption is not necessarily valid for sulphate, but as found earlier (Kaasik & Sõukand, 2000), sulphate in the precipitation affected by the emissions of oil shale combustion is most likely of mineral origin and follow nearly 1:1 mass ratio with calcium ion. Spheroidal fly ash particles constitute a specific product of high-temperature combustion, formed due to incomplete burning or melting of mineral matter in the furnace. Spheroidal particles in lake and bog sediments are found to be a sensitive and chemically inert marker of fossil fuel combustion (Alliksaar et al., 1998). In this study the microscopically determined particle deposition fluxes (number of particles larger than 5 μm) are applied. Compared to chemical markers spheroidal particles seem to be more sensitive for identifying the oil shale combustion products (Table 1, 2).

Table 1. Precipitation amounts and airborne deposition fluxes (average and standard deviation), December 2 – 14, 2002.

Site type (number of samples)	Precipitation, mm		Ca^{2+}, mg/m^2 per day		SO$_4^{2-}$, mg/m^2 per day		Particles, 10^3 no./m^2 per day	
Woodland (6)	7.2	±1.2	6.37	±0.99	6.88	±1.22	49.2	±8.8
Open land (6)	8.8	±1.7	7.53	±3.25	7.83	±3.06	70.4	±19.8
Woodland, background (2)	2.4	±0.6	0.22	±0.03	0.42	±0.02	0.6	±0.2
Open land, background (2)	4.7	±0.9	0.42	±0.01	0.58	±0.21	0.9	±0.3

Table 2. Airborne deposition fluxes, August 2 – 12, 2002 (no precipitation registered).

Sample No.	Site type	Ca^{2+}, mg/m^2 per day	SO$_4^{2-}$, mg/m^2 per day	Particles, 10^3 no./m^2 per day
1	Bog forest	10.9	4.7	3.9
2	Open bog	14.7	5.5	2.6
3	Bog forest	11.6	2.4	3.6
4	Open bog	14.3	4.8	1.9
6	Bog forest (background)	13.7	-	0.9
5	Open bog (background)	14.8	-	0.3

No systematic difference between woodland (incl. bog forest, tall pine forest, bushwood) and open (incl. bog with sparse low trees, lake ice) sites was found in December. In August the deposition fluxes of spheroidal particles under the forest seem systematically (nearly twice) higher than on the open land. It is remarkable that in this case the calcium flux does not indicate clear difference neither between forest and open, nor for polluted and background sites. The summertime sampling occurred during long-lasting dry anticyclonal weather with severe forest fires and soil erosion around. Therefore high background concentrations are explicable. Different behaviour of Ca and spheroidal particles in woodland versus open land is a matter of question. Different size of particles may be a reason. Obviously, most of summertime Ca is not of high-temperature combustion origin.

3.2. Meteorological background data

All meteorological data were extracted from the online database of NOAA Air Resources Laboratory (http://www.arl.noaa.gov/ready.html). The analysis product FNL contains global surface and three-dimensional data with 6 hours time step.

The wind speed at 10 m reference height (surface wind) was used to determine u_*. Let us point out that due to coarse resolution of global data set the surface wind represents an average dynamically adjusted value rather than actual wind over the actual terrain. We can apply it for different underlying surfaces despite small topographical elements and zero-displacement height, which can make problems when applying on-site wind measurements.

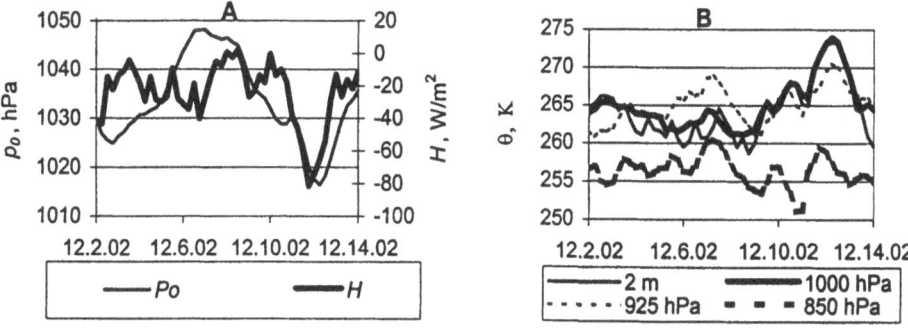

Figure 2. Time series of surface air pressure and heat flux (A) and potential temperatures at different levels (B) during December 2 – 14, 2002, eastern Estonia.

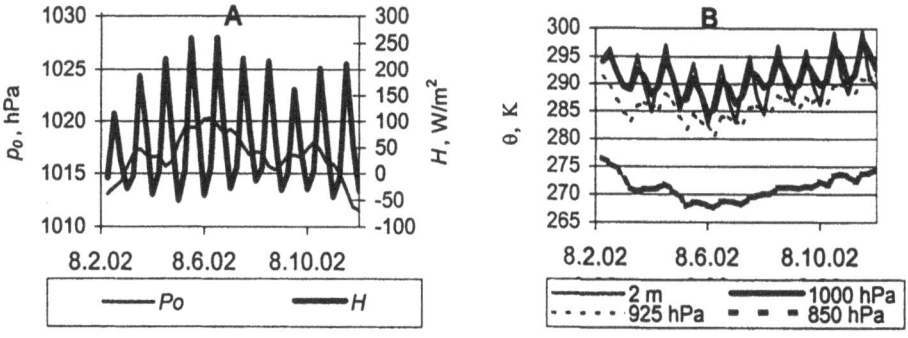

Figure 3. Time series of surface air pressure and heat flux (A) and potential temperatures at different levels (B) during August 2 – 11, 2002, eastern Estonia.

Surface heat flux H is directly reported in FNL data set (not regarding micro-scale details, however). Temperatures, heights and wind speeds were extracted at 1000, 925 and 850 hPa levels. To calculate the Brunt-Väisällä frequency N, potential temperatures at 925 and 1000 hPa (approximately 800 and 200 – 400 m) levels were used. Average wind direction of 1000 and 925 hPa level was used to determine the periods of favourable wind, blowing from the power plants (stack heights 150 – 250 m, highly buoyant plumes) to sampling sites (±30° from exact direction).

During the wintertime sampling the long-lasting surface inversion under the cold high-pressure area definitely dominated (Fig. 2), although surface heat flux data are not always consistent with stratification (extremely strong downward flux together with low-pressure outbreak and warming at Dec. 13). Problem can be related to low vertical resolution of FNL data set. During summertime sampling the boundary-layer conditions were highly convective daytime, but FNL suggest rather strong inversions at night hours.

4. DEPOSITION FLUXES *VS.* THEORETICAL DEPOSITION VELOCITIES

Until the parameterisation schemes (Sec. 2) are not included into an air pollution transport and dispersion model, we cannot reproduce the adequate deposition fluxes theoretically. But we can compare the theoretical ratios of observed fluxes and time-averaged deposition velocities over forested and open land.

By its definition the deposition velocity is proportional to the deposition flux at each time moment: $I = v_d C$, where C is concentration in the air. This assumption is not strictly valid for time-averaged deposition velocity and flux, but we can apply it as first approximation. Selection of favourable wind directions (Subsec. 3.1) enhances the reliability of this approach due to excluding of probable very low particle concentrations in the air. For wintertime series the friction v_d was calculated separately with two sets of empirical coefficients C_{u1}, C_{u2} in Eq. (7), see end of Sec. 2. The nighttime summer inversions were treated according to the classical Monin-Obukhov theory, Eq. (5).

For both summer- and wintertime the calculated deposition velocities for lowest roughness lengths are close to the gravitational settling velocity. During summertime, however, there does not exist $z_0 \approx 0.01$ m like a snow-covered open land in winter. More realistic values for summertime open bog are in the range of 0.05 – 0.1 m. Assuming the roughness of forested bog about 0.5 – 1 m, we get about 2 – 3 times difference in the deposition velocity between forested and open bog sites under favourable wind directions, which is rather consistent with measured deposition fluxes (Table 2). The estimated wintertime deposition velocities depend on selected set of parameters C_{u1}, C_{u2} (Zilitinkevich et al, 1998 or Zilitinkevich & Galanca 2000). The later one gives lower deposition velocities, especially for favourable wind direction and low roughness lengths. In this case the gravitational settling velocity (0.095 m/s) is dominating up to roughness about 0.1 m. At "forest-size" roughness lengths, however, the deposition velocity is growing rapidly, which is inconsistent with indifferent or even lower measured deposition flux for the forest (Subsec. 3.1.).

Thus, we found that qualitatively the wintertime landscape-insensitive deposition might be due to the domination of gravitational settling in very stable surface layer, but according to Eq. (3), (4) and (7) the criterion for domination ($u*$ 0) is extremely sensitive to the reference wind speed, Brunt-Väisällä frequency and not well-known

496

empirical coefficients. Both wind and temperature profiles are most probably oversmoothed by the techniques synoptical of analysis. Thus, too strong surface-layer winds and too small Brunt-Väisällä frequencies are expected, both leading to the underestimation of stability.

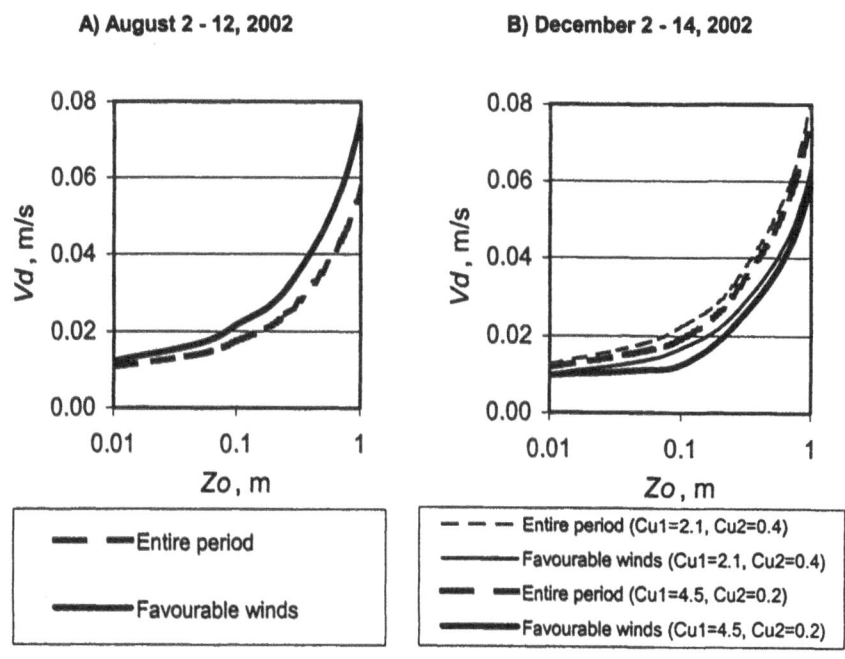

A) August 2 - 12, 2002   **B) December 2 - 14, 2002**

Figure 4. Computed average deposition velocities for summer- (A) and wintertime (B) sampling periods. Time with favourable winds constitutes 52% of summer and 43% of winter time-series. Wintertime deposition velocities are calculated separately for two sets of empirical parameters (see Sec. 2).

Based on experimental deposition fluxes and computed deposition velocities, we can estimate the concentrations of fly ash in the air. Keeping in mind the definition of deposition flux and assuming that the Ca flux is 7 mg/m^2 per day and v_d=0.01 m/s (mainly gravitational settling), we get 8 mg/m^3 Ca during the wintertime sampling. Assuming that oil-shale fly ash contains about 23% Ca (Pets et al, 1985), the estimated average fly ash concentration in the air is 35 mg/m^3. Analogical calculations for August 2002 (assuming Ca flux 13 mg/m^2 per day and average v_d=0.03 m/s) yield about 5 mg/m^3 Ca in the air. Unfortunately, we cannot conclude anything about fly ash concentrations because of high background values of Ca. In this context we have to stress that summertime measurement sites were located further away from Narva power plants than wintertime ones (15 – 20 and about 45 km respectively). Therefore smaller impact is expected. No reliable data about the concentration of spheroidal particles in emitted fly ash is available. Thus, despite higher sensitivity, deposition data of spheroidal particles don't have "absolute calibration" and cannot be applied for tracing the fly ash concentrations yet. Need additional research is recognised.

5. CONCLUSIONS

Relying on both surface-layer theory and deposition measurements, the deposition process of coarse solid particles is highly sensitive to the surface roughness. The sensitivity is less in deep inversion, when deposition, except gravitational settling, vanishes.

For application of surface-layer theory to the particle deposition in patchy sub-boreal bog landscape the earlier indirectly estimated values of empirical coefficients C_{u1}, C_{u2} (Zilitinkevich et al., 1998) fit slightly better with field measurement data than the later ones (Zilitinkevich & Galanca, 2000), but for the work with analysed (not on-site measured) profiles probably a third set of parameters should be established.

To enhance the quality of comparison, the need for both experimental (more fieldwork, aerosol measurements) and theoretical (application of above surface-layer scaling in a dispersion-advection model) updates is realised. More detail meteorological analysis is needed to decrease the uncertainty of input parameters (possible application of the HIRLAM model data in future). The application of surface-layer scaling (Sec. 2) is planned in the AEROPOL model and the Estonian meso-scale air pollution model (under development).

ACKNOWLEDGEMENTS

This study was funded by the Estonian Science Foundation (grant 5002). Special thanks to J. Ivask and E. Taevere for kind assistance.

REFERENCES

T. Alliksaar, P. Hörstedt, I. Renberg, Characteristic fly-ash particles from oil-shale combustion found in lake sediments, Water, Air, and Soil Pollution, **104**, 149-160 (1998).

L. Aunela, E. Häsänen, V. Kinnunen, K. Larjava, A. Mehtonen, T. Salmikangas, J. Leskelä, J. Loosaar, Emissions from Estonian oil shale power plant, *Oil Shale*, **12**, 5, 165-177 (1995).

EEA, *Corine Land Cover, A European Community Project*, European Environmental Agency, Copenhagen, Denmark (1992, dataset on CD).

EMEP *EMEP manual for sampling and chemical analysis.* NILU, Kjeller, 97 (1996).

M. Kaasik, R. Rõõm, O. Røyset, M.Vadset, Ü. Sõukand, K. Tõugu, H. Kaasik, Elemental and base anions deposition in the snow cover of north-eastern Estonia. The impact of industrial emission., *Water, Air, and Soil Pollution*, **121**, 324-366 (2000).

M. Kaasik, Ü. Sõukand, Balance of alkaline and acidic pollution loads in the area affected by oil shale combustion. *Oil Shale*, **17**, 2, 113-128 (2000).

T. A. McMahon, P. J. Denison, Empirical atmospheric deposition parameters – a survey. *Atm. Environ.*, **13**, 571 – 585 (1979).

L. I. Pets, P. A. Vaganov, I. Knoth, Ü. Haldna, H. Schwenke, C. Schnier, R. Juga, Micro-elements in the ashes of the Baltic Thermoelectric Power Plant, *Oil Shale*, **2**, 4 379-390 (1985, in Russian).

W. Ruijgrok, C. I. Davidson, K. W. Nicholson, Dry deposition of particles. Implications and recommendations for mapping of deposition over Europe. *Tellus*, **47B**, 587 – 601 (1995).

S. Zilitinkevich, P.-E. Johansson, D. V. Mironov, A. Baklanov, A similarity-theory model for wind profile and resistance law in stably stratified planetary boundary layers, *J. of Wind Engineering and Industrial Aerodynamics*, **74-77**, 209-218 (1998).

S. Zilitinkevich, P. Galanca, An extended similarity theory for the stably stratified atmospheric surface layer, Q. J. R. Meteorol Soc., **126**, 1913-1923 (2000).

DISCUSSION

J. BARTNICKI   You have not included gravitational settling velocities in your parameterization of dry deposition velocities. Why?

M.KAASIK   Gravitational settling velocity is of same order of magnitude as other deposition velocity components, therefore I cannot neglect it. (I misunderstood the question: I thought, he is asking, why I included the gravitational settling velocity. Actually, gravitational settling velocity was included, but Dr. Bartnicki did not realise that.)

THERMODYNAMICAL AND AIRFLOW CONDITIONS WITHIN THE LOWER TROPOSPHERE DURING THE TROPOPAUSE FOLD EVENT LEADING TO ELEVATED SURFACE OZONE CONCENTRATIONS

Zvjezdana Bencetić Klaić, Danijel Belušić and Ivana Herceg Bulić[*]

1. INTRODUCTION

We investigated one episode of a winter stratospheric ozone intrusion over Zagreb, Croatia on February 6, 1990, when unusually high ozone concentrations were measured at two sites in the greater Zagreb area. Lisac et al. (1993) already studied the same episode. They argue that the air must be of stratospheric origin, since photochemistry cannot generate large amounts of ozone at this time of year. Their conclusion is also supported by the spiky character of the ozone level behavior, which was recorded on both stations, and which suggests rapid downward transport related to cross-tropopause exchange (Schuepbach et al., 1999). Lisac et al. confirmed the intrusion by means of analysis of available routine surface and upper air data. Finally, they roughly estimated a probable three-dimensional trajectory of an intruded air parcel. The trajectory starts on February 5 at 01 LST (Zagreb local time) at 300 hPa surface above the Baltic sea, and due to its anticyclonic curvature arrives in Zagreb on February 6 at 13 LST from the east. The above study was limited by the rough spatial and temporal resolution of the routine data, and, it did not offer a clear explanation of the time delay in the ozone peak, which was recorded at one of the measuring sites, which, we believe, is attained in the present study.

In order to investigate the above event, we used a nonhydrostatic prognostic mesoscale model, where we focused on the lower troposphere. Our results enabled a very detailed insight into the vertical structure of a tropopause fold, found in the lower troposphere. Additionally, they suggested a slight modification of the trajectory of the intruded parcel proposed by Lisac et al.

2. OZONE EPISODE ON FEBRUARY 6, 1990

According to Lisac et al. (1993) an episode of high ozone concentration was recorded at two measuring sites in the greater Zagreb area on February 6, 1990 (Fig. 1., left). One site, Ruđer Bošković Institute (RBI), is situated in the northern residential part of Zagreb at 180

[*] Andrija Mohorovičić Geophysical Institute, Faculty of Science, University of Zagreb, Croatia.

Air Pollution Modeling and Its Application XVI, Edited by
Borrego and Incecik, Kluwer Academic/Plenum Publishers, New York, 2004

m ASL. Due to clearly observable typical urban diurnal variation of ozone concentration, RBI is representative for an urban area with occasional photosmog situations. The other site, Puntijarka (P), is on the ridge of Medvednica mountain at 980 m about 10 km north of RBI. It is isolated from major pollution sources, and, considering its ozone concentration behavior it exhibits a rural character. Therefore, it may be regarded as a representative of regional ozone concentration levels. During the winter months, hourly ozone volume fraction at site P is on the average 35 ppbv (Cvitaš et al., 1997).

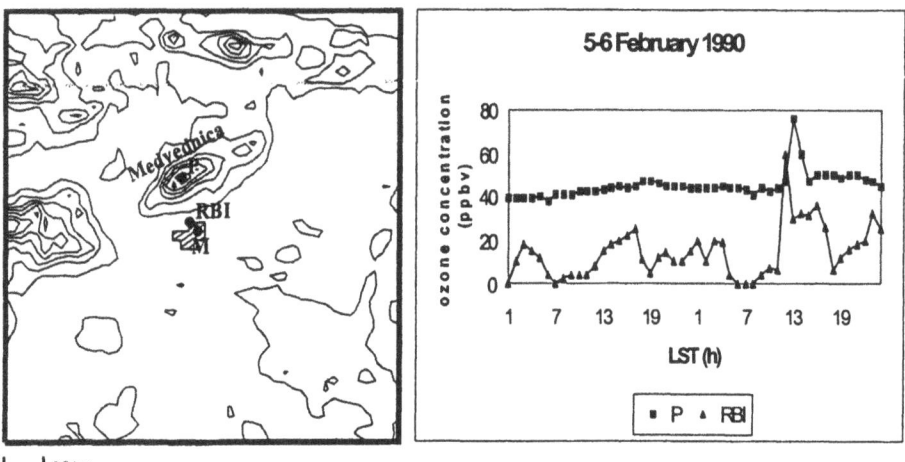

Fig. 1. Left - monitoring sites in the greater Zagreb area where elevated ozone concentrations were recorded: Puntijarka observatory (P, on the top of Medvednica mountain at 980 m ASL, 45°54' N, 15°58' E) and Ruđer Bošković Institute (RBI, in residential part of Zagreb at 180 m ASL, 45°50' N, 16° E). M is position of Zagreb-Maksimir aerological observatory (128 m ASL, 45°49'N and 16°02'E). The densely urbanized part of Zagreb is hatched. Topography contours are given for every 100 m. Right: measured hourly mean concentrations for 5 - 6 February 1990.

A sudden ozone concentration spike from 2 ppbv to 100 ppbv was at first recorded at RBI at 11.45 LST (10.45 UTC). About one hour later, a rise from 40 ppbv to above the maximum measurable value of the instrument (i.e. higher than 100 ppbv) was recorded at the mountain site P. These sharp peaks resulted in elevated hourly mean concentrations at both sites, where 60 ppbv and 76 ppbv corresponded to RBI at 12 LST and P at 13 LST, respectively (Fig. 1., right). During the following hours hourly mean concentrations decreased noticeably at both stations. It is worth noting that variations of daily ozone volume fractions in Vienna were almost identical to RBI values during the period of study (Lisac et al., 1993).

3. SYNOPTIC CONDITIONS

The synoptic conditions were analyzed by means of standard diagnostic surface and upper level charts. A particular aim was to recognize synoptic features that might induce tropopause fold, and associated processes of stratosphere-troposphere exchange.

As illustrated in Fig. 2., on February 5 at 00 UTC a low-pressure center over Iceland was found. At the same time, a stable anticyclone dominated, southeast of the cyclone, over the major part of Europe. A large-scale disturbance was present at upper isobaric levels, and it extended throughout the whole troposphere. It was characterized with an upper-level ridge over central Europe between two troughs: one placed over the Atlantic (western trough) and the other one over Europe (eastern trough).

On the western side of the ridge, a confluence and consequent SW-NE oriented jet stream is found. Both, the entrance (confluent) and the exit (diffluent) regions of the jet stream are favorable for the development of the transverse secondary circulations (Andrews et al., 1987; Uccellini and Kocin, 1987; Reed, 1990; Cox et al., 1995), which might cause the stratosphere-troposphere exchange. Stratospheric air can penetrate into the troposphere only by sinking motions. These are established by secondary circulations in the right front (right-diffluent) and the left rear (left-confluent) quadrants of the jet stream, respectively. However, the latter could not lead to the transport of air towards Zagreb.

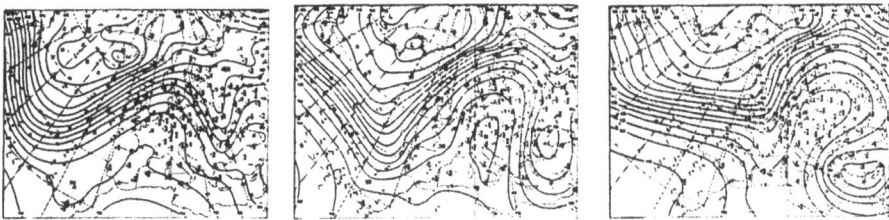

Fig. 2. Diagnostic 300 hPa charts for 00 UTC over Europe for February 5, 6 and 7, 1990 (left, center and right, respectively).

In time, the confluence of the geopotential contours at 300 hPa isobaric level strengthened, as well as the jet stream located in the upper atmosphere between the UK and Scandinavia. Therefore, consequent transverse secondary circulations should also intensify. As the upper level disturbance propagated toward the east, the ridge extended towards the north, while the eastern through deepened and extended southward. Relative topography RT 500/1000 indicated a cold air trough that coincided with the geopotential trough over eastern Europe. Eventually, the southern part of the cold trough had cut from the main stream and had formed COL (cut-off low) system, which was on February 6 at 00 UTC placed over the Black Sea. The cyclonic circulation extended from the earth's surface throughout the entire troposphere.

On the western part of COL a secondary jet stream with an axis oriented in NNE-SSW direction at 300 hPa is found. Maximal winds for February 6 at 00 UTC are found approximately 600 km east of Zagreb at latitudes of approximately $\varphi = 45°N$, while confluent and diffluent regions are found northward and southward, respectively. Therefore, a secondary-circulation-induced sinking motion should be found on February 6 at 00 UTC found southeast of Zagreb.

On February 7 at 00 UTC the COL moved southward toward the Mediterranean Sea, while its secondary jet stream disappeared. At the same time, the strength of the jet stream above northwestern Europe decreased.

The above facts suggest several processes responsible for the sinking of the stratospheric air deep into the troposphere, and at the same time, its gradual advection towards Zagreb. According to the synoptic conditions, sinking started on February 5 in the right front region of the jet stream found above northwestern Europe. It intensified over time together with the jet stream. Due to established synoptic formations, parcels of stratospheric origin once found in troposphere over northern Europe, were deflected southward (initially towards SE, later towards SSW, and finally towards SW). While

503

passing through the right front region of secondary jet stream (i.e. southwestern part of COL), parcels are, due to established subsidence, again forced to sink. In the same region another tropopause fold might occur as a result of the COL system. In addition, sinking of the air westward from COL system (i.e. over Central Europe) is maintained due to the stable anticyclone.

4. NUMERICAL SIMULATION OF THE CONDITIONS WITHIN THE LOWER TROPOSPHERE

Meteorological conditions within the lower troposphere were simulated by the MEsoscale MOdel (MEMO) version 6.0. More details on this Eulerian nonhydrostatic prognostic mesoscale model could be found in studies of Moussiopoulos (1994, 1995) and Kunz and Moussiopoulos (1995). Model performance over the various European regions is described in several papers (Moussiopoulos, 1995; Moussiopoulos et al., 1997a, 1997b; Klaić et al., 2002; Klaić and Nitis, 2001-2002).

Table 1. Specification of domains.

Grid	Domain size (km^2)	Horizontal resolution $\Delta x = \Delta y$ (km)	Integration time step (s)
Coarse	1080 x 1080 (Fig. 3)	9	30
Medium	250 x 250	5	20
Fine	100 x 100 (Fig. 1 left)	2	10

Simulation was performed for the period from 4 to 8 February 1990, where the first day was used for pre-run purposes. The model was driven with the routinely measured surface and radiosonde data for Budapest, Belgrade, Brindisi, Pratica Di Mare, Milan, Munich and Zagreb. The double-nested grid option was employed with a coarse, medium and fine grids as specified in Table 1. Fine and the coarse grids are also shown in Fig 1. left and Fig. 3, respectively. Medium grid (not shown here) was employed only to enable the fine grid calculations.

The model top was set at 6000 m above the sea level. In the vertical, 25 layers were selected with a finer resolution at lower altitudes. Layer depths gradually increased from 20 m at the bottom to about 900 m at the top of the modeling domain. Apart from the roughness length for forest (z_0 = 0.4 m) and urban surfaces (z_0 = 0.8 m), values of other model parameters are taken as in the study of Klaić et al. (2002).

5. RESULTS

Figure 3 illustrates modeled surface winds and potential vorticities at isentropic surface θ = 30.0°C for the coarse domain on February 5 at 12 LST. Potential vorticity PV is calculated from $PV = -g\,(\partial\theta/\partial p)(\zeta + f)$, where g is acceleration due to gravity, ζ is relative vorticity on the considered isentropic surface, f is Coriolis parameter and θ is potential temperature.

The above isentropic surface is selected since it is high enough, so that air parcels of stratospheric origin may still retain increased PV values. Yet, it is low enough to enable air parcels descend down to the ground without significant changes in their chemical structure. During the period of study, selected isentropic surface was, according to the model results, above Zagreb found at heights between 4.7 and 5.1 km ASL.

On February 5 at 12 LST there are two distinct regions of increased PV values (over $1.2 \cdot 10^{-6}$ K m Pa^{-1} s^{-3}) in the lower troposphere (Fig. 3). One, which is above northern Italy southward from the Alps, is beyond the scope of this study. The other is found northwest of Zagreb. Its horizontal scale is of the order of 100 km.

The air with the high PV values moves southward over time. Simultaneously, PV values gradually decrease. At 20 LST the region of the highest PV ($\leq 1.2 \cdot 10^{-6}$ K m Pa^{-1} s^{-3}), is still north of Zagreb, and it passes above Zagreb on February 6 at 00 LST. At later times, only the fragments of $PV \leq 1.0 \cdot 10^{-6}$ K m Pa^{-1} s^{-3} are found southeast of Zagreb until February 6 at 12 LST.

Figure 4 illustrates two north-south vertical cross-sections through Zagreb for February 5, in the coarse grid domain. For February 5 at 12 LST it clearly shows a sloping streamer of increased PV stretching throughout almost the whole investigated lower 6 km of the troposphere. (The same was also found at previous times, starting from February 5 at 00 LST.) Therefore, it seems reasonable to assume that the streamer extends higher. However, a simulation of whole troposphere would substantially prolong required CPU time, if the spatial and temporal resolution remained as fine as in this study.

Due to the northern airflow in the free troposphere, the streamer is sloped toward the south. The horizontal width of the streamer is of the order of magnitude of 100 km, which agrees with other studies (Andrews et al., 1987; Appenzeller and Davies, 1992). Further, the vertical structure of the streamer, as obtained for example for February 5 at 12 LST, shows a train of horizontally stretched vortices. The horizontal dimensions of the individual vortices vary from 30 to 70 km, while vertical dimensions are between 500 to 1500 m. These vortices are characterized with very large PV gradients, particularly in the vertical direction, where changes of PV up to $1.5 \cdot 10^{-6}$ K m Pa^{-1} s^{-3} / 100 km and $0.2 \cdot 10^{-6}$ K m Pa^{-1} s^{-3} / 100 m are found in horizontal and vertical direction, respectively. Additionally, the lower part of the streamer is characterized with closely spaced isentropes.

The wave-like distribution of the individual vortices is also accompanied by vertical convergence of the wind flow (not shown here). In other words, in the border regions between the two individual vortices subsidence of the air from above and the upward movement of the air from below are found.

In the course of time, PV values within the vortices gradually decrease, vortices continue to sink, and they are more and more split apart. On February 6 at 00 LST the initial streamer is broken into two distinct parts. One, with PV values up to 1.6·10-6 K m Pa^{-1} s^{-3} is found approximately above Vienna. It stretches from the ground up to about 3 km ASL. The other, which is found above Zagreb, is more likely 'stratospheric residue' with PV values up to $1.0 \cdot 10^{-6}$ K m Pa^{-1} s^{-3}. This residue moves gradually southward, and therefore, starting from February 6 at 4 LST it is found south of Zagreb.

Figure 5 shows modeled surface winds for the fine grid domain, i.e. within the greater Zagreb area for February 6. One may note that winds in the southeast quadrant of the domain are predominantly northward or northeastward. This, together with the fact that stratospheric air was at times after February 6 at 00 LST found aloft, south and/or southeast of Zagreb, explains why increased ozone levels on February 6 were first recorded at RBI, and afterwards, about one hour later, at measuring site P which is about 10 km north of RBI. More so, such a scenario would require an average northward wind

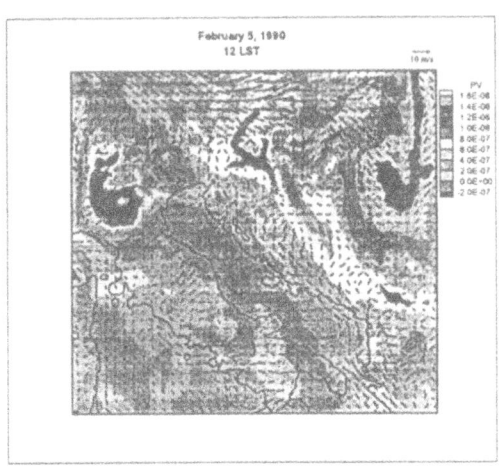

Fig. 3. Modeled potential vorticity (K m Pa^{-1} s^{-3}) at isentropic surface $\Theta = 30.0°$ and surface winds (10 m above the ground) for the coarse grid domain for February 5 at 12 LST. Wind vectors are given at resolution of 27 km. Point M denotes the position of Zagreb-Maksimir aerological observatory.

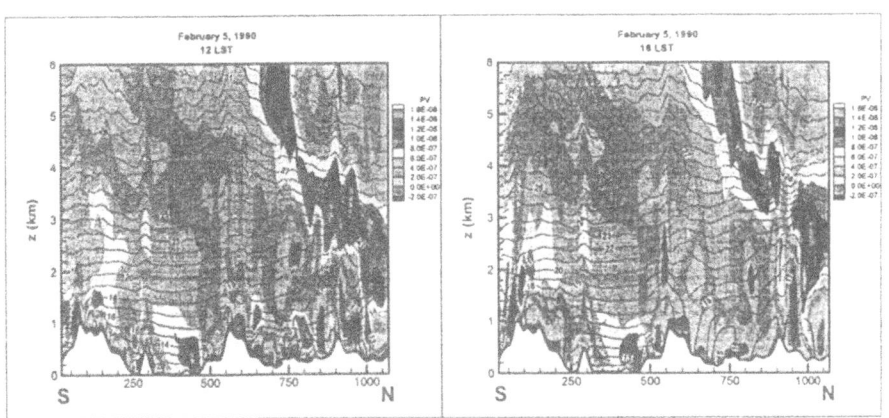

Fig. 4. North-south vertical cross section trough the Zagreb-Maksimir point M (see Fig. 1 left) for the coarse grid domain. Point M is indicated with the black circle. Modeled potential vorticity (*PV*, K m Pa^{-1} s^{-3}) and potential temperature field (lines, in °C) are shown. The distance in abscissa is given in km and *z* is a height above the sea level.

component of about 2.8 m s^{-1}, which agrees well with modeled surface winds in the southeast quadrant of the domain.

Finally, as seen from Fig. 5, an up and down slope local wind circulation established on south-facing slopes of Medvednica mountain, as expected under anticyclonic synoptic conditions. This supports the assumption proposed by Lisac et al. that daytime up-slope winds enhanced ozone transport toward the mountaintop, resulting in the higher ozone peak at the site P on February 6 compared to that at the site RBI.

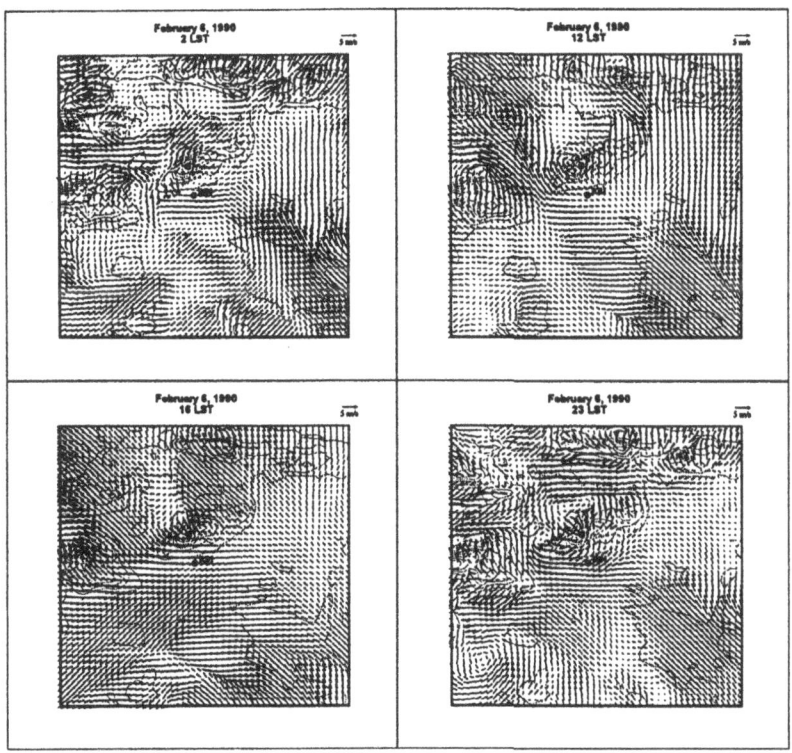

Fig. 5. Modeled surface winds (10 m above the ground) over the greater Zagreb area (fine grid domain). Wind vectors are given at resolution of 2 km. Positions of points P and RBI are indicated with circles. Topography contours are given for every 100 m.

6. CONCLUSIONS

The employed mesoscale model confirmed the very deep wintertime tropopause fold, which was accompanied with elevated ozone values. The event was associated with a jet stream and COL development. In times from 1.5 to 1 day prior to the elevated ozone event in Zagreb, the calculated PV field in the first 6 km of the troposphere shows a streamer stretching almost throughout the whole lower troposphere, where PV values up to more than $1.6 \cdot 10^{-6}$ K m Pa^{-1} s^{-3} are found. The horizontal scales of the modeled streamers agree with the values suggested by other authors.

The model reproduced the significant fine-scale features of the streamer. The streamer was composed of the wave-like train of vortices characterized by particularly large vertical PV gradients (up to $\Delta PV = 0.2 \cdot 10^{-6}$ K m Pa^{-1} s^{-3} / 100 m). The horizontal dimensions of the individual vortices varied from 30-70 km, while vertical dimensions were between 500-1500 m. Further, the model well captured the streamer's deformation (i.e. slope) due to synoptic scale airflow, its splitting apart, and its final irreversible mixing with the adjacent tropospheric air.

Predominantly southward movement of the PV anomaly coincided with the predominantly southward movement of the upper level trough and COL system.

Our results suggest a slight modification of the final stage of the trajectory of the intruded parcel, which was estimated by Lisac et al. (1993). Based on both, modeled positions of the PV anomaly residuals and the surface wind fields at a resolution of 2 km, we conclude that ozone rich air should reach the surface southeast of Zagreb. The same also explains why elevated ozone levels were at first measured at the southern measuring site.

We conclude that in the examined case stratospheric air found in the troposphere retained elevated ozone levels about one day longer than its thermodynamic signature (i.e. PV values and closely spaced isentropes).

Finally, we would like to emphasize that such events may limit performance of photochemical models of the atmospheric boundary layer.

ACKNOWLEDGMENTS

We are grateful to Dr. Inga Lisac of Andrija Mohorovičić Geophysical Institute for providing the hourly mean ozone concentrations at ozonometric monitoring sites. Dr. Rainer Kunz of Grundlagenentwicklung Draegerwerk AG, Luebeck, Germany kindly gave us useful information regarding MEMO model source code. Dr. Darko Koračin of Desert Research Institute, Reno, NV, USA provided topography and landuse coarse domain data.

REFERENCES

Andrews, D. G., Holton, J. R., Leovy, C. B., 1987: *Middle Atmosphere Dynamics*. Accademic Press, Inc., San Diego, 489 pp.

Appenzeller, C., Davies, H. C., 1992: Structure of stratospheric intrusions into the troposphere. *Nature*, **358**, 570-572.

Cox, B. D., Bithell, M., Gray, L. J., 1995: A general circulation model study of a tropopause-folding event at middle latitudes. *Q. J. R. Meteorol. Soc.*, **121**, 883-910.

Cvitaš, T., Kezele, N., Klasinc, L., 1997: Boundary-layer ozone in Croatia. *J. Atmos. Chem.*, **28**, 125-134.

Klaić, Z. B., Nitis, T., 2001-2002: Application of mesoscale model (MEMO) to the greater Zagreb area during summertime anticyclonic weather conditions. *Geofizika*, **18-19**, 31-43.

Klaić, Z. B., Nitis, T., Kos, I., Moussiopoulos, N., 2002: Modification of the local winds due to hypothetical urbanization of the Zagreb surroundings. *Meteorol. Atmos. Phys.*, **79**, 1-12.

Kunz, R., Moussiopoulos N., 1995: Simulation of the wind field in Athens using refined boundary conditions. *Atmos. Environ.*, **29**, 3575-3591.

Lisac, I., Marki, A., Tiljak, D., Klasinc, L., Cvitaš, T., 1993: Stratospheric ozone intrusion over Zagreb, Croatia, on February 6, 1990. *Meteorol. Zeitschrift*, N.F. **2**, 224-231.

Moussiopoulos, N. (ed.), 1994: *The EUMAC Zooming model: model structure and applications*. Garmish-Partenkirchen: EUROTRAC special publication, ISS, 266 pp.

Moussiopoulos, N., 1995: The EUMAC Zooming model, a tool for local-to-regional air quality studies. *Meteorol. Atmos. Phys.*, **57**, 115-133.

Moussiopoulos, N., Sahm, P., Karatzas, K., Papalexiou, S., Karagiannidis, A., 1997a: Assesing the impact of the new Athens airport to urban air quality with contemporary air pollution models. *Atmos. Environ.*, **31**, 1497-1511.

Moussiopoulos, N., Sahm, P., Kunz, R., Vögele, T., Schneider, C., Kessler C., 1997b: High-resolution simulations of the wind flow and the ozone formation during the Heilbronn ozone experiment. *Atmos. Environ.*, **31**, 3177-3186.

Reed, R.J., 1990: Advances in knowledge and understanding of extratropical cyclones during the past quarter century: and overview. In *Extratropical cyclones. The Erik Palmén memorial volume.* (Newton, C. and Holopainen E.O., eds.), American Meteorological Society, Boston, 167-191.

Schuepbach, E., Davies, T. D., Massacand, A. C., 1999: An unusual springtime ozone episode at high elevation in the Swiss Alps: contributions both from cross-tropopause exchange and from the boundary layer. *Atmos. Environ.*, **33**, 1735-1744.

Uccellini, L. W., Kocin, P. J., 1987: The interaction of jet streak circulations during heavy snow events along the east coast of the United States. *Wea. Forecasting*, **1**, 289-308.

DISCUSSION

G.KALLOS

In your simulation, did you take into account condensation processes? If not, how do you expect to have your results modified from the ones you presented here?

Z.B.KLAIC

MEMO version 6.0 does not incorporate condensation processes. However, during the investigated period the modeling domain was within the high-pressure field, which is characterized by subsidence. Consequently, condensation might occur only in the shallow ground based layer during the nighttime radiative cooling of the earth. Therefore, I expect similar results if condensation processes were taken into account.

D. STEYN

I would like to see some more evidence that the modelled high PV air is also high O_3 air. Your model output shows a volume of high PV air moving down Italy. Was there a coincident increase in O_3 at the high altitude station in the Northern Appenines?

Z.B.KLAIC

Unfortunately, ozone concentration data for Italy were not available. However, more evidence of stratospheric intrusion over Zagreb is given in the previous study of Lisac et al. (1993), where measured data were analyzed. This analysis showed that increase of ozone concentrations had occurred simultaneously with the decrease of relative humidity and the increase of potential vorticity (both determined from radiosonde data for Zagreb). Additionally, elevated ozone concentrations were also recorded in Vienna, Austria, and K-puszta, Hungary.

EVALUATION OF AURORA SIMULATED BENZENE CONCENTRATIONS FOR THE URBAN AREA OF ANTWERP

Filip Lefebre, Koen De Ridder, Nicolas Lewyckyj, Liliane Janssen, Jozef Cornelis, Françoise Geyskens and Clemens Mensink[*]

1. INTRODUCTION

Urban air quality is a major topic in the European environmental policy because more than 70 % of the European inhabitants and a major part of the working places are located inside an urbanized area. This leads to increased levels of human exposure to harmful air pollutants such as ground-level ozone, nitrogen dioxide, volatile organic components and fine particles. In particular, the road traffic emitted BTEX-aromatics (Benzene-Toluene-Ethylbenzene and Xylene) are characteristic for local urban air pollution. Moreover, benzene is highly carcinogenic and therefore an annual averaged limit value of 5 μg m^{-3} has been set in the framework of the Air Quality Framework Directives 96/62/EC and 2000/69/EC (WHO, 2000). This limit value should be obtained in 2010. Traffic related benzene concentrations are controlled by the petrol composition, the car park composition and traffic speed as well as the atmospheric conditions.

Until now, modeling studies have mostly focused on the determination of the concentration inside street canyons where the highest values can be found (Fenger et al., 1998). Nonetheless, people living in residential neighborhoods should not be exposed to urban background concentrations exceeding the critical levels either.

In this study, the urban air quality model AURORA (Air quality modeling in Urban Regions using an Optimal Resolution Approach) has been used to simulate the background benzene concentrations for the whole Antwerp urban area. The simulations have been done for a week in March 1998 during which an intensive benzene measurement campaign took place. The measurements were part of the MACBETH (Monitoring of Atmospheric Concentrations of Benzene in European Towns and Homes) project (Geyskens et al., 1999). The measurements have been used to verify the model

[*] Filip Lefebre, Koen De Ridder, Nicolas Lewyckyj, Lilane Janssen, Jozef Cornelis, Clemens Mensink and Françoise Geyskens, Flemish Institute for Technological Research (Vito), Boeretang 200, B-2400 Mol, Belgium. Corresponding author: Filip.Lefebre@vito.be , Tel: + 32 14 33 68 47, Fax: + 32 14 32 27 95.

accuracy in simulating the urban background benzene concentrations over a wider urban region.

In the next section, the AURORA model that has been used is this study will be shortly described. Afterwards, the model results will be compared with the available benzene measurements. Finally, results from two sensitivity experiments in which the influence of the lateral boundary conditions and the impact of the ring road emissions have been studied, will be shown.

2. THE INTEGRATED AURORA MODEL

2.1 General description

The Eulerian air quality model AURORA has been designed to calculate urban air quality at different scales. Its field of application ranges from the whole urban area (about 100 x 100 km) down to the street canyon level (Mensink et al., 2001; 2002). Lateral boundary conditions are provided by large-scale simulations with an area of about 850×850 km through successive nesting. In this study, a one-way three-level nested approach (resolutions of 16, 4 and 1 km) has been used. The three levels respectively correspond with the N3-, N2- and the N1-domain.

2.2 Emission modeling

Two different emission modules account for the calculation of the benzene emission fluxes. A first module calculates the large-scale emissions (domains N1 and N2) while the second one takes into account the fine-grid emissions (N3-domain).

The large-scale emissions are based on CORINAIR 1985 (EC, 1995), CORINAIR 1990 (EEA, 1995) and CORINAIR 1994 (EEA, 1997) emissions that are updated with more recent EMEP data (Mylona, 1999). Benzene emissions are calculated from the NMVOC-emissions by using average benzene fractions (Mc Innes, 1996).

The industrial emissions for the inner high-resolution domain are based on the Flemish Emission Inventory (VMM, 2001) that is constructed from individually reported stack emissions by the companies themselves. High-resolution emissions for spatial heating are also delivered by the Flemish Environmental Agency. They are calculated based on total fuel consumption and statistical data from a national inquiry on public housing (VMM, 2001). Lastly, the derivation of the traffic emissions is based on a traffic flow model for the whole Flanders region and the city of Antwerp in combination with a transport emission model (Mensink et al., 2000; Lewyckyj et al., 2002). The traffic emission model computes hourly benzene emissions from passengers cars, light duty vehicles, heavy duty vehicles and busses in function of road type, vehicle type, fuel type, traffic volume, vehicle age, trip length distribution and the actual ambient temperature. Cold start emissions and evaporation losses are included in the model. The emission factors used in the model are derived from the COPERT-III methodology. Figure 1 shows an example of a benzene emission flux pattern for the Antwerp urban area.

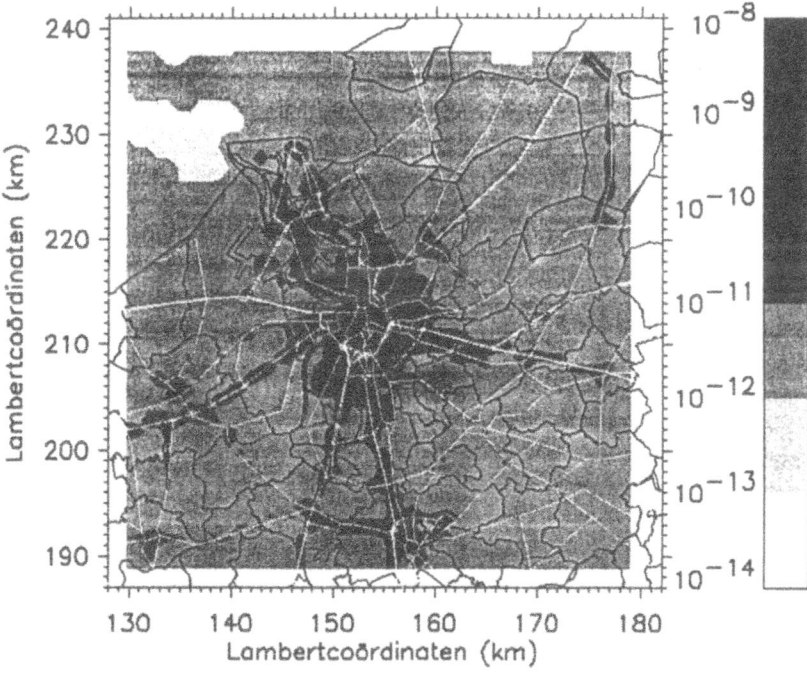

Figure 1. Benzene emission pattern for the urban area of Antwerp (in kg m^{-2} s^{-1}).

2.3 Atmospheric advection and diffusion

The meteorological fields used in the calculation of the atmospheric advection and vertical diffusion are simulated with the non-hydrostatic meteorological model ARPS (Xue et al., 1995) that is nested into ECMWF (European Centre for Medium Range Weather Forecasts) analysis. In this way, high-resolution atmospheric features such as surface inversions and meso-circulations are accurately resolved. To be consistent with the fine atmospheric resolution at which ARPS is applied, high-resolution topography, vegetation and land use data are being used.

Afterwards, the modelling of the atmospheric advection and diffusion of the benzene concentrations was done using state of the art routines (De Ridder et al., 2001). Removal of benzene through dry deposition has been implemented with a deposition velocity of 1 mm s^{-1}.

3. BENZENE MEASUREMENTS

Benzene measurements are available from the European LIFE project called MACBETH "Monitoring of Atmospheric Concentrations of Benzene in European Towns and Homes" (Geyskens et al. (1999)). During MACBETH, data have been collected for four periods of five days in 1998 by means of passive diffusive samplers (Radiello tubes). The Vito-contribution concentrated on Antwerp. Therefore, passive samplers were installed at 101 locations inside Antwerp. During the observational period in March 1998 (23-27/03/1998), 91 urban samples could be analyzed. In a first step, a careful screening of all 91 locations has been done by using high-resolution city maps in order to check the representativeness of the locations. Only 42 locations were classified as urban background locations. The remaining stations were, or located inside a street-canyon, or located too close to streets (for example on street corners), or could not be pointed out on high-resolution maps.

4. MODEL RESULTS

4.1 Reference experiment

The AURORA model has been applied for the week in March 1998 during which the benzene measurements took place. Weekly average concentrations have been calculated from the model output and the model grid value closest to each observation site has been used to analyze the model results. The highest concentrations are found over the Antwerp urban area and in the harbor area located north-west of the city centre (Figure 2). The high concentrations in the urban area are due to intense traffic. The simulated maximum in the harbor area is related to 3 important diffusive benzene sources (loading and unloading activities) located inside the Antwerp harbor area, but unfortunately no MACBETH measurements exist to verify this modelled maximum.

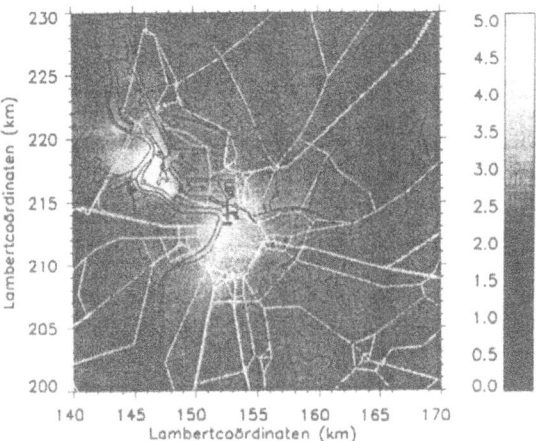

Figure 2. Average modelled benzene concentrations (μg m^{-3}). The shown area is a zoom within the whole model domain shown in figure 1. White lines correspond with the major road network.

However, in this context it is worthwhile to mention the work reported by Kerremans and Geukens (1995) who analyzed 136 half-hourly samples taken inside the Antwerp harbour area between January and July 1994. Strong variations in function of the prevailing wind direction as well as maximum average values per location up to 30 μg m^{-3} were found.

The agreement between the average values for all observations is satisfying but becomes much better when the averaging is restricted to the urban background stations (Table 1). Also the root-mean-square-error between the observations and the model decreases when only urban background stations are used. Overall, simulated concentrations do not deviate more than 50 % from the observed values (Figure 3).

Table 1. Model results against observations

	# stations	Observed mean (μg m^{-3})	Simulated mean (μg m^{-3})	RMSE between Obs en Model (μg m^{-3})
All observations	91	3.13	2.90	0.94
Urban background	42	2.81	2.83	0.73

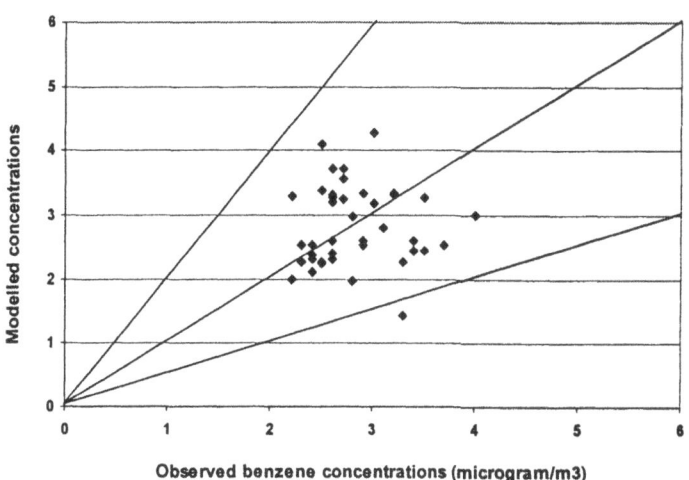

Figure 3. Comparison between observed and simulated benzene concentrations for all urban background observations.

4.2 Sensitivity experiments

4.2.1 Influence of remote benzene emissions on urban benzene concentrations

The influence of remote (outside the N3-domain) benzene emissions on urban benzene concentrations has been analyzed by putting the lateral benzene conditions to zero in case of inflow wind conditions (Table 2).

Table 2. Time averaged benzene concentrations for the reference simulation and the sensitivity to the lateral boundary conditions ($\mu g\ m^{-3}$).

Area	Reference experiment	No lateral boundary conditions
All domain	1.530	0.461 (- 69 %)
Within city ring	3.461	2.373 (- 31 %)
Outside city ring	1.476	0.408 (- 72 %)

Emissions from outside the N3-domain contribute for about 70 % to the benzene levels outside the city ring. On the contrary, inside the city ring remote emissions are responsible for only 30 % of the benzene concentrations. Therefore, effective local emission reductions can be applied to reduce benzene air pollution inside the city ring.

4.2.2 Impact of ring road pollution

Benzene is principally emitted by cars and other motorized road transport activities. Therefore, the combined use of a traffic flow model and a traffic emission model allows the assessment in terms of air quality caused by changed traffic flows. The latter can be due to speed limitations or modifications in the road network such as introduction of one-way streets, construction of tunnels as well as altered city structures. For the moment, we do not dispose of a traffic flow model and therefore we have introduced a more drastic change in the emissions by cutting traffic emissions from parts of the ring road. This choice is motivated by the idea of roofing parts of the ring road and the lay out of some public green space facilities on top of it. These experiments suppose that the emitted traffic emissions are filtered before their release in the atmosphere above the roofed part of the ring.

Because of the very localized nature of this experiment, an additional refinement nesting-level N4 with a horizontal resolution of 250 m has been used. Due to computing resources limitations, the N4 model domain was restricted to 20 x 20 km. Figure 4 shows, for a sub-region of N4, the difference in average benzene concentration between the reference experiment and the sensitivity experiment in which all ring road traffic emissions were put to zero. Significant reductions ranging between 0.3 and 0.8 μg m^{-3} are found in the immediate vicinity of the ring road. Expressed as a percentage change, reductions of up to 25 % are encountered. Even though we are quite confident in the simulated *average* concentration decrease, it may well be that very close to the ring (less than 50 m) the decrease is larger, and a bit farther away (but still within 1 or 2 grid cells) the decrease may be lower than the *average*.

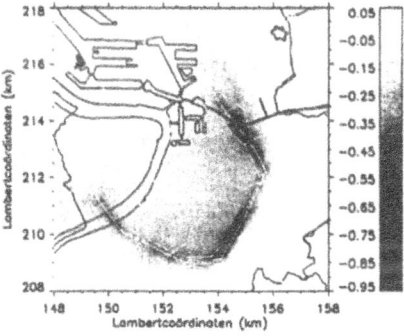 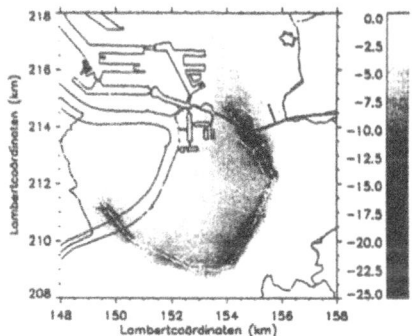

Figure 4. Difference in benzene concentration between the sensitivity experiment (no ring road emissions) and the reference experiment. Differences are expressed in $\mu g \ m^{-3}$ (left panel) and in % reductions compared to the reference experiment (right panel).

5. CONCLUSIONS

A 3-dimensional nested Eulerian air pollution model called AURORA has been applied during a week in March 1998 in order to simulate the corresponding benzene concentration field. City averaged simulated benzene concentrations agree favorably with the available observations in case only urban background measurements are used to verify the model with. Overall, simulated benzene concentrations are within an error bound of 50 % of the observed values when looked at the individual measurements. The simulated benzene field also indicates twice as high benzene concentrations over the Antwerp urban area as compared to the rural surroundings.

Afterwards, the model has been used to determine the level of contribution of remote benzene emissions on urban benzene levels. Emissions from outside the city of Antwerp are found to be related to 30 % of the benzene concentrations inside the city ring but outside the city ring, their importance amounts to more than 70 %.

Finally, a sensitivity experiment has been conducted in which all emissions due to ring road traffic have been pulled out of the system. It was found that locally in a band centered on the ring road, benzene concentrations were reduced by 0.3 to 0.8 $\mu g \ m^{-3}$ which corresponds with relative reductions of at its maximum 25 %.

6. ACKNOWLEDGMENTS

This work is a contribution to the EUROTRAC-2 subproject SATURN and has been supported by the EU 5th framework programme project BUGS (Benefits of Urban Green Space, http://www.vito.be/bugs).

Atmospheric simulations were done using the Advanced Regional Prediction System (ARPS) developed by the Centre for Analysis and Prediction of Storms, University of Oklahoma, http://www.caps.ou.edu/ARPS (Xue et al., 1995).

REFERENCES

De Ridder, K. and Mensink, C., 2001. Improved algorithms for advection and vertical diffusion in AURORA, *Proceedings of the 25th NATO/CCMS International Technical Meeting on Air Pollution Modelling and its Applications*, Louvain-la-Neuve (15-19 Octobre 2001), Belgium, pp. 289-294.

EC (1995) CORINAIR, Inventaire des émissions de dioxyde de soufre, d'oxydes d'azote et de composés organiques volatils dans la Communauté européenne en 1985, Corine, *EC publication EUR 13232 FR*, Luxembourg.

EEA (1995) CORINAIR 1990 summary tables: Fifth Set, European Environmental Agency, Copenhagen, August 1995, Copenhagen.

EEA (1997) CORINAIR 1994, Copenhagen.

Fenger, J., Hertel, O. and Palmgren, F. (Eds.), 1998. Urban Air Pollution – European Aspects. Kluwer Academic Publishers.

Geyskens, F., Bormans, R., Lambrechts, M. and Goelen, E., 1999. Measurement campaign in Antwerp for quantification of personal exposure to representative volatile organic compounds, *Vito report 1999/DIA/R/048*, Mol (in Dutch).

Kerremans, W. and Geukens, M., 1995. Measurement campaign air pollution in the northern harbor-area (Meetcampagne luchtverontreiniging in het noordelijk havengebied), City of Antwerp, Centrum for air- and water pollution (in Dutch).

Lewyckyj, N., Colles, A., Cornelis, J., Janssen, L., Cornelis, E. and De Vlieger, I., 2002. Uitbouw milieu-impactmodule gekoppeld aan multi-modale verkeers-en vervoersmodellen, *Vito report 2002/TAP/R/019*, Mol (in Dutch).

Mc Innes G. (ed.), 1996. Atmospheric emission inventory guide book, A joint EMEP/CORINAIR Production, European Environmental Agency, Copenhagen, B710/9-11.

Mensink, C., De Vlieger, I. and Nys, J., 2000. An urban transport emission model for the Antwerp area, *Atmospheric Environment*, **34**, 4595-4602.

Mensink, C., De Ridder, K., Lewyckyj, N., Delobbe, L., Van Haver, Ph. and Janssen, L., 2001. Air quality modelling in Urban Regions using an Optimal Resolution Approach (AURORA), *Proceedings of the 7th International Conference on Harmonisation within Atmospheric Dispersion Modelling for Regulatory Purposes*, Belgirate, pp. 459-463, European Commission, Joint Research Centre, Ispra, May 2001.

Mensink, C., Lewyckyj, N. and Janssen, L., 2002. A new concept for air quality modelling in street canyons. *Water, Air and Soil Pollution* (in press).

Mylona, S., 1999. EMEP EMISSION DATA, *Status Report 1999 Emep MSC-W*, Report 1/99, Oslo.

VMM, 2001. Lozingen in de lucht 1980-2000, (in Dutch).

WHO, 2000. Air Quality Guidelines for Europe, 2nd Edition, *WHO Regional Publications*, Copenhagen, in press.

Xue, M., Droegemeier, K.K., Wong, V., Shapiro, A., and Brewster, K.. 1995: ARPS Version 4.0 User's Guide, 380pp, Available from CAPS, University of Oklahoma, Norman OK 73019, USA.

518

DISCUSSION

D. G. STEYN   Your Figure 3 and Table 1 show quite clearly that the model matches observations only to within 100%. This means that the model configures only domain-wide averages and net spatial variability. Could you not achieve this level of agreement with a simple box model?

F. LEFEBRE   Regional grouping of observations and model output (not in the abstract but shown during the presentation) has shown that the model is still able to obtain sub-area variability over the city. This would not have been possible with a box model. The model complexity is also needed for future model studies in which land use changes will be applied in order to obtain different city patterns.

J. BRECHLER   Is the ARPS model capable to describe flow deformations caused by the city structure (buildings) on due resolution 250 m?

F. LEFEBRE   The ARPS model does not allow to describe flow deformations caused by each individual building. However at 250m, the overall effect of buildings on the wind fields (reduced wind speed) is accounted for by the use of a larger roughness length compared to the surrounding rural areas.

MODELLING OF ODOROUS EMISSIONS
Case study for the city of Izmir-Turkey

Aysen Muezzinoglu[*], Tolga Elbir

1. INTRODUCTION

Concentrations and emission fluxes of sulfur containing gaseous pollutants emitted from the polluted stream deltas in the city of Izmir were determined during summer of 2001 (Muezzinoglu, 2003). The streams were acting like open sewers collecting industrial and domestic wastewaters until August 2001. Many of these malodorous gas emissions were due to anaerobic decomposition of organic water pollutants brought by the creeks into the river deltas. Impacts of the engineering projects to collect and send the wastewaters to a central treatment plant were first notable in September 2001 as the measured emission rates of the malodorous gases were minimized. Emission flux rates from polluted water surfaces (mg $m^{-2} h^{-1}$) and emission concentrations (mg m^{-3}) of the studied gases which consisted of H_2S, dimethylsulfide, 2-propane thiol, 2-butane thiol, thiophene and diphenylsulfide were submitted in another publication (Muezzinoglu, 2003).

Ambient levels of the odorous gases can be predicted with dispersion models at selected locations in an urban area having odor problems. In Izmir among the odorous gases H_2S was selected for dispersion estimations using suitable dispersion models. The selection of this odorous compound was based on two factors; firstly H_2S was the only compound among the malodorous sulfur gases with default dispersion coefficients in the CALPUFF model, and

[*] Department of Environmental Engineering, Dokuz Eylul University, Kaynaklar Campus, Buca, 35160-Izmir,Turkey.

secondly it was noted that this gas has linearly varied with the perceived odors in the city between June and September 2001. During June measurement campaign H_2S ranged between $16.86 - 1463.50$ mg m^{-3} in the emissions from the river segments. However, during September 2001 campaign after the clean-up and diversion of the wastewater discharges, odors were diminished as indicated by H_2S levels approaching zero. Calculated H_2S emission rates from the water surfaces were assumed to be spatially homogenous over the malodorous stream sections and in the lagoons formed at the regulated river mouths. Thus total emissions per hour were calculated by multiplying these rates by the surface area of the odorous segments.

2. MATERIALS AND METHODS

Dispersion models are used to predict perceived odors or odorous matter concentrations in the emissions and ambient air that are presumably related to each other. However, models that can be used for calculating odorous emissions are under discussion by several researchers. In several cases such models were developed to estimate the emission rates either from measured ambient levels of odorants (Smith, 1995; Sarkar and Hobbs, 2003) or odor emissions from point industrial sources were dispersed in order to allow safe separation distances to residential areas (Schauberger et al., 2001; Piringer and Schauberger, 1999; Schauberger et al., 2000). It is generally accepted that odors and odorous compounds behave similar to other air pollutants during dispersion. In other words they are distributed according to Gaussian plume dispersion models at crosswind dimensions (Smith, 1995).

Recently a critical discussion has arisen on the issue of averaging times in such models. In air pollution dispersion models averaging times are normally much longer than the time scales of odor perception (Venkatram, 2002; Mussio et al., 2001). Necessity of use of very short averaging times in the order of a few seconds to minutes (Mussio et al., 2001) in odor modeling arises from the perception characteristics of odors detected by humans (Sarkar and Hobbs, 2002). Therefore, the methods of estimations of short averaging time odor concentrations from long-term estimations are under theoretical discussion. In some regulatory air pollution models such as Industrial Source Complex (ISC) short-term concentrations representing perceived odors or odorous matter as averaged over a time period of less than a few minutes are predicted from 1-hour estimates. In CALPUFF for determining such short-term peak concentrations the same approach is used. This conversion is based on the power law equation of the form:

$$C_s = C_h (T_h / T_s)^p \qquad (1)$$

where C_s is the required short-term average odor concentration, C_h is the hourly average concentration calculated by the dispersion model, T_h is 1 hour period expressed in minutes or seconds, T_s is the short term period in minutes or seconds. Exponent (p) varies between 0.2-0.5 according to the literature (Venkatram, 2002). Instruction manual for CALPUFF model (Scire et al., 2000) advices use of $1/5^{th}$ power which corresponds to p=0.2 in equation (1). Venkatram (2002) criticized the use of power law equations such as (1) for predicting the odor concentrations from longer averaging time estimates, and proposed another model which takes into account the standard deviations of time-averaged concentrations in both of the two different averaging times rather than the concentrations themselves. He proposed use of a model utilizing the of short-term (T_s) odor concentration standard deviation of probability distribution (σ_c^s), and the long term (T_h) concentration standard deviation of probability distribution (σ_c^h):

$$\sigma_c^s / \sigma_c^h = (T_h / T_s)^{1/2} \qquad (2)$$

As the total duration of added up "odor-on" incidences approach the longer period of time, the probability function approaches a normal distribution. Venkatram (2002) has indicated that as the total probability of sensing the odor during the longer interval is equal to 40 instantaneous sniffs or more, binomial distribution of frequencies approach a normal distribution. In an on-going study standart deviations of field data consisting of odor measurements at short and long durations are being tested to see the validity of the model proposed in (2).

In an earlier study measured organic sulfur compound emission fluxes such as dimethyl-sulfide, 2-propane-thiol, 2-butane-thiol, hydrogen sulfide (H_2S), thiophene and diphenyl-sulfide were measured at five creek surfaces (Arap, Bostanli, Meles, Manda and Ilica) and in the wastewater treatment plant in the city of Izmir. Study area is shown in Fig.1. These fluxes were measured by using a flux chamber as emitted from highly odorous polluted water surfaces in June and September 2001. Emissions were found especially high during the June period that was before the engineering project application to stop the wastewater discharges into the creeks (Muezzinoglu, 2003).

Keeping in mind that the odor pulses of the stimulants of 2001 study in Izmir were continuously emitted from polluted water surfaces, normal distribution of concentrations at crosswind and vertical dimensions could be assumed. This enabled the use of Gaussian models to calculate the airborne levels of odorants. Keeping this in mind, total emissions from polluted creeks and treatment plant surfaces were estimated for malodorous sulfur containing gases in Izmir (Muezzinoglu, 2003). Emissions of odor-causing sulfur gases

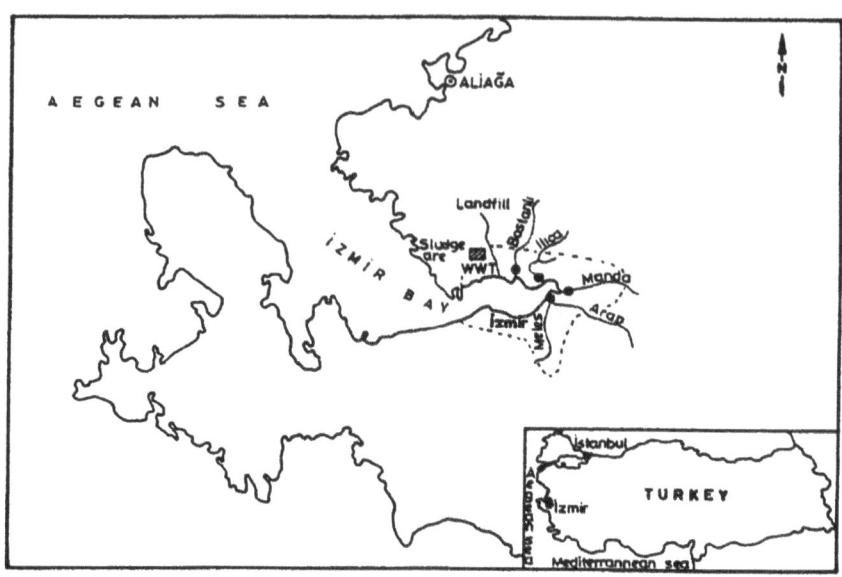

Figure 1. General layout of the study area

were summed up on the basis of sulfur concentrations and evaluated as total odorous sulfur in the air in equivalents of H_2S. Results are plotted into a map of the city of Izmir indicating different categories of H_2S concentrations. These emissions could then be dispersed in order to predict the ambient concentrations of odorous gases downwind of the complaint areas. Calculated H_2S emissions were dispersed using the CALMET module and its Lagrangian, multi-layer and gridded non-steady-state puff dispersion model CALPUFF. Three minutes averaging time for emissions of the odorous compound were derived from hourly average results with the assumption made in accordance with equation (1). Hourly average meteorological data were obtained from the meteorology service and used as input of the model. Model was run for two representative days corresponding to monthly average meteorological conditions in June and September 2001.

Model results distinctly showed that two main areas in the city were important odor sources. These were the Izmir wastewater treatment plant and the delta of Manda creek that was highly polluted at that time. Maximum ground level H_2S concentrations were measured as 250 mg m^{-3} around the treatment plant in June while maximum H_2S concentration was 27 mg m^{-3} at the creek deltas. In September, all H_2S levels in the creek site were very low, while treatment plant odors persisted with rather high H_2S. Equal odor isophlets were drawn on a GIS map of the area. It is assumed that other odorous gases coming from the same area sources show very similar dispersions and the resulting pollution maps have similar odor distributions. These maps showed the success of the engineering project during summer of 2001.

3. RESULTS AND DISCUSSION

CALPUFF model results for three minutes averaging time, H_2S concentrations near the ground level are shown in odor maps. Fig.2 belongs to the predictions carried out for the last week of June 2001 and indicates the odor distribution around the two distinct odor emission source clusters. At the left hand side in this figure Izmir wastewater treatment plant and its associated sludge management area is indicated. At the right hand side major creek discharge zones are shown. Fig. 3 corresponds to the September 2001 odor levels at the same sites. Wastewater treatment plant impacts are not indicated in Fig.3 as it is nearly the same as in Fig.2. However by comparing creek emission dispersions in Fig.2 and Fig.3 it can be seen that the odorous gas dispersion has been much better in September 2001. Detailed H_2S isophlets are separately shown within larger circles in both figures.

It must be added that H_2S concentrations predicted in ambient air before and after the municipal engineering project that took place in August 2001 were found by modeling. It does not include chemical transformations of sulfurous gases in the air, which are known to be rather rapid. These concentrations are given in (mg m^{-3}) units that are typical for air pollutants. For odorous substances, however, odor units (OU) are defined as the number of treshold values of specific compounds, such as H_2S in this study. Odor treshold level of H_2S is 0.21 µg m^{-3} (Cha and Turk, 2000) and dividing the predicted concentrations given by the model Fig.2 and Fig. 3 were drawn. Fig.3 shows the the odor isophlets in OU in the city after engineering project application in summer of 2001. Therefore this is the same odor map representing the conditions in the city today. A comparison of Fig.2 and Fig. 3 indicates that the sewerage system project made a difference in the odor levels of the city of Izmir if H_2S would be a conservative gas. It is known however, that sulfur containing organic gases are rapidly oxidized to sulfur oxides as shown in other research work (Müezzinoğlu, 2003). Therefore although the OU levels drawn for the city are still above the sensory limits according to model results, actual situation has largely improved. According to measurement results concurrent with the model operation, H_2S was not detectable in the air of Izmir.

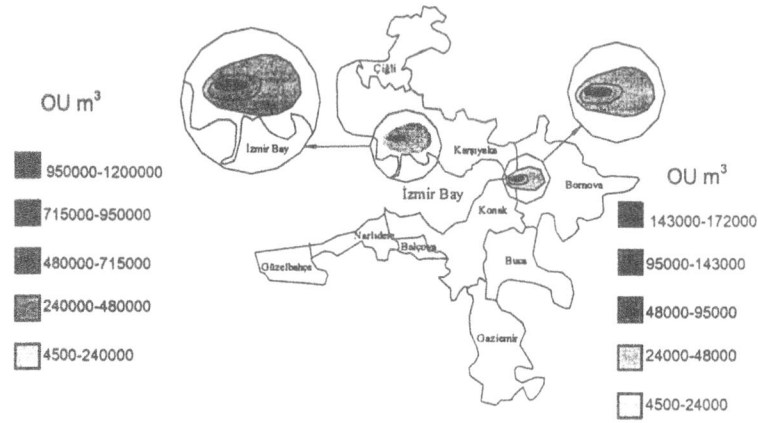

Figure 2. Predicted H$_2$S dispersions on June 2001 (before cleaning the streams)

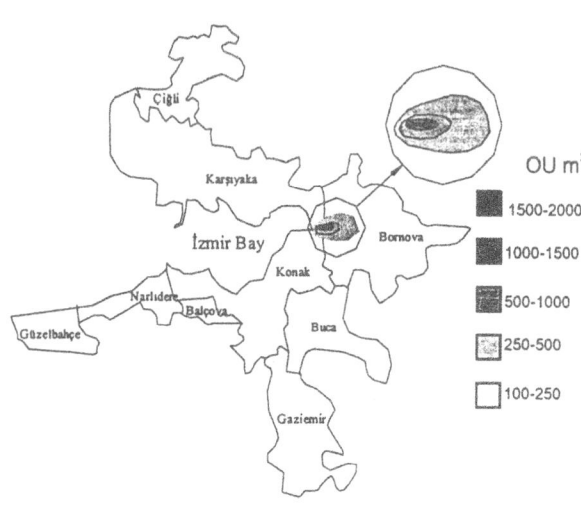

Figure 3. Predicted H$_2$S dispersions on September 2001 (after cleaning)

4. CONCLUSIONS

Odor distribution maps for high and low emission cases during June and September 2001 measurement campaigns in Izmir were drawn in order to note the distribution of perceived odors in the city. This aimed at assessing the success level of an engineering project for cleaning the streams and creeks carrying domestic and industrial pollutants into the sea. During August 2001 these wastewaters were fully diverted into the sewerage system and the beds of the rivers were cleaned.

As the odor maps were drawn over a GIS-based mapping system by direct transfer of the dispersion model outputs from the gridded area, it was quite possible to relate the odor concentrations with several different land use areas in the city. Odor complaint distribution is confined to a small area for the moment. From another research project aiming at measurement of H_2S levels that most of the urban area is already below the odor perception level of 3-5 odor units. In the past before the wastewater collection system was completed almost all of the urban area was above annoyance level. So the engineering project proved to be successful in abating with the odor complaints in the city of Izmir.

5. REFERENCES

Cha, S.S., Turk, A., 2000, Odor control, in: Air Pollution, D.H.F. Liu and B.G. Liptak (eds), Lewis Publishers, Boca Raton, pp. 23-24.

Muezzinoglu, A., 2003, A study of volatile organic sulfur emissions causing urban odors, *Chemosphere*, **51**(4): 245-252.

Mussio, P., Gnyp, A.W., Henshaw, P.F., 2001, A fluctuating plume dispersion model for the prediction of odour-impact frequencies from continuous stationary sources, *Atmospheric Environment*, **35**: 2955.

Piringer, M., Schauberger, G., 1999, Comparison of a Gaussian diffusion model with guidelines for calculating the separation distance between livestock farming and residential areas to avoid odour annoyance, *Atmospheric Environment*, **33**: 2219.

Sarkar, U., Hobbs, S.E., 2002, Odour from municipal solid waste (MSW) landfills: a study on the analysis of perception, *Environment International*, **27**: 655.

Sarkar, U., Hobbs, S.E., 2003, Landfill odour: assessment of emissions by the flux footprint method, *Environmental Modelling and Software*, online edition.

Schauberger, G., Piringer, M., Petz, E., 2000, Diurnal and annual variation of the sensation distance of odour emitted by livestock buildings calculated by the Austrian odour dispersion model (AODM), *Atmospheric Environment*, **34**: 4839.

Schauberger, G., Piringer, M., Petz, E., 2001, Separation distance to avoid odour nuisance due to livestock calculated by Austrian odour dispersion model (AODM), *Agriculture, Ecosystems and Environment*, **87**: 13.

Scire, J.S., Strimatis, D.G., Yamartino, R.J., 2000, A User's Guide for the CALPUFF Dispersion Model, Earth Tech. Inc., Concord, MA, pp..521-522.

Smith, R.J., 1995, A Gaussian model for estimating odour emissions form area sources, *Mathematical Computation and Modelling,* 21(9): 23-29.

Venkatram, A., 2002, Accounting for averaging time in air pollution modelling, *Atmospheric Environment,* **36**: 2165.

DATA FROM AN URBAN STREET MONITORING STATION AND ITS APPLICATION IN MODEL VALIDATION

Michael Schatzmann, D. Grawe, B. Leitl, and W.J. Müller[*]

1. INTRODUCTION

The quality of CFD model results depends largely on the quality of parameterisations employed within the computer codes. Such parameterisations are needed in order to close the equations and to describe the effects of sub-grid-scale processes on the development of the flow. These parameterisations are essentially empirical. To justify the assumptions involved and to determine the values of the constants the parameterisations contain appropriate data sets with which the model results can be compared are needed.

In the subsequent analysis we restrict ourselves to obstacle resolving, micro-scale meteorological models developed to predict traffic-generated urban air pollution. The emission source in such models is usually assumed to be a line source since it is not yet feasible to simulate individual moving vehicles. For momentum-free line sources and under conditions as subsequently specified, it is to be expected that the concentration C $[g/m^3]$ (in excess above background) at any point in the vicinity of the source is proportional to the source strength Q/L $[g/(s \, \forall \, m)]$ and inverse proportional to the wind velocity u [m/s] measured at a reference height well above the buildings. Dimensional reasoning suggests the introduction of a normalized concentration c^* [-] which depends on the following non-dimensional variables

$$c^* = \frac{C \cdot u \cdot H}{(Q/L)} = f\left(DD, \frac{l_i}{H}, Re, \frac{L_M}{H}, TIT\right) \qquad (1)$$

with the additional parameters:

H = characteristic building height (in m), here H = 25 m,
DD = wind direction (in degree),

---

[*] Michael Schatzmann, D. Grawe, B. Leitl, Meteorological Institute of the University of Hamburg, and W.J. Müller, Lower Saxony State Agency for Ecology (NLÖ), Hanover, Germany.

Air Pollution Modeling and Its Application XVI, Edited by
Borrego and Incecik, Kluwer Academic/Plenum Publishers, New York, 2004

l_i/H = multiple length scales (normalized by H) describing all details of the urban geometry,

Re = Reynolds number (Re = $H \, \forall \, u \, / \, v$),

L_M/H = Monin-Obukhov length (normalized by H) which characterizes the density stratification,

TIT = an appropriately defined dimensionless parameter describing the traffic induced turbulence.

Chemical reactions are excluded from the analysis, i.e. only passive tracer dispersion is considered here. If the wind speed is sufficiently high, the Reynolds number takes on an above-critical value and the turbulence within the canopy layer is dominated by shear turbulence. This means that effects of stratification or traffic induced turbulence should be of minor importance. Then, for a given urban landscape (H and all l_i/H-values are fixed), c^* is a function of the wind direction alone

$$c^* = \frac{C \cdot u \cdot H}{(Q / L)} = f(DD) \tag{2}$$

Practitioners seldom care for the limiting conditions mentioned so far. Those who carry out the field measurements either provide the raw data or group the excess concentration mean values from a particular measurement campaign according to Eq. (2). Numerical modellers use the processed data, believing that they represent the "truth" and tune their models accordingly. Using the example of field data from an urban monitoring station it will be demonstrated how dangerous this practice can be.

2. FIELD EXPERIMENTS

For more than a decade, the Lower Saxony State Agency for Ecology (NLÖ, 1995) has operated a monitoring station at the pedestrian walkway in a busy street canyon (Goettinger Strasse in Hanover, Germany) with a load of up to 30 000 vehicles/day. Based on automated traffic counts and information on the composition of the German vehicle fleet, reasonable estimates of pollutant emission rates are available. Concentrations of NO, NO_2, Benzene and other pollutants are continuously measured and stored in the form of 30 min or 60 min average values. The above-roof wind and background concentrations are also monitored. More information on the site can be found in Mueller et al (2002), a pictorial impression is given by Fig. 1.

Figure 1: Photography of the field test site 'Goettinger Straße'.

3. DATA AQUISITION AND DATA PROCESSING

The subsequent discussion is based on the assumption that all data is collected with state of the art equipment, and that all possible care has been taken to assure the quality of the measurements. In case of the Goettinger Strasse data this condition is without doubt fulfilled.

A permanent monitoring station like that operated by NLÖ in Hanover provides the opportunity to analyse data collected over long periods of time. This creates the possibility that particular dispersion situations are met several times and that the statistical significance of measured results can be assessed.

Fig. 2 shows c^*-concentrations of NO_x from a whole year (1994), presented acc. to Eq. (2) as a function of wind direction only. NO_x, calculated from measured NO and NO_2, can be regarded as a passive tracer for the short dispersion time periods of interest here. Each of the approximately 12 500 dots in Fig. 2 represents a half-hourly mean value, and the curve the average over all 30 min values.

The scatter of data points in Fig. 2 is striking. In order to decrease the spread, the data points are subsequently re-analysed and all those points are eliminated which do not properly meet the conditions underlying Eq. (2). This is done in a step by step procedure by scrutinizing the individual parameters which enter the equation.

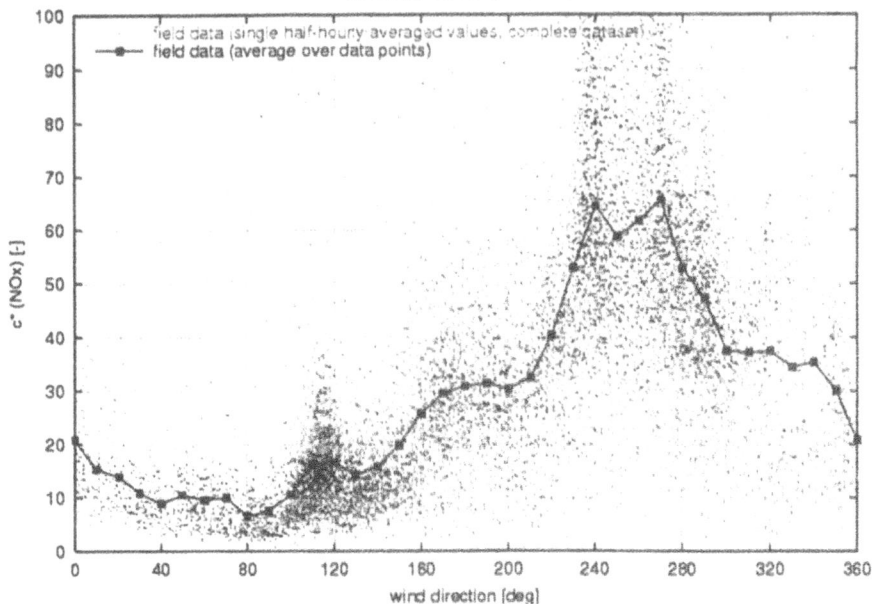

Figure 2: Normalized half-hourly mean concentration values as a function of wind direction measured over a period of one year at the street monitoring station 'Goettinger Strasse' in Hanover/Germany.

Determination of C: In the definition of c*, C is the concentration excess above ambient, which means the background concentration C_b needs to be subtracted from the value C_m measured at the monitoring station. In a city environment with numerous sources and large local concentration differences it is not easy to determine a meaningful background. In case of the Goettinger Strasse the background C_b is measured on top of the NLÖ-building (left hand side in Fig.1) about 30 m above street level. Although for certain wind directions the background is likely to be increased by pollutants flushed out of the street canyon, this can not be quantified. Thus, C_b as measured is assumed to be correct.

Determination of wind direction and wind velocity: The wind vector is also measured on top of the NLÖ-building at a mast 10 m above the highest elevation of the building complex and 42 m above street level. The wind directions used in Fig. 2 are those which were directly measured. From the velocity u_{42} a representative free-stream reference wind speed u (100 m above ground) was calculated assuming the existence of a power law wind profile above the urban canopy with an exponent of n = 0.3. Wind directional changes between 42 m and 100 m height were neglected. The present practice is similar to that frequently applied in CFD modelling. The velocity u (=u_{100}) corresponds to the wind speed at the top of the numerical domain.

Although the velocity is measured 10 m above roof level, free stream conditions are most likely not yet met. It remains unknown to what degree the flow is disturbed by surrounding obstacles or the NLÖ-building itself. If such disturbances occur, they are surely different for different wind directions.

Line source approximation: The c^*-equations (1) and (2) are valid only for line sources. The question arises of what traffic rate must be exceeded before the line source approximation holds. To find an answer, all data were grouped according to the traffic rate and plotted according to Eq. (2). As Fig. 3 shows, there is large scatter between the different curves. However, with the exception of the lowest traffic category (< 60 vehicles/30 min), all curves have a similar shape. For safety reasons, not only the lowest but also the second lowest class will be neglected in the subsequent analysis, i.e. only half hourly values with more than 119 vehicles/min will be taken into account. The elimination of low traffic data points corresponds to the elimination of some of the night-time measurements.

Determination of the source strength: The determination of the emission rate per unit length Q/L is usually somewhat of a problem. To obtain an estimate as reliable as possible, automated traffic counters are used at the Hanover site. These counters register not only the number of vehicles per time interval and per lane, they also discriminate between passengers cars and trucks. In combination with knowledge of the composition of the German vehicle fleet in the year 1994, the prevailing driving pattern alongside the monitoring station and emission factors for specific vehicle types, Q/L-values can be estimated. The present study uses the emission model MOBILEV which is recommended by the German Environmental Agency. It should be noted, however, that there are several sets of emission factors in use. Depending on which one is chosen, the whole ensemble of measured points in Fig. 2 moves up or down by about 50 %, the degree of scatter is not affected.

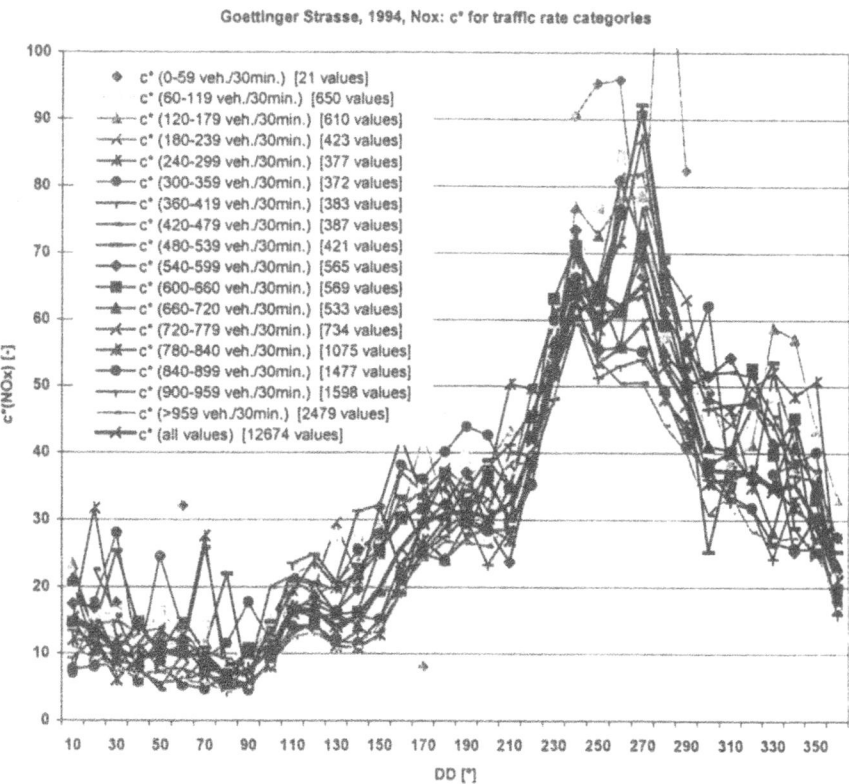

Figure 3: Concentration c^* as a function of wind direction and traffic rate.   533

Minimum wind velocity: In the derivation of Eq. (1) it was made clear that the c*-concept is not applicable to low wind situations. Only if the wind speed rises above a certain minimum value it can be assumed that

(1) the critical Reynolds number is exceeded,
(2) stability effects inside the canopy layer are negligible, and
(3) the dispersion is governed by wind generated rather than by traffic induced turbulence.

In order to determine the minimum wind speed, the data were split into 9 velocity classes. As Fig. 4 shows, at low wind speeds c* appears to be rather independent of the wind direction. This suggests that traffic induced turbulence is the major mixing mechanism. With increasing velocity the wind seems to form a secondary flow inside the canyon with the consequence of higher concentrations for westerly than for easterly winds. It appears that the curves take on a similar form for wind velocities $u_{100} > 3.9$ m/s which corresponds to wind speeds in unobstructed terrain and at standard anemometer height of about $u_{10} > 2$ m/s.

The street canyon has an approximate north-south orientation. Winds from 77° or 257° would be exactly perpendicular to the canyon. The monitoring station is positioned at the walkway west of the traffic lanes. The c*-values show a maximum for westerly winds which is in line with expectation. The exact shape of the individual curves and the fact that they still show some dependency on wind speed even for the highest velocity classes is not fully understood.

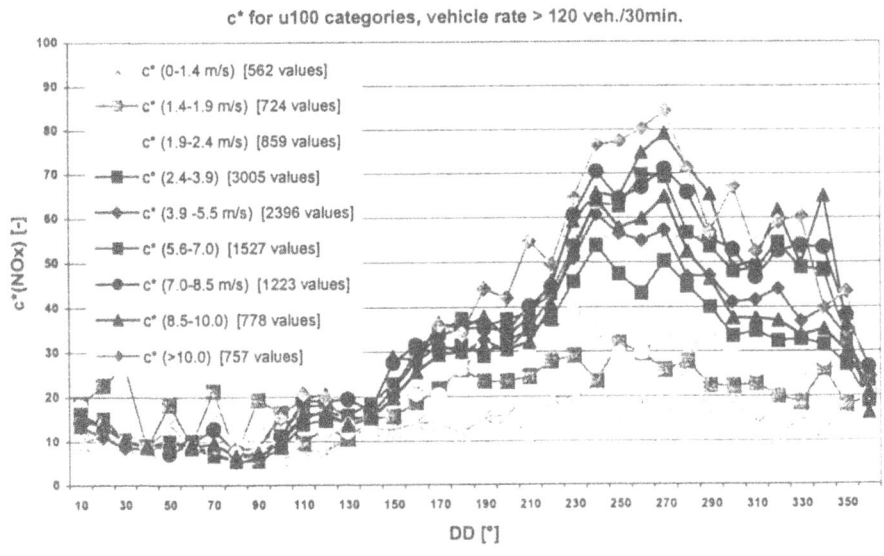

Figure 4: Concentration c* as a function of wind direction and wind velocity interval.

4. CONCLUSIONS

Fig. 5 replicates the data presentation of Fig. 1 but comprises, from the original 12424 values, only those 6562 values which survived the filtering process explained in section 3. Compared to Fig. 1 the spread of data points is somewhat reduced but still significant.

The large scatter of data points supports the physical notion which has been put forward in a paper by Schatzmann and Leitl (2002). They showed that high resolution concentration measurements carried out within the urban canopy layer provide highly fluctuating and intermittent signals. Periods of zero excess concentrations are interspersed with non-zero, rapidly varying concentrations. Even under constant meteorological conditions long averaging times would be required in order to obtain repeatable time mean concentrations. It is to be expect that the commonly used 30 min measurement cycles are simply too short, and that they have the character of random samples only. Depending on the wind direction the variability between seemingly identical cases can be large. To simply increase the averaging time would not solve but worsen the problem since over periods longer than 30 min a systematic trend in meteorological conditions due to the diurnal cycle must be expected.

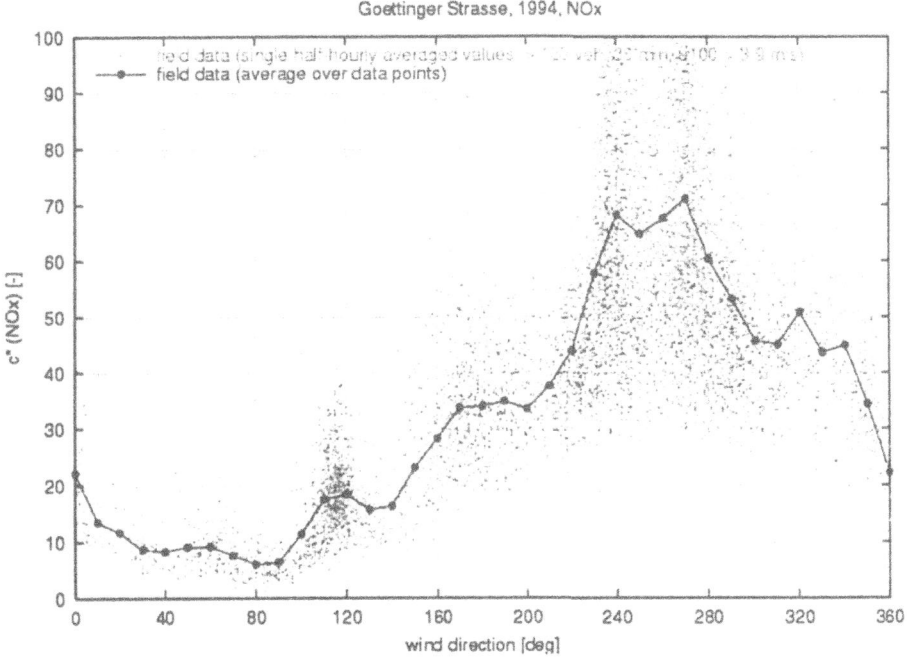

Figure 5: Normalized half-hourly mean concentration as a function of wind direction. Data points not in line with the assumptions underlying Eq. (2) were eliminated.

Fig. 5 furthermore indicates that single measurements cannot be representative for locations exposed to highly fluctuating and intermittent concentrations. Each individual data point in the figure is likewise justifiable. It must be concluded that only long duration measurement campaigns within which similar dispersion episodes occur several times are meaningful. Such experiments allow representative ensemble mean values to be determined (curve in Fig. 5). The usefulness of data generated in episodic field campaigns for model validation purposes is clearly limited.

Finally the numerous and somewhat arbitrary decisions made when processing the raw data need to be reviewed (section 3). Although reasons were given for the elimination of parts of the data base or the selection of a particular set of emission factors, other similarly plausible choices could have been made. The widespread belief that the uncertainty of field data sets is solely related to the inaccuracy of the instruments is surely fiction.

4.1 Acknowledgements

Financial support from the German Federal Ministry for Education and Research within the Atmospheric Research Programme AFO 2000 and from the EU project FUMAPEX is gratefully acknowledged.

5. REFERENCES

Müller, W.,J., Heits, B., and Schatzmann, M. (2002): A Prototype Station for the Collection of Urban Meteorological Data. Proceedings, 8th International Conference on Harmonisation within Atmospheric Dispersion Modelling for Regulatory Purposes, Sofia, Bulgaria, October 14-17.

NLÖ, 1995: Lufthygienisches Überwachungssystem Niedersachsen – Standortbeschreibung der NLÖ Stationen. Bericht, Niedersächsisches Landesamt für Ökologie, Göttinger Str. 14, 30449 Hannover (in German).

Schatzmann, M, and Leitl, B. (2002): Validation and Application of Obstacle Resolving Dispersion Models. Atmospheric Environment, Vol.36, pp. 4811-4821.

DISCUSSION

S. -E. GRYNING   You say that your data cannot be trusted. I think your data are good and can be trusted. But it is simply so that what you want you don't measure and what you measure you don't want.

M. SCHATZMANN   Let us be more precise, (1) these are not my data and (2) I do believe that the data have been taken with all possible care. My point is that half-hour mean values taken within the urban canopy layer have a random character and lack representativeness. If the same measurement would be repeated under exactly identical meteorological and source conditions one would measure something else. Prove was given in my slide with the dimensionless presentation of street level concentrations versus wind direction for data from a whole year (Fig.5). The data points should fall on a single curve, but quite obviously they do not show this behaviour.

Therefore, it is not appropriate to tune a numerical model to match a particular measurement. Since individual data points have the character of snapshots only, we need long series of field measurements which allow ensemble averaging (the red line in Fig.5). An alternative would be to use data measured under controlled conditions in boundary layer wind tunnels, but that is another story.

E. GENIKHOVICH   I agree with Sven-Erik that your measurement data reflect reality and I think that they are perfectly applicably for validation purposes. However from my point of view the models that predict only mean value of concentration at given receptor points cannot properly reproduce real situations which are characterized by the high level of stochasticity. That is why I was presenting in my talk different approaches to modeling based on calculations upper percentiles, PDFs and ensembles.

SENSITIVITY OF AIR TRAJECTORIES IN SEA BREEZE ENVIRONMENT TO TURBULENCE AND SURFACE FORMULATIONS

Aurore Porson[1], Guy Schayes[1]

1. INTRODUCTION

Using models to represent the behaviour of air and its pollutants necessitates several approximations. Two important ones are the turbulence closure and the representation of the ground characteristics. In most meso-scale models, the effect of turbulence is represented through eddy exchange coefficients determined by mixing lengths. In the experiments presented here, the sensitivity to mixing length formulation is analysed with a meso-scale model using a 1.5 order turbulent closure scheme and two associated mixing lengths: the length lk for the turbulent diffusion coefficient and the length le for the dissipation term. The two turbulence formulations we compare are those of Therry-Lacarrère (1983) and of Bougeault-Lacarrère (1989).

The goal of this study is to show the possible importance of the closure and ground characteristic assumptions on tracer experiments, particularly when used in a complex terrain such as a costal environment surrounded by large topography gradients.

Following that way, first of all, through idealised sea-breeze simulations in 2D, we depicted the main differences between both formulations using a simple situation : a vertical plane perpendicular to the coast line, extending 200 km over the sea and 200 km over uniform flat land. We especially focused on the sea breeze development and extinction.

From these simplified simulations, we can put in evidence the major differences between the behaviour of the two formulations. We notice mainly that simulations with Bougeault-Lacarrère (B-L) leads to less turbulent energy, lower BL depths, but higher

[1] Institut d'Astronomie et de Géophysique G. Lemaître, Université catholique de Louvain
B-1348 Louvain-la-Neuve, Belgium

mixing, which favours the stationary development of the sea breeze circulation. Therry-Lacarrère (T-L) formulation leads to more turbulent kinetic energy production during the day and subsequently higher boundary layer depths.

Secondly, these behaviours were also observed in the more complex 3D simulations over the Escompte campaign domain, south of France. This area includes notably complex topography surrounding the northeast of the city of Marseille. On the western side, the presence of Etang de Berre allows possible interactions between the Mediterranean sea and the lake breezes. This site choice is also motivated by the importance of the sources of pollutants emitted by the industrial pole of Etang de Berre and by the large city of Marseille (see figure 3).

Since the dynamical aspects (sea breeze front, boundary layer depth, wind speed) seemed to be strongly influenced by the different mixing length formulations, we used the use a tracer to compute trajectories and to emphasise the sensitivity analysis.

Finally a last sensitivity analysis was performed by modifying the strength of the surface heat fluxes by changing the evaporation rate. This change also alters strongly the intensity of the breeze system and has therefore a strong influence on the trajectories.

These elements represent boundary conditions on which we have usually little information before making routine meso-scale runs in preparation of air quality simulations. They represent inherent weaknesses for operational use of air quality models in complex coastal environments.

2. MODEL DESCRIPTION AND SET UP

The Topographic Vorticity Meso-scale model in non-hydrostatic mode (TVMnh) is used. Prognostic variables are the two vorticity components (allowing to avoid the role of pressure and density in dynamical equations), potential temperature and specific humidity. Surface temperature and moisture values are computed using surface energy and moisture balance equations and the Deardoff force-restore formulation.

Turbulence closure scheme is 1.5 order using a prognostic equation on turbulent kinetic energy (tke). Boundary layer height Hi is estimated as the height where the tke value falls under 10% of its local surface value. In the tke budget equation, the dissipation term ε and the turbulent diffusion coefficients K depend on the mixing length formulations lk and le as shown in the following equations, with C_1 and C_2 constant.

$$K = C_1 \, l_k \, (tke)^{1/2} \tag{1}$$

$$\varepsilon = C_2 \, \frac{(tke)^{3/2}}{l_e} \tag{2}$$

In the formulation by Therry-Lacarrère (T-L), the mixing lengths are functions of surface stability parameters (as L, the Monin-Obukhov length) and boundary layer height Hi. The equations for le and lk are similar but with different coefficients.

In the formulation of Bougeault-Lacarrère (B-L), two mixing lengths lup and $ldown$ are first determined in function of the shape of the temperature and tke profiles.

540

At each vertical level, the two *lup* and *ldown* are calculated as the largest distance a parcel can travel in the vertical direction due to the negative buoyancy effect on the kinetic energy. The mixing lengths *le* and *lk* are combinations of the two lengths up and down.

As it will be seen, once a stable layer appears above or under a level, the length *lk* is reduced rapidly to small values. This feature seems to be the main difference between both formulations. This is a very critical topic and we will see to which extent it affects the 2D sea breeze and 3D simulations over a complex terrain.

3. SENTIVITY TO TURBULENCE CLOSURE SCHEME

3.1.   2D Simulations

In these idealised simulations, during the night, mixing lengths and *tke* values are observed to be smaller for the Bougeault-Lacarrère simulation. Moreover, for this formulation, due to the presence of thermal stability, the *tke* budget is restricted close to the surface. Evolution during the day shows that the *tke* budget remains lower for this B-L formulation despite the rapid growth and extension of the mixing lengths *lk*. Nevertheless, higher diffusion coefficients appear during the day for the B-L case. This causes a greater homogeneity of the horizontal wind components inside the boundary layer depth and consequently higher wind speeds close to the surface.

This simple 2D configuration is aimed to find the major differences between the behaviour of the two selected mixing lengths formulations. The vertical plane is oriented in the North-South direction; the horizontal extent is 400 km and the vertical one is 4 km high. The domain is divided in 200 km of sea (southward) and 200 km of land. In the example shown, the geostrophic forcing wind (3 m/s) is offshore (thus against the day breeze).

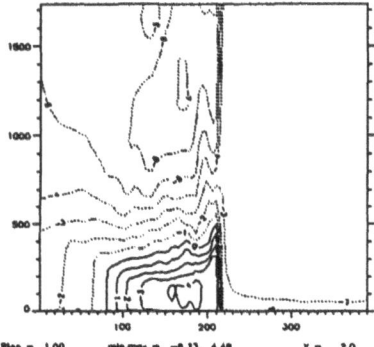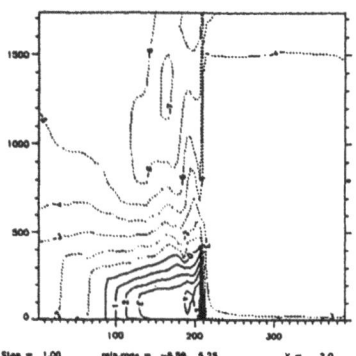

Figure 1 Horizontal wind component (m/s) along the vertical plane perpendicular to the coast, at 17.00 LST. Sea surface covers the first 200 km and land surface, the last 200 km : a) T-L formulation and b) B-L formulation.

Figure 1 presents the two sea breeze circulations obtained at 1700 LST for the two formulations. At 17.00 LST, the sea breeze front looks more vigorous for the B-L formulation than for the T-L one. Indeed, as we said earlier, horizontal wind speeds are stronger for the B-L formulation. This higher horizontal wind convergence is evidently responsible for a more vigorous sea breeze front formation. This is confirmed by stronger vertical motions at the front for the B-L case but also by a less rapid inland progression.

If we examine the front progression later, at 21.00 LST, when convective activity has significantly decreased, the layout of remaining *tke* is much different: nothing persists on land except at the front location for both formulations.

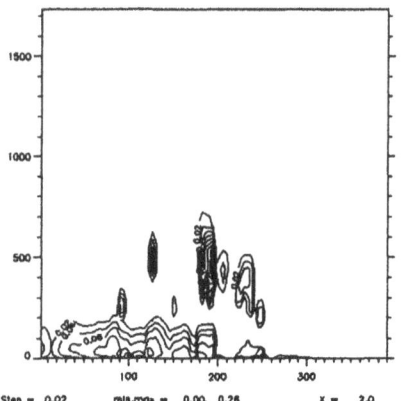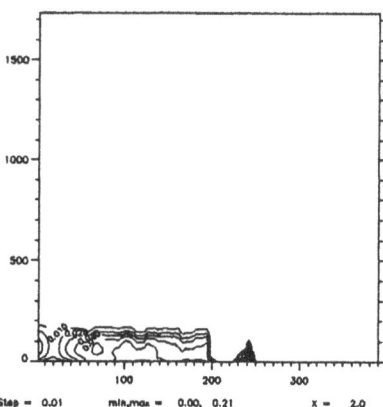

Figure 2 TKE in a YZ plan at 21.00 LST. Sea surface covers the first 200 kilometers and terrain surface covers the last 200 kilometers : a) T-L formulation, b) B-L formulation.

For the T-L formulation, as seen on figure 2, significant *tke* formation takes place in the return flow of the sea breeze circulation. That is another example which amplifies the differences between the formulations. Indeed, in the T-L formulation, the sink due to the stability increase is not strong enough to counteract the production linked with the wind shear. By contrast, in the B-L formulation, the stability is a strong sink, so no *tke* remains in the return current and over land.

To sum up, the idealised 2D simulations have shown that boundary layer depths are higher in the T-L formulation and that the sea breeze front is more stationary in the B-L simulation.

3.2. 3D Simulations

We turn now to the 3D simulations for the particularly complex terrain of the ESCOMPTE domain. This domain has large topography gradients. Some mountains reach more than 1000 m only 15 km from the shore. On figure 3, the topography used to run the model is detailed.

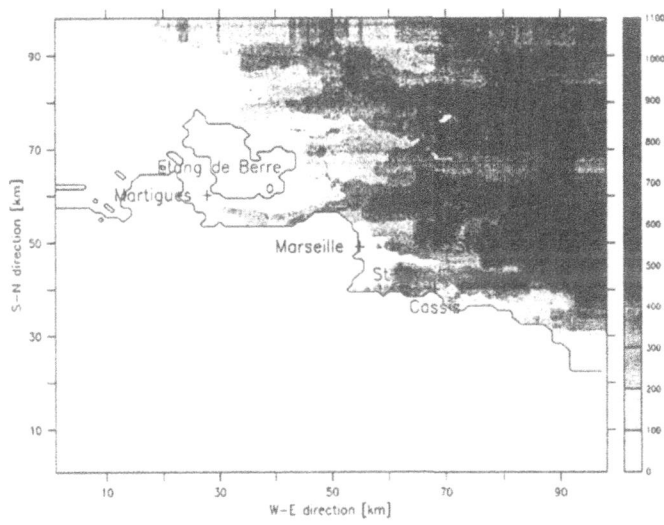

Figure 3 : Topography used in the model initialisation. The St-Baume culminates at 1147m, the Ste-Victoire mountain at 1011m; the Etoile and St-Cyr reach around 600 m.

The model is run with a horizontal grid size of 1 km and extends 100 km in both directions ; about one third of the domain is covered by the Mediterranean Sea.

The simulation is initialised with the conditions of the 30 June 2000 at 20.00 LST as measured by the radio sounding at Martigues. (see figure 3, south of Etang de Berre). The geostrophic wind forcing is 6 m/s in a south-easterly direction. Lateral boundary conditions are left open. Eight land types were adopted to describe the diversity of the domain. We mainly find forests in the west and over the Ste-Baume mountain, shrubs in the east. Elsewhere, cropland is dominant.

Differences between the two simulations appear from the begin of the day, around 8.00 LST. The T-L simulation already starts to produce slight sea breeze circulation along the coast, whereas the southwards wind transport still dominates in the B-L simulation. Progressively in the afternoon, a large sea breeze circulation formation tends to prevail for the T-L simulation : the lake breeze is strongly reduced south of Etang de Berre but is also strongly reinforced by the northwards sea breeze progression as presented in figure 4a presents.

Actually, the B-L formulation, as shown in figure 4b, leads to boundary layer maxima on the western boarder of the lake indicating convergence areas between the main geostrophic flow, the lake breeze and the sea breeze. Lower boundary layers for the formulation of T-L around the lake area can also be explained by the important sea breeze inland progression. Then we can conclude that sea breeze circulations are responsible for noticeable differences in boundary layer heights between the two formulations. One can also observe higher boundary layers above the sea for the T-L simulation linked to *tke* formation in the sea breeze return circulation, as already mentioned in the 2D simulations.

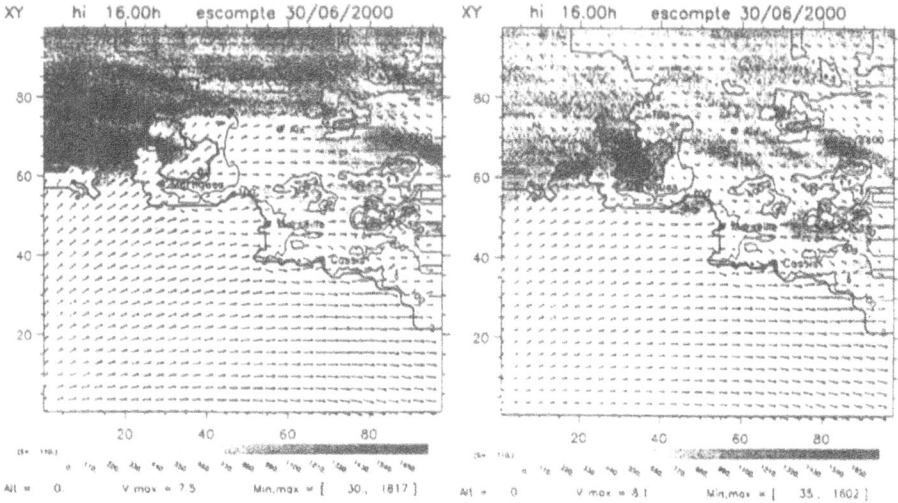

Figure 4 Boundary layer heights (m) and wind surface components (m/s) over the Escompte campaign domain at 16.00 LST : a) for the simulation of Therry-Lacarrère, b) for the simulation of Bougeault-Lacarrère.

Finally, tracer trajectories calculated for both simulations lead to very different results caused by the various sea breeze inland progressions as we highlighted earlier. An example is illustrated on figure 5.

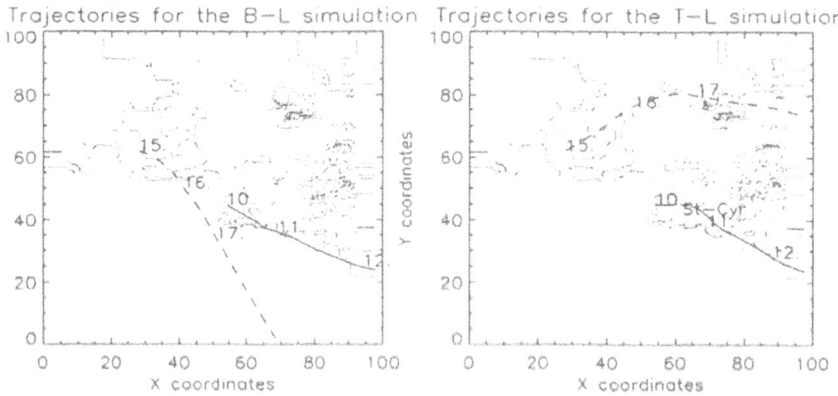

Figure 5 Trajectories of particles emitted over the Marseille area and over the industrial site of Etang de Berre. Particles are launched at 10.00 LST from the Marseille area and at 15.00 LST for the industial site located Etang de Berre area: a) for the B-L simulation and b) for the T-L simulation.

For particles emitted from Marseille at 10.00 LST, we notice that the south-eastwards motion is more rapid for the B-L simulation. This could be for two reasons: first, as we explained by the 2D simulations, horizontal winds are effectively more vigorous for B-L than for T-L simulation. Secondly, the south-eastwards transport is delayed by a sea breeze circulation appearing later over Marseille for the T-L simulation.

544

Because these experiments do not include dispersion effects, we do not expect high correlations of tracer heights with boundary layer heights calculations. Tracer heights can actually only be related to vertical motions. These ones are enhanced particularly in sea-lake breeze fronts convergence and over mountains, so that differences in trajectories heights exist : at 11.00 LST, the particles released from Marseille are submitted to a strong subsidence motion down the St-Cyr mountain for the B-L simulation. Indeed, the remaining presence of a thin stable surface layer at that location contributes to a downslope flow.

Large trajectory differences are also observed for particles launched next to Etang de Berre at 15.00 LST, as also shown on figure 5. Sea breeze circulation enhanced by the lake breeze, for the T-L simulation, controls clearly the transport northwards, whereas a seawards transport prevails for the B-L simulation. This difference underlines the critical location of the emission site of Etang de Berre related to the inland progression of the sea-land breeze fronts.

4. SENSITIVITY TO GROUND CHARACTERISTICS

The land types are distinguished by the values of albedo, emissivity, rugosity, resistance to evaporation and thermal capacity. In one first experiment, we tested the effect of cutting by half the surface resistance values to evaporation for the simulation of Therry-Lacarrère. The sea and the lake breezes are obviously affected by the induced reduction of sensible heat fluxes; then, the wind transport tends to prevail in this last simulation. This is illustrated on figure 6 where we can observe the trajectories of particles released from an industrial site at 15.00LST next to Etang de Berre.

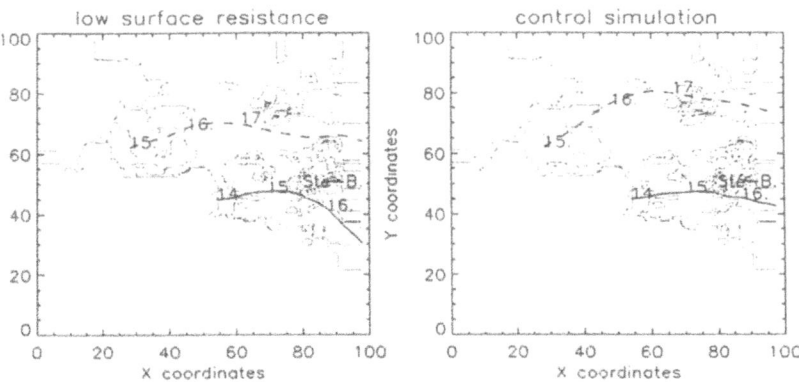

Figure 6 Trajectories of particles emitted over the Marseille area and an industrial site of Etang de Berre. Particles are launched at 14.00 LST for the Marseille area and at 15.00 LST for the Etang de Berre area : a) for the surface resistance cut by half and b) for the control run.

Figure 6 also shows particles released from Marseille at 14.00LST. We see that these trajectories follow an important veering at the approach of the mountain Ste-Baume. This area is quite critical due to very steeped slopes. In this experiment, the reduction of the surface sensible heat fluxes attenuates partly upslope movements. This is what we observed by comparing the vertical velocities between the control run and this last experiment. Actually, for that last simulation, ascending motions are slowed down and subsidence motions enhanced so that the large veering observed in the particles trajectories could be due to stronger subsidence motions down the mountain.

5. CONCLUSION

Mixing length formulations/turbulence closure schemes and surface parameter values are two elements which infer quite well on the results of simulations.

Firstly we analysed the sensitivity to turbulence closure schemes with the mixing length formulations of Therry-Lacarrère (1983) and Bougeault-Lacarrère (1989). Through idealised 2-D simulations, we highlighted the differences between the behaviour of the two formulations (B-L producing less *tke*, lower BL depths and more stationary sea breeze fronts). Then, the two closure schemes were applied to the simulation of the situation of the pollution episode of 30th June 2000 near Marseille showing noticeable differences in sea breeze inland progressions. This was emphasised by computing air particles trajectories emitted from the most polluted areas over the domain (Etang de Berre area and the city of Marseille). Large differences of the simulated trajectories do not only appear horizontally but also along the vertical. Secondly, results reveal also a high dependence on surface characteristics values. One of these features is the role played by the surface resistance to evaporation controlling the sensible heat flux budget, hence sea breeze progressions and also slope movements.

Results developed in this paper suggest a higher consideration of these elements in future interpretation of data, comparison with observations and subsequent chemical simulations; they also point out important weaknesses inherent to most meso-scale models used in complex situations.

6. ACKNOWLEDGEMENTS

This work was supported by the Belgian FRIA. We also would like to thank H. Goosse and X. Fettweis for their constructive comments.

7. REFERENCES

Bougeault, P., Lacarrère P., 1989, Parameterisation of orography-induced turbulence in a mesobeta-scale model. *MWR*, **117**, 1872-1890.

Schayes G., Thunis T., Bornstein R., 1996, Topographic vorticity-mode mesoscale-β (TVM) model. Part I: formulation. *Journal of Applied Meteorology*, **35**, 1818-1823.

Simpson J..E., 1994, *Sea Breeze And Local Winds*.

Therry G., Lacarrère P., 1983, Improving the eddy kinetic energy model for planetary boundary layer description, *Boundary Layer Meteorology*, **25**, 63-88

Thunis P., Clappier A., 2000, Formulation and evaluation of a nonhydrostatic mesoscale vorticity model (TVM), *MWR*, **128**, 3236-3250.

DISCUSSION

E. GENIKHOVICH   Because of high inhomogeneity of your flow and its complexity, one could expect that small perturbations in the wind flow and/or location of the release point could result in noticeable sensitivity.

A. PORSON   Concerning the locations, we launched particles at several places located close to each other in order to examine both flows behaviour, especially around the city of Marseille. The resulting trajectories showed most of the time similar results except when the sea breeze cirulation is being established. The way we used to perform the 3D simulations through idealised 2D simulations can give us a strong insight about the different inland progressions in both formulations. Further examination can be achieved by simply using only grass and sea and check up if the observed sensitivity would still be significant.

SHORT-TERM PREDICTON OF NOx LEVELS IN İSTANBUL USING ARTIFICIAL NEURAL NETWORK APPROACH

R. Tansel Coşkuner, Gülen Güllü, and Gürdal Tuncel*

1. INTRODUCTION

Neural networks are very useful tools for prediction and classification applications and they have been used for a wide variety of applications where statistical methods are employed. Some applications in environmental engineering area are; atmospheric corrosion modeling (Pintos et al., 2000), short-term predictions of hourly surface ozone concentrations and SO_2 concentrations (Gardner and Dorling, 2000; Mok and Tam, 1998), identification of unknown air pollution sources (Reich et al., 1999), pH modeling (Moatar et al., 1999), optimal CO_2 control in a greenhouse (Linker et al., 1998) and prediction of water resources variables (Maier and Dandy, 2000).

Recently, the use of neural networks, and in particular the multilayer perceptron (MLP), has been shown to be effective alternatives to more traditional statistical techniques. Unlike other statistical techniques the MLP neural networks makes no prior assumptions concerning the data distribution. It can model highly non-linear functions and can be trained to accurately generalize when presented with new, unseen data (Gardner and Dorling, 1998). Due to the highly non-linear nature of atmospheric data, and its complex relationship with meteorological data, MLP has many applications in the atmospheric sciences (e.g., Yi and Prybutok, 1996; Jorquera et al., 1998; Boznar et al., 1993; Mok and Tam, 1998; Gardner and Dorling, 1999; Kolehmainen et al., 2001).

The objective of this study is to develop a short-term prediction model for nitrogen oxides levels in the ambient atmosphere of İstanbul. Multilayer perceptron (MLP) neural network structure with back propagation learning algorithm was used for short-term prediction of NO_x levels at the Kadıköy Monitoring Station in İstanbul. The model was developed using the monitoring data of air quality and key meteorological parameters from Göztepe Meteorological Station in İstanbul during 1997-1999.

---

* R. Tansel Coşkuner, Middle East Technical University, Ankara, Türkiye. Gülen Güllü, Hacettepe University, Ankara, Türkiye. Gürdal Tuncel, Middle East Technical University, Ankara, Türkiye.

Air Pollution Modeling and Its Application XVI, Edited by
Borrego and Incecik, Kluwer Academic/Plenum Publishers, New York, 2004

549

2. DATA

İstanbul is the most densely populated city in Turkey. The city of İstanbul covers an area of 5,712 km² and based on the last census conducted in October 2000 by the State Institute of Statistics, its population is declared as over 10 million. İstanbul can be defined as a highly industrialized urban area. The traffic load in the city is considerably high as it is the connection point between the Asia and the Europe. The air pollutant emissions are considerably high in İstanbul due its dense population and its traffic load.

The hourly average temperature ranges between 5.5°C and 9.9°C. The average relative humidity during these periods changes between 65% and 85%. Most of the time during these data periods the atmospheric conditions are neutral or stable.

The air quality data used in this study were obtained from an air quality monitoring program, which was funded and conducted, by the İstanbul Greater Metropolitan Municipality, Directorate of Environmental Protection and Improvement.

The pollutant parameters monitored were, ozone, NO_x, NO_2, NO, SO_2, PM, CO, total hydrocarbons, methane (CH_4) and non-methane hydrocarbons (nMHC). The NO_x data used in this study is taken from the Kadıköy Monitoring station and the data covers a period between January 1, 1997 and July 31, 1999. There were missing measurements during this 31 months of duration and the total number of hourly NO_x data covered approximately 75.8 % of this period. Average NO_x concentrations are given in Table 1. The minimum and maximum NO_x concentrations observed during the measurement period are 2.0 µg/m³ (1.06 ppb) and 1707.0 µg/m³ (909.16 ppb), respectively. The average NO_x concentration during this period is 104.27 µg/m³ (55.54 ppb).

The hourly climatological data and the mixing height values calculated based on the radiosonde measurements of Göztepe Meteorological Station operated by the General Directorate of the Turkish State Meteorological Service. The Göztepe Meteorology Station is located at 40.58° N latitude and 29.05° E longitude and it is close to the Kadıköy Station where the air pollution monitoring was performed. The meteorological variables used in this study are; temperature, atmospheric pressure, cloud cover, relative humidity, wind direction, wind speed, base of lowest cloud, solarity, mixing height, Pasquill stability classes.

Hourly data is only available for temperature, atmospheric pressure, wind direction, wind speed and solarity parameters. The cloud cover, relative humidity and base of lowest cloud observations are performed three times a day at 07:00 a.m., 14:00 p.m. and 21:00 p.m. The data for these three parameters were interpolated to hourly values.

The "no cloud conditions" for the base of lowest cloud parameter were substituted by a value of 7000 m, which is higher than the maximum value observed.

Hourly Pasquill Stability Classes were calculated from meteorological surface observations using PCRAMMET, a pre-processor program developed by U.S. EPA. Pasquill stability classes are expressed with an index of A to G (or numerically 1 to 7) where "A", "B" and "C" corresponds to unstable conditions, "D" to neutral and "E", "F" and G" corresponds to stable conditions.

Twice daily mixing height data is present only for the training data set period. Mixing height data for the period of the validation data set and the test data set does not exist. Therefore for these periods the mixing height data from the corresponding time of the year 1997 was used for replacement. The interpolation of these twice daily mixing height values to hourly values were performed by using PCRAMMET.

Table 1. Statistical summary of NO, NO_2 and NO_x in İstanbul

Parameter	Minimum (μg/m³)	Maximum (μg/m³)	Average (μg/m³)
NO	1.0	1677.0	59.4
NO_2	1.0	448.0	48.3
NO_x	2.0	1707.0	104.3

In this study JavaNNS was used as neural network simulator under Windows 2000 environment, which is freely available via internet and can be downloaded from the http://www-ra.informatik.uni-tuebingen.de.

3. METHODOLOGY

3.1. Multilayer Perceptron

The MLP neural networks developed in this study were trained by using the backpropagation algorithm, which is implemented in JavaNNS. After preprocessing of the input data, pattern sets containing missing observations were eliminated, as the neural network simulator use only complete sets of input patterns. Once the complete sets of input patterns are prepared, they were changed into ASCII format according to the input file grammar used by the simulator.

3.2. Activation Functions and Data Normalization

In this study, the logistic sigmoid function was used as the activation function for the development of the artificial neural network model and data were normalized to a range of [0, 1]. Data normalization process was carried out by determining the maximum and minimum values of each variable over the whole data set and scaling was performed using Eq. (1).

$$\hat{x} = (x - x_{min}) / (x_{max} - x_{min}) \tag{1}$$

where;
$\hat{x}$ = normalized value
x_{min} = minimum value
x_{max} = maximum value

3.3. Network Structure

Network structure is generally defined by the number of hidden layers and the number of nodes in each of these layers. The use of one hidden layer and two hidden layers were investigated, however the use of two hidden layers did not improve the efficiency of the neural network model. As the number of connection weights increased, the training process slowed down significantly. Therefore, one hidden layer was used in this study.

Hecht-Nielsen (1987) suggests Eq. (2) as the upper limit for the number of hidden layer nodes in order to ensure that ANNs are able to approximate any continuous function.

$$N^H \leq 2N^I + 1 \tag{2}$$

where;

N^H = number of hidden layer nodes
N^I = number of input layer nodes

Rogers and Dowla (1994) recommends considering the relationship between the number of training samples and network size. To satisfy this criterion Eq. (3) can be used as the upper limit for the number of hidden layer nodes.

$$N^H \leq N^{TR} / (N^I + 1) \tag{3}$$

where;

N^{TR} = number of training samples

Equation (2) and (3) were used to define the upper limit for the number of the hidden layer nodes during this study, and different node numbers lower than these upper limits were tried to determine the best network structure. In this study, the number of input layer nodes is 11 and the number of training patterns is 705. Using formula (2) and (3) the upper limit was found as 23 and 59, respectively. For the safe of the study the upper limit of the number of hidden layer nodes is taken as 20.

3.4. Pattern Sets

The minimum, maximum and average NO_x values for the selected training, validation, and test data sets are summarized in Table 2.

Table 2. The training, validation, and test pattern sets

Properties	Training	Validation	Test
Starting Date of the Period	01.01.1997	06.12.1998	25.01.1999
Ending Date of the Period	30.01.1997	14.01.1999	07.03.1999
Number of Patterns	705	949	995
Average NO_x Concentration ($\mu g/m^3$)	126.75	98.23	144.24
Minimum NO_x Concentration ($\mu g/m^3$)	6	6	6
Maximum NO_x Concentration ($\mu g/m^3$)	1707	1221	1610

The validation set error was used as the stopping criterion during the training process. After the training of the network is completed, the test data set was used for the verification of the generalization ability of the trained neural network model.

4. RESULTS AND DISCUSSION

4.1. Initial Model Development Process

Initial model development process was performed using all the meteorological parameters and the previous hour NO_x concentrations. A neural network model with 10 hidden layer nodes was found to give the best prediction results for all the training, validation and test data sets. The generalization ability of the selected neural network model was checked using the test data set that is not introduced to the model during the training step. The r^2 and the correlation coefficient for the prediction results of the test data sets were 75% and 86%, respectively.

4.2. Alternative Model Structures

Then alternative model structures to increase the prediction time of the NO_x concentrations were analyzed, and it is tried to develop a prediction model for NO_x concentrations for two hours (T2) and 24 hours (T24) in advance. The performance of these neural network models is given in Table 3. The model performance of the two hours in advance and 24 hours in advance were compared with the neural network model for prediction of one hour (T1+2) in advance.

There is a decrease in prediction accuracy of the model for two hours in advance compared to the selected neural network model for prediction of one-hour in advance. The r^2 of the model is 0.48. The peak NO_x were significantly underestimated. Increasing the prediction period even by one hour significantly decrease the generalization ability of the selected neural network model. However it should be noted that, the general pattern is predicted fairly well, which is important in terms of regulatory purpose, but exact NO_x concentrations at maxima are not predicted as accurately.

There is a significant decrease in prediction accuracy of the model for prediction of 24-hour in advance. The r^2 for this model is 0.14, and the correlation coefficient is 0.37. The model failed in catching the variation pattern of the NO_x concentrations. As a result of all these model improvement attempts, it is found that the NO_x concentrations are highly dependent on the previous hour concentrations.

Table 3. Neural Network Models Developed to Increase the Prediction Time

Network ID	Training Cycles	Training Set		Validation Set		Test Set	
		r^2	Corr. Coeff.	r^2	Corr. Coeff.	r^2	Corr. Coeff.
T2	4640	0.6	0.81	0.5	0.72	0.4	0.70
T1+2	2730	0.7	0.87	0.7	0.85	0.7	0.85
T24	1350	0.4	0.68	0.1	0.42	0.1	0.37

4.3. Implementation of a Model Performance Improvement Method

A model performance improvement method, which is carried out by repeating the peak hour concentrations at several times, was implemented. For this purpose two days of the training data containing the highest peak hours were repeated several times. The results showed that the repetition of the peak hours did not improve the generalization ability of the selected neural network model.

4.4. Refinement of the Neural Network Model

Finally, input parameter reduction is performed in order to ease the use of the model for future prediction purposes, and to ease the preparation and supply of the parameters required for prediction process. For this purpose a detailed analysis of the correlations between the input parameters and the output parameter was performed in order to eliminate the parameters showing weak correlations with the output parameter, NO_x. The input parameters, having high inter-correlation were also assessed for the refinement of the selected neural network model.

The input parameters were eliminated one by one from the selected neural network model based on the correlation analysis, and the generalization ability for each refined prediction model were analyzed in order to select an optimum neural network prediction model. Previous hour NO_x concentrations, temperature, wind speed, solarity, cloud height, wind direction, played an important role for the prediction of the NO_x concentration for one hour in advance.

The scatterplot of predicted and observed NO_x concentrations versus patterns for the optimum neural network model is given in Figure 1.

5. CONCLUSION

A short-term prediction model for nitrogen oxides levels in Kadıköy, İstanbul was developed in this study. For this purpose multilayer perceptron (MLP) neural network structure with back propagation learning algorithm was used. The model was developed using the monitoring data of air quality and key meteorological parameters from Göztepe Meteorological Station in İstanbul during 1997-1999. The results indicate that accurate prediction can be made for the next hour NO_x levels.

Attempts can be made to improve the modeling performance for predicting the NO_x concentrations in longer terms. Counts on the traffic inter-section points, as input parameter might be helpful in improving the model performance for prediction of longer terms. Since the neural network models very much rely on the quality of the input data, obtaining data with improved quality would help to decrease the error levels of the models. ANN algorithms other than the back propagation algorithm can be used in future studies, in order to improve the model performance. Short–term prediction models for other air pollutants, such as for ozone, SO_2, or PM, can be developed using ANN methodology.

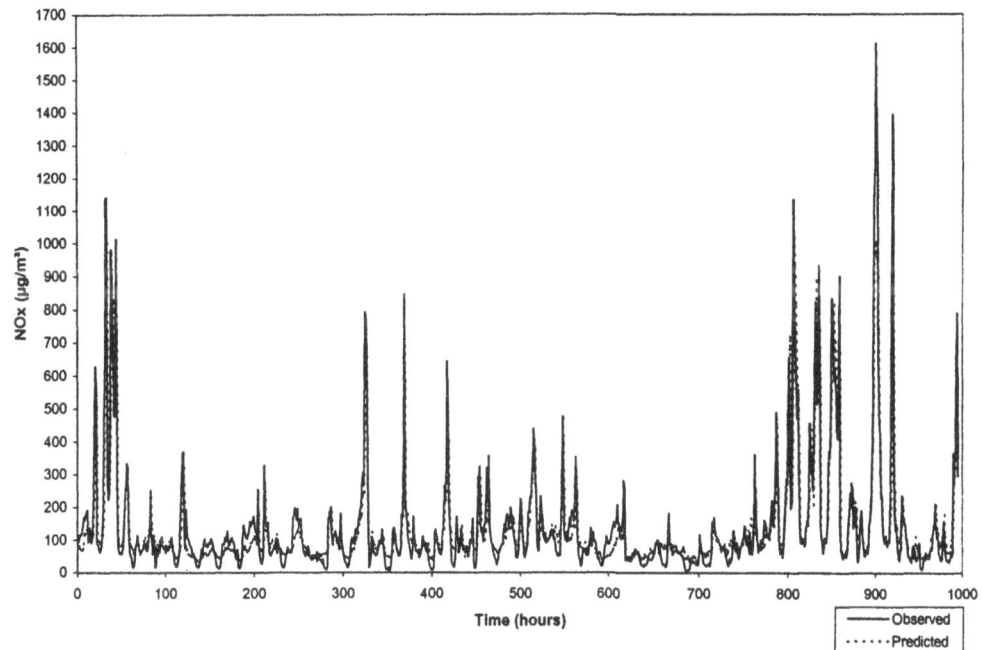

Figure 1. Time series plot of observed and predicted NO$_x$ concentrations for the test data set using the optimum network

6. ACKNOWLEDGMENTS

The authors are grateful to Prof. Dr. Mustafa Öztürk for providing the air pollution data used in this study.

7. REFERENCES

Boznar, M., Lesjak, M., and Mlakar, P., 1993, A neural network based method for short-term predictions of ambient SO$_2$ concentrations in highly polluted industrial areas of complex terrain, *Atmospheric Environment.* **27B(2)**: pp. 221-230.

Gardner, M.W. and Dorling, S.R., 1998, Artificial neural network (multilayer perceptron) a review of applications in the atmospheric sciences, *Atmospheric Environment.* **32(14/15)**: pp. 2627-2636.

Gardner, M.W. and Dorling, S.R., 1999, Neural network modeling of hourly NO$_x$ and NO$_2$ concentrations in urban air in London, *Atmospheric Environment.* **33(5)**: pp. 709-719.

Gardner, M.W. and Dorling, S.R., 2000, Statistical surface ozone models: an improved methodology to account for non-linear behaviour, *Atmospheric Environment.* **34**: pp. 21-34.

Hecht-Nielsen, R., 1987, Kolmogorov's mapping neural network existence theorem, *First IEEE International Joint Conference on Neural Networks*: pp. 11-14.

Jorquera, H., Perez, R., Cipriano, A., Espejo, A., Letelier, M.V., and Acuna, G., 1998, forecasting ozone daily maximum levels at Santiago, Chile, *Atmospheric Environment.* **32(20)**: pp. 3415-3424.

Kolehmainen, M., Martikainen. H., and Ruuskanen. J., 2001, Neural networks and periodic components used in air quality forecasting, *Atmospheric Environment.* **35**: pp. 815-825.

Linker, R., Seginer, I., and Gutman, P.O., 1998, Optimal CO_2 control in a greenhouse modeled with neural networks, *Computers and Electronics in Agriculture.* **19**: pp. 289-310.

Maier H.R. and Dandy, G.C., 2000, Neural networks for the prediction and forecasting of water resources variables; a review of modeling issues and applications, *Environmental Modeling and Software.* **15**: pp. 101-124.

Moatar, F., Fessant, F., and Poirel, A., 1999, pH modeling by neural networks, application of control and validation data series in the Middle Loire River, *Ecological Modeling.* **120**: pp. 141-156.

Mok, K.M. and Tam, S.C., 1998, Short-term prediction of SO_2 concentration in Macau with artificial neural networks, *Energy and Buildings.* **28**: pp. 279-286.

Pintos, S., Queipo, N.V., Rincon, O.T., Rincon, A., and Morcillo, M., 2000, Artificial neural network modeling of atmospheric corroison in the MICAT project, *Corroison Science.* **42**: pp. 35-52.

Reich, S.L., Gomez, D.R., and Dawidowski, L.E., 1999, Artificial neural network for the identification of unknown air pollution sources, *Atmospheric Environment.* **33**: pp. 3045-3052.

Rogers, L.L. and Dowla, F.U., 1994. Optimization of groundwater remediation using artificial neural networks with parallel solute transport modeling, *Water Resources Research.* **30**(2): pp. 457-481.

Yi J. and Prybutok, R., 1996, A neural network model for prediction of daily maximum ozone concentration in an industrialized urban area, *Environmental Pollution.* **92**(3): pp. 349-357.

DISCUSSION

F.LEFEBRE   Did you try to use predicted mixing heights (forecasted) instead of using measured mixing heights?

T. COSKUNER

P.BUILTJES   I have the impression that the NO_2 - conc. are mainly determined by local emissions. What is your opinion?

T. COSKUNER

D.SYRAKOV   Have you used the t-24 NO_x values in case of 24-hour forecast?

T. COSKUNER

O.D.YAY   Since you include the previous hour NO_x concentration as an input, may be the model could work for other locations too. Did you try that? Does it work for other locations or only for Kadıköy?

T. COSKUNER

P.SUPPAN   Have you tried to split NOx and check the learning process of NO and NO_2. I assume the learning process for NO will be more easier than for NO_2.

T. COSKUNER

J.BARTNICKI   Does your model works well for all days of the week e.g. Saturdays and Sundays when traffic is low?

T. COSKUNER

ADAPTIVE NEURO-FUZZY BASED METHOD FOR DAILY ESTIMATION OF SO₂ CONCENTRATION IN CITY OF ZONGULDAK

Yilmaz Yildirim[*], Mahmut Bayramoglu, Lokman H. Tecer and Gultekin Yalcin

Abstract: Air pollution continues to be a major problem in many countries. Mathematical models are useful in relating emissions to air quality under a variety of meteorological conditions and source emission concentrations over an urban area. Meanwhile, the forecasting capability of sophisticated models is limited to very large and complex terrains. In this study, adaptive neuro- fuzzy logic method has been proposed to estimate the impact of meteorological factors on SO₂ pollution levels. The model forecasts satisfactorily the trends in SO₂ concentration levels, with performance between 78-90%.
Key words: SO₂ pollution, neural networks, fuzzy logic, modeling

1. INTRODUCTION

The importance of the preventing air pollution has been increasing in recent years, due to increasing knowledge of polluting sources and their pollution levels. SO₂ is one of the environmentally important air pollutants that have been associated with urban air quality problems during winters in Turkey as well as in other parts of the world. Air pollution phenomenon takes place within the atmospheric planetary boundary layer under the combined effects of meteorological factors, earth surface topographic features and the release of air pollutants from various sources.

In recent years, Artificial Intelligence (AI) based techniques have been proposed as alternatives to traditional statistical ones in many scientific disciplines. Artificial neural

[*] Yilmaz Yildirim and Lokman H. Tecer, Zonguldak Karaelmas University, Engineering Faculty, 67100 Zonguldak/Turkey. Mahmut Bayramoglu, Gebze Institute of Technology, Engineering Faculty, 41420 Kocaeli/Turkey. Gultekin Yalcin, Turkish State Meteorological Service, Ankara/Turkey.

networks (ANN), one of the most popular AI methods, are considered to be simplified mathematical models of brain-like systems. A summarized review of the applications of ANN in the atmospheric sciences has been carried out by Gardner and Dorling (1998). ANN models have been studied by various investigators for SO_2 (Boznar et al., 1993; Perautonis et al., 1994; Mlakar and Boznar, 1997; Reich et al., 1999; Andretta et al., 2000, Perez, 2001), for NO, NO_2 and NOx (Gardner and Dorling, 1999; Perez and Trier, 2001), ozone (Ryan, 1995; Jorquera et al, 1998; Gardner and Dorling, 2000) and PM2.5 (Perez et al., 2000; Perez and Ryes, 2001) concentration forecasting

On the other hand, new AI techniques have been developed, such as "Soft computing", which aims at integrating such powerful artificial intelligence methodologies as neural networks and fuzzy inference systems. While fuzzy logic performs an inference mechanism under cognitive uncertainty, neural networks offer exciting advantages, such as learning, adaptation, fault-tolerance, parallelism and generalization. To enable a system to deal with cognitive uncertainties in a manner more like humans, one may incorporate the concept of fuzz logic into the neural networks. The resulting hybrid system is called neuro-fuzzy network (Fullér, 1995). A literature survey revealed that this method was not fully investigated for the prediction of SO_2 or other pollutants concentration.

This study aims to estimate SO_2 pollution levels depending on meteorological parameters such as relative humidity, wind speed, precipitation and temperature, by using an artificial intelligence method known as ANFIS which is claimed as a universal approximator to represent highly non-linear functions more powerfully than conventional statistical methods (Jang et al., 1997) and the most reliable forecasting method comparing to neural network and time series (Jorquerra et al. 1998).

2. THEORETICAL SURVEY

2.1. Fuzzy Inference and Fuzzy Modelling

Fuzzy inference is a method that interprets the values in the input vector and assigns values to the output by means of some set of fuzzy " IF-THEN" rules;

IF x is A THEN y is B,

where A and B are labels of fuzzy sets, e.g, "low", "high". Each fuzzy set is characterized by appropriate membership functions that map each element to a membership value between 0 and 1. The *IF* part (antecedent) and *THEN* part (consequent) of a rule can have multiple parts linked by Boolean operators *(AND, OR)* .

A fuzzy inference system consists of a rule base containing fuzzy rules, a database defining the membership functions of the fuzzy sets used in the fuzzy rules, and a reasoning mechanism which performs the inference procedure. A suitable fuzzy inference system for sample-data based fuzzy modelling is Sugeno's system ; its output is crisp, so, without the time consuming defuzzification operation, it is by far more suitable and it lends itself to the use of adaptive techniques (Sugeno and Kang, 1988 and Takagi and Sugeno, 1985). In

Sugeno's system, the output of each rule is a predetermined function of input variables. For example, in a first-order model with two inputs, the i.th rule is described as:

$$IF\ x\ is\ X_i\ AND\ y\ is\ Y_i,\ THEN\ \ f_i = p_{i,0} + p_{i,1}x + p_{i,2}y \qquad (1)$$

where lowercase variables (x, y) denote the inputs, uppercase variables (X, Y) stand for the fuzzy sets corresponding to the domain of each linguistic label, and p_i is a set of adjustable parameters.

2.2. Anfis: Adaptive Neuro-Fuzzy Inference System

A neural network structure consists of a number of nodes connected through directional links. Each node is characterized by a node function with fixed or adjustable parameters. The training phase of a neural network is a process to determine optimum parameters values to sufficiently fit the training data. The basic learning rule is the well-known back-propagation method which seeks to minimize some measure of error, usually sum of squared differences between a network's outputs and desired outputs. Meanwhile, "over-training" diminishes the forecasting capability of the network due to its structure excessively adapted to the training data. The model performance is always checked by means of distinct test data, and relatively good fitting is expected especially in this testing phase.

3. MATERIAL AND METHODS

Zonguldak is a coastal city located in the Western Black Sea region of Turkey, situated on a coast surrounded by mountains to the south, east and west. It has a current population of about 108,000. Although the city is located on the shore, the hilly landscape surrounds the city centre from SE and SW. Around the city is mainly forest.

Air pollution measurements carried out by the Zonguldak Public Health Center for the last 12 years have shown that there has been a high level of pollution in the city during winter season between November and March (Yildirim and Uzun, 2000). In citywide, two air quality measurement stations were established by local authority to observe air quality trends. While station 1 was set around the hospital, houses and some social clubs, station 2 was set at the city's busiest traffic road and around the other activities such as schools, and private and government offices. The acidimetric method was applied for analysis of sulfur dioxide (WHO, 1976). The daily values of SO_2 concentrations used in the model as training and testing data were collected from station 1 as seen in Figure 1..

4. RESULTS AND DISCUSSION

Model building, training and testing are performed by means of graphical user interface supplied in "MATLAB Fuzzy Toolbox". Fuzzy rules set of ANFIS structure is generated by subtractive clustering method. First order Sugeno model is preferred as inference system for its simplicity. Neural networks are trained by hybrid method as suggested by Jang et al. (1997). By considering the statistical aspect of the air pollution modeling, Gaussian type membership functions are used in this study, described by the following equation:

$$\mu_{i_i}(x) = \exp\left\{-\left(\frac{x-c_i}{a_i}\right)^2\right\}$$

(2)

where a_i, and c_i are memberships function parameters.

Figure 1. Topographic structure of City of Zonguldak and location of pollution measurement stations and meteorological station..

In order to evaluate the performance of the neural fuzzy model, two statistical indices shall be employed: the root-mean-square error (RMSE) and the index of agreement (IA), defined as

$$RMSE = \sqrt{\frac{1}{N}\sum_{i=1}^{N}(o_i - p_i)^2}$$

(3)

$$IA = 1 - \frac{\sum_{i=1}^{N}(o_i - p_i)^2}{\sum_{i=1}^{N}\left[|o_i'| + |p_i'|\right]^2}$$

(4)

where oi and pi are the observed and forecast SO_2 concentration values on day i, N is the number of days in the test set, $p_i' = p_i - o_m$ and $o_i' = o_i - o_m$, with o_m the average observed SO_2 concentration. The index of agreement is a dimensionless index bounded between 0 (showing no agreement at all) and 1 (perfect agreement of the time series). Both indices make assessments of the global performance of the model; in order to check the forecast accuracy, the number of successful forecast and the number of false positives shall also be used (Jarquerra et al., 1998).

Approximately 150 data of each winter season were used for model estimation. Preliminary tests showed that approximately 50 training epochs are sufficient for good training and testing performances.

This study is primarily focused on two important questions; 1) what is the ideal set of input parameters? and 2) how can we use large data accumulated over past years to train ANFIS so that good forecasting performance is obtained?

4.1. Input Variable Selection

The criteria considered for the model input variables was decided using cross-correlation coefficient employed by many investigators (i.e. Perez, 2001). Cross-correlation coefficient is relation between pollutant and meteorological variables such as wind speed, temperature, relative humidity etc. The cross-correlation coefficient is described as

$$C_{x,j} = \frac{\langle x(t)y_j(t)\rangle - \langle x\rangle\langle y_j\rangle}{\sqrt{(\langle x^2\rangle - \langle x\rangle^2)(\langle y_j^2\rangle - \langle y_j\rangle^2)}} \qquad (5)$$

where $\langle\ \rangle$ means average over whole series, $x(t)$ is the SO_2 series and $y_j(t)$ is the series for meteorological variable ($j=1$ wind speed, $j=2$ precipitation, $j=3$ temperature and $j=4$ is relative humidity) and t in Equation (5) runs through the whole series. C_{xj} varies between 1 (total correlation) and -1 (anti-correlation). A value zero indicates no correlation at all.

In this study cross-correlation was calculated between SO_2 as a pollutant and some of the meteorological values such as wind speed, temperature, relative humidity and precipitation. For 1996-97 winter season data, it was found as $C_{x1}=$ -0.35, $C_{x2}=$ -0.09, $C_{x3}=$ -0.35 and $C_{x4}=$ 0.18 and it was similar for the other winter seasons. These results indicate that there are anti-correlation between SO_2 pollution and wind speed, temperature and precipitation, and positive correlation between SO_2 and relative humidity.

In addition to meteorological parameters which are naturally included in the input set, the previous day's SO_2 concentration is also taken into account: it is obvious that pollution emission is a continuous process, and in this way, it is intended to reflect this fact into the model which becomes equivalent to a second order ARX model (Yildirim et al., 2002).

Table 1. Effect of input set on the training and test performances of ANFIS

Model input set	Training error	Average test error
RH, WS, P, T, $SO_{2, j-1}$	15	16
WS, P, T, $SO_{2, j-1}$	17	27
RH, WS, P, T	18	26
WS, P, T	19	27

RH; relative humidity, WS; wind speed, P; precipitation, T; temperature
$SO_{2, j-1}$; previous day's SO_2 concentration

In order to find an optimal input set which minimizes test error given by equation (3), all of the combinations of input variables have been considered. For this purpose, the accumulated past year winter season's data has been used for training, and each of the following years winter season's data (1997-98, 1998-99, 1999-2000, 2000-2001) has been separately used for testing purposes. Typical results have been represented in Table 1. It is clear that wind speed, relative humidity, temperature and previous day's SO_2 concentration are required parameters for an acceptable model performance. The input set consists of five parameters, namely; wind speed, temperature, relative humidity, precipitation and previous day's SO_2 concentration. An ANFIS structure with five input variables is given in Figure 2; each input variable is characterized by 8 gaussian membership functions and the rule base contains 8 rules of first order Sugeno type.

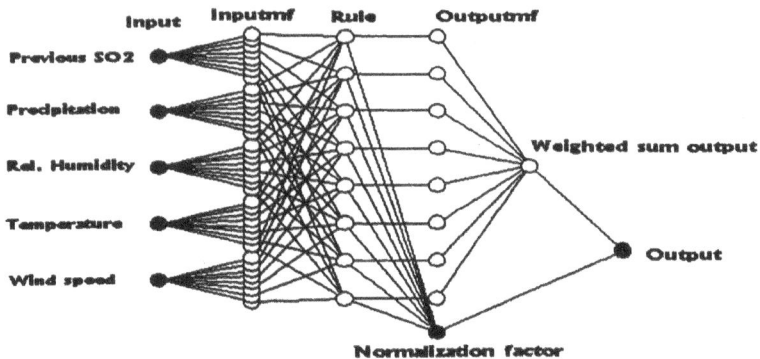

Figure 2. ANFIS Model structure used in training and testing of model. Five input parameters, 8 Gaussian membership function and 8 rules of first order Sugeno type.

4.2. Training ANFIS

Data accumulated over past years offers great potential as training data to obtain a well-trained ANFIS. It is clear that when more data is used in the training phase, ANFIS is more adapted to nonlinear functional dependency between input variables and the output. Various alternatives exist for beneficiating the large number of data. The obvious way is to use accumulated data of past years for training and the following year's data for testing purposes. The results obtained are shown in Table 2. Training and test performances are close in magnitudes, which means clearly that the ANFIS structure is not over-trained, and obviously optimally trained.

Table 2. Use of yearly progressive training sets and related performances.

Training sets	Test sets	Training Error	Test Error
1996-1997	1997-1998	16	16
1996-1997, 1997-1998	1998-1999	17	18
1996-97,1997-98, 1998-99	1999-2000	18	18
1996-97,1997-98, 1998-99, 1999-2000	2000-2001	17	15

On the other hand, test errors which asses the variance between measured and predicted values according to Equation (3), is of the same order of magnitude when compared with the standard error of the analytical method used in SO_2 concentration measurement. Thus, it may be concluded that, statistically ANFIS modeling is a valid approach and variations between model outputs and measured values result greatly from measurement errors of random nature. To reinforce this conclusion, model outputs and measured data are given in Figure 3, as time-trends for 2000-20001 winter season. Table 3 represents the statistical evaluation of the neural fuzzy model for station 1, indicating acceptable forecasting limits between 78-90%. RMSE and IA assessment parameters given in Table 3, show that the forecasting capability of the model increases gradually. As expected; primarily, the model forecasts successfully the time-trends, and secondly, magnitudes of pollutant concentration are also estimated in acceptable limits.

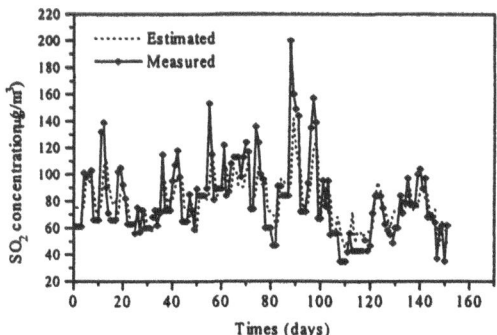

Figure 3. Time plot of measured and predicted SO_2 concentration values for 2000-2001-winter season. Training data: 1996-97, 1997-98, 1998-99 and 1999-2000-winter season; 601 data for each variable. Testing data: 2000-01-winter season; 150 data for each variable

Table 3. Statistical evaluation of the neural fuzzy model for station 1

Test Period N=153	RMSE ($\mu g/m^3$)	IA (0-1)	Number of successful forecasts	Number of false positives
1997-1998 winter season	20.3	0.685	134	19
1998-1999 winter season	25.5	0.695	128	25
1999-2000 winter season	20.1	0.713	130	23
2000-2001 winter season	19.9	0.787	138	15

5. CONCLUSION

In this study, a new methodology based on neural fuzzy method has been proposed to estimate daily SO_2 pollution concentrations over an urban area. Effective input variables in the model may be ranked as; temperature, SO_2 concentration of previous day, wind speed, relative humidity and precipitation. It is seen that temperature and previous day's SO_2 concentration are indispensable parameters for an acceptable model performance.

When the different combination of data sets has been examined from test performance point of view, it is found that cumulated input sets of five years gave the best statistical results. Measured and estimated SO_2 pollutant concentrations show peak points together within 15% testing error limits. With a better set of training patterns, one would be able to predict SO_2 concentration with high accuracy.

ACKNOWLEDGEMENTS

This study was supported by Z. Karaelmas University Research Grant 2002-45-10-16.

6. REFERENCES

Andretta, M., Eleuteri, A., Fortezza, F., Manco, D., Mingozzi, L., Serra, R and Tagliaferri, R., 2000. Neural Networks for Sulphur Dioxide Ground Level Concentrations Forecasting, *Neural Computing and Applications*, 9, 93-100.

Boznar, M., Lesjack, M and Mlakar, P., 1993. A Neural Network Based Method for Short-term Predictions of Ambient SO_2 Concentrations in Highly Polluted Industrial Areas of Complex Terrain, *Atmospheric Environment*, vol.270-B, no:2, 221-230.

Donald, A.W., 1986. A Guide to Expert Systems, Addision-Waley, Reading, M.A.

Fuller, A.D., 1995. Neural Fuzzy Systems, Abo Akademi University, Abo.

Gardner, M.W. and Dorling, S.R., 1998. Artificial Neural Network (The Multi Layer Perception)- A Review of Applications in the Atmospheric Sciences, *Atmospheric Environment*, 32, 2657-2636.

Gardner, M.W. and Dorling, S.R., 1999. Neural Network Modeling and Prediction of Hourly NOx and NO_2 Concentration in Urban Air in London, *Atmospheric Environment*, 33, 709-719.

Gardner, M.W. and Dorling, S.R., 2000. Statistical Surface Ozone Models: An Improved Methodology to Account for Non-linear Behavior, *Atmospheric Environment*, 34, 21-34.

Jang, J.S.R, Sun, C.T and Mizutani, E., 1997. Neuro-fuzzy and Soft Computing: A Computational Approach to Learning and Machine Intelligence, Prentice Hall, N.J.

Jorquera, H., Perez, R., Cipriano, A., Espejo, A., Letelier, M.V and Acuna, G., 1998. Forecasting Ozone Daily Maximum Levels at Santiago, Chile, *Atmospheric Environment*, 32, 3415-3424.

Mlakar, P and Boznar, M., 1997. Perceptron Neural Network-Based Model Prdicts Air Pollution, Proceeding on Intelligent Information System, IIS'97, IEEE Computer Society, 0-8186-8218-3, pp.345-349.

Perez, P., 2001. Prediction of Sulfur Dioxide Concentration at a Site Near Downtown Santiago, Chile, *Atmospheric Environment*, 35, 4929-4935.

Perez, P and Ryes, J. 2001. Prediction of particulate air pollution using neural techniques, *Neural Computing and Applications*, 10, 165-171.

Perez, P and Trier, A., 2000. Prediction of PM2.5 concentrations several hours in advance using neural networks in Santiago, Chile, *Atmospheric Environment*, 34, 1189-1196.

Perez, P and Trier, A., 2001. Prediction of NO and NO2 Concentrations Near a Street with Heavy Traffic in Santiago, Chile, *Atmospheric Environment*, 35, 1783-1789.

Sugeno, M and Kang, GT., 1988. Structure Identification of Fuzzy Model, Fuzzy Sets and Systems, 28, 15-33.

Tagaki, T and Sugeno, M. 1985. Fuzzy Identification of Systems and its Application to Modelling and Control, IEEE Transactions on Systems, Men and Cybernetics, 15, 116-132.

WHO, 1976. Selected Methods of Measuring Air Pollutants, World Health Organization, Geneva, No. 24, pp. 17-43.

Yildirim, Y and Uzun, N, 2000. An investigation of combustion efficiency in domesting housing heating systems: Zonguldak case, Fifth National Symposium on Combustion and Air Pollution Control, , 18-21 June 2000, Firat University, Elazig, Turkey, p13.

Yildirim, Y., Demircioglu, N., Kobya, M and Bayramoglu, M., 2002. A mathematical modeling of sulphur dioxide pollution in Erzurum City, *Environmental Pollution*, 118(3), 411-417.

566

DISCUSSION

B. KLAIC
Did you investigate cross-correlation with air pressure (higher pressure should result in the higher surface concentrations)

Y. YILDIRIM
I have investigated cross-correlation with air pressure, but it was found not much an effective parameter along the period empyloyed in the study. It may be effective for some other periods.

J. BARTNICKI
Your model is independent on SO_2 emission sources. What happens if there is a rapid and significant change in the emmission sources can you still use the model or do you have to start from the beginning?

Y. YILDIRIM
The model base is artificial intelligent, training and testing. We train the computer with sufficient and suitable data, we get the results (forecasting). If we train computer for the significant change in the emission source with sufficient data, we could forecast changes; i.e. If we train the computer with traffic related data, we can forecast the pollutions from traffic sources.

M.Z. BOZNAR
Only one short comment : If you use only linear correlation (to decide whether to use a parameter as an input to the model or not) you can losse some good inputs because "neural networks" and neuro fuzzy models can learn highly non-linear relationships. There are other methods based on trained perceptron neural networks that perform significantly better.

Y. YILDIRIM
I have employed input parameters without and with cross-correlation in preliminary study. I have found that input parameters are same for both case in forecasting. I did not use trained perceptron neural networks.

TRANSITION METAL IONS IN CLOUD CHEMISTRY

L. Deguillaume, M. Leriche, A. Marinoni, and N. Chaumerliac[*]

1. INTRODUCTION

Cloud, fog and rain chemistry have an important effect on both regional and global scales (Lelieveld and Crutzen, 1991; Jacob, 2000). Actually, there are still some remaining questions about processes in the atmospheric liquid phase related, in particular, to the role of transition metal ions, to the presence of VOCs (Volatile Organic Compounds), and to particulate matter that can act as cloud nuclei (Facchini, 2002).

TMI are incorporated into tropospheric liquid phase via aerosols, which often contain metal oxide particles. This metal oxide can turn into soluble metals by undergoing thermal and photochemical processes at the surface of the particles (Zuo and Hoigné, 1992; Erel et al, 1993; Siefert et al., 1994; Faust et al., 1994; Hoigné et al, 1994; Sulzberger et al., 1994; Sulzberger and Laubscher, 1995). When aerosol particles act as CCN (Cloud Condensation Nuclei), soluble metals dissolve into cloud droplets.

Various field campaigns report concentrations of dissolved trace metals into rainwater (Jickells et al., 1984; Ross, 1987; Lim et al., 1993) and in cloud water samples (Anastasio et al., 1994; Sedlak et al., 1997). Laboratory measurements also demonstrate the transfer of trace metals from the solid to the liquid phase (Spokes et al. 1994, Desboeufs et al., 1999). Iron (Fe), manganese (Mn) and copper (Cu) are the most abundant transition metals in the atmospheric liquid water as well as in aerosols. The major effects of transition metals are principally linked to the homogeneous aqueous phase chemistry.

In the atmospheric liquid phase, the speciation of dissolved Fe, Cu and Mn is still subject of major uncertainties. Especially, in the case of Mn, no measurement is available. While most of the Fe(II) is thought to be present as the free Fe^{2+} ion at pH<3, most of the Fe(III) is complexed with OH^-, SO_4^{2-}, and organic ions such as oxalate (Faust, 1994). Copper appears to be present as organic complexes (Spokes et al, 1996; Nimmo and Fones, 1997) which are less reactive towards HO_2/O_2^- radicals than towards free ions (von Piekowski et al., 1993). Key uncertainties for Cu and Fe are the complexation state of the metals and the kinetic reaction constants involving the metal complexes.

[*] Laurent Deguillaume, Maud Leriche, Marinoni Angela and Nadine Chaumerliac, LaMP/OPGC, Université Blaise Pascal, CNRS, 24 avenue des Landais, 63177, Aubière, France.

In this paper, we present new developments of the fully explicit multiphase chemistry model M2C2 (Model of Multiphase Cloud Chemistry) from Leriche et al. (2000; 2001) including the incorporation of the transition metal ions chemistry and variable actinic flux in the aqueous phase. The model is then applied to chemical conditions already described in Herrmann et al. (2000) who simulated three different chemical scenarios (urban, rural and marine). The ratio of Fe(II)/Fe(III) exhibits a diurnal variation with values in agreement with the few measurements of Fe speciation available.

2. DESCRIPTION OF THE MULTIPHASE BOX MODEL

The M2C2 is the result of the coupling (Leriche et al., 2001) between a multiphase chemistry model described in Leriche et al. (2000) and a quasi-spectral microphysics model based upon the parameterisation of Berry and Reinhardt (1974). For the purpose of this study, M2C2 is used with prescribed microphysics in order to compare results with Hermann et al. (2000) simulations.

The chemistry included in the chemical module is explicit. The gas-phase mechanism includes the oxidation of methane, the chemistry of NO_y and ammonia; it is derived and has been updated after Madronich and Calvert (1990). The exchange of chemical species between the gas phase and the aqueous phase is parameterised with the mass transfer kinetic formulation developed by Schwartz (1986). The aqueous phase chemistry includes the detailed chemistry of HO_x, chlorine, carbonates, NO_y and sulphur and the oxidation of organic volatile compounds (VOCs) with one carbon atom. This aqueous phase chemical mechanism has been recently updated (Leriche et al., 2003). The droplet pH is calculated at each time step by solving a simplified ionic balance equation.

The transition metal ions considered are iron, manganese and copper because these species are known to play a major role on HO_x and sulphur chemistry. This mechanism includes 106 reactions describing the chemistry of TMI with HO_x, sulphur, NO_y and VOCs.

This chemical scheme is updated after the chemical mechanism of Herrmann et al. (2000) and improvements in the chemical mechanism are listed and discussed in the following.

In the iron mechanism, the ferryl ion FeO^{2+} and its reactivity have been added. Until recently, the oxidation of ferrous ion (Fe(II)) by ozone was considered as producing OH radicals and Fe(III). However, Logager et al. (1992) provided final evidence that this reaction between Fe(II) and ozone produces Fe(IV) (the ferryl ion) and oxygen. Rate constants and activation energies for reactions of the ferryl ion with selected inorganic and organic compounds present in atmospheric water have been measured by Jacobsen et al. (1997a, 1998b). In all Fe(IV) reactions, Fe(III) is formed. Fe(III) reacts with HO_x, sulphite and VOCs (formaldehyde and formic acid). The reaction between the ferryl ion and Fe^{2+} ion branches into two reactions forming Fe^{3+} or dimer $Fe(OH)_2Fe^{4+}$. The formation of dimer $Fe(OH)_2Fe^{4+}$ is favoured at higher temperature. Afterwards, this species produces also Fe^{3+} but with a very slow rate (Jacobsen et al., 1997a). For a pH around 3, the ferryl ion plays a role mainly as a temporary OH radical sink whereas for more acidic solution, the ferryl ion is more likely to react as a distinct species (Jacobsen et al, 1998b).

Another update of the Herrmann et al. (2000) mechanism concerns the addition of the reaction between Fe(II) and peroxyl radicals which forms a transient intermediate $FeCH_3O_2^{2+}$, which later decomposes into Fe^{3+} and CH_3OOH (Khaikin et al., 1996).

In M2C2 model, the equilibrium between sulphate plus Fe^{3+} and the iron-sulphato-complex $[Fe(SO_4)]^+$ is considered, while the reaction between Fe^{2+} and SO_4^- is not taken into account because there is no evidence that this reaction takes place in the cloud phase (Mc Elroy and Waygood, 1990). Moreover, the reactions between HO_2/O_2^- radicals and $[Fe(SO_4)]^{2+}$ are considered (Rush and Bielski, 1985).

Copper chemistry is very similar to the mechanism presented in CAPRAM2.3.

Manganese complexes, as well as the manganyl ion, have been added in M2C2. Looking at Mn(II) oxidation by ozone, available information about the kinetic and the mechanism of this reaction is contradictory. Sheng (1993) has advanced that the reaction between Mn^{2+} and ozone produced Mn(III) and free OH radical based upon a theoretical study. Jacobsen et al. (1998a) described a plausible mechanism based upon laboratory investigations: Mn(II) reacts with ozone forming the manganyl ion MnO^{2+} (Mn(IV)) without formation of free OH radical. Then, MnO^{2+} rapidly reacts with Mn^{2+} to form Mn(III). This mechanism is similar to the one describing the reactivity of the ferryl ion.

In M2C2, the interaction of HO_2/O_2^- radical with Mn^{2+} is described by two equilibriums with the formation of Mn(II) superoxide complex MnO_2^+ (Jacobsen et al. (1997b). This complex is destroyed by its self reaction and by the reaction with HO_2 which leads to the formation of H_2O_2 and Mn^{2+}.

The equilibriums leading to the formation of hydroxocomplexes $[Mn(OH)]^{2+}$ and $[Mn(OH)_2]^+$ are also considered in M2C2.

Finally, the Mn^{4+} ion formed in the reversible self-reaction of Mn(III) (Rosseinsky, 1963) reacts with H_2O_2 leading to Mn(II) as suggested by Jacobsen et al. (1997b).

Photolysis frequencies are calculated using the Tropospheric Ultraviolet-Visible Model (TUV version 4.1) developed by Madronich and Flocke (1999), which has been extended to include calculations of photolysis frequencies in cloud droplets. The conditions used to calculate actinic flux are: 51°N, 0° longitude, 1000hPa and 288.15K on June 1997, the 24th. The actinic flux is obtained every 15 minutes after running the TUV model. To calculate the photolysis coefficients inside the droplets, we multiply values of clear sky actinic flux by 1.6 (Ruggaber et al., 1997) and we use available laboratory data (cross sections and quantum yields).

A main difference between M2C2 and CAPRAM2.3 is the accounting of ozone (O_3), nitrate radical (NO_3), ferric ion (Fe^{3+}), methyl hydroperoxide (CH_3OOH) and hydroxymethyl hydroperoxide ($OHCH_2OOH$) photolysis in droplets. For these two last compounds, since no data are available, photolysis coefficients in the aqueous phase are calculated following Madronich and Flocke (1999) who estimated that the actinic flux inside a droplet is increased by a factor of 1.6 compared to the droplet surrounding.

3. RESULTS AND VALIDATION OF M2C2

In order to evaluate the role of the transition metal ions on cloud chemistry, three simulations have been performed with different initial chemical conditions and are compared with results from CAPRAM2.3 model (Herrmann et al., 2000). These three chemical scenarios correspond to marine, averaged continental (remote) and polluted continental (urban) conditions and are described in details in Herrmann et al. (2000).

Same conditions are also used for prescribed microphysics: the liquid water content (0.3 g/m³), the temperature (288.15 K), the pressure (1000 hPa) and the droplet radius (1 μm) are set constant during the simulation, which lasts 48 h. This small value of radius was chosen to demonstrate the maximum effect of the aqueous phase reactivity on tropospheric chemistry. The simulated values of the pH are 2.8, 3.2, 3.45 for the urban, the remote and the marine scenarios respectively.

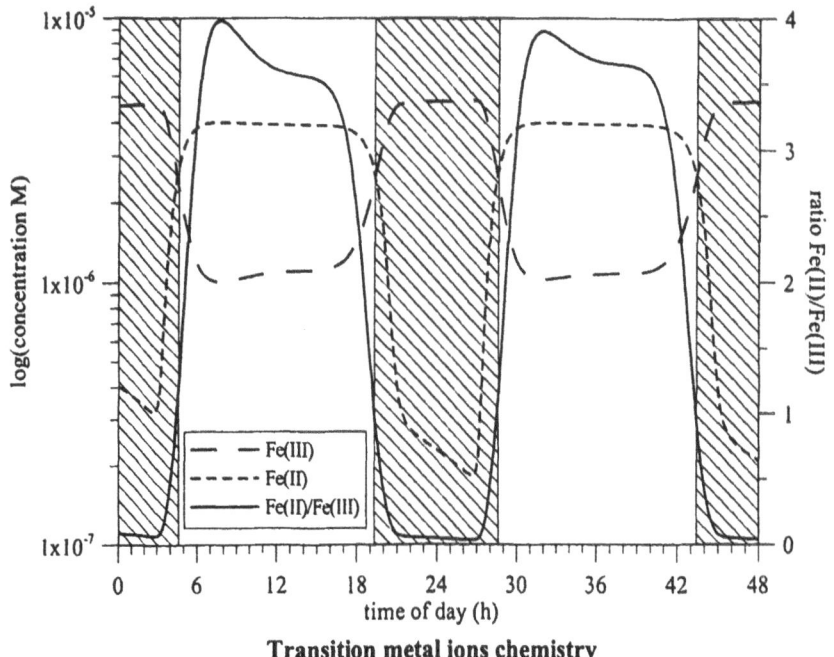

Transition metal ions chemistry

Figure 1. Time evolution of Fe(III) and Fe(II) concentrations and of the ratio Fe(II)/Fe(III) for the urban scenario. Shaded areas correspond to night-time periods.

The transition metals are initialized as Fe^{3+}, Cu^+ and Mn^{3+}. In the urban case, iron initial concentration is 10 and 100 times larger than respectively in the remote case and in the marine case. The initial concentrations of manganese and copper are the same and the ratio Fe/(Mn or Cu) is equal to 20 in the two continental cases and to 50 in the marine scenario. For manganese and copper, initial concentrations in the urban case are 10 and 250 times larger than, respectively, in the remote case and in the marine case. All these conditions are similar to the scenarios described by Herrmann et al. (2000).

Figure 1 represents the time evolution of the Fe(II) and Fe(III) concentrations and of the Fe(II)/Fe(III) ratio in the urban conditions for which iron concentrations are particularly important. During the day, Fe(III) is converted into Fe(II) and a strong dependency of Fe(II)/Fe(III) ratio on the photolysis is observed. The same behaviour is obtained for the two other scenarios.

To go into more details, Figure 2 shows the relative contribution of iron redox reactions at noon and at midnight in the urban case.

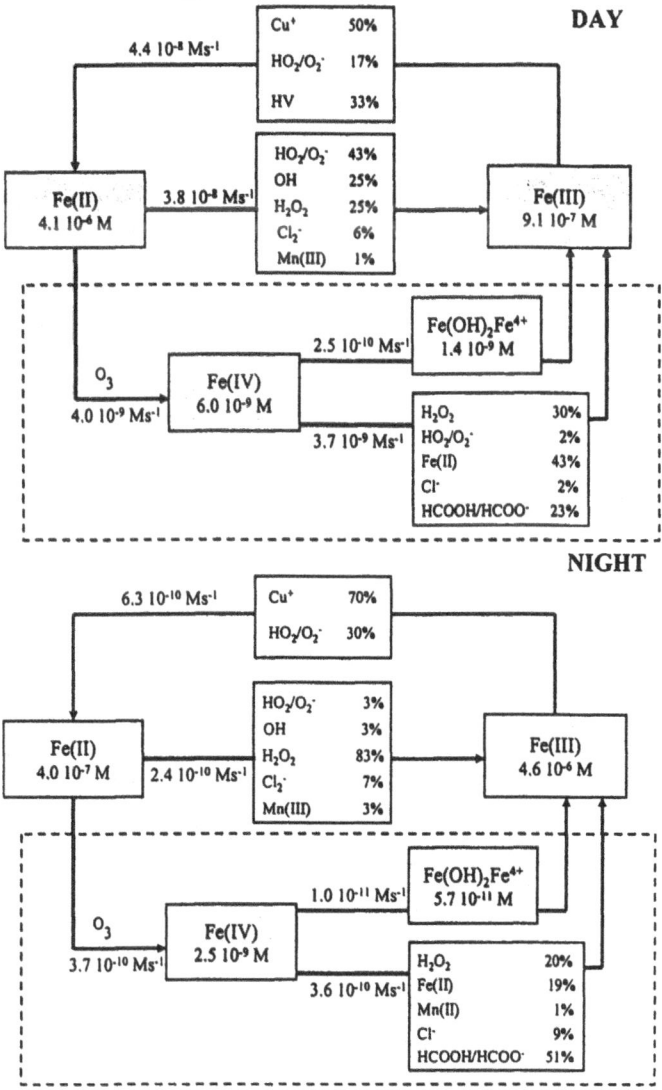

Figure 2: Iron redox reactions during the day and at night for the urban scenario.

Apart from the dominant contribution of photolysis processes in the total production of Fe(II), the main differences between day and night are due to the redox reactions between Cu(I) and Fe(III) and to the reactions between Fe(III) and HO_2/O_2^- radicals. Concerning the production of Fe(III), reactions between Fe(II) and radicals play a major role during the day. These reactions are negligible during the night and the most important source for Fe(III) is the reaction of Fe(II) with O_3 producing the ferryl ion Fe(IV) (Logager et al., 1992). Reactions of the ferryl ion with selected inorganic and organic compounds (especially with formic acid, Fe(II) and H_2O_2 during the night) lead to the production of Fe(III) (Jacobsen et al., 1997a, 1998b). The diurnal behaviour of the Fe(II)/Fe(III) ratio is not reproduced by Herrmann et al. (2000), who found no variation.

Looking now at manganese, it is completely reduced to Mn(II) and the only interaction between manganese species and photolysis activities occurs via reactions of manganese with HO_2/O_2^- radicals.

Mn(III) reacts with hydrogen peroxide to form $Mn(II)_{aq}$-superoxide complex MnO_2^+. The stability of this complex is governed by the two equilibriums between Mn^{2+} plus HO_2/O_2^- radicals and MnO_2^+ (Jacobsen et al., 1997b). These equilibriums are shifted towards Mn^{2+} production and mainly contribute to the global reduction of Mn(III) to Mn^{2+} (60% of total destruction of Mn(III) in the urban case, 96% in the remote case and 97% in the marine case). In the urban case, the reaction between Fe(II) with Mn(III) in the form of $[Mn(OH)]^{2+}$ complex contributes in a significant amount to the destruction of Mn(III) (37%) whereas it is negligible in the two other cases. This is explained by the higher initial concentration of iron and because Fe(II) represents 80% of dissolved Fe. In addition, the contribution of the equilibrium between Mn^{2+} plus HO_2/O_2^- radicals and MnO_2^+ is less important in the urban case because of the lower concentration of HO_2/O_2^-. The reaction of Mn(III) with hydrogen peroxide plays a role in the production of MnO^{2+} but is negligible in the total destruction of Mn(III) in the three cases.

In Herrmann et al. (2000), the equilibrium between Mn^{2+} plus HO_2/O_2^- radicals and MnO_2^+ is not considered. Moreover, in their urban simulation, the reduction of Mn(III) is driven by the reaction with H_2O_2 (90%) whereas the contribution of Fe(II) is less important (10%). This is due to the difference between the manganese mechanisms considered in M2C2 and in CAPRAM2.3: the complex $[Mn(OH)]^{2+}$ (Mn(III) form) is neglected in CAPRAM2.3˙even if this form predominates for the acidic pH calculated in the scenarios. In the present study, the reaction between $[Mn(OH)]^{2+}$ and Fe^{2+} is faster than the reaction between $[Mn(OH)]^{2+}$ and H_2O_2.

In the three scenarios, we found that copper is totally oxidised to Cu(II) by dissolved oxygen. In CAPRAM2.3, this reaction is not considered and only 60% of copper is oxidised by Fe(III) during day time.

3.2. Aqueous phase OH radical

OH is one of the most important radical in both gas and aqueous phases. Figure 3 represents the relative contribution of the corresponding sources and sinks for OH radical for the three cases at noon, the second day. In the remote and marine cases, the transfer from the gas phase is the most important source for OH radicals (80 and 97 % respectively). In the urban case, it only represents 41 % of the total OH production. Indeed, the photolysis of Fe(III) complexes (contributes around 30% to the production of the OH radical in this scenario. The Fenton-type reactions of metals ions (Fe^{2+} and Cu^+) also lead to the production of OH in significant amount in the continental cases (respectively 27% and 11% in the urban and remote cases). These results are directly related to the initial concentrations of TMI. In the urban case, Fe(III) initial concentration is ten times larger than in the remote scenario and one hundred times larger than in the maritime case. Concerning the OH destruction, the oxidation of organic species are an important sink, especially in the continental cases where initial concentrations of organic species in the gas phase are one order of magnitude larger than in the marine case. Another important destruction of OH radical occurs via the equilibrium between OH plus Cl^- and $ClOH^-$ which is shifted towards production of $ClOH^-$. This contribution is more important in the marine case than in the two continental cases because, in the marine case, initial concentration of chlorine in the gas phase and in the aqueous phase are larger than in the continental cases. Reaction between OH and Fe(II) also contributes to the destruction of OH in the continental cases where initial TMI concentrations are larger.

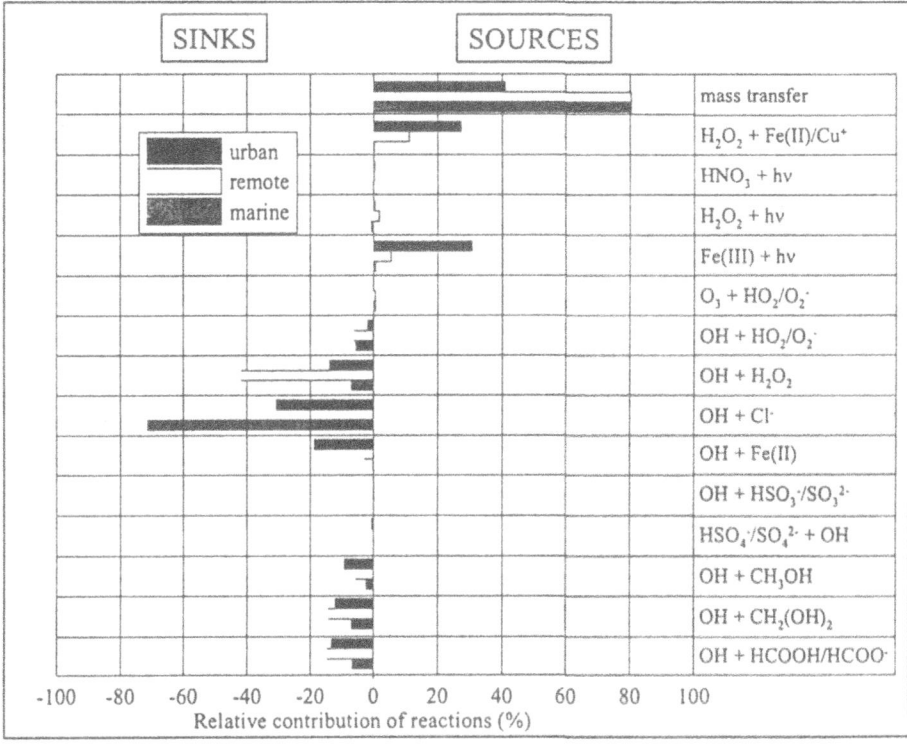

Figure 3: Sinks and sources for OH_{aq} for urban, remote and marine conditions at noon 2nd day. Values of production and loss fluxes are identical and are in [M s^{-1}]: urban $5.7 \ 10^{-8}$, remote $3.1 \ 10^{-8}$, marine $2.4 \ 10^{-8}$.

Concentrations for OH radicals computed in M2C2 seem quite reasonable since they are in the range of previous modelling studies (Herrmann et al., 2000; Warneck, 2000; Jacob et al., 1989). Going into details of the scenarios simulated by CAPRAM2.3, one can observe that OH is more important in the urban case while Herrmann found it more important in the marine case. This has to be related to the contribution of Fe(III) complexes photolysis in the OH production which did not appear as an OH source in CAPRAM2.3. Indeed, in CAPRAM2.3, the reaction between Cu(I) and O_2 is not considered whereas in M2C2 this reaction prevails on reaction of Cu(I) with Fe(III) in the conversion of Cu(I) to Cu(II). Therefore, in Herrmann et al. (2000), even if photolysis of Fe(III) complexes is considered, it cannot contribute significantly to $OH_{(aq)}$ production, since Fe(III) is totally reduced by Cu(I). In the present work, there is a diurnal variation of Fe(II)/Fe(III) ratio with a destruction of Fe(III) during the day by photolysis processes.

4. DISCUSSION

The original features of the updated mechanism introduced in M2C2 are now reviewed in order to highlight new findings and to discuss them not only in the context of Herrmann et al. (2000) scenarios but also against other available model results and field measurements.

The analysis of TMI chemistry in M2C2 shows a diurnal variation of the Fe(II)/Fe(III) ratio due to the Fe(III) photolysis, which was not observed by Herrmann et al. (2000). Some measurements of the iron speciation in atmospheric liquid phase are available and exhibit a very high variability in the Fe(II)/Fe(III) ratio. This information about speciation is important because it is directly linked to the efficiency of atmospheric redox cycles in atmospheric liquid phase. The results obtained with M2C2 model for the percentage of Fe(II) towards dissolved Fe for the three cases are given in Table 1 as well as field measurements in clouds found in the literature. Measurements indicate higher values during the day than during the night. This particular feature is quite well reproduced by M2C2 model, contrary to CAPRAM 2.3, even if the model seems to underestimate night time values. In conclusion, results from M2C2 on the Fe(II)/Fe(III) ratio agree quite favourably with the available measurements and is able to reproduce the natural diurnal variability of this ratio among various scenarios.

The analysis of the major sinks and sources for OH in the aqueous phase shows the importance of the TMI chemistry in the concentration levels of the OH radical, especially due to the photolysis of Fe(III) complexes and to the Fenton-type reactions of metal ions (Fe^{2+} and Cu^+) in the continental scenarios.

	day			night		
	M2C2	CAPRAM2.3	measured	M2C2	CAPRAM2.3	measured
urban	80%	99.7%	90% [*] [a] 70-90% [b] 40-80% [d] 62% [f]	5%	99.7%	20-60% [*] [a] 0[#]-40% [d] 45-60% [e] 50% [f]
remote	80%	98%	76% [c]	20%	98%	26-55% [c]
marine	80%	85%	60-80% [S] [c]	30%	85%	no value

[*] percentage of Fe(II) out of total amount of Fe
[#] Fe(II) is below detection limits
[S] marine air mass with anthropogenic influence
[a] Behra and Sigg. 1990. [b] Pehkonen et al.. 1992. [c] Erel et al.. 1993. [d] Sedlak et al.. 1997.
[e] Schwanz et al.. 1998. [f] Deutsch et al.. 2001.

Table 1. Percentage of Fe(II) out of dissolved Fe during the day and during the night for three scenarios and from available measurements

5. CONCLUSION

This study presents new developments of the multiphase model of cloud chemistry M2C2 including the incorporation of transitions metal ions and the capability to use variable photolysis frequencies in aqueous phase. This new version of the model has been applied to three different chemical scenarios in order to evaluate the impact of transitions metal ions on multiphase chemistry. These three scenarios represent urban, remote and marine conditions following Herrmann et al. (2000).

The chemistry of transition metal ions, of OH radical is analyzed in details with special focus on new developments which leads to differences with Herrmann et al. (2000) model. The main difference concerns the Fe(II)/Fe(III) ratio, which is compared directly with available measurements and shows that the M2C2 model is able to represent the natural diurnal variability of this ratio.

6. REFERENCES

Anastasio, C., Faust, B.C., and Allen, J.M., 1994, Aqueous phase photochemical formation of hydrogen peroxide in authentic cloud waters, *J. Geophys. Res.* 99: 8231-8248.

Behra, P., and Sigg, L., 1990, Evidence for redox cycling of iron in atmospheric water droplets, *Nature* 344: 419-421.

Berry, E.X., and Reinhardt, R.L., 1974, An analysis of cloud drops growth by collection., *J. Atmos. Sci.* 31: 1814-2135.

Desboeufs, K.V., Losno, R., Vimeux, F., Cholbi, S., 1999, The pH-dependent dissolution of wind-transported Saharan dust, *J. Geophys. Res.* 104-D17: 21287-21299.

Deutsch, F.; Hoffmann, P., Ortner, H.M., 2001, Field Experimental Investigations on the Fe(II)- and Fe(III)-Content in Cloudwater Samples, *J. Atmos. Chem.* 40-1: 87-105.

Erel, Y., Pekhonen, S.O., and Hoffmann, M., 1993, Redox Chemistry of Iron in fog and stratus clouds, *J. Geophys. Res.* 98: 18423-18434.

Facchini, M.C., Cloud, 2002, Atmospheric Chemistry and Climate, *IGACtivities Newsletter*, 26.

Faust, B.C., 1994, Photochemistry, fogs, and aerosols, *Environ. Sci. Technol.* 28-5: 217.

Herrmann, H., Ervens, B., Jacobi, H.–W., Wolke, R., Nowacki, P., and Zellner, R., 2000, CAPRAM2.3: A chemical aqueous radical mechanism for tropospheric chemistry, *J. Atmos. Chem.* 36: 231-284.

Hoigné, J., Zuo, Y., Nowell, L., 1994, Photochemical reactions in atmospheric waters: role of dissolved iron species, In: Helz, G.R., Zepp, R.G., Crosby, D.G. (Eds), *Aquatic and Surface Photochemistry.*, Boca Raton, Lewis Publishers, pp.75-84, and references cited therein.

Jacob, D.J., Gottlieb, E.W., Prather, M.J., 1989, Chemistry of polluted cloudy boundary layer, *J. Geophys. Res.* 94: 12975-13002.

Jacob, D.J., 2000, Heterogeneous chemistry and tropospheric ozone, *Atmos. Environ*, 34: 2131-2159.

Jacobsen, F., Holcman, J., and Sehested, K., 1997a, Activation parameters of ferryl ion reactions in aqueous acid solutions, *Int. J. Chem. Kinet.* 29: 17-24.

Jacobsen, F., Holcman, J., Sehested, K., 1997b, Manganese (II) superoxide complex in aqueous solutions, *J. Phys. Chem.* 101: 1324-1328.

Jacobsen, F., Holcman, J., Sehested, K., 1998a, Oxidation of manganese (II) by ozone and reduction of manganese (III) by hydrogen peroxide in acidic solution, *Int. J. Chem Kinet.* 30: 207-214.

Jacobsen, F., Holcman, J., Sehested, K., 1998b, Reaction of the ferryl ion with some compounds found in cloud water, *Int. J. Chem. Kinet* 30: 215-221.

Jickells, T.D., Knap, A.H., Church, T.M., 1984, Trace metals in Bermuda rainwater, *J. Geophys. Res.* 89-D1: 1423-1428.

Khaikin, G.I., Alfassi, Z.B., Huie, R.E., and Neta, P., 1996, Oxidation of ferrous and ferrocyanide ions by peroxyl radicals, J. Phys. Chem. 100: 7072-7077.

Lelieved, J., and Crutzen,1991, P.J., The role of clouds in tropospheric photochemistry, *J. Atmos. Chem.* 12: 229-227.

Leriche, M., Voisin, D., Chaumerliac, N., Monod, A., and Aumont, B., 2000, A model for tropospheric multiphase chemistry: Application to one cloudy event during the CIME experiment, *Atmos. Environ.* 34: 5015-5036.

Leriche, M., Chaumerliac, N., and Monod, A., 2001, Coupling quasi-spectral microphysics with multiphase chemistry: A case study of a polluted air mass at the top of the Puy de Dôme mountain (France), *Atmos. Environ.* 35: 5411-5423.

Leriche, M., Deguillaume, L., Chaumerliac, N., 2003, Modeling study of strong acids formation and partitioning in a polluted cloud during wintertime, *submitted to J. Geophys. Res.*

Lim, B., Jickells, T.D., Colin, J.L., and Losno, R., 1993, Solubilities of Al, Pb, Cu, and Zn in rain sampled in the marine environment over the North Atlantic Ocean and Mediterranean Sea, *Global Biogeochemical Cycles* 8: 349-362.

Logager, T., Holcman, J., Sehested, K., and Petersen, T, 1992, Oxidation of ferrous ions by ozone, *Inorg. Chem.* 31: 3523-3529.

Mc Elroy, W. J., Waygood, S.J., 1990, Kinetics of the reactions of the SO_4^- radical with SO_4^-, $S_2O_8^{2-}$, H_2O and Fe^{2+}, *J. Chem. Soc. Faraday Trans.* 86: 14, 2557-2564.

Madronich, S. and J.G. Calvert, 1990, The NCAR Master Mechanism of the gas phase chemistry, *NCAR technical Note*, TN-333+SRT, Boulder Colorado.

Madronich S., and Flocke., S.,1999, The role of solar radiation in atmospheric chemistry. in *Handbook of Environmental Chemistry* (P. Boule, ed.), Springer-Verlag, Heidelberg, pp.1-26.

Nimmo, M., Fones, G.R., 1997, The potential pool of Co, Ni, Cu, Pb and Cd organic complexing ligands in coastal and urban rain waters, *Atmos. Environ.* 31-5: 693-702.

Pehkonen, S.O., Erel, Y., Hoffmann, M.R., 1992, Simultaneous spectrophotometric measurement of Fe(II) and Fe(III) in atmospheric water, *Environ. Sci. Technol.* 26-9: 1731-1776.

Piechowski, M. von, Nauser, T., Hoignè, T., and Buhler, R. E., 1993, O_2^- decay catalysed by Cu^{2+} and Cu^+ ions in aqueous solutions : a pulse radiolysis study for atmospheric chemistry, *Ber. Bunsenges. Phys. Chem.* 97: 762-771.

Ross, H.B., 1987, Trace metals in precipitation in Sweden, *Water Air and Soil Pollution* 36: 349-353.

Rosseinsky, D.R., 1963, The reaction between Mercury(I) and Managanese(II) in aqueous perchlorate solution, *J. Chem. Soc.*: 1181-1186.

Ruggaber, A., Dlugi, R., Bott, A., Forkel, R., Hermann, H., and Jacobi, H.-W., 1997, Modelling of radiation quantities and photolysis frequencies in the aqueous-phase in the troposphere, *Atmos. Environ.* 31: 3137-3150.

Rush, J.D., and Bielski, B.H.J., 1985, Pulse radiolytic studies of the reactions of HO_2/O_2^- with Fe(II)/Fe(III) ions. The reactivity of HO_2/O_2^- with ferric ions and its implication on the occurrence of the Haber-Weiss reaction, *J. Phys. Chem* 89: 5062-5066.

Schwanz, M., Warneck, P., Preiss, M., Hoffmann, P., 1998, Chemical speciation of iron in fog water, *Contr. Atmos. Phys.* 71-1: 131-143.

Schwartz, S.E., 1986, Mass-transport considerations pertinent to aqueous phase reactions of gases in liquid water clouds, in: *Chemistry of Multiphase Atmospheric Systems*, Jaeschke, W. Eds., NATO ASI Series, G6, Spinger-Verlag, pp. 415-471.

Sedlak, D.L., Hoignè, J., David, M.M., Colvile, R.N., Seyffer, E., Acker, K., Wiepercht, W., Lind, J.A., and Fuzzi, S., 1997, The cloudwater chemistry of iron and copper at Great Dun Fell, U.K., *Atmos. Environ.* 31: 2515-2526.

Sheng, H.L., 1993, *J. Chem. Tech. Biotechnol.*, 56: 163-167.

Siefert, R.L., Pehkonen, S.O., Erel, Y., Hoffmann, M.R., 1994, Iron photochemistry of aqueous suspensions of ambient aerosol with added organic acids, *Geochimica et Cosmochimica Acta* 58-15: 3271-3279.

Spokes, L.J., Jickells, T.D., and Lim, B., 1994, Solubilisation of aerosol trace metals by cloud processing : a Laboratory Study, *Geochimica and Cosmochimica Acta* 58-15: 3281-3287.

Spokes, L.J., Lucia, M., Campos, A.M., and Jickells, T.D., 1996, The role of organic matter in controlling copper speciation in precipitation, *Atmos. Environ.* 30: 3959-3966.

Stuglik, Z., Zagorski, Z.P., 1981, Pulse radiolysis of neutral iron(II) solutions: oxidation of ferrous ions by OH radicals, *Radiat. Phys. Chem.* 17: 229.

Sulzberger, B., Laubscher, H., Karametaxas, G., 1994, In: Helz, G.R., Zepp, R.G., Crosby, D.G. (Eds), *Aquatic and Surface Photochemistry.*, Lewis Publishers, Boca Raton, pp. 53-73, and references cited therein.

Sulzberger, B., Laubscher, H., 1995, Reactivity of various types of iron (III)(hydr)oxides towards light-induced dissociation, *Marine Chemistry* 50: 103-115.

Warneck, P., 2000, *Chemistry of the natural atmosphere*, International Geophysics Series, Vol. 71, Academic Press, pp. 927.

Zuo, Y., and Hoignè, J., 1992, Formation of hydrogen peroxide and depletion of oxalic acid in atmospheric water by photolysis of iron(III)-oxalato compounds, *Environ. Sci. Technol.*, 26: 1014-1022.

DISCUSSION

D. MICHELANGELI   You have selected a particle radius of 1μm, which is small for cloud drops. Did you do any sensitivity studies to investigate the effect of changing drop size?

L DEGUILLAUME   For this particular study on TMI, droplet radius is set up to 1μm in order to get the maximum efficiency for aqueous phase reactivity.
In a previous study (Leriche et al., 2001) sensitivity tests were performed and concluded to a very high dependency of aqueous chemistry upon drop radius. Finally, the chemistry model presented is actually coupled with microphysical model in which the radius of the cloud droplet is varying following microphysical conversions (Leriche et al., 2003).

POSTER SESSION

MODELLING PLUME RISE AND DISPERSION OF POWER PLANT FLUE GASES DISCHARGED THROUGH COOLING TOWERS

Öznur Oğuz, Coşkun Yurteri and Gürdal Tuncel[*]

1. INTRODUCTION

Conventionally, treated flue gases of fossil-fueled thermal power plants are discharged through specially designed stacks in order to increase atmospheric dispersion. One of the innovative technologies involves the discharge through cooling towers instead of conventional tall stacks. The so-called "cooling tower discharge" (CTD) technology has proven to provide better dispersion of the power plant plume in the atmosphere (Schatzmann et al., 1984; Glamse et al., 1989; Ernst et al., 1986). Owing to its natural buoyancy and large heat content, the cooling tower plume mixed with the treated flue gas reaches considerably greater heights in the atmosphere. The CTD plume can even penetrate through the inversion layers (Ernst et al., 1986). The CTD technology, therefore lead a considerable decrease in the GLC values of pollutants by providing a better dispersion in the atmosphere (Ernst et al., 1986; Schatzmann et al., 1987).

The purpose of this study is to evaluate influence of the CTD application on air quality impacts of the Afşin-Elbistan B Thermal Power Plant.

2. METHODOLOGY

A real case that is the proposed Afşin-Elbistan B Thermal Power Plant (AEBTPP) having a CTD application was evaluated in this study. AEBTPP was proposed to construct with a generating capacity of 4x350 MW in order to utilize lignite reserves in the Afşin and Elbistan districts, which are located in Kahraman Maraş Province, Türkiye. Proven lignite deposit in the region is 3,357 million tons. The Afşin-Elbistan A Thermal Power Plant (AEATPP) already locates in the region with four units each with an installed capacity of 340 MW.

---

[*] Öznur Oğuz and Prof. Gürdal Tuncel, Middle East Technical University, Department of Environmental Engineering 06531 Ankara, Türkiye. Prof. Coşkun Yurteri, ENVY Energy and Environmental Investments Inc., 06450 Ankara, Türkiye.

Air Pollution Modeling and Its Application XVI, Edited by
Borrego and Incecik, Kluwer Academic/Plenum Publishers, New York, 2004

583

Mathematical modeling studies were carried out in order to assess the impacts of AEBTPP on the regional air quality and to investigate the effectiveness of the CTD technology to reduce the ground level concentrations (GLC) of sulfur dioxide (SO_2), nitrogen oxides (NO_x) and particulate matter (PM). The well-known "Industrial Source Complex, Short-Term" (ISCST) dispersion model was utilized for this purpose. The ISCST is one of the most developed computer models used for the estimation of hourly, daily and yearly GLC values of various pollutants with real-time meteorological data. The model uses three different categories of data; namely, (i) meteorological data; (ii) receptor data; and (iii) source parameters. An additional model was used for the cooling tower plume rise calculations. The S/P model is an integral hydrodynamic model for cooling tower plume rise calculations, which was developed by Schatzmann and Policastro (1984).

Modeling studies were conducted under different scenario conditions. In addition to the CTD application, FGD technology and building downwash effects considering cumulative as well as sole impacts of the plants (i.e., AEATPP and AEBTPP) were also evaluated in the scenarios.

3.   RESULTS AND DISCUSSION

The effect of the CTD application at the AEBTPP is thoroughly assessed under different scenario conditions. The results reveal that the use of CTD technology can result in 60% reduction in the GLC values of pollutants emitted from the AEBTPP. Additionally, the maximum GLC values occur away from the plant, on the northeast of the study area by the influence of the prevailing winds.

Cumulative impacts of the AEATPP and AEBTPP were also evaluated. The results of the simulations for this condition have demonstrated that when the existing plant is equipped with an FGD unit, the predicted annual averages of the GLC values of pollutants at the grid points do not exceed "long term limits" in the Turkish Air Quality Control Regulation; however, short-term limits are not complied. Additional runs demonstrated that the reduction of more than 95% in the GLC values of all the modeled pollutants due to the combined effect of taller stack and the FGD unit that are installed in the AEATPP is observed.

The building downwash algorithm of the model has known to have some deficiencies (EPRI, 1997). Thus several runs were carried out to evaluate affect of building downwash algorithm on the predicted GLC values of pollutants. Results demonstrated that the model runs without using building downwash algorithm resulted in much close agreement with the field measurements.

4. CONCLUSION

The effectiveness of the CTD technology to reduce air quality impacts of the power plants was investigated for the AEBTPP case. Mathematical modeling was used as the tool. A well-known dispersion model ISCST and plume rise model S/P were utilized for this purpose. Results of the modeling studies carried out for different scenario conditions demonstrated that the CTD application provides remarkable decrease in the predicted GLC values of pollutants emitted from AEBTPP.

5. REFERENCES

Glamser, J., Eikmeier, M., and Petzel, H.K., 1989, Advanced Concepts in FGD Technology: The Shu Process with Cooling Tower Discharge, *JAPCA*, **39** (9), 1262-1267.

Ernst, G., Leidinger, B.J.G., Natusch, K., Petzel, K.H., and Scholl, G., 1986, Cooling Tower and flue Gas Desulfurization Plant of the Model Power Station Volklingen, *Brennstoff Warme Kraft*, 38(11), 510-511.

Schatzmann, M., Lohmeyer A. and Ortner, G., 1987, Flue Gas Discharge from Cooling Towers: Wind Tunnel Investigation of Building Downwash Effect on Ground-Level Concentrations, *Atmospheric Environment*, **21**(8), 1713-1724.

Schatzmann, M. and Policastro, A.J., 1984, An Advanced Integral Model for Cooling Tower Plume Dispersion, *Atmospheric Environment*, **18**(4), 663-674.

EPRI, 1997, *Plume Rise and Downwash Modeling Project*, Conducted by Electrical Power Research Institute.

BULGARIAN EMERGENCY RESPONSE SYSTEM: STRUCTURE, DESCRIPTION, PERFORMENCE

Dimiter Syrakov, Maria Prodanova, Kiril Slavov[*]

A PC-oriented Emergency Response System (BERS) is developed and works in the Bulgarian National Institute of Meteorology and Hydrology (NIMH). Creation and development of BERS was highly stimulated by the European project ETEX.

BERS comprises of two main parts - the operational and the accidental parts, which are applied to region "Europe"- EU, and region "Northern Hemisphere" - NH. The operational part runs automatically every 12 hours, after new meteorological information is available via GTS of WMO (DWD "Global model" 3-days forecast is used). This module performs the following tasks: preparation of the necessary input meteorological information for both trajectory and dispersion models; creation of archives with meteorological data; trajectory calculations for specific NPP; visualization of results; and uploading the trajectory maps in the NIMH web site for use by interested institutions.

The accidental part is activated manually when a real radioactive release occurred or during emergency exercises. In both cases the source information is provided by the user. Two Bulgarian dispersion models - LED and EMAP are the cores of the accidental part. LED (**L**agrangean-**E**ulerian **D**iffusion) is a typical puff-model [2]. The action of each continuous pollution source is presented as releases of successive clouds (puffs), which are transported and dispersed independently. The movement description is Lagrangean (Euler-forward 3D scheme), and the diffusion, transformation and removals - Eulerian (proper analytical solutions). EMAP (**E**ulerian **M**odel for **A**ir **P**ollution) is realized using the time splitting approach applied both for both processes and directions [1]. The semi-empirical diffusion-advection equations for scalar quantities are solved in terrain-following co-ordinates. Non-equidistant grid spacing is set in vertical. The numerical solution is based on discretization applied on staggered grids. Conservative properties are fully preserved within the discrete model equations. Advective terms are treated with the TRAP scheme [1]. Displaying the same simulation properties as the Bott scheme (explicit, conservative, positive definite, transportable, limited numerical dispersion), it is several times faster. The advective boundary conditions are zero for income and "open boundary" for outcome flows. Turbulent diffusion is described by means of the simplest

---

[*] National Institute of Meteorology and Hydrology, 66 Tzarigradsko chaussee Bulvd.,
Sofia 1784, Bulgaria, E-mail: Dimiter.Syrakov@meteo.bg

Air Pollution Modeling and Its Application XVI, Edited by
Borrego and Incecik, Kluwer Academic/Plenum Publishers, New York, 2004

587

schemes – explicit in horizontal and implicit in vertical. The bottom boundary condition for the vertical diffusion equation is the dry deposition flux; the top boundary condition is optionally "open boundary" and "hard lid" type. The lateral boundary conditions for diffusion are "open boundary" type. The wet deposition is also taken into account by means of the simplest decay scheme. In the surface layer a parameterization is applied permitting to have the first computational level at the top of SL [3]. It provides a good estimate for the roughness level concentration and accounts also for the action of continuous sources on the earth surface. The outcome from these models is the ground-level concentration and accumulated deposition; concentration and deposition maps being upload to specialized ftp-site for use by the interested institutions.

BERS took part in RTMOD project and now is actively participating in the ENCEMBLE project (http://ensemble.ei.jrc.it) supported by the European Commission DG-RTD Nuclear Fission Program (Contact point <stefano.galmarini@jrc.it>). Examples of BERS performance (the best and the worst ones) are shown in Fig.1 and 2 as captured by ENSEMBLE web-sight. The FMS (Figure of Merit in Space) estimate gives the part of overlapping area to all polluted area.

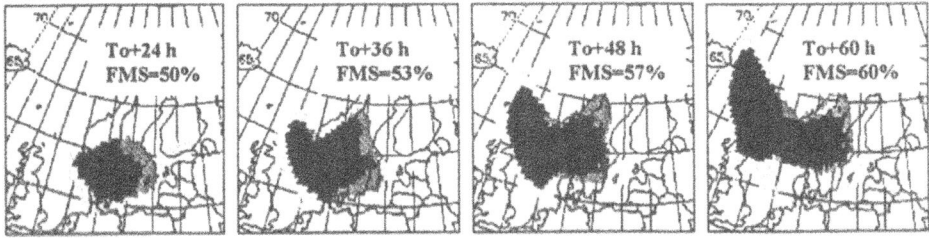

Figure 1. BERS simulation of ENSEMBLE dry run No.5 – release from Stockholm. Light gray - 9 EU models average prediction; dark gray - BERS prediction; black – overlapping. Concentration threshold - 0.01 Bq/m³.

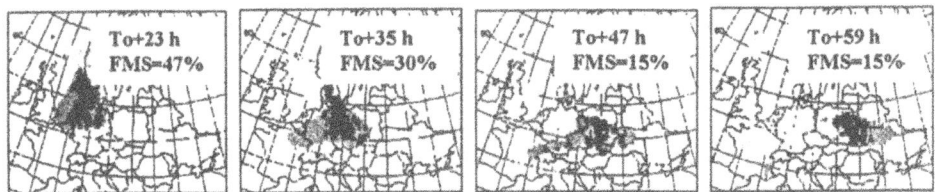

Figure 1. As in Fig. 1 but for ENSEMBLE dry run No.5.

The presented results prove that good emergency response can be produced on the base of personal computers using meteorological information, distributed via GTS. This can be useful for warning-system's development in the east European countries without own numerical weather forecast.

REFERENCE

1. Syrakov, D., 1995, On a PC-oriented Eulerian Multi-Level Model for Long-Term Calculations of the Regional Sulphur Deposition, in *Air Pollution Modelling and its Application XI*, **21**, Gryning S.E. and Schiermeier F.A., eds., Plenum Press, N.Y. and London, pp. 645-646.
2. Syrakov, D., Djolov, D. and Yordanov, D., 1983, Incorporation of planetary boundary layer dynamics in a numerical model of long-range air pollution transport, *Boundary Layer Meteorology*, 26, pp. 1-13.
3. Syrakov, D. and Yordanov, D., 1997, Parameterization of SL Diffusion Processes Accounting for Surface Source Action, *Proc. of 22nd NATO/CCMS International Technical Meeting on Air Pollution Modelling and its Application*, 2-6 June 1997, Clermont-Ferrand, France, pp. 111-118.

AIR QUALITY IMPACT OF INDUSTRIAL PLANTS (TEAP)

Vidmantas Ulevicius, Kestutis Senuta, Kristina Plauskaite[*]

1. INTRODUCTION

Environmental problems are growing in complexity and scope. Local management solutions alone can no longer address many of today's problems. Awareness of the significant influence that human activities have on natural environment and ecosystems is increasing.

The air quality impact of the industrial emissions is one of the most important areas of research in the last years due to public interest of environmental issues and also from the view of industrial approach. Knowing of different air concentrations in the area where an industrial plants is located in real-time and forecasting opens new possibilities for the public, environmental authorities and industrial managers.

The TEAP project is focusing on developing an integrated software tool to assist industrial plants on assessing the air quality impact of their emissions on the surrounding areas and also to optimize the cost/effective balance and also the production processes based on the air quality forecasts (Byun et al., 1998).

It was developed basic modules of an air quality impact management system for Lithuanian industrial plants by taking into consideration the characteristics of these industrial plants on emissions and processes. The system is using meteorological data from satellite, MM5 (mesoscale meteorological model) and CMAQ (community multiscale air quality modelling system) and working in almost on real time and allow to take appropriate actions on industrial production to avoid the exceedences on pollution concentrations when the main reason is due to industrial activities and no other alternative method is applied.

2. RESULTS AND DISCUSSION

[*] Vidmantas Ulevicius, Kesutis Senuta, Kristina Plauskaite, Institute of Physics, Vilnius, Lithuania LT-2053.

Air Pollution Modeling and Its Application XVI, Edited by
Borrego and Incecik, Kluwer Academic/Plenum Publishers, New York, 2004

589

In order to confirm the applicability and reliability of the system an exemplary application to the industrial plant was performed. The system was applied for analyzing and evaluating the impact of various air quality regulations concerning technical measures for the reduction of SO_2, NO_X, and VOC emissions from the most important emission sources of the area. These legislative modulations have already been implemented or will be implemented by 2010. Apart from the "business as usual" scenario which assumes full compliance with the European legislation until the year 2010, three hypothetical situations involving 50% reduction of SO_2, NO_X, VOC emissions on top of the "business as usual" scenario were simulated. The assessment was performed for the period 1990-2010, considering the year 1990 as the base case scenario to serve as the reference for the evaluation of the proposed measures.

The maps of exceedances of critical loads for sulphur and nitrogen compounds are also were built. Critical loads of sulphur and nitrogen compounds have been mapped for forest ecosystems, using recently available calculation methods. The results of calculation showed that critical loads of nitrogen compounds lay in the range from 1.7 to 3.0 $gN/m^2 \cdot yr$ with the lowest values in western and southeastern parts of plant. The range of critical loads of sulphur compounds was found to be from 0.9 to 1.5 $gS/m^2 \cdot yr$ with the lowest values in southern and northeastern parts of plant. Comparison of the calculated critical loads with modelling data indicated that exceedances of critical loads of sulphur and nitrogen compounds can be as high as 1 gS (or N)/$m^2 \cdot yr$.

3. ACKNOWLWDGMENTS

The research described in this paper was supported partially by the Lithuanian Foundation for Science and Education. The authors gratefully thank for this assistance.

4. REFERENCES

Byun, D. W., Young, J., Gibson, G., Godowitch, J., Binkowski, F., Roselle, S., Benjey, B., Pleim, J., Ching, J. K. S., Novak, J., Coats, C., Odman, T., Hanna, A., Alapaty, K., Mathur, R., McHenry, J., Shankar, U., Fine, S., Xiu, A., and Jang, C., 1998, Description of the Model-3 Community Multiscale Air Quality (CMAQ) model, *Proc. of the American Meteorological Society 78th Annual Meeting.* (Phoenix, AZ, Jan 11-16, 1998), pp. 264-268.

NUMERICAL SIMULATION OF PHOTOCHEMICAL EPISODES IN ALPINE VALLEYS

Guillaume Brulfert, Jean-Pierre Chollet, Charles Chemel, and Marie-Aurélie Kerbiriou*

1. INTRODUCTION

Alpine valleys are sensitive to air pollution due to emission sources (traffic, industries, individual heating), morphology (narrow valley surrounded by high ridge), local meteorology (temperature inversions and slope winds). Such situations are rarely investigated with specific research programs taking into account detailed atmospheric chemistry coupling gas and aerosol phases. Following the accident under the Mont Blanc tunnel on March 24th, 1999, international traffic between France and Italy was stopped through the Chamonix Valley (France). The heavy-duty traffic (about 2130 trucks per day) has been diverted to the Maurienne Valley, with up to 4250 trucks per day. The program POVA (Pollution des Vallées Alpines) started in May 2000, with the objective to analyse air quality and develop atmospheric modelling in each of the two valleys, in order to study impact of traffic and local development scenarios.

2. MODELS

Atmospheric prediction model ARPS 4.5.2 (Advanced Regional Prediction System), developed at CAPS (Center for Analysis and Prediction of Storms) (Xue, 2000), enables to resolve atmosphere dynamics above complex terrain. A large (250 * 300 kms) field is driven by the ALADIN verticals of the French Meteorological Office and forces the field of interest (25 * 25 kms). This model is coupled off-line to the TAPOM 1.5.2 code of atmospheric chemistry (Transport and Air POllution Model) developed at the LPAS of the EPFLausanne (Clappier, 1998; Gong and Cho, 1993). TAPOM uses the Regional Atmospheric Chemistry Modeling (RACM) scheme (Stockwell et al.,1997). 300-meters grid cells to calculate dynamics and reactive chemistry make possible to represent accurately dynamics in the valley (slope winds) (Anquetin et al., 1999) and to process chemistry at fine scale.

*G. Brulfert, J.P. Chollet, C. Chemel, and M.A. Kerbiriou, Laboratoire des Ecoulements Géophysiques et Industriels, Université Joseph Fourrier, Institut National Polytechnique de Grenoble et Centre National de la Recherche Scientifique, BP 53, 38041 Grenoble Cedex 9, France. brulfert@hmg.inpg.fr

Air Pollution Modeling and Its Application XVI, Edited by
Borrego and Incecik, Kluwer Academic/Plenum Publishers, New York, 2004

3. SIMULATION OF PHOTOCHEMICAL EPISODES IN CHAMONIX VALLEY

The emission inventory is based on the CORINAIR methodology and SNAPS's codes, uses a 100x100m grid and includes information (land use, population, traffic, industries,..) gathered from administrations and field investigations. The yearly inventory is turned into hourly emissions for a set of RACM data in specific cases (summer/winter, week day/week end/holidays, ...) to be used in scenarios (Middleton et al., 1990). Then, different scenarios are tested in the valley of Chamonix with various sources: with and without road traffic, with and without heavy vehicles, for different meteorological cases. Sunny days, without synoptic wind, explain the contribution of local sources of pollution with mixing by slope winds. Days with a constant synoptic wind in the bottom of the valley gives an insight into the pollution background and boundary conditions species. Dispersion of a passive tracer enables to describe how primary pollutants are either mixed inside the valley or transported beyond the limits of the valley. Preliminary results, without comparison with campaign data, show that typical species of secondary pollutants (HNO3, PAN) depend on road traffic and heavy –vehicles.

Several sensitivity tests are being performed, based on real variation of the gases emissions and background. The first results are very encouraging, when compared to fields measurements.

4. CONCLUSION

The objective of the program is now to estimate the respective impacts of the various emissions sources present in the valleys and evaluate the variability in air concentrations according to changes in environmental conditions. Then the same study will begin for the other valley of POVA program (Maurienne valley) and the impact of road traffic will be evaluated.

5. ACKNOWLEDGEMENTS

The program POVA is supported by Région Rhône Alpes, ADEME, METL, MEDD. Meteorological data are provided by Météo France, traffic data by STFTR, ATMB, DDE Savoie et Haute Savoie. Computations were done on MIRAGE (plate-forme grenobloise de modélisation numérique de l'environnement et du climat).

6. REFERENCES

Anquetin, S., Guilbaud, C., Chollet, J.P., 1999. Thermal valley inversion impact on the dispersion of a passive polluant in a complex mountainous area, *Atmospheric Environnement 33*, pp. 3953-3959.

Clappier, A., 1998, A correction method for use multidimensionnal time splitting advection algorithms : application to two and three dimensional transport, *Monthly Weather Revue 126*, pp. 232-242.

Gong, W. and Cho, H-R (1993) A numerical scheme for the integration of the gas phase chemical rate equations in a three-dimensional atmospheric models, *Atmospheric Environment 27A(14)*, pp. 2147-2160.

Middleton, P., Stockwell, W.R., Carter, W.P.L., 1990. Aggregation and analysis of volatile organic compound emissions for regional modelling, *Atmospheric Environnement 24A*, pp. 1107-1133.

Stockwell, R., Kirchner, F., Kuhn, M., Seefeld, S., November 27, 1997. A new mechanism for atmospheric chemistry modeling, *Journal of Geophysical Research*, vol. 102, No. D22, pp. 25,847-25,879.

Xue, M., Droegemeir, V., Wong, V., 2000. The Advanced Regional Prediction System (ARPS)- A multi-scale nonhydrostatic atmospheric simulation and prediction model. Part I : Model dynamics and verification, *Meteorology and atmospheric physics*, volume 75, Issue 3/4, pp. 161-193.

AIR FLOWS AND POLLUTION TRANSPORT IN THE SOFIA VALLEY UNDER SOME TYPICAL BACKGROUND CONDITIONS

K. Ganev, R. Dimitrova and N. Miloshev[*]

1. INTRODUCTION

The present study demonstrates some specific characteristics of the local flow system and pollution transport in the region of Sofia valley. The city of Sofia is situated in the bottom of a deep valley between high surrounding mountains. The study of topography induced mesoscale effects is interesting itself and this, together with the great social and economic importance of the region, is the motivation for this work.

2. NUMERICAL SIMULATION

A tree-dimensional quasi-hydrostatic mesoscale dynamic model[1,2] is used for studying the mesoscale flow systems under typical synoptic conditions.

The IMSM[3] (a 3D Eulerian model) is applied for the pollution transport simulations. The calculations are carried out in a domain with horizontal resolution of 1.5km.

A case with very stable background atmosphere, in which the mesoscale effects are very well displayed, is chosen as an example. The blocking mountain effects can be seen in the case of south background wind (Figure 1). At both the levels the north (opposite to the background wind) wind components dominate over the whole valley. The zone of the calm conditions can be seen in front of a north slope of the Vitosha mountain.

Real emission data for the major high point sources, situated in the region and for area sources in the city of Sofia is used as an input to the IMSM. The SO_2 concentration fields that correspond to the circulation from Figure 1 are shown in Figure 2. The tendency of pollution convergence in front of the north Vitosha slope is the reason why the sulfur pollution in that region is almost as big as around the large point sources in Pernik and Kremikovci (Figure 2.a). A very well known climatic phenomenon – flow channeling along the Vladaia canyon (wind rotation around the west mountain slope and directing of the flow from Pernik towards Sofia) can be seen in both the wind (Figure 1) and air pollution (Figure 2) fields.

[*] K. Ganev, R. Dimitrova, N. Miloshev, Institute of Geophysics, Bulgarian Academy of Sciences, Acad. G. Bonchev str., bl.3, Sofia 1113, Bulgaria

Air Pollution Modeling and Its Application XVI, Edited by
Borrego and Incecik, Kluwer Academic/Plenum Publishers, New York, 2004

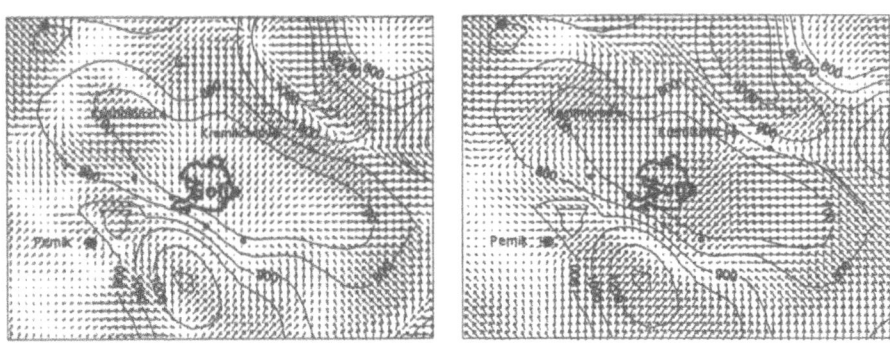

Figure 1. The simulated wind field at the level of 10m (a) and 300m (b) under S (2m/s) background wind.

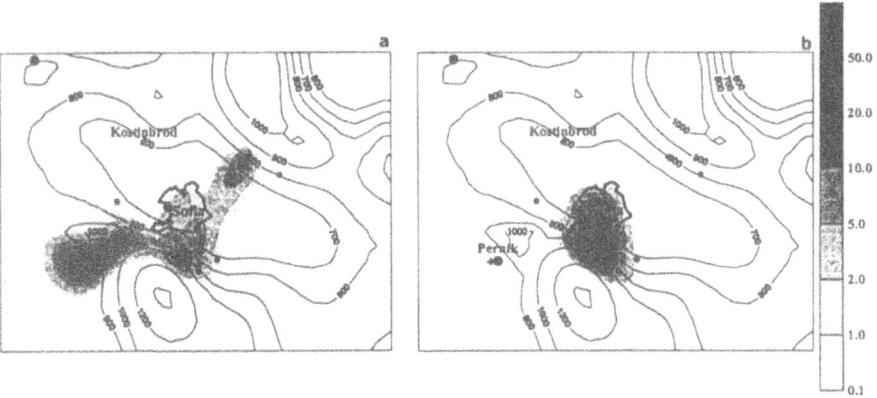

Figure 2. The vertically averaged concentration fields [$\mu g/m^3$] from high point (a) and area (b) sources.

3. CONCLUSIONS

The numerical experiments performed demonstrate quite a good qualitative agreement of the simulated wind fields and the results from the laboratory and the field experiments. In most of the cases the pollutants from the major sources and those from the Sofia surface sources settle in front of the north Vitosha slope. The mesoscale effects dominate the formation of the pollution pattern in the Sofia Valley.

4. REFERENCES

1. K. Ganev, Numerical study of the local flow systems in the "Kozloduy" NPP region - some preliminary results, *Bulg. Geophys. J.*, **XIX**, 9-23 (1994).
2. R. Dimitrova, Air flows and pollution transport in the Sofia Valley under highly stable background conditions, *Bulg. Geophys. J.*, **XXVII**, 98-110 (2001).
3. E. Syrakov and K.Ganev, On the accounting for some sub-scale effects in the long range air pollution modelling over a complex terrain, *XXIII International Technical Meeting on Air Pollution Modelling and its Applications, 28.09-02.10.1998, Varna, Bulgaria,* Kluwer Academic/Plenum Publ. Corp., 107-115 (1998).

AN APPLICATION OF SCHWARZ-CHRISTOFFEL TRANSFORMATION TO GRID GENERATION IN 2-D ATMOSPHERIC FLOW MODELING

H.Erdun[*], S. İncecik[**], M.O.Kaya[***] and İ.Özkol[***]

1. INTRODUCTION

Numerical modeling for dispersion of air pollution over a complex topography is achieved through usage of non-orthogonal grid structure. This stems mainly from the fact that a non-orthogonal system (for example, terrain-following height coordinate system) instead of an orthogonal one is more suitable for development of a computer code.

The use of coordinate transformation for the air pollution applications has been limited for the simplest types of transformations. The only vertical coordinates are transformed to coincide with the lower boundary.

In order to generate well fitted two-dimensional grid mesh that is most suitable to represent atmospheric domain with irregular lower boundary due to topography, it is possible to use conformal mapping techniques (Sharman, R.D., et all, 1988). In the conformal transformation, the domain is transformed into one of the simple geometries (i.e. rectangle) on which the solution is easy. In this paper we preferred to use one of the well-known conformal mapping techniques, the Schwarz-Christoffel Transformation (SCT).

2. DEFINITION OF TOPOGRAHY AND GRID GENERATION

A coordinate generation method for use in the atmospheric flow modeling is given here which is much simpler, more accurate, and more flexible than currently existing methods. This approach is based on numerical integration of SCT for general curved lines or surfaces. In addition, this method directly provides the two-dimensional incompressible potential flow solution for flow past complex shapes including flows with free streamlines. In the atmospheric flow modeling, the usage of SCT is introduced by Erdun et al., (1996).

---

[*] BNP-AK-Dresdner Bank AŞ, Büyükdere Cad. 1.Levent Plaza, No: 173, 1.Levent, Istanbul, Turkey
[**] Istanbul Technical University, Department of Meteorology, Faculty of Aeronautics and Astronautics, 34469 Maslak, Istanbul, Turkey
[***] Istanbul Technical University, Department of Aeronautics, Faculty of Aeronautics and Astronautics, 34469 Maslak, Istanbul, Turkey

Air Pollution Modeling and Its Application XVI, Edited by
Borrego and Incecik, Kluwer Academic/Plenum Publishers, New York, 2004

595

SCT for polygons is well known in literature (e.g. Carrier et al., 1966) The extension to curved surfaces is less known and is absent from most texts.

The most familiar form of the SC Theorem is given by

$$\frac{dy}{d\lambda} = M \prod_{i=1}^{n} (\lambda - a_i)^{\alpha_i/\pi}$$

This formula represents a conformal mapping of the upper half of the· λ-plane onto a region into the interior of the polygon shown in the y-plane (Erdun, et al., 1996). The quantity n relates to the number of corners, M is complex constants to be determined, and the values of points a_i's are real constants also to be determined ($i=1...n$). The α_i's are the outside angles of the corners, defined positive for clockwise rotation when the curve is followed in the direction of the enumeration of the corners, where the enumeration of the a_i in the λ-plane is taken from left to right.

3. AN APPLICATION OF SHWARZ-CHRISTOFFEL TRANSFORMATION

Conformal mapping with SCT can be viewed as a coordinate generation as well as a boundary integral method. Generated coordinates for several types topographical shapes are shown in Fig 1.

SCT can be used in all 2-dimensional models with various scales from micro to planetary scale. However, in micro scales with phenomenon where topographical effects are more important, SCT plays a critical important role in order to get better model results.

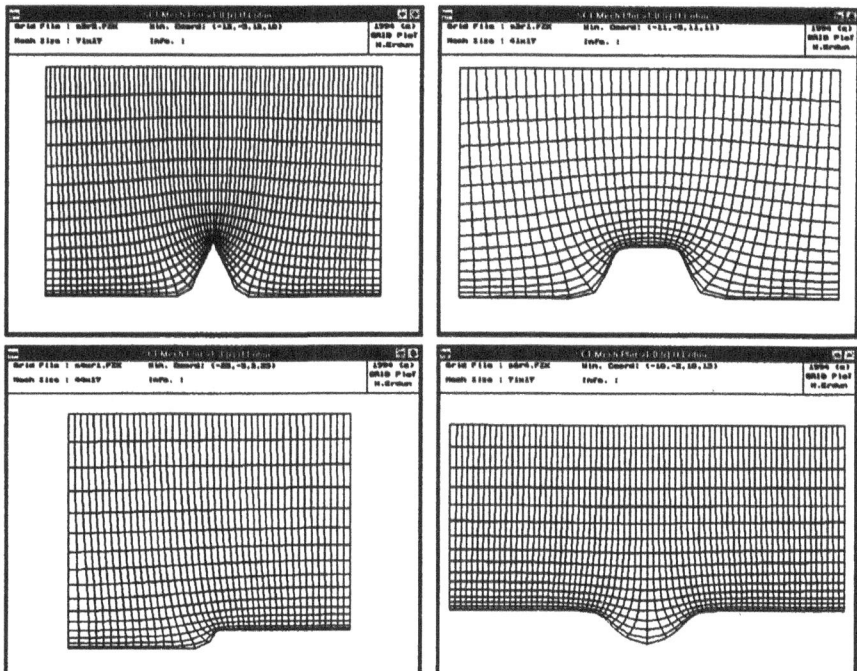

Fig 1. Generated grid meshes by using SC transformation.

In this study, we have used the mean momentum, continuity and heat equations for shallow convection system. Since the model is designed for two-dimensional, the local derivatives in y-direction are removed from the model

596

equations. The basic dynamical model equations can be solved on a finite grid. The details on model equations are given in Erdun, 1996.

In order to test the model a modified form of *"Witch of Agnesi"* mountain profile (e.g., Bacmeister and Schoeberl, 1989) as topography is used. This mountain is steeper than original profile. The results are shown in Fig 2.

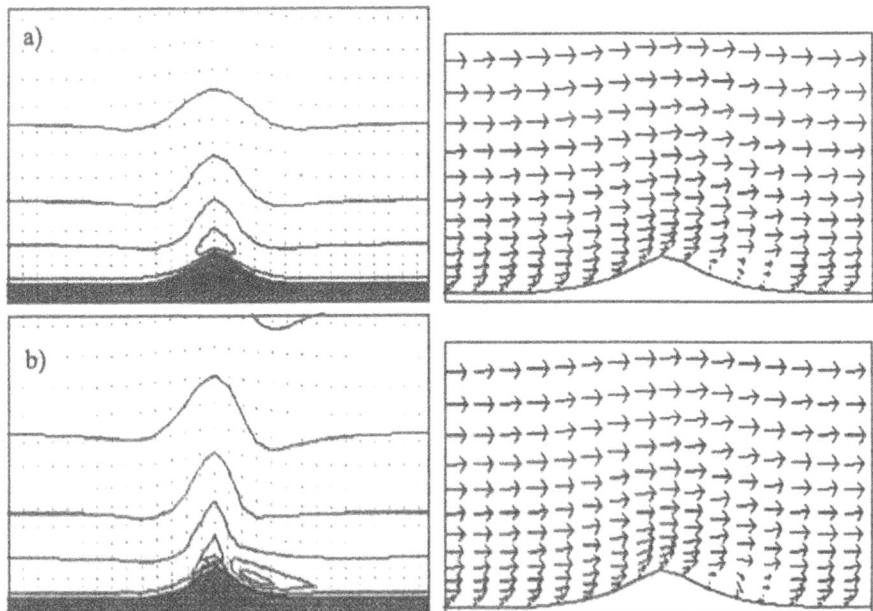

Fig 2. Calculated mixing coefficient fields and wind fields at time step 2 (a) and 1250 (b).

4. CONCLUSIONS

SCT can be used to generate a well-fitted grid-mesh for the two-dimensional flows over complex topography that has any shapes with any slope and size. SCT has a capability for generating a grid-mesh without singularities in the vicinity of very sharp edges of the topography. SCT can generate higher resolution near the ground, and coarse resolution in the upper part of the model domain. SCT can be applied to all scales of atmospheric flow modeling.

Having these capabilities make this SC technique more powerful than other similar techniques and especially commonly used the terrain-following coordinate system in the numerically simulation of the atmospheric flow for dispersion of air pollution.

5. REFERENCES

Bacmeister. J.T., and M.R. Schoeberl, (1989), Breakdown of vertically propagating two-dimensional gravity waves forced by orography, 46, 14, 2109-2131.

Carrier, G.F., M.Krook and C.E.Pearson, (1966), Functions of a complex variable, McGraw-Hill Book Company.

Erdun, H., S.Incecik, M.O.Kaya and I.Ozkol, (1996), Realistic approach to the 2-D flow modeling for complex topography, 4th Workshop on Harmonization Within Atmospheric Dispersion Modeling for Regulatory Purposes}, VITO, Oostende, Belgium, 275-284.

Erdun, H, (1997), Two-dimensional atmospheric flow modeling by using Schwarz-Christoffel transformation over complex topography, Ph.D. thesis, ITU.

Sharman.R.D., T.L.Keller and M.G.Wurtele, (1988), Incompressible and anelastic flow simulations on numerically generated grids, M.Weather Review, 116, 1124-1136.

A MODELING STUDY TO ESTIMATE THE SOURCES OF ATMOSPHERIC NITROGEN DEPOSITION IN THE LAKE TAHOE BASIN

Jülide K. Koračin, Leland W. Tarnay, Alan W. Gertler[*]

1. INTRODUCTION

Lake Tahoe, which is located between the California-Nevada borders, is famous for its clarity; however, decline in Lake Tahoe's water clarity was reported as 0.25 m per year[1]. Nutrient loading (phosphorus and nitrogen (N)) in last decades has been shown as the primary cause for this decline, and nearly half of total N input to the lake was attributed to atmospheric deposition[2].

Identification of nutrient sources is crucial in order to develop effective management strategies to restore the water clarity in the lake. Location of the Lake Tahoe basin downwind of the highly polluted urban areas in California increases the possibility that regional pollutants can be transported into the basin. Local emissions associated with recent urbanization and the use of motor vehicles need to be considered since their impact may be significant. While there had been several measurement studies in the basin for quantification of atmospheric N, to date no study identified the contributions from in-basin vs. out-of-basin sources of atmospheric N concentrations.

In this study, we used advanced meteorological and chemical transport/dispersion models to predict the atmospheric N impact for selected days in August 2000. This period coincided with measurements of ambient nitric acid (HNO_3), which was shown as the major contributor to N deposition in the Lake Tahoe basin [3].

---

[*] Julide K. Koračin and Alan W. Gertler, Desert Research Institute, 2215 Raggio Parkway, Reno, Nevada 89512, USA. Leland W.Tarnay, NPS, 45 MacArthur Blvd., N.W. Washington, D.C. 20007, USA.

Air Pollution Modeling and Its Application XVI, Edited by
Borrego and Incecik, Kluwer Academic/Plenum Publishers, New York, 2004

2. METHOD AND RESULTS

The modeling framework included CALMET/CALPUFF diagnostic meteorological and chemical transport/dispersion models[4,5] as well as prognostic meteorological Mesoscale Model 5 (MM5)[6], which was used as initial guess field in CALMET. The modeling domain encompassed the significant industrial areas in California and included the Sacramento valley and the San Francisco Bay area. Due to the complex terrain features, horizontal grid cells (500x530) were set up at 500 m. To determine the optimum model parameters, a series of sensitivity studies were also performed. Mixing heights simulated by CALMET was adjusted based on the performance comparisons of MM5 and CALMET vs. available upper air observations in the domain.

An hourly gridded (4 km) emission inventory was obtained from the California Air Resources Board (CARB) for the regional sources. In the basin, local mobile source emissions were apportioned spatially and temporally to the routes around the lake according to traffic volumes and vehicle miles traveled. All sources were implemented as point sources in CALPUFF.

Three days (August 10, 16, and 22) were selected for simulations based on prevailing westerly winds and the measured maximum and minimum ambient HNO_3 concentrations in the basin. Overall model performance was evaluated using statistical methods which showed that model results were not significantly different ($p > 0.05$) from the measurements for the two out of three days. Simulated average HNO_3 concentrations (with estimated background values added; 0.3 $\mu g\ m^{-3}$) in the basin comprised approximately 60% of the measured ambient concentrations. Local sources were responsible for 90 % of these simulated concentrations. Using the existing emission inventory, our results indicate that the impact of out-of-basin sources to the Lake Tahoe basin appears to be of secondary importance. Based on these results, in-basin sources of atmospheric N need to be controlled as part of the restoration strategies for Lake Tahoe and other environmental management programs in the region.

3. REFERENCES

1. Reuter, J.E., Miller, W.W., in: Air Quality/*Lake Tahoe Watershed Assessment*, edited by D.M. Murphy and S.M. Knopp Technical Report USDA Forest Service, Berkeley, 2000, pp. 215-377.
2. Cliff, S.S., Cahill, T.A, in: Aquatic Resources/*Lake Tahoe Watershed Assessment*, edited by D.M. Murphy and S.M. Knopp Technical Report USDA Forest Service, Berkeley, 2000, pp. 131-209.
3. Tarnay, L.W., Gertler, A.W., Blank, R.R., Taylor, G.E., Preliminary measurements of summer nitric acid and ammonia concentrations in the Lake Tahoe basin airshed: implications for dry deposition of atmospheric nitrogen. *Environmental Pollution* 113, 145-153 (2001).
4. Scire, J.S., Robe, F.R., Fernau, M.E., Yamartino, R.J., *A User's Guide for the CALMET Meteorological Model*. Earth Tech., Concord, 2000, p. 332.
5. Scire, J.S., Strimaitis, D.J., Yamartino, R.J., *A User's Guide for the CALPUFF Dispersion Model*. Earth Tech., Concord, 2000, p. 521.
6. Grell, G.A., Dudhia, J., and Stauffer, D. R., *A Description of the Fifth-Generation Penn State/NCAR Mesoscale Model (MM5)*. National Center for Atmospheric Research Techn. Note TN-398, Boulder, 1995, p. 122.

URBAN EFFECTS ON AIR POLLUTANT DISPERSION IN VERY COMPLEX TERRAIN: THE ATHENS CASE.

Alberto Martilli, Yves-Alain Roulet, Martin Junier, Frank Kirchner, Mathias W. Rotach and Alain Clappier[1].

1. INTRODUCTION

The presence of a city can strongly modify mesoscale circulations induced by very complex topography and pollutant dispersion. A series of mesoscale numerical simulations are carried out for one day of the MEDCAPHOT campaign in Athens (Greece, Ziomas, 1998), in order to evaluate the impact of the city and the sensitivity of model results to the precision in the parameterization of the urban effects.

2. METHODOLOGY

The mesoscale model used is boussinesq, anelastic, non-hydrostatic (FVM, Clappier et al. 1996) with a detailed urban parameterization (Martilli et al. 2002) taking into account the impact of horizontal (streets and roofs) and vertical (walls) urban surfaces on heat and momentum fluxes. Three simulations are carried out: one with the detailed urban parameterization (*Urban*), a second without any urban parameterization (*Rural*) and a third with a simpler and less detailed urban parameterization (modification in roughness length and heat soil capacity, *Trad*). Meteorological fields are, then, passed to a eulerian photochemical model (TAPOM) to compute the dispersion of reactive pollutants.

3. METEOROLOGY.

The three simulations reproduce the main characteristics of the land-sea breeze cycle. However, *Urban* reproduces the observed nocturnal Urban Heat Island, while the others two simulations do not. As a consequence, *Urban* has weaker land breezes. Furthermore, during daytime, compared to *Urban*, *Rural* has moister soil, and lower

---

[1] Alberto Martilli, University of British Columbia, Vancouver, BC V6T 1Z4 Canada, Yves-Alain Roulet, Martin Junier, Frank Kirchner, Alain Clappier, EPFL Lausanne, Switzerland, Mathias W. Rotach, ETHZ, Zurich, Switzerland.

surface sensible heat fluxes (inducing a delay in sea breeze formation), while *Trad*, storing less energy in the ground, has higher sensible heat fluxes, stronger sea breezes and higher PBL.

4. AIR POLLUTION

The differences in meteorological fields modify the pollutant distribution. The morning peak of NOx in the downtown area (Fig. 1), which is strongly linked with the rate of increase of PBL height, is overestimated by *Rural*, and underestimated by *Trad*, while *Urban* agrees better. For a secondary pollutant like O_3 *Trad* and *Rural* underestimate the peak by 15-20 ppb, and, again, *Urban* matches better.

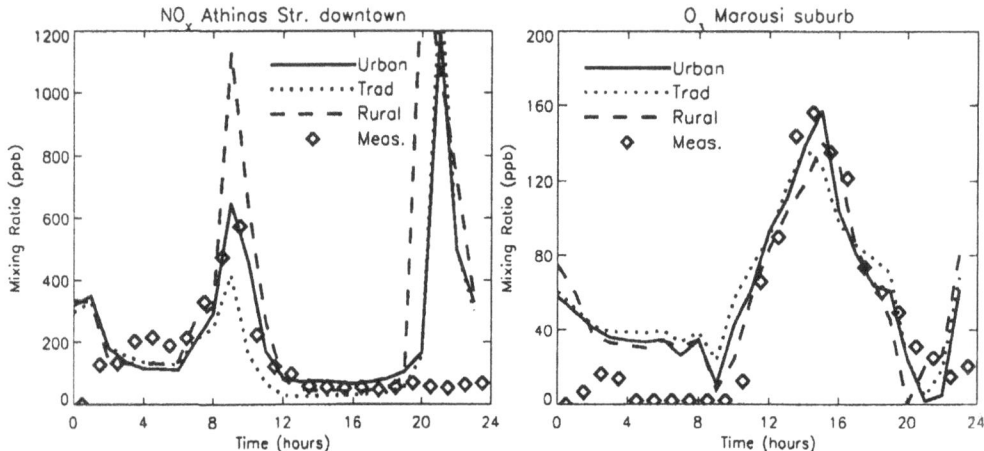

Figure 1. Comparison between measurements and the three simulations for NO_X in the city center, and O_3 north of the city

The daytime O_3 spatial distribution for *Urban* shows one maximum over the sea, where pollutants emitted during night are transported by land breezes, and one in the north-east part of the city, at the sea breeze convergence. Since sea breezes are stronger in *Trad*, pollutants are transported back over land quicker, and the maximum over the sea is lower. Moreover PBL is higher over land, and the second maximum is reduced. On the other hand, in *Rural*, sea breezes start later and pollutants stay longer over the sea, resulting in a stronger maximum over the sea. The chemical regimes are also modified: in *Trad* the extension of the NO_x saturated region is 192 km^2, in *Urban* is 276 km^2, while in *Rural* is 424 km^2

References

Clappier, A., Perrochet. P., Martilli, A., Muller, F. and Krueger, B. C.:,1996, A new non-hydrostatic mesoscale model using a CVFE (control volume finite element) Discretisation technique, in *Proceedings of EUROTRAC Symposium '96.*, Editors: P. M. Borrell et al., Computational Mechanics Publications, Southampton, pp 527-531.

Martilli, A., Clappier, A. and Rotach, M. W., 2002, An urban surface exchange parameterisation for mesoscale models, *Boundary-Layer Meteorol.*, 104, 261-304.

Ziomas, C., 1998, The Mediterranean Campaign of Photochemical Tracers-Transport and Chemical evolution (MEDCAPHOT-TRACE): an Outline. *Atmos. Environ.*, 32, 2045-2053.

NESTING THE EMEP REGIONAL MODEL

Peter Wind and Leonor Tarrasón[1]

The EMEP models have been traditionally used for the analysis of the transport, chemical transformations and deposition of air pollutants at the regional level (EMEP, 2002). The Unifed EMEP Eulerian model has been recently developed to allow a flexible choice of the model domain and model resolution, with size of gridcells ranging from 50x50 km² and down to 5x5km² (Wind et al., 2002). Calculations at different scales are coupled through one-way nesting. This allows the EMEP model to link regional long range transport to urban and local pollution, and in this way determine the influence of background contributions to urban air pollution levels. The description of intercontinental exchange of pollution in a hemispheric scale is also facilitated within the same framework.

The EMEP model may now be used with meteorological data from several different numerical weather prediction models: HIRLAM50, HIRLAM20, MM5 and ECMWF model will be tested. In addition, the Norwegian meteorological institute (met.no) has developed a new interpolation system that ensures the mass conservation properties of the atmospheric flow (Holstad and Lie, 2002). These different meteorological driver data and interpolated fields have been used here to study ozone formation and NO_2 levels in Oslo and surrounding areas.

The initial results presented here are part of a more comprehensive study where the influence of different input parameters to local scale simulations are analysed separately. In particular, we are interested in the differences in local scale simulations of particulate matter, NO_2 and ozone when using:

- Different emission data at different scales (aggregated or interpolated)
- Different meteorological drivers (with refined topography or interpolated fields)
- Different boundary conditions and extension of the model domain

---

[1] Norwegian Meteorological Institute (MET.NO), N-0313 Oslo, Norway

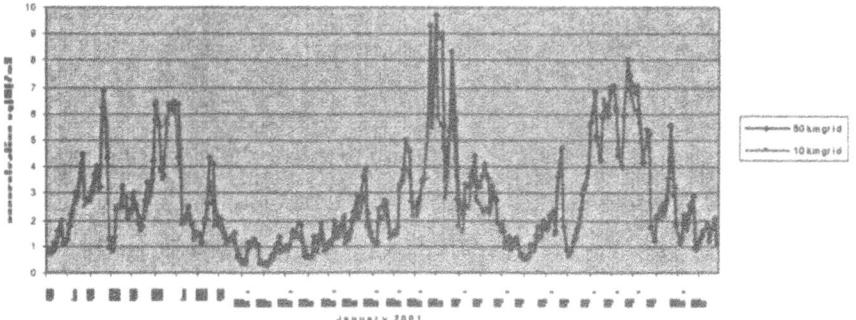

The first figure shows numerical differences in the evolution of NO_2 concentrations in Oslo in January 2001 using the same model, meteorological input fields (HIRLAM50) and emissions (EMEP), but using two different grid resolutions. The curve with diamonds is calculated using a grid with 50 km resolution and the squares are the results obtained using a grid with 10 km resolution nested into the 50 km grid. The calculation in the 10 km grid uses interpolated emissions and meteorology. As expected the results are very similar for the two simulations. This shows that the nested model is operational and can be used to study the influence of different input parameters.

NO2 concentration in Berlin

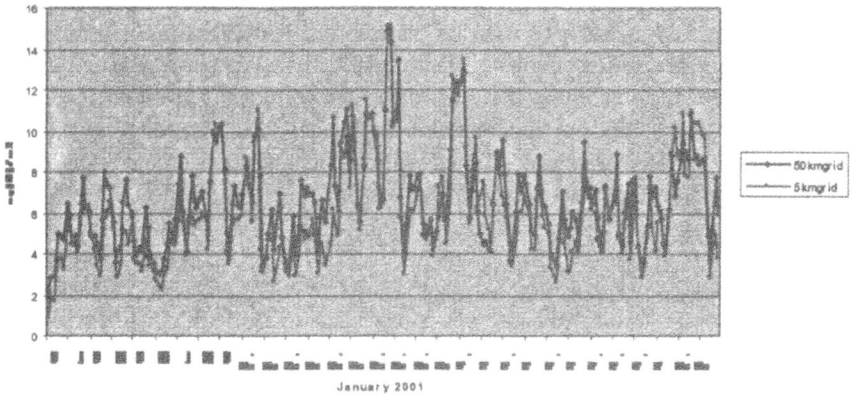

The second figure shows calculated NO_2 concentrations close to Berlin, January 2001 using two different grid resolutions. The experiment has been designed to characterise the effect of refined emission information in the calculated fields. The calculation in the fine 5 km resolution grid is nested in a large grid with 50 km resolution. The emissions are given in a fine 5 km resolution grid around Berlin (CityDelta) and aggregated for the coarse grid calculation. The meteorology in the fine grid is interpolated from the coarse grid meteorology. There are visible differences between the two simulations, due to the effect of refined emission fields. These will be studied in further details in the course of this project.

Acknowledgments

This work has been financed by the Norwegian Ministry of Fisheries (FID), the Norwegian Ministry of the Environment (MD) and the Norwegian Pollution Control Authority (SFT).

References

EMEP, 2002, Transboundary Acidification, Eutrophication and Ground Level Ozone in Europe, Norwegian Meteorological Institute, Oslo, Norway, EMEP/MSC-W Report 1 and 2/2002.

EMEP: Co-operative Programme for Monitoring and Evaluation of the Long Range Transmission of Air Pollutants in Europe; http://www.emep.int

Holstad, A. and Lie, I. , 2002 , On the computation of mass conservative wind and vertical velocity fields, Research Report No. 141, Norwegian Meteorological Institute, Oslo, Norway.

Wind, P., Tarrason, L., Berge, B., Slørdal, L. C., Solberg, S., Walker, S.-E., 2002, Development of a modelling system able to link hemispheric-regional and local air pollution, Norwegian Meteorological Institute, Oslo, Norway, EMEP/MSC-W Note 5/2002.

CityDelta: http://rea.ei.jrc.it/netshare/thunis/citydelta

MODELLING THE UK NH_x, S, NO_x BUDGETS FOR 2010 WITH AN ATMOSPHERIC TRANSPORT MODEL

N. Fournier, A.J. Dore and M.A. Sutton *

1. INTRODUCTION

The Gothenburg Protocol to Abate Acidification, Eutrophication and Ground-level Ozone was adopted on 30 November 1999. The Protocol sets emission ceilings for 2010 for four pollutants: SO_2, NO_x, VOC_s and NH_3. Emissions reduction of SO_2, NO_x and NH_3 for the time period 1990 to 2010 have been estimated over Europe by (UN/ECE, 1999). To consider the effect of these changes on the British Isles, the FRAME model is applied here to model deposition of sulphur and nitrogen considering the implementation of the Gothenburg Protocol scenario of emissions for 2010 throughout the British Isles and Europe.

2. RESULTS

An atmospheric transport model, FRAME (Fine Resolution Atmospheric Multi-species Exchange), is used to assess the NH_x, S, NO_x deposition over the British Isles (BI : UK and Ireland) for 1996 and 2010 (Fournier *et al.*, 2001). The 2010 Gothenburg scenario corresponds to 70 %, 40 % and 16 % reduction in the 1996 SO_2, NO_x and NH_3 emissions, respectively, over the BI.

The following Table shows the UK reduced nitrogen budget obtained with the FRAME model for the 2010 Gothenburg Protocol scenario. The first column indicates the results obtained for 1996 with FRAME. The reduction in both import (-21 %) and emissions (-16 %) of reduced nitrogen in 2010 led to less deposition to the UK than in 1996. The total deposition declined by 14 %. However, because of a reduced rate of formation of NH_4^+ in 2010 (NH_3 emissions will not decline as much as for SO_2 and NO_x), concentrations of NH_3 will not decrease as much as

*N. Fournier, A.J. Dore, M.A. Sutton, Centre for Ecology and Hydrology, Edinburgh Research Station, EH26, UK.

NH_3 emissions. This has the result that dry deposition of NH_3 is only expected to decrease by 7 % in 2010 compared with 1996.

UK NH_x kt N yr^{-1}	FRAME 1996	FRAME 2010	Reduction (%)
Import	52	42	-21
Emission	291	245	-16
Dry deposition	103	96	-7
Wet deposition	105	83	-21
Total deposition	208	179	-14
Export	135	108	-21

Concerning the UK sulphur budget obtained with FRAME for 2010, the reduction in both import (-82 %) and emissions (-70 %) led to less deposition to the UK than in 1996. The total deposition declined by 68 % from 329 kt to 106 kt. However, the sulphur wet deposition shows a lower decline of only 64 % between 1996 and 2010. This indicates that there are non-linearities in the relationship between emission and wet deposition patterns. This is caused at least in part by the increased neutralisation of ammonia. Similarly, the reduction in both import (-53 %) and emissions (-40 %) of oxidised nitrogen in 2010 is estimated to lead to less deposition to the UK than in 1996. The total deposition declines by 44 % from 216 kt to 121 kt. This does not exhibit significant non-linearities between emission and deposition. The decline of both dry and wet deposition roughly follows the 40 % amplitude in NO_x emissions reduction.

3. CONCLUSION

This modelling study shows that the application of European policies to reduce further emissions of SO_2, NO_x and NH_3 by 2010 would be beneficial, in terms of deposition of nitrogen and sulphur compounds, for the UK in 2010. Considering these targets of emissions reductions in 2010 in FRAME led to decreases in the total deposition of sulphur, oxidised and reduced nitrogen to the UK by 68, 44 and 14 %, respectively from 1996 to 2010. However, non-linearities in the relationship between the emissions and wet deposition patterns of sulphur have been detected. Ammonia, for which emissions will not decline as much as for SO_2 and NO_x, has an increasing role and the relative contribution of nitrogen compounds to the total deposition has increased from 1996 to 2010.

REFERENCES

Fournier N., Pais V.A., Sutton M.A., Weston K.J., Dragosits U., Tang S.Y. and Aherne J. (2001) Parallelisation and application of an atmospheric transport model simulating dispersion and deposition of ammonia over the British Isles. *Environmental Pollution* 116(1), 95-107.

UN/ECE (1999) *Protocol to the 1979 Convention on Long-range Transboundary Air Pollution to Abate Acidification, Eutrophication, and Ground-level Ozone.* United Nations Economic Commission for Europe, Geneva, Switzerland.

ATMOSPHERIC WET DEPOSITION OF SOLUBLE MACRO-NUTRIENTS IN THE CILICIAN BASIN

Türkan Özsoy*

1. INTRODUCTION

Pulses of mineral dust from North Africa into the Mediterranean region could deposit macro-nutrients (e.g. PO_4^{3-} and NO_3^-) (Bergametti et al., 1992) at the sea surface. and possibly contribute to marine biological production. Phosphate, PO_4^{3-} is generally known to be the limiting nutrient for phytoplankton production in the Eastern Mediterranean, typically having a high N:P ratio (>20:1) near the sea surface (Krom et al., 1991). Most studies in the western (Rimmelin et al., 1999) and the eastern (Herut et al., 1999) Mediterranean regions have been directed at the determination of macro-nutrient contents of wet deposition. Particular attention has been given to inorganic nitrogen for its major role in anthropogenic air pollution related to industrial/agricultural activities or traffic emissions, which are partly responsible for increased acidity of rainwater. In order to estimate wet deposition atmospheric fluxes of macro-nutrients, soluble inorganic phosphate (PO_4^{3-}), nitrate (NO_3^-) and nitrite (NO_2^-) concentrations in precipitation (from February 1996 to June 1997) have been measured at a coastal sampling site, Erdemli, Turkey. Soluble inorganic P, a reactive, bioavailable, limiting macro-nutrient in the oligotrophic waters of the eastern Mediterranean was studied with respect to its contribution to biological productivity.

The sampling site, (36°33'54" N, 34°15'18" E) located on the south-eastern Mediterranean coast of Turkey can be accepted as a remote area. A total of 87 precipitation samples on an event basis have been collected by a wet/dry sampler analyzer, between Feb. 1996 –June 1997. Soluble macro-nutrient analysis were performed using a Technicon Model two-channel autoanalyzer (Grasshoff et al., 1983). The analytical precision was found to be approximately 7 % for PO_4^{3-} and 8 % for both TON and NO_2^- anions. Additional to pH and conductivity, Aluminum concentrations have been measured as an indicator of dust in the particulate fraction of the precipitation samples. In order to identify the source of dust transportation, three dimensional, 3-days backward trajectories for air masses arriving at the sampling point on 12 00 UT of the

*Mersin University, Science and Literature Faculty, Chemistry Department, Çiftlikköy Campus, Mersin, 33342, Turkey

Air Pollution Modeling and Its Application XVI, Edited by
Borrego and Incecik, Kluwer Academic/Plenum Publishers, New York, 2004

sampling day at levels of 900, 850, 700 and 500 hPa were calculated according to the trajectory model of the ECMWF in Reading, England.

2. RESULTS AND DISCUSSION

Particulate Al measurements of membrane filters and air mass back trajectory analysis of the corresponding rainy days revealed 12 dust transport episodes, associated with 18 "red rain" events. No differences were found between the mean reactive P and TON concentrations of red rain and normal rain events. Most likely as a result of low solubility of crustal phosphorus (Lepple, 1971), dust episodes were not found to be important sources of reactive P, in terms of wet deposition.

Table. Statistical results of; pH, conductivity ($\mu S\ cm^{-1}$), particulate Al (Al_{par}), PO_4^{3-}-P and TON concentrations (μM) in the precipitation samples at Erdemli. The number of the samples are given in parenthesis.

Parameter	Arit. Mean	Geo. Mean	VWM	Min.-Max.
pH (87)	5.6±0.9	5.5	4.95	3.5-7.6
Conductivity (41)	74.6±79.6	51.4	-	12.9-391.0
Al_{par} (84)	88.77±247.50	15.28	56.31	0.37-1843.0
PO_4^{3-}-P (61)	1.09±2.94	0.19	0.51	BDL-16.09
TON (61)	45.09±50.31	28.13	26.50	BDL-247.40

VWM: Volume Weighted Mean
BDL: Below Detection Limit

The calculated annual wet deposition flux of particulate Al, 942 mg m^{-2} yr^{-1} was found to be the highest value among reported values from various stations around the world (Özsoy and Saydam, 2000). The annual wet deposition fluxes of reactive P and TON into the Cilician Basin were respectively estimated to be 0.010 g P m^{-2} yr^{-1} and 0.23 g TON m^{-2} y^{-1}, which are comparable to the fluxes from land-based sources in the NE Mediterranean. The atmospheric deposition of bioavailable P and TON in these amounts could lead to corresponding new production values of 0.40 g C m^{-2} yr^{-1} and 1.31 g C m^{-2} yr^{-1}, accounting respectively for 3.3 % and 11 % of the estimated mean annual new production, 12 g C m^{-2} yr^{-1} (Ediger, 1995) in the NE Mediterranean .

3. REFERENCES

Bergametti, G., Remoudaki, E., Losno, R., Steiner, E. and Chatenet, B., 1992, Source, transport and deposition of atmospheric phosphorus over the northwestern Mediterranean. *J. Atmos. Chem.*, 14:501-513.

Ediger, D., 1995, *Interrelationships Among Primary Production, Chlorophyll and Environmental Conditions in the Northern Levantine Basin*, PhD Thesis, IMS-METU, p.178.

Grasshoff, K., Ehrhardt, M., Kremling, K., 1983, Determination of nutrients, in: *Methods of Seawater Analysis*, 2nd edition, Verlag Chemie, Weinheim, pp.125-188.

Herut, B., Krom, M. D., Pan, G., Mortimer, R., 1999, Atmospheric input of nitrogen and phosphorus to the Southeast Mediterranean: Sources, fluxes and possible impact, *Limnol. Oceanogr.*, 44 (7):1683-1692.

Krom, M. D., Kress, N. and Brenner, S., 1991, Phosphorus limitation of primary productivity in the eastern Mediterranean, *Limnol. Oceanogr.*, 36:424-432.

Lepple, F. K., 1971, *Eolian Dust Over the North Atlantic Ocean*, PhD Thesis, University of Delaware.

Özsoy, T. and Saydam, A. C., 2000, Acidic and Alkaline precipitation in the Cilician Basin, northeastern Mediterranean Sea, *Sci. Tot. Environ.*, 253 (1-3):93-109.

Rimmelin, P., Dumon, J-C., Burdloff, D. and Maneux, E., 1999, Atmospheric deposits of dissolved inorganic nitrogen in the southwest of France, *Sci. Tot. Environ.*, 226:213-225.

NEW VERSION OF THE MODEL FOR LONG RANGE TRANSPORT OF POLLUTANT SUBSTANCES IN THE ATMOSPHERE OF IGCE AND HMC OF RUSSIA

Dr. A.Degtiarev , N. Shtyreva

This model (MLTT2) was created for carried out of a long range and transboundary transport of harmful emitions in the atmosphere and their deposition on the surface. The model enables to settle an invoice for forecasting and monitoring distribution and dissipation of industrial pollutants, chemical and nuclear emergencies.

Applications of this model are emergency planning, scientific researches and air quality assessments. The model calculates gases and aerosols concentrations of pollutants in the atmosphere and their deposition on the surface.

The account of distribution of polluting substances from many sources can be carried out in the model. Sources can be persistence, instant or intensity of sources can vary. The MLTT2 simulates an account of individual qualities for each source of pollution and the given polluting substance. The source can emit one or more polluting substance with given properties, for example, various pollutant gases or a polydispersive structure of emission. MLTT2 uses a regional-to-continental or global spatial scale. The model uses complex meteorology in spherical (latitude-longitude) coordinates and 8-vertical sigma levels. MLTT2 is based on a numerical decision of the advective-duffusion equation system of transport of pollutants in the atmosphere. The Lagrangian approach for an advective transport of polluting substances is realized in the model. For the decision of a system of transport of polluting substances is realized in the model. For the decision of a system of significances of wind velocity in the point, where particular particles of pollution are located, are calculated with the help 3-dimension interpolation. It should note that the interpolation in the model is the most frequently executed procedure; therefore it should satisfy to rigid computing requirements. The used procedure is based on a method of construction of digital-linear function, conterminous in the point of a grid by the given function.

A.   Degtiarev, Institute for global climate and ecology; N. Shtyreva, Hydrometeocentre of   Russia,   Bolshoy   Predtechenskiy   per.,   9-13,   Moscow,   Russia

The Monte-Carlo method for account of turbulent diffusion in the atmosphere is realized in the model. The Monte-Carlo method is used for the approximation of diffuse members. Parameterisation of vertical and horizontal turbulent flows plays an essential role for the correct decision of the given task. Now in the model the new and modified circuit parameterisation vertical turbulent, dry and wet depositions of polluting substances on a surface are used. Coefficient of vertical turbulence in surface layer is defined with the help of the account of flows of quantity of movement and heat on a surface. The speed gravitational deposition of aerosol particles of pollution in MLTT2 is determined under the formula Stoks-Kanningem. The speed dry deposition of particles of pollution in surface layer is determined from a condition of proportionality of speed dry deposition of particles at height of 1 m. The speed wet deposition is calculated depending on intensity of deposits and relation of concentration of polluting substance in deposits to concentration of polluting substance in air.

The model uses a mechanism of secondary migration of polluting substances from behind wind intercepting from a surface. This parameterisation also based on a Monte-Carlo method. The probability of secondary migration of a particular particle of pollution on the surface depends on such factors as a module of surface wind, type of surface, roughness, and soil moisture.

The variant of the MLTT2 in a spherical coordinate system is to the utmost set - up on use of meteorological information of the operative model of the intermediate term forecast of Hydrometeocentre of Russia. The data of components of wind velocity, temperature, humidity, surface pressure and a vertical diffusion coefficient of this model can be directly used by the model of polluting substances transport.

The distinctive features of the MLTT are: 1) the model of transport is combined with the computing model of intermediate term forecasts of the weather by Hydrometeocentre of Russia (or another forecasting model of the similar scale), that it is important for operative accounts, 2) in the model the data of the main meteorological elements on the boundary layer of the atmosphere are taken into account, where the main transport of anthropogenic pollution occurs, it is enough in details (at the 4-th vertical accounting levels), and, as is known, wind and temperature structures in the low troposphere are rather changed, 3) the model permits to settle an invoice as for the whole Northern hemisphere, as for any given region, that permits essentially to speed up calculations.

The MLTT2 was used for examination of possible air pollution of construction of the Baltic Pipe-line System in Russia. The model was used for accounts of distant carry sandy aerosols from Northern Africa through the Mediterranean Sea in area of Northern Caucasus, distribution of pollution from fires the Moscow Region and smog in Moscow in 1999, and also for revealing consequences of wood fires on Far East in 2001.

INVESTIGATION OF THE VARIABILITY OF SO₂ AND PM IN THE LARGE CITIES OF TURKEY

Mete Tayanç[1], Nuriye Garipağaoğlu[2] and Bülent O. Akkoyunlu[3]

1. INTRODUCTION

This study presents the analysis of SO_2 and PM concentrations that were sampled in 14 large cities of Turkey during the period of 1990-2000. In the atmosphere of these cities higher SO_2 and PM levels exist than those of the other cities in Turkey. Results indicate that heating season (october-march) average concentration of SO_2 for all these cities ranges between 37-404 $\mu g/m^3$ and PM ranges between 72.1-158.4 $\mu g/m^3$. The main reason of high SO_2 and PM concentration is the low quality lignite consumption. The other reasons can be the industrial activities and the topography of the urban areas. However in 9 cities, decreasing SO_2 concentrations are detected in the late 1990s.

2. RESULTS AND CONCLUSIONS

SO_2 and PM concentration levels were investigated for 14 cities including three large and industrialized cities to assess air pollution during the heating seasons between 1990 and 2000. Seasonal average SO_2 concentration levels for 9 cities, showing the significant decreases in air pollution, are depicted in Figure 1. Erzurum has the highest SO_2 concentration in 1993-1994 heating season with a value of 404 $\mu g/m^3$. Other seasonal average maximum SO_2 concentrations are found in Sivas and İstanbul as 402 $\mu g/m^3$ in 1990-1991 and 379 $\mu g/m^3$ in 1991-1992, respectively. It is intuitively clear in Figure 1 that for these 9 cities, maximum SO_2 concentration is found in the first two heating seasons. With progressing years SO_2 concentration for all cities decreased considerably, especially, İstanbul, Kocaeli and Ankara showed the sharpest decreases. The decreases are calculated by taking the ratio of the concentrations of 1990-1991 season to 1999-2000 season. The change in the SO_2 concentration for the above three cities are calculated as almost the same, -%83. Owing to the increasing air pollution problems in Ankara towards the late 1980s, a program of switching coal to natural gas usage has been initiated. Afterwards in many cities of Turkey natural gas consumption for heating has increased significantly and coal has been subjected to quality control via municipalities.

---

[1] Marmara University, Dept. of Environmental Eng., Göztepe, İstanbul, Turkey
[2] Marmara University, Dept. of Geography, Göztepe, İstanbul, Turkey
[3] Marmara University, Dept. of Physics, Göztepe, İstanbul, Turkey

Air Pollution Modeling and Its Application XVI, Edited by
Borrego and Incecik, Kluwer Academic/Plenum Publishers, New York, 2004

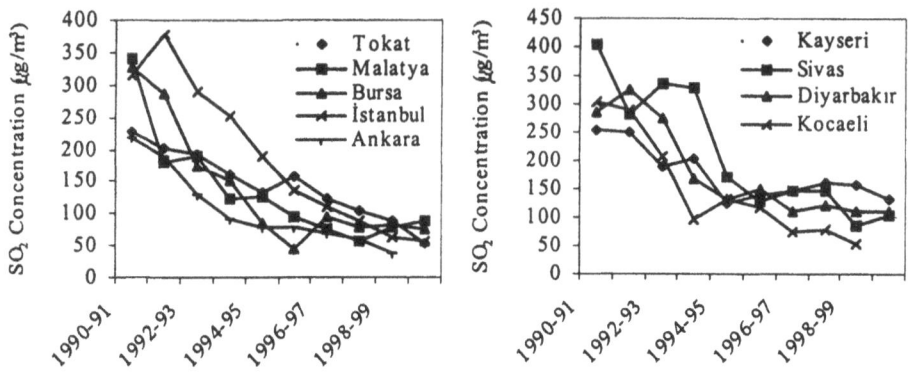

Figure 1. Temporal variation of heating season SO_2 for selected cities of Turkey from 1990 to 2000.

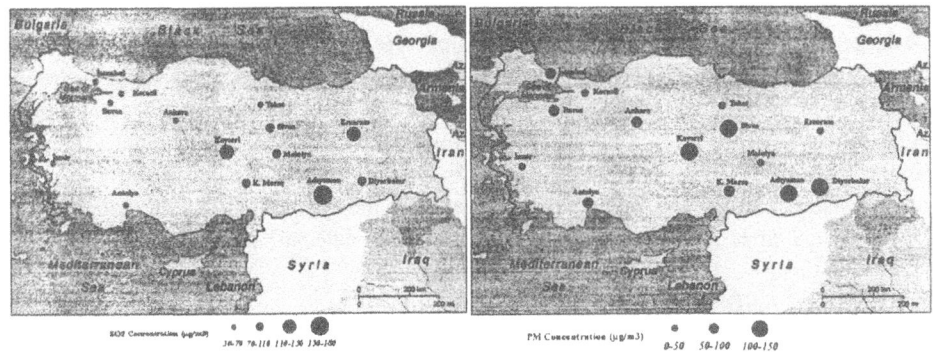

Figure 2. Spatial distribution of SO_2 and PM over Turkey in 1999-2000 heating season

The spatial distribution of SO_2 and PM of 14 cities in 1999-2000 heating season is shown in Figure 2. Adıyaman, Kayseri and Erzurum have the highest SO_2 concentrations and Sivas, Kayseri, Adıyaman and Diyarbakır have the highest PM concentrations, over 100 $\mu g/m^3$ as the long term average. During recent years, it is obvious that mid-east large cities of Turkey generally have higher pollution levels than the other cities. SO_2 concentration in İstanbul was decreased from 315 $\mu g/m^3$ in 1990-1991 heating season to only 57 $\mu g/m^3$ in 1999-2000 leading to a total 82% decrease. On the other hand, SO_2 concentration in Adıyaman was decreased from 193 $\mu g/m^3$ in 1990-1991 heating season to only 168 $\mu g/m^3$ in 1999-2000, leading to a total 13% decrease. Antalya has the lowest heating seasonal average SO_2 concentration as 62 $\mu g/m^3$, lower than the limits and the lowest standard deviation as 17.6. Correlation coefficients between the monthly mean concentration values of SO_2 and PM between 1991 and 1993 years are calculated and found to range from 0.47 to 0.91 in these 14 cities. These considerably high correlation coefficients suggest that SO_2 and PM are behaving alike in the environment.

614

A METHODOLOGY FOR ENVIRONMENTAL URBAN PLANNING IN BURSA (TURKEY)

Tugba Albayrak, Fahrettin Çeliktaş, Mihnet Tekinay[1], M. Peronaci, N. Colonna, S. Racalbuto, P. Picini, M. De Cassan[2]

1. INTRODUCTION

The Metropolitan Municipality of Bursa (Turkey) and ENEA (Italian Agency for New Technology, Energy and the Environment) are cooperating within the framework of the MEDA Programme, funded by the European Commission – Directorate "EuropAid Cooperation Office (AIDCO).

Scope of this cooperation is to provide Bursa local Authorities with a methodological tool for energy and environment urban planning, by implementing a demo of an Environment Atlas with energy and environment indicators and by developing a test case on the air quality of the Bursa Metropolitan area in relation to the human activities.

The methodological tool is made up by an air emission inventory, a numerical model to evaluate the main pollutants concentrations at ground level and a G.I.S.

The activities can be summarized as follows:

- Identification of the more suitable socio-economic and environmental indicators;
- Implementation of an air emission inventory related to the traffic and house heating pollution, using the COPERT III methodology;
- Gathering of the data of the main pollutant emissions from the major industrial chimneys;
- Gathering of the relevant socio-economic data describing the Metropolitan area;
- To use a Gaussian multi-source numerical model developed by ENEA to evaluate the main pollutants concentrations at ground level due to the industrial sector, the

---

[1] Bursa Metropolitan Municipality – Environment Dept.
[2] Environment Dept. - ENEA (Italian Agency for New Technology, Energy and the Environment) - Italy

Air Pollution Modeling and Its Application XVI, Edited by
Borrego and Incecik, Kluwer Academic/Plenum Publishers, New York, 2004

traffic and the house heating; the model also allows to assess the impact on the population to the evaluated exposure to the main pollutants;

• To implement a G.I.S. to be used as tool for environment urban planning.

2. BACKGROUND INFORMATION AND FUTURE NEEDS

The City of Bursa is the fourth biggest city of Turkey with about 1.300.000 inhabitants; since the 1960s an intensive industrial growth caused many different problems mainly in the sectors of air pollution, waste management, transportation: therefore there is an urgent need to develop appropriate tools for urban planning.

With regard to the air pollution issue, the Bursa Metropolitan Municipality has been developing a monitoring programme since 2001. Nowadays four Air Pollution Monitoring Stations (AIRNET) are operating, to continuously monitor SO_2, NO_X, CO, HC, PM and O_3.

Due to the actual situation, there is a need to better evaluate the different contributions from the industrial sector, the traffic and the house heating to the total air pollution, so to better plan future policies and technical responses to these driving forces.

A robust criteria to implement the existing AIRNET, both in terms of new stations and more representative locations, is needed as well.

3. THE AIR EMISSION INVENTORY AND THE NUMERICAL MODEL

An air emission inventory related to the traffic and house heating pollution, has been implementing in ACCESS, using the COPERT III methodology.

In addition the emissions of SO_2, NO_X, CO, PM pollutants from 22 main chimneys in the Bursa Metropolitan area have been collected as well.

A numerical model to evaluate atmospheric dispersion of pollutants in limited areas (10 to 30 km far from the main sources of pollutants) characterized sufficiently by homogeneous and stationary atmospheric conditions has been developed by ENEA and applied to the Bursa Metropolitan area.

The model (Gaussian multi-source model) allows to carry out both "short term" and climatologic "long term" simulations. This Gaussian-type solution takes into consideration the actual height of the emissions for hot sources; it also consider the lateral and vertical dispersion of plumes computed on the bases of the atmospheric stability described in terms of the Pasquill-Turner classes.

Climatologic elaborations are based on the use of the Joint Frequency Functions for a number of representative stations. The model can consider both point sources and areal sources.

MODELLING AEROSOL EVOLUTION IN AN INDUSTRIAL PLUME- 1. MODEL DEVELOPMENT

Sun Hee Cho and Diane V. Michelangeli[*]

1. INTRODUCTION

Aerosols in the atmosphere are recognized as being a major health concern and causing damage to vegetation, reduced visibility, and interference with economic development, thus necessitating more detailed studies of processes leading to their formation and evolution. Therefore, a numerical model to simulate realistically the evolution of aerosols from an anthropogenic source, such as an industrial plume, is necessary. In the past there have been various attempts to improve the understanding of the basic mechanisms modifying the plume particulate phase, to elucidate how this processes depends on source variables and background conditions, and to determine how there importance varies in different parts of the plume (Seigneur, 1989). In this work, a Community Aerosol & Radiation Model for Atmospheres (CARMA), which is a 3-dimentional aerosol microphysics model, is used.

2. METHODS

The present microphysical model calculates the time dependent particle size distributions in an industrial region together with changes in gas phase mixing ratios of sulfuric acid vapor (H_2SO_4), assuming an initial size distribution of pre-existing particles, and SO_2 emission. The CARMA model was modified with the inclusion of gas phase chemistry, Gaussian dispersion, binary homogeneous nucleation of sulfuric acid vapor (H_2SO_4) with water vapor (H_2O), and heterogeneous nucleation (adsorption) of pre-existing particles with sulfuric acid vapor for the formation of new particles. Growth (condensation/evaporation), coagulation and sedimentation were already included in the model. Changes in the coagulation kernel can influence the form of the particle size

[*] Sun Hee Cho and Diane V. Michelangeli, York University, 4700 Keele Street, Toronto, Ontario, M3J 1P3, Canada

distribution. We have added van der Waals forces to the Brownian coagulation kernel and have included the turbulent coagulation kernel. The goal is to improve the microphysical description of aerosols with the industrial plume for further comparison with measurement data.

3. RESULTS AND DISCUSSION

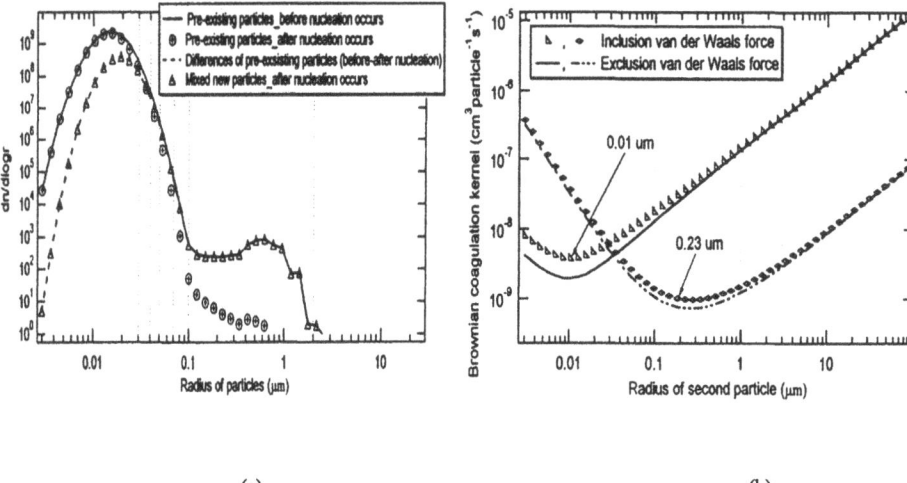

(a)   (b)

Figure 1. (a). The result of nucleation process as a function of particles size. (b) Brownian coagulation kernel comparison as a function of all particle sizes.

Figure 1. (a) shows that the new mixed particles (symbol: Δ) are formed by nucleation of sulfuric acid on the pre-existing particles (solid line). The differences between the pre-existing particles before (solid line) and after (symbol : $\oplus$) nucleation is shown as the dash-dot line, and is, as expect, identical to the curve of mixed new particles (symbol: Δ). The comparison between the Brownian coagulation kernel with and without van der Waals froces is shown in Figure 1. (b). Since the enhancement factor is maximum when particles are the same size, we find the largest impact of van der Waals froces when the particles are the same size.

REFERENCES

Jacobson, M.A., and R. Turco, 1994. Modeling coagulation among particles of different composition and size. *Atmos. Environ.*, Vol. 28, 1327-1338.

Turco, R.P., P. Hamill, O.B. Toon, R.C. Whitten and C.S Kiang, 1979 a, b. The NASA-Ames research center stratospheric aerosol model: I physical processes and computational analogs. *NASA Tech. Publ.*, 1362, iii-94.

Seigneur C. 1982. A model of sulfate aerosol dynamics in atmospheric plumes. *Atmos. Environ.*, Vol. 16, 2207-2228.

DEVELOPMENT AND VALIDATION OF A PREDICTIVE MODEL TO ASSESS THE IMPACT OF COASTAL OPERATIONS ON URBAN SCALE AIR QUALITY

Alan W. Gertler, Darko R. Koracin, Julide Koracin, John M. Lewis, Menachem Luria, John C. Sagebiel, and William R. Stockwell[*]

1. INTRODUCTION

One of the most pervasive air quality problems is the high level of urban/regional O_3. Approximately 90 urban areas in the U.S., containing some 30% of the U.S. population, still exceed the O_3 standard, with little immediate prospect of attainment.[1] All the major urban areas in California are classified as non-attainment for O_3.[2] Control of O_3 in the troposphere is complicated by the fact that it is a secondary pollutant. Particulate matter (PM) is also a serious environmental issue. In 1999 all but four counties in California failed to meet the state PM_{10} standard.[2] Recognizing the health impacts of fine particulates ($PM_{2.5}$), the U.S. EPA has proposed new standards for $PM_{2.5}$. Unlike PM_{10}, a significant fraction of ambient $PM_{2.5}$ is produced by secondary reactions.

Many of the urban areas classified as non-attainment for O_3 or PM_{10} and facing non-attainment for $PM_{2.5}$ are located along the east and west coasts of the U.S. and are home to major U.S. Department of Defense (DoD) facilities. These operations can be significant sources of the O_3 and $PM_{2.5}$ forming precursors, direct $PM_{2.5}$ and PM_{10} emissions, and emissions of toxic species. Much of the uncertainty in developing an understanding of the causes of reduced air quality in urban areas is due to uncertainty on the emissions inventories;[1] however, in coastal areas the situation is confounded by the complex meteorology associated with the land/sea interface. In order to address this

---

[*] Alan W. Gertler, Darko R. Koracin, Julide Koracin, John M. Lewis, John C. Sagebiel, and William R. Stockwell, Desert Research Institute, Division of Atmospheric Sciences, Reno, Nevada 89512. Menachem Luria, The Hebrew University, P.O. Box 1255, Jerusalem, Israel 91094.

Air Pollution Modeling and Its Application XVI, Edited by
Borrego and Incecik, Kluwer Academic/Plenum Publishers, New York, 2004

issue, the primary objective of this work is to develop and validate a prognostic modeling system capable of assessing the impact of emissions in coastal areas on air quality.

2. MODELLING APPROACH

We are currently in the process of linking state-of-the art meteorological, transport, and chemical modules into a hybrid model for the prediction of emissions, transport, transformation, and deposition in coastal regions. This approach couples the advantages of a Lagrangian random particle dispersion model with the advantages of a Eulerian chemical model. The San Diego area located in southern California has been chosen as the test case for model development and validation.

The modeling system will be composed of three integrated components:

- An emissions processing system that allows for area sources, stationary sources, and mobile sources.
- A prognostic meteorological component that uses operationally available forecasts for the domain of interest.
- A prognostic hybrid Lagrangian random particle dispersion component-model[3] coupled with a Eulerian chemical component-model[4] that uses output from the meteorological component to simulate the transport, dispersion, and the chemical transformations of pollutants emanating from specified emission sources.

To date we have developed an emissions inventory incorporating hourly gridded emissions coupled with two research grade inventories (SCOS97[5] and Barrio Logan[6]) to provide initial model input. Using tracer data from the 2001 Barrio Logan microscale experiment, we have evaluated transport and dispersion on a very small domain (0.433km). This result will be used to evaluate the performance of the Lagrangian random particle model. For the Eulerian chemical component, we have used CAMx to develop a baseline for comparison and are exploring the possibility of further developing the HYSPLIT-Chem model[7] to incorporate more advanced chemistry. As part of the model validation, an aircraft study is planned for July 2003 to obtain real-world data for model validation.

3. REFERENCES

1. Seinfeld, J.H., et al., *Rethinking the Ozone Problem in Urban and Regional Air Pollution* (National Academy Press, Washington, D.C., 1991)
2. Alexis, A., P. Gaffney, C. Garcia, M. Nystrom, and R. Rood, *The 1999 California Almanac of Emissions and Air Quality* (California Air Resources Board, Sacramento, CA, 2000).
3. Koracin, D., V. Isakov and J. Frye, A Lagrangian particle dispersion model (LAP) applied to transport and dispersion of chemical tracers in complex terrain. Presented at the Tenth Joint Conference on the Applications of Air Pollution Meteorology, Phoenix, AZ, 11-16 January 1998.
4. Stockwell, W.R., F. Kirchner, M. Kuhn, and S. Seefeld, A new mechanism for regional atmospheric chemistry modeling, *J. Geophys. Res.*, **102**, 25847-25879 (1997).
5. Fujita, E.M. and D. Lawson. *SCOS97-NARSTO 1997 Southern California Ozone Study and Aerosol Study: Final Report* (California Air Resources Board, Sacramento, CA, 1999)
6. Isakov, V., T. Sax, A. Venkatram, D. Pankratz, J. Heumann, and D. Fitz. Near field dispersion modeling for regulatory applications, *J Air & Waste Manage. Assoc*, in press (2003)
7. Stein, A.F., Lamb D., and R.R. Draxler, Incorporation of detailed chemistry into a three-dimensional Lagrangian-Eulerian hybrid model: Application to regional tropospheric ozone, *Atmospheric Environment*, **34**, 4361-4372 (2000).

VARIATIONS OF FIRST GUESS AND ASSIMILATION DATA IN MM5 SIMULATIONS FOR AIR QUALITY MODELING IN SWITZERLAND

Johannes Keller, Sebnem Andreani-Aksoyoglu and Andre S.H. Prevot[*]

1. INTRODUCTION

In the next future the Laboratory of Atmospheric Chemistry (LAC) will use the meso-scale model MM5 as meteorological driver for air quality modeling in Switzerland and its adjacent countries. The model is initialized by data of the "alpine model" (aLMo) of MeteoSwiss. aLMo is a non-hydrostatic model operational at MeteoSwiss since April 2001. It is based on the Local Model (LM) developed in the frame of COSMO (COnsortium for Small scale MOdelling of the five national weather services of Germany, Switzerland, Italy, Greece and Poland). The model runs in a configuration of 385x325 grid points with a horizontal grid mesh of 7 km. It has 45 levels up to 23 km a.s.l., whereof 19 layers being within the first 2 km. Since January 2002, hourly aLMo outputs are available both as forecasts and as analyses, the latter being assimilated with soundings and surface measurements (wind only from stations below 100 m a.s.l.).

In a preliminary study, we simulated wind fields in Switzerland with MM5 for May 13, 2002. Two two-way nested domains centered at Zurich were selected. We specified the dimensions as 132 km x 108 km (domain 1) and 70 km x 46 km (domain 2) with grid resolutions of 3 and 1 km, respectively. The current version of MM5 uses 25 sigma pressure levels. The thickness of the bottom layer at 950 hPa corresponds to about 40 m.

First, MM5 was run without data assimilation. The model was initialized by aLMo forecasts (S1) and analyses (S2). In the third case (S3), S2 was supplemented with 4D nudging. There is an option to include surface observations and to generate a gridded surface field for the assimilation. The assimilation interval was 6 hours for the first guess data and 2 hours for the assimilation of surface data provided by the ANETZ network of MeteoSwiss. The last case (S4) is similar to S3, but the time step of the surface nudging was increased to 6 hours. The model was initialized at 6:00 UTC (7:00 CET) and was run for 12 hours. In the following, wind fields at 18:00 UTC are compared.

---

[*] Laboratory of Atmospheric Chemistry, Paul Scherrer Institut, CH-5232 Villigen PSI, Switzerland

Air Pollution Modeling and Its Application XVI, Edited by
Borrego and Incecik, Kluwer Academic/Plenum Publishers, New York, 2004

2. RESULTS

At the time of initialization (6:00 UTC), horizontal surface wind speeds below 5 m/s were forecasted in most locations of the coarse domain. The aLMo analyses show comparable wind velocities, whereas the wind direction differed by roughly ±40° compared to the forecast. At 18:00 UTC wind speeds are higher, particularly at mountain grid points. Analysis winds of aLMo mostly exceed the forecasted values. There are substantial discrepancies between aLMo and ANETZ data mainly at elevated locations.

At 18:00 UTC, surface wind fields simulated with MM5 do not change much at altitudes below roughly 1000 m a.s.l., if first guess data are taken from aLMo analysis (S2) instead of forecast data (S1). At higher altitudes, however, substantial differences between aLMo wind fields and MM5 results were observed, for both the forecast (S1) and the analysis data (S2). The left panel of Figure 1 shows aLMo analysis and MM5 windfields.

If 4D and surface assimilations are applied (S3), the MM5 wind field of the coarse domain 1 is modified mainly in the area of domain 2 (right panel of Figure 1). It is evident that nudging induces a southward flow through the valleys west of Zurich. However, further analysis of the space and time dependence of these results is needed to interpret the wind patterns. Finally, the increase of the surface assimilation time step from 2 to 6 hours (S4) affect the results of domain 1 only at a few locations in the area of domain 2.

In the near future, the patterns will be investigated for various parameterizations of MM5 and for different meteorological situations. Other domains and spatial resolutions will be selected. For instance, regions located in the alpine valleys of southern Switzerland will be included.

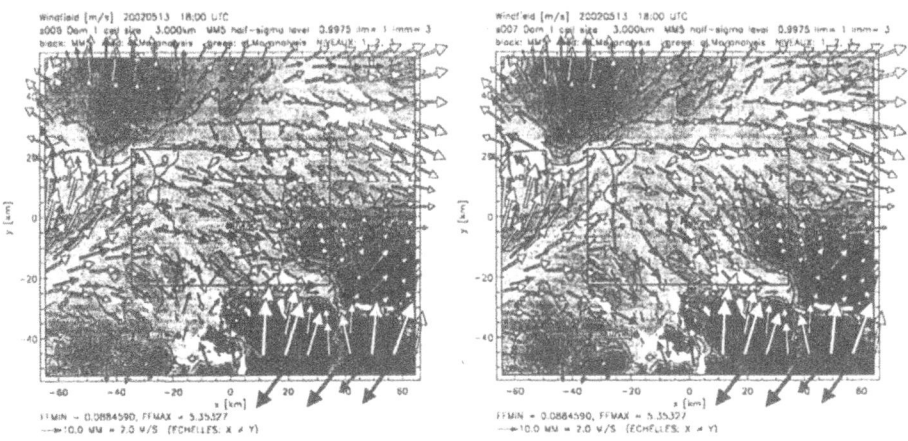

Figure 1. Wind fields of the lowest vertical layer of coarse domain 1 at 18:00 UTC (17:00 CET) on May 13, 2002. The MM5 simulation started at 06:00 AM. aLMo analysis data were used as a first guess. Left: MM5 simulation without 4D nudging; right: MM5 simulation with 4D nudging (every 6 hours) and surface nudging (every 2 hours). Open arrows: aLMo analysis; solid arrows: MM5 output.

3. ACKNOWLEDGEMENTS

We thank MeteoSwiss for providing aLMo and ANETZ data.

PARAMETERIZATION OF PBL USING SURFACE DATA FOR APPLICATION IN DISPERSION MODELLING

Dimiter Yordanov, Dimiter Syrakov, and Maria Kolarova[*]

The aim of the present work is to determine the vertical profiles of wind velocity, temperature and turbulent exchange coefficient in the Planetary Boundary Layer (PBL) using data collected from the automatic meteorological stations. The automatic meteorological stations (recently introduced in the NIMH network) provide hourly measurements of the wind velocity at 10 m and atmospheric stability classes based on measurements of the wind pulsations.

Two simple two-layer models of PBL are applied: YORDAN - for stable and neutral conditions and YORCON a CPBL model - for convective cases (Yordanov et al., 1997). Both models are developed in accordance with the similarity theory and consist of a Surface Layer (SL) and an Ekman layer over it. The PBL models are used to obtain the vertical profiles of the temperature, wind velocity and the turbulent exchange coefficient in PBL from the surface wind measurements and atmospheric stability data. As input to these models the internal to PBL parameters are needed. The internal parameters can be obtained from the experimental data applying two approaches: first one (called "top-down" approach) uses data for the geostrophic wind and the potential temperature and was described by Yordanov et al. (1997). The second one (called "bottom-up" approach) uses data from the surface meteorological observations. The description of this new approach was presented in details by Yordanov et al. (2002a; 2002b). The proposed approach determines the external to PBL parameters from the meteorological data measured in the surface layer on the basis of the Similarity theory.

To relate the M.O. length scale L to the roughness length z_o the empirical curves of Golder are used fitted by power low functions:

---

[*] Dimiter Yordanov, Geophysical Institute, Bulgarian Academy of Sciences (BAS), Sofia, Bulgaria. Dimiter Syrakov, National Institute of Meteorology and Hydrology (NIMH), BAS, Sofia, Bulgaria. Maria Kolarova National Institute of Meteorology and Hydrology (NIMH), BAS, Sofia, Bulgaria.

$$L^{-1} = a\,z_0^b,\qquad(1)$$

where the constants a and b depend on the stability class.

Having the stability class experimentally determined every hour from the automatic measurements, the corresponding Monin-Obukhov length scale L can be calculated at given roughness from Eq. (1).

The friction velocity u_* can be determined from the measurements of the wind velocity at $10m$ and the surface profile functions taken as log-linear for stable and neutral stratification, applying the PBL model YORDAN:

$$u_* = \frac{\kappa\,u_a}{f_u(\zeta_a,\zeta_0)},\qquad(2)$$

where κ is fon Karman constant, u_a is the wind velocity at the anemometer height, which is $z_a = 10m$, f_u is a function of the dimensionless height $\zeta_a = z_a/L$, and the dimensionless roughness $\zeta_0 = z_0/L$. All parameters in the right part of Eq. (2) are measured by the automatic stations.

For unstable stratification the universal profiles of Businger are used, applying the CPBL model YORCON. Under convective conditions the evolution of the convective PBL height (mixing layer height) at conditions of horizontal homogeneity is calculated from the M.-O. length scale L and u_* following the approach described by Yordanov et al. (1997).

Applying the relationship between the external and internal to PBL parameters given by the resistance and heat exchange laws the geostrophic wind - $|V_g|$, the cross isobaric angle - $|\alpha|$, and $\delta\theta$ - the difference between the potential temperature at PBL height and at the ground can be determined as shown by Yordanov et al. (2002a; 2002b).

Applying the similarity theory and resistance lows we can generally determine the surface turbulent fluxes defined by L and u_* from the external to PBL parameters ($|V_g|$, $|\alpha|$ and $\delta\theta$), often determined from the numerical weather prediction. The proposed approach solves the inverse problem that consists of determination of the external to PBL parameters and the vertical profiles of the temperature, wind velocity and the turbulent exchange coefficient (given by the PBL models) from the ground station meteorological measurements of the wind and atmospheric stability. The proposed parameterization was successfully applied in different practical tasks concerning air pollution modeling.

REFERENCES

Yordanov, D., Syrakov, D., and Kolarova, M., 1997, On the parametrization of the PBL of the atmosphere, *The Determination of the Mixing Height-Current Progress and Problems*, EURASAP Workshop Proc., 1-3 Oct. 1997, RISO Nat. Lab., Roskilde, Denmark, eds. S.-E. Gryning, F.Beyrich, E. Batchvarova, Riso-R-997(EN), pp. 117-120.

Yordanov, D., Syrakov, D., and Kolarova, M., 2002a, Parameterization of PBL from the surface wind and stability class data, *NATO ARW "Air Pollution Processes in Regional Scale"*, 13-15 June 2002, Halkidiki, Greece, pp. 71-82.

Yordanov, D., Syrakov, D., and Kolarova, M., 2002b, Parameterization of convective PBL using surface data for the wind and stability classes, Proc. 8th Int. Conf. on Harmonization within Atmospheric Dispersion Modelling for Regulatory Purposes, Sofia, Bulgaria, 14-17 Oct. 2002, eds. E. Batchvarova and D. Syrakov, pp.220-225.

Dispersion of pollutants from an elevated source in the Residual Layer: the influence of the convective decaying turbulence in the ground-level concentration

Davidson Moreira, Marco T. Vilhena, Antônio Goulart, Gervásio Degrazia, Domenico Anfossi, Jonas Carvalho and Paulo Ferreira Neto[*]

1. INTRODUCTION

About half hour before sunset over land, the surface heat flux (positive during the day) begins to decrease and then, during night-time, becomes negative and, consequently, a stable boundary layer (SBL) develops near the ground. Above this SBL, the convective boundary layer (CBL) starts to decay. In this study, employing simulations with a Eulerian analytical diffusion model, the influence of this decaying convective turbulence in the dispersion of contaminants is investigated. The important case of pollutants released from a tall stack during the characteristic sunset transition time (about one hour) is considered.

2. MODEL

In our model, the decaying convective turbulence in the residual layer and shear dominated turbulence in the SBL are parameterized. A SBL with a height of 100m was considered, whereas above of this shear dominated SBL a decaying convective turbulence is present. The dispersion model used in this study was derived by Moreira et al. (1999). The eddy diffusivity which describes the decaying convective turbulence is derived from a budget equation for the turbulent kinetic energy (Goulart et al. 2002):

$$\frac{K_z}{z_i w_*} = \frac{0.079}{\sqrt{1 + 2t_*^{1.7}}} \tag{1}$$

On the other hand, the eddy diffusivity which describes the shear dominated turbulence is derived from the classical statistical diffusion theory (Degrazia et al., 1996):

[*] Davidson Moreira, Jonas Carvalho and Paulo Ferreira Neto, ULBRA – PPGEEAM, Canoas, Brazil. Marco T. Vilhena, Universidade Federal do Rio Grande do Sul, Instituto de Matemática, Porto Alegre, Brazil. Antonio Goulart, URI – Departameto de Ciências Exatas e da Terra, Santo Ângelo, Brazil. Gervásio Degrazia, UFSM – Departamento de Física, Santa Maria, Brazil. Domenico Anfossi, CNR – ICG, Turin, Italy.

$$\frac{K_\alpha}{u_* h} = \frac{2\sqrt{\pi}\,0{,}64\,p_i^2\left(1-z/h\right)^{9_1}(z/h)X\left[2\sqrt{\pi}\,0{,}64(z/h)+8p_i(fm)_i\left(1-z/h\right)^{9_1/2}X\right]}{\left[2\sqrt{\pi}\,0{,}64(z/h)+16p_i(fm)_i\left(1-z/h\right)^{9_1/2}X\right]^2} \qquad (2)$$

3. RESULTS

The figure 1 shows that the released contaminants in the decaying convective boundary layer produce pronounced ground-level concentrations. This result is a manifestation of the turbulent diffusion provocated by the decaying of the convective energy-containing eddies. We conclude that the parameterization of this decaying eddies, during the transition between the CBL and the SBL, is of fundamental importance to calculate the ground-level concentrations originated from elevated pollutants sources.

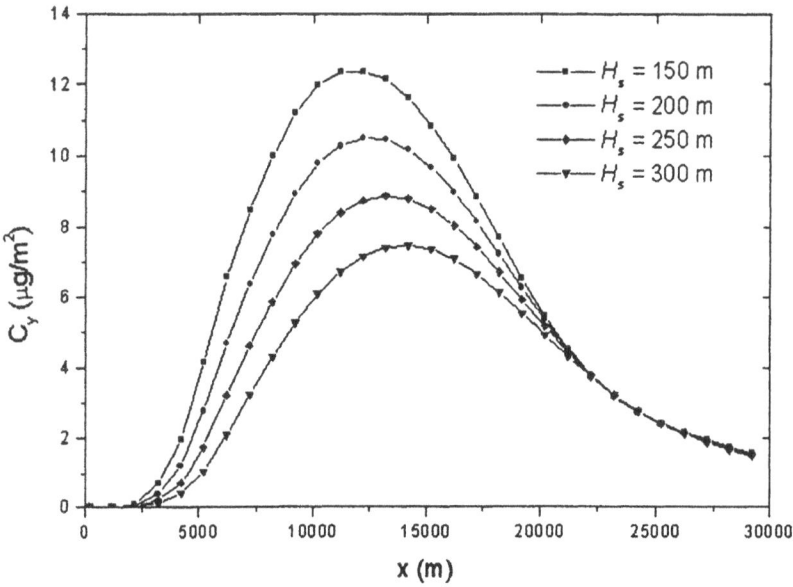

Figure 1 - The crosswind integrated ground-level concentration for different sources height H_s.

3. REFERENCES

Goulart, A., Degrazia, G.A., Anfossi, D. and Acevedo, O., 2002. Modelling and eddy diffusivity for convective decaying turbulence in the residual layer. *15th Symposium on Boundary layers and turbulence*, American Meteorological Society, 267-268.

Degrazia, G.A., Vilhena, M.T., Moraes, O.L.L., 1996. An algebraic expression for the eddy diffusivities in the stable bondary layer: a description of near-source diffusion. *Il Nuovo Cimento*, **19**:.399-403.

Frisch, U., 1995, Turbulence, ed. Cambridge University Press, pp. 296.

Hinze J.O., 1975, Turbulence, Ed. Mc. Graw Hill. pp. 790.

Moreira, D.M., G.A. Degrazia and M.T. Vilhena, 1999. Dispersion from low sources in a convective boundary layer: an analytical model. *Il Nuovo Cimento*, **22**C: 685-691.

Nieuwstadt, F.T.M. and Brost R.A., 1986, The decay of convective turbulence, *J. Atmos. Sci.* **43**: 532.

Applications of Rams and Aermod models to evaluate pollution dispersion in a coastal valley

R. Cocci Grifoni, G. Passerini, S. Tascini[*]

1. INTRODUCTION

AERMOD is an air pollution model that possibly incorporates all the current understanding of dispersion and micrometeorology to model the impact of sources at short distances. Amongst all its capabilities, it can make use of the more advanced meteorological information such as that produced by a mesoscale model, namely RAMS.

RAMS is capable to model weather systems such as land/sea breezes and mountain circulations and it is suitable to model meteorological conditions in a complex coastal area. In this paper we compare the results of a set of AERMOD runs using RAMS data as input with the results of AERMOD runs made using surface and upper air data from a rather distant Airbase Station.

2. APPLICATION OF THE MODELS

The main purpose of this paper is to evaluate the use of the mesoscale model, RAMS (Pielke et al., 1992), in preparing meteorological input values for the pollutant dispersion model AERMOD (EPA, 1998). We have developed a procedure to extract data from the mesoscale model RAMS, ingest it into AERMOD meteorological preprocessor AERMET, and then run AERMOD to obtain pollutant concentration levels in a quite wide area.

The coast of Marche Region (Italy) was selected as study area due to the presence of different types of topographically influenced flow patterns. Results from a series of sensitivity experiments (Latini G. et al, 2002) indicated significant topographical forcing although the synoptic forcing still remains quite strong in this valley. RAMS was run with two and three nested grids being the finest grid of 1-km horizontal spacing. AERMOD was run to predict short-term, one-hour-average concentrations at each receptor using three months of winter meteorology, namely November 2001,

[*] Dipartimento di Energetica, Università Politecnica delle Marche, 60131 Ancona, Italy

Air Pollution Modeling and Its Application XVI, Edited by
Borrego and Incecik, Kluwer Academic/Plenum Publishers, New York, 2004

December 2001 and January 2002. A discrete Cartesian receptor grid network containing 400 receptors with a resolution of 500 meters was selected.

3. RESULTS AND CONCLUSIONS

As an example, here we present a typical result of our simulations, namely related to 3-4 December 2001. The location of the source and the predicted 1-hour maximum-impact locations are depicted in Figures 1a and 1b. For the first simulation (Figure 1a) AERMOD modelling system has been initialised using the wind fields predicted by RAMS. AERMOD was then used to calculate 1-hour-average values for each day of the selected period. For the second simulation AERMOD was run using upper air data from the closest Airbase Station, namely Pratica di Mare Airport. Unfortunately, this station is about 200 km far from our area. Hourly surface data came from a station located nearby the emission source. The source configuration was the same for both the experiments.

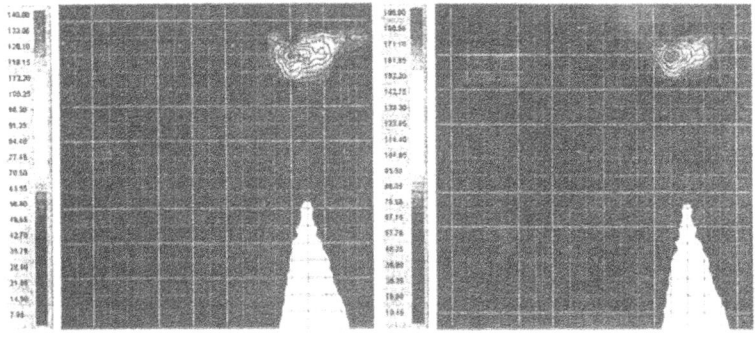

Figure 1a,1b. AERMOD and RAMS-AERMOD maximum predicted 1-hr concentrations.

From these results it is readily apparent that both model configurations predicted very similar short-term-average concentrations, when using identical input data. The locations of concentration maximums are almost the same but the values predicted using RAMS-AERMOD are slightly higher. Based on our experience we can postulate that RAMS-AERMOD values are the most likely to be exact. This could be due to the evident capability of RAMS to model PBL structure and parameters.

4. REFERENCES

1. U.S. EPA, 1998. User's Guide for the AERMOD. Office of Air Quality Planning and Standards, Research Triangle Park, NC.
2. Latini, G., Cocci Grifoni, R., Passerini, G., 2002: The Optimal Choice of AERMOD Input Data in Complex Areas. Air Pollution Modelling and its Applications XV, eds. Borrego and Schayes, Kluwer Academic/Plenum Publishers, New York, 513-514
3. Pielke, R. A., W. R. Cotton, R. L. Walko, C. J. Tremback, W. A. Lyons, L. D & Grasso, M. E. Nicholls, M. D. Moran, D. A. Wesley, T. J. Lee, and J. H. Copeland, 1992: A comprehensive meteorological modeling system--RAMS. Meteor. Atmos. Phys., 49, 69-91.

MODELLING ACTIVITY IN THE FRAMEWORK OF THE NATIONAL PROJECT "TRANSFORMATION OF AIR-POLLUTION, MODELLING ITS TRANSPORT AND DISPERSION"

Tomas Halenka[1], Josef Brechler, Jan Bednar[*]

1. INTRODUCTION

In the framework of the presented activity a spatial distribution of tropospheric ozone is modelled. An overview of the ozone distribution will be given for local modelling study with very high resolution. Model results will be tested against measured data that will be collected both from field campaigns and from existing monitoring networks. Within this modelling activity results of some formerly and simultaneously solved projects can be also used.

2. CASE STUDY

Modelling activity in the framework of the mentioned project mainly deals with distribution of ground concentration of photooxidative air-pollution that can be represented with tropospheric ozone O_3. Chemical submodel SMOG solving this kind of task has been developed at the Department of Meteorology and Environment Protection, Faculty of Mathematics and Physics, Charles University, Prague, see Bednar et al.,2001, for detailes.

For the first test the region of Cervenohorske sedlo (Northern Moravia) which is covered by measurements quite well was selected to be model area. There was interesting episode in June 2000. Basic inputs for modelling study contain mainly well detailed data concerning emission sources (technical detailes and emissions of NO_x and organics) from REZZO1, information concerning the sources located in the area of interest from REZZO2 and area sources of REZZO3 devided with respect to technology (combustion processes, ventilation of organics) as well as emissions from transportation (REZZO4). Biogenic emissions are not involved in the above mentioned information.

---

[1] Regular associate of the Abdus Salam ICTP, Trieste, Italy

[*] Tomas Halenka, Josef Brechler, Jan Bednar, Charles University in Prague, Faculty of Math. and Physics, Prague, Czech Republic

Air Pollution Modeling and Its Application XVI, Edited by
Borrego and Incecik, Kluwer Academic/Plenum Publishers, New York, 2004

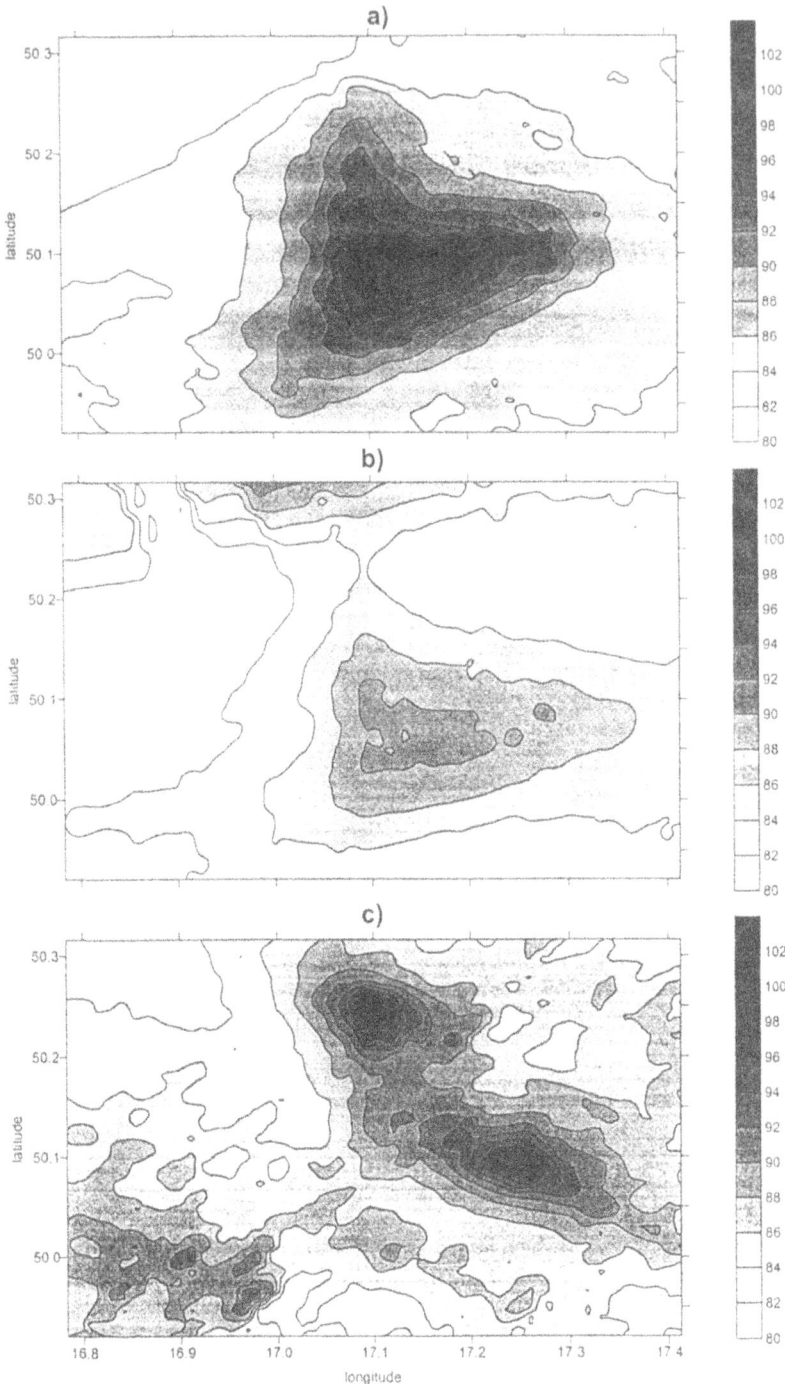

Figure 1. Surface concentration of O_3 ($\mu g.m^{-3}$).

630

Three typical groups of synoptic patterns were analyzed in June 2000 for which surface concentrations of NO, NO_2 and O_3 were computed in area of interest. Those are:

a)   weak anticyclonic situation, WSW advection with wind of about 5m/s (850 hPa), with surface temperature about 27°C, cloud cover about 40% (12 days)

b)   cyclonic situation, W advection with wind of about 9m/s, with surface temperature about 23°C, cloud cover about 70% (6 days)

c)   typical anticyclonic situation, weak SE advection with wind of about 3m/s, with surface temperature about 30°C and cloudness about 10% (9 days)

Days with precipitation were not included. The case a) is generally considered to be with quite good conditions for formation of photochemical smog, whereas the case b) is rather opposite. In case c) we could expect exceptional conditions for photochemical smog formation, but, unfortunately, as working only with emission data from Czech Republic and having in mind the location of the model area, limited number of emission sources could be taken into account in this case. Despite of this, and as a consequence of the importance of local sources in formation of photochemical smog, our results represent this case to be with the highest concentrations of O_3.

The results presented in Fig. 1 are validated against the monitoring stations. In first guess, this study served to obtain rough estimate of potential of our model SMOG, originally developed rather for photochemical smog modelling in urban environment, in case of rather remote areas. Based on comparison with real data it seems to make a sense to continue with more detailed experiment with the simulation day by day with nesting of the SMOG model into meteorological model to have more detailed description of flow field and some other parameters for the model computation. Work in this direction is now in progress, trajectory computation with all necessary parameters being computed by the meteorological model as a pre-processor.

ACKNOWLEDGMENT

This work has been performed in the framework of the project VaV/740/2/01 funded by the Ministry of Environment of the Czech Republic and in the framework of projects 205/01/1120 and 205/02/1488 funded by the Grant Agency of the Czech Republic.

REFERENCES

1. Machálek, P., 2001: Trends in air-pollution in the years 1990 – 1999. In Proc.: *Atmosphere 2001*, Brno, 2001 (in Czech).
2. Leníček, J.,. Sekyra, M., Kociánová, S., Novák, J., Beneš, I., 1997: Determination of VOC in some locations of the Czech Republic, *Air Protection*, **2**, 10 – 12 (in Czech).
3. Mesinger, F., Janjić, Z., Nickovič, S., Gavrilov, D., Deaven, D.G.,1988: The step-mountain coordinate: Model description and performance for cases of Alpine lee cyclogenesis and for a case of an Appalachian redevelopment. *Mon. Wea. Rev.*,**116**, 1493 – 1518.
4. Bednář, J., Brechler, J., Halenka, T. and Kopáček, J., 2001: Modelling of Summer Photochemical Smog in the Prague Region, *Phys. Chem. Earth (B)*, **6**, 129 – 136.
5. Grell, G.A., Dudhia, J., Stauffer, D.R., 1994: A Description of the Fifth-Generation Penn/Srate /NCAR Mesoscale Model (MM5), NCAR Technical Note, NCAR/TN-398 + STR.
6. ENVIRON International Corp., 1998: User's Guide: Comprehensive Air Quality Model with Extension (CAMx). Version 2.00., Novato, California 94945-5010.
7. Schlünzen, K. H., Bigalke, K., Lenz, C.-J., Lüpkes, Ch., Niemeier, U., von Salzen, K., 1996: Concept and Realuization of the Mesoscale Transport and Fluid Model 'METRAS', Universität Hamburg, 1996.

A METHOD OF ESTIMATING UNCERTAINTY IN SIMULATED TRANSPORT AND DISPERSION OF POLLUTANTS IN COMPLEX TERRAIN

Darko Koračin and Anna Panorska[*]

1. INTRODUCTION

Accurate prediction of the transport and dispersion of atmospheric pollutants and tracers in complex terrain represents a significant research challenge. The differences between simulations and measurements are caused by model simplifications and assumptions, uncertainty in measurements, and variability of natural processes that have a random component. The uncertainties in the model and measurements could be reduced to certain extent, while the natural variability cannot be fully accounted for. In this study, we present a method of estimating uncertainties in the simulated dispersion as a residual term in the equation for the fractional error.

2. METHOD AND RESULTS

The main objective of the study is to provide a rigorous statistical model for the partitioned error of an atmospheric dispersion model. We define that error as a fractional error:

$$E_f = (C_0 - C_m)/C_0 \tag{1}$$

where C_0 is the observed tracer concentration and C_m is the concentration predicted by the model. We hypothesize that the model fractional error is a linear combination of the main normalized error components due to:

---

[*] Darko Koračin, Desert Research Institute, 2215 Raggio Parkway, Reno, Nevada 89512, USA. Anna Panorska, University of Nevada, Campus Box 084,Reno, Nevada 89557, USA.

- Emissions (E_e)
- Measurements (E_m)
- Atmospheric and dispersion models (E_{ad})
- Stochastic processes (E_s).

Furthermore, using results from the studies by Koračin et al. (2000a,b), we can estimate independently the error due to the atmospheric model. Using the results from these studies, one can separate the total error due to both the atmospheric and dispersion models (E_{ad}) into two separate errors: E_a and E_d for the atmospheric and dispersion model errors, respectively. In this case, the model fractional error is:

$$(C_0 - C_m) / C_0 = E_e + E_m + E_a + E_d + E_s \qquad (2)$$

The errors due to emissions, measurements, and the atmospheric model can be represented by normal distributions with known standard deviations. The error due to stochastic processes can be also represented by a normal distribution with unknown standard deviation. The dispersion model error represents an unknown residual term. Using the Maximum Likelihood Estimators (MLE) technique, the standard deviation of the stochastic error and the dispersion model error can be computed.

The method has been demonstrated by using observations and modeling results from a tracer experiment in the complex terrain of the southwestern US (Green, 1999; Koračin et al., 2000a; Podnar et al., 2002). Mesoscale Model 5 (MM5) (Grell, 1995) was used to provide atmospheric fields that were input to the dispersion simulation. A Lagrangian random particle dispersion model (Koračin et al., 1999) with three optional turbulence parameterizations has been used as a testbed for the method application. The dispersion model yielded high correlation coefficients with measurements ranging from 0.7-0.8. The dispersion error and the variance of the stochastic component have been compared for model runs with various turbulence schemes. Improvement of the dispersion model success by using higher vertical resolution in the atmospheric model as compared to the base run has been also evidenced by using the developed method. These preliminary results indicate that the normalized dispersion error appears to be strongly influenced by the choice of a turbulence scheme while the increased vertical resolution only slightly improves the dispersion error estimates for all turbulence schemes.

3. REFERENCES

Green, M. C., 1999, The Project MOHAVE tracer study, Study design, data quality, and overview of results. *Atmos. Environ.*, 33, 1955-1968.

Grell, G.A., Dudhia, J., and Stauffer, D. R., 1995, *A Description of the Fifth-Generation Penn State/NCAR Mesoscale Model (MM5)*. National Center for Atmospheric Research, Techn. Note TN-398, 122 pp.

Koračin, D., J. Frye, and V. Isakov, 2000a, A method of evaluating atmospheric models using tracer measurements. *J. Appl. Meteor.*, 39, 201-221.

Koračin, D., V. Isakov, and J. Frye, 2000b, A method of evaluating atmospheric models using tracer measurements: Main algorithms. *Int. J. Environment and Pollution*, 14, Nos. 1-6, 89-97.

Koračin, D., V. Isakov, D. Podnar and J. Frye, 1999, Application of a Lagrangian random particle dispersion model to the short-term impact of mobile emissions. Proceedings of the Transport and Air Pollution conference, Graz, Austria, 31 May - 2 June 1999.

Podnar, D., D. Koračin, and A. Panorska, 2002, Application of artificial neural networks to modeling the transport and dispersion of tracers in complex terrain. *Atmos. Environ.*, 36, 561-570.

EVALUATION OF THE ADAPTIVE GRID
AIR POLLUTION MODEL

M. Talat Odman and Maudood N. Khan[*]

1. INTRODUCTION AND METHODOLOGY

The adaptive grid air pollution model developed by Odman et al. (2002) is used to simulate the July 9-17, 1995 ozone episode in the Tennessee Valley. The model is evaluated using observations from a surface network and compared to its own static grid version where the adaptive grid feature is turned off. In the static grid simulation, the grid is uniform with 8-km resolution. The adaptive grid employs the same grid structure but dynamically refines the grid in places where the gradients in NO_x concentrations change rapidly. This may result in grid sizes as small as 200 m locally downwind of major sources but the total number of grid cells does not change.

Meteorological fields and all emissions other than those from point sources were processed on a uniform grid with 8-km resolution. While the static grid model uses these input data directly, the adaptive grid model interpolates them to the new grid locations after each adaptation. The point source emissions are dumped into the grid cell containing the stack. Finding this cell requires a search procedure in the adaptive grid simulation.

The modeled ozone levels were compared to the observations at 69 surface stations within the domain. A normalized station error is calculated using all available hourly ozone observations at that station. This error is the average of the absolute values of the differences between the modeled and observed values divided by the observed values.

3. RESULTS

The average of the normalized station errors is 44% with the static grid and 41% with the adaptive grid model. The normalized error with the adaptive grid model is smaller at 80% of the stations. The decrease in error is as much as 24% with an average of about 5%. At the remaining 20% of the stations the normalized error with the adaptive grid model increases as much as 6% with an average of about 3%.

---

[*] Talat Odman, Georgia Institute of Technology, Atlanta, Georgia, 30332-0512, talat.odman@ce.gatech.edu, Telephone: 770-754-4971, Fax: 770-754-8266.

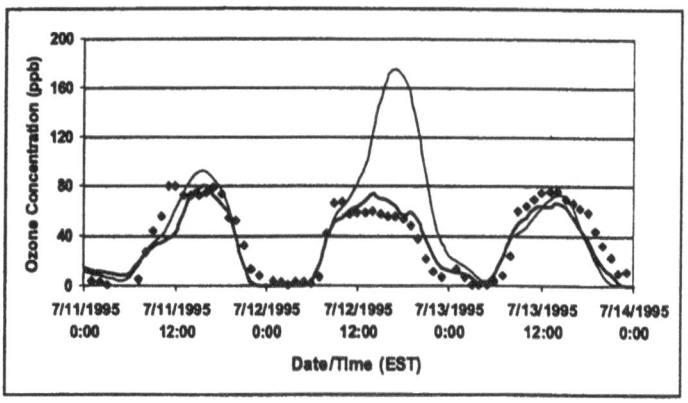

Figure 1. O$_3$ at Livingston Co., KY: observations ($\bullet$), static grid (—) and adaptive grid (—) modeling results.

Large differences between the two models ozone results are observed when there is a major point source plume near the monitoring sites. Figure 1 illustrates one such case. Investigation of this case revealed a very complex source-receptor relationship. On July 11, northeasterly winds dispersed the plume from Cumberland power plant towards Midwestern Tennessee. The plume is less dispersed on the adaptive grid simulation with a more distinct core further downwind. During the night, a divergence in the wind field led to completely different trajectories. The puff traveled west in the adaptive grid while it veered northwest in the static grid simulation. The next day, southeasterly winds carried the puff into Kentucky in the static grid simulation. This resulted in high ozone estimates at Livingston County, KY. This was not the case in the adaptive grid simulation, which estimated ozone levels in better agreement with the observations.

4. CONCLUSION

The ozone simulation skill of the adaptive grid model was evaluated. The NO$_x$ fields showed gradients with a level of detail that is likely unprecedented in regional scale simulations. Compared to the static grid model, the adaptive grid model estimated ozone concentrations more consistent with observations at stations downwind of power plants and resolved some very complex long-range source-receptor relationships.

5. ACKNOWLEDGEMENT

This research is supported by U.S. EPA STAR Program, Grant No. R 827028-01-0.

6. REFERENCE

Odman, M. T., Khan, M. N., Srivastava, R. K. and McRae, D. S., 2002. Initial application of the adaptive grid air pollution model, in: *Air Pollution Modeling and its Application XV*, C. Borrego and G. Schayes, eds., Kluwer Academic/Plenum Publishers, New York, pp. 319-328.

STATISTICAL MODELING OF WINTER AIR-POLLUTION STUDY IN URBAN AREA OF SIVAS, TURKEY.

Lokman H. Tecer[*], Ali Yılmaz[**], and [*]Yılmaz Yıldırım

ABSTRACT

In this study air pollution level in the city center of Sivas for a 10 years' period was analyzed. Winter season' and annual averages of sulfur dioxide and particulate matter measurements have been carried out in central area of Sivas by City Health Management. Specifically the winter seasons' concentrations are presented and their causes and their change with respect to meteorological parameters are discussed. The statistical relations between pollutants and meteorological parameters have been found to be significant.

Key words: Urban Air Pollution, Meteorology, and Statistic.

1. INTRODUCTION

Sivas where the terrestrial climate effects are seen and whose winters are cold and long, sometimes experiences air pollution problem in winter. The air pollution in city causes from the cool that is being used for heating, negative meteorological conditions, topographic structure and unsystematic urbanization (TCDIE,1992).

In this study air pollution level in the city center of Sivas for 10 years period includes from 1990 to 2000 are analyzed. Air pollution and meteorological factors have been evaluated together and air pollution predict models have been developed using statistical models (Simpson and Layton, 1983; Brown and Bilger, 1998).

2. METHODOLOGY

The measurements of air pollutants (SO_2, PM) began measured at 4 fixed stations of city center. Each parameter was measured as daily averages. The data that chosen as predictor variables in the predict models were obtained from The Turkish State

---

[*] Karaelmas University, Eng. Fac. Environmental Eng. 67100, Zonguldak, Turkey
[**] Cumhuriyet University, Eng. Fac. Environmental Eng. 58140, Sivas, Turkey

Meteorological Service. The 10 years values of wind speed, temperature, cloudiness, total solar radiation, the time of solar radiation, humidity and precipitation were used.

Using the most suitable variables, air pollution predict models have been developed using statistical models. Model's statistical parameters were estimated by using SPSS 8.0 statistical software (SPSS(a), 1998; Kinnear and Gray, 1996).

3. RESULTS AND DISCUSSION

Sivas city's winter period air pollution trend have shown a decreasing. By 1995-1996 winter period there is serious pollution problem in city. The cause of the air pollution is coal and fuel oil that used for approximately 6 months for heating (Polat, 1990). High sulphur ratio coal that using in city is air pollutant by oneself. Also negative meteorological conditions are another effect that causes the accumulation of the pollutants in Sivas.

From observation results we can understand especially until 1995-1996 winter period there was SO_2 and PM concentrations in high levels. December, January, February is the most polluted months. Decreases in air pollution after 1995-1996 winter attributed to increase in using high quality and import coal (Beyazıt, and Bali, 1996). From 1996 to 2000 this decrease in SO_2 and PM pollution have continued but still can't reach air quality criteria that recommended by WHO.

4. CONCLUSSION

In this study SO_2 and PM concentrations were modeled by using meteorological parameters of last 10 years. Statistical SO_2 and PM model's determined coefficients were found as R=0,805 and R=0,725 by using multiple linear regression method, respectively. The coefficients of variables, that effect accumulation and/or dispersion of pollutants as wind, temperature, cloud ness, total solar radiation are significant at the level %99. The possible pollution levels can be determined by using the result of the pollutants predict models at some meteorological conditions which effect negatively the pollutants.

SO_2 trend observed in city is explained with previous day's SO_2 concentration, maximum speed of wind and temperature. SO_2 concentrations have opposite proportion between temperature and speed of wind. PM trend changes like this too.

The basis cause of SO_2 and PM pollution is less calorie and high sulphur coal using for heating and other fuels. Using high quality and import coals after 1993 decreases pollution to important levels. In spite of these in December, January and February there is still air-pollution in high levels. Negative meteorological conditions and topographic structure raises effects of pollution in these months.

5. REFERENCES

Beyazıt, N., Bali, U., 1996. Sivas'ta Hava Kirliliği ve Meteorolojik Parametrelerle İlişkisinin Araştırılması, I. Uludağ Çevre Mühendisliği Sempozyumu, 24-26 Haziran, Bursa, Turkey.
Brown, R.J. and Bilger, R.W., 1998. Experiments on a Reacting Plume – 2. Conditional Concentration Statistics. Atmospheric Environment, 9, 417 – 423.
Kinnear, P.R. and Gray, C.D. (1996). SPSS for Windows Made Simple, Erlbaum (UK) Taylor and Francis, UK.
Polat, H., 1990. Sivas'ta Hava ve Su Kirliliği. Çevre ve İnsan, 52.
Simpson, R.W. and Layton, A.P., 1983. Forecasting Peak Ozone Levels. Atm. Env., 17, 9,1649 – 1654.
SPSS(a), (1998) Base 8.0 for Window, User Guide, SPSS Inc. USA.
SPSS, (1998). SPSS Base 8.0, Applications Guide, SPSS Inc. U.S.A.
T.C. Devlet İstatistik Enstitüsü. (1992). Çevre İst., Hava Kirliliği 1984-1991. DIE matbası, Ankara, Turkiye.

PARTICIPANTS

The 26th NATO/CCMS International Technical Meeting on Air Pollution Modelling and Its Application, Istanbul, Turkey, May 26-30, 2003

BELGIUM

Lefebre F.   Vito-Flemish Institute for Technological Research
                      Boeretang 200 B-2400 Mol
                      e-mail: Filip.Lefebre@vito.be

Porson A.   Université catholique de Louvain
                     Institut d'Astronomie et Geophysique
                     2, Chemin du Cyclotron B-1348 Louvain-La-Neuve
                     e-mail: porson@astr.ucl.ac.be

Schayes G.   Université catholique de Louvain
                     Institut d'Astronomie et Géophysique
                     2, Chemin du Cyclotron B-1348 Louvain-La-Neuve
                     e-mail: schayes@astr.ucl.ac.be

BULGARIA

Batchvarova E.   National Institute of Meteorology and Hydrology
                        66 Blvd. 'Tzarigradsko Chaussee' 1784 Sofia
                        e-mail: Ekaterina.Batchvarova@meteo.bg

Dimitrova R.   Bulgarian Academy of Sciences
                        Institute of Geophysics Acad.
                        G. Bonchev Str. Bl. 3 Sofia 1113
                        e-mail: rendim@geophys.bas.bg

Ganev K.   Bulgarian Academy of Sciences
                        Institute of Geophysics Acad.
                        G. Bonchev Str. Bl. 3 Sofia 1113
                        e-mail: kganev@geophys.bas.bg

Kolarova M.   Bulgarian Academy of Sciences
                        Institute of Geophysics Acad.
                        G. Bonchev Str. Bl. 3 Sofia 1113
                        e-mail: Maria.Kolarova@meteo.bg

Miloshev N.   Bulgarian Academy of Sciences
                        Institute of Geophysics Acad.
                        G. Bonchev Str. Bl. 3 Sofia 1113
                        e-mail: miloshev@geophys.bas.bg

Petrova Prodanova M.   National Institute of Meteorology and Hydrology
                            66 'Tzarigradsko Chaussee' Blvd. 1784 Sofia
                            e-mail: Maria.Prodanova@meteo.bg

Slavov K.   Bulgarian Academy of Sciences
                            Institute of Geophysics Acad.
                            G. Bonchev Str. Bl. 3 Sofia 1113
                            e-mail: Kiril.Slavov@meteo.bg

Syrakov D.   National Institute of Meteorology and Hydrology
                            66 'Tzarigradsko Chaussee' Blvd. 1784 Sofia
                            e-mail: dimiter.syrakov@meteo.bg

Yordanov D.   Bulgarian Academy of Sciences
                            Institute of Geophysics Acad.
                            G. Bonchev Str. Bl. 3 Sofia 1113
                            e-mail: kganev@geophys.bas.bg

CANADA

Ainsie B.   University of British Colombia
                            Atmospheric Science Programme
                            6339 Store Road UBC Vancouver
                            e-mail: bainslie@eos.ubc.ca

Cho S.   York University
                                4700 Keele St. Toronto Ontario M3J 1P3
                                e-mail: sunnycho@yorku.ca

Gong W.   Meteorological Service of Canada
                                4905 Dufferin St. Downsview Ontario M3H 5T4
                                e-mail: wanmin.gong@ec.gc.ca

Martilli A.   University of British Colombia
                                Vancouver BC V6T 1Z4
                                e-mail: amartilli@eos.ubc.ca

Michelangeli D.   York University
                                4700 Keele St. Toronto Ontario M3J 1P3
                                e-mail: dvm@yorku.ca

Norman A.-L.   The University of Calgary
                                Department of Physics and Astronomy
                                2500 Calgary Alberta T2N 1N4
                                e-mail: annlisen@phas.ucalgary.ca

Sloan J.   University of Waterloo
                                Department of Chemistry
                                e-mail: sloanj@uwaterloo.ca

Steyn D. G.　　　　　　　University of British Colombia
　　　　　　　　　　　　Atmospheric Science Programme
　　　　　　　　　　　　Vancouver B.C.
　　　　　　　　　　　　e-mail: dsteyn@eos.ubc.ca

CROATIA

Klaic Z. B.　　　　　　　University of Zagreb
　　　　　　　　　　　　Faculty of Science
　　　　　　　　　　　　e-mail: zklaic@rudjer.irb.hr

CZECH REPUBLIC

Brechler J.　　　　　　　Charles University
　　　　　　　　　　　　Department of Meteorology and Env. Protection
　　　　　　　　　　　　e-mail: josef.brechler@mff.cuni.cz

Bubnik J.　　　　　　　　Czech Hydrometeorological Institute
　　　　　　　　　　　　e-mail: bubnik@chmi.cz

Halenka T.　　　　　　　Charles University
　　　　　　　　　　　　Department of Meteorology and Env. Protection
　　　　　　　　　　　　e-mail: tomas.halenka@mff.cuni.cz

DENMARK

Geels C.
National Environmental Research Institute
Department of Atmospheric Environment
Frederiksborgvej 399   DK-4000 Roskilde
e-mail: cag@dmu.dk

Gryning S.-E.
Risø National Laboratory
Wind Energy Department
DK-4000 Roskilde
e-mail: Sven-Erik.Gryning@risoe.dk

ESTONIA

Kaasik M.
University of Tartu
Institute of Environmental Physics
Ulikooli 18 50090 Tartu
e-mail: mkaasik@ut.ee

FINLAND

Siljamo P.
Finnish Meteorological Institute
PO Box 503 Vuorikatu 24
FIN 00101 Helsinki
e-mail: pilvi.siljamo@fmi.fi

Sofiev M.

Finnish Meteorological Institute
PO Box 503 Vuorikatu 24 Helsinki
e-mail: mikhail.sofiev@FMI.FI

FRANCE

Brulfert G.

Université Joseph Fourrier
Lab. des Ecoulements Géophys. et Industriels
BP 53 38041 Grenoble Cedex 9
e-mail: guillaume.brulfert@hmg.inpg.fr

Chaumerliac N.

Laboratorie de Meteorologie Physique/OPGC
Université Blaise Pascal CNRS
24 avenue des Landais 63177 Aubiére Cedex
e-mail: N. Chaumerliac@opgc.univ-bpclermont.fr

Deguillaume L.

Laboratorie de Meteorologie Physique/OPGC
Université Blaise Pascal CNRS
24 avenue des Landais 63177 Aubiére Cedex
e-mail: deguilla@opgc.univ-bpclermont.fr

GERMANY

Flemming J.
Freie Universitaet Berlin
Karl-Heinrich-Becker Weg 10
D-12165 Berlin
e-mail: flemming@zedat.fu-berlin.de

Reimer E.
Universitaet Berlin
Karl-Heinrich-Becker Weg 10
D-12165 Berlin
e-mail: reimer@zedat.fu-berlin.de

Renner E.
Institute for Tropospheric Research
Permoserstr. 15 D-04303 Leipzig
e-mail: renner@tropos.de

Schatzmann M.
University of Hamburg
Meteorological Institute
Bundesstr. 55 D-20146 Hamburg
e-mail: schatzmann@dkrz.de

Suppan P.
Institute of Meteorology and Climate Research
Environmental Atmospheric Research IMK-IFU
Research Centre Karlsruhe
D-82467Garmich-Partenkirchen
e-mail: peter.suppan@imk.fzk.de

646

Wolke R.

Institute for Tropospheric Research
Permoserstr. 15 D-04303 Leipzig
e-mail: wolke@tropos.de

GREECE

Boucouvala D.

Department of Applied Physics
Laboratory of Meteorology, Build. Phys-V,
University of Athens, 15784 Athens
e-mail: dbouc@atlas.uoa.gr

Kallos G.

University of Athens
School of Physics AM&WF Group
Bldg PHYS-5 15784 Athens
e-mail: kallos@mg.uoa.gr

Lisaridis I.

Aristotle University
Department of Environmental Physics
e-mail: hralys@auth.gr

Moussiopoulos N.

Aristotle University
Lab. of Heat Transfer and Env. Engineering
Box 483 GR-54124 Thessaloniki
e-mail: moussio@vergina.eng.auth.gr

Pastras G.   Aristotle University
                               Department of Environmental Physics
                               e-mail: pastras@panafonet.gr

ISRAEL

Haikin N.   Nuclear Research Centre-Negev
                               PO Box 9001 Beer Sheva 84190
                               e-mail: nitsah@nrcn.org.il

Terliuc B.   Nuclear Research Centre-Negev
                               PO Box 9001 Beer Sheva 84190
                               e-mail: terliuc@zahav.net.il

ITALY

Agostini E.   DIMNP-Pisa University
                               e-mail: eleonora.agostini@ing.unipi.it

Anfossi D.   Istituto di Scienze dell'Atmosfera e del Clima
                               Corso Fiume 4 Torino
                               e-mail: anfossi@to.infn.it

D'Allura A.   University of Milano
                               BICOCCA
                               e-mail: alessio.dallura@unimib.it

Grandoni G.   ENEA-CASACCIA
                               e-mail: giovanni.grandoni@casaccia.enea.it

Thunis P.   Institute for Environment and Sustainability
                               Joint Research Centre
                               TP 280 21020 Ispra
                               e-mail: philippe.thunis@jrc.it

Racalbuto S.   ENEA-CASACCIA

Tascini S.   Universita Politecnica delle Marche
                               Dipartimento di Energetica
                               60131 Ancona

Trini Castelli S.   ISAC CNR
                               Home 4 10133 Torino
                               e-mail: trini@to.infn.it

KOREA

Park S.-U.   Seoul National University
                               School of Earth and Environmental Sciences
                               151-747 Seoul
                               e-mail: hjin@snupbl.snu.ac.kr

LITHUANIA

Ulevicius V.   Institute of Physics
                             Savanouri 231 Vilnius LT-2053
                             e-mail: ulevicv@ktl.mii.lt

NETHERLANDS

Builtjes P.   TNO-MEP Apeldoorn
                             e-mail: p.j.h.builtjes@mep.tno.nl

Dosio A.   Wageningen University
                             Meteorology and Air Quality
                             Duivendaal 2 6701 AP
                             e-mail: dosio@hp1.met.wau.nl

Fournier N.   CEH-Centre of Ecology and Hydrology
                             Iaan van Roos En Doorn 7 2514 BC Den Haag
                             e-mail: nicolas@onetelnet.nl

Velders G. J. M.   Netherlands Env. Assessment Agency RIVM
                             PO Box 3720 BA Bilthoven
                             e-mail: guus.velders@rivm.nl

NORWAY

Bartnicki J.

Norwegian Meteorological Institute
PO Box 43 Blindern N-0313 Oslo
e-mail: jerzy.bartnicki@met.no

Oyvind S.

University of Oslo
Department of Geophysics
PO Box 1022 Blindern N-0315
e-mail: oyvind.seland@geofysikk.uio.no

Saltbones J.

Norwegian Meteorological Institute
PO Box 43 Blindern N-0313 Oslo
e-mail: jorgen.saltbones@met.no

Wind P.

Norwegian Meteorological Institute
N-0313 Oslo
e-mail: peter.wind@met.no

POLAND

Juda-Rezler K.

Warsaw University of Technology
Institute of Environmental Engineering Systems
Nowowiejska 20 00-653 Warsaw
e-mail: katarzyna.juda-rezler@is.pw.edu.pl

PORTUGAL

Borrego C.   University of Aveiro
                              Department of Environment and Planning
                              3810-193
                              e-mail: borrego@ua.pt

Carvalho A.   University of Aveiro
                              Department of Environment and Planning
                              3810-193
                              e-mail: avc@dao.ua.pt

Ferreira J.   University of Aveiro
                              Department of Environment and Planning
                              3810-193
                              e-mail: jferreira@dao.ua.pt

Miranda A. I.   University of Aveiro
                              Department of Environment and Planning
                              3810-193
                              e-mail: aicm@dao.ua.pt

RUSSIA

Genikhovich E.

Voeikov Main Geophysical Observatory
194121 St. Petersburg
e-mail: ego@main.mgo.rssi.ru

Ryaboshapko A.

EMEP Meteorological Synthesizing Centre "EAST"
Arhitektor Vlasov Str. 51 Moscow 117393
e-mail: alexey.ryaboshapko@msceast.org

SLOVENIA

Boznar M. Z.

Jozef Stefan Institute and AMES
Jamova 39 SI-1000 Ljubljana
e-mail: marija.boznar@ames.si

SPAIN

Baldasano J.

Universitat Politéchnica de Catalunya
Department of Engineering Projects
Avd. Diagonal 647 Barcelona
e-mail: jose.baldasano@upc.es

San Jose R.

Technical University of Madrid
Boadilla del Monte 28660 Madrid
e-mail: roberto@fi.upm.es

SWITZERLAND

Aksoyoglu S. A.   Paul Scherrer Institut
CH-5232 Villigen PSI
e-mail: sebnem.andreani@psi.ch

Keller J.   Paul Scherrer Institut
CH-5232 Villigen PSI
e-mail: johannes.keller@psi.ch

TURKEY

Atımtay A.   Middle East Technical University
Environmental Engineering Department
06531   Ankara
e-mail: aatimtay@metu.edu.tr

Coskuner, T.   Middle East Technical University
Ankara

Erdun, H.   BNP-AK Dresdner Bank
Buyukdere St. No: 173 Levent Istanbul

Incecik S.   Istanbul Technical University
Department of Meteorology
Maslak 34469 Istanbul
e-mail: incecik@itu.edu.tr

Muezzinoglu A.   Dokuz Eylul University
                          Department of Environmental Engineering
                          Buca Izmir
                          e-mail: aysen.muezzin@deu.edu.tr

Oguz O.   Middle East Technical University
                          Department of Environmental Engineering
                          06531 Ankara
                          e-mail: oznuro@metu.edu.tr

Tayanc, M.   Marmara University
                          Department of Environmental Engineering
                          Goztepe Istanbul
                          e-mail: mtayanc@eng.marmara.edu.tr

Tecer L. H.   Karaelmas University
                          Engineering Faculty
                          67100 Zonguldak
                          e-mail: yildirim61@yahoo.com

Topcu S.   Istanbul Technical University
                          Department of Meteorology
                          Maslak 34469 Istanbul
                          e-mail: stopcu@itu.edu.tr

USA

Aneja V. P.   North Carolina State University
                           Dept. of Marine, Earth and Atmospheric Sciences
                           Raleigh NC 27695-8208
                           e-mail: viney_aneja@ncsu.edu

Bornstein R.   San Jose State University
                            e-mail: pblmodel@hotmail.com

Dupont S.   US EPA
                            Research Triangle Park NC
                            E243-03 NERL/AMD/AMDB
                            e-mail: dupont@hpcc.epa.gov

Gertler A.   Desert Research Institute
                            2215 Raggio Parkway
                            Reno Nevada 89512
                            e-mail: alang@dri.edu

Hogrefe C.   Atmospheric Sciences Reserch Center
                            University of Albany
                            Albany NY 12222
                            e-mail: chogrefe@dec.state.ny.us

Kang D.   University Corporation for Atmospheric Research
                          Boulder CO 80301
                          e-mail: kang@hpcc.epa.gov

Koracin J.   Desert Research Institute
                          2215 Raggio Parkway
                          Reno Nevada 89512
                          e-mail: jkahya@dri.edu

Koracin D.   Desert Research Institute
                          2215 Raggio Parkway
                          Reno Nevada 89512
                          e-mail: darko@dri.edu

Odman M. T.   Georgia Institute of Technology
                          Atlanta 30332-0512
                          e-mail: talat.odman@ce.gatech.edu

Russell T.   Georgia Institute of Technology
                          e-mail: trussell@ce.gatech.edu

YUGOSLAVIA

Grsic Z.   Institute of Nuclear Sciences Vinca
                          PO Box 522 SCG-11001 Belgrade
                          e-mail: grsa@rt270.vin.bg.ac.yu

AUTHOR INDEX

SUBJECT INDEX

The manufacturer's authorised representative in the EU is Springer
Nature Customer Service Centre GmbH, Europaplatz 3, 69115 Heidelberg,
Germany. If you have any concerns regarding our products, please
contact ProductSafety@springernature.com

Printed and bound by CPI Group (UK) Ltd, Croydon, CR0 4YY
23/04/2026
02095628-0009